Proceedings of the

16th International Conference

on

Defects in Semiconductors

Proceedings of the

16th International Conference

on

Defects in Semiconductors

Part 3

Lehigh University, Bethlehem, Pennsylvania
22 - 26 July 1991

Edited by
Gordon Davies, Gary G. DeLeo and Michael Stavola

Trans Tech Publications
Switzerland — Germany — UK — USA

This work was partially supported by U.S. Air Force Grant AFOSR-91-0217 issued by the U.S. Air Force Office of Scientific Research.

This work was partially supported by the Department of The Army Grant DAAL03-91-G-0113 issued by the Army Research Office. The views, opinions, and/or findings contained in this report are those of the authors and should not be construed as an official Department of the Army position, policy, or decision, unless so designated by other documentation.

This work relates to Department of Navy Grants N00014-91-J-1761 and N00014-91-J-1912 issued by the Office of Naval Research. The United States Government has a royalty-free license throughout the world in all copyrightable material contained herein.

Copyright © 1992 Trans Tech Publications Ltd

ISBN 0-87849-628-9
(Pts. 1, 2 & 3)

Volumes 83-87 (Pts. 1, 2 & 3) of
Materials Science Forum
ISSN 0255 - 5476

Distributed in the Americas by

Trans Tech Publications
c/o Gower Publishing Company
Old Post Road
Brookfield VT 05036
USA

Fax: (802) 276 - 3837

and worldwide by

Trans Tech Publications Ltd
Segantinistr. 216
CH-8049 Zürich
Switzerland

Fax: (++41) 1 342 05 29

CONTENTS — PART 3

14. NEW TECHNIQUES

Atomic defect configurations identified by nuclear techniques Th. Wichert (Invited Plenary Paper)	1081
Combination of deep level transient spectroscopy and transmutation of radioactive impurities M. Lang, G. Pensl, M. Gebhard, N. Achtziger and M. Uhrmacher	1097
Identification of band gap states in silicon by deep level transient spectroscopy on radioactive isotopes J.W. Petersen and J. Nielsen	1103
Microscopy of Frenkel pairs in semiconductors by nuclear techniques R. Sielemann, H. Häßlein, M. Brüßler and H. Metzner	1109
Modern muon spin spectroscopic methods in semiconductor physics R.L. Lichti, C.D. Lamp, S.R. Kreitzman, R.F. Kiefl, J.W. Schneider, Ch. Niedermayer, K. Chow, T. Pfiz, T.L. Estle, S.A. Dodds, B. Hitti and R.C. DuVarney	1115
Muon stopping sites in semiconductors from decay-positron channeling H. Simmler, P. Eschle, H. Keller, W. Kündig, W. Odermatt, B.D. Patterson, I.M. Savić, J.W. Schneider, B. Stäuble-Pümpin, U. Straumann and P. Truöl	1121
Polarized spectroscopy of complex luminescence centers S.S. Ostapenko and M.K. Sheinkman	1127
ONP spectroscopy of defects in silicon N.T. Bagraev and I.S. Polovtsev	1135
Nuclear spin polarization by optical pumping of nitrogen impurities in semiconductors K. Murakami, K. Hara and K. Masuda	1141
On the analysis of digital DLTS data C.A.B. Ball and A.B. Conibear	1147
A re-evaluation of electric-field enhanced emission measurements for use in type and charge state determination of point defects W.R. Buchwald, H.G. Grimmeiss, F.C. Rong, N.M. Johnson, E.H. Poindexter and H. Pettersson	1153
X-ray spectroscopy following neutron irradiation of semiconductor silicon A.J. Filo, A. Meyer, C.C. Swanson and J.P. Lavine	1159

Spin dependent recombination at deep centers in Si — electrically detected magnetic resonance 1165
P. Christmann, M. Bernauer, C. Wetzel, A. Asenov, B.K. Meyer and A. Endrös

Excited defect energy states from temperature dependent ESR 1171
C. Kisielowski, K. Maier, J. Schneider and V. Oding

Dislocation associated defects in gallium arsenide by scanning tunneling microscopy 1177
O. Siboulet, S. Gauthier, J.C. Girard, W. Sacks, S. Rousset and J. Klein

15. DEFECTS IN SiC AND DIAMOND

Transition metals in silicon carbide (SiC) : vanadium and titanium 1183
K. Maier, H.D. Müller and J. Schneider (Invited)

Photoluminescence excitation spectroscopy of cubic SiC grown by chemical vapor deposition on Si substrates 1195
J.A. Freitas, P.B. Klein and S.G. Bishop

Paramagnetic defects in SiC based materials 1201
O. Chauvet, L. Zuppiroli, J. Ardonceau, I. Solomon, Y.C. Wang and R.F. Davis

Acceptors in silicon carbide : ODMR data 1207
P.G. Baranov and N.G. Romanov

Luminescence and absorption of vanadium (V^{4+}) in 6H silicon carbide 1213
A. Dörnen, Y. Latushko, W. Suttrop, G. Pensl, S. Leibenzeder and R. Stein

Impurity-defect reactions in ion-implanted diamond 1219
A.A. Gippius

16. DEFECTS IN II-VI SEMICONDUCTORS

Native defect compensation in wide-band-gap semiconductors 1225
D.B. Laks, C.G. Van de Walle, G.F. Neumark and S.T. Pantelides (Invited)

ODMR investigations of the A-centres in CdTe 1235
D.M. Hofmann, P. Omling, H.G. Grimmeiss, D. Sinerius, K.W. Benz and B.K. Meyer

Picosecond energy transfer between excitons and defects in II-VI semiconductors 1241
R. Heitz, C. Fricke, A. Hoffmann, and I. Broser

Luminescence of a 5d-centre in ZnS 1247
R. Heitz, P. Thurian, A. Hoffmann and I. Broser

Interstitial defects in II-VI semiconductors : role of the cation d states — 1253
J.T. Schick, C.G. Morgan-Pond and J.I. Landman

PAC study of the acceptor Li in II-VI semiconductors — 1259
H. Wolf, Th. Krings, U. Ott, U. Hornauer and Th. Wichert

Generation of metastable shallow donors under cooling in hexagonal II-VI semiconductors — 1265
N.E. Korsunskaya, I.V. Markevich, E.P. Shulga, I.A. Drozdova and M.K. Sheinkman

17. HETERO-EPITAXY AND STRAINED LAYERS

Strain relief in thin films : can we control it? — 1271
F.K. LeGoues (Invited)

Composition modulation effects on the generation of defects in $In_{0.54}Ga_{0.46}As$ strained layers — 1285
F. Peiró, A. Cornet, A. Herms, J.R. Morante, S. Clark and R.H. Williams

Electron-trapping defects in MBE-grown relaxed n-$In_{0.05}Ga_{0.95}As$ on gallium arsenide — 1291
A.C. Irvine, L.K. Howard and D.W. Palmer

Atomic ordering in (110)InGaAs and its influence on electron mobility — 1297
O. Ueda and Y. Nakata

Dopant diffusion in $Si_{0.7}Ge_{0.3}$ — 1303
D. Mathiot and J.C. Dupuy

18. DISLOCATIONS

Characterisation of dislocations in the presence of transition metal contamination — 1309
V. Higgs, E.C. Lightowlers, C.E. Norman and P. Kightley

Correlation of the D-band photoluminescence with spatial properties of dislocations in silicon — 1315
K. Weronek, J. Weber, A. Höpner, F. Ernst, R. Buchner, M. Stefaniak and H. Alexander

Photoluminescence and electronic structure of dislocations in Si crystals — 1321
Yu. Lelikov, Yu. Rebane, S. Ruvimov, D. Tarhin, A. Sitnikova and Yu. Shreter

Characterization of point defects in Si crystals by highly spatially resolved photoluminescence — 1327
M. Tajima, H. Takeno and T. Abe

Theoretical study on the structure and properties of dislocations in semiconductors — 1333
K. Masuda-Jindo

19. SUPERLATTICES

Solid state processes at the atomic level A. Ourmazd, F.H. Baumann, M. Bode, Y. Kim and J.A. Rentschler (Invited)	1339
Theory of Zn-enhanced disordering in GaAs/AlAs superlattices Q.-M. Zhang, C. Wang and J. Bernholc	1351
Spatial partition of photocarriers trapped at deep level defects in multiple quantum well structures D.D. Nolte, R.M. Brubaker, Q.N. Wang and M.R. Melloch	1357
Picosecond dynamics of exciton capture, emission and recombination at shallow impurities in center-doped AlGaAs/GaAs quantum wells C.I. Harris, H. Kalt, B. Monemar, P.O. Holtz, J.P. Bergman, M. Sundaram, J.L. Merz and A.C. Gossard	1363
Spectroscopy of shallow donor impurities in GaAs/GaAlAs multi-quantum wells J.L. Dunn, E. Pearl and C.A. Bates	1369
Excitons bound at shallow impurities in GaAs/AlGaAs quantum wells with varying doping levels P.O. Holtz, B. Monemar, M. Sundaram, J.L. Merz, A.C. Gossard, C.I. Harris and H. Kalt	1375

20. DEFECTS AT SURFACES AND INTERFACES AND IN LOW-DIMENSIONAL STRUCTURES

Scanning tunneling microscopy studies of semiconductor surface defects J.E. Demuth (Invited)	1381
The atomic and electronic structure of ordered buried B(2×1) layers in Si(100) M. Needels, M.S. Hybertsen and M. Schluter	1391
Negative U systems at semiconductor surfaces G. Allan and M. Lannoo	1397
Ab initio calculations on effect of Ga-S bonds on passivation of GaAs surface - a proposal for new surface treatment T. Ohno and K. Wada	1403
Two-dimensionally localized vibrational mode due to Al atoms substituting for Ga one-monolayer in GaAs H. Ono and T. Baba	1409
O surrounding of P_b defects at the (111) Si/SiO$_2$ interface A. Stesmans	1415

^{17}O hyperfine study of the P_b center 1421
J.H. Stathis, S. Rigo, I. Trimaille and M.S. Crowder

Defects induced by high electric field stress and the trivalent silicon defects at the Si-SiO_2 interface 1427
D. Vuillaume, A. Mir and D. Goguenheim

21. PROCESSING-INDUCED DEFECTS

Interstitial defect reactions in silicon processed by reactive ion etching 1433
J.L. Benton, M.A. Kennedy, J. Michel and L.C. Kimerling

Anomalous damage depths in low-energy ion beam processed III-V semiconductors 1439
S.J. Pearton, F. Ren, T.R. Fullowan, R.F. Kopf, W.S. Hobson, C.R. Abernathy, A. Katz, U.K. Chakrabarti and V. Swaminathan

Photoluminescence characterisation of the silicon surface exposed to plasma treatment 1445
A. Henry, B. Monemar, J.L. Lindström, T.D. Bestwick and G.S. Oehrlein

An analysis of point defect fluxes during SiO_2 precipitation in silicon 1451
W.J. Taylor, T.Y. Tan and U.M. Gosele

Deep states associated with copper decorated oxidation induced stacking faults in silicon 1457
M. Kaniewska, J. Kaniewski and A.R. Peaker

Electrical properties of oxidation-induced stacking faults in n-type silicon 1463
J. Kaniewski, M. Kaniewska and A.R. Peaker

Study of internal oxide gettering for CZ silicon : effects of oxide particle size and number density and assessment of thermal stability of gettering for copper and nickel 1469
Z. Laczik, G.R. Booker, R. Falster and P. Török

Morphology change of oxygen precipitates in CZ-Si wafers during two-step heat-treatment 1475
M. Hasebe, J.W. Corbett and K. Kawakami

Annealing of damage in GaAs and InP after implantation of Cd and In 1481
W. Pfeiffer, M. Deicher, R. Kalish, R. Keller, R. Magerle, N. Moriya, P. Pross, H. Skudlik, Th. Wichert, H. Wolf and ISOLDE Collaboration

Ion implantation induced sheet stress due to defects in thin (100) silicon films 1487
J. Yuan, A.J. Yencha and J.W. Corbett

Observation of a trivalent Ge defect in oxygen implanted SiGe alloys M.E. Zvanut, W.E. Carlos, M.E. Twigg, R.E. Stahlbush and D.J. Godbey	1493
Comparison between defects introduced during electron beam evaporation of Pt and Ti on n-GaAs F.D. Auret, G. Myburg, L.J. Bredell, W.O. Barnard and H.W. Kunert	1499
Enhanced-diffusion in electron-beam doping of semiconductors T. Wada, M. Takeda and T. Kondo	1503
High temperature defect-free rapid thermal annealing of III-V substrates in metallorganic controlled ambient A. Katz, A. Feingold, S.J. Pearton, M. Geva, S. Nakahara and E. Lane	1509

22. EFFECTS OF DEFECTS ON DEVICES

The properties of individual Si/SiO_2 defects and their link to 1/F noise M.J. Uren (Invited)	1519
Hydrogen induced defects and defect passivation in silicon solar cells B.L. Sopori	1531
A study of radiation induced defects in silicon solar cells showing improved radiation resistance J. Peters, T. Markvart and A. Willoughby	1539
Defects and Schottky barrier formation : a positive proof for epitaxial Al on AlGaAs Schottky diodes J.M. Langer and P. Revva	1545
Recombination-enhanced diffusion of Be in GaAs M. Uematsu and K. Wada	1551
Role of the diffusivity of Be and C in the performance of GaAs/AlGaAs heterojunction bipolar transistors F. Ren, T.R. Fullowan, J. Lothian, P.W. Wisk, C.R. Abernathy, R.F. Kopf, A.B. Emerson, S.W. Downey and S.J. Pearton	1557
Effects of the substrate-epitaxial layer interface on the DLTS spectra in MESFET and HFET devices M. Spector, M.L. Gray, J.D. Yoder, A.M. Sergent and J.C. Licini	1563
The study of interfacial traps of InP metal-insulator-semiconductor structure L. Lu, J. Zhou, W. Qu and S. Zhang	1569

Atomic Defect Configurations Identified by Nuclear Techniques

Thomas Wichert

Technische Physik, Universität des Saarlandes, 6600 Saarbrücken, FRG

Abstract

The impact of nuclear techniques as analytical tools for the identification of atomic defect configurations is discussed. Perturbed γγ angular correlation and Mössbauer spectroscopy, similar to the magnetic resonance techniques, obtain their local information with respect to a probe atom via electric or magnetic hyperfine fields that originate from the probe's microscopic surrounding. The strength of nuclear techniques is their ability of supplying information on the chemical nature rather than on the electrical properties of a defect. By employing radioactive isotopes low dopant and defect concentrations (10^{16} cm^{-3}) can be investigated. Both methods are qualified for the study of individual dopant atoms and their reactions with other defects, like impurity atoms or native lattice defects. Recent investigations of atomic defect configurations in elemental, III-V, and II-VI semiconductors will be discussed.

1. Introduction

Besides the electrical and optical properties of a defect, the chemical nature of the constituents and the resulting atomistic structure are of interest. In this respect, significant contributions have been made by "nuclear techniques", i.e. by techniques that arise from the field of nuclear physics and/or use properties of the atomic nucleus. Some of the recent results, particular those obtained by the perturbed γγ angular correlation technique (PAC) and by the Mössbauer spectroscopy (MS) will be discussed in this paper. Thereby, the probe atoms are used for watching the behavior of dopant atoms, in particular, their interactions with other defects. The PAC technique will be emphasized because it is comparatively new in the field of defects in semiconductors. The reader will find several other nuclear techniques in this conference which are not discussed here; such as, muon-spin-rotation, β-NMR, or positron annihilation. - In PAC and MS local information is obtained with reference to a probe atom which in a PAC or in a MS "source" experiment is always a radioactive atom whereas in a MS "absorber" experiment a stable isotope is used. But, just the use of radioactive probe atoms guarantees the often needed sensitivity for the detection of defects in semiconductors. Thereby, the local electronic and geometrical structures about the probe atoms are observed via the hyperfine interactions which results from the interaction of the electromagnetic fields of the solid with the nuclear moments of the probe atoms. Thus, both techniques are similar to the well known ESR/ENDOR techniques which measure mainly the magnetic interactions of paramagnetic centers. For the interpretation of the experimental results obtained with help of radioactive probe atoms, however, it should be taken into account that the defect complex is formed with the probe atom but the resulting hyperfine interaction is measured, after a β-decay, at the daughter isotope which in most cases is chemically different from the probe atom. We shall start with a brief summary of the parameters used in both techniques; for more details see[1,2,3].

Perturbed γγ Angular Correlation: The parameter used for the characterization of defects is the electric field gradient tensor (EFG) which is the second derivative of the electrostatic potential of the solid. The three components of the traceless tensor are measured during the radioactive decay of the probe atom which leads to the nuclear ground state of the daughter nucleus via an isomeric state; this process is associated with the emission of two γ rays bracketing this isomeric state. The electric quadrupole moment Q of this state interacts with the EFG and effects a spin dependent energy shift $\Delta E = \hbar\omega$. This energy shift modulates in time the probability to detect both γ rays in coincidence, whereby the maximum possible time difference elapsed between the detection of the two γ rays is determined by the lifetime of isomeric state. The resulting PAC time spectrum contains the following information: The fundamental modulation frequency is determined by the strength V_{zz} of the EFG tensor which is expressed via the coupling constant $\nu_Q = eQV_{zz}$. The ratio of two modulation frequencies yields the asymmetry parameter $\eta = (V_{xx}-V_{yy})/V_{zz}$ of the tensor with $0 \leq \eta \leq 1$. The amplitudes of the modulations are determined by the fraction of probe atoms involved in the defect complex and by the orientation of the defect complex with respect to the host lattice. A possible relaxation of the amplitudes is indicative of either a distribution of different EFGs or a change of an EFG during the nuclear lifetime of the isomeric state; the latter process gives information on dynamic effects of the defect complex.

Mössbauer Spectroscopy: The parameter mainly used for the characterization of defects is the isomer shift (IS) which is determined by the electron density at the site of the nucleus. The associated energy shift ΔE offsets the energy E_o of the γ quantum which is emitted during the radioactive decay of the probe atom and leads from the isomeric nuclear state to the ground state of the daughter nucleus. This tiny offset can be determined by the resonance-like energy dependence of the absorption probability when the γ quantum interacts with the identical (daughter) nucleus that is located in an absorber. The Mössbauer spectrum is measured as a function of the Doppler velocity v which is used to change the energy E_o by $v/c \cdot E_o$ in order to achieve the nuclear resonance absorption of the emitted γ rays. From the required resonance velocity v_{res} the defect induced IS is obtained. In a similar way, values v_{res} are obtained which result from the energy shifts of a defect induced EFG. The area of the resonance line is determined by the fraction of probe atoms involved in the respective defect and, via the Debye Waller factor (DWF), by the local dynamics of the defect complex. Because of the DWF the area of the resonance line becomes temperature dependent and decreases at higher temperatures. The width of the resonance line plays the role of the relaxation of the modulation amplitude in the PAC experiment. The lower limit of the observable relaxation rate is given by the inverse nuclear lifetime of the isomeric nuclear state.

2. Passivation by Hydrogen Atoms

The electrical deactivation of shallow acceptors in Si, first observed by Sah et al. in MOS devices[4] and by Pankove et al. after plasma etching[5], pointed to an interaction between the acceptor atoms and a defect in Si which was identified as the H atom, hitherto. Here, a good opportunity was offered to employ nuclear techniques for the analysis of possible reactions of the acceptor atoms on a microscopic scale. The alternative technique, the Electron Spin Resonance (ESR), that is often used for a chemical identification of a defect in

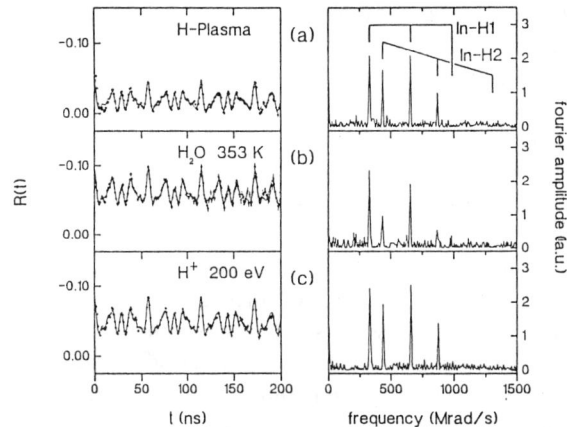

Figure 1. PAC time spectra and their Fourier transforms measured at the probe atom ^{111}In/Cd in Si. Visible are the different oscillation frequencies connected with the two EFGs that characterize the In-H1 and In-H2 pair.

question failed because of the absence of paramagnetic centers, in this case. Since the observed passivation extended only to a depth of a few μm, radioactive ^{111}In atoms were introduced at a depth of 0.16 μm by implantation with 350 keV which at the same time served as a shallow acceptors and as a probe atom for the PAC experiment. After having annealed the implantation induced radiation damage and subsequently etched the surface with a H plasma the PAC time spectrum shown in Fig. 1 (a) is obtained[6,7]. It unequivocally shows the presence of a defect at a well defined lattice site of the immediate neighborhood of the acceptor through the strength of the EFG which falls off with the third power of the distance acceptor-defect and through the sharpness of the associated frequencies which are visible in the Fourier transform of the PAC time spectrum (Fig. 1 (a), right).

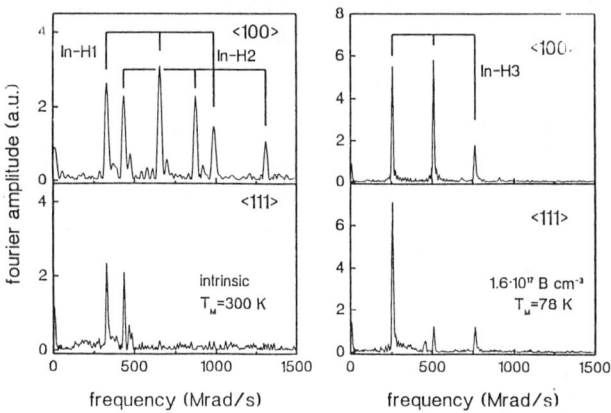

Figure 2. Fourier transforms of PAC time spectra with the γ detectors oriented along <100> and <111> lattice directions. The corresponding variations of the Fourier components show the <111> orientations of the three EFG tensors.

Actually, two different EFGs are observed in this experiment, as is visible by the presence of two frequency triplets, whereby for each triplet always holds that the third frequency is the sum of the two lower ones (see also Fig. 2, top panels). The reason for the occurrence of two EFGs will be explained in section 5 where the dynamics of defects is discussed. The here used local acceptor concentrations ranging between 1 and $5 \cdot 10^{16}$ cm^{-3} matches those used in the experiment by Pankove so that a direct comparison between the observed In-defect pairing and the increase of the electrical resistivity can be made. Thus, the passivation of the acceptor atoms is obviously accompanied by the formation of close pairs. From the symmetry and the orientation of the defect specific EFG tensors it is concluded that the formed pairs possess axial symmetry and are oriented along one of the four equivalent <111> directions of the Si lattice. Fig 2 (left) shows the results of such an orientation experiment:[6,7] The axial symmetry of the complex is obtained from the ratio of the fundamental to the first harmonic frequency which for both frequency triplets is 1:2. Then, the orientation of the symmetry axis is obtained from the amplitudes of the three frequencies when the γγ-coincidences are recorded with the γ-detectors positioned along different lattice directions of the Si wafer, here along <100> and <111>; the strong reductions of the amplitudes of the higher harmonics observed in the latter case show the <111> orientations of the formed In-defect complexes[2].

Although the EFGs characterizing different defect configurations are experimentally well known within about 1 %, an identification of the trapped defects is not possible solely based on the EFGs because of the insufficient accuracy of the theoretical calculation of defect induced EFGs in covalently bonded semiconductors. The fact, however, that PAC measurements are feasible under a wide range of different sample conditions still allows the identification of the defect in question by a suited combination of experiments. Thus, the information from the experiments by Tevendale et al.[8] was used to see whether boiling of Si in water would also create the same In-defect complex that was observed after the plasma etch. The PAC spectrum in Fig. 1 (b) immediately reveals the identity of the formed In-defect complexes. Thus, the presence of lattice damage possibly introduced during the plasma treatment is already excluded as a source for the trapped defect. For the final identification of the defect, mass separated ion implantation at low energies of about 100 eV has turned out as a successful tool for it represents an element specific doping procedure which almost completely avoids the uncontrolled creation of radiation damage. The result in Fig. 1 (c), obtained after implanting 10^{14} H$^+$ cm^{-2} with 200 eV finally unmasks the trapped defect as the H atom. As listed in Table I, three different EFGs characterize the formed In-H complex pointing to different electronic and/or structural configurations of the formed complex. It is visible from Fig. 2 (right panel) that also the third EFG (In-H3) has the same <111> axial symmetry as observed for the other two EFGs. - Here, it should be mentioned that the hyperfine signature of the trapped H atoms is very rich what has turned out to be typically for the trapping of interstitial defects at the probe ^{111}In/Cd in Si (see Tables I and II).

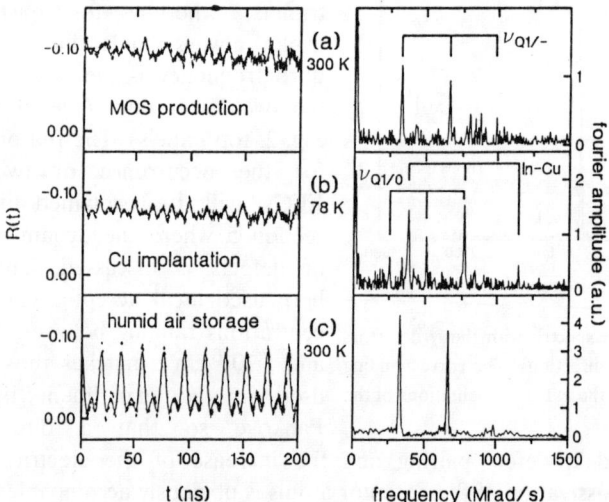

Figure 3. Three examples of unintentional incorporation of H into Si causing the formation of In-H pairs. Note, $\nu_{Q1/-}$ and $\nu_{Q1/0}$ correspond to In-H1 and In-H3 in Table I, respectively.

After knowing the hyperfine parameters of the In-H pairs, PAC can be used to pursue possible contamination pathways for the introduction of H into Si. Three results of such investigations, which often were obtained more or less accidentally, are collected in Fig. 3:[9] (a) During evaporation of a Pd layer on top of a SiO$_2$/Si substrate, needed for the preparation of a MOS contact, H was released, possibly from the SiO$_2$ layer, and diffused into the ^{111}In doped Si crystal forming In-H1 pairs (characterized by $\nu_{Q1/-}$). (b) Low energy (200 eV) Cu$^+$ implantation effected in addition to the formation of In-Cu pairs (see section 3) also the formation of In-H3 pairs ($\nu_{Q1/0}$); here, possible sources for the contamination with H were water or hydrocarbon compounds adsorbed on the Si surface. In an electrical resistivity experiment, In-H and In-Cu, would be indistinguishable because both complexes passivate acceptors and contribute to a resistivity increase whereas by PAC both types of complexes are easily distinguished because of their different hyperfine parameters. Also implantation of p-Si with Ar^+, Ni^+, or Li^+ ions lead to the formation of In-H pairs in agreement with corresponding observations by Seager for B in Si^{10}. (c) The last example shows that after a shallow implantation of In with 60 keV about 50 % of the In atoms were passivated by H when the doped Si wafer was exposed to humid air at 30° C for about 1 day.

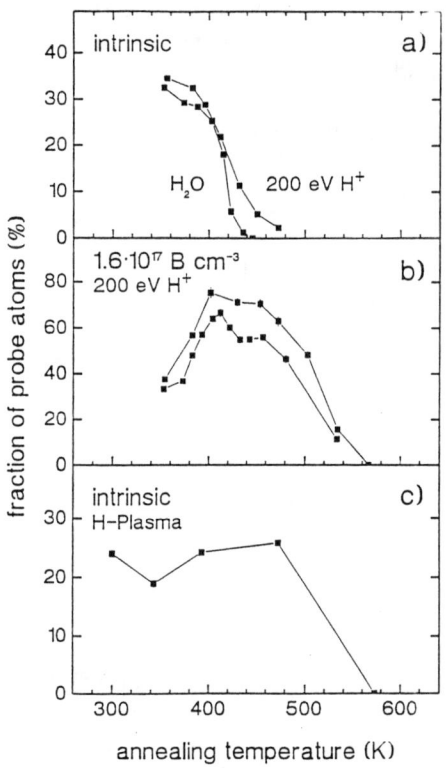

Figure 4. Thermal stabilities of the In-H pairs in Si observed under different sample conditions.

The reduction in the fraction of In-H pairs measured during an isochronal annealing experiment allows to determine the thermal stability of the In-H pairs. Fig. 4 compares the results for different sample conditions.[11] From the amplitude of the modulation in the PAC spectrum the fraction of passivated In atoms is directly obtained. Obviously, Figs. 4 (a) and (c) show quite different thermal stabilities for the In-H pairs which differ by about 100 K. In panel (a) the dominant traps are the implanted radioactive ^{111}In atoms whose peak concentration is around $5 \cdot 10^{16}$ cm^{-3}. Because of the low trap concentration and with that low probability of retrapping of H the lowest thermal stability of the In-H pairs is observed corresponding to a dissociation energy E_D(In-H) = 1.3 eV if a single step dissociation process and an attempt frequency of 10^{13} s^{-1} is assumed. - Under the same assumptions most of the other dissociation energies E_D listed in Table I were deduced. - In contrast, in panel (c) the increased stability of the pairs is connected with the presence of the plasma induced intrinsic defects, being absent in the case of low energy H implantation or boiling in water, which also represent traps for the H atoms; because of frequent dissociation and retrapping processes caused by the higher trap concentration H remains until much higher temperatures in the Si sample as compared to the first case. In panel (b) the artificial increase in the thermal stability is effected by co-doping the sample with B acceptors which also interact with H atoms. The here first observed increase in the fractions of In-H pairs, occurring at 380 K, is caused by the dissociation of B-H pairs. This process feeds the fraction of In-H pairs via the release of H atoms which are then available for additional trapping by isolated In atoms. From this type of co-doping experiments the following set of dissociation energies was obtained by PAC: E_D(B-H) = 1.2 eV < E_D(Ga-H) < E_D(In-H) = 1.3 eV.

From the data in Fig. 4 it is evident that dissociation energies reported for acceptor passivation in Si critically depend on the H-loading conditions and the respective acceptor and H concentrations. Thus, Pearton et al. reported dissociation energies of 1.1, 1.6, 1.9, and 2.1 eV for B, Al, Ga, and In, respectively,[12] and Stavola et al. dissociation temperatures of

450, 480, and 580 K for B, Al, and Ga respectively[13]. In those experiments H was introduced via plasma loading and the acceptor concentration was well above 10^{17} cm^{-3}; consequently, the observed stabilities of the acceptor-H pairs were shifted to higher values. The problem of retrapping of H by the acceptor atoms was excluded in experiments by Weber and Zundel[14,15] who performed their stability studies in the depletion region of a biased Schottky contact; their thermal stabilities, determined for various acceptor-H pairs, agree well with those obtained by PAC. - That the In-H pairing observed by PAC really corresponds to the microscopic reaction that takes place during the electrical passivation as observed in resistivity experiments was recently demonstrated by PAC and electrical measurements performed at the identical In-doped Si sample; it was shown that the number of In-H pairs exactly accounts for the number of deactivated In atoms.[16]

Table I. Acceptor-hydrogen pairs formed with the acceptors ^{111}In/Cd or ^{111m}Cd/Cd.

COMPLEX	v_Q (MHz)	T (K)	η	<ijk>	E_D (eV)	Ref.
Silicon						
In-H1	360	78	0.	<111>	1.3	2
In-H2	480	78	0.	<111>	1.3	
In-H3	270	78	0.	<111>	1.3	
Germanium						
In-H	436	78	0.	<111>	1.1	17
GaAs						
Cd-H1	476	78	0.	<111>	1.35	18, 19
Cd-H2	495	78	0.	<111>	1.35	
InP						
Cd-H	484	295	0.	<111>	1.8	18, 20
GaP						
Cd-H	525	295	0.			21
InAs						
Cd-H1	427	295	0.	<111>		21
Cd-H2	581	295	0.	<111>		
InSb						
Cd-H	357	295	0.			21

In Table I, the hyperfine parameters for different acceptor-H pairs observed so far by PAC in several semiconductors are collected. In Si[2] and Ge[17], the pairs are formed with the probe atom ^{111}In and are detected at the daughter isotope ^{111}Cd. In the III-V compounds GaAs[18,19], InP[18,20], GaP, InAs, and InSb[21], the pairs are formed with the probe ^{111m}Cd and are detected also at the element Cd whereby the identical isomeric state of ^{111}Cd is used as in the case of ^{111}In. The microscopic nature of the complexes Cd-H2 observed in GaAs and in InAs, respectively, is not yet completely identified; it can not be excluded that these EFGs are caused by trapping of a second H atom or even by an intrinsic lattice defect. - For the II-VI semiconductors no pairing with H atoms has been reported, so far.

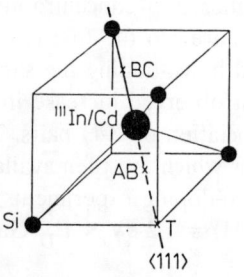

Figure 5. Distinct interstitial sites in the Si lattice with reference to the probe atom ^{111}In/Cd. BC: bond center, AB: antibonding, T: tetrahedral interstitial.

In all cases, the listed complexes are axially symmetric (η = 0) and the symmetry axis is along a <111> lattice direction. This means, all lattice sites proposed, so far, for H next to an acceptor[22,12] would be in agreement with these PAC results. Of course, it would be tempting to explain the occurrence of the three different EFGs for the In-H pair in Si, among others, by the population of different lattice sites, e.g. the bond center (BC) and antibonding (AB) site in Fig. 5[6,23] what, however, has to be speculative at this point. An experimental solution to the problem of determining the actual H site at the acceptor atom is, in principal, offered by the channeling technique. By substituting D for H the lattice site of D at the presence of the acceptor B was determined via the angular dependence of the reaction yield of the D(^{3}He,p)^{4}He nuclear reactions. Summarizing the available ion channeling experiments, it turns out that the results for D in Si are much in favor for a BC site next to the acceptor B[24,25]; but, there is also evidence for an AB site[26]. In interpreting these channeling results, however, it should be taken into account that i) the analyzing beam influences the apparent D site, ii) D and B concentrations of typically 10^{19} cm^{-3} were required, and that iii) the technique does not supply direct information on the number and the population of different D sites.

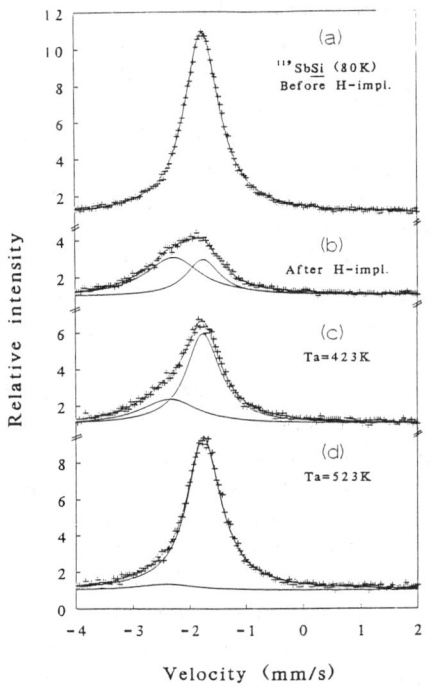

Figure 6. Conversion electron Mössbauer spectra measured at the probe ^{119}Sb/Sn in Si. The two resonance lines correspond to Sb-H pairs (left) and isolated substitutional Sb atoms (right).

Using the Mössbauer probe ^{119}Sb/^{119}Sn, it was possible to look also for donor-H pairing in Si which was proposed to occur first by Johnson et al.[27]; here, the formation of the complex Sb-H is of interest [28,29]. For the Mössbauer source experiment, first Si was doped with the donor ^{119}Sb by implantation and subsequently annealed; the substitutional incorporated ^{119}Sb atoms are characterized by the isomer shift IS_s = 1.80 mm/s, as is visible in the conversion electron Mössbauer spectrum in Fig. 6 (top). After low energy H implantation (200 eV) this line is reduced and a second resonance line characterized by IS_H = 2.34 mm/s becomes visible. It disappears after annealing at 423 K and 523 K in favor of the substitutional Sb line. The extracted dissociation energy of E_D = 1.4 eV agrees well with the values obtained by infrared spectroscopy[30]. Supported by the low DWF of the H associated site and because of the absence of a defect induced EFG for the trapped H defect the more distant AB site has been proposed that belongs to the Si atom neighboring the ^{119}Sb probe atom. The temperature dependence of the DWF also points to a possible tunneling motion of the H atom in the neighborhood of the daughter isotope ^{119}Sn.

3. Reactions between Dopant Atoms and Metallic Impurities

In many respects, the behavior of the metallic impurities is similar to that of the above discussed H atoms; in particular, they tend to migrate also via interstitial lattice sites and exhibit strong interactions with dopant atoms. It has been this similarity that brought up the suggestion that the species which passivates acceptors in Si after chemomechanical polishing[31] might also consists of H atoms[12,22]; possibly, in a different, metastable state because i) the passivation observed in resistivity experiments extends across the whole wafer of a few mm thickness instead of reaching only a depth of a few μm and ii) the thermal

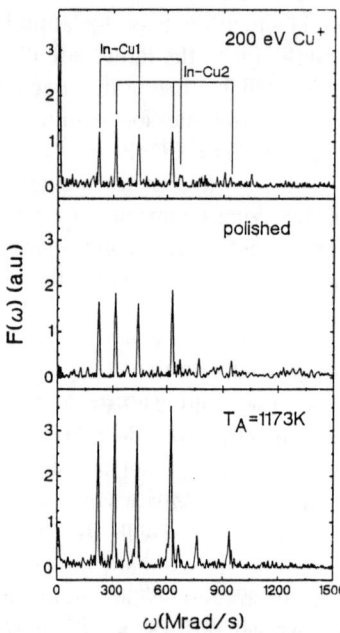

Figure 7. Formation of ^{111}In-Cu pairs in Si under different experimental conditions.

stability of the passivation is significantly lower. By implanting ^{111}In acceptors into the backside of Si wafers which subsequently were chemomechanically polished on their front it was straightforward to show by PAC that this species X does not form any of the already known In-H pairs[32,33] because the X defect gives rise to three new EFGs - two of them are shown in Fig. 7 (middle panel). By implanting low energy Cu$^+$ ions (200 eV) into Si and comparing the resulting EFGs of both experiments (top and middle panel of Fig. 7) the EFGs are shown to correspond to the formation of In-Cu pairs[34] (see Table II). Since exclusively implantation of Cu$^+$ ions but not of other ions, created the three EFGs that characterize the trapped X defect an assignment of this defect to an interstitial Si atom[35] was dismissed. Further studies in p-Si revealed that as-delivered (unpolished) Si wafers, implanted with 350 keV ^{111}In$^+$, have to contain already Cu as precipitates[33]: Then, it can be explained that as soon as the wafers reach a temperature at which the solubility of the Cu atoms[36] becomes comparable to the local ^{111}In concentration of about $5 \cdot 10^{16}$ cm^{-3}, In-Cu pairs are visible what is illustrated in Fig. 7 (bottom panel). The fractions of In-Cu pairs measured at 78 K as a function of annealing temperature of the Si wafer are plotted in Fig. 8[33]. They strongly increase above 973 K and reach a maximum around 1173 K. At still higher temperatures, the high Cu concentration seems to favor the Cu-Cu interaction over the acceptor-Cu interactions so that the In-Cu fraction decreases. For comparison, Fig. 8 shows the increase of the electrical resistivity measured at the identical sample used for PAC; the same behavior of the resistivity is obtained at p-type Si wafers that were not implanted with ^{111}In$^+$ before. The increase in resistivity, which corresponds to the formation of passivating acceptor-Cu pairs, nicely parallels the fraction of In-Cu pairs measured by PAC. It should be noted that the high sensitivity of the PAC technique in this case relies mainly on the fact that Cu precipitates are known to form close to the Si surface[37], just where the ^{111}In probe atoms reside after implantation with 350 keV.

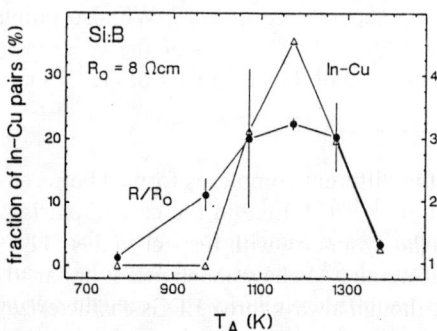

Figure 8. Comparison of the fraction of In-Cu pairs (left scale) and the electrical resistivity (right scale) measured at the same sample as a function of annealing temperature of Si.

Fig. 9 shows the thermal stability of the In-Cu pairs following different treatments of the Si wafers[9]. Depending on the procedure employed the thermal stability varies by as much as 200 K if one compares the dissociation after Cu$^+$ implantation and after chemical polishing. Since in the case of the low energy implantation only $2 \cdot 10^{14}$ Cu$^+$ cm^{-2} are introduced into the near surface region whereas the polishing procedure is known of generating very high concentrations of Cu atoms across the whole wafer the different observed stabilities seem to be caused by the different concentrations and depth distributions of the Cu atoms which lead to different apparent stabilities of the In-Cu pair

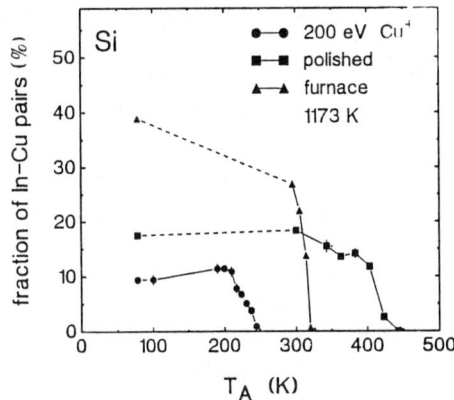

Figure 9. Thermal stability of In-Cu pairs in Si observed under different sample conditions.

because of the different retrapping probabilities. Taking the lowest observed stability the dissociation energy $E_D = 0.7$ eV is obtained which is in good agreement with corresponding values reported by Zundel et al.[35,38].

The identification of the X defect, induced by polishing, with Cu atoms was in seemingly contradiction to the much too high diffusivity of the X defect, determined by Zundel et al. at 250 K[35], when compared with the diffusivity of Cu, determined by Hall and Racette around 800 K[39]. The fact, however, that in both experiments the acceptor concentrations differed by 5 orders of magnitude along with the shown tendency of Cu atoms to form pairs with acceptor atoms reconciled the two different results with each other and led to a correction of the published diffusion energy of 0.43 eV for Cu in Si to the new value of 0.15 eV[34]. Thus, it is obvious from these examples that the use of thermodynamical parameters can become problematic for proving or disproving the identity of a particular defect complex.

Table II. Pairing between the acceptors ^{111}In/Cd or ^{111m}Cd/Cd and interstitial metal atoms.

COMPLEX	v_Q(MHz)	T (K)	η	<ijk>	E_D (eV)	Ref.
Silicon						
In-Li1	172	78	0.	<111>	1.0	2, 17
In-Li2	260	78	0.	<111>	1.0	
In-Li3	323	78	0.	<111>	1.0	
In-Na1	170	78	0.			20
In-Na2	249	78	0.			
In-Na3	274	78	0.			
In-Cu1	237	78	0.	<111>	0.7	34
In-Cu2	334	78	0.	<111>	0.7	
In-Cu3	408	78	0.	<111>	0.7	
In-Fe1	378	295	0.	<311>	<1.4	43
In-Fe2	432	295	0.32		<1.4	
In-Fe3	473	295	0.	<111>	<1.4	
In-Cr1	370	30	0.	<111>	<1.5	43
In-Cr2	473	295	0.	<111>	<1.5	
Germanium						
In-Li	163	78	0.	<111>	0.85	17, 20
GaAs						
Cd-Li	86	295	0.86			20
InP						
Cd-Li	90	295	0.85			20

Along with their hyperfine parameters, Table II lists the different complexes formed between interstitial metallic impurities and the acceptors ^{111}In or ^{111m}Cd. Like in the case of pairing with H, also here most of the observed pairs show axial symmetry with respect to a <111> lattice direction. Further, it turns out that in Si the In-Cu pairs[34] behave much more similar to the In-Li[17,2] and In-Na pairs[20] than to the In-H pairs, though always three EFGs characterize the respective pairs. Since Li bound to an acceptor atom is known to reside at or close to the tetrahedral interstitial site[40] a near tetrahedral site is also assumed for Cu trapped by the acceptor In. Such an assignment would be in agreement with calculations by Estreicher[41] who proposes the BC site for H, in agreement with van de Walle[42], and the

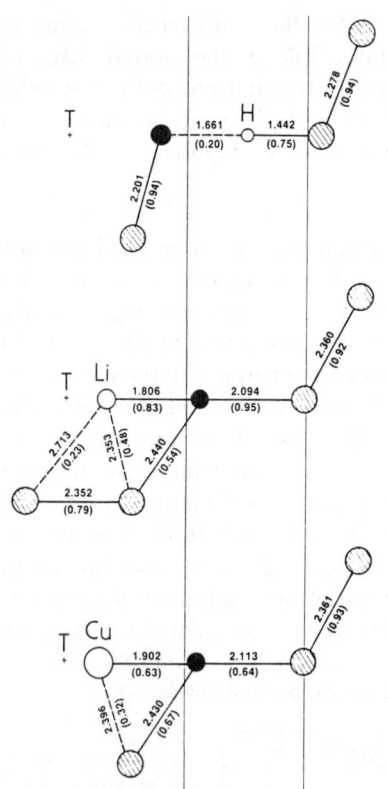

Figure 10. Lowest energy configurations calculated for pairing between the acceptor B (full circle) and three different interstitial impurities in Si. Given are the bond length in 0.1 nm and the "degree" of bonding (values in brackets).

antibonding site close to the tetrahedral interstitial site for Li and Cu when associated with the acceptor B in Si (see Fig. 10). The dissociation energies E_D listed in Tables I and II can be thought off as the sum of the binding energy E_B of the complex and of the migration energy E_M of the interstitial impurity atom. Taking into account that these complexes are coulombic bound the estimate of $E_B = 0.5$ eV for the binding energy is plausible. With this value good agreement with corresponding migration energies reported in the literature is obtained[36]. In particular, for In-Cu the value $E_M = 0.2$ eV is obtained which agrees well with the corrected diffusion energy for Cu in Si of 0.15 eV, mentioned above.

Of especial interest are the results by Reislöhner and Witthuhn[43] who reported the formation of In-Fe and In-Cr pairs which are also visible in ESR experiments[44,45]. The fact, that for two by PAC observed In-Fe pairs the symmetry axis of the EFG is not along a <111> direction would be in agreement with the <100> orientation observed by ESR for the In-Fe complexes. In addition, Chantre and Kimmerling[46] concluded from DLTS experiments that the formation of acceptor-Fe pairs should be charge controlled leading to bistability effects which might become observable in future PAC experiments.

4. Intrinsic Defects

The identification of complexes formed between radioactive probe atoms and intrinsic defects is much more complicated than of the complexes discussed above, in spite of the well defined, unique EFGs which characterize each of the complexes listed in Table III. The reasons are: The lack of reliable calculations of EFGs created by defect complexes in semiconductors and the difficulty of creating a single type of intrinsic defect.

For the three In-defect complexes observed by PAC in Si[2,47,48], Fig. 11 shows their formation probabilities as a function of annealing temperature T_A. After implanting 350 keV ^{111}In$^+$ ions into n-type Si at 295 K, at 400 K the formation of In-De1 is observed which reaches its maximum just below 600 K and, at higher temperatures, is followed by a new complex, In-De2, that is stable up to 1000 K. At the same time, the fraction of In atoms on unperturbed lattice sites (crosses in Fig. 11) which is identified via its zero EFG steadily increases to 100 % what indicates the incorporation of the implanted In atoms as electrically active dopants. When the completely annealed sample was post-implanted with Si$^+$ or P$^+$ ions at 100 K the complex In-De3 was formed around 250 K; the formation of this complex is more pronounced after irradiation with MeV electrons or protons. The behavior of the three

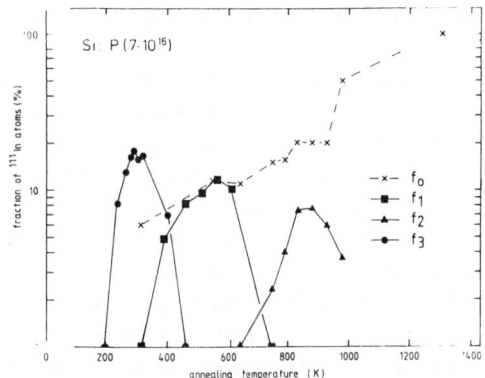

Figure 11. Dependences of different ^{111}In-defect fractions on the annealing temperature which were created by ^{111}In$^+$ and P$^+$ or Si$^+$ implantation. Shown are isolated In atoms (f_o) and the complexes In-De1 to In-De3 (f_1 to f_3).

complexes as a function of annealing temperature in Fig. 11 suggest a formation of successively larger defects at the In site. On this basis, In-De3 has been assumed to contain the smallest number of defects and has been identified as a close In-vacancy pair what is supported by its easy production through electrons and the measured <111> orientation of the complex. Although a convincing interpretation of the other two complexes is difficult to offer, the unique EFGs of all complexes prevent that they are mixed up with the other complexes, which e.g. are listed in Tables I and II. Thus, the two EFGs characterizing the hydrogen complexes In-H1 and In-H2 were originally assigned to complexes formed between In and an intrinsic defect[49]; on the basis of the experiments shown in Fig. 1, however, this assignment could be corrected[6].

For Ge, the EFGs of two types of In-defect complexes are known.[50,51,52,53]. In-De1 is locally created through the recoil of a neutrino which originates from the β-decay of the isotope ^{111}Sn to ^{111}In and transfers a recoil energy of 29 eV to the ^{111}In probe atom.[50] As a consequence, an isolated Frenkel pair can be created either with or in the next vicinity of the ^{111}In probe. The complex has been assigned to an In-vacancy pair created through the <111> displacement of the In atom from its substitutional site.[51,52] The corresponding EFG is exclusively observed in p-type Ge. The complex In-De2 was created by electron irradiation[51] and is similar in its properties to complex In-De3 in Si. Therefore, it has been also assigned to a close In-vacancy pair oriented along a <111> lattice direction[51]. Since the EFG is exclusively observed in n-type Ge it is assumed that both, In-De1 and In-De2, represent an In-vacancy pair which can occur in two different charge states depending on the position of the Fermi level and, in turn, gives rise to the two different observed EFGs. - Also listed in Table III are In-defect pairs found in several II-VI compounds. As discussed at this

Table III. Pairing between the dopant atom ^{111}In/Cd and intrinsic lattice defects.

COMPLEX	v_Q(MHz)	T (K)	η	<ijk>	E_D (eV)	Ref.
Silicon						
In-De1	28.	295	0.	<111>	2.0	2, 47
In-De2	142.	295	0.42		3.0	2, 47
In-De3	451.	78	0.	<111>	1.3	2, 48
Germanium						
In-De1	54.0	78	0.	<111>	1.3	50, 51
In-De2	420.	78	0.	<111>	1.1	50, 52
ZnS						
In-V	81.5	295	0.16			54
CdS						
In-V1	72.4	295	0.35			54, 55
In-V2	78.7	295	0.21			54, 55
CdTe						
In-V	60.0	295	0.10			54, 56 57

conference[54], the trapped defects are identified a the vacancy of the metallic (group II) sublattice, forming with the donor In so-called A centers in ZnS[54], CdS[54,55], and in CdTe[54,56,57].

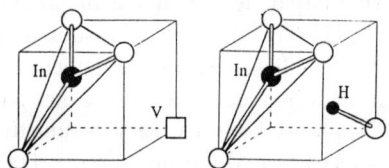

Figure 12. Microscopic models of (left) In-vacancy pairs In-De3 (Si) and In-De2 (Ge) and (right) In-hydrogen pairs In-H2 (Si) and In-H (Ge), emphasizing the defect induced relaxation of the acceptor atom In.

Whereas H and the metallic interstitial atoms in Si always create three different EFGs at the site of the probe ^{111}In/Cd this is not the case for the intrinsic defects in Si. In this respect, the intrinsic defects behave similar the substitutional donor atoms P, As, Sb, and Bi trapped by In acceptor atoms; also their pairing is not characterized by three different EFGs[58]. A second point to note is the close similarity of the EFGs of the In-vacancy defect and of the In-H complex in Si and in Ge. Besides the identical symmetries and orientations of the EFGs, in Si the corresponding coupling constants are 451 MHz (In-De3) and 480 MHz (In-H1); in Ge they are still closer with 420 MHz (In-De2) and 436 MHz (In-H). Thus, for Ge, where at 295 K the frequencies of both defect complexes are identical within 2 MHz these similarities led to some temporary confusion which only through PAC experiments under varying sample conditions were resolved[59,17]. In addition to the similar EFGs, which arise from locally similar electric charge distributions about the probe atom, also the observed thermal stabilities of the In-H and the In-vacancy pair are almost identical for each semiconductor. An explanation might be that a H atom and a vacancy exert a similar effect on the neighboring acceptor atom (see Fig. 12): The H atom trapped at the bond center site just saturates one covalent bond of the neighboring Si or Ge atom thereby leaving the In acceptor only threefold coordinated[42,22]; similarly, the trapped vacancy leaves the acceptor threefold coordinated because of the missing lattice atom. The resulting relaxation and the threefold coordination of the acceptor atom occurring for both the trapped H or vacancy seem to be responsible for the very similar EFGs and dissociation energies of these two types of complexes in Si and in Ge. - Of course, there are many results on intrinsic defects in semiconductors obtained by Mössbauer spectroscopy[60,3] which have not been discussed at this place.

5. Dynamic Effects

Whenever changes in the local geometrical or electronic structure of the probe atom occur during the lifetime of the isomeric state of the daughter nucleus which is in the range of 100 ns the rate of change of these dynamic effects can be determined.

Diffusion processes can represent one possible reason for such a dynamic process as shown for Fe in Si where in-beam Mössbauer spectroscopy succeeded in determining the diffusion of interstitial Fe_i atoms between 500 K and 700 K.[61] In these experiments, the 140 ns isomeric state of the daughter nucleus ^{57}Fe is populated by Coulomb excitation instead of by the β-decay of the parent ^{57}Co nucleus and the excited ^{57}Fe ion is recoil implanted into n-Si. In this way, which requires the use of a heavy ion-accelerator but manages with concentrations of 10^{15} Fe cm^{-3} the problem of the low solubility of Fe in Si has been taken into account.

The Fe_i atoms are characterized by the isomer shift IS = -0.84 mm/s and the corresponding resonance line is visible in the right part of the first spectrum shown in Fig. 13, which is

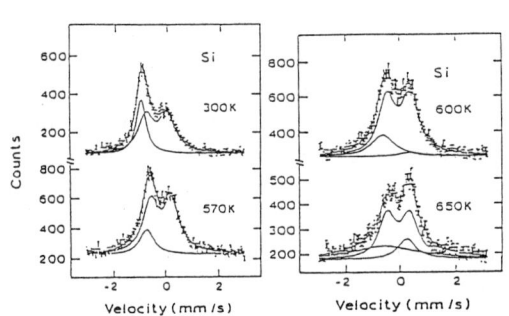
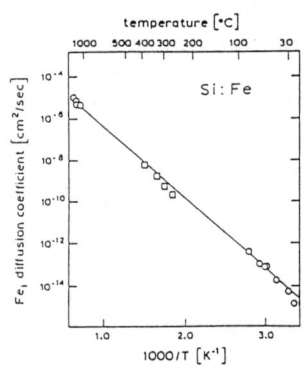

Figure 13. Mössbauer spectra recorded at temperatures between 300 K and 680 K using Coulomb excited ^{57}Fe that is implanted into Si.

Figure 14. Diffusion coefficients of Fe in Si. Squares: data extracted from the Mössbauer resonance line at IS = -0.84 mm/s; circles: data from the literature.

recorded at a sample temperature of 300 K. With increasing temperature this resonance line starts to broaden what is a manifestation of the onset of diffusion of the Fe$_i$ atoms on the time scale of the observing nuclear probe atom. The line broadening $\Delta\Gamma$ is connected to the diffusion coefficient D via $\Delta\Gamma = 12 \cdot \hbar \cdot D \cdot l^{-2}$ where l is the elementary jump length. Taking l = 0.235 nm, the distance between two neighboring tetrahedral interstitial sites, and assuming the Arrhenius law $D = D_o \exp(-E_a/kT)$ for the temperature dependence of the diffusion coefficient the diffusion data plotted in Fig. 14 (squares) can be extracted from the Mössbauer spectra. The fact that the new data lie on the interpolating line that connects the already known low- and high-temperature diffusions data for Fe in Si (circles) justifies this interpolation proposed by Weber[36] and shows the same mechanism to govern the diffusion of Fe$_i$ atoms between 300 K and 1500 K. - Local diffusion of Cu[62] and H[63] atoms about the probe atom ^{111}In/Cd in Si was observed by PAC. Again, the corresponding dynamics that is associated with fluctuating EFGs characterizing the In-Cu or In-H pairs becomes observable as soon as the time for a single reorientation of the ^{111}Cd-Cu or ^{111}Cd-H pair becomes comparable to the nuclear lifetime of the probe's daughter nucleus ^{111}Cd.

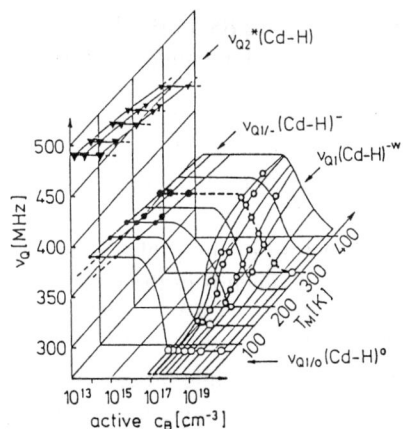

Figure 15. The EFGs of ^{111}In/Cd-H pairs in Si (expressed by their coupling constants ν_Q) as a function of the electrically active B concentration c_B and sample temperature T_M. Note, the coupling constants $\nu_{Q1/-}$, $\nu_{Q1/0}$, and ν_{Q2} correspond to In-H1, In-H3, and In-H2 in Table I, respectively.

An example where the respective EFGs characterizing a probe atom-defect pair depend on the electrical properties of the semiconductor is shown for ^{111}In/Cd-H pairs in Si which, as mentioned above, are characterized by three different EFGs. In Fig. 15. the value of the measured EFG or the coupling constant ν_Q is shown to depend on both the concentration of electrically active acceptors c_B and the temperature T_M of the p-Si sample. The data show that two of the three EFGs, In-H1 and In-H3 listed in Table I, actually represent two different charge states of the same ^{111}In/Cd-H pair. When the Fermi level is close to the valence band in p-type Si, i.e. at high c_B ot low T_M, the corresponding EFG observed at the daughter nucleus ^{111}Cd is $\nu_{Q1/0}$ (In-H3) and its value is determined by the neutral complex (Cd-H)^0 - note that Cd forms a double acceptor in Si. If c_B is decreased or T_M is increased the Fermi level moves towards the center of the band gap and the negative (Cd-H)$^-$ charge state is observed characterized by $\nu_{Q1/-}$ (In-H3). In between, the coupling constant is described by $\nu_{Q1} = w \cdot \nu_{Q1/-} + (1-w) \cdot \nu_{Q1/0}$, where w is the probability for the population of the negatively charged state which is determined by the position and sharpness of the Fermi distribution. When in the intermediate region of c_B and T_A the charge fluctuation happens within the nuclear time window relaxation effects are observable in the PAC spectra. A similar effect is reported to occur for pairs in Si formed between the probe ^{111}In/Cd and substitutional donor atoms, like P or As.[64]

Independent of the coupling constant ν_{Q1} whose actual value is controlled by the charge state of the Cd-H complex the presence of the third EFG, characterized by ν_{Q2} (In-H2 in Table I), shows that a second type of ^{111}In/Cd-H pair exists which was assigned to a structurally different configuration of Cd-H. The population of ν_{Q2} is qualitatively represented by the size of the closed triangles in Fig. 15. This state is observed at low acceptor concentrations c_B and temperatures T_M and its population competes with that of $\nu_{Q1/-}$. Here, a structural transition, probably already taking place at the ^{111}In-H complex and marked by $\nu_{Q1/-}$ and ν_{Q2} occurs when the hole concentration in the region of the probe atom decreases. This interpretation has been confirmed by additional PAC experiments that were performed in the space charge region of a Ti/Si Schottky contact which allows to control the hole concentration without a change of dopant concentration or temperature of the sample.[23]

6. Concluding Remarks

The presented examples should have demonstrated how a chemical identification of defects in semiconductors is possible if radioactive probe atoms are used that are suited for PAC or Mössbauer experiments. It should be pointed out that radioactive atoms are also used in combination with other techniques which, in part, are represented in this conference. Among them, there are two techniques which should be mentioned: The use of radioactive atoms allows to perform lattice location experiments via the blocking/channeling effect for charged particles at much lower defect concentrations than it is possible in conventional ion channeling experiments.[65] This possibility meets a requirement that is of importance in many defect studies in semiconductors. Second, the measurement of DLTS at samples doped with radioactive instead of with stable impurities allows to combine the high electrical sensitivity of DLTS with the possibility of a chemical identification of the observed electronic levels.[66,67,68] The intensity of the DLTS signal that is related to the radioactive impurity atoms will decay or grow according to the nuclear lifetime of the respective impurity atom.

In summary, nuclear techniques will rarely be suited for systematic studies of defects in semiconductors because of their special requirements; rather, they are a tool for exemplary investigations in prudently selected systems whose results can be transferred to similar systems, as well. If applicable, nuclear techniques deliver information on defects that are often, if at all, difficult to obtain by other techniques.

Acknowledgements

It is a great pleasure to thank M. Deicher, R. Keller, Th. Krings, R. Magerle, W. Pfeiffer, E. Recknagel, H. Skudlik, M.L. Swanson and H. Wolf for their experimental support and clarifying discussions and, in particular, Jörg Weber for his help and advice during the first stages of the PAC experiments in semiconductors. This work has been financially supported by the Deutsche Forschungsgemeinschaft and the Bundesminister für Forschung und Technologie.

References

1. G. Langouche, *Submicroscopic Investigation of Defects in Semiconductors* (Elsevier, Amsterdam, 1991) in press.
2. Th. Wichert, M. Deicher, G. Grübel, R. Keller, N. Schulz and H. Skudlik, Appl. Phys. A 48, 59 (1989).
3. A. Nylandstedt Larsen, J. W. Petersen and G. Weyer, Mater. Sci. Forum 38-41, 1137 (1989).
4. C.-T. Sah, J. Y.-C. Sun and J. J.-T. Tzou, Appl. Phys. Lett. 43, 204 (1983).
5. J.I. Pankove, D.E. Carlson, J.E. Berkeyheiser and R.O. Wance, Phys. Rev. Lett. 51, 2224 (1983).
6. Th. Wichert, H. Skudlik, M. Deicher, G. Grübel, R. Keller, E. Recknagel and L. Song, Phys. Rev. Lett. 59, 2087 (1987).
7. H. Skudlik, M. Deicher, R. Keller, R. Magerle, W. Pfeiffer, P. Pross, E. Recknagel and Th. Wichert, Phys. Rev. B, submitted.
8. A.J. Tavendale, D. Alexiev and A.A. Williams, Appl. Phys. Lett. 47, 316 (1985).
9. R. Keller, M. Deicher, W. Pfeiffer, H. Skudlik, D. Steiner and Th. Wichert, *Identification of Process-Induced Defects in Silicon by PAC*, in: *Defect Control in Semiconductors*, ed. K. Sumino (North-Holland, Amsterdam, 1990) p. 377.
10. C.H. Seager, R.A. Anderson and J.K.G. Panitz, J. Mater Res. 2, 96 (1987).
11. H. Skudlik, M. Deicher, R. Keller, W. Pfeiffer, D. Steiner and Th. Wichert, *Stability of Acceptor-Hydrogen Complexes in Semiconductors*, in: *Defect Control in Semiconductors*, ed. K. Sumino (North-Holland, Amsterdam, 1990) p. 413.
12. S.J. Pearton, J.W. Corbett and T.S. Shi, Appl. Phys. A 43, 153 (1987).
13. M. Stavola, J.S. Pearton, J. Lopata and W.C. Dautremont-Smith, Phys. Rev. B 37, 8313 (1988).
14. T. Zundel and J. Weber, Phys. Rev. B 39, 13549 (1989).
15. J. Weber and T. Zundel, this conference.
16. H. Skudlik, M. Deicher, R. Keller, R. M. W Pfeiffer, P. Pross and Th. Wichert, Nucl. Instrum. & Methods B, in press.
17. M. Deicher, R. Keller, W. Pfeiffer, H. Skudlik, D. Steiner, E. Recknagel and Th. Wichert, Mater. Sci. Engineer. B4, 25 (1989).
18. A. Baurichter, M. Deicher, S. Deubler, D. Forkel, H. Plank, H. Wolf and W. Witthuhn, Appl. Phys. Lett. 55, 2301 (1989).
19. W. Pfeiffer, M. Deicher, R. Keller, R. Magerle, E. Recknagel, H. Skudlik, Th. Wichert, H. Wolf, D. Forkel, N. Moriya and R. Kalish, Appl. Phys. Lett. 16, 1751 (1991).
20. M. Deicher, Nucl. Instrum. & Methods B, in press.
21. A. Baurichter, M. Deicher, S. Deubler, D. Forkel, J. Meier, H. Wolf and W. Witthuhn, Appl. Surf. Sci. 50, 165 (1991).
22. S.J. Pearton, M. Stavola and J.W. Corbett, Mater. Sci. Forum 38-41, 25 (1989).
23. H. Skudlik, M. Deicher, R. Keller, R. Magerle, W. Pfeiffer, D. Steiner, E. Recknagel and Th. Wichert, Phys. Rev. B, submitted.
24. A.D. Marwick, G.S. Oehrlein, J.H. Barret and N.M.Johnson, Mat. Res. Soc. Symp. Proc. 104, 259 (1988).
25. B. Bech Nielsen and J. U. Andersen, and S. J. Pearton Phys. Rev. Lett. 60, 321 (1988).
26. Th. Wichert, H. Skudlik, H.D. Carstanjen, T. Enders, M. Deicher, G. Grübel, R. Keller, L. Song and M. Stutzmann, Mat. Res. Soc. Symp. Proc. 104, 265 (1988).
27. N.M. Johnson, C. Herring and D.J. Chadi, Phys. Rev. Lett. 56, 769 (1986).
28. Z.N. Liang and L. Niesen, Hyperfine Interactions 60, 749 (1990).
29. Z.N. Liang and L. Niesen, Nucl. Instrum. & Methods B, in press.
30. K. Bergman, M. Stavola, S.J. Pearton and T. Hayes, Phys. Rev. B 38, 9643 (1988).

31. A. Schnegg, H. Prigge, M. Grundner, P.O. Hahn and H. Jacob, Mat. Res. Soc. Symp. Proc. 104, 291 (1988).
32. M. Deicher, G. Grübel R. Keller, E. Recknagel, N. Schulz, H. Skudlik, Th. Wichert, H. Prigge and A. Schnegg, Inst. Phys. Conf. Ser. 95, 155 (1988).
33. Th. Wichert, R. Keller, M. Deicher, W. Pfeiffer, H. Skudlik and D. Steiner, Mat. Res. Soc. Symp. Proc. 163, 245 (1990).
34. R. Keller, M. Deicher, W. Pfeiffer, H. Skudlik, D. Steiner, and Th. Wichert Phys. Rev. Lett. 65, 2023 (1990).
35. T. Zundel, J. Weber, B. Benson, P.O. Hahn, A. Schnegg and H. Prigge, Appl. Phys. Lett. 53, 1426 (1988).
36. E. R. Weber, Appl. Phys. A 30, 1 (1983).
37. L.T. Canham, M.R. Dyball and K.G. Barraclough, J. Appl. Phys. 66, 920 (1989).
38. T. Prescha, T. Zundel and J. Weber, Mater. Sci. Engineer. B4, 79 (1989).
39. R.N. Hall and J.H. Racette, J. Appl. Phys. 35, 379 (1964).
40. M. Cardona, S.C. Shen and S.P. Varma, Phys. Rev. B 23, 5329 (1981).
41. S.K. Estreicher, Phys. Rev. B 41, 5447 (1990).
42. P.J.H. Denteneer, C.G. Van de Walle and S.T. Pantelides, Phys. Rev. Lett. 62, 1884 (1989).
43. U. Reislöhner and W. Witthuhn, Mater. Sci. Forum 65-66, 281 (1990).
44. W. Gelhoff, P. Emanuelson, P. Omling and H.G. Grimmeiss, Phys. Rev. B 41, 8560 (1990).
45. P. Emanuelsson, P. Omlin, H.G. Grimmeiss, W. Gelhoff and J. Kreissl, this conference.
46. A. Chantre and L.C. Kimmerling, Mater. Sci. Forum 10-12, 387 (1986).
47. D. Forkel, F. Meyer, W. Witthuhn and H. Wolf, Hyperfine Interactions 35, 715 (1987).
48. D. Forkel, S. Deubler, U. Reislöhner, K. Spörl, W. Witthuhn and H. Wolf, Mater. Sci. Forum 38-41, 1251 (1989).
49. G.J. Kemerick and F. Pleiter, Phys. Lett. 121, 367 (1987).
50. M. Brüssler, H. Metzner and R. Sielemann, Mater. Sci. Forum 38-41, 1205 (1989).
51. U. Feuser, R. Vianden and A.F. Pasquevich, Hyperfine Interactions 60, 829 (1990).
52. R. Sielemann, *Perturbed Angular Correlation*, in: *Submicroscopic Investigation of Defects in Semiconductors*, ed. G. Langouche, Chapt. 2.5 (Elsevier, Amsterdam, 1991).
53. R. Sielemann, H. Hässlein, M. Brüssler and H. Metzner, this conference.
54. H. Wolf, T. Krings, U. Ott, U. Hornauser and Th. Wichert, this conference.
55. R. Magerle, M. Deicher, U. Desnica, R. Keller, W. Pfeiffer, F. Pleiter, H. Skudlik and Th. Wichert, Appl. Surface Science, in print (1991).
56. R. Kalish, M. Deicher and G. Schatz, J. Appl. Phys. 53, 4793 (1982).
57. D. Wegner and E.A. Meyer, J. Phys.: Condensed Matter 1, 5403 (1989).
58. Th. Wichert and M.L. Swanson, J. Appl. Phys. 66, 3026 (1989).
59. M. Deicher, R. Keller, W. Pfeiffer, H. Skudlik and Th. Wichert, Physica B 170, 335 (1991).
60. G. Langouche, *Characterization of semiconductors by Mössbauer spectroscopy*, in: *Mössbauer spectroscopy applied to inorganic chemistry*, eds. G. J. Long and F. Grandjean, Vol. 3, Chapt. 10 (Plenum Publishing Corporation, 1989) p. 445.
61. P. Schwalbach, S. Laubach, M. Hartick, E. Kankeleit, B. Keck, M. Menningen and R. Sielemann, Phys. Rev. Lett. 64, 1274 (1990).
62. R. Keller, M. Deicher, R. Magerle, W. Pfeiffer, H. Skudlik and Th. Wichert, Nucl. Instrum. & Methods B, in press.
63. M. Gebhard, B. Vogt and W. Witthuhn, this conference.
64. N. Achtziger, *Perturbed Angular Correlation*, in: *Submicroscopic Investigation of Defects in Semiconductors*, ed. G. Langouche, Chapt. 2.3 (Elsevier, Amsterdam, 1991).
65. H. Hofsäss and G. Lindner, Physics Report 201, 121 (1991).
66. M. Lang, G. Pensl, M. Gebhard, N. Achtziger and M. Uhrmacher, Appl. Phys. A 53, 95 (1991).
67. M. Lang, G. Pensl, M. Gebhard, N. Achtziger and M. Uhrmacher, this conference.
68. J. W. Petersen and J. Nielsen, this conference.

COMBINATION OF DEEP LEVEL TRANSIENT SPECTROSCOPY AND TRANSMUTATION OF RADIOACTIVE IMPURITIES

MANFRED LANG[1], GERHARD PENSL[1], MARION GEBHARD[2], NORBERT ACHTZIGER[2] AND MICHAEL UHRMACHER[3]
[1] Institut für Angewandte Physik, Universität Erlangen-Nürnberg
 Staudtstraße 7, D-8520 Erlangen, Germany
[2] Physikalisches Institut, Universität Erlangen-Nürnberg,
 Erwin-Rommel-Straße 1, D-8520 Erlangen, Germany
[3] II. Physikalisches Institut, Universität Göttingen,
 Bunsenstraße 7-9, D-3400 Göttingen, Germany

ABSTRACT

Deep level transient spectroscopy (DLTS) is used to sensitively probe deep centres containing one or more radioactive isotopes. The number of the particular probe atoms that participate in the centres observed is determined from the decay and formation laws of the corresponding defect concentrations. The method is applied to silicon samples implanted with radioactive ^{111}In*; one of the samples is further doped with iron by means of diffusion. The indium ions decay into stable cadmium ions with mean lifetime τ of 4.08 d. We have identified a series of Cd-related centres whose concentrations exponentially increase with τ. The growth of Cd-related centres considerable deviates from the expected growth laws in samples that contain interstitial iron atoms (Fe_i). The deviations are due to slowly diffusing Fe_i atoms which form CdFe pairs; the level observed at E_v+485 meV is attributed to these pairs.

1. Introduction

The application of deep level transient spectroscopy (DLTS) to defect centres that contain radioactive isotopes combines the advantage of high sensitivity to low concentrations with the opportunity also to identify the chemical nature and number of particular probe atoms that make up the observed centres [1,2,3]. The transmutation of unstable parent nuclei into their stable daughter isotopes leads to a decrease or increase of specific defect centres. If these centres are observable in the DLTS spectra, an analysis of their peak heights (corresponding to their defect concentrations), determined at various decay times, clearly reveals the number of parent or daughter nuclei that are incorporated.

In this paper, we give the different decay and formation laws for centres containing one or two radioactive nuclei and demonstrate the schemes for silicon samples that are implanted with radioactive ^{111}In* ions. ^{111}In* decays into stable ^{111}Cd with a mean lifetime τ of 4.08d. In addition, we have diffused iron into one ^{111}In*-implanted Si sample. The advantage of our scheme is that all the incorporated impurities can be detected by DLTS and that the value of the In mean lifetime allows the observation of defect concentrations in a wide range within a reasonable period. We have identified the singly and doubly charged state of the Cd double acceptor, two Cd-related complexes containing one Cd atom each and, finally, a donor-like complex that is attributed to CdFe pairs.

2. Deep level transient spectroscopy by means of radioactive impurities (DLTS-RI)

The DLTS-RI method is based on the observation of electrically active defects which contain one or more radioactive nuclei. The radioactive parent nuclei decay into stable daughter isotopes according to the mean lifetime τ. This transmutation may be accompanied by a decrease or an increase of a certain number of electrically active defects. Their concentrations are monitored by DLTS as a function of the decay time. Fitting the DLTS data with the growth laws appropriate for one or two probe atoms, we are able to determine the size of the complex. In the following, we show the decay and formation laws for defect centres containing one or two radioactive isotopes.

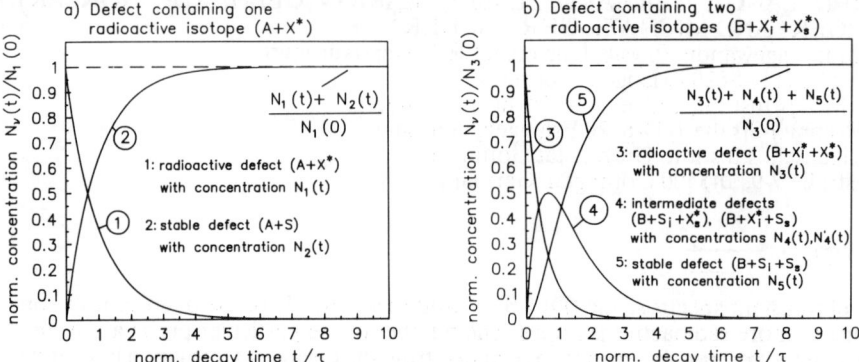

Fig. 1. The calculated curves show the normalized defect concentration $N_v(t)/N_{1/3}(o)$ vs. normalized decay time t/τ. $N_1(o)$ and $N_3(o)$ are the starting concentrations of radioactive defects at decay time t=0, τ is the mean lifetime of the parent isotope, and the subscript v runs from 1 to 5. Curves 1,2,3,4 and 5 are calculated with Eq. (1), (2), (3), (4), and (5), respectively.

The curves in Fig. 1a are calculated for the case that only one probe atom X^* participates in the defect centre; generally, this centre may also contain an additional stable partner A. The decay of defect $(A+X^*)$ according to

$$(A+X^*) \xrightarrow{\tau} (A+S)$$

is given by an exponential law

$$N_1(t)/N_1(o) = \exp(-t/\tau) \qquad (1)$$

where S is the stable daughter nucleus.
The increase of the stable defect (A+S) is described by

$$N_2(t) = N_1(o)[1 - \exp(-t/\tau)] \qquad (2)$$

Fig. 1b considers the case that two probe atoms X^*_i, X^*_s and possibly a further stable partner B form a defect complex. X^*_i and X^*_s are chemically identical. The subscripts i and s indicate that they may differ in any of their physical properties, e.g. may reside on crystallographically inequivalent lattice sites. The stable complex $(B + S_i + S_s)$ is formed via the intermediate states $(B + S_i + X^*_s)$ and $(B + X^*_i + S_s)$ as indicated in the following decay scheme

$$2(B + X^*_i + X^*_s) \xrightarrow{\tau/2} (B + S_i + X^*_s) + (B + X^*_i + S_s) \xrightarrow{\tau} 2(B + S_i + S_s)$$

Taking into account that the radioactive decay of the two probe atoms is statistically independent and assuming that the intermediate states are not distinguishable, the foregoing decay scheme is governed by the following decay and formation equations which are represented in Fig. 1b (see curves 3,4, and 5).

$$N_3(t) = N_3(o) \cdot e^{-2t/\tau} \qquad (3)$$

$$[N_4(t) + N'_4(t)] = N_3(o) \cdot 2e^{-t/\tau}(1 - e^{-t/\tau}) \qquad (4)$$

$$N_5(t) = N_3(o)(1 - e^{-t/\tau})^2 \qquad (5)$$

In the present paper, the DLTS-RI method is used to investigate defects in silicon which are introduced by implantation of radioactive ^{111}In*. This isotope decays into stable ^{111}Cd with a mean lifetime τ of 4.08d.

$$^{111}\text{In}^* + e^- \xrightarrow{4.08d} {}^{111}\text{Cd} + \gamma_1 + \gamma_2 + \nu_e \qquad (6)$$

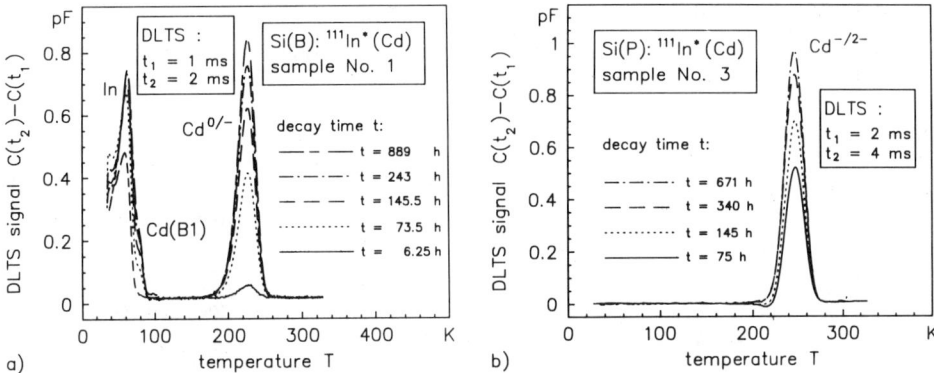

Fig.2. DLTS spectra of ^{111}In*-implanted Si samples taken at various decay times. a) B-doped sample No. 1, b) P-doped sample No. 3

where e⁻ is the captured electron and γ_1, γ_2, and ν_e are the two emitted γ-quanta and the emitted neutrino, respectively.

Both In and Cd form deep acceptors in silicon which can be probed by DLTS. The system Si:In has been thoroughly investigated [4] whereas the data published on the Cd acceptor in Si provide an ambiguous view (see [3]).

3. Experimental

For this investigation, we employ Cz-grown n- and p-type Si wafers supplied by Wacker Chemitronic. The data relating to the starting material and the processing steps are summarized in Table I. The implanted ^{111}In* dose is judged to have an uncertainty of ± 50%. Gold/titanium Schottky contacts (⌀=0.7 mm) are fabricated on the front side of the n/p-type Si samples; large area ohmic contacts (Al) are evaporated on the back side. The Fe diffusion is carried out in an evacuated quartz ampule. The DLTS equipment used is described in Ref. [5].

Table I. Data of the starting Si material and of the processing parameters of the investigated samples

Sample No.	Material	Processing parameters
1	Cz-Si(B), <111> [B]=6·10^{16}cm^{-3}	implantation: ^{111}In* energy: 400keV dose: ≈10^{11}cm^{-2} subsequent annealing: T=1000°C/t=90 min.
2	Cz-Si(Al) <111> [Al]=8·10^{16}cm^{-3}	
3	Cz-Si(P), <111> [P]=3·10^{16}cm^{-3}	
4	Cz-Si(B), <111> [B]=6·10^{16}cm^{-3}	implantation: ^{111}In* energy: 400keV dose: ≈10^{11}cm^{-2} subsequent annealing combined with Fe-diffusion: T=1020°C/T=30min rapidly cooled down second annealing after 400h: T=100°C/T=60min.

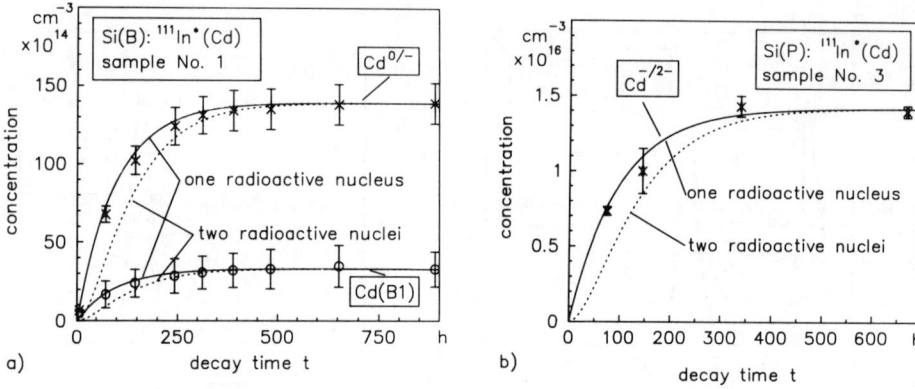

Fig.3. Concentration of stable defects vs. decay time of the ^{111}In* parent atoms. The solid curves are least-squares fits of Eq.(2) to the experimental data (crosses, circles). A time constant of (95±5)h is obtained from all three fits. The dashed curves are calculated with Eq. (5) using the same values of the saturation concentration and mean lifetime as for the calculation of the solid curves. a) B-doped sample No.1, b) P-doped sample No. 3.

4. DLTS results for ^{111}In*-implanted samples

Figs. 2a, b show DLTS spectra of the B-doped sample No. 1 and the P-doped sample No.3 which were both subjected to an ^{111}In* implantation of comparable dose. The spectra are taken at various decay times, whereby the zero point of the indium decay time is set at the end of the annealing process since the heat treatment at around 1000°C results in an out-diffusion of those Cd atoms that have already been produced. The DLTS spectra of the B-doped sample (see Fig. 2a) display three peaks. The peak at approx. 70K decreases with increasing decay time; it is attributed to the substitutional In acceptor (for detailed discussion, see [3,4]). The Cd(B1) peak at T=78K overlaps the In peak on the high temperature side and a third peak (Cd$^{0/-}$) is observed at T=200K. The Cd(B1) and Cd$^{0/-}$ peak heights increase with increasing decay time. The DLTS spectrum of the Al-doped sample No. 2 (not shown) also displays the In and Cd$^{0/-}$ level and a third peak at T=78K, termed Cd(Al1), that corresponds to Cd(B1) in the B-doped sample No.1. The DLTS spectrum of the P-doped sample No.3 (see Fig. 2b) shows one sharp peak (Cd$^{-/2-}$) at T=240K that strongly increases with the decay time.

Table II. DLTS results for ^{111}In*-implanted samples No. 1,2,3 and for the ^{111}In*-implanted and Fe-diffused sample No. 4

Defect centre	Sample No.	Mean peak temperature <T> [K]	Mean electric field <F> 10^5 [V/cm]	Energy level E^a[meV]	Energy level E^b[meV]	Capture cross section σ^a [cm^2]
Cd$^{0/-}$	1,4	216	7.5	E_v+455^c	E_v+510^c	4·10^{-15}
Cd$^{-/2-}$	3	240	5.5	E_c-430	E_c-470	5·10^{-16}
Cd(B1)	1,4	78	8.0	E_v+190^c	E_v+210^c	7·10^{-14}
Cd(Al1)	2	78	8.5	E_v+195^c	E_v+215^c	4·10^{-14}
CdFe	4	260	0.5	E_v+485	E_v+530	1·10^{-14}

^{a} Evaluated for temperature-independent capture cross-section σ=const.
^{b} Evaluated for temperature-dependent capture cross-section σ=T^{-2}
^{c} Corrected for the Poole-Frenkel effect at <F> given in column 4 (see [3])

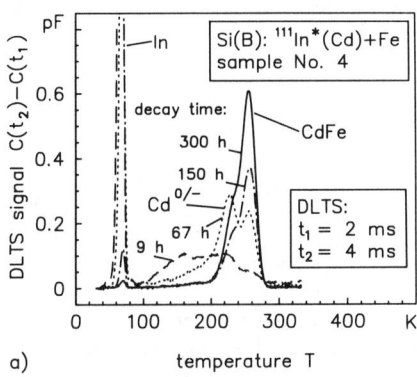
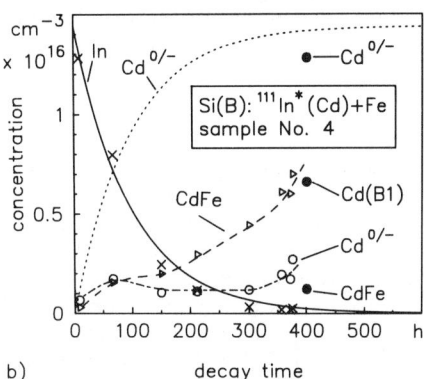

Fig. 4.a) DLTS spectra of the ^{111}In*-implanted and Fe-diffused Si sample No.4 taken at various decay times. b) Concentration of stable defects v. decay time; the concentrations are determined from the corresponding peak heights in Fig.4a). The solid curve is a fit of Eq.(1) to the measured ^{111}In* concentrations. The dotted curve is calculated with Eq.(2) using $N_1(o)=1.45 \cdot 10^{16}$cm^{-3} and $\tau=98$h. The full circles are measured after a decay time of 400h and a second anneal was performed at 100°C for 60 min.

Ionization energies ΔE_i and capture cross sections σ of the observed centres are derived from an Arrhenius plot $\ln (\tau_{p/e} / 1\mu s)$ vs. T^{-1}, where $\tau_{p/e}$ is the hole/electron emission time constant.
Since the temperature dependence $\sigma(T)$ is unknown for these centres, we define a range in which the ionization energies are likely to be; we do this by assuming two possible cases: σ=const. and $\sigma \approx T^{-2}$ (see Table II, column 5). In order to obtain the zero-field ionization energies, the emission time constants $\tau_{p/e}(T)$ (available from the DLTS measurements) have to be corrected for the Poole-Frenkel effect [3,6] that takes into account the enhancement of the emission rate from an attractive coulomb center in the presence of a high electric field. The resulting energy levels and capture cross sections are summarized in Table. II (column 5 and 6).
In Figs. 3a,b, the defect concentrations of centres Cd$^{0/-}$, Cd(B1), and Cd$^{-/2-}$ determined from the DLTS peak heights are depicted as function of the indium decay time (see crosses and circles). The solid curves are least squares fits of Eq. (2) to the experimental data. $N_1(o)$ is given by the saturation concentrations of the various centres at decay times >600h. All three solid curves result in the same time constant $\tau=(95\pm5)$h; this value agrees within the error bar with the value of the mean lifetime of the ^{111}In* probe atoms ($\tau=98$h). The same value of τ is obtained for the formation of the Cd(Al1) concentrations. The agreement of the two time constants clearly demonstrates that one Cd atom is incorporated in each of the centres (Cd$^{0/-}$, Cd$^{-/2-}$, Cd(B1), Cd(Al1)). For comparison, we have calculated the defect formation assuming that two probe atoms are incorporated in each centre (see dashed curves in Figs. 3a,b). The calculation is based on Eq.(5) using the same parameters as for the solid curves. The dashed curves deviate appreciably from the experimental data; moreover, an intermediate state given by Eq. (4) cannot be observed. The following experimental results strongly support the supposition that the Cd$^{0/-}$ and Cd$^{-/2-}$ levels belong to the singly and doubly charged state of the isolated Cd double acceptor:
(1) from the fit (Eq.(2)) to the Cd$^{0/-}$/Cd$^{-/2-}$ concentrations, it is known that these centres contain one Cd atom each,
(2) the measured concentrations ($\sim 10^{16}$cm^{-3}) are the same for both centres (at same implantation dose),
(3) the analysis of the Poole-Frenkel effect results in acceptor-like states and predict the charge states o/- and -/2- for the two levels.

We propose that the isolated Cd double acceptor generated by transmutation of radioactive In atoms resides on a lattice site like the In acceptor itself [4]. The maximum energy of 1eV that can be transferred during the ^{111}In* decay (Eq.(6)) is much smaller than the displacement energy of approx. 20eV necessary to kick a Si atom out of its lattice site. At the present time, it is not clear whether centres Cd(B1) and Cd(A11) are identical or consist of different partners. Both centres contain one Cd atom. They act as acceptors and their ionization energies are equal within our error bar of ± 10 meV.

5. DLTS results for the ^{111}In*-implanted and Fe-diffused sample

DLTS spectra of sample No.4 are shown in Fig.4a; the spectra are taken at various In decay times. The ^{111}In* level at T=70K rapidly decreases with decay time. At short times (t=9h), a broad peak around 180K appears in the DLTS spectrum. After a decay time of 67h, primarily two levels grow out of the broad peak, namely the Cd$^{0/-}$ and a so far unknown level that is attributed to a CdFe complex ($\Delta E_i = 507$ meV); its electrical behavior is donor-like. The CdFe complex becomes dominant in the DLTS spectrum with increasing decay time. Fig. 4b depicts the measured defect concentrations as a function of the decay time. The ^{111}In* level decays exponentially shown by the DLTS data (crosses) and by the solid curve calculated with Eq.(1) (parameters: $N_1(o) = 1.45 \cdot 10^{16} cm^{-3}$, $\tau = 98$h). By contrast to sample No.1, the Cd$^{0/-}$ level (open circles) does not show the expected exponential increase that is indicated by the dotted curve also labelled Cd$^{0/-}$; the Cd(B1) level is not at all observed. Instead the CdFe level appears with slowly increasing concentration. It is assumed, that interstitial Fe atoms slowly diffuse in the Si lattice at room temperature and form pairs with isolated Cd atoms; similar pairing has already been reported for acceptors of the third column of the periodic table [7]. The full circles in Fig. 4b denote Cd-related defects (Cd$^{0/-}$, Cd(B1), and CdFe) that are generated after a decay time of 400h and by a second annealing step. This heat treatment leads to a redistribution of Cd-related defects. The Cd$^{0/-}$ and Cd(B1) level now reach values comparable to those of sample No. 1 and the concentration of the CdFe level is strongly reduced.

6. Summary

Deep level transient spectroscopy using radioactive impurities (DLTS-RI) is a defect specific analysis method; it provides high sensitivity to low concentrations and additionally gives the opportunity to identify the chemical nature and number of particular probe atoms that participate in the observed defect centre. The DLTS-RI method is demonstrated for ^{111}In*-implanted Si samples. We have identified the singly and doubly charged state of the isolated Cd double acceptor, and two Cd-related complexes Cd(B1)/Cd(A11) that contain one Cd atom each. The ^{111}In*-implanted and Fe-diffused Si sample reveals that the growth kinetics of stable daughter defects may significantly deviate from the expected growth laws if additional reactions - like the slow diffusion of interstitial Fe - proceeds in parallel with the radioactive decay.

Acknowledgements: The authors would like to thank Prof.M.Schulz for his continued interest. Appreciation is also expressed to Prof.W.J.Choyke, University of Pittsburgh, for stimulating discussions and critical reading of the manuscript.

REFERENCES

1. M.Lang, G.Pensl, M.Gebhard, M.Uhrmacher: Verhandl. DPG (VI) 25, 824 (1990)
2. J.W.Petersen, J.Nielsen: Appl.Phys.Lett. 56, 1122 (1990)
3. M.Lang, G.Pensl, M.Gebhard, N.Achtziger, M.Uhrmacher: Appl.Phys. A53, 94 (1991)
4. R.Helbig: In Festkörperprobleme (Advances in Solid State Physics) Vol. XXV, ed.by P.Grosse (Vieweg, Braunschweig 1985) p. 573
5. K.Hölzlein, G.Pensl, M.Schulz, P.Stolz: Rev. Sci. Instrum. 57, 1373 (1986)
6. J.L.Hartke: J.Appl.Phys. 39, 4871 (1968)
7. A.Chantre, L.C.Kimerling: Defects in Semiconductors, ed. by H.J. von Bardeleben, Materials Science Forum Vol. 10-12 (1986), p. 387

IDENTIFICATION OF BAND GAP STATES IN SILICON BY DEEP LEVEL TRANSIENT SPECTROSCOPY ON RADIOACTIVE IMPURITIES

JON WULFF PETERSEN AND JACOB NIELSEN

Institute of Physics, University of Aarhus, DK-8000 Aarhus C, Denmark,
and the ISOLDE Collaboration, CERN, CH-1211 Geneva 23, Switzerland

ABSTRACT

The deep level transient spectroscopy technique has been applied to silicon doped with radioactive isotopes of the elements Au, Pt, Ir, and Os. The disappearance or appearance of features in the spectra, when the incorporated radioactive atoms transmute, identifies unambigously an impurity involved in the observed centers. A possible generalization of the method is discussed.

1. Introduction

The identification of defects giving rise to band gap levels is a major problem in semiconductor physics. The introduction of Deep Level Transient Spectroscopy (DLTS) led to drastically improved sensitivities for detecting such states [1], but did not solve the problem of identifying the constituents of the observed centres. Here it is demonstrated that in some cases this problem can be solved by combining the DLTS technique with the use of radioactive dopants. The cases of Au, Pt, Ir, and Os were chosen since there is considerable fundamental as well as applied interest in the 5d elements as intentional as well as unintentional dopants in silicon. Furthermore, strong and very clean ion beams of the Hg mother isotopes could be obtained at the ISOLDE on-line mass separator at CERN [2].

We argue at the end of the paper that the method can be generalized to solve other identification problems by combining the use of radioisotopes with other sensitive semiconductor techniques, like e.g. EPR and photoluminescence.

2. Experimental results and discussion.

Radioactive ^{195}Hg, ^{193}Hg, and ^{189}Hg isotopes were produced as spallation products from irradiations of a molten lead target with 600 MeV protons. The isotopes are delivered from the ISOLDE isotope separator [2] as mass-analyzed 60 keV ion beams with a fluence of roughly 10^9 ions/s. The ions were implanted into standard n- and p-type float zone <100> Si (from Wacker Chemitronic) with a resistivity of 0.5 - 1.0 Ω cm. It is of crucial importance to the feasability of the experiment that the radioactive beams were not contaminated with stable isotopes of the elements investigated. The samples were annealed at temperatures between 1050 and 1300 K to remove the radiation damage and diffuse the implanted isotopes into the substrate. Au, Pt, or Ti films were evaporated on to the surface of the samples to produce Schottky barriers. The DLTS measurements were performed with a commercial instrument (Semitrap).

The decay sequences for the three radioactive isotopes are as follows:

$$^{195}Hg \xrightarrow{10\ h} {}^{195}Au \xrightarrow{183\ d} {}^{195}Pt$$

$$^{193}Hg \xrightarrow{4\ h} {}^{193}Au \xrightarrow{17.5\ h} {}^{193}Pt \xrightarrow{50\ yr} {}^{193}Ir$$

$$^{189}Hg \xrightarrow{8\ m} {}^{189}Au \xrightarrow{29\ m} {}^{189}Pt \xrightarrow{10.9\ h} {}^{189}Ir \xrightarrow{13\ d} {}^{189}Os$$

Fig.1 shows a temperature scan of a ^{189}Ir diffused n-Si sample. The spectrum is dominated by two peaks of equal intensities. The low temperature peak has an activation energy of 0.27 eV and a majority capture cross section of $2.6\ 10^{-14}\ cm^2$.

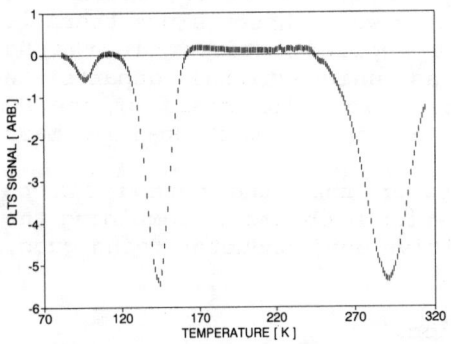
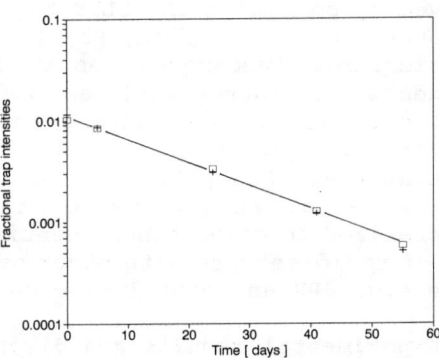

Fig.1. DLTS temperature scan from a ^{189}Ir containing n-Si sample. The repetition frequency was 250 Hz.

Fig.2. Time evolution of the .27 eV (□) and the .50 eV (+) peak intensities. The results are normalized to the time zero value of the .50 eV peak.

The high temperature peak has an activation energy of 0.51 eV and a cross section of 6.7×10^{-15} cm^2. As the ^{189}Ir isotopes transmute into ^{189}Os isotopes, the intensities of the two main peaks in Fig.1 disappear, but no new features appear in the spectrum. The disappearance of the two peaks as a function of time is shown in Fig.2. The decay constant closely matches that of the radiactive decay of ^{189}Ir.

The results demonstrate unambigously that Ir is involved in the centre(s) giving rise to these signals. The fact, that the intensities of the two components - although varying in different samples and diodes by orders of magnitude - were always equal, is a strong indication that they originate from the same center. They might be the signature of two charge states of the substitutional Ir impurity. We shall return to this point shortly. The fact that no new features show up in the spectra, when the ^{189}Ir isotopes decay into ^{189}Os strongly suggests that Os in this lattice position does not produce any levels, which act as electron traps, in the "visible" part of the upper half of the band gap of Si (the "visible" part being roughly E_c - .1 to E_c - .6 eV). We intend to extend the search to the other half of the band gap in a future experiment.

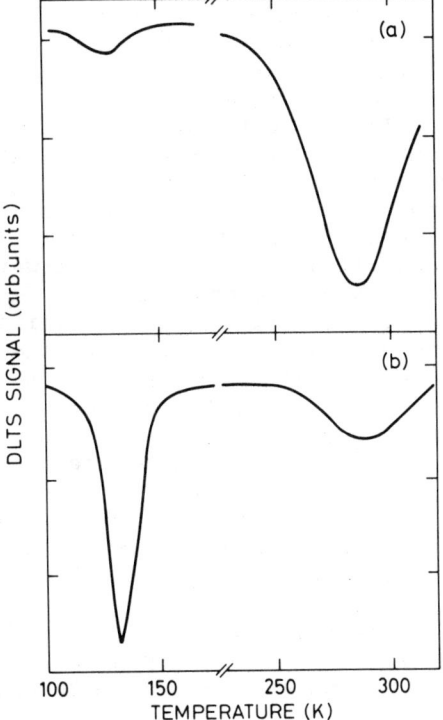

Fig.3. DLTS scans of a ^{195}Au containing sample measured at a time interval of one year. The repetition frequncy was 250 Hz.

Previously, a brief report of the results of our first experiments concerned with Si doped with ^{195}Au and ^{193}Pt has been published [3]. Here an example will be shown, and the results will be very briefly reviewed.

Fig.3 shows two temperature scans of a n-Si sample doped with ^{195}Au, measured at a time interval of one year (corresponding to two halflives of ^{195}Au). The spectra contain two main peaks, of which the high temperature one disappears, when ^{195}Au decays to ^{195}Pt, and the low temperature peak grows in accordingly - in a one-to-one correspondance. The two peaks have activation energies of 0.55 eV and 0.22 eV, respectively. The measurements demonstrate unambigously that Au is involved in the center giving rise to the 0.55 eV level and that this center - when a Pt atom has replaced the Au atom - gives rise to the 0.22 eV level.

Detailed measurements have shown [3] that the four levels normally assigned to Au and Pt in n- and p-Si [4] indeed contain Au and Pt, respectively. Furthermore, the measurements strongly suggest that all four levels arise from the same center, which in the case of Pt is known to be a distorted substitutional site [5]. The results thus suggest that the "well-known" Au levels [6] also originate from isolated, substitutional atoms, since the nuclear transmutation is unlikely to lead to a site change of the daughter atom. It should be stressed that the nuclear decays do not impart sufficient energy (E^R < 1 eV) to reimplant the daughter isotopes, but do, on the other hand, permit electronic rearrangements and eventually distortions.

The results obtained for Ir does not allow a definite identification of the center giving rise to the signals observed. However, by looking at the decay chain of ^{189}Hg it is clear that in a future experiment the annealing could be carried out before the decay of ^{189}Pt, thus exploiting that the lattice position of Pt is known. It is even straightforward to do mixed implantations where one or more isotopes serve as "guards" and built-in calibrations.

Quite generally, the availability at ISOLDE of many isotopes of one element, decaying in one or more generations through other elements, allow sophisticated identifications of deep levels induced by several elements, if only the behaviour of one element in the decay chain (like here Pt) is well understood. We intend to exploit this to study the deep levels induced by a series of 3d, 4d, and 5d impurities in Si, Ge, and SiGe alloys in the future.

In a second generation of experiments, studies of the complexing behaviour of impurities in semiconductors can be envisaged by making mixed implantations of more than one radioactive isotope.

Several types of identification problems in semiconductor physics can in our opinion be solved by using radioisotopes in conjunction with sensitive techniques, which do not require a large number of centers to obtain a signal, like e.g. DLTS, EPR, and photoluminescence. Radioisotopes of suitable halflives (days to a year) are readily available for some thirty elements, and can be made available for for at least twenty more elements, covering all groups of the periodic system.

This work was supported by the Danish Accelerator Council and by the Danish Natural Science Research Council.

References.

1. D.V.Lang, J.Appl.Phys. 45, 3023 (1974).
2. H.L.Ravn, Phys.Rep. 54, 201 (1979).
3. J.W.Petersen, J.Nielsen, Apll.Phys.Lett. 56, 1122 (1990).
4. S.D.Brotherton, and J.E.Lowther, Phys.Rev.Lett. 44, 606 (1980).
5. C.A.J.Ammerlaan, A.B.van Oosten, Phys.Scr. T25, 342 (1989), and references cited therin.
6. D.V.Lang, H.G.Grimmeiss, E.Meijer, M.Jaros, Phys.Rev. B22, 3917 (1980); L.-A.Ledebo, Zhan-Guo Wang, Appl.Phys.Lett. 42, 680 (1983); J.Utzig, W.Schroter, Appl.Phys.Lett. 45, 761 (1984).

MICROSCOPY OF FRENKEL PAIRS IN SEMICONDUCTORS BY NUCLEAR TECHNIQUES

R. SIELEMANN, H. HÄßLEIN, M. BRÜßLER, AND H. METZNER

Hahn-Meitner-Insitut Berlin, Glienicker Str. 100, D-100 Berlin 39, Germany

ABSTRACT

Monoenergetic neutrino-recoil atoms which originate from electron-capture decays can be used as primary knock-on atoms for the production of isolated single Frenkel pairs in semiconductors. This production process combined with perturbed angular correlations of γ-rays (PAC) gives a microscopic experimental access to Frenkel pairs from the standpoint of the primary knock-on atom in otherwise undamaged material. Experiments on germanium and III-V semiconductors are reported and the role of complimentary defect-trapping experiments after defect production by electron irradiation is described.

1. Introduction

Perturbed angular correlations of γ-rays (PAC) has in recent years become an important tool to study defects in materials, in particular in metals. Unique features of the technique are high sensitivity (radioactive detection) and the capability of obtaining structure information on an atomic scale. In semiconductors, the charge state of the PAC probe atoms will strongly influence the measured electric field gradient which renders the techniques simultaneously sensitive to structural and electronic properties. In a 'standard' PAC experiment the probe atom detects defects by a thermally activated trapping process. This trapping process involves migration of defects and requires an attractive potential between probe and defect which in semiconductors may be supplied by either elastic or Coulombic interaction or both. A general overview on PAC in semiconductors is given by Wichert at this conference [1].

We have recently introduced a technique which yields a strictly microscopic picture of Frenkel pair production and thermal behavior [2,3]. Instead of taking the 'standard' PAC probe ^{111}In as parent activity the radioactive precursor ^{111}Sn is used. ^{111}Sn decays to ^{111}In via electron capture thereby emitting a neutrino with an energy of 2.5 MeV. This neutrino emission imparts the tiny recoil energy of 29 eV to the resulting ^{111}In which in this way in addition to its function as probe atom becomes the primary knock-on atom (PKA) in defect production. The low recoil energy of 29 eV is just above threshold for Frenkel pair formation in many solids and one expects production of single Frenkel pairs with both Frenkel partners in close vicinity to the probe atom. In this sense we have a truly local experimental technique which should give insight into the Frenkel pair production process and the thermal behavior of closely spaced vacancies and interstitials. Information concerning type and structure of a defect may be obtainable from the low energy collision process in a solid and can be combined with information from trapping experiments.

2. Experimental Realization

The neutrino recoil technique starts with the parent activity ^{111}Sn. The decay ^{111}Sn to ^{111}In via electron capture gives a recoil energy $E_R = Q^2/2Mc^2 = 29\,eV$ to the ^{111}In PAC probe, where $Q/c^2 = 2.5\,MeV/c^2$ is the mass difference between ^{111}Sn and ^{111}In and M is the mass of ^{111}In. The corresponding nuclear decay scheme is shown in fig. 1. ^{111}Sn is produced by a heavy-ion induced nuclear reaction

and deeply implanted (µm range) into any desired material. After implantation all
radiation damage is completely annealed before the neutrino recoil process occurs.
The deep implantation is favorable for annealing procedures on the one hand and
on the other hand allows work with low probe concentrations which can be kept
below 10^{14} cm^{-3}. Thus investigations of semiconductors with doping levels above
10^{15} cm^{-3} can be performed without significantly altering the Fermi level by the
probe atom incorporation.

The short half life of ^{111}Sn ($T_{\frac{1}{2}}=35$min) followed by the long half-life of the 'standard' activity ^{111}In imposes a special experimental schedule which is described in
[3].

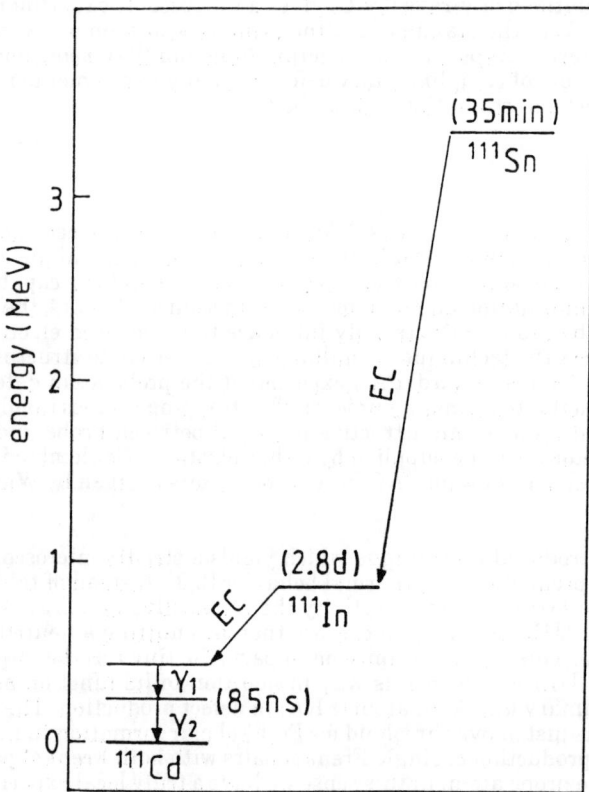

Fig. 1: Simplified decay scheme of ^{111}Sn which is produced and implanted by the nuclear reaction ^{93}Nb(^{22}Ne, p3n) ^{111}Sn. The neutrino recoil occurs in the EC decay of ^{111}Sn to ^{111}In.

3. Experimental Results and Discussion

3.1 Experiments in Germanium

A large variety of PAC experiments have been performed to study intrinsic defects
in Germanium. These include studies with the neutrino recoil technique, trapping
experiments after electron irradiation and studies of the correlated damage after

probe atom implantation. Some preliminary results have been published [4,5] and a publication drawing conclusions from all experiments is presently being prepared [6].

3.1.1 Experiments in p-Germanium

The material used was doped with 6×10^{17} Ga cm^{-3}. The PAC probe concentration obtained from heavy-ion recoil implantation was below 10^{14} cm^{-3}, thus no shift of the Fermi level is expected. Results of the neutrino recoil process are shown in fig. 2. The spectrum measured at 4 K immediately following the recoil process consists of two fractions: fraction (1) with about 12% of the probes is decorated with a defect characterized by the quadrupole interaction frequency $v_Q = e^2Qq/h = 54$ MHz ($\eta = 0$); a second fraction (2) displays a spectrum which is characterized by a frequency distribution, which, however, is strongly temperature dependent and changes reversibly with temperature. A control experiment was performed with ^{111}In instead of ^{111}Sn which shows that fraction (1) with $v_Q = 54$ MHz is due to the neutrino recoil process. This defect signal vanishes irreversibly at $T \geq 200$ K and can thus uniquely be ascribed to an intrinsic defect.

In addition to these recoil experiments trapping experiments were also performed. For this purpose the samples were doped with ^{111}In as described above, furnace annealed and then irradiated with 2×10^{16} cm^{-2} electrons at 1.1 MeV energy. Irradiation temperature was 77 K. Fig. 3 shows the result. Immediately after irradiation no defect can be detected, for $T \geq 250$ K, however, the defect $v_Q = 54$ MHz, already known from the recoil experiment, is trapped at the probe. In this case the defect configuration is stable up to 400 K and then completely vanishes.

From the recoil and the complementary trapping experiments one can draw the following conclusion: as the threshold for Frenkel pair production in Ge is in the region 15-25eV [7], the defect observed after recoil must be one constituent of a single Frenkel pair. Since the stability of this defect after trapping is maintained up to 400 K, the disappearance in the recoil experiment already at $T \geq 200$ K must be due to correlated recombination between the Frenkel partners. The question, whether the 54 MHz defect represents a vacancy or an interstitial associated with the PAC probe cannot be answered in a straightforward way: the microscopic mechanism by which a Frenkel pair is created is not sufficiently well known for crystals of the diamond structure. In the close packed fcc structure Frenkel pairs are produced by replacement collision sequences leading to a situation where the PKA, in our case ^{111}Sn/^{111}In, comes to rest adjacent to a vacancy. This has been shown for Cu [3]. Though the situation is not as clear for Ge, available theory and experiments suggest that after neutrino recoil the ^{111}In probe must sit beside a vacancy [8]. Hence, the results in p-Ge can best be described by vacancies freely migrating around 250 K, the temperature at which they are trapped at ^{111}In. The bond to the PAC probe is broken around 400 K. Correlated recombination takes place around 200 K where the (not visible) interstitial annihilates the vacancy.

3.1.2 Experiments in n-Germanium

The material used was doped with 1×10^{16} Sb cm^{-3}. It is remarkable that the results in n-Germanium completely differ from those in p-Ge which directly shows the influence of the Fermi level. We obtained the following defect signals after neutrino recoil and electron irradiation:

ν recoil: $v_Q = 44$ MHz, $\eta = 0.6$;
electron irradiation: $v_Q = 423$ MHz, $\eta = 0$.

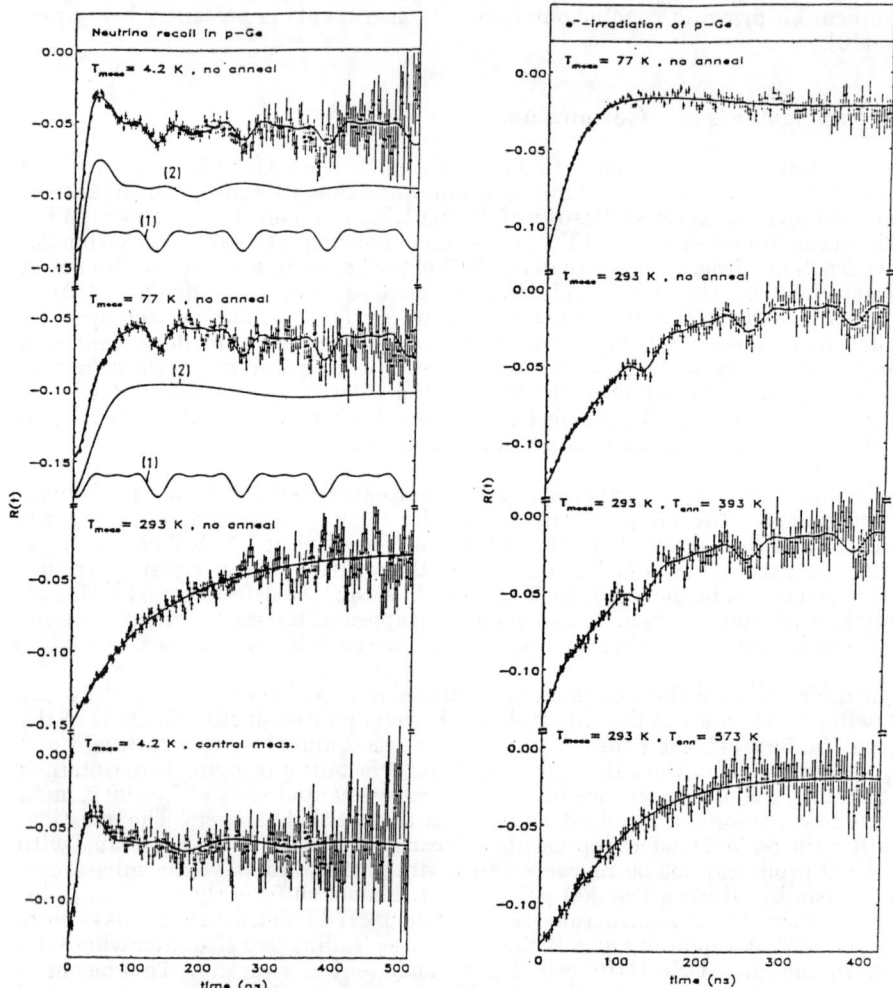

Fig. 2 (left): PAC spectra after neutrino recoil in p-Ge at various measuring temperatures. Fraction 1 is the defect signal $v_Q = 54$Mhz, $\eta = 0$. The bottom spectrum measured finally at 4.2 K shows that fraction 1 has disappeared.

Fig. 3 (right): PAC spectra in p-Ge after electron irradiation. The frequency $v_Q = 54$Mhz is also observed after neutrino recoil (fig. 2).

The 44 MHz defect has a rather weak signal and vanishes between 77 K and room temperature. The 423 MHz defect is not seen after neutrino recoil but appears with a large fraction up to 30% upon annealing above 200 K. Since in this case recoil and trapping techique yield different results, the interpretation is more difficult. It appears, however, from the strong similarity in trapping temperature (between 200 K and 250 K) and thermal breakup (betwenn 300 K and 400 K) in both p- and n-Ge, that the 423 MHz defect is also a vacany-probe complex resulting from trapping of a more negatively charged vacancy. The 44 MHz defect resulting from neutrino recoil might be due to a direct displacement of the probe to an interstitial site. This, however, is presently only speculative. More detailed arguments are given in [6].

3.1.3 Defects from Correlated Implantation Damage

In these experiments PAC was measured immediately after implantation of the probes in both p- and n-Ge. Implantation and PAC measurements were perfomed at room temperature. The PAC in both cases show that the crystals are heavily damaged, but in p- as well as in n-type material small fractions of the 54MHz and the 423 MHz defect appear simultaneously in sharp contrast to the experiments described above. This is a result similar to the result recently communicated by Feuser et al. [9], where in all cases high ^{111}In concentrations and heavy radiation damage were present. We assume that the strong inhomogeneities in the implantation experiments can account for the simultaneous appearance of the differently charged probe-vacancy pairs.

3.2 Experiments in III-V Semiconductors

Rather intensive research has been carried out to produce Frenkel defects in the III-V semiconductors InP and InSb by the neutrino recoil technique. Conventional trapping experiments with ^{111}In cannot be carried out in these compounds since the probe atom In is a constituent of the compounds and therefore does not represent a trap for defects. Threshold energies given in the literature are around 6-10 eV [8], so one expects Frenkel pair production by the 29 eV recoil energy imparted to ^{111}In in the recoil experiment. However, in none of these materials regardless of the type of conduction a neutrino recoil induced defect could be observed so far. One possilbe explanation is that, starting from the In-sublattice where the ^{111}In and also ^{111}Sn are expected to reside prior to neutrino recoil, no stable defect configuration exists. Further research devoted to this problem is in progress.

4. Conclusion

The neutrino recoil in combination with PAC yields insight into production and thermal behavior of Frenkel pairs at an atomic level which includes information on both vacancy and interstitial. Its capability to help identifying defects as intrinsic is of special importance in semiconductors. The technique of introducing radioactive probes by nuclear reactions and recoil implantation permits low probe concentrations (less than 10^{14} cm^{-3}). Detailed information on point defects is obtained when the neutrino recoil technique is combined with the 'standard' technique of trapping irradiation induced defects at the nuclear probes. It is concluded that two different defects detected by PAC in germanium under various experimental conditions result from the accociation of the probe with differently charged vacancies.

5. References

[1] Th. Wichert, this conference.
[2] H. Metzner, R. Sielemann, R. Butt, S. Klaumünzer, Phys. Rev. Let. 53, 290 (1984).
[3] H. Metzner, R. Sielemann, S. Klaumünzer, E. Hunger, Phys. Rev. B36, 9535 (1987).
[4] M. Brüßler, H. Metzner, R. Sielemann, Materials Science Forum 38-41, 1205 (1989).
[5] M. Brüßler, H. Metzner, R. Sielemann, Hyperfine Interactions 60, 809 (1990).
[6] M. Brüßler, H. Häßlein, H. Metzner, R. Sielemann, to be published.
[7] F. Poulin, J.C Bourgoin, Revue Phys. Appl. 15, 15 (1980).
[8] J.C Bourgoin, M. Lanoo, in *Point Defects in Semiconductors II* (Springer, Berlin 1983).
[9] U. Feuser, R. Vianden, A.F. Pasquevich, Hyperfine Interactions 60, 829 (1990).

MODERN MUON SPECTROSCOPIC METHODS IN SEMICONDUCTOR PHYSICS

R.L. LICHTI[1], C.D. LAMP[1], S.R. KREITZMAN[2], R.F. KIEFL[2],
J.W. SCHNEIDER[2], Ch. NIEDERMAYER[2], K. CHOW[2], T. PFIZ[3],
T.L. ESTLE[4], S.A. DODDS[4], B. HITTI[4], and R.C. DuVARNEY[5]

[1] Department of Physics, Texas Tech University, Lubbock, TX 79409, USA
[2] TRIUMF and University of British Columbia, Vancouver, BC V6T 2A3, Canada
[3] Max Planck Institut fuer Metallforschung, Institut fuer Physik, Stuttgart, Germany
[4] Department of Physics, Rice University, Houston, TX 77251, USA
[5] Department of Physics, Emory University, Atlanta, GA 30322, USA

ABSTRACT

The application of muon level-crossing resonance (μLCR) and RF muon-spin resonance (RF-μSR) in investigations of the defect states and reactions of muonium in semiconductors is discussed. Spectra from μLCR were crucial in identifying the bond-centered position as the stable neutral defect state for muonium in GaAs and Si. RF-μSR has yielded reaction rate parameters for muonium site-change and charge-state reactions in Si.

1. INTRODUCTION

The muonium defect center in semiconductors is expected to be identical to the hydrogen center, except for vibrational effects due to the mass difference. Indeed, much of the current experimental knowledge of isolated H impurities is from investigation of the muonium analog. Traditional time-differential muon-spin-rotation (TD-μSR) measurements[1] have established two neutral isolated muonium centers in tetrahedrally bonded semiconductors: an isotropic state, known as Mu in the μSR literature, assigned to tetrahedral interstices (T); and a near bond-center state (BC), labeled as anomalous muonium Mu*. Here we discuss two modern muon spectroscopic methods which have contributed significantly to our understanding of muonium centers in semiconducting materials. These are muon level-crossing resonance (μLCR),[2] which was crucial in identifying Mu* as Mu(BC),[3,4] and an RF resonance technique (RF-μSR)[5] recently applied to the study of muonium charge-state reactions.

As applied at TRIUMF, both of these techniques use integral decay positron counting methods to measure variations of the time-integral muon decay asymmetry (proportional to the time-averaged muon-spin polarization) as a function of a control parameter, typically the magnetic field. A generic experimental arrangement for such measurements is depicted in figure 1. The muon beam is spin polarized opposite to the momentum and the (longitudinal) magnetic field is applied parallel to the initial polarization direction. The backward and forward positron counters record decay rates for positrons emitted parallel and antiparallel to the initial muon polarization. The asymmetry is then the normalized difference of these two muon decay rates. For typical longitudinal field situations the muon polarization is retained except under technique-specific resonant conditions discussed below. Neither μLCR nor RF-μSR is dependent on retention of muon-spin phase coherence, and are therefore much more sensitive to final states of any reaction than are TD spin-precession measurements.

2. LEVEL-CROSSING RESONANCE: Identification of Mu(BC)

Results from μLCR in GaAs[3] and in Si[4] have been crucial in the identification of the BC state[6,7] as the stable state for isolated muonium (and hydrogen) in these systems. We have subsequently applied μLCR to the study of both Mu(BC) and Mu(T) states in a number of semiconductor hosts. This method takes advantage of state mixing which occurs to avoid the crossing of two energy levels, in particular it measures the loss of polarization when muon spin states are mixed in avoiding a crossing. In muonium spectroscopy the $|m_\mu, m_e\rangle = |++\rangle$ and the $a|+-\rangle + b|-+\rangle$ states are involved in the avoided crossing. When interactions of muonium with a neighboring nucleus are included,

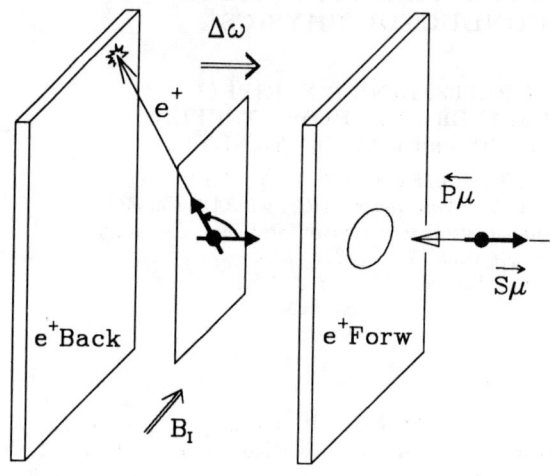

Figure 1. Geometry for an integral μSR experiment. After the incoming muon stops in the sample it precesses in the effective field until it decays into a positron. The positron emerges preferentially along the spin direction of the muon thereby allowing the polarization of the muon to be measured along the axes of the e^+ counters. The field in which the muon precesses comes from the varying *off resonance* contribution of the control parameter $\Delta\omega$ and the driving field B_I, which arises from hyperfine interactions in the LCR case, or an external RF field for RF-μSR.

each level is split by the nuclear spin states and associated hyperfine and quadrupole terms. The level crossings for the combined system are avoided by state mixing, leading to reduced muon polarization and decay asymmetry for the specific magnetic fields where the zero'th order levels would have crossed. The positions of the LCR lines, thus, carry information on the nuclear hyperfine interaction between the muonium and coupled neighboring nuclei. Line positions are also shifted by the nuclear quadrupole interaction, which helps in identifying the nucleus. Information obtainable from μLCR is, therefore, very similar to that from a double resonance experiment such as ENDOR. For an anisotropic center like Mu*, LCR spectra taken with a single crystal sample, aligned with the field along each of several high-symmetry directions, allows full determination of the nuclear hyperfine and quadrupole tensors. In turn, these yield the geometry of the defect state and the associated electronic spin density on each nucleus, including the muon.

Figure 2 shows a set of μLCR spectra for Mu* in GaAs.[3] A flip-coil field modulation technique was used, giving the typical derivative lineshapes. The dominant μLCR lines are all from one As and one Ga position along the $\langle 111 \rangle$ symmetry axis of Mu* with 83% of the spin density on these two atoms, conclusively confirming the near BC model for this center. A simultaneous fit of all observed near-neighbor lines yielded the parameters listed in table 1. Careful TD-μSR measurements reveal resolved hyperfine spin-precession spectra in excellent agreement with the μLCR parameters, further verifying this assignment. A similar series of μLCR measurements on Mu* in Si yield proof of the BC model in that case, as well.[4] The resulting parameters for bond-centered Mu* in Si are also listed in table 1.

Muon level-crossing spectroscopy has proven to be invaluable in identifying the geometries of isolated muonium defect centers in semiconductors. In a slightly different form, μLCR spectra from the interaction of muon spin states and impurity nuclear quadrupole levels are proposed as a means of observing muonium-impurity pair formation.[8]

3. RF MUONIUM SPECTROSCOPY: Charge-state reactions

RF spectroscopic methods,[5] similar to cw NMR, have been applied at TRIUMF to the study of isolated muonium centers in semiconductors, most recently to muonium charge-state reactions in Si and Ge. The apparatus as currently operated for semiconductor experiments uses a high power (1kW) broadband (10-250MHz) RF amplifier driving a specially designed non-resonant coil matched to a 50 Ohm transmission line. The coil design allows full use of the muon beam intensity onto a sample of up to 2.5×3cm cross section. The broadband features allow maximum flexibility in either the constant-frequency, swept-field mode, or with swept RF frequency. Figure 3 shows the three RF signals for silicon, used to monitor the Mu, Mu*, and diamagnetic ionized (μ^+) states.

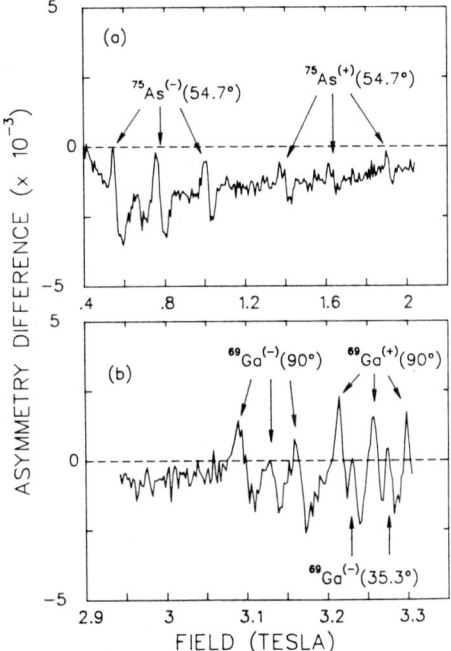

Figure 2. The level-crossing-resonance spectra of Mu* in GaAs for **B** applied along the (a)⟨100⟩ and (b)⟨110⟩ directions. The resonances are labeled by the nucleus involved, the sign of M_s, and Θ, the angle between the symmetry axis and the magnetic field.

In previous studies of muonium in silicon,[1] reactions have been identified with major changes in the TD-μSR signals as a function of temperature. The Mu* relaxation rate increases rapidly above 120K and the signal disappears at about 150K. This fraction of the signal strength is recovered in the μ^+ signal by about 200K. Since both signals are identified as coming from the BC site,[4,9] this Mu* → μ^+ reaction is assigned to Mu(BC)→Mu$^+$(BC) ionization. A fit to the Mu* relaxation increase using a thermally activated process gives an energy of 0.168eV and a prefactor of 9.5×10^{11}s^{-1}. A second charge-state reaction is seen in the disappearance of the Mu signal near 270K and the subsequent recovery of that fraction of the intensity near 600K. Attempts to obtain rate parameters for this reaction yield sample-dependent results, suggesting that the rate-limiting step may be dependent on some sample property such as impurity content or strain. This reaction most likely involves a site change as well as ionization. The TD relaxation data for the isotropic state Mu in the sample used for the RF-measurements indicate a slow process with a prefactor of 6.5×10^8s^{-1} and a 0.114eV barrier. The results from TD measurements in Ge are slightly different with both muonium signals disappearing near 100K and the recovery of intensity in the diamagnetic signal showing a single step, centered between 220 and 300K depending on the sample.[1]

RF methods detect all the muons in a given state, while TD methods require that the spin polarization and phase be retained during any reaction into that state. Given these differences and the TD data, the RF μ^+ intensity should show steps near 150 and 270K. A few μ^+ data points obtained during RF development runs[10] indicate that the Mu* intensity is recovered by 185K and that by 317K all the muons contribute to the diamagnetic signal.

Table 1. Hyperfine and quadrupole parameters for Mu(BC) in GaAs and Si (from references 3 and 4 respectively), and the resulting spin densities on each nucleus.

State	Nucleus	$A_{\parallel}/h$ (MHz)	$A_{\perp}/h$ (MHz)	Q/h (MHz)	Spin Densities	
					s	p
Ga-Mu-As	μ^+	218.54(3)	87.87(5)	...	0.0392	...
	^{75}As	563.1(4)	128.4(2)	6.26(7)	0.0186	0.434
	^{69}Ga	1052(2)	867.9(3)	0.36(1)	0.0761	0.303
Si-Mu-Si	μ^+	-16.82	-92.59	...	0.0151	...
	^{29}Si	-137.5(1)	-73.96(5)	...	0.0207	0.186

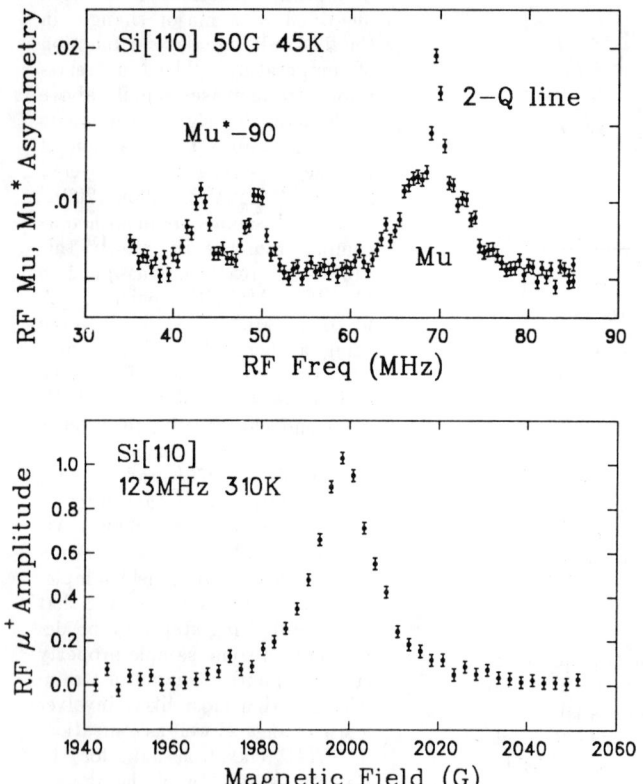

Figure 3. RFμSR spectra in Si. The upper diagram shows lines of the two paramagnetic centers, taken with swept frequency, while the lower curve shows a typical diamagnetic field scan. The very narrow 2-quantum line of the Mu center is due to a simultaneous electron-muon flip and it sits atop the two single quantum transitions.

Figure 4 displays the RF-μSR intensity of μ^+ as a function of temperature in both Si and Ge. The origin of the lowest temperature step for Si is not yet understood since no changes in the neutral muonium signals are correlated with this feature. Above 80K the data are more or less as expected, and the appropriate reductions in muonium signals are seen for the higher temperature μ^+ features. Relative step sizes indicate that the Mu* fraction is recovered by 150K and the 150-300K increase represents the initial Mu fraction. However, this latter increase appears to be more complicated than anticipated from the TD data, and may require two processes. Table 2 lists the rate parameters from a preliminary fit of the RF-μSR data in Si using a single process for each feature.

In modeling the temperature dependence above 80K we start from the simplest picture consistent with earlier data, using only neutral and positively charged states at the T and BC sites. A more complete picture involving the negative charge state (stable at the T site) is used in explaining the recent E3' DLTS annealing data of Holm et al.[11] With the simpler picture we have reproduced the observed μ^+ temperature dependence assuming all reaction paths into the Mu$^+$(BC) state are active and each of the three local minima, Mu(T), Mu(BC) and Mu$^+$(BC), are initially populated in ratios found from low temperature TD measurements. These reaction paths are: 1) Mu(BC)→Mu$^+$(BC)+e_c, 2) Mu(T)→Mu(BC) followed by rapid ionization, and 3) Mu(T)→Mu$^+$(T)+e_c followed by a site change. A simulation using the rate parameters from the Mu* TD results reproduces the 150K step in the μ^+ RF intensity reasonably well. However, a fit to this region yields a barrier which is 50% higher with a correspondingly higher prefactor. On the other hand, the disappearance of the Mu* RF-signal, which is clearly correlated with the 150K μ^+ step, gives a much lower barrier. Our preliminary rate parameters for this μ^+ step are characteristic of an optical process, consistent with simple ionization. The higher-temperature, broad μ^+ intensity increase can be roughly reproduced by a single slow process giving the parameters in the table. When two processes are used, interaction between parameters leads to fit instabilities. If one rate is kept slow to be consistent with the TD data for the Mu signal, the other process is fast in keeping with reaction paths 2) and 3) above. The density of the current RF data is insufficient to cleanly separate the Mu→ μ^+ transition into two well defined reaction paths, and additional investigation is planned. We have also fit the disappearance of the RF signal for Mu, which is associated with the broad μ^+ feature, and obtain a very low energy barrier. The disappearance of both neutral muonium features give much lower barriers than the associated increase in the diamagnetic signal strength. We do not have a satisfactory explanation for this result at present. There are now several separate data sets yielding information on the dynamics of charge-state reactions of muonium in silicon. The general

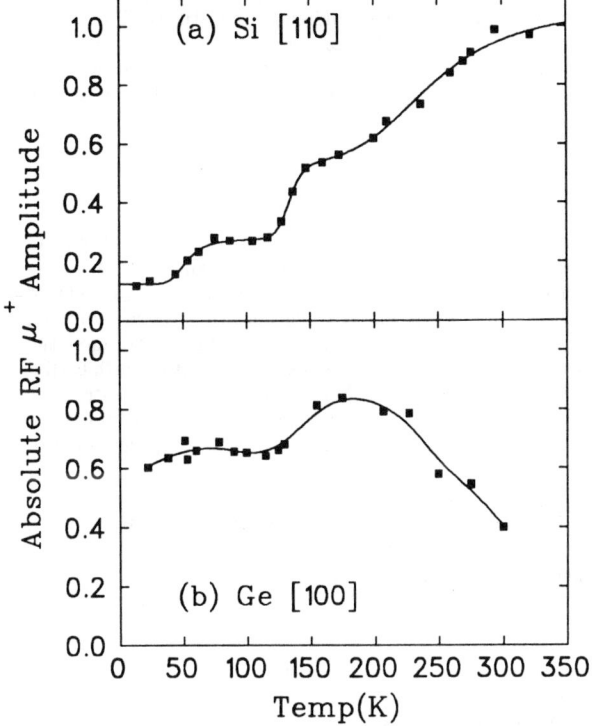

Figure 4. RF diamagnetic amplitudes in high resistivity Si and Ge. The curve through the Si data is a fit to three activated processes, while that through the Ge points is simply a guide to the eye. The RF data here is indicative of *final* state occupation the diamagnetic state.

features remain consistent with the $Mu^* \to \mu^+$ and $Mu \to \mu^+$ assignments from the TD data. A detailed description of these reactions will have to take into account the differences in measurement techniques and resulting rate parameters. A simple comparison of parameters suggests the presence of intermediate steps, such as the T→BC site change ($Mu \to Mu^*$) inferred from the slow Mu-related processes.

The main difference between our picture and that of Holm et al[11] is the presence of $Mu^-(T)$ in their model. While all of the existing μSR data are consistent with the two charge-state model, the diamagnetic muon signal could come from either Mu^+ or Mu^- in any site. In assigning this signal to $Mu^+(BC)$ we rely heavily on positron channeling results[9] and theoretical models which give the BC site as the stable state for H^+. It is possible that some of the μ^+ signal could be from $Mu^-(T)$. The easiest way to incorporate this into our picture would be for the direct $Mu(T)$ ionization to be an electron capture process; a site change would then not be expected to follow. The RF-μSR data for μ^+ and the DLTS annealing data yield very similar energy parameters, however the reaction assignments are quite different. At present a consistent picture combining the hydrogen and muonium results in silicon is not obvious, thus, we have chosen to interpret the muonium results using a simple, internally-consistent model.

Table 2. Rate parameters from preliminary fits to RF-μSR data in Si and assignment of the rate-limiting step of each process.

Reaction Step	RF Feature	Temperature (K)	Prefactor (s^{-1})	Barrier (eV)
Unassigned	μ^+ step	50	6×10^9	0.032
$Mu(BC) \to Mu^+(BC)$	μ^+ step	135	5.2×10^{16}	0.284
	Mu^* decrease	140		0.101
$Mu(T) \to Mu(BC)$	μ^+ increase	250	6×10^9	0.153
	Mu decrease	250		0.029

The RF intensity of the μ^+ signal in Ge is also shown in figure 4. The most striking difference between the Ge and Si data is the drop in diamagnetic signal strength above 240K seen for Ge. This is a clear indication of a reaction out of the ionized state at higher temperatures. This result represents a major difference in the dynamics of muonium in Ge as compared to Si, and was not anticipated from the TD measurements. Since many hydrogen-impurity passivation reactions are thought to occur via Coulomb capture processes, the disappearance of ionized states of muonium below room temperature in Ge offers a possible explanation of the reduced effectiveness of hydrogen passivation in Ge as compared to Si.

4. CONCLUSION

In conclusion, we have demonstrated the usefulness of μLCR and RF-μSR integral methods in the study of muonium states and reactions in semiconductors. These methods will continue to be developed and modified, yielding detailed information on the muonium analog of the hydrogen-related defects in these materials, and have a particularly significant role with respect to isolated hydrogen and the early stages of the development of hydrogen-related complexes. We have examined the charge-state reactions of isolated muonium in Si and Ge using RF methods in conjunction with earlier TD measurements, and expect to expand such investigations to other muonium reactions. Progress to date establishes integral RF-μSR methods as a means to examine the final states of muonium reactions in semiconductors, yielding information on reaction dynamics which is complimentary to that from traditional time-differential techniques.

This work was supported in part by grants from the Robert A. Welch Foundation (C-1048, D-1053, and D-1167), the United States National Science Foundation (DMR-8917639), the National Research Council and the National Science and Engineering Research Council of Canada.

REFERENCES

1. For a review of muonium in semiconductors see:
   B.D. Patterson, Rev. Mod. Phys. **60**, 69 (1988).
2. R.F. Kielf and S.R. Kreitzman, in *Perspectives in Muon Science*, Ed. T. Yamazaki, K. Nakai, and K. Nagamine. To be published.
3. R.F. Kielf *et al.*, Phys. Rev. Lett. **58**, 1780 (1987).
4. R.F. Kielf *et al.*, Phys. Rev. Lett. **60**, 224 (1988).
5. S.R. Kreitzman, Hyperfine Interactions **65**, 1055 (1990).
6. S.J.F. Cox and M.C.R. Symons, Chem. Phys. Lett. **126**, 516 (1986).
7. T.L. Estle, S. Estreicher, and D.S. Marynick, Phys. Rev. Lett. **58**, 1547 (1987).
8. Dj.M. Maric *et al.*, J. Phys. C (to be published).
9. H. Simmler *et al.*, Hyperfine Interactions **64**, 535 (1990).
10. S.R. Kreitzman *et al.*, Hyperfine Interactions **64**, 561 (1990).
11. B. Holm *et al.*, Phys. Rev. Lett. **66**, 2360 (1991).

MUON STOPPING SITES IN SEMICONDUCTORS FROM DECAY-POSITRON CHANNELING

H. SIMMLER, P. ESCHLE, H. KELLER, W. KÜNDIG, W. ODERMATT,
B.D. PATTERSON, I.M. SAVIĆ, J.W. SCHNEIDER, B. STÄUBLE-PÜMPIN,
U. STRAUMANN, and P. TRUÖL
Physics-Institute, University of Zurich, Schönberggasse 9, CH-8001 Zurich, Switzerland

Abstract

The lattice positions of implanted positive muons (μ^+) in intrinsic semiconductors (Si, GaAs, InP) have been investigated by μ-decay positron channeling at temperatures ranging from 95 K to 400 K. In high purity float-zone (FZ) Si a metastable μ site is observed: Below 200 K, the pattern is consistent with a fraction of 40% near a BC (bond-center) site and 60% near a T (tetrahedral) site. Above 200 K, the T-like fraction undergoes a transition to the BC-like site, where virtually all muons are located above 300 K. By comparison with muon-spin-rotation (μSR) measurements, these sites can be associated with the known paramagnetic muonium (μ^+e^-) states observed in numerous semiconductors: the metastable site corresponds to the isotropic state (Mu), the BC-like configuration is the stable site for both the anisotropic state (Mu*) at low temperatures as well as the final ionized state ('μ^+') at higher temperatures. In GaAs, there is evidence for a similar metastability. In InP, a stable near-BC configuration is observed throughout the temperature range investigated. Thus a BC-like configuration is found to be most stable in all measurements.

1 Introduction

During the thermalization process of low energy (4 MeV) positive muons (μ^+) in solids radiation damage is very low, yielding stopping sites far away from induced defects [1]. Thus the μ-decay

$$\mu^+ \xrightarrow{2.2\mu s} e^+ + \nu_\mu + \overline{\nu_e}. \qquad (1)$$

allows in principle a one-by-one observation of an isolated single charge in an otherwise unperturbed lattice environment. Since bound states μ^+e^- (muonium) in semiconductors were discovered [2], detailed information on the hyperfine structure of muonium centers has been obtained by means of the muon-spin-rotation (μSR) technique [1]. In spite of the conceptual simplicity, these centers turned out to be rather complex: At low temperatures, two paramgnetic states coexist in a number of semiconductors (e.g. Si, Ge, GaAs, GaP): a so called "normal", isotropic state (Mu) and an "anomalous", strongly anisotropic state (Mu*) with trigonal symmetry. In the normal state the small hyperfine coupling (e.g. 45% of the vacuum value in Si) indicates a substantial delocalization of the unpaired electron spin density. In the anomalous state the electron spin density at the muon is very close to zero. The diamagnetic fraction is usually small at low temperatures. At higher temperatures, the paramagnetic signals disappear and subsequently the diamagnetic fraction increases to almost 100%.

During the last decade, numerous theoretical and computational studies of a single charge in a diamond or zincblende structure were performed. However, many inconsistent results were obtained. Only in the last few years the calculations (see, e.g. [3]) tend to confirm a model based on qualitative arguments by Cox and Symons [4]: it is generally agreed that Mu* is located near the bond center (BC) of two adjacent host ions, and Mu is trapped in a T (tetrahedral) cage. Experimentally, a near-BC site in Si is supported by a μSR/LCR measurements [5], and a T-like site is found for the two Mu states observed in CuCl [6]. However, details of the local geometry are still unclear, and the quantitative agreement between calculations and experiments is incomplete.

Beside the fundamental interest in the investigation of a simple defect, the problem of hydrogen in semiconductors is of great technological interest because of its strong interaction with defects, such as passivation of shallow and deep impurities [7]. With standard experimental techniques the observation of isolated paramagnetic hydrogen centers in semiconductors is difficult. However, except for the dynamical properties, muonium is expected to be an almost equivalent hydrogen substitute (the reduced mass differs by only 0.4%).

In addition to the study of hyperfine interactions, the μ-decay positrons can be used for blocking experiments yielding direct information on the muon lattice sites. In the following, we describe the experimental setup for μ-decay positron blocking experiments and present results obtained for intrinsic semiconductors.

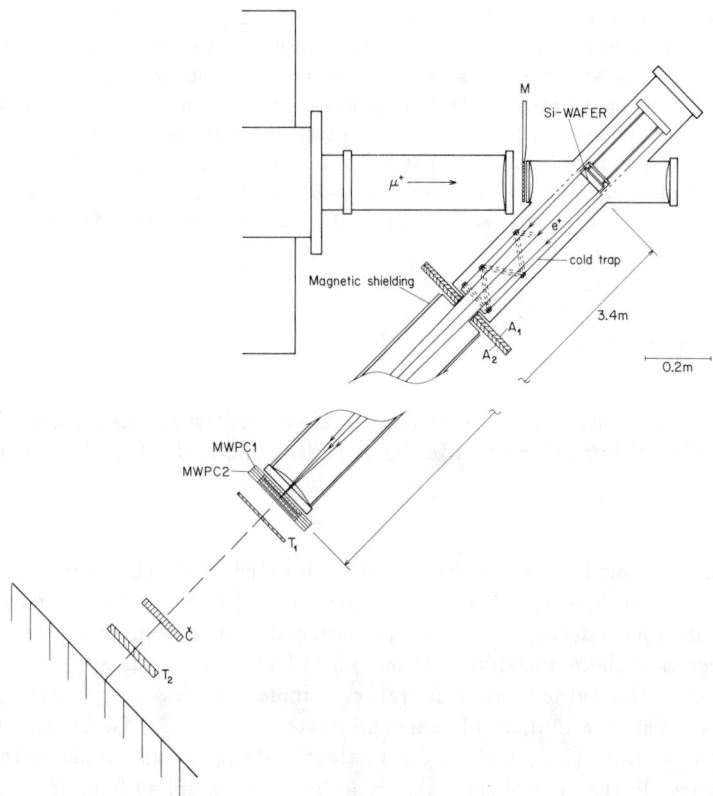

Figure 1: Schematic representation of the μ/e channeling experiment. A significant event is defined by the trigger condition $\overline{(A_1 + A_2)} \cdot T_1 \cdot \check{C} \cdot T_2$. The direction of the positron path compared to the decay channel is determined by two successive 2D multiwire proportional chambers (MWPC1 and MWPC2).

2 Experimental setup and data analysis

The experimental setup is shown schematically in Fig. 1. The experiments were done at the Paul-Scherrer-Institute (PSI) in Switzerland, using a high intensity low energy muon beam ($E_\mu \simeq 4$ MeV, rate $\simeq 8 \times 10^6$ s^{-1}) for the implantation. Because of a critical dechanneling length of only about 30 μm in the sample, the muons were carefully degraded to a mean range of about 20 μm. In order to use the whole beam spot area (≈ 10 cm^2) without loosing resolution, the sample wafers were spherically bent with a 3.4 m radius of curvature. At the center of the sphere, the decay positrons were recorded with a 2D multi-wire proportional chamber. Scintillation counters were used to define the observed solid angle and to trigger the chamber readout. To reduce multiple scattering in a surface condensate, the target region and the positron path were evaporated and surrounded by cryogenic shields kept at about 80 K. The sample temperature could be varied between 95 K and 400 K.

The obtained 2D histograms were smoothed and projected along the crystal-plane–chamber intersection line in question. The amplitude of channeling/blocking-effects observed was about 5% in maximum. Obviously several sources contribute to a reduction of the channeling effect and to a smooting of the experimental dip/peak-shape: multiple electron scattering of the positrons along its path between planes, scattering at other crystal defects, the non perfect spherical shape of the samples, multiple scattering outside the sample, etc.

The experimental one-dimensional particle flux histograms were compared with simulated planar profiles. The simulations done in this work are based on (i) a multiplane continuum approximation of the potentials, and (ii) the assumption of statistical equilibrium in the transverse phase space [8]. The nonrelativistic transverse motion of the relatively heavy μ-decay positron ($\overline{\gamma} \simeq 73$) is assumed to be classical. Based on these assumptions the relative positron yield (for equally spaced planes) is

$$P_r(E_\perp) = L_0 \int_{L_0} dx f(E_\perp, x) p_\mu(x), \qquad (2)$$

where L_0 is the plane distance and $p_\mu(x)$ is the muon position distribution. The equilibrium distribution of particles with energy $E_\perp$ in the transverse space is given by $f(E_\perp, x) \propto 1/\sqrt{E_\perp - V(x)}$, which is evaluated using thermally averaged planar Doyle/Turner potentials $V(x)$ [9]. The final yield of positrons with kinetic energy E_0 and tilt angle ψ relative to the planes is obtained by integrating over the distribution of transverse energies $g(E_\perp, E_0 \psi^2, z)$:

$$\chi(E_0 \psi^2, z) = \int dE_\perp g(E_\perp, E_0 \psi^2, z) P_r(E_\perp), \qquad (3)$$

$$g(E_\perp, E_0 \psi^2, 0) = L_0^{-1} \frac{dL}{dE}(E_\perp - E_0 \psi^2), \qquad (4)$$

where $L(E)$ is the accessible interval of a positron with transverse energy E. Dechanneling inside the sample has been included by modifying $g(E_\perp, z)$ with increasing depth according to the increase of the mean planar angular spread due to multiple electron scattering. The result is

$$g(E_\perp, z) = \int dE'_\perp \frac{g(E'_\perp, 0)}{\sqrt{8\pi E_d(z) E_\perp}} \left[\exp -\frac{(\sqrt{E'_\perp} - \sqrt{E_\perp})^2}{2E_d} + \exp -\frac{(\sqrt{E'_\perp} + \sqrt{E_\perp})^2}{2E_d} \right]. \qquad (5)$$

The dechanneling energy $E_d(z)$ is proportional to the path length. In the Bethe-Bloch approximation we estimate $dE_d/dz \approx 0.7$ eV/μm in Si. Since other perturbations mentioned above are not included, we expect to reproduce the characteristics, but not the magnitude and details of the line shape. To avoid very time-consuming calculations, further assumptions are made: (i) a fixed positron path length of 15 μm, and (ii) a fixed positron energy of 37 MeV.

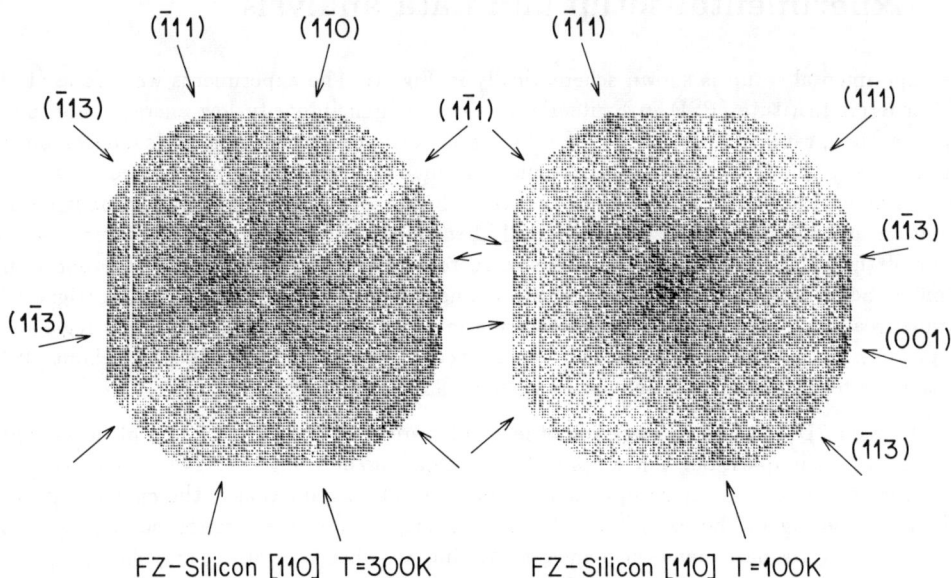

Figure 2: Intensity plot of the μ-decay positron flux emerging from FZ silicon around a $\langle 110 \rangle$ direction (dark and light features correspond, respectively, to high and low counting rates). The observed muon site transition occurs between 200 K and 300 K.

3 Results and Discussion

Some preliminary results of the experiments presented here were already reported in [10]. In Fig. 2 positron flux histograms for float-zone (FZ) silicon are shown. At 300 K blocking of the (111)-planes is clearly visible. Minor features are due to the (110)- and (113)-planes. At 100 K the effect of the (111)-planes is opposite: the positrons are now channeled along these planes, indicating a different muon site. Small effects come from the (001)- and (113)-planes. The site transition occurs between 200 K and 300 K.

In comparison to the spectroscopic data the positional transition is correlated to the disappearance of Mu. We conclude that Mu is a bistable defect in Si: the charge state transition Mu $\rightarrow$ (Mu)$^{\pm}$ [11] is related to a site transition. At lower temperatures, the muon sites are not affected during the charge changing reaction Mu* $\rightarrow$ (Mu*)$^{\pm}$ [11] . We conclude that the Mu*-location is stable for both the neutral and an ionized charge state.

In order to identify the different sites, the temperature dependence of the different muonium fractions states have to be taken into account. A good qualitative agreement (Fig. 3) is obtained with the following configurations: (i) a superposition of a near-BC fraction (40%) and a near-T fraction (60%) at 100 K, and (ii) a single near-BC site at 300 K. In detail, the T-site is displaced towards the anti-bondig (AB) site by 0.56 Å with a gaussian distribution width of 0.7 Å. This result confirms recent calculations [12] which found a local minimum of the adiabatic energy surface at a displaced T site (simultaneously the observed large reduction of the electron-spin density was obtained). The measurements also qualitatively agree with recent calculations finding the global minimum of the total energy at the BC site [13], for both the neutral and the positive charge state. In these (and most other) calculations the muon is well localized at the center of a strongly relaxed Si–Si bond. However, our measurements clearly indicate a motion of the muon in the mid plane between two adjacent Si hosts, a possible configuration also discussed in [4]. In our simulation (Fig. 3) the BC site is displaced by 0.5 Å perpendicular to the bond and a rotation around the bond is assumed.

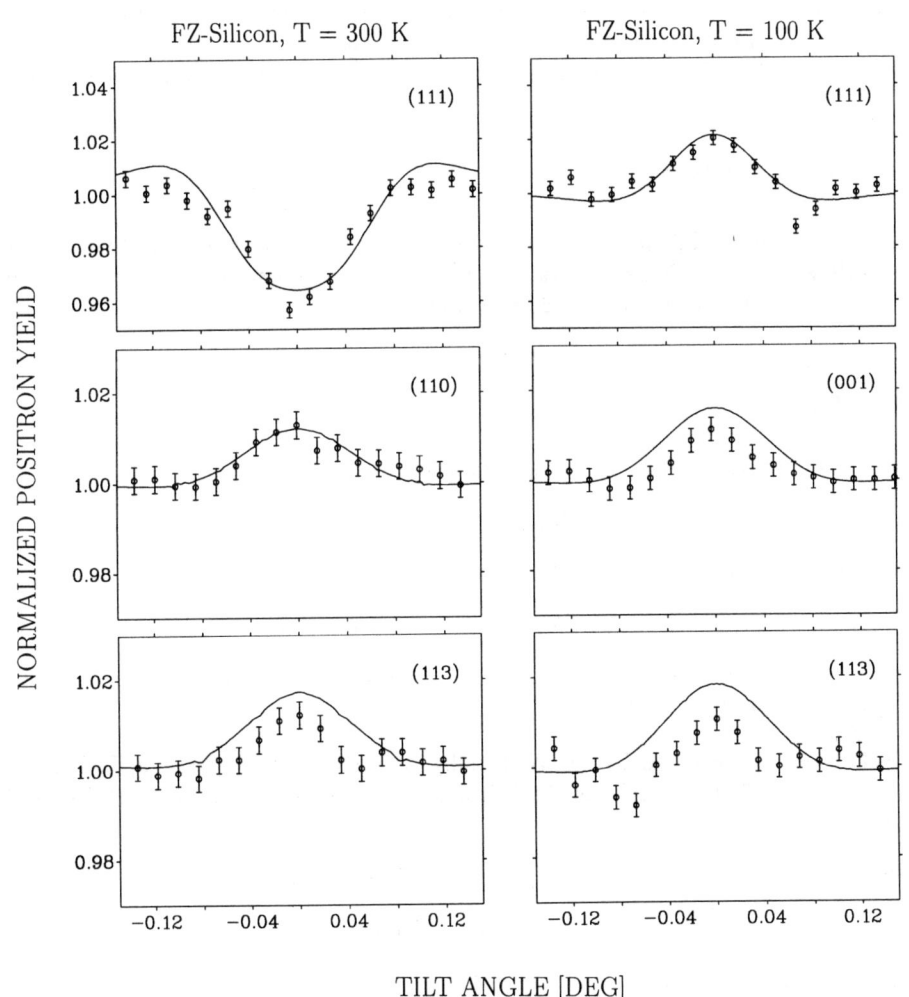

Figure 3: Comparison of measured and simulated planar μ-decay positron yields in Si: at 300 K the simulation corresponds to a single near-BC site, at 100 K a near-BC (40%) and a near-T (60%) component are superimposed (see text).

In GaAs also a metastable T-like site could be observed up to room temperature, in agreement with the existence of Mu. At higher temperatures a similar transition from channeling to blocking along (111)-planes is observed. However, the interpretation is not as clear as in Si: the site transition seems to be incomplete and occurs at higher temperatures than the spectroscopic transition. Furthermore the strong relaxations of the transverse field precession signals are not clearly due to charge state transitions. In InP no indication of a metastable site was found, in agreement with the lack of Mu precession signals. A BC-like site was observed in the temperature range investigated. The charge states at this site are unknown since only a diamagnetic signal could be detected by standard μSR.

4 Conclusions

In summary, the measurements showed the existence of a metastable T-like site for μ^+ in GaAs and Si. In InP, GaAs and Si a BC-like configuration was found to be most stable. For the latter host, a consistent description of μSR and channeling results can be given: At low temperatures two neutral states Mu (near-T) and Mu* (near-BC) coexist. On ionization at higher temperatures, the BC-configuration remains stable, while $(Mu)^+$ is trapped at BC sites.

Acknowledgements

We are grateful to K. Kotsuki, Nippon Mining (U.K.) for supplying us with specially oriented, high purity InP wafers. We thank the staff of PSI for technical assistance. This work was supported by PSI and the Swiss National Science Foundation.

References

[1] See, e.g., B.D. Patterson, Rev. Mod. Phys. **60** (1988).

[2] J.I. Friedmann, V.L. Telegdi, Phys. Rev. **105**, 1681 (1957).

[3] S.K. Estreicher, Phys. Rev. B **36**, 9122 (1987).

[4] S.F.J. Cox, M.C.R. Symons, Chem. Phys. Lett. **126**, 516 (1986).

[5] R.F. Kiefl, M. Celio, T.L. Estle, S.R. Kreitzman, G.M. Luke, T.M. Riseman, E.J. Ansaldo, Phys. Rev. Lett. **60**, 224 (1988).

[6] J.W. Schneider, M. Celio, H. Keller, W. Kündig, W. Odermatt, B. Pümpin, I.M. Savić, H. Simmler, T.L. Estle, C. Schwab, R.F. Kiefl, and D. Renker, Phys. Rev. B **34**, 1474 (1990).

[7] S.J. Pearton, M. Stavola, J.W. Corbett, Mat. Sci. Forum **38-41**, 25 (1989).

[8] J. Lindhard, K. Dan. Vidensk. Selsk., Mat.-Fys. Medd. **34**, No. 14 (1965).

[9] P.A. Doyle, P.S. Turner, Acta. Cryst. **A24**, 390 (1968).

[10] H. Simmler, P. Eschle, H. Keller, W. Kündig, W. Odermatt, B.D. Patterson, B. Pümpin, I.M. Savić, J.W. Schneider, U. Straumann, P. Truöl, Hyperfine Int. **64**, 535 (1990).

[11] S.R. Kreitzman, T. Pfiz, S. Sun-Mack, T.M. Riseman, J.H. Brewer, D.L. Williams, T.L. Estle, Hyperfine Int. **64**, 561 (1990).

[12] Dj.M. Maric, S. Vogel, P.F. Meier, S.K. Estreicher, Hyperfine Int. **64**, 573 (1990).

[13] C.G. Van de Walle, Phys. Rev. Lett. **64**, 669 (1990).

POLARIZED SPECTROSCOPY OF COMPLEX LUMINESCENCE CENTERS

S.S.OSTAPENKO AND M.K.SHEINKMAN
Institute of Semiconductors of Ukrainian Academy of Sciences, pr.Nauki 45, 252650, Kiev, USSR

ABSTRACT

Polarization diagram method is theoretically developed for three various paths: two-dipole problem, distributed pair centers (DA-pairs), dipole centers in hexagonal type crystals. The results of calculations are applied to evaluate the symmetry and to propose the models of two complex centers in GaP and CdS.

Introduction

Impurity atoms as well as native point defects can form the complex luminescent centers (CLC) composed of two or more particles. Peculiarity of photoluminescence (PL) band attributed to definite CLC is that of one part of complex center participate in absorption of excitation light while the luminescence transition is linked to another part of center. The symmetry of CLC is lower then that of the crystal lattice basically due to the fact of space correlation of both parts of CLC. As the result the anisotropy of electron transitions in optical absorption and emission spectra are observed. The methods of the polarization optical spectroscopy, in particular, the method of the luminescence polarization diagrams (PD) [1] can be applied to reveal the symmetry of CLC. The PD method in the case of a cubic crystals is based on measurement and theoretical analysis of the polarization degree of PL band induced by linear polarized excitation light. The theory of PD method was extended to three various paths: i) two-dipole problem of CLC in cubic crystals [2]; ii) polarized PL of angular distributed donor-acceptor pairs [3]; iii) polarization of CLC in hexagonal-type crystals [4].
In this paper we demonstrate the effectiveness of the PD method to investigate the properties and to reveal the structure of two specific CLC in GaP and CdS.

Method of polarization diagrams

Following to PD method anisotropic PL centers are excited by linear polarized light specified by angle φ between electric field vector of excitation beam and some of crystal axis, e.g <100> of cubic lattice or C axis of hexagonal lattice. Orthogonal scheme of the PL excitation is suitable for these measurements: the direction of excitation light and that of PL recording are perpendicular to one another. The PL intensity demonstrates linear polarization due to removal of orientation degeneracy of anisotropic centers. The angular dependence of PL polarization degree defines the PD curve:

$$P(\varphi) = \frac{I_{\parallel}(\varphi) - I_{\perp}(\varphi)}{I_{\parallel}(\varphi) + I_{\perp}(\varphi)} \qquad (1)$$

$I_{\parallel}$ and $I_{\perp}$ are the intensities of two orthogonally polarized PL components when the analyzer is respectively parallel and perpendicular to the chosen crystal axis.
In two-dipole case different PD curves were classified with respect

to the mutual orientations of absorbing and emitting dipoles [2]. Each dipole is oriented along <100>, <111> or <110> cubic crystal axis. These calculations are applied to analyze the PD curves of various CLC in GaAs [5] and GaP [6]. Another problem to be solved is that of symmetry of DA-pairs randomly distributed in a cubic lattice. Such CLC are characterized by arbitrary orientation of optical dipole directions specified by the unit vector <a,b,c>. It was found the relation between DA-pair coordinates and of the PD extreme points [3]

$$a^2b^2 + a^2c^2 + b^2c^2 = \frac{P(0)}{3P(0) + 2P(45°)} \quad (2)$$

The results of calculations were used for the DA-pairs in AlSb [7]. As to employment of PD method in hexagonal lattice (CdS) one should take into account the effect of crystal birefringence and also the peculiarity of symmetry equivalent orientations in such a lattice. The dipole center can be described by angle Ψ with respect to C axis of crystal and phenomenological parameter of β ($0 \leq \beta \leq 1$) attributed to superposition of π and σ dipoles in optical transitions. Due to calculations [4] the value of Ψ is expressed via intensities of polarized PL

$$\cos^2\Psi = \frac{1 \pm S_1}{1 \pm 3S_1} \; ; \quad S_1^2 = 2 \cdot \frac{I_\parallel(0) \cdot I_\parallel(90°) - I_\perp(0) \cdot I_\perp(90°)}{[I_\perp(0) - I_\parallel(0)] \cdot [I_\perp(90°) - I_\parallel(90°)]} \quad (3)$$

The information available from the PD method is the experimental proof of the optical anisotropy of PL center, orientation of optical dipoles evaluated from theory and, finally, the center symmetry. The symmetry of CLC provides the background to propose its model.

Polarized photo- and thermoluminescence of deep centers in GaP

Specific CLC are observed in n-GaP contaminating N impurity. It is argued that for one of them NN_1-pair forms the complex center with the deep acceptor of Cu_{Ga} as shown on fig.1.

1. Spectroscopy, kinetics and temperature measurements.

Bulk Czochralski grown n-GaP single crystals doped with S or Te of $10^{17} cm^{-3}$ are investigated. The contamination of residual N is found

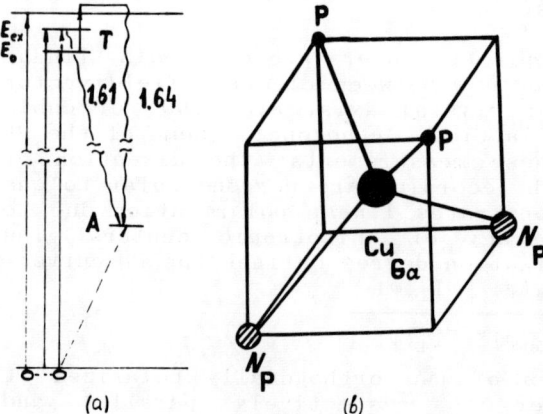

Fig.1. Scheme of electron transitions (a), the model of CLC (b).

of order of $10^{16} cm^{-3}$. Three PL bands are observed with maxima of $h\nu_A$ = 1.55 - 1.58eV (A), $h\nu_B$ = 1.42eV (B) and $h\nu_C$ = 1.72eV (C)- fig.2, curve 1. The following is concerning to A-band and the data for B- and C-bands are summarized in Table. All samples could be separated as two various groups. The following is specific for the group one of crystals. 1) A-band is excited preferentially in "impurity" spectral region of 2.2 - 2.3eV; 2) the kinetics of PL demonstrates slow decay (I_s)

with $\tau_s \sim 10^2$s at 77K (insert on Fig.2). The PL band of $I_s(h\nu_{lum})$

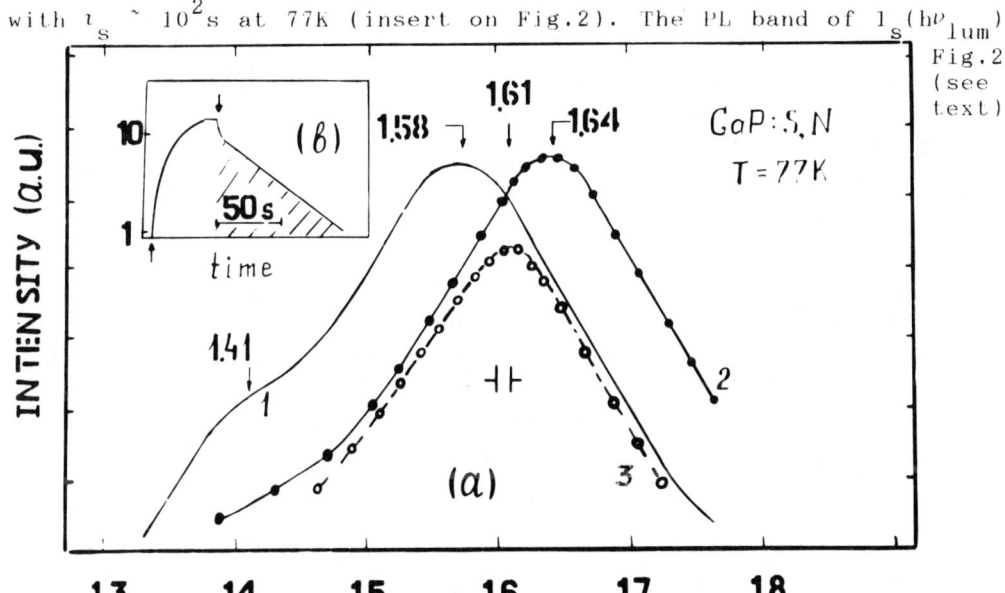

Fig.2 (see text)

possesses the maximum of 1.61eV (fig.2, c.3); 3) the value of τ_s exponentially decreases when the temperature is increased with the activation energy of $\Delta=0.095$eV. This is accompanied by a spectral shift of A-band maximum up to 1.64eV (fig.2, c.2); 4) the thermoluminescence (TL) peaks with the maxima of $TL_A=97$K, $TL_B=118$K and $TL_C=103$K are observed. The luminescence spectrum of the TL_A-peak coincides with A-band of PL at T>100K and, hence, the same deep center participats both in PL and TL. We notice that TL_B- peak and TL_C- peak are linked to another PL bands (Table). The characteristics 1) - 4) are specific for the group one of GaP crystals and are not attributed to the crystals of group two.
We carried out detailed temperature, spectral and kinetic measurements of TL_A-peak. It was found: i) the energy of the electron trap related to the peak is $E_c - (0.13\pm0.01)$eV;

ii) the threshold of the "impurity" excitation is given by $h\nu_{th} = 2.20 \pm 0.02$eV;

iii) spectrum of IR-quenching of the TL peak is related to the hole photoionization from the deep acceptor of E_v +0.51eV and coincides with that of 1.65 eV PL band in GaP:Cu.

2. *Polarized photo and thermoluminescence*
To reveal the symmetry of the deep center we used the method of PD. The $P(\varphi)$ curve of the PL and of the TL lightsum are shown in

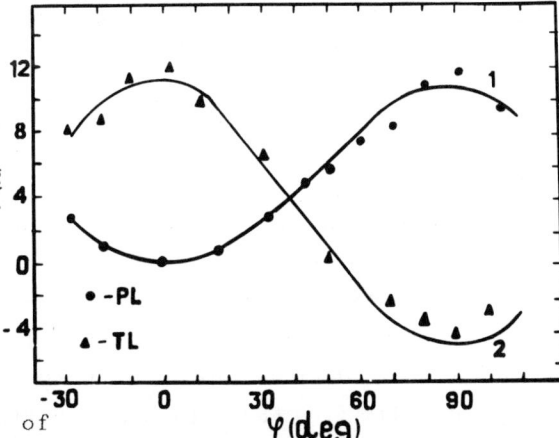

Fig.3. Polarization diagrams of A-band and TL-peak.

fig.3. The observation of the induced polarization of luminescence directly demonstrates the optical anisotropy of corresponding center. The prominent feature of the polarized TL is that of PD curve is essentially different from the diagram attributed to the PL. Actually, these curves can not be adjusted to one another by the parameters of the PD theory. As the consequence, the mutual orientations of absorbing and of emitting optical dipoles calculated are various for PL and TL. These dipoles are oriented along <110> - <111> axes in the case of PL and are directed along <110> - <001> for TL (Table).

3. Discussion

To explain the peculiar features of the PL and TL as well as polarization data concerning to A-band and TL_A peak we propose the new model of the CLC (fig.1). This center is composed of deep acceptor (A-center) with the level of $E_v + 0.51eV$ and of a deep electron trap (T-center) with the energy of the ground state $E_c - 0.13eV$ (E_o) and that of the excited state $E_c - 0.035eV$ (E_{ex}).

In fig.1a the scheme of the electron transitions given rise to PL, TL and their excitation is shown. The "impurity" excitation of both A-band and of TL_A-peak are caused by indirect excitation process. The electron is optically transfered from v-band to T-level (either E_o or E_{ex}), free hole is captured by the A-center within the same CLC. The subsequent recombination of the electron thermally released to c-band with the bound hole leads to 1.64eV band of PL and TL. The slow thermally activated PL kinetics (I_s) is resulted from the center-to-center recombination of the electron bound to T-trap and the hole bound to A-center. This electron recombines via the excited state which is thermally populated with the barrier $\Delta = E_0 - E_{ex}$.

A specific feature of the TL is the observed polarization under polarized indirect excitation. Our data contradict to the trivial scheme of polarized TL, when the lightsum storage is resulted from the direct photoionization of anisotropic deep acceptor following by the capture and release of electrons to/from distant traps. Actually, this is demonstrated by the facts: 1) PD curves for PL and TL are essentially different; 2) "impurity" excitation are caused by electron transitions v-band ⇒ T-level, i.e. related to indirect excitation process; 3) each TL peak is linked one-to-one to the specific luminescence band (Table). These facts as well as polarization characteristics of PL and TL are naturally explained if T and A centers are nearest neighbours in GaP lattice and form the CLC. The strong localization of wave functions both of the captured electron in the ground state of T-trap and of the bound to A-center hole explains the storage of the lightsum within such a complex center at low temperatures and possibility to observe TL.

4. The model of CLC

The components of CLC are identified as Cu_{Ga} - deep acceptor (A) and NN_1-pair - isoelectronic trap (T).

A-center. The deep acceptor of $E_v+0.51eV$ in GaP was identified with Cu_{Ga} impurity in [8]. In our crystals the concentration of T-A complexes increases after diffusion of Cu at $950°C$. The orientation of emitting optical dipole attributed to c-band ⇒ A-center transition is linked to <100> crystal axis. This orientation is unambiguously related to A-center symmetry and corresponds to the symmetry of Cu_{Ga} complex center observed by ODMR [9].

T-center. The identification of T-trap as NN_1 is argued by the following: 1) the energy positions of the ground and excited states are identical with those of NN_1 [10]; 2) the orientation of the optical dipoles concerning to v-band → T-level transition is <110> that coincides with the data of uniaxial stress measurements for NN_1-pair [11]; 3) A-T complexes are observed only in crystals contaminating nitrogen.

5. Conclusions

We notice that the centers demonstrating polarized TL are fairly typical for bulk GaP. Various luminescent bands as well as TL peaks linked to them are shown in Table. We are not able presently to propose substantiated models for the rest of centers. However, it is clear from the polarization data that each of them is related to the complex of deep acceptor and of electron trap. We believe that such centers represent a new system of the crystal defects being able locally accumulate the lightsum within a crystal.

Table

Energy position of luminescence, eV	TL-peak maximum, K	Deep acceptor energy, eV	Electron trap energy, eV	Orientation of optical dipoles
1.64	97	E_v + 0.51	0.13	<110>-<001>
1.42	118	E_v + 0.93	0.17	<110>-<1$\bar{1}$0>
1.72	103	E_v + 0.51	0.15	<110>-<111>

Light stimulated PL metastability in electron irradiated CdS

The origin of metastability for deep centers in a semiconductors is still the point of current technological and physical interest. Only a few of cases are known that provide well supported microscopical model of centers involved in metastable transitions [12]. We present the summerized results of comprehensive study of the new effect attributed to the transient behavior of the "red" (PL) band ($h\nu_{max}$=1.68eV) in electron irradiated CdS. The mechanism for PL metastability is developed within the model of a three-particle CLC.

1. Experiment

Czochralski grown bulk n-CdS single crystals are studied. The crystals are irradiated by 1.2 MeV electrons at 200K with the dose of 10^{16}-10^{18} cm^{-2}. As the result of electron irradiation the intensity of the "red" PL band of $h\nu_{max}$=1.68eV demonstrates the stimulation, ΔI, and subsequent decline with a time of intensity (fig.4,c.2). This transient behavior is observed under PL excitation in a spectral region of stimulation light $h\nu_{st}$=2.5eV and unambiguously related to the preliminary illumination of a sample with the pumping light of $h\nu_p$= 2.0-2.4eV (fig.4,c.1).

The concentration of centers of the stimulated-"red"-luminescence is gradually increased with an irradiation dose and reaches the value of $10^{13} cm^{-3}$. Thus, the effect of PL stimulation is directly

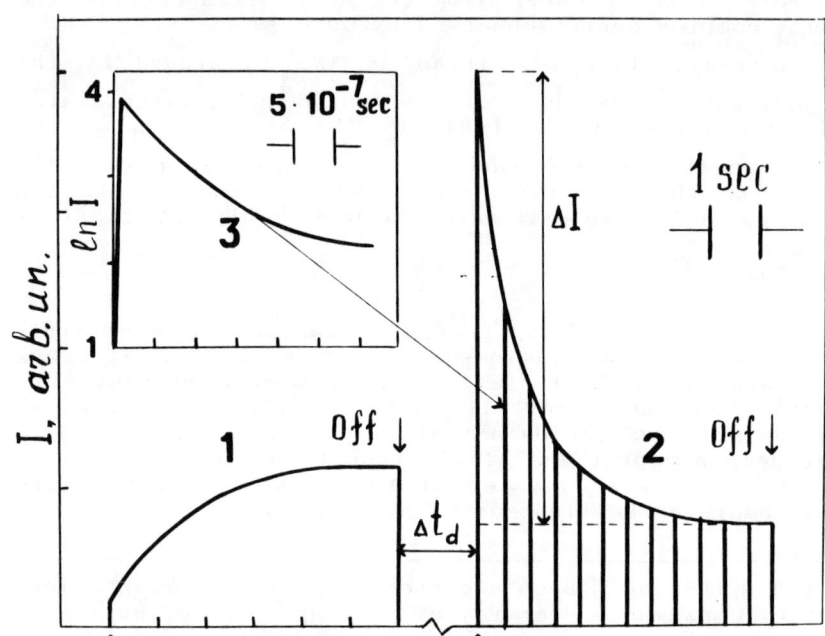

Fig.4 Kinetics of stimulated PL under pumping (1) and stimulated(2) light. On insert-pulse excitation.

related to e-beam treatment of CdS.
The following data are essential reasons to evaluate the mechanism for the effect observed. 1) stimulation of ΔI is not accompanied with stimulation of the photocurrent in n-type samples. It means that stimulated PL is resulted from the recombination of localized electrons. 2) The kinetic of ΔI measured with the pulse excitation (fig.4,c.3) is much slower then that of a free hole capture. Thus, stimulated PL is originated from the bound-to-bound recombination. 3) The value of ΔI is quenched when the temperature is raising up with an activation energy of $\Delta = 0.08$ eV. This energy is linked to the gap between the ground state ($E_c - 0.12$ eV) and excited state ($E_c - 0.04$ eV) of the peculiar donor center (D). The level of D-center can be directly filled with electrons by pumping light as demonstrated by TSC measurements. After switching of the pumping light off the ΔI decays in a dark following to relation: $\Delta I \sim \exp(-\Delta t_d / \tau_r)$ with $\tau_r = 5 \cdot 10^3$ s at 77K. This process is the result of thermal release of electrons from D-center. 4) Besides of D-center, the level of a shallow acceptor (A) is observed in pumping and stimulation spectra. Its energy position is evaluated and given by $E_v + 0.14$ eV. 5) The low symmetry of center being the consequence of anisotropy of deep acceptor K with the level of $E_v + 0.68$ eV is demonstrated by the PD method [14]. We emphasis the coincidence of the polarization characteristics of stationary, I, and stimulated, ΔI, "red" PL. In particular, polarization degrees of PL and PL excitation as well as their polarization diagrams are identical for both bands. The later means the same orientation of the optical dipoles within the centers (equality of angle Ψ) and identity of symmetry of both centers. 6) stimulated PL is annealed at $T_{an} > 160°C$ with the activation energy of $(0.43 - 0.49)$ eV. The annealing mechanism is identified as diffusion of mobile S_i atoms in CdS lattice and their recombination with D-component of CLC [15]. Thus, we observe in experiments the complex triple PL center consisted of deep acceptor K, deep donor D and shallow acceptor A.

2. Mechanism of metastability

By analyzing the data it was concluded that no one mechanism of metastability available can be fitted to experiments [16]. The new mechanism for the PL metastability was proposed within the model of a triple D-A-K center (fig.5,a-c). According to this mechanism the components of a triple center can be recharged as a result of the pumping light illumination. This is a two-step process: i) intercenter A $\Rightarrow$ D optical electron transition (a); ii) hole tunneling A $\Rightarrow$ K (b). The total charge state of a complex is, evidently, unchanged after pumping, but recharging of D and K levels due to localization of electron and hole is taken place (c). We suggest that D $\Rightarrow$ K recombination has a low probability after pumping as a result of: i) strong localization of the electron and the hole wave functions of the D and K centers (both deep) and separation of D and K in a lattice; ii) the existence of the repulsive barrier of negatively charged A-center near- neighboring to K. The barrier is removed under stimulated light due to photoionization of A-center that makes D $\Rightarrow$ K transition allowed (c).

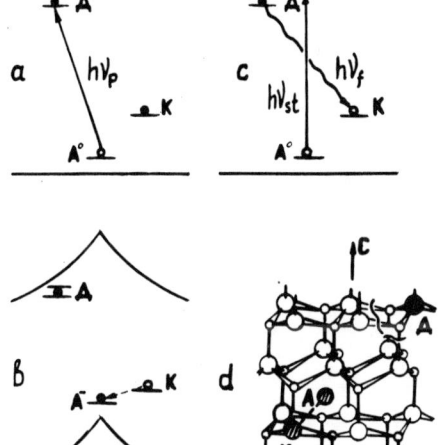

Fig.5. On mechanism of PL metastability (a-c); K-Cu_{Cd}, A-S_i, D- V_S (d).

The model of a triple center is developed in [14]. It is argued that A and D are the components of Frenkel pair, respectively, S_i and V_S, created by electron irradiation in the vicinity of deep acceptor (K). The later was identified with substitutional Cu_{Cd} impurity.

References
1. P.P.Feofilov, The Physical Basis of Polarized Emission,1961,NY.
2. I.A.Buyanova,E.I.Oborina and S.S.Ostapenko,Semicond.Sci.Technol, 4, 797 (1989).
3. S.S.Ostapenko, J.Phys.C (1991), be published.
4. S.S.Ostapenko, M.A.Tanatar and M.K.Sheinkman, Opt.Spectr.(USSR), 48, 430 (1980).
5. I.A.Buyanova,S.S.Ostapenko and M.K.Sheinkman, Sov.Phys.Solid State, 27, 461 (1985).
6. I.A.Buyanova,S.S.Ostapenko and M.K.Sheinkman, Sov.Phys. Semicond., 20, 1123 (1986).
7. S.S.Ostapenko, G.Hofmann, Solid State Commun.,74, 447 (1990).
8. H.G.Grimmeiss, B.Monemar, Phys.Stat.Sol.(a), 19, 505 (1973).
9. H.P.Gislason,et al.:Phys.Rev.B, 26, 827 (1982).
10. E.Cohen, M.D.Sturge, Phys.Rev.B, 15, 1039 (1977).
11. B.Gil, et al. Phys.Rev.B, 33, 2690 (1986).
12. G.D.Watkins, Materials Science Forum 38-41, 39 (1989).
13. N.S.Bogdanyuk, A.P.Galushka, S.S.Oatapenko and M.K.Sheinkman, Sov.Phys.Semicond. 18, 189 (1984).
14. N.S.Bogdanyuk, A.P.Galushka, S.S.Oatapenko and M.K.Sheinkman, Sov.Phys.Solid State, 27, 1155 (1985).
15. N.S.Bogdanyuk, S.S.Ostapenko, Phys.Stat.Sol.(a), 96,621 (1986)
16. S.S.Ostapenko, Semicond.Sci.Technol. (1991), be published.

ONP SPECTROSCOPY OF DEFECTS IN SILICON

Nickolay T. Bagraev and Igor S. Polovtsev

A. F. Ioffe Physico-Technical Institute, Leningrad, 194021, U.S.S.R.

ABSTRACT

Optical Nuclear Polarization (ONP) technique was used to study metastable properties of the gold donor center in silicon. A model of a deep defect's symmetry changing $C_{3v} \to C_{1h} \to D_{2d}$ with its charge state ($D^- \to D^0 \to D^+$) is proposed to account for observed optically induced quenching and regeneration of Au^0 centers.

1. Introduction

The fundamental discovery of optical pumping by A. Kastler [1], paralleled by major advances in magnetic resonance spectroscopy, gave rise to investigations of optical nuclear polarization (ONP) in semiconductors [2-5].
The ONP technique requires the creation of an electron system in spin-nonequilibrium, in which state the orientation of lattice nuclei is governed by hyperfine interaction (HFI). It has been established [3] that ONP does not involve conduction electrons and is due, instead, to the HFI of optically oriented electrons captured on defects with neighboring lattice nuclei. Hence, the ONP technique provides an important tool for investigation of shallow/deep impurity centers [3,4], thermal [3] and radiation [4] defects, magnetic impurity centers [3] and dislocations [3,4] in semiconductors.
Metastability of deep centers, a feature of light-stimulated reactions involving defects in semiconductors, is here considered as a further example of the ONP technique's great investigatory potential.

2. The ONP technique

Electrons and lattice nuclei magnetic isotopes are known to form interrelated spin systems in crystals. If the electrons are prevented - by some external force - from reaching thermodynamic equilibrium, they will necessarily engage - due to their spin-nonequilibrium state - in a hyperfine interaction with lattice nuclei, causing the crystal's nuclei to undergo dynamic polarization:

$$P_n - P_{n0} = \xi \cdot (P_e - P_{e0}) \qquad (1)$$

where P_e and P_n are the nonequilibrium polarization states of electrons and nuclei respectively, and P_{e0} and P_{n0} are the respective Boltzmann's equilibrium polarization states for electrons and nuclei. $P_n = (n_+ - n_-)/(n_+ + n_-)$; $P_e = (N_+ - N_-)/(N_+ + N_-)$; N_+, N_-, n_+ and n_- are the respective populations of the states with the electron and the nuclear spin projections; $m_S = +1/2$, $m_S = -1/2$ and $m_I = +1/2$, $m_I = -1/2$; and ξ is the term accounting for the relative contributions made to nuclear polarization by the contact and the dipole-dipole HFI modes.
One way to accomplish nonequilibrium polarization of electrons

($P_e \neq P_{e0}$) in a semiconductor crystal is by optical pumping. By this method, illumination is carried out with circularly polarized light using a longitudinal magnetic field, to create a strong polarization for electrons through their photoexcitation to the conduction band, so that when subsequently captured at impurity centers or lattice defects, they can polarize the neighboring lattice nuclei ($P_n \approx P_e$) via hyperfine interaction [3,4]. The ONP process begins with the formation of a sphere of polarized lattice nuclei around a defect: $P_{nm} = \xi(\delta) \cdot P_e$. In the next stage, ONP spreads from this sphere of radius δ over the entire crystal via nuclear spin diffusion at a rate which is governed for defects uniformly distributed in the crystal bulk by an exponential law of the form [3]:

$$P_n = P_{nm}(1 - \exp(-t/T_1)) \qquad (2)$$

Here T_1 is the spin diffusion dependent nuclear spin-lattice relaxation time: $1/T_1 = 4\pi N D \delta$; where D is the nuclear spin diffusion coefficient, δ is the nuclear spin diffusion radius [3], and N is the concentration of defects that are involved in the ONP process.

The two ways of recording electrons and nuclei in the polarized state are optical detection of magnetic resonance [5] and classical NMR [3,4]. Using experimental values of P_n obtained for specific relaxation times (2), magnetic field strengths, light intensities etc., it is possible to establish various P_{nm} and T_1 values, and thus gain information on the nature, concentration, and distribution of impurities and defects that are present in a given semiconductor crystal [3,4]. Below, the capabilities of the ONP technique are illustrated in the instance of n-type silicon single crystals containing gold donor centers.

The samples used in the experiment were n-type silicon with initial resistivity 1.0 $\Omega \cdot$cm. Doping with gold was done using high-temperature (1200°C) diffusion technique, followed by quenching in oil. Spectroscopic data revealed the presence of a level in the gap at $E_v + 0.35$ eV, known to correspond to an Au^0 donor center resulting from the $Au^+ \rightarrow Au^0$ transition. Illumination of the samples was carried out at 77 K using circularly polarized light from a 1 kW incandescent lamp, with the magnetic field (0.7 Oe) being oriented along the light path [3]. Following illumination, the samples were heated to room temperature and placed in a wide-line NMR radiospectrometer where ONP is measured using fast adiabatic passage technique [3,4]. Nuclear magnetization values were taken in the samples ranging with regard to illumination time from 5 min to 7 hours. Extrapolating the results of measurements gives the degree of ONP, P_{nm}, and the relaxation time T_1 for the ^{29}Si nuclei under investigation.

3. Results

The observed ONP in n-type silicon doped with gold can be described in terms of the following model [3]. A neutral Au^0 donor center captures a hole, generated by interband pumping, from the valency band: $Au^0 + h \rightarrow Au^+$, and becomes an effective trap for optically polarized electrons. Addition of an electron

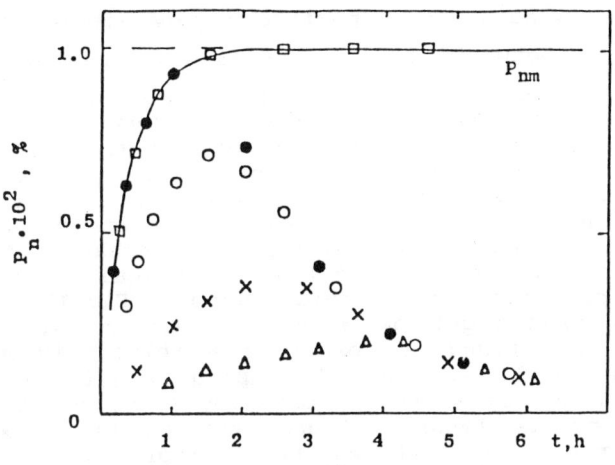

Fig.1
Kinetics of the ONP degree in silicon doped with gold; (□)monochromatic interband circularly polarized light; (•)broad spectrum circularly polarized light; (○,×,Δ) broad spectrum circularly polarized light after prior illumination at hυ=0.9 eV for 2h (○), 3h (×) and 4h (Δ). Solid line - calculated dependence from Eq.(2).

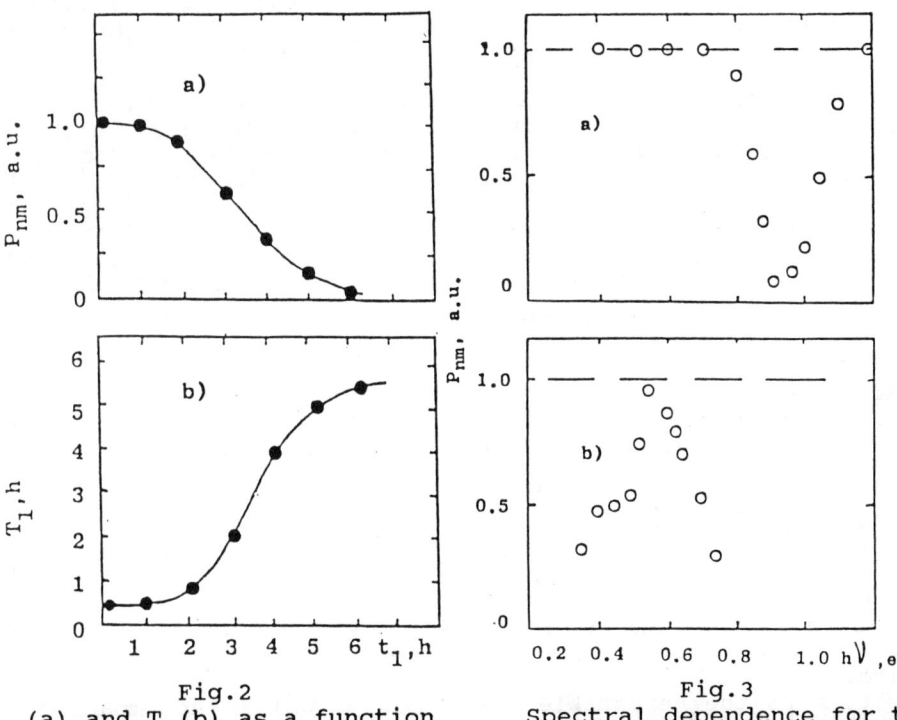

Fig.2
P_{nm} (a) and T_1 (b) as a function of prior illumination time at hυ=0.9 eV in silicon doped with gold

Fig.3
Spectral dependence for the effect of prior illumination on the ONP degree in silicon doped with gold; pumping time 5h; (a) quenching, (b) regeneration

from the conduction band to Au^+ turns it into a polarized Au^0 center: $Au^+ + e^* \rightarrow Au^{0*}$. The hyperfine dipole-dipole interaction that must arise between the polarized Au^{0*} centers and the surrounding ^{29}Si nuclei results in these nuclei becoming polarized [3,4], with the polarized state eventually setting up over the whole of the crystal because of nuclear spin diffusion (see Eq.(2)). The ONP degree, however, was found varying, depending on the properties of light used for optical pumping. Whereas illumination with monochromatic interband circularly polarized light produced an ONP degree that behaved strictly in accordance with the above described classical mechanism for lattice nuclei polarization in semiconductors (Fig.1), the use of circularly polarized light of a broad spectrum (including the impurity light) was found to result in a relatively much lower ONP degree under prolonged optical pumping (Fig.1). In order to understand the role played in the observed anomalous behavior of the ONP degree by the impurity light, the kinetical experiments on the ONP degree were performed using prior illumination with monochromatic impurity light. Figures 1-3 show that as quenching on the ONP degree decreases with prior illumination time, the nuclear spin-lattice relaxation time (T_1) increases (Figs.1 and 2). This is evidence of a reduction in the concentration of paramagnetic Au^0 centers, the carriers of the ONP. On impurity light shutdown, the gold donor center system continued in a metastable state for an indefinitely long period of time as long as the temperature was kept at 77 K. The concentration of the Au^0 centers regained its initial level only on heating the samples to room temperature. The spectral dependence of the ONP degree quenching is shown in Fig.3. A similar quenching event was observed for interband photoconductivity. Taken together, these results suggest that the observed anomaly in the behavior of the ONP degree is due to metastability of the isolated gold center in silicon; an appropriate model is presented in Fig.4. The model postulates that because of a nonmonotonic dependence of the electron-vibrational constant on the number of electrons present at the defect, different charge states of a deep defect would occupy different positions in the lattice and would accordingly have dissimilar symmetry (Fig.4)[6].. While Au^- would tend to C_{3v} and Au^+ to D_{2d}, the paramagnetic state Au^0 is most likely to assume the lowest symmetry - C_{1h} [7] which lies closest to the lattice site.

Careful examination of the effect produced by prior illumination with $h\nu=I_1$ light will show that, in fact, we have an optical analog of the negative-U reaction:

$Au^0 + h\nu(I_1) \rightarrow Au^+$

$$2Au^0 + h\nu(I_1) \rightarrow Au^- + Au^+ \quad (3)$$

$Au^+ + 2e \rightarrow Au^-$

leading to a slow charge exchange process as the gold centers tunnel between the positions of different symmetry in the silicon lattice.

Transition to the metastable state ($Au^- + Au^+$) is accompanied on the one hand by a reduction in the concentration of Au^0 donor centers, and on the other, by a sharp drop in the lifetime of photoexcited electrons because of the photo-induced recombination

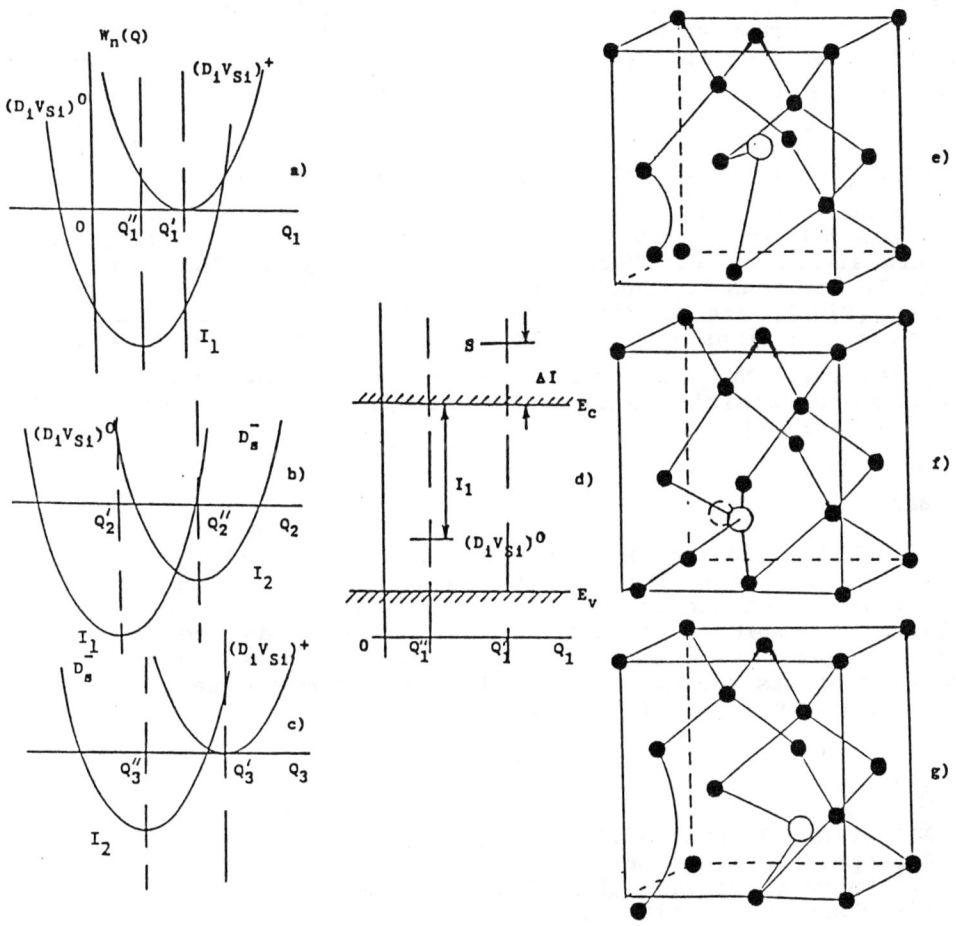

Fig.4
Two-electron adiabatic potentials (a,b,c) and equivalent one-electron band diagram (d) for the gold donor center in n-type silicon; $I_2=I_1+\Delta I_0$. Model of the gold donor center in silicon (e) $Au^-=D_s^-$, (g) $Au^0=(D_1V_{Si})^0$, (f) $Au^+=(D_1V_{Si})^+$; Configuration coordinates Q_1, Q_2 and Q_3 approximately correspond to [100], [111] and [$\bar{1}1\bar{1}$] directions in the order given.

corresponding decline in the ONP degree, as P_e is proportional to the rate of filling the Au^0 donor centers by photoexcited electrons, which is a function of their lifetime. An adequate test of the proposed model was provided by the observation of ONP regeneration over the spectral range corresponding to the reverse reaction of $Au^+ + Au^- + h\nu(E_g-I_1) \rightarrow 2Au^0$ (Figs.3b and 4). The regeneration of the ONP degree is additionally aided by photo-ionization experienced by induced Au^- centers (Fig.4b): $Au^- + h\nu \rightarrow Au^0$. It is of interest that the highest probability for this process occurs at pumping light energies corresponding to transition of electrons to the L valley of the conduction band. It may thus be concluded that the reconstructed Au^- and Au^+ states (Fig.4) are of different symmetry and formed mainly from the wave functions of L and X valleys of the conduction band, respectively. The unreconstructed state Au^0 is probably in a low-symmetry position and is due to the wave functions of the valency band [8].

Summary

The capability of the ONP technique as a tool to study defects in semiconductors has been demonstrated in its application to investigate metastable properties of gold donor centers in silicon. A model of an isolated gold center has been described which suggests that each of the defect's charged states has a symmetry of its own and is formed mainly from the wave functions of different valleys of the conduction band.

References

1. A.Kastler, J.Phys.Rad.<u>11</u>, 255 (1950).
2. G.Lampel, Phys.Rev.Lett.<u>20</u>, 491 (1968).
3. N.T.Bagraev, Physica <u>116B</u>, 236 (1983).
4. L.S.Vlasenko, Physica <u>116B</u>, 281 (1983).
5. G.Hermann, G.Lampel, V.I.Safarov, Ann.Phys.Fr.<u>10</u>, 1147 (1985).
6. N.T.Bagraev, V.A.Mashkov, Sol.St.Commun.<u>65</u>,1111 (1988).
7. M.Höhne, phys.stat.sol.(b) <u>119</u>, K117 (1983).
8. I.G.Atabaev, N.T.Bagraev, V.A.Mashkov, M.S.Saidov, U.Sirozhev, A.Yusupov, Sov.Phys.Semicond.<u>23</u>, 525 (1989).

Nuclear Spin Polarization by Optical Pumping of Nitrogen Impurities
in Semiconcuctors

Kouichi Murakami, Kazusato Hara, and Kohzoh Masuda
Institute of Materials Science, University of Tsukuba
Tsukuba Academic City, Ibaraki 305, Japan

ABSTRACT

We have investigated optical pumping effects of off-center nitrogen (N) in Si, on-center N in hexagonal single crystal SiC (6H-SiC), and off-center N in diamond. It is found that three physical conditions, i.e. (i) nuclear spin memory, (ii) no electron spin memory, and (iii) large spin flip-flop relaxation rate w_2, are necessary for dynamical nuclear spin polarization induced by unpolarized light illumination.

1. Introduction

Metastability and bistability of defects in semiconductors have recently been studied with interest from fundamental and technical points of view.[1] There are several examples in silicon. Light element impurities such as hydrogen, carbon, nitrogen, and oxygen show metastability/bistability in the forms of bond-center and tetrahedral interstitial H [1], a pair of interstitial C and a substitutional C [1], off-center N [2], and a vacancy-interstitial O pair [3], respectively. Among many defects and impurities with metastable and/or bistable configurations, the off-center N is one of defects that have geometrically simple structures.[2] For study of meta-stability/bistability of defects we think motional effects are significantly important. In a previous paper [4], we reported for the first time an electron spin resonance (ESR) measurement of the off-center N in Si revealed the motional effects among four equivalent off-center sites and one on-center site and an optical pumping effect yielding dynamical spin polarization of nitrogen nuclei. This nuclear spin polarization may be a useful probe for the study on motional effects and metastabilty/bistability, as well as hyperfine interaction, of defects and impurities in semiconductors.

In order to explain the mechanism of optical-pumping nuclear spin polarization of nitrogen impurities with hyperfine (hf) interactions in semiconductors, we have investigated off-center N in Si, on-center N in 6H-SiC, and off-center N in diamond. The off-center N centers in Si and diamond have bistable and metastable configurations [2,5], so that optical excitation induces local motion among off-center sites and an on-center site even at low temperatures, whereas the on-center N in SiC [6] shows no local motions. In this paper, we report physical conditions for paramagnetic nitrogen impurities to show nuclear polarization by unpolarized light illumination.

2. Experimental Procedure

Off-center N in Si was introduced by ion implantation and subsequent pulsed laser annealing.[2] The concentration was $\sim 1 \times 10^{18}$/cm^3 within a surface layer of ~ 100 nm. The energy level was evaluated to be E_c-0.33eV.[7] On-center N in 6H-SiC with a band gap of 3.0eV was doped during subrimation crystal growth. The concentration was obtained by Hall effect measurements at room temperature to be 3.1×10^{17}/cm^3. The energy level of shallow donor N in 6H-SiC was reported to be approximately 0.1 eV.[8] Off-center N center in diamond with a band gap of 5.5 eV forms a level of E_c-1.7 eV.[9]

The ESR measurements were mainly made at a temperature of 4 K with an X-band spectrometer, using a continuous-flow liquid helium cryostat. Unpolarized light illumination was performed through a window of ESR cavity (TEM_{011} mode). Background illumination (300 K and visible) was not eliminated. A Nd:YAG laser with a wavelength of 1060 nm or a He-Ne laser with 1150 nm were used for Si and SiC samples, and a Xe-lamp or monohromatic light with energies higher than 2.7 eV for diamond samples. Microwave powers and light intensities were varied for each ESR measurement to investigate optical pumping effects.

3. Results and Discussion

3.1. Optical Pumping Effects

Typical ESR spectra of N impurities are shown in Fig. 1. Nitrogen, ^{14}N (99.63% natural abundance), has a nuclear spin of 1, so that three hyperfine (hf) lines are observed if g and hyperfine tensors are isotropic.[2,5,6] Significant changes in each intensity of the hf lines of off-center N in Si by optical pumping are seen in Fig. 1(a) and (b). The three hf lines decreases in intensity at 4 K by light illumination at relatively low microwave powers. The degree of decrease is larger in the order of hf lines at high, intermediate, and low magnetic fields that corresond to M_I (quantum number of N nuclear spin) = -1, 0, and +1, respectively. At high microwave powers ESR saturation occurs and the hf lines of M_I = +1 are enhanced by light illumination whereas hf lines of M_I = -1 are supressed.

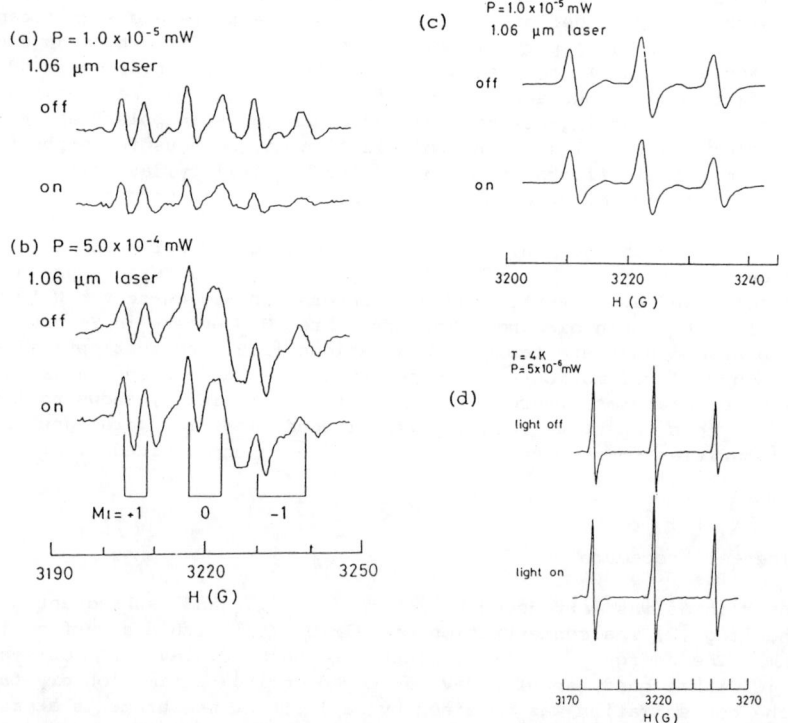

Fig.1 ESR spectra taken at 4 K without (off) and with (on) light illumination for (a) and (b) off-center N in Si (H // <110>), (c) on-center N in SiC (H // <100>), and (d) off-center N in diamond (H ⊥ c-axis).

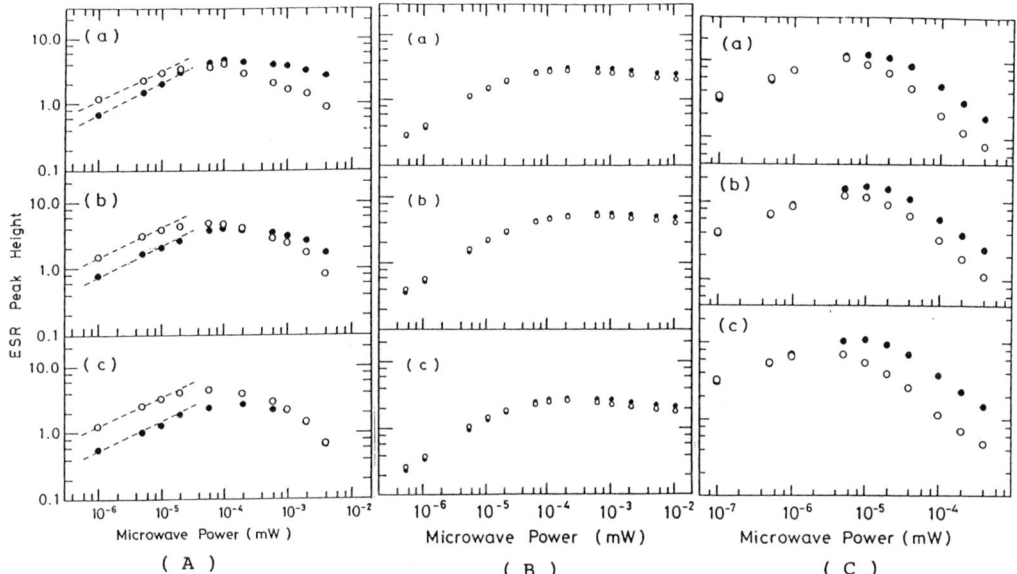

Fig.2 Microwave power dependence of ESR hf lines (●: under light illumination, o: without light illumination) for (A) off-center N in Si, (B) on-center N in SiC and (C) off-center N in diamond. The top (a), middle (b), and bottom (c) figures correspond to $M_I=+1$, 0, and -1, respectively.

For hf lines of N in SiC and diamond, no significant changes are observed in Fig. 1(c) and (d). The relative amplitudes of ESR for M_I = +1, 0, and -1 are plotted as functions of microwave power without (off) and with (on) light illumination in Fig. 2. As reported in the previous paper [4], ESR changes of off-center N in Si by light illumination is caused by optical pumping effect resulting in dynamical nuclear spin polarization. Such an optical pumping effect can not be seen in ESR spectra of SiC and diamond. In these cases light illumination affects only the spin-lattice relaxation process, i.e. shortening of the relaxation time.

3-2. Physical Conditions for Nitrogen Nuclear Spin Polarization

For paramagnetic nitrogen impurities with a hf interaction between an unpaired electron of N atom and the N nucleus in semiconductors, energy levels in a strong magnetic field and transitions are shown in Fig. 3(a). Light illumination induces excitation (process U) of unpaired electrons to the conduction bands, and then the excited electrons are captured by N^+ ion (process R). For off-center N centers the illumination leads to local motion of N atoms at low temperatures, as shown in Fig. 3(b). The Fermi contact hf interaction between the nuclear and electron spins leads to a strong modification of the state of the nuclear spin system under certain severe conditions. In the complete excitation and relaxation cycle, three physical conditions are found to be important and necessary for dynamical nuclear spin polarization of N impurities in semiconductors.

The first is nuclear spin memory during the optical pumping cycle, otherwise nuclear polarization does not take place.[10] This condition is generally satisfied in semiconductors at low temperatures, since coupling between the nuclei and lattice is very weak.[10,11]

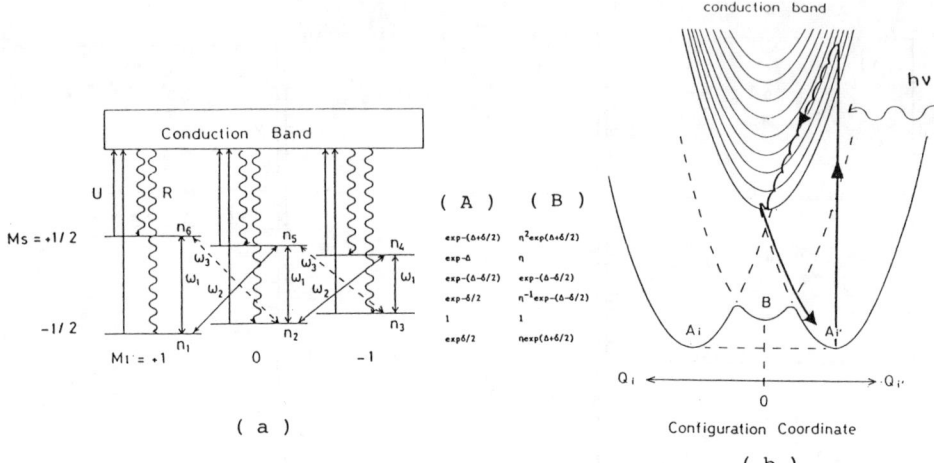

Fig.3 (a) Energy levels and transitions for N impurities in semiconductors. Columns (A) and (B) indicate populations in thermal equilibrium and in optical pumping state, respectively. (Δ and δ represent the zeeman energy $g\beta H/kT$ and hyperfine energy A/kT.)
(b) Configuration coordinate potential energy curve for off-center N centers in Si and diamond.

The second is no electron spin memory. This condition should be satisfied for the electron spins to be strongly out of equilibrium by unpolarized light illumination. Before any illumination the relative population n_i (i= 1,2,..,6) are in equilibrium as indicated in column (A) of Fig. 3(a) and, therefore, the nuclear spin polarization p is approximately 2.6×10^{-5} at 4 K for off-center N in Si. The spin-orbit interaction becomes larger in the order of Si, SiC and C. Therefore, for off-center N in Si, excited electrons is thought to lose spin memory through the spin-orbit interaction in the conduction band in contrast to diamond with a much smaller spin-orbit interaction. This condition yields changes in population at each levels. Thus, the population ratio η ($=n_6/n_1$, n_5/n_2, n_4/n_3) between the two electron spin levels leads to increase; i.e., optical pumping occurs. Figure 4 shows laser intensity dependence of the optical pumping effect of off-center N in Si. At higher intensities the temperature of excited electrons is relatively higher than the lattice temperature, so that spin memory is more easily lost through the spin-orbit interaction, i.e., enhancement in hf line changes is observed.

Due to the form $A\vec{I}\cdot\vec{S} = A/2(S_+I_- +S_-I_+ +S_zI_z)$ of the isotropic hf interaction term, the nuclear are dynamically polarized.[10,11] This is because the term $(S_+I_- +S_-I_+)$ allows simultaneous reversal of a nuclear spin and an electron spin (flip-flop). Consequently the third is relaxation process w_2 arising from modulation of the Fermi contact hf interaction which is caused by lattice vibration. This should not be too small compared with the spin-lattice relaxation rate w_1 and be much larger than w_3 mainly originating from modulation of dipole-dipole hf interaction.[10] The Fermi contact hf interaction leads to a strong enhancement of the nuclear spin polarization, if the above-mentioned conditions are fulfilled. For the off-center N centers in Si and diamond and on-center N in SiC, the Fermi contact hf interaction is much larger than the dipole-dipole hf interaction [2,5,6], so that $w_2 \gg w_3$. Figure 3 shows temperature dependence of the optical pumping effect of off-center N in Si. Significant changes are seen below 50 K, but there were no changes at temperatures above 50 K. Assuming populations shown in column (B) of Fig.3(a), p is evaluated from the result of Fig.2(A) to be strongly enhanced up to 4.2×10^{-2} at 4 K, and η is estimated to be 0.93 that is larger than the value of 0.89 in thermal equilibrium. The spin-

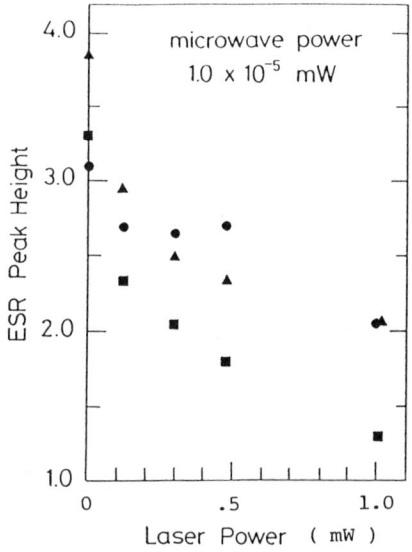
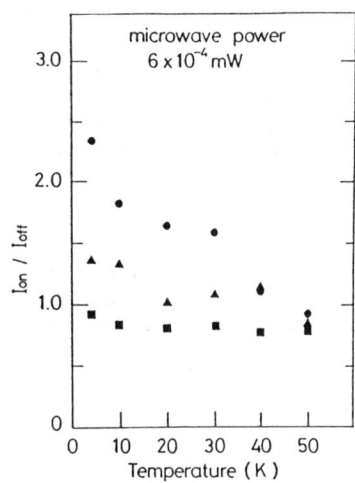

Fig.4 Laser intensity dependence of the amplitude of three ESR hf lines of off-center N in Si. Symbols ●, ▲, and ■ indicate result obtained for M_I=+1, 0, and -1, respectively.

Fig.5 Temperature dependence of the relative value of ESR hf lines during light illumination to that without light. Symbols ●, ▲, and ■ indicate values obtained for M_I=+1, 0, and -1, respectively.

lattice relaxation rate is known to rapidly increase with increasing temperature. Because of $w_1 \gg w_2$ above 50 K, no optical pumping effects take place. Also for on-center N in SiC, w_1 is large as compared with those of Si and diamond, as seen from Fig. 2. Therefore it is difficult to observe dynamical nuclear spin polarization for on-center N in SiC by light illumination.

In the previous paper we proposed an idea that enhancement of w_2 may occur by time dependent $A(t)(S_- I_+ + S_+ I_-)$ caused by local motion of N at off-center sites induced by light illumination.[4] The result that on-center N in SiC show no dynamical spin polarization is consistent with the idea. It is inconsistent that off-center N in diamond also shows no dynamical polarization, however, this comes from unfulfilling the second condition due to a small spin-orbit interaction. In order to moreover investigate the possibility, we have tried to measure the Overhauser effect in ESR spectra. For off-center N in Si and on-center N in SiC the Overhauser effect were not detected at 4 K in the condition of microwave absorption saturation, but only off-center N in diamond showed the effect due to w_2 process. These results confirm that local motion of metastable/bistable off-center N in Si induced by light illumination makes important contribution to dynamical nuclear spin polarization. More studies are anticipated to clarify the effect of local motion of defects and impurities.

4. Summary

We have investigated dynamical nuclear spin polarization of nitrogen impurities in semiconductors (Si, 6H-SiC, and diamond) by optical pumping. Physical conditions for N nuclear spins to dynamically polarize at low temperatures by unpolarized light illumination were found as follows;

(1) The nuclear spin memory is retained in the overall optical pumping cycle,
(2) electron spin memory is lost in the conduction band through the spin-orbit interaction, and
(3) the spin flip-flop relaxation rate w_2 is not too small as compared with the spin-lattice relaxation rate w_1 and much larger than w_3.
A possibility is comfirmed that local motion of off-center ^{3}N in Si induced by light illumination enhances the spin flip-flop relaxation through modulation of the Fermi contact hf interaction.

Acknowledgement

We would like to thank Dr. T. Niina's group (Semiconductor Lab., Sanyo Electric Co., Ltd.) for supplying high quality 6H-SiC crystals doped with various concentrations of N impurity.

References

1. G.D. Watkins, Proc. 15th Int. Conf. Defects in Semiconductors (Trans Tech. Pub., 1989) p.39.
2. K. Murakami, H. Kuribayashi, and K. Masuda, Phys. Rev. B 38, 1589 (1988).
3. A. Chantre and D. Bois, Phys. Rev. B 31, 7979 (1985).
4. K. Murakami, H. Kuribayashi, K. Hara, and K. Masuda, Proc. Int. Conf. Defect Cotrol in Semiconductors (North-Holland, 1990) p.447.
5. W.V. Smith, P.P. Sorokin, I.L. Gelles, and G.J. Lasher, Phys. Rev. 115, 1546 (1959).
6. H.H. Woodbury and G.W. Ludwig, Phys. Rev. 124, 1083 (1961).
7. K. Murakami, H. Kuribayashi, and K. Masuda, Jpn. J. Appl. Phys. 27, L1414 (1988).
8. M. Ikeda, H. Matsunami, and T. Tanaka, J. Luminescence 20, 111 (1979).
9. J. Koppitz, O.F. Schirman, and M. Seal, J. Phys. C: Solid State Phys. 19, 1123 (1986).
10. C.D. Jeffries, Electron Paramagnetic Resonance, ed. S. Geschwind (Plenum Press, NY-London, 1972) p.217.
11. P. Paget and V.L. Berkovits, Optical Orientation, eds. F. Meier and B.P. Zakharchenya (North-Holland, Amsterdam, 1984) p.381.

ON THE ANALYSIS OF DIGITAL DLTS DATA

C.A.B. BALL and A.B. CONIBEAR
Department of Physics, University of Port Elizabeth
P.O. Box 1600, Port Elizabeth, 6000, South Africa.

ABSTRACT

A digital storage oscilloscope is used to capture capacitance transients for the DLTS experiment. Two novel data analysis methods are presented, one using a double rate window and another a digital simulation of the analogue process in a lock in amplifier. These methods provide considerably more data than earlier real-time analogue techniques and can reduce the time needed for the experiment. They further constitute a sensitive test as to whether the capacitance decay contains more than one exponential component. An experimental example shows that trap parameters should be treated with caution if obtained by conventional analogue methods using spectral peak positions only.

I. INTRODUCTION

The DLTS experiment as originally proposed by Lang[1] required several temperature scans for the construction of a single Arrhenius plot. One way of reducing the redundancy inherent in many of the conventional analogue methods is to use digital means to capture the capacitance-time data and then to subject this transient to appropriate analysis.

In this paper we describe a system in which a digital storage oscilloscope is used to capture capacitance transients during temperature scanning. Two methods of determining time constants of decay curves are shown; one a very simple calculation using a double rate window (DRW) and the other essentially a digital simulation of the analogue process in a lock in amplifier (SLIA). Both techniques allow many Arrhenius points to be obtained with a single temperature scan. They also constitute a sensitive test of whether a DLTS peak may be identified with a single trap with temperature independent energy and capture cross section.

II. EXPERIMENT

The data acquisition is done by a Nicolet 4094B digital oscilloscope with a Nicolet 4570 plug in. The large range of times per point available on the plug in (100ns to 10s) allows the determination of the time constant of the capacitance decay over a wide range of temperatures. Once a capacitance transient has been obtained it is dumped to a computer and the curve is digitally reduced to 32 points by a parabolic averaging technique. These points along with other relevant data are stored on disk for further processing and archive purposes.

III. THEORY

The basic problem of the DLTS experiment is to determine the time constant $\tau(T)$ of the decay of the capacitance transient, assumed to be

$$C(t) = C_o - \Delta C_o \exp\left(-\frac{t}{\tau(T)}\right), \tag{1}$$

at each temperature T. The original method[1] used was to calculate the DLTS signal, $C(t_1)-C(t_2)$, for a given rate window defined by two instants t_1 and t_2 measured from the end of the filling pulse and to determine at which temperature a maximum occurs. At this maximum temperature the time constant is a simple function of t_1 and t_2.

The rate window can be varied to obtain different DLTS spectra and the temperature at which a maximum occurs for each rate window may be found. From these data the Arrhenius curve of $\log(T^2\tau)$ against $1/T$ may be plotted, and the slope and y-intercept are used to calculate the trap energy and the capture cross section.[1] A similar method may be used with a lock in amplifier data acquisition system.[2,3]

In the case of digital systems, one set of capacitance transients for a range of temperatures is all that is necessary to obtain the Arrhenius plot. Two methods were found useful, giving different insights. These will be referred to as the double rate window (DRW) technique and the simulated lock in amplifier (SLIA) technique.

A. DOUBLE RATE WINDOW TECHNIQUE.

Refer to figure 1. Assuming that the capacitance transient is of the form of equation (1), one obtains

$$C_2 - C_1 = \Delta C_o \exp\left(-\frac{t}{\tau}\right)\left[1 - \exp\left(-\frac{\Delta t}{\tau}\right)\right] \tag{3}$$

with a similar expression for $C_3 - C_2$.
It follows that

$$\frac{1}{\tau} = \frac{1}{\Delta t} \ln\left(\frac{C_2 - C_1}{C_3 - C_2}\right) \tag{4}$$

and the time constant may be obtained for any given decay curve. If the capacitance decay is exponential this should be constant for any choice of t and Δt. This is often not the case however, and the variation of the time constant with different values of t and Δt for a given transient gives an indication of the deviation from exponentiality.

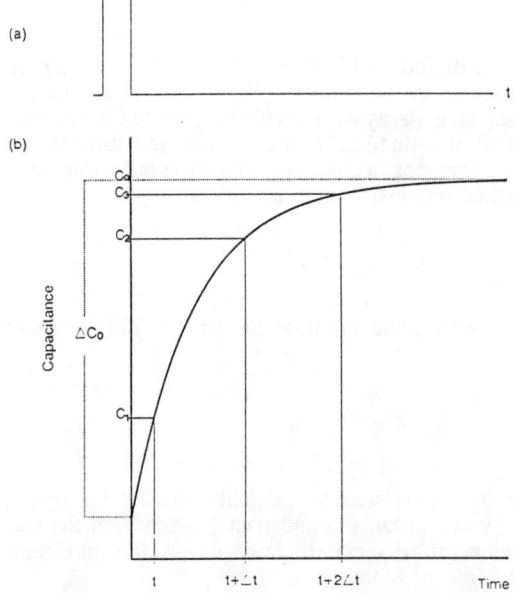

Figure 1
(a) Bias pulse.
(b) Schematic diagram of capacitance transient with double rate window.

B. SIMULATED LOCK IN AMPLIFIER TECHNIQUE

In the SLIA technique the DLTS signal is the integral

$$S = \int_{t_d}^{t_d+T_o} C(t)w(t)dt \qquad (5)$$

which is calculated numerically. Figure 2 indicates the meanings of T_d and T_o. The interval T_o is arbitrary within the limits of the measured transient and variation of T_o can be used as a test for exponentiality.

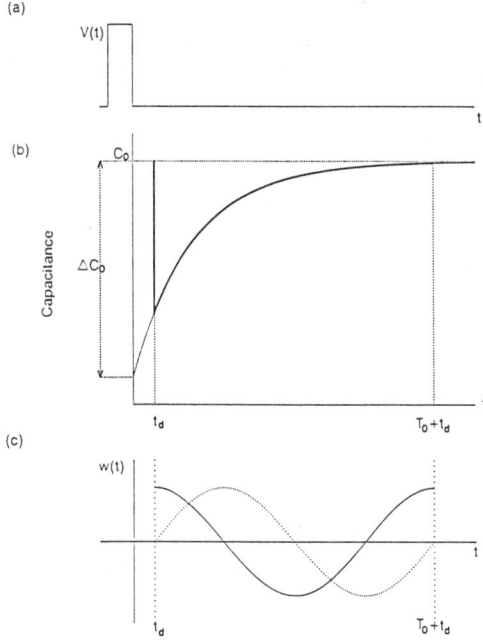

Figure 2
(a) Bias pulse.
(b) Schematic diagram of capacitance transient with t_d and T_o for SLIA method.
(c) Weighting function.

We have used weighting function

$$w(t) = \sin(2\pi\{t-t_d-T_o/2\}/T_o) \qquad (6)$$

and the cosine of the same argument. Since the integral of w(t) over the given interval is zero, these weighting functions eliminate the influence of the constant C_o in (1).

A theoretical curve of $S_{th}(T)$ for the DLTS signal may be obtained by substituting equations (1) and (6) into (5). The result is:

$$S_{th}(\tau) = \frac{\Delta C_o 2\pi\tau^2 \exp(-t_d/\tau)\{1-\exp(-T_o/\tau)\}}{T_o^2 + 4\pi^2\tau^2}, \quad (7)$$

for the sine weighting function with a similar expression for the cosine. The DLTS signal given by equation (5) is of course in effect the same signal as would be obtained with a lock in amplifier[2]. In lock in amplifier DLTS, a time constant and corresponding temperature are determined from the signal maximum, necessitating several scans at different frequencies to generate an Arrhenius plot. With our SLIA method the determination of the time constant is not limited to the signal maximum and so the Arrhenius plot is in principle generated from a single temperature scan.

Suppose capacitance transients have been obtained at temperatures T_i, $i=1,2...N$. The algorithm to obtain the corresponding time constants τ_i is as follows:

1. Calculate $S_{exp}(T_i)$ numerically from equation (5) using the experimental capacitance transient captured at T_i.

2. Find by interpolation the temperature T_{max} at which S_{exp} is a maximum.

3. Using equation (7) find the time constant τ_{max} at which $S_{th}(\tau)$ is a maximum.

4. For each temperature T_i solve the equation

$$S_{th}(\tau) = S_{th}(\tau_{max}) \frac{S(T)}{S(T_{max})} \quad (9)$$

for τ_i. $S_{th}(\tau_i)$ is given by equation (7), $S_{exp}(T_i)$ has been calculated in step 1 above, $S_{exp}(T_{max})$ in step 2 and $S_{th}(\tau_{max})$ in step 3.

Note that there are two solutions for any given T_i (except for T_{max}) and the ambiguity is removed by noting that if $T_i > T_{max}$ then $\tau_i > \tau_{max}$. The solution of equation (9) is illustrated schematically in fig. 3.

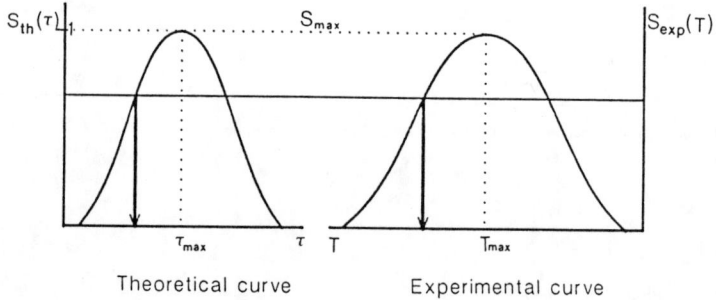

Figure 3
Schematic diagram illustrating solution of equation 9.

The weighting function, w(t), used should not influence the value of τ thus obtained if the capacitance decay is exponential. However, if the decay is the sum of more than one exponential term, the cosine weighting function will emphasize the initial part of the decay transient which will be more dependent on the shorter time constants in the sum, while the sine weighting function tends to emphasize the longer time constants.

The SLIA technique is preferable to the DRW technique for the calculation of trap energies and capture cross sections since an integral over every point on the C(t) curve in the interval (t_d, t_d+T_o) is used to obtain the DLTS signal, reducing the noise very considerably. However as a test for non-exponentiality the DRW method is more sensitive and has the advantage of a considerably reduced computation time.

IV. EXPERIMENTAL EXAMPLES AND DISCUSSION

The much studied EL2 trap in GaAs has been chosen to illustrate the use of the SLIA and DRW methods. Results of analysis done on an OMVPE n-type GaAs sample are presented below.

Fig. 4 shows a set of three Arrhenius lines, one generated by the DRW method and the other two by the SLIA method using a sine and a cosine weighting function. The three curves are not co-linear, indicating non-exponentiality of the transient, with the DRW method being the most sensitive indicator, followed by the cosine weighting function.

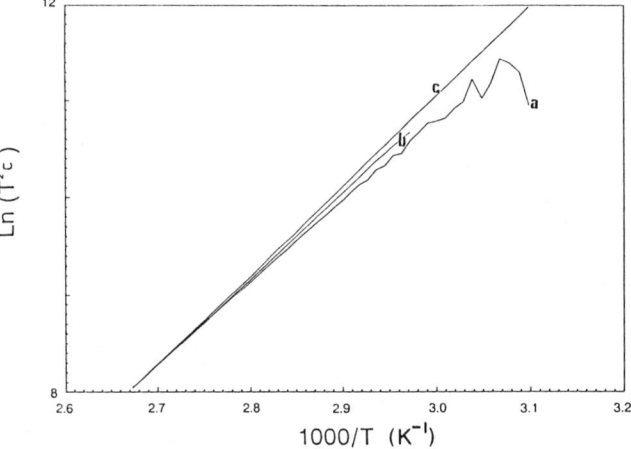

Figure 4
Arrhenius lines for the EL2 trap in GaAs.
a - DRW method
b - SLIA method, cosine weighting function.
c - SLIA method, sine weighting function.

The non-exponentiality was magnified when the pulse height was small, irrespective of the value of the constant reverse bias. This would seem to indicate that it was a property of the boundary between the depletion region and the neutral semiconductor. This was confirmed by using alternating pulse heights (3.2 V and 2 V) with a subtraction procedure[4] which made it possible to sample a region away the boundary. The analysis of this data was consistent with the EL2 being a single trap of energy $E_t = 0.82 \pm 0.02$ eV with respect to the conduction band and capture cross section 3E-13 ± 2E-13 cm²; the rather error margin being attributed to uncertainty in temperature measurements.

Using only DLTS peak maxima to draw Arrhenius curves can give rise to erroneous trap parameters, especially in the case where traps in a sample are close in energy. Fig. 10 shows Arrhenius curves for various T_o generated by the SLIA method for an OMVPE grown $Al_{0.19}Ga_{0.81}As$ sample. The inset is a conventional DLTS spectrum for the same temperature range. The crosses on the graph correspond to the points generated from the DLTS peak maxima for each curve.

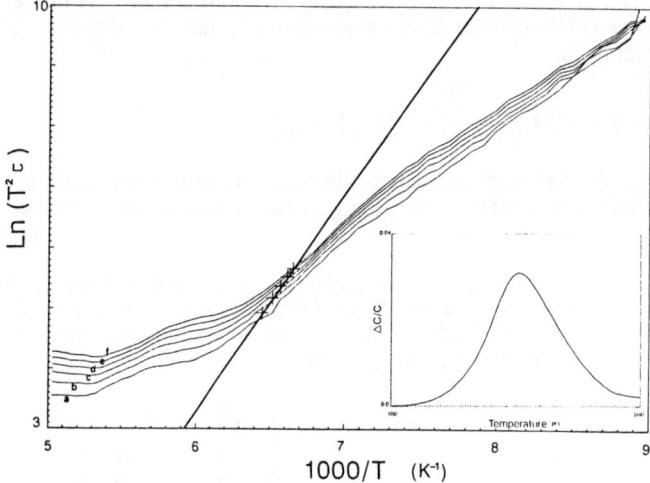

Figure 5
Arrhenius lines for $Al_{0.19}Ga_{0.81}As$.
(SLIA method with $t_d = 0.002$ s and $T_0 = 0.08$s, 0.012s, 0.016s, 0.020s, 0.024s and 0.028s for curves a-f respectively).
Crosses correspond to DLTS peak maxima.
Inset: Conventional DLTS spectrum

Although these points are relatively co-linear, the resulting trap parameters are very different to those that would result from an analysis using all the Arrhenius points. Use of very short pulse widths showed that the DLTS peak contains contributions from at least two traps, a fact supported by the shape of the Arrhenius curves which show two approximately linear regions in the $1000/T$ ranges 5.3 to 5.8 and 6.1 to 7.5 K^{-1}.

V. CONCLUSION

We have shown two digital methods for acquiring and processing DLTS data. Such methods give considerably more insight into the nature of the capacitance transients being analysed, and in particular into non-exponential behaviour. The DRW method, while relatively sensitive to noise in the data, is also most sensitive to non-exponentiality and has the advantage of requiring very little computation in its implementation. The SLIA method involves more computation, but because it uses every data point in the transient, gives a more stable result over a large temperature range.

ACKNOWLEDGEMENT

This work was supported in part by the South African Foundation for Research and Development. The authors would like to acknowledge Mr A. Venter and Prof. F.D. Auret for the use of their samples and to thank Prof. Auret for helpful discussions.

REFERENCES

1. D.V.Lang, J.Appl.Phys. **45**, 3023 (1974).
2. F.D Auret, Rev.Sci.Instrum. **57**, 1597 (1986).
3. G.L.Miller, D.V.Lang and L.C.Kimerling, Ann.Rev.Mater.Sci. **7**, 377 (1977).
4. H.Lefevre and M.Schulz, Appl.Phys. **12**, 45 (1977).

A REEVALUATION OF ELECTRIC-FIELD ENHANCED EMISSION MEASUREMENTS FOR USE IN TYPE AND CHARGE STATE DETERMINATION OF POINT DEFECTS

W.R. Buchwald[1], H.G. Grimmeiss[2], F.C. Rong[3], N.M. Johnson[4], E.H. Poindexter[1], H. Pettersson[2]

[1] Electronics Technology and Devices Laboratory, Fort Monmouth NJ, USA, 07703-5601
[2] Department of Solid State Physics, University of Lund, S-22007, Lund, Sweden
[3] GEO-Centers, Inc., Electronic Technology and Devices Lab., Fort Monmouth NJ, USA, 07703-5601
[4] XEROX Palo Alto Research Center, Palo Alto, CA, USA, 94304

ABSTRACT

Deep level transient spectroscopy and single-shot junction space charge capacitance transient measurements have been used to measure the effects of electric-field on the thermal emission rates of trapped carriers in both GaAs and Si. These measurements reveal inconsistencies when compared to existing field enhanced emission (i.e. Poole-Frenkel effect) theory, which casts doubt on the validity of using these measurements for the determination of defect type, (i.e. donor or acceptor-like transitions), or charge state of the emitting center. Space charge measurements performed on S and Se doped Si reveal field dependencies inconsistent with existing theory for the well known charge state of these centers as determined by IR absorption measurements. DLTS measurements performed on a defect previously observed in MOCVD grown GaAs show unambiguously an electric-field dependent emission rate, but differing quantitative results depending on the method used to extract relevant field dependent parameters. In the case of GaAs, the apparent inconsistencies can be resolved if a temperature dependent field-effect is assumed; but it is shown that these types of measurements are not accurate enough for the unambiguous determination of either defect type or charge state.

1. INTRODUCTION

The observation of field-enhanced, thermal emission, of trapped carriers has been used in the past for the determination of the type of trapping center present (i.e donor or acceptor)[1]. These defect type determinations have been primarily based on the Poole-Frenkel effect model[2]. The Poole-Frenkel effect models the potential experienced by an emitted carrier as Coulombic, giving rise to a reduction in the measured thermal activation energy that is proportional to square-root of the applied electric field. Based on this model, it has been assumed that if the defect site acquires a net charge upon emission of a trapped carrier, electric field enhanced emission should be observed. Theoretical calculations also model certain defect sites that do not acquire a net charge upon emission of trapped carriers (i.e. neutral acceptors) as potential wells that could also have a field effected emission process, however, most of these are proportional to the applied field to powers other then one half.[3]. From this it seems that the observation of field assisted thermal emission would be enough to determine, not only the type of defect present, but also its charge state.

This paper will report that even careful field effect measurements on well studied defects in Si[4] (i.e. S and Se) do not necessarily give results consistent with the known type of the defects studied. Field enhanced emission data is also presented for a defect in GaAs which not only shows these types of measurements are not of an accuracy suitable to determine defect type for this defect, but also that the method used to acquire the relevant field dependent parameters effects the acquired values. It is also shown that the ambiguity with respect to the method used to obtain field enhanced emission parameters can be resolved, at least in the case of GaAs, by assuming a temperature dependent, field assisted, emission process.[5]

2. EXPERIMENTAL AND THEORETICAL DETAILS

The Si samples used in this work were sulfur and selenium diffused p^+n diodes similar to those previously studied[6-8]. Special care was taken to make sure that only the isolated impurities were studied. For the case of the GaAs samples, Pt Schottky diodes were fabricated on MOCVD grown epi-layers as previously described[9]. The GaAs defect reported in this work has not been identified, but was found previously to have a field dependent emission rate, and an activation energy for thermal emission of electrons of E_c-0.61eV when the sample was reversed biased at -2.5V.

In order to investigate any field effect on the Si samples of this work, single shot transient measurements were performed[10]. In order to have a precise value for the electric-field effecting the emitting center, a narrow range in the space charge region was selected by taking the difference between two transients recorded at two slightly different fill pulses V_1 and V_2 while keeping the reverse bias constant. The sample was held at a fixed temperature, the transients recorded, the transient subtraction performed and then the results were plotted semi-logarithmicaly to obtain the emission rate as a function of temperature. In the case of the GaAs samples used in this work, the technique of double deep level transient spectroscopy (DDLTS) was used to define a narrow observation window to allow for a precise determination of the applied electric field[11]. The standard rate window technique, and Arrhenius plots were used to determine the emission rates of the GaAs samples.

The standard Poole-Frenkel effect model treats the trap energy associated with the thermal emission of trapped carriers as electric-field dependent. The emission rate of trapped carriers, e^t_n, is be given by:

$$e^t_n(F,T) = A\, T^2 \exp[-\Delta E_M(F)/kT] \quad (1)$$

where the measured trap depth, ΔE_M, is given by,

$$\Delta E_M(F) = \Delta E_M(0) - \Delta E_{th}. \quad (2)$$

In the above equations F is the field effecting the emitting center, T is Kelvin degrees, k is Boltzmanns constant. δE_{th} is the reduction in the thermal activation energy due to the applied electric-field. The constant A is independent of temperature and given by

$$A = \sigma^0\, v_{th}\, N_c\, T^{-2} \exp(\Delta S/k) \quad (3)$$

where $\sigma^0 = \sigma \exp(-E_c/kT)$ is the capture cross section, ΔS is the change in enthalpy, v_{th} the average thermal velocity and N_c the density of conduction band states. A simple treatment for a one dimensional Coulombic potential shows the field dependent decrease in the activation energy to be given by[2]

$$\Delta E_{th} = c\, F^{1/2} \quad (4)$$

From the above equations it seems that if DLTS measurements are used to obtain defect emission rates, plots of ΔE_M versus the square root of the applied field should give the zero field trap depth as well as the Poole-Frenkel coefficient, c. It is also easily shown that by combining equations (1),(2) and (4) above, the emission rate e^t_n can be written as

$$e^t_n(F,T) = e^t_n(0,T) \exp[c\, F^{1/2}/kT] \quad (5)$$

In this case, if a temperature independent Poole-Frenkel coefficient is assumed, then plots of $\ln(e^t_n)$ versus the square root of the electric field should also give the Poole-Frenkel coefficient, c. Regardless of the method used (i.e. plotting ΔE_M vs. $F^{1/2}$ or $\ln(e^t_n)$ vs. $F^{1/2}$) identical values of c should be obtained.

3. EXPERIMENTAL RESULTS

The isolated S and Se doped defects in Si are extremely well studied and are known to act as double donors[12-14]. Therefore, according to the classic, one dimensional Poole-Frenkel effect model, both the neutral, (S^0, Se^0, i.e. filled with two electrons), and charged, (S^+, Se^+ i.e. filled with one electrons) centers should show field enhanced emission upon release of their trapped electrons. Figure 1 shows the $\ln(e^t_n)$ versus $F^{1/2}$ plots for the neutral S and Se in Si at the indicated temperature. As can be seen, the neutral S^0 center shows virtually no field enhanced emission while the neutral Se^0 center shows a field dependence that is not proportional to $F^{1/2}$. Similar results were obtained at different temperatures[4]. It is now pointed out that the effect of electric-field on the thermal activation energy can also be calculated in three dimensions and in turn, leads to a much more complicated expression for ΔE_{th}[15]. This, however, does not alter the result that neutral S in Si, a known donor, shows no field effect. It will also be shown later in this work, that due to the inaccuracy of junction space charge techniques, and the fact that theoretical models for acceptors exist that predict electric-field dependent emission for these centers, it is irrelevant which form of ΔE_{th} is used. At this point it is clear however, that at least for these well known donor centers, the Poole-Frenkel effect is not apparent in one case, and in the other, the one dimensional Poole-Frenkel effect model does not adequately describe the field assisted emission. The reason for no field effect for neutral S is unclear at the moment, however, it has been suggested that donor centers that have associated with them a temperature dependent capture cross section, implying an energy barrier to carrier capture, could have their potentials perturbed to such an extent that any long range Coulombic effects, (such as the Poole-Frenkel effect), could be suppressed[16]. The neutral S and Se centers in Si are known to have temperature dependent capture cross sections, therefore their field assisted emission processes could be suppressed.

One problem associated with field enhanced emission experiments is obtaining an accurate value for the field effecting the emitting center. The trap filling parameters can be adjusted to define a narrow region in the space charge region, but unless the zero field trap depth is known the calculated value of the electric-field is imprecise. For the GaAs samples of this work, the following method was used to overcome this problem. Initially a guess was made of the zero field trap depth based on the low field trap depth obtained via DLTS measurements. Next, equation (2) above

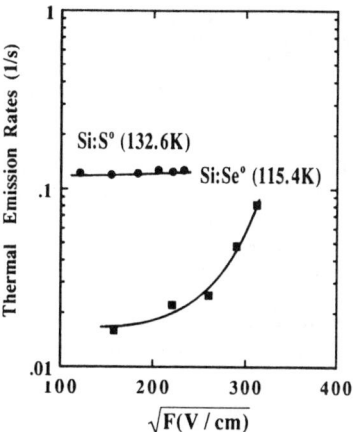

Figure 1: Thermal emission rates versus the square root of applied electric field for sulfur and selenium doped Si.

was used and these ΔE_{DLTS} values were plotted versus the square root of the field. The y-intercept should then be the zero-field trap depth. If the y-intercept was not equal to the initial value used for the zero-field trap depth, the electric fields were recalculated using this new extracted value. This procedure was continued until the y-intercept matched the zero-field trap depth by 1%. In this manner the electric fields calculated were at least consistent with the extracted zero field trap depth. Figure 2 shows the thermal emission rates obtained from DLTS measurements plotted versus the square root of the applied field for the GaAs samples. Based on Equation (5) above, the slopes of the curves in Figure 2 should give the Poole-Frenkel coefficient, c. Figure 3 also plots the actual ΔE_{DLTS} versus the square root of the applied field. This plot was made based on equation (2) above and again the slope should give the Poole-Frenkel coefficient.

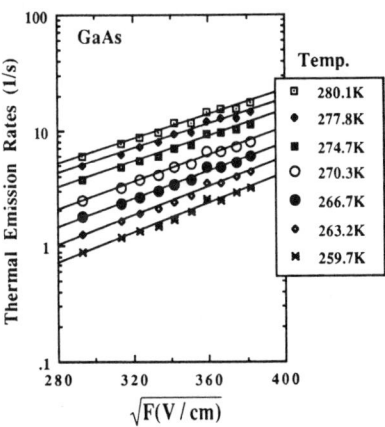

Figure 2: Thermal emission rates versus the square root of applied field.

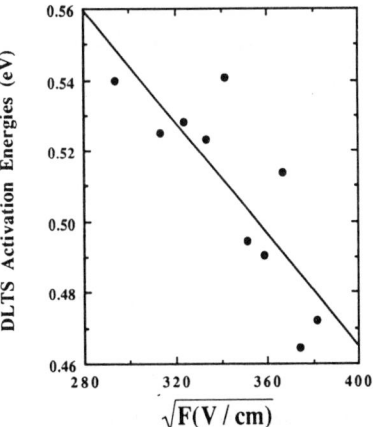

Figure 3: DLTS activation energies versus the square root of the applied field.

What becomes immediately apparent from these plots is that the Poole-Frenkel coefficients are not equal for the two methods used. From Figure 2 the value of c appears to be temperature dependent with a value for c of $2.95 \times 10^{-4} ev/(V/cm)^{1/2}$ at 280K and $3.32 \times 10^{-4} ev/(V/cm)^{1/2}$ at 260K. The value of c obtained from Figure 3 gives a value of $4.8 \times 10^{-4} eV/(V/cm)^{1/2}$ which is different from that obtained via Figure 2. At this point it is tempting to make the assumption that even though the two methods used to extract the coefficient c give differing values, the unambiguous observation of a field assisted emission process is enough to at least determine the donor like nature of this defect. Theoretical work however, on neutral acceptor centers, seems to indicate that for certain potentials, such as a polarization potential, there should still be a field assisted emission process but it should be proportional to the field to the four-fifths power. Figure 4 is a plot of the thermal emission rates versus the applied field to the four-fifths[3]. It is clear from this figure that the accuracy of the measurement technique is not sufficient to determine the difference between the square root of the field, or, of the field to the four-fifths power. Therefore, even the observation of a field assisted emission process is not enough to determine the donor like nature of an emitting center.

From the analysis of this work it would appear that the Poole-Frenkel coefficient is temperature dependent. If it is assumed that c takes the form

$$\Delta E_{th} = (a - bT) F^{1/2} \qquad (6)$$

then, the Poole-Frenkel coefficients obtained from Figure 2 are now the temperature dependent Poole-Frenkel coefficients. Figure 5 shows the plots of these temperature dependent coefficients versus inverse temperature. As can be seen by this figure, a straight line is obtained implying a linear temperature dependence for the Poole-Frenkel coefficients of the GaAs sample used in this work. It is also pointed out, that assuming a temperature dependent Poole-Frenkel coefficient, the apparent discrepancies obtained when obtaining c from different analysis can be easily resolved. The value obtained for c from Figure 3 is the temperature dependent Poole-Frenkel coefficient at T=0K. This value was found to be $4.8 \times 10^{-4} eV(V/cm)^{-1/2}$ and is in reasonable agreement with the T=0K value of $5.2 \times 10^{-4} eV(V/cm)^{-1/2}$ extracted from Figure 5.

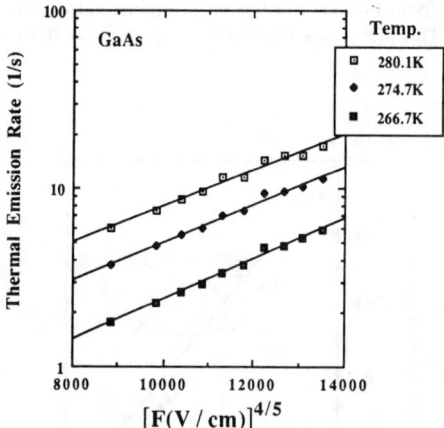

Figure 4: Thermal emission rates versus the applied field to the four-fifths power.

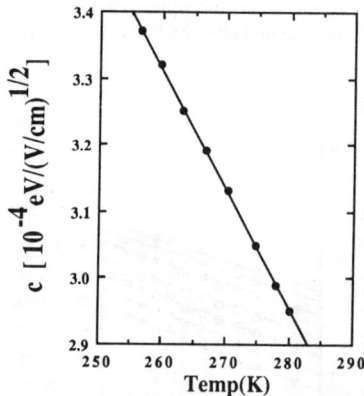

Figure 5: Poole-Frenkel coefficients, c, versus temperature.

4. DISCUSSION AND CONCLUSION

In conclusion, due to the inaccuracy of junction space charge measurement techniques, and the fact that both Coulombic and certain non-Coulombic traps are theorized to have field enhanced emission associated with them, the observation of field enhanced emission is not sufficient to unambiguously determine the charge sate or the donor like nature of an emitting center. Experimental evidence also suggests that the absence of field enhanced emission is also not sufficient to unambiguously determine the acceptor like nature of a defect center. It is also shown that based on the GaAs measurements of this work, certain inconsistencies exist with respect to the extracted Poole-Frenkel coefficients and the analysis method used. These inconsistencies, at least in this work, can be resolved by assuming a temperature dependent field enhanced emission process.

5. REFERENCES

1) L.C. Kimmerling, J.L. Benton, Appl. Phys. Lett. **39**, 410 (1981)
2) J. Frenkel, Phys. Rev. **54**, 647 (1938)
3) P.A. Martin, B.G. Streetman, K. Hess, J. Appl. Phys., **29**, 1907 (1984)
4) H. Pettersson, H.G. Grimmeiss, Phys. Rev. B **42**, 1381 (1990)
5) H.G. Grimmeiss, W.R. Buchwald, F.C. Rong, E.H. Poindexter... (to be published)
6) H.G. Grimmeiss, E. Janzen, B. Skarstam, J. Appl. Phys. **51**, 3740 (1980)
7) H.G. Grimmeiss, E. Janzen, B. Skarstam, J. Appl. Phys. **51**, 4214 (1980)
8) H.G. Grimmeiss, E. Janzen, H. Ennen, O. Schirmer, J. Schneider, R. Worner, C. Holm, E. Sirtl, P.Wagner, Phys. Rev. B,**24**,4571 (1981)
9) W.R. Buchwald, N.M. Johnson, L.P. Trombetta, Appl. Phys. Lett.**50**, 1007 (1987)
10) H.G. Grimmeiss, C. Ovren, J. Phys. E. Sci. Instrum.,**14**, 1032 (1981)
11) H. Lefevre, M. Schulz, Appl. Phys. Lett.**12**, 45 (1977)
12) E. Janzen, R. Stedman, G. Grossman, H.G. Grimmeiss Phys. Rev. B **29**, 1907   (1984)
13) M. Kleverman, K. Bergman, H.G. Grimmeiss Phys. Rev. B **31**, 8000 (1986)
14) M. Kleverman, K. Bergman, H.G. Grimmeiss, Semicond. Sci. Technol. **1**, 49   (1986)
15) J.L. Hartke J. Appl. Phys. **39**, 4871 (1968)
16) W.R. Buchwald, N.M. Johnson J. Appl. Phys. **64**, 958 (1988)

X-RAY SPECTROSCOPY FOLLOWING NEUTRON IRRADIATION OF SEMICONDUCTOR SILICON.

A. J. Filo, A. J. Meyer, C. C. Swanson
Analytical Technology Division, Eastman Kodak Company, Rochester, N. Y. 14650-02155
J. P. Lavine
Image Acquisition Products Division, Eastman Kodak Company, Rochester, N. Y. 14650-02008

ABSTRACT

Transition metals form defects in silicon that lead to reduced device performance and yield. Thus, it is necessary to improve techniques that detect metals in silicon. Neutron activation analysis has long been used to identify metals and determine their concentration through the measurement of their characteristic gamma rays. However, lower energy gamma rays and x-rays (<120 keV) are also emitted by metals that have been irradiated in silicon and it is advantageous to detect these x-rays. This paper describes the use of a low-energy photon spectrometer (LEPS) system and the gains in metal detection that are possible.

X-rays from elements such as iron, chromium, copper, zinc, and germanium have been detected in silicon wafers with the aid of the LEPS detector. When these elements are irradiated by neutrons, they undergo a neutron capture event to form an unstable isotope that decays by electron capture to a nucleus with one less proton. An example of this would be $^{63}Cu\,(n,\gamma)\,^{64}Cu \rightarrow\,^{64}Ni$. The Ni atom then emits the x-rays as a result of the rearrangement of the electron orbitals. The near-surface region of the silicon wafer is probed since the low-energy x-rays detected by the LEPS are attenuated within the sample. This allows the location of the metal atoms to be isolated to either the front or back surface of the wafer. Advantages for the technique are the high resolution of the detector, the reduction in background due to the high-energy gamma-ray processes, and the fact that there are fewer peaks associated with the spectrum than there are with a gamma-ray spectrum.

I. Introduction

The introduction of contamination, especially transition group metals (Cu, Fe, Co, Ni, etc), during the fabrication of integrated circuits has been a continuous problem for the semiconductor industry. The presence of these metals has had a negative impact on the silicon device performance through the formation of microdefects in the Si lattice during heat treatments and the reduction of the minority carrier lifetimes through the formation of generation-recombination centers. The concentrations of the contaminants needed to degrade the lifetimes are very small [1] and are certainly beyond the reach of most conventional analytical chemical techniques. Secondary ion mass spectrometry (SIMS) and total reflectance x-ray fluorescence (TXRF) with their ability to quantify and profile impurities both in the bulk and at the surface have been invaluable tools in the semiconductor industry. Neutron activation analysis (NAA), following reactor neutron irradiation, is a versatile tool capable of determining the total impurity concentration within the wafer with no information on the depth distribution. The use of γ ray spectroscopy with high-resolution, high-efficiency Ge detectors offers very high sensitivity for most elements in the Si matrix. Unfortunately, when dealing with contamination studies as a result of processing, there exists a fundamental problem with any of these techniques. This problem is that they all assume a uniform distribution of the contaminants. This is not always the case as observed by the Haze technique [2,3] or by autoradiography [4] following reactor neutron irradiation. Haze and autoradiography both have shown that the contamination introduced during device fabrication may be localized on the wafer.

In general, when a material is irradiated with neutrons, the radionuclides formed decay by β emission which is followed by the emission of γ rays. The resulting γ ray spectrum tends to be rather complex depending upon the contaminants present. The presence of low-level contaminants may be obscured by spectral interferences and high background due to bremsstrahlung radiation and Compton scattering. X-ray spectrometry using a Si(Li) detector or a low energy photon spectrometer (LEPS), which is a planar high-purity Ge detector, may be used to detect the x-rays or low energy γ rays emitted following the decay of some radionuclides. The advantages of using the LEPS detector are that the x-ray spectra tend to be less complex so overlaps of x-ray peaks are minimized, and there is unambiguous identification of the elements since the x-ray energies are well known. The detector has high resolution with a lower background since the contribution from the γ rays is minimized due to a reduced response to higher energy photons. Spectral interferences that may occur are due to fluorescent x-rays from the detector and bremsstrahlung radiation from the β particles. The latter effect may be overcome by use of absorbers or by a magnetic field [5,6] set to deflect the β particles. Self-absorption of the low energy x-rays may be a problem for some matrices but it may be used to some advantage in differentiating between front and back surface contamination.

II. Experimental

Liquid spectrometric standard solutions from the NIST and Alpha Products SPECPURE Plasma Emission Standards were used to determine the sensitivities for each element. The x-rays and low energy γ rays were measured on an Ortec LEPS detector with an energy resolution of 195 eV FWHM at 5.9 keV. The γ ray spectra were measured on a Canberra Ge detector with a relative efficiency of 27%

and an energy resolution of 1.8 keV FWHM at 1332 keV. The samples were irradiated at the Cornell University Triga Reactor for 1 hour at a thermal neutron flux of 1×10^{12} n/cm^2/sec, an integrated epithermal neutron flux of 1×10^{10} n/cm^2/sec, and a fast neutron flux of 8×10^{10} n/cm^2/sec. A volume of 150 µl of the irradiated solution was used to make a point source standard with the activity measured in a fixed position. The same sample was used to measure both the x-ray and γ ray spectra.

The Si wafers analyzed were 4-inch diameter n-type Czochralski wafers with a (100) orientation and a resistivity of 30 ohm-cm. The Si wafers were irradiated at the University of Missouri Research Reactor for 75 hours at a thermal neutron flux of 5×10^{13} n/cm^2/sec, an integrated epithermal neutron flux of 1×10^{12} n/cm^2/sec, and a fast neutron flux of 8×10^{12} n/cm^2/sec. The samples were cleaned in a dilute aqua regia solution and allowed to decay for at least 46 hours prior to data acquisition. The concentrations of the contaminants were determined using the parametric method of NAA using the fundamental nuclear parameters [7,8]. The cross sections, resonance integrals, and isotopic abundances used were from [9], the γ ray intensities and half-lives from the tables by Erdtmann [10], and the x-ray specific parameters were obtained from the *Table of the Isotopes* [11].

III. Discussion

NAA has been used for the analysis of Si and the characterization of contamination introduced during processing [12,13] for many years. Si is an ideal matrix for contamination studies due to its nuclear properties (short half-life, low neutron capture cross section, and low γ ray intensity). When a material is irradiated by neutrons, all isotopes have a probability of capturing a neutron and forming a radionuclide. The radionuclide formed may then decay by ß emission to a stable nuclide that is in an excited state (daughter nuclide). This excited daughter nuclide then de-excites via the emission of γ rays to the ground state. In addition, there are several modes of ß decay that give rise to the emission of x-rays. These modes of decay are: the internal conversion (IC) process following an isomeric transition (IT), orbital electron capture (EC), and ß decay accompanied by IC. The origin of the x-rays detected for the IC process is the parent nuclide with atomic number Z, the x-rays emitted from EC are from the daughter nuclide with atomic number Z-1, and the x-rays from the ß decay plus IC process are also from the daughter nuclide but with the atomic number Z+1. There are approximately 60 elements that may be detected by measuring their x-rays but many of them have half-lives that are relatively short so that measurement is difficult unless they are done at the reactor. There are approximately 30 elements that have half-lives that are sufficiently long so that they are readily measured. The measurement of x-rays following neutron irradiation has been employed for the measurement of the rare earth elements in geological material as well as U and Th [14-18] but nothing has appeared with respect to semiconductor material. Some of the more common elements of interest to the semiconductor industry are listed in Table I. The mode of decay, the x-rays detected and an enhancement factor, which is a ratio for the x-ray to the γ ray sensitivities, are also listed. As we can see from Table I, the elements listed all have an advantage of measuring the x-ray over their associated γ rays and would indicate a preference for their measurement by x-ray spectrometry.

Table I. Elements which decay by x-ray emission following neutron irradiation. Mode of decay, origin of the x-rays, and enhancement factors are listed.

Element	Isotope	Decay	X-ray emitted	Enhancement Factor (EF)
Cr	^{51}Cr	EC	V	2.9
Fe	^{55}Fe	EC	Mn	no γ rays
Cu	^{64}Cu	EC	Ni	150
Zn	^{65}Zn	EC	Cu	3.3
Ge	^{71}Ge	EC	Ga	no γ rays
Pt	^{191}Pt	EC	Ir	12.6
	^{195m}Pt	IT	Pt	2.3

Note: Fe and Ge have no enhancement factors (EF) since there are no γ rays associated with the decay from the isotope emitting the x-rays. If we were to compare the response from the two radionuclides of Fe and Ge, the EF for Fe would be 3.5 and 285.7 for Ge.

The detection of Cu and Ge in silicon may only be unambiguously determined by the measurement of their x-rays, unless they are present in substantial quantities. Their only γ-emitting isotopes have short half-lives and are not generally observed. Ge does have a long-lived isotope but it only decays with the emission of x-rays. Table II contains the sensitivity or detection limits for the elements of interest in terms of total number of atoms detected. The detection limits are listed this way in order to eliminate the dependence on volume and sample size. The energies of the x-rays are sufficiently low that they undergo scattering and absorption in the host matrix more then the higher energy γ rays and the measured activity must be corrected for this absorption and scattering. There is essentially no absorption effects for the γ rays emitted in a Si wafer. Typical ranges for the x-rays will vary from about 160 and 350 μm for the Cr and Cu x-rays to over 650 μm for Ge. All of the elements above Ga would have ranges in excess of the thickness of the Si wafer.

Table II. Detection limits for the x-ray emitting elements following neutron irradiation. Limits are listed as atoms in a Si matrix.

Element/Isotope	X-ray Detection Limit (atoms)	γ Ray Detection Limit (atoms)
Cr/^{51}Cr	7.3x10^{10}	2.1x10^{11}
Fe/^{55}Fe	1.1x10^{13}	-
Fe/^{59}Fe	-	3.8x10^{13}
Cu/^{64}Cu	3.3x10^{10}	5.0x10^{12}
Zn/^{65}Zn	6.1x10^{11}	2.0x10^{12}
Ge/^{71}Ge	2.1x10^{10}	-
Ge/^{77}Ge	-	6.0x10^{12}
Pt/^{191}Pt	2.3x10^{11}	2.9x10^{12}
Pt/^{195m}Pt	1.4x10^{11}	3.2x10^{11}

Note: The detection limits were determined assuming a thermal neutron flux of 5x10^{13} n/cm^2/sec, an epithermal neutron flux of 1x10^{12} n/cm^2/sec, an irradiation time of 75 hours, and a decay time of 46 hours.

An example of x-ray spectrometry on an unprocessed starting substrate is tabulated in Table III. The results of the x-ray measurement are presented for both the polished front surface and the back surface. The irradiation conditions are the same as those listed for Table II. The measured activity of the sample has been corrected for absorption within the Si matrix using the photon cross sections by Storm and Israel [19].

Table III. Analysis of the emitted x-rays from a polished Si wafer following neutron irradiation. The results are listed as atoms/cm^2.

Sample	Cu	Ge	Fe
Front Side	9.3x10^{12}	2.9x10^{12}	3.3x10^{14}
Back Side	1.7x10^{13}	2.6x10^{12}	4.8x10^{14}

It should be no surprise to see the level of Fe and Cu contamination present in this sample. The presence of Ge has not been reported in the past, although one would expect to see it since its chemistry is similar to Si, and it is very difficult to determine following γ ray spectroscopy. However, as stated earlier, x-ray spectroscopy is a viable way to detect Ge.

In conclusion, the use of x-ray spectroscopy has some advantages over γ ray spectroscopy as stated in the introduction. When used in conjunction with γ ray spectroscopy, it aids in the determination of some of the elements of interest in the semiconductor industry. The determination of Cr, Fe, and Zn are all enhanced and the detection limit for Cu and Ge is reduced approximately two orders of magnitude. The low energy x-rays also allow us to differentiate between the contamination present on the front and back sides of the wafer.

IV. References

1) L. E. Katz, P. F. Schmidt, C. W. Pearce, J. Electrochem. Soc. 128(3), 620-624 (1981).
2) M. Domenici, G. Ferrero, P. Malinverni, Semiconductor Processing, ASTM STP-850, 257-271(1984).
3) W. McColgin and R. Andrus, Unpublished data.
4) E. W. Haas, R. Hofmann, Solid-State Elec. 30(3), 329-337 (1987).
5) S. Amiel, M. Mantel, Z. B. Alfassi, J. Radioanal. Chem. 37, 189-193 (1977).
6) M. Mantel, Z. B. Alfassi, S. Amiel, Anal. Chem. 50(3), 441-444 (1978),
7) P. F. Schmidt, D. J. McMillan, Anal. Chem. 48(13), 1962-1969 (1976).
8) P. F. Schmidt, J. E. Riley, Jr., D. J. McMillan, Anal. Chem 51(2), 189-196 (1979).
9) S. F. Mughabghab, M. Divadeenam, N. E. Holden (eds), Neutron Cross Sections, Academic Press, New York (1981).
10) G. Erdtmann, W. Soyka, The Gamma-Rays of the Radionuclides, vol 7, Verlag Chemie, New York (1979). G. Erdtmann, Neutron Activation Tables, vol 6, Verlag Chemie, New York (1976).
11) C. M. Lederer, V. S. Shirley (eds), Table of the Isotopes, 7th edn., Wiley, New York (1978).
12) P. F. Schmidt, C. W. Pearce, J. Electrochem. Soc. 128(3), 630-637 (1981).
13) T. Bereznai, F. DeCorte, J. Hoste, Radiochem. Radioanal. Letters 17(3), 219-228 (1974).
14) C. Shenberg, J. Gilat, H. L. Finston, Anal. Chem. 39, 780 (1967)
15) K. K. S. Pillay, N. W. Mitter, J. Radioanal. Chem. 2, 97-107 (1969).
16) M. Mantel, S. Amiel, Anal. Chem. 45(14), 2393-2399 (1973).
17) H. M. Lau, M. Sakanour, K. Komura, Nuclear Instru. Meth. 200, 561-565 (1982).
18) M. Mantel, S. Amiel, J. Radioanal. Chem 16, 127-137 (1973).
19) E. Storm, H. I. Israel, Nuclear Data Tables A7, 565-681 (1970).

Spin dependent recombination at deep centers in Si - electrically detected magnetic resonance

P. Christmann, M. Bernauer, C. Wetzel, A. Asenov, B. K. Meyer and A. Endrös[*]
Physik - Department, Technical University of Munich, E 16, D-8046 Garching, F. R. G.
[*] Siemens AG, Abt. ZFE ME FKE 42, Otto Hahn Ring 6, D-8000 Munich, F.R.G.

ABSTRACT

Electrically detected magnetic resonance (EDMR) experiments showing spin dependent recombination in commercial p-n diodes are presented. In Si:Pt pin-diodes for the first time an axial defect is observed by EDMR ($g_{||}$ =1.97, $g_{\perp}$ =2.04). Calculations of the recombination current taking into account the energy levels, capture cross sections and concentrations of the Pt defects as obtained by Deep Level Transient Spectroscopy (DLTS) show that the spin dependent recombination occurs at the Pt donor level. In 1 N4007 p^+-n-diodes the isotropic resonance at g=2.004 often attributed to dangling bond centers (P_b) is studied in detail by EDMR and DLTS. Isochronal and isothermal annealing experiments show that the defect behaves very similar as the divacany in bulk Si.

1. Introduction

The detailed knowledge about the the microscopic structure of deep and shallow impurities in Si has mainly be supported by conventional electron paramagnetic resonance (EPR). While in the late fifties shallow donors and acceptors were of current interest [1], in the sixties the role of transition metal elements [2] and vacancy type defects [3] in Si were investigated in great detail. EPR allowed to conclude about site (interstitial, substitutional) and symmetry (isolated, complexed, Jahn-Teller distorted) of the point defects. While EPR is best suited for volume samples, it is often limited for applications in thin, epitaxial samples and especially in active devices. Detection of electrically active defects by EPR in a space charge region of a Si diode could bring together two important characterisation techniques, i.e. the space charge techniques like DLTS and EPR. DLTS gives the information about the number of defects, their capture cross section and energetical position of the level in gap, informations not always easy to obtain by EPR, but, more serious, cannot be obtained on the same sample.

2. Experimental

The EDMR experiments were performed at room temperature on commercially available Si p-n diodes (1N 4007, BY 448). By forward biasing the diode, with a current flow in the range from 1 nA up to 1 mA (recombination current regime), synchronous changes in the voltage under constant current (Keithley 225 Current Source) by on/off modulation of the microwaves were detected. The sample was placed in a cylindrical TE_{011} resonator with a loaded Q of 12000. The microwave system operates in the 9 GHz region (X-band). A maximum power of 200 mW was delivered by a frequency stabilized klystron (Varian V-262). DPPH was used as a g-marker. Narrow band lock-in detection (PAR 124A) was essential.

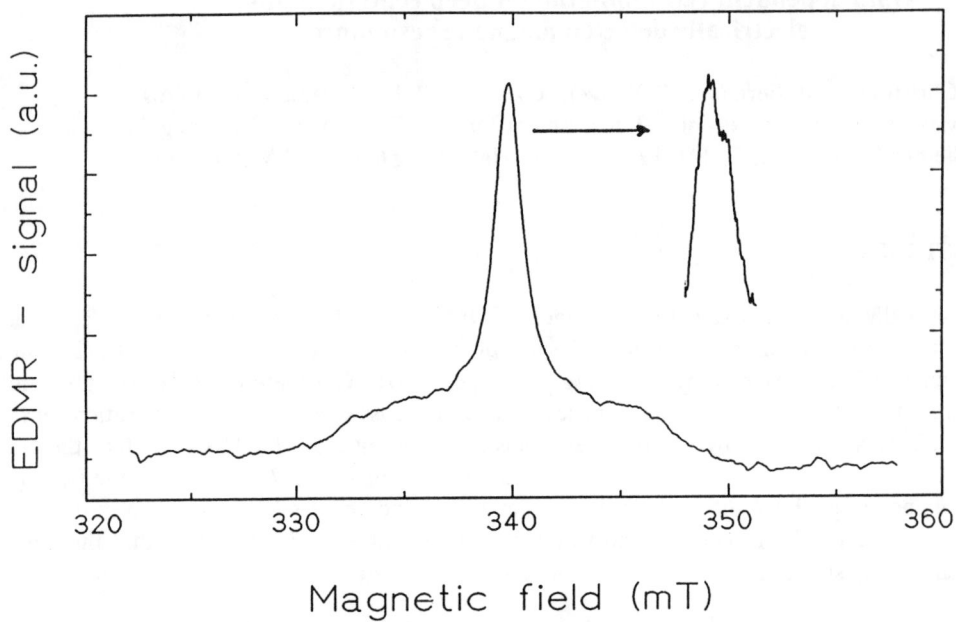

Fig.3 : Electrically detected magnetic resonance observed in the diode 1 N4007 (inset: measurement under high resolution)

Si/SiO$_2$ interface [8], which is in conflict with our result that the resonance is observable in the recombination current in the space charge region of the diode. Therefore defects present in the space charge region must account for the resonance. DLTS experiments revealed the presence of majority carrier traps at E_v + 0.25 ± 0.005 eV and E_v + 0.3 ± 0.05, additionally a miniority carrier trap is found at appr. E_c - 0.35 ± 0.05 eV. Neither energy levels can be identified with the metal elements Au and Pt often used as dopants. By a comparison with reported energy levels for intrinsic vacancy type defects, two of them fall into the energy range where the levels of the divacancy [11] have been observed (G7: E_c - 0.3 eV; G6: E_v + 0.24 eV). The DLTS experiments give for the concentrations of all observed traps 10^{10} to 10^{11} cm^{-3}. To support the idea that the divacancy has been observed we studied the behaviour of the EDMR and DLTS signals under isochronal and isothermal annealing steps. For 15 min annealing times from 180 °C up to 350 °C the decay of the EDMR signal is shown in fig.4 together with the kinetics of G7 and G6 taken from [10]. The isothermal annealing experiments at 310 and 330 °C showed thermally activated behaviour with an activation energy of 1.45 ± 0.05 eV, the pre-exponential factor was determined to be (2 ± 1) · 10^9 s^{-1}. Both results correlate closely with the well established kinetics of the divacancy [11,12]. The charge state of the divacancy, which is responsible for the EDMR resonance signal, the singly ionised donor and acceptor charge states of the divacany are paramagnetic, remains undetermined at the moment. Low temperature measurements are needed to compare our results with conventional ESR investigation, which were restricted to temperatures below 120 K.

Vacancy type defects in Si are produced by particle irradiation (electrons, neutrons etc). The device fabrication also includes processes where vacancies can be formed (e.g. implantation, metallisation). If after the last fabrication step no proper annealing has been carried out (that might be plausible for such a low cost device) vacancy type defects might still be present in the Si pn-diodes.

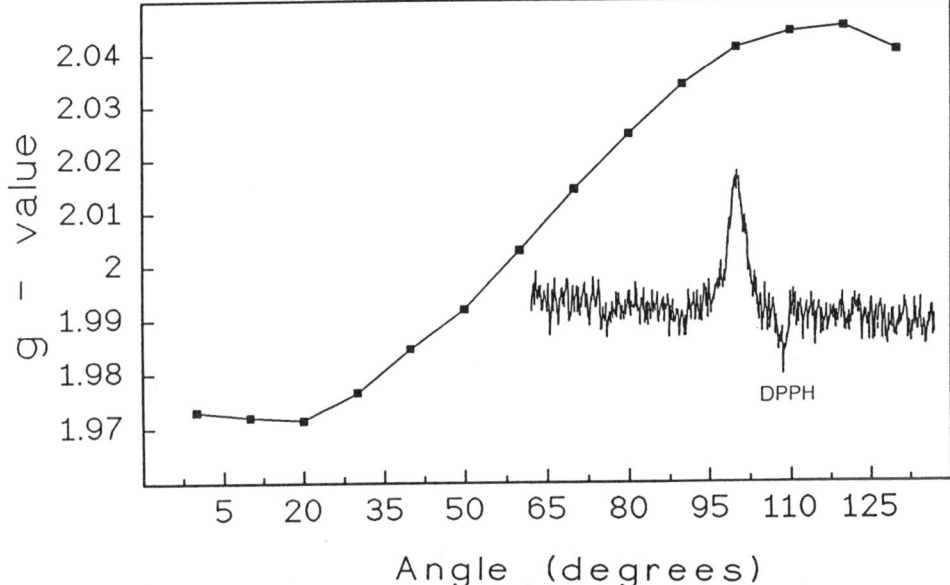

Fig.1 : Angular dependence of the electrically detected magnetic resonance line at room temperature, $\nu = 9.51$ GHz. The inset shows the resonance together with the g-marker (negative signal).

3. Experimental Results and Discussion

a) Si:Pt p-n-diodes

Fig.1 (see inset) shows the EDMR signal obtained as microwave induced change in the voltage, when the diode is forward biased under a constant current of 500 nA (the negative signal is due to the g-marker DPPH). The signal is observed from a few nA up to mA with a maximum at 500 nA in the recombination current range of the diode. However, in contrast to other experiments, the EDMR signal is not isotropic: We observe the resonance position for $B_0 \| (111)$ at $g = 1.97 \pm 0.01$ and for $B_0 \perp (111)$ at $g = 2.04 \pm 0.01$, the measured angular dependence is periodic in 180 degrees and resolves the axial symmetry of the defect (see fig.1). The observed g-values point to a deep center in Si.

EDMR detects the spin resonance of the defects in the space charge region of the pn-diode. This gives the possibility to use the same device for capacitance spectroscopy, DLTS. A complete analysis of the DLTS experiments revealed the presence of only one dominating majority carrier trap at $E_c - 0.23 \pm 0.005$ eV with a electron capture cross section (σ) of $5 \cdot 10^{-14}$ cm^{-2} and a concentration of appr. $6 \cdot 10^{13}$ cm^{-3} (the shallow background doping was estimated from C-V measurments to be appr. $4 \cdot 10^{14}$ cm^{-3}). These results identify the center as the Pt 0/- acceptor level in Si [4]. DLTS also revealed the presence of two additional traps at $E_c - 0.55$ eV and $E_c - 0.3$ eV but in concentrations 10^{-4} below the background doping. Under forward bias the Pt related donor level at $E_v + 0.32$ eV was observed as a minority carrier trap. The DLTS results are in line with informations obtained from the manufacturer of the diodes. Pt was introduced into the diodes as a life time killer resulting in faster switching times [5].

The question which charge state of Pt is involved in the spin dependent recombination, can be answered when we examine the relative contributions of the Pt donor and acceptor level in the Shockley-Read-Hall recombination of the diode. We use the numerical 1 dimensional

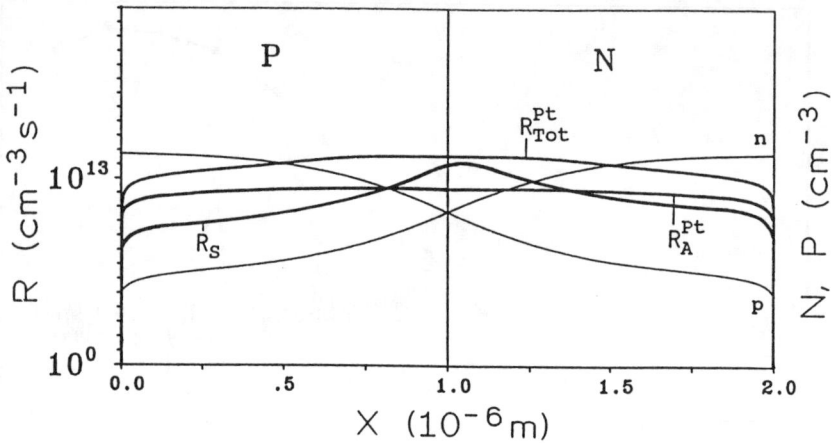

Fig.2 : Numerical simulation of the different recombination rates for the diode BY 448 Si:Pt

solution of the stationary semiconductor equation (poisson equation, continuity equations for electrons and holes) [6]. The doping concentrations on both sides of the p-n junction are taken to be $N_A = N_D = 5 \cdot 10^{14}$ cm^{-3}. For the calculation the voltage was set to $V_D = 0.1$ V. In the numerical simulation the recombination rate R_s through mid gap centers with concentration dependent electron and hole lifetimes is compared with the recombination rate R_{Pt} through the Pt acceptor and Pt donor level. The parameters for the Pt centers, i.e. binding energies and cross sections are taken from the DLTS experiments. For the Pt 0/- acceptor level we used $E_c - 0.23$ eV and $\sigma_A = 5 \cdot 10^{-14}$ cm^2, for the donor +/0 level the values $E_v + 0.32$ eV and $\sigma_D = 1 \cdot 10^{-14}$ cm^2, both with equal concentrations of $7.5 \cdot 10^{13}$ cm^{-3}. The electron and hole lifetimes are taken to be equal, their respective values for the acceptor and donor centers were estimated to be $\tau_A = 6 \cdot 10^{-7}$ s and $\tau_D = 1 \cdot 10^{-7}$ s. The result of the simulation is shown in fig.2. The Pt related recombination is higher than the typical Si (R_s) recombination. The difference, which is about one order of magnitude for $V_D = 0.1$ V (the corresponding current is 65 nA), increases further with increasing applied voltage. One further important aspect is, that the contribution of the Pt acceptor level to the total recombination rate R_{tot} is negligible compared to the contribution of the donor level. This recombination rate does not differ from R_{tot} shown in fig.2 (the simulation also shows that contributions from the shallow donor level Pt$^{++/+}$ appr. 70 meV above the valence band [7] have no influence on the recombination). Pt$^+$ is expected to be paramagnetic and our calculations of the recombination current indicate that the donor level is dominating. We therefore tentatively identify the EDMR resonance to Pt$^+$. However, no EPR results for a comparison are available so far.

b) Si:p-n diodes

The first successful experiment using EDMR experiments in commercial 1 N4007 Si p-n diodes dates back till 1976 [8]. No clear microscopic structure identification of the defect being responsible for the spin dependent recombination has been given, similarities with the P_b centers have been pointed out [9,10]. Fig. 3 shows the EDMR signal obtained as microwave induced change in the voltage, when the diode is forward biased under a constant current of 500 nA. The EDMR signal is within the experimental resolution isotropic with a g-value of g=2.004, under higher resolution and signal averaging we note that the it consists of two lines. Recent experiments attributed the resonance to the P_b centers located at the

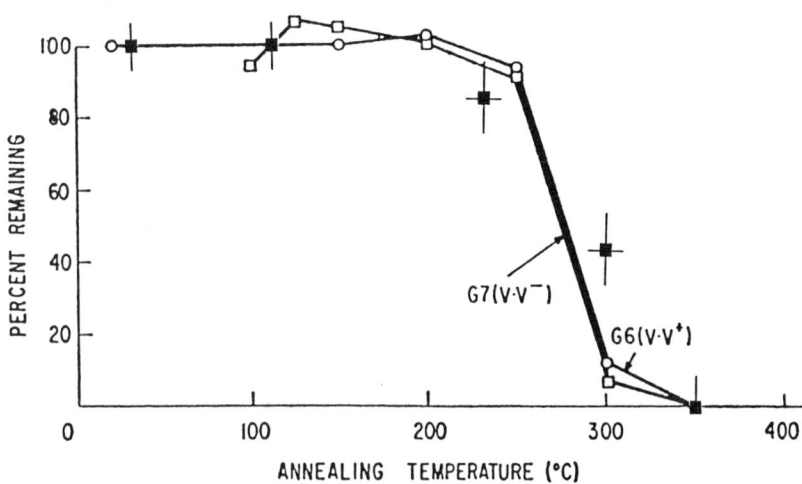

Fig. 4: Isochronal (15min) annealing experiments (resonance intensity as a function of annealing temperature, full squares) performed on the diode 1 N4007 in comparison with the results reported in ref. 11 (open symbols).

In summary, electrically detected magnetic resonance investigations in Si p-n diodes showed spin dependent recombination with deep centers. We present experimental evidence that a Pt-related center and the divacancy defect are involved. The EDMR technique can be an important step forward to apply the great potential of magnetic resonance to the defect characterisation in active devices.

References

[1] G. Feher, Phys. Rev. **114**, 1219 (1959)
[2] G.W. Ludwig and H.H. Woodbury, Solid State Phy. **13**, 223 (1962)
[3] G.D. Watkins, in Radiation Damage in Semiconductors (Dunod, Paris, 1964) p. 97.
[4] W. Stöffler and J. Weber, Phys.Rev.B **33**, 8892 (1986)
[5] J. Van Tiggelen, Philips, Eindhoven, private communication
[6] S.M. Sze, "Physics of Semiconductor Devices", 2nd edition, Wiley, N.Y., p.37 (1981)
[7] R. Zimmermann and H. Ryssel, Appl. Phys. Lett. **56**, 5 (1991)
[8] I. Solomon, Solid State Comm. **20**, 215 (1976)
[9] D.J. Lepine, Phys. Rev.B **6**, 436 (1972)
[10] For recent review see Semic.Sci.Technol. Vol.4 Nr.12 (1989)
[11] G.D. Watkins and J.W. Corbett, Phys. Rev. **138**, A543 (1965)
[12] A.O. Evwaraye and E. Sun, J. Appl. Phys. **47**, 3776 (1976)

EXCITED DEFECT ENERGY STATES FROM TEMPERATURE DEPENDENT ESR

C. KISIELOWSKI[1], K. MAIER[1], J. SCHNEIDER[1] AND V. ODING[2]

[1] FRAUNHOFER INSTITUT FÜR ANGEWANDTE FESTKÖRPERPHYSIK,
TULLASTR. 72, D-7800 FREIBURG, GERMANY
[2] A.F. IOFFE PHYSICO-TECHNICAL INSTITUTE, 194021 LENINGRAD, USSR

ABSTRACT

We report on the determination of valley-orbit splittings of nitrogen in different SiC polytypes by exploiting the temperature dependence of the ESR signal amplitudes. The effect is shown to be caused by an Orbach spin-lattice relaxation process and, therefore, enables us to measure relaxation times. The reliability of the method was tested by detection of the 1s ground state splitting of phosphorus and arsenic in silicon and it is applied to other defects in Si:P as well as to the Ga_{As} antisite related defect in GaAs. New results on hyperfine splittings are presented, too.

1. INTRODUCTION

The electrons of group V donors in silicon are known to exhibit an exponential spin-lattice relaxation at sufficiently high temperatures and the energy involved, ΔE, is the valley-orbit splitting of the 1s ground state[1]. ΔE determines the spin-lattice relaxation time T_1 and its measurement in the interesting temperature range usually requires saturation experiments and/or the observation of line broadening by Electron-Spin-Resonance[1].

Binding energies of the ground - and excited states are nowadays extracted from optical - or Raman spectroscopy because of their enormous energy resolution. However, additional informations are required to assign a spectrum to a particular impurity at a certain lattice site and some of these methods are much less sensitive than ESR spectroscopy. In addition the valley-orbit splitting of the ground state together with the hyperfine splitting observable by ESR give useful informations for corrections to the effective-mass theory which usually predicts well the binding energy of excited states but not the central-cell potential.

It is the purpose of this paper to show that ground state splittings can be simply determined by the detection of an anomalous decrease of the ESR line amplitude with increasing temperature which is caused by the exponential temperature dependence of T_1. The phenomenon is observed on donor resonances in different substances as well as on an acceptor resonance. Compensation is suspected to influence the data.

2. MODEL

We assume inhomogeneously broadened ESR lines of Gaussian shape composed of Lorentzian spin-packets with half-widths $\Delta H^g_{1/2}$ and $\Delta H^l_{1/2}$, respectively. At low temperatures the spin-spin relaxation time T_2 determines $\Delta H^l_{1/2}$. However, at high temperature T_1 rapidly decreases and equals T_2 and, thereby, determines the width of the Lorentzian spin-packets because $T_1^{-1} = T_2^{-1} = \gamma * \Delta H^l_{1/2}$ (γ = gyromagnetic ratio) as shown in ref. 1.

Computer simulation of this prosess was done by multiplication of i Lorentzian line amplitudes centered at H_{li} with a Gaussian distribution centered at H_0 on a finite magnetic field scale $i*H_l$. For a given half-width $\Delta H^g_{1/2}$ we typically used $H_{l(i+1)} - H_{li} = 1/100 * \Delta H^g_{1/2}$ and i=1000. Increasing the half-widths of the Lorentzian lines we kept constant the area A' under each Lorentzian absorption curve and added up all contributions of the other lines to a particular position H_{li}. This simulated the transition from a Gaussian absorption curve ($\Delta H^l_{1/2}/\Delta H^g_{1/2} << 1$) to a Lorentzian one ($\Delta H^l_{1/2}/\Delta H^g_{1/2} >> 1$). The absorption derivative of the resulting line with amplitude Y'_1, and its integral value A' were determined numerically to check up the influence of different modulation amplitudes and the constancy of the total intensity which represents the number of spins.

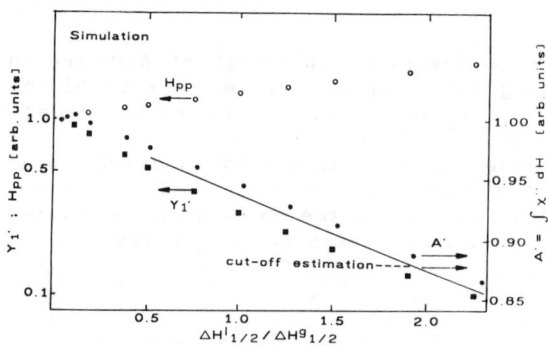

FIGURE 1: Simulated dependence of the peak-to-peak width Hpp, the integral absorption A' and the amplitude of abs. derivative Y'_1 on $R = \Delta H^l_{1/2}/\Delta H^g_{1/2}$. Note different scales. Estimation from ref.2.

Figure 1 shows the dependence of A', Y'_1 and the peak-to-peak width Hpp of the absorption derivative on the ratio $R = \Delta H^l_{1/2}/\Delta H^g_{1/2}$. For R < 0.1 A' changed only within 2% with a tendency to increase with larger R. The drop of A' for R > 0.1 can well be explaind by cut-off effects which for Lorentzian lines can be estimated by: $A' \sim A_0 * (1 - \Delta H^l_{1/2} / \Pi*a)$; with a being half of the finite field width[2]. Usually, the increase of Hpp (by a factor of 2 in this example) is exploited[3] to determine T_1 even though Y'_1 changes in same range by one order of magnitude.

Within an experiment the decay of the line amplitudes will even be sharper because of the natural temperature dependence of the absorption derivative amplitude Y' which is assumed to be given by the Curie law: $Y' \sim C/T$. (It is noted that other dependencies may be observed but it is easy to extend the results to such cases). Thus, to compare with the simulation we have $Y'_1 = Y'*T$ and the deviation from the Curie law in case of line broadening will be given by $C - Y'*T$.

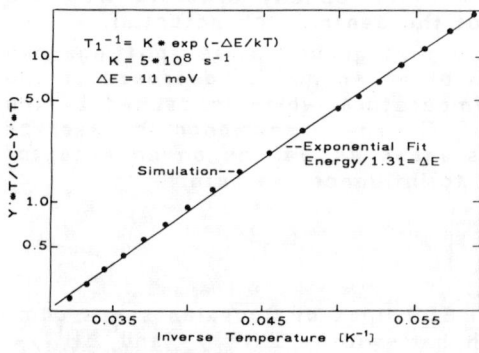

FIGURE 2: Approximation of simulated and normalized amplitudes by an exponential function. Numerical values from ref.3.

Numerical values for the exponential decay of T_1 in Si:P were taken from ref.3 (see fig.2). They couple the Lorentzian line widths with a particular temperature because $T_1^{-1} = \gamma * \Delta H^l_{1/2}$. By use of fig.1 the relation $Y'*T(T)$ is, therefore, fixed if we choose C=1 and $\Delta H^g_{1/2}(P) \approx 2.2G$. One expects $Y' \sim C/T * (1-f(T))$ with f(T) being an unknown function describing the deviation from the Curie law. f(T) must be rather complicated because the resonance lines are neither Lorentzian nor Gaussian. However, we show in fig. 2 that the simulated ratio $Y'*T/(C-Y'*T)$ can be approximated very well by $B*exp(\Delta E/mkT)$. From comparison with the input value $\Delta E=11$meV we find m=1.31.

Several possible dependencies were controlled by simulation and they are as follows: 1) m does not depend on the modulation amplitude H_m if $H_m <$ 0.5 H_{pp}. Otherwise it decreases and needs to be corrected. 2) The plot of Fig.2 is sensitive against a proper choice of C. Wrong Curie constants produce S-shaped curves. 3) If two parallel Orbach relaxation processes are present with different energies, $|\Delta E_1 - \Delta E_2|$ must be larger than $\approx$ 5meV to recognize a bending in the curve. Otherwise a mean energy is determined. 4) In case of signal superposition the plot is dominated by the largest signal and the highest energy. 5) For $\Delta H^l_{1/2} > 2.3\ \Delta H^g_{1/2}$ the plot deviates from linearity.

3. APPLICATIONS

ESR spectra were recorded with a Bruker X-band spectrometer between 2K and 293K. The absolute temperature calibration was 1K. Different samples were investigated and they are listed in table I together with a summary of the experimental results. Small samples were used to avoid disturbences of the cavity by conducting samples.

Figure 1a shows ESR spectra of the phosphorus donor. With increasing temperature the line widths change by a factor of ≈ 2 and the decay of the signal amplitudes is enormous. The line shape changes from Gaussian to Lorentzian with higher detection temperature and the central hyperfine splitting A decreases. Also, this is exactly the temperature range where one would expect to find line broadening and decay on basis of the data of fig.2 were we used values for T_1 measured on P in Si[3]. Thus, Orbach relaxation to the excited valley-orbit states causes the effects. We attribute the reduction of A to an admixture of the E and/or T_2 states (which have a node at the nucleus site) into the A_1 ground state by transitions which do not flip the spin. No resonance of electrons in the conduction band can be seen because excitation into the conduction band is still very weak at this temperature. Similar observations were made on the arsenic spectra with the exception that we did not record a reduction of A. In fig.4a we show the Arrhenius plots of the signal amplitudes. Using m=1.31 the extracted energies are both by a common factor of 1.28 larger than literature values for the A_1 to T_2 splitting of P and As in

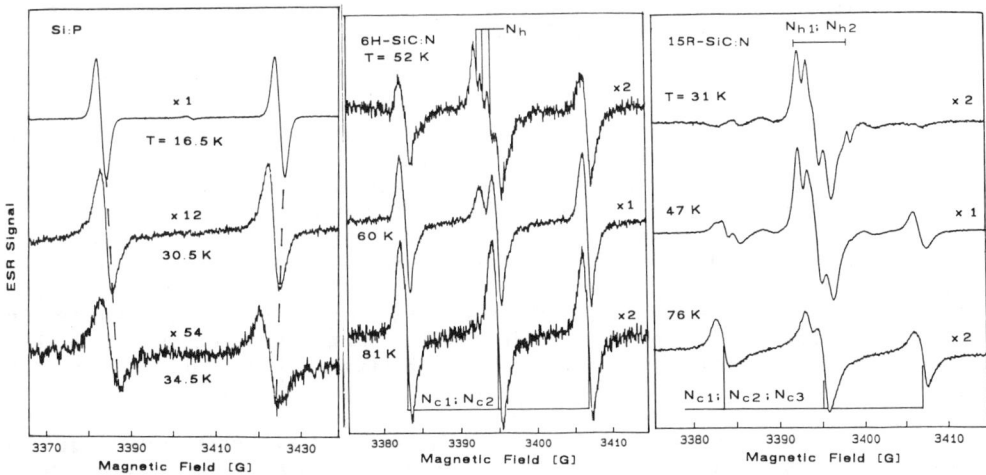

FIGURE: 3a   3b   3c
ESR spectra of the indicated systems. Hyperfine splitting is caused by the nuclear spins $I_P = 1/2\hbar$ and $I_N = 1\hbar$. Scaling factors and detection temperatures are indicated. N_c, N_h nitrogen on cubic or hexagonal site. 6H-SiC: two cub.- and one hex. site; 15R-SiC: three cub.- and two hex. sites

TABLE I:
Materials properties and summary of the experimental results:
g= g-value of the dopant; T_1= spin-lattice relaxation time;
ΔE = energy from Orbach process; experimental error of $\Delta E \approx 15\%$;
ΔE_{lit}= valley-orbit splitting from literature[4,6];
*= energy value adjusted with m=1.67; += signal superposition
A = hyperfine splitting; $A_m(T)$ = minimum detected at temperature T;

Substance	Doping [cm-3]	g	$T_1(T)$ [s(K)]	A [G]	$A_m(T)$ [G(K)]	ΔE [meV]	ΔE_{lit} [meV]
Si:P	10^{15}	1.998	$3.2\ 10^{-7}(26)$	42.0	37.7(34)	11.8*	11.7
Si:As	10^{15}	1.998	$2.2\ 10^{-7}(26)$	71.0	-	21.0*	21.1
Si:P,diff.	inhomog.	1.998	$6.0\ 10^{-7}(34)$	-	-	11.3	-
PD-Si:P	10^{15}	1.996 anisotr.	$2.9\ 10^{-7}(27)$	-	-	3.4; 12.6	-
3C-SiC:N	$\geq 10^{18}$	2.005	$3.2\ 10^{-7}(23)$	-	-	4.7	-
4H-SiC:Nc	$\approx 10^{17}$	2.004	$4.0\ 10^{-7}(82)$	18.0	17.5(109)	45.5	-
6H-SiC:Nh polycryst.	$\approx 10^{17}$	2.005	$6.7\ 10^{-7}(19)$	1.1	1.0(35)	3.8; 13.0	13.0
6H-SiC:N $\approx 10^{16}$ Nc+:		2.003	$1.0\ 10^{-6}(81)$	11.5	-	61.1	60.3 62.6
Nh:		2.005	$1.0\ 10^{-6}(44)$	1.0	0.7(56)	12.3	13.0
15R-SiC:N $\approx 10^{16}$ Nc+:		2.003	$1.0\ 10^{-6}(81)$	10.8-11.8	-	58.8	54.9 50.6 46.0
Nh1+:		2.005	$1.2\ 10^{-6}(33)$	-	-	8.1	7.7
Nh2+:		2.005	$0.8\ 10^{-6}(45)$	≈ 1.3	-	12±4	11.6
PD-GaAs: undop. Ga+ As (fourlines)		≈ 2.05	$\approx 10^{-8}(12)$	-	-	3±2	-
Singlet+ (backgr.)		≈ 2.05	$\approx 10^{-9}(47)$	-	-	21±10	-

silicon (11.7 and 21.1meV)[4]. This suggests that the algorithm used in the simulation underestimates the drop of Y'_1 (R) in fig.1. The same conclusion comes out from comparison of absolute T_1 values: from fig 1 one extracts that $H^1_{1/2}$ equals $0.5H^9_{1/2}$ for $0.52*Y_1'$. From this condition we determined T_1 from the amplitudes and listed them in table I ($3.2\ 10^{-7}$ and $2.2\ 10^{-7}$ s for P and As at 26K, respectively). Compared with T_1 measurements of ref. 3 ($1\ 10^{-7}$ and $8\ 10^{-8}$ s for P and As at 26K) the agreement is already satisfactory but our values are systematically larger suggesting again a steeper decay of Y'_1 (R) in fig.1. Thus, we fixed m to be 1.67 from the experiments on As and P in Si (table I) and used this value later on.

Nitrogen in SiC polytypes exhibits different valley-orbit splittings depending on its local environment and the polytype[5-7]. In figure 3b we show X-band spectra of 6H-SiC where we could resolve the hyperfine splitting of nitrogen on the hexagonal lattice site N_h. The two cubic sites N_{c1}, N_{c2} of this polytype can partly be resolved in other orientations of the crystal with respect to the magnetic field. Determination of the energy

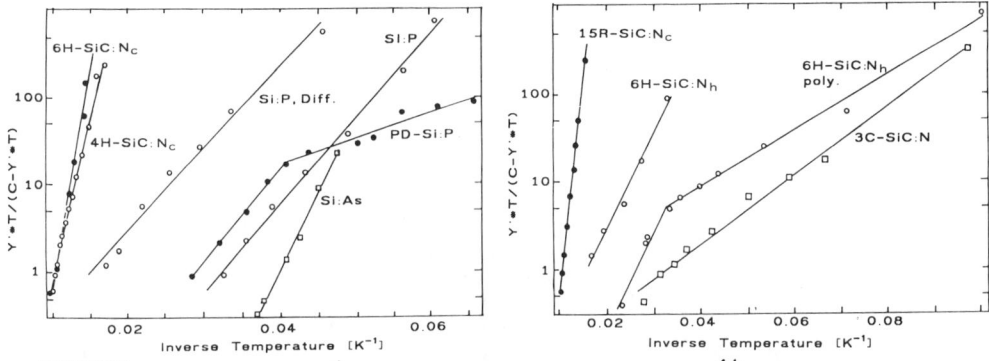

FIGURE: 4a 4b
Normalized ESR line amplitudes and their approximation by an exponential fit versus inverse temperature. For results see table I.

related to an Orbach process is shown in figures 4a and 4b. The results are listed in table I: measured energies are in good agreement with literature data of valley-orbit splittings in 6H- and 15R-SiC [6] but different values of the cubic lattice sites could not be observed because of the signal superposition; the hyperfine interaction of N_h depends on temperature as in case of Si:P which is why the signal degenerates to a singlet at 60K in fig 3b; T_1-values of N on cubic and hexagonal lattice sites are very different. In the <u>15R-SiC</u> spectra of fig.3c the 6 lines of N on the two hexagonal lattice sites interfere with three lines of N on cubic sites at 3395 G which makes measurements of line amplitudes difficult. However, the N_h signal decays very similar to that in 6H-SiC (12.3 meV) with the exception of the isolated line at 3398 G which is nearly absent from the spectrum at 47K because ΔE= 8.1 meV. Literature values for valley-orbit splittings of the hexagonal lattice sites are: 7.7 and 11.6 meV [6] and they suggest that the isolated line is part of one N_h spectrum. The three spectra of N on cubic lattice sites superimpose and can partly be resolved in fig.3c at 47 K and 3383 G. From table I it can be seen that ΔE is slightly smaller than in 6H-SiC but the superposition causes deviation from the established values. In <u>4H-SiC</u> the hyperfine splitting of cubic nitrogen is the largest of all polytypes but ΔE is smallest.

In fig.4b we compare the results on 6H-SiC: N_h ,6H-SiC:N_h(polycrystalline) and 3C-SiC:N (with higher doping the triplet structure of N in cubic SiC vanishes and is replaced by a singlet [8] which we observed). For the 6H-SiC polycrystalline sample it can be seen from fig.4b that it exhibits a very shallow activation (3.8 meV) at low temperatures before the valley-orbit splitting shows up. In case of 3C-SiC:N we measured a comparable small activation energy (4.7meV) throughout the temperature range. This makes us doubt to attribute this energy to a valley-orbit splitting. From experiments on Si:P it is known [9] that T_1 shortens in case of compensation. Since a similar effect is observed in the PD-Si:P sample (fig.4a) which we know to be compensated [10] it is tempting to ascribe such low activation energies to hopping processes between occupied and unoccupied donor sites within an impurity band but other interpretations are still possible.

The ground state splitting of different donors in silicon is characteristic [4] and may help to identify impurities. This is useful in cases where no hyperfine splitting is observable by ESR for some reason. In plastically deformed PD-Si:P a new anisotropic and dislocation related signal was observed [10] showing no hyperfine splitting. Also, P diffusion in Si can give rise to anisotropic, unsplit line at g=1.998 in spite of the low P concentrations in the sample ($< 10^{16} cm^{-3}$). In fig. 4a and table I we show that the valley-orbit splittings of both defects are very similar to that of isolated P in Si which suggests that phosphorus is the impurity causing the spectra.

An application to an acceptor concerns fig.5 which shows the As_{Ga} antisite defect and a new four line spectrum detectable in plastically deformed PD-GaAs[11]. From the huge similarity of the new spectrum to one in PD-GaP it was recently concluded that it is due to the Ga_{As} antisite defect[12] which should be a double acceptor if it is isolated. It can be seen from fig.5 that the four line structure around 3000 G degenerates to a broad line as the temperature increases. Between 40 and 80K the broad line disappears, too, and an almost undisturbed As_{Ga} antisite spectrum is detectable. The inset of fig.5 shows that simulation of a temperature dependent broadening of the four lines ($H_{PP}\sim$ 150G) on a background line ($H_{PP}\sim$ 500G) explains the observation. Broadening of single lines is hardly observable because of signal superposition but from the amplitude drop of the four lines, 3±2 meV could be extracted while the background line decays with 21±10meV at 60±20K. Since T_1 must be short enough to broaden also the background line, which requires high temperatures, the result suggests that the four lines and the background are due to one defect with T_1 showing a shallow and steep decay like PD-Si:P in fig.4a. Compensation is apparent due to the presence of the As_{Ga} signal and acceptor excited states are important. Table I shows that T_1 is very short at 12K which agrees with the fact that the spectrum is hardly saturable even at 5K with 0.5W of microwave power. Very similar saturation behaviour is observed on the Ga_P-spectrum in PD-GaP[12] which stresses the identity of the spectra and suggests that the resolution of Ga antisite related defects in GaAs and GaP is limited by T_1 broadening.

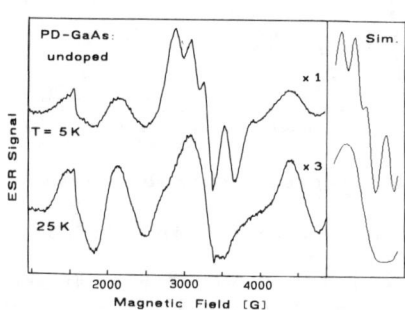

FIGURE 5: ESR signals in plastically deformed GaAs: As_{Ga} and Ga_{As} antisite defects. The inset shows the simulated decay of the Ga_{As} lines due to broadening.

ACKNOWLEDGEMENTS
One of the authors (C.K) appreciates stimulating discussions with G.D. Watkins and he acknowledges that P.Pirouz, W.Zulehner and W.Lerch/N. Stolwijk supplied the 3C-SiC:N , 6H-SiC (poly.) and Si:P (diff.) samples.

REFERENCES:
1) T.G.Castner, Jr., Phys. Rev. 130, 58 (1963)
2) C.P.Poole, Jr. in "Electron Spin Resonance" (Wiley Interscience, NY 1967), p.810
3) T.G.Castner, Jr., Phys. Rev. Let., 8, 13 (1962)
4) R.Sauer in "Landolt-Börnstein" 22, ed. M. Schulz (Springer, Berlin 1989), p.326
5) O.V.Vakulenko, O.A.Guseva, Sov. Phys. Semicond., 15, 886 (1981)
6) P.J.Colwell, M.V.Klein, Phys. Rev. B, 6, 498 (1972)
7) E.N.Kalabukhova, N.N.Kabdin, S.N.Lukin, E.N.Mokhov, B.D.Shanina, Sov. Phys. Solid State, 378 (1989)
8) H.Okumura, M.Yamanaka, M.Shinohara, E.Muneyama, S.Kuroda, K.Endo, I.Nashiyama, H.Daimon, E.Sakamura, S.Misawa, S.Yoshida in "Amorphous an crystalline Silicon Carbide", eds. G.L. Harris, C.Y.-W.Yang, Springer Proc. in Physics 34, 1989, p.107
9) G.Feher, A.Gere, Phys. Rev. 114, 1245 (1959)
10) C.Kisielowski, J.Palm, B.Bollig, H.Alexander, Phys. Rev. B (1991), in press
11) M.Wattenbach, J.Krüger, C.Kisielowski-Kemmerich, H.Alexander, Mat. Science Forum, 38-41, 73 (1989)
12) J. Palm, C. Kisielowski-Kemmerich, H.Alexander, Appl. Phys. Let. 58, 68 (1991)

DISLOCATION ASSOCIATED DEFECTS IN GALLIUM ARSENIDE BY SCANNING TUNNELING MICROSCOPY

O. SIBOULET, S. GAUTHIER, J.C. GIRARD, W. SACKS, S. ROUSSET and J. KLEIN
Groupe de Physique des Solides, Universités Paris VI et VII, Tour 23, 2 place Jussieu, 75251 Paris Cédex 05.

ABSTRACT

STM is used to analyse volume defects of GaAs. Very high defect concentration areas are observed. The defects are identified to be mainly gallium multivacancies. The rôle of dislocations in the clustering of these vacancies is shown. The inhomogeneity we identified in our sample could account for the discrepancy which is commonly observed between DLTS and positron annihilation in the determination of vacancy concentration.

Scanning tunneling microscopy (STM) was developed by Binnig, Rohrer, Gerber and Weibel [1]. In this technique, a tip is scanned over a surface and the distance between tip and sample is monitored by piezoelectric ceramics so as to keep the tunneling current constant. Atomic resolution can be achieved on many samples, such as the 7x7 reconstruction on silicon (111) [2] or graphite [3]. The tunneling current is due to the overlap of the wave functions of the tip and sample. If the rôle of the tip is neglected [4], the images can be interpreted in terms of local density of states of the surface [5]. Our microscope is similar to the one developed by Gerber [6]. The operating pressure in the vacuum chamber is below 10^{-10} torr. The time between cleaving and imaging is less than 20mn. Prior to the experiment, the tip is heated *in situ* by electron bombardment. The whole system allows us to obtain atomic resolution regularly.
The sample we deal with in this paper is an as grown HB wafer, Si doped at a concentration of $10^{18} cm^{-3}$.
The cleaved surface of Gallium arsenide - (110) - was first observed by STM by Feenstra and Fein [7]. According to the polarization of the tunneling junction Vt, taken to be the sample bias relative to the tip, electrons tunnel from the tip to the sample (Vt>0) or from the sample to the tip (V_t<0). Then, on semiconducting samples, STM reveals conduction or valence states, according to the applied bias. In gallium arsenide, the local density of filled (valence) states is higher on arsenic atoms, whereas the local density of empty (conduction) states is higher on gallium atoms. This tendency is stronger on the surfaces, since electrons of the gallium dangling bonds transfer to the arsenic dangling bonds [8]. This is why it can be said that negative bias shows arsenic atoms whereas positive bias shows gallium atoms. These two types of measurement can be carried out simultaneously by inverting bias between each scan [9].

Point defects were resolved by Stroscio *et al* [10] as well as by Cox [11]. Cox observed arsenic and gallium point defects, which he attributes mainly to vacancies.

We present our results in two steps. To begin with, a very high defect density area is shown, and then a smaller area with smaller defects. The analysis of these two results allows us to draw out conclusions.

We observe very high defect concentration areas. Figure 1 is an area of 33nmx26nm, taken at negative bias and showing about a hundred [1-10] rows of arsenic atoms. 12 dark spots, about 2nm large, can be seen. But we are mostly concerned by areas extending in the [1-10] direction, about 2nm wide, where the position of the tip during scanning shows instabilities. These areas are indicated by "D". Now, the only extended defects in crystals are dislocations. Besides, the [1-10] direction is the intersection of the (110) cleavage plane with one of the (111) or (11-1) slip planes of GaAs dislocations. We conclude that these areas of tip instability can only be related to dislocations.

The reason why dislocations appear in the cleavage-plane is that they are split : their stacking faults extend in the (111) or (11-1) planes.

Because of the dislocations, one expects to see steps or extra atomic rows. In figure 1, the black to white scale corresponds to 0.2 nm, which is the height of a monoatomic step on a [110] face of GaAs. The presence of steps in the area can be ruled out, because they would give rise to a strong contrast. Since there are no steps, Burgers vectors have to belong to the cleavage-plane. This reduces the possibilities to the vectors 1/2[1-10] or -1/2[1-10]. Thorough analysis by Fourier transforms did not reveal any extra [001] row or stacking fault area required by dislocations with such Burgers vectors. It can be inferred that dislocations are not present when we examine the sample. What we see are not the dislocations, but the point defects which clustered during growth on the stacking fault planes.

We attribute them to gallium vacancies: instabilities of the tip are very probably due to scanning over arsenic atoms, linked to gallium atoms among which many are missing. In fact, a zoom of another area of the sample helps to determine the nature of these defects.

Before doing so, we can obtain a rough estimate of the defect concentration in figure 1. Since there is about one (111) (or (1-11)) ex-stacking fault plane out of ten crystallographic planes, and that these stacking faults have attracted approximately one defect monolayer within 3 or 4 interatomic distances, the average defect concentration in this area is about 10%, which is also $2 \times 10^{21} cm^{-3}$. Such high values are not unexpected [12].

As we said, we need to carry out the analysis of smaller defects to confirm that we observe gallium vacancies. Figure 2a and 2b are images of another area taken simultaneously at two different voltages: figure 2a shows arsenic atoms and figure 2b shows gallium atoms. *Although they look different from the ones in figure 1, we shall see that defects in figure 2 have the same origin: dislocations.* Figure 2c and 2d show cross-sectional cuts displayed at the same surface location. We analyse the defect on the right of 2a and 2b to be a gallium multivacancy.

In figure 2b we see atoms missing on the gallium rows. On the 2c cross-sectional cuts, which are near the defect, Z(2.2V) shows local maxima evenly

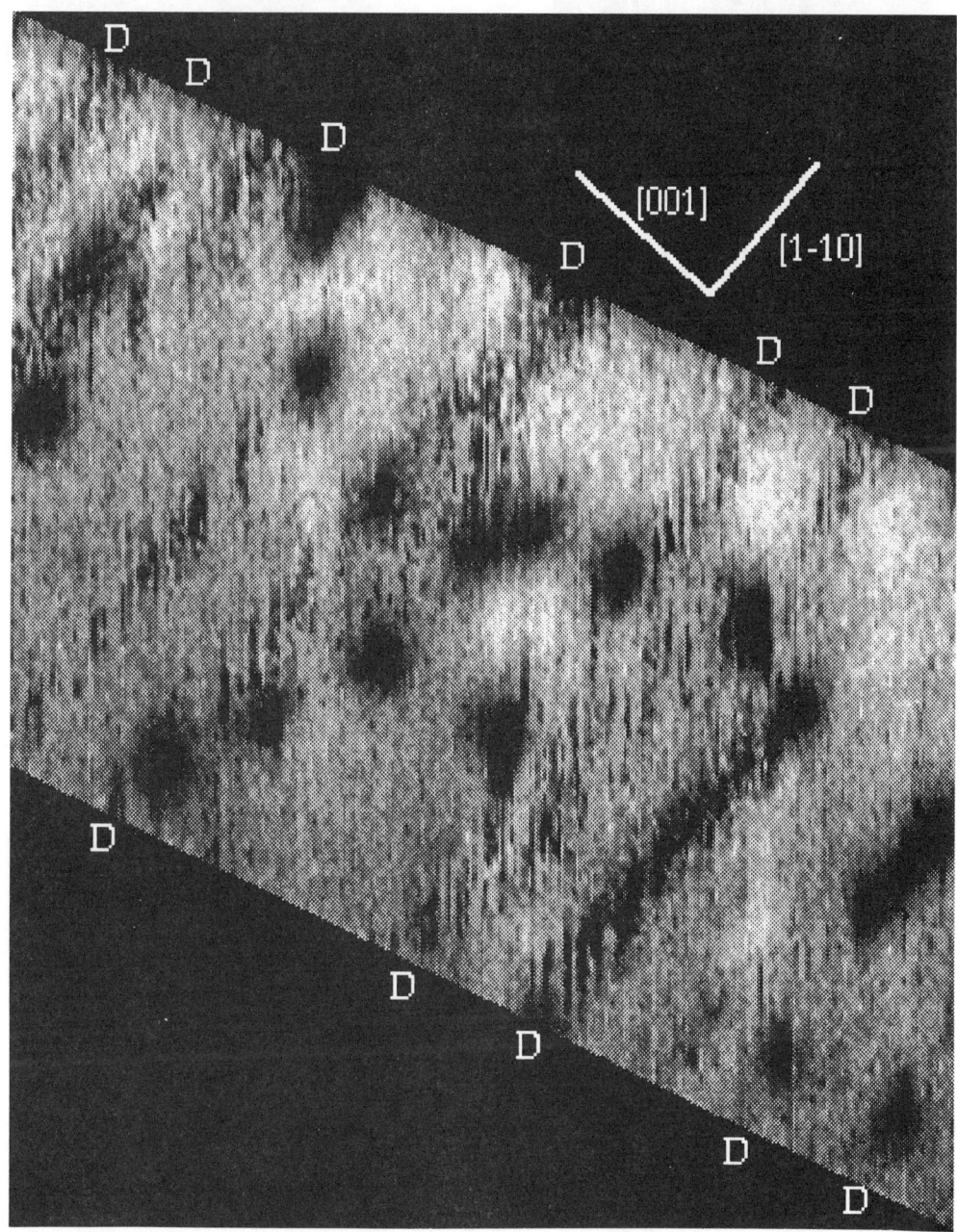

Figure 1 : a high defect concentration area.
n+ (Si) HB sample. Size: 26nmx33nm. 0.2nm from black to white. Tunneling bias: –1.9V. As atoms are seen.
About 12 dark spots can be seen. Besides, areas extending in the [1-10] direction, about 3 or 4 atomic rows wide, indicated by "D", produced tip instabilities during scanning. They are due to gallium vacancies clustered on split dislocations.

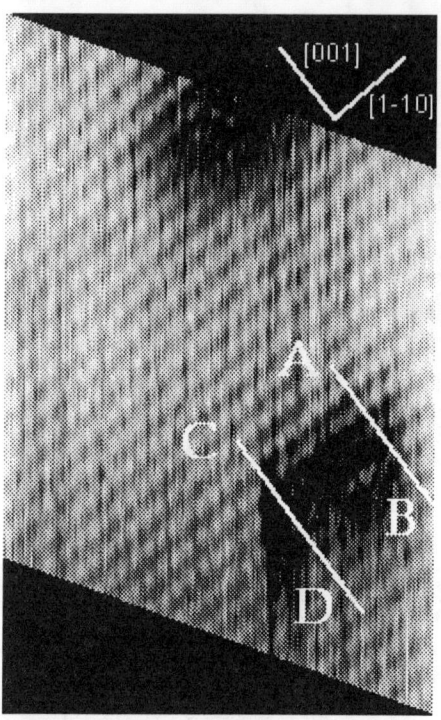

Figure 2 a : gallium multivacancies. n+ (Si) HB sample. Size: 11nmx17nm. Tunneling bias: -1.5V. **As** atoms are seen.

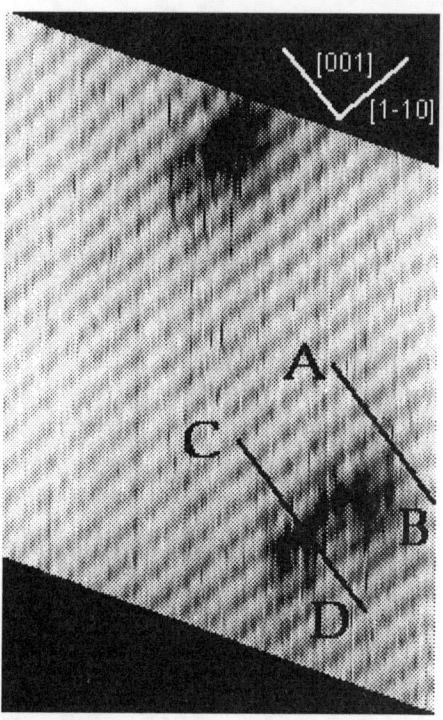

Figure 2 b : the same area taken simultaneusly at a tunneling bias of 2.2V. **Ga** atoms are seen.

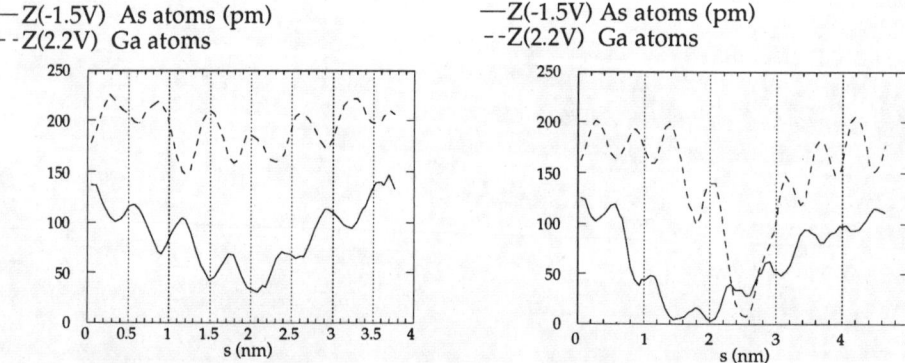

Figure 2 c : AB

Figure 2 d : CD

2 c and 2 d are cross-sectionnal cuts taken at two polarities between A and B or C and D. Positive polarity shows gallium atoms, whereas negative polariry shows arsenic atoms. In 2d, the missing *local* maximum reveals the absence of gallium atoms, whereas no arsenic atom is missing.

spaced, whereas in 2d, a gallium atom is clearly missing around the position 2.5nm.

2a is more difficult to analyse, because extra negative charge produced by the defect induces band-bending and repels electrons from the valence band. When the tip is scanned over this area at negative bias, the control loop lowers it so as to keep the current constant. This is a purely electronic effect. In figure 2a, *local* maxima can be seen on the location of each arsenic atom. The impression one can get from observation of the image is confirmed by thorough examination of the cross-sectional cuts Z(-1.5V). In figure 2d, maxima on arsenic atom locations can be distinguished, just as in 2c, which is taken near the defect.

Figure 2 shows that only gallium positions are concerned by the defect. More difficult to rule out is the possibility of arsenic antisite clustering or dopant (Si) clustering. We observe many defects ranging from the one in figure 2 to the ones showing tip instabilities in figure 1, so that they can with no error be attributed to the same origin. If they were any kind of atom clusters, they would not show such tip instabilities.

Other methods confirm this hypothesis. X-ray intensity measurements have shown that that the regions surrounding dislocations have a higher arsenic concentration [13]. Positron annihilation (PA) reveals the existence of high vacancy concentrations, of the order of 10^{18}cm^{-3} [14]. The defects are attributed to arsenic vacancies because gallium vacancies, which are acceptors, should trap free electrons and decrease conductivity in n-type material. However, conductivity measurements are not relevant for highly inhomogeneous materials.

The [1-10] direction of the defect is strong evidence that it was created by a dislocation. In larger images, we can see that many of the defects such as those in figure 2, although several nanometers distant, are often aligned together in the [1-10] direction. Besides, clustering of gallium vacancies is unexpected, since without the presence of dislocations, negatively charged defects should repel each other. The rôle of dislocations in the clustering of gallium vacancies in figure 2 is clear.

A discrepancy between PA and deep level transient spectroscopy (DLTS) can be accounted for by the inhomogeneity of the vacancy concentration. The DLTS is insensitive to areas where defect concentration is too high, whereas positron annihilation is not. If inhomogeneity in vacancy density is established, it is normal the DLTS should give a lower apparent defect concentration (10^{16}cm^{-3}) than PA (10^{18}cm^{-3}). Actually, we confirm that there exist areas of very high vacancy concentration in an as grown HB wafer.

We have used STM to study native defects in an as grown HB sample. Gallium multivacancies are identified, and the rôle of split dislocations in the production of such defects is shown. Areas with concentrations of defects, mainly vacancies, up to 2×10^{21}cm^{-3} are observed. Since no technique can be more local than STM, it is normal that our experiments should yield concentrations higher than what is usually obtained. So far, STM has only been used as a surface analysis technique. But in any kind of cleavable conductive or semiconducting sample, *volume defects* can also be analysed.

The only restriction is the high concentration required: above $10^{18} cm^{-3}$. But STM will provide structural information, which is of paramount importance in the determination of defects.

---

[1] G. Binnig, H. Rohrer, CH. Gerber and E. Weibel, Phys. Rev. Lett. 49, 57 (1982)

[2] G. Binnig, H. Rohrer, CH. Gerber and E. Weibel, Phys. Rev. Lett. 50, 120 (1983)

[3] I.P. Batra, N. García, H. Rohrer, H. Salemink, E. Stoll and S. Ciraci, Surf. Sci. 181, 126 (1987)

[4] W. Sacks and C. Noguera, J. Vac. Sci. Technol. B9, 488 (1991)

[5] J. Tersoff and D.R. Hamann, Phys. Rev. B31, 805 (1985)

[6] Ch. Gerber, G. Binnig, H. Fuchs, O. Marti and H. Rohrer, Rev. Sci. Inst. 57, 221 (1986)

[7] R.M. Feenstra and A.P. Fein, Phys. Rev. B32, 1394 (1985)

[8] S.Y. Tong and W.N. Mei, J Vc. Sci. Technol. B2, 393 (1984)

[9] R.M. Feenstra, J.A. Stroscio, J. Tersoff and A.P. Fein, Phys. Rev. Lett. 58, 1192 (1987)

[10] J.A. Stroscio R.M. Feenstra, D.M. News and A.P. Fein, J Vac. Sci. Technol. 6, 499 (1988)

[11] G. Cox, K.H. Szynka, U. Poppe and K. Urban, Vacuum 41, 591 (1990)

[12] J.C. Bourgoin, H.J. von Bardeleben and D. Stiévenard, J. App. Phys. 64, R65 (1988)

[13] I Fujimoto, Japanese J. App. Phys. 23, L287 (1984)

[14] G. Dlubek, O. Brümmer, F. Plazaola and P. Hautojärvi, J. Phys. C19, 331 (1986)

Transition Metals in Silicon Carbide (SiC):
Vanadium and Titanium

Karin Maier, Harald D. Müller and Jürgen Schneider

Fraunhofer Institut für Angewandte Festkörperphysik,
Tullastrasse 72, D-7800 Freiburg, Germany

ABSTRACT

Vanadium, substituting for silicon in SiC-polytypes, has been identified as an amphoteric deep level defect. Electron Spin Resonance (ESR) is observed for the neutral state $V^{4+}(3d^1)$, $S=1/2$ and for the A^--state $V^{3+}(3d^2)$, $S=1$. By photo-ESR, the position of the $(0/+)$ donor level is found to occur near midgap in *6H*-SiC. Near-infrared, 1.3 - 1.5 μm, photoluminescence and absorption, arising from internal 3d-shell transitions, $^2T_2 \leftrightarrow {}^2E$, are observed for the neutral state $V^{4+}(3d^1)$. These sharp-line spectra were found to be very specific for a given SiC-polytype. Isoelectronic, electrically inactive, *titanium* impurities have been found, by ESR, to complex preferentially with shallow nitrogen donors. The resulting Ti_{Si}-N_C pair then acts as deep donor, $E_C - 0.6$ eV, in *6H*-SiC. The isolated titanium defect forms a deep acceptor state in *4H*-SiC, but not in the *6H*-polytype.

Introduction

The understanding of deep level defects in silicon carbide (SiC) is still in its infancy. For example very little is known about transition metals in SiC, as their location in the lattice (substitutional or interstitial?), or their presumed electrical activity as deep level impurities. This lack of knowledge is surprising, and challenging, in view of the fact that transition metals, in particular *titanium* and *vanadium*, are practically unavoidable contaminations in Lely-grown SiC-crystals [1]. Obviously, identification and control of transition metal deep level impurities is prerequisite for satisfactory performance of electronic and opto-electronic devices fabricated from silicon carbide. We point out that SiC currently attracts renewed interest because of its potential use in high-power, high-speed, high-temperature and high-radiation resistant devices [2].

1. Amphoteric Vanadium

By electron spin resonance (ESR) we have recently detected [3],[4] that omnipresent vanadium contaminations in Lely-grown SiC crystals act as amphoteric deep level impurities, substituting the various silicon sites in the lattice. Three charge states of vanadium were found to exist in SiC: $V^{3+}(3d^2)$, $V^{4+}(3d^1)$, and $V^{5+}(3d^0)$, their relative occurrence depending on the position of the Fermi level. In n-type *4H-* and *6H-*SiC, the ESR-spectrum of the ionized acceptor (A⁻) charge state $V^{3+}(3d^2)$, S=1 is dominant. No vanadium-related ESR signals are detectable in *p*-type SiC, indicating the occurrence of vanadium in its pentavalent diamagnetic charge state $V^{5+}(3d^0)$, which corresponds to that of an ionized donor (D⁺). Thus vanadium in SiC acts as an electrically amphoteric impurity, introducing two levels, $D^°/D^+ \triangleq V^{4+}/V^{5+}$ and $A^-/A^° \triangleq V^{4+}/V^{3+}$ in the bandgap. The role of vanadium as a minority-carrier lifetime killer in SiC-based optoelectronic devices has been suggested from these results.

$V^{4+}{}_{Si}$ $(3d^1)$, S=½

ESR of the neutral charge state $V^{4+}(3d^1) \triangleq D^°$, $A^°$, S=1/2 is preferentially observed in strongly compensated material. The ESR-spectrum shown in Fig.1, recorded under H//c, results from the superposition of three hyperfine octets of the isotope ^{51}V (I=7/2, 99.8%), arising from $V^{4+}(3d^1)$ on the hexagonal ($\alpha \triangleq h_1$) and the two quasicubic ($\beta, \gamma \triangleq k_1, k_2$) lattice sites in 6H-SiC. In addition the ESR-signal of the neutral nitrogen (^{14}N) donor is apparent, showing that the sample is weakly *n*-type.

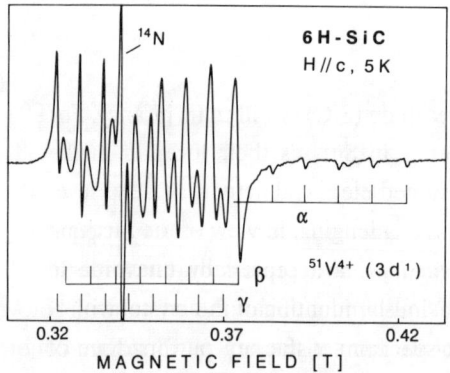
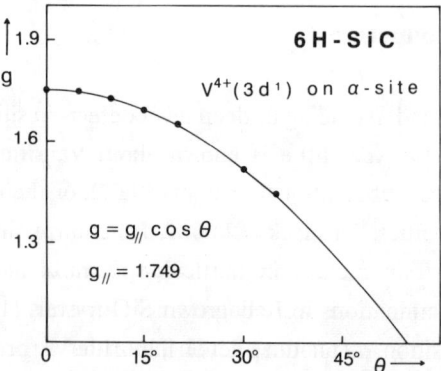

Fig.1. ESR spectrum of $V^{4+}(3d^1)$ in strongly compensated 6H-SiC. The three ^{51}V hyperfine octets arise from vanadium on the hexagonal (α) and the two quasicubic (β, γ) lattice sites. f=9.54 GHz.

Fig.2. Angular dependence of the g factor of $V^{4+}(3d^1)$ on the hexagonal α site in 6H-SiC. The solid line is calculated according to $g=g_{//} \cos \vartheta$, with $g_{//}=1.749$.

Upon rotating the magnetic field H away from the high symmetry orientation H//c, the β- and γ-signals show a weak angular dependence, which will not be discussed here. In contrast, the ^{51}V hyperfine octet α rapidly shifts to higher field under such rotation, as illustrated in Fig.2. Under H⊥c, the g-factor approaches zero. Such behaviour is expected for a ^2E-state in a strong trigonal (C_{3v}) crystalline electric field, as characteristic for the α-site in 6H-SiC. In this case, the ^2E-ground state splits into two close lying Kramers doublets, $\Gamma_{5,6}$ and Γ_4, by the combined action of the trigonal field (~Δ') and spin-orbit coupling (~λ'), as illustrated in Fig.3. In zeroth order, both Kramers doublet states have the same g-factors, with $g_{//}=2$ and $g_\perp=0$. A further negative g-shift of $g_{//}$, as observed experimentally, can be accounted for by higher order perturbations.

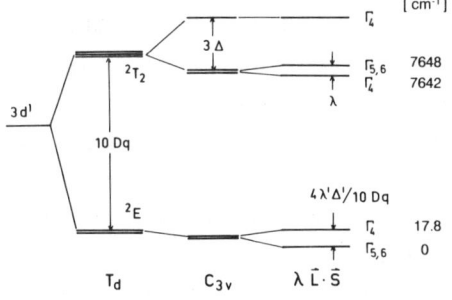

Fig.3. Crystal field and spin-orbit splitting of a $3d^1$ configuration in a trigonally distorted tetrahedral coordination, as appropriate for V^{4+} on the α site in 6H-SiC. The energetical positions of the levels indicated were determined by IR spectroscopy.

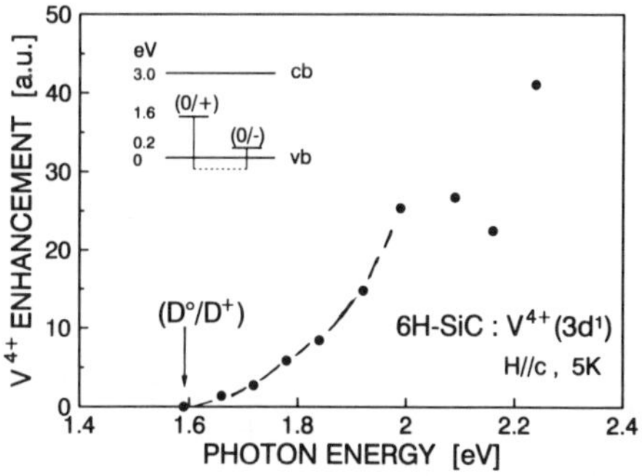

Fig.4. Photo-enhancement of the ESR of $V^{4+}(3d^1)$ in weakly p-type 6H-SiC. The inset illustrates the hole transfer between the deep ionized vanadium donor and an ionized shallow aluminum acceptor.

The position of the donor level $V^{4+}/V^{5+} \hat{=} D^\circ/D^+$ in the bandgap of 6H-SiC could be determined by photo-ESR experiments, performed on a weakly p-type 6H-SiC:Al crystal [4]. In this sample all vanadium donor impurities were ionized, V^{5+} ($3d^\circ$). Consequently, in the dark, only the ESR-spectra of neutral aluminum acceptors on the three silicon lattice sites were detectable. However, after in situ illumination of the crystal with below bandgap photon energies, hν≥1.6eV, the ESR-spectrum of the neutral vanadium donor, $V^{4+}(3d^1)$, could be observed. This ESR-enhancement is caused by the photo-neutralisation process $V^{5+}(3d^\circ) + h\nu(\geq 1.6eV) \rightarrow V^{4+}(3d^1) + h^+$, or $D^+ + h\nu \rightarrow D^\circ + h^+$. The free hole h^+ created in this reaction is subsequently trapped by ionized acceptors, as evidenced by some enhancement of the Al-acceptor ESR after such illumination. At temperatures above 200 K the trapped hole is thermally released from the aluminum acceptors and diffuses back to the neutral vanadium donors, $V^{4+}(3d^1) + h^+ \rightarrow V^{5+}(3d^0)$. The photo-neutralisation spectrum of the ionized vanadium donor on the quasi-cubic β and γ-sites in *6H*-SiC is shown in Fig.4. The photon threshold at 1.6 eV fixes the energetical position of the β-site and γ-site donor levels V^{4+}/V^{5+} at $E_V+1.6$ eV. The corresponding value for the α-site donor has not been determined, because of the rather low intensity of its ESR-spectrum, see Fig.1.

Infrared Luminescence and Absorption

We have discovered that *all* SiC crystals so far investigated by us exhibit a characteristic infrared emission in the 1.3-1.5 μm spectral range [3]. It is anticipated that this luminescence very likely arises from intra-3d-shell transitions, $^2T_2 \rightarrow {}^2E$, of $V^{4+}(3d^1)$.

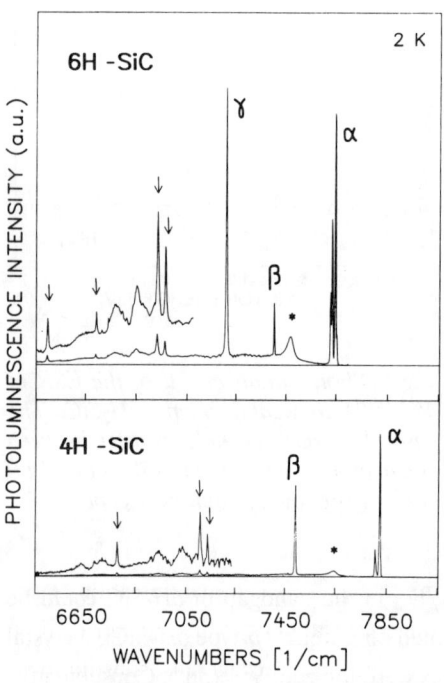

Fig.5. *Low-temperature photoluminescence spectrum of vanadium in 6H-SiC. The zero phonon lines α, β, and γ arise from $^2T_2 \rightarrow {}^2E$ transitions of $V^{4+}(3d^1)$ occupying the three substitutional silicon sites in 6H-SiC. Intrinsic and vanadium-induced phonon sidebands are observed at lower photon energies. Note that only two zero-phonon lines, α and β, are observed for the 4H-SiC polytype. After ref. [3].*

A high-resolution spectrum taken at 2 K on an *n*-type Lely-grown *6H*-SiC crystal is shown in Fig.5. Identical spectra were observed for *p*-type or compensated *6H*-SiC. Above ~40 K the sharp line features in Fig.5 start to broaden and merge into a broad emission band extending from 1.3 to 1.5 μm at 300 K. In Fig.5, three different spectrally narrow sets of zero-phonon lines (ZPLs) can be distinguished for *6H*-SiC. We label them α, β and γ and assign them to substitutional vanadium impurities occupying the hexagonal ($h_1 \triangleq \alpha$) and the two quasi-cubic ($k_1, k_2 \triangleq \beta, \gamma$) lattice sites in the *6H*-polytype.

The above assignment is nicely confirmed by observing that only *two* sets of ZPLs are found in *4H*-SiC (see Fig.5), since in this SiC polytype, one hexagonal (α) and only one quasi-cubic (β) substitutional lattice site exist. On the other hand, more than three sets of ZPLs have been observed in *15R* polytype crystals, where two hexagonal and three quasi-cubic substitutional sites exist. In mixed polytype SiC crystals, a superposition of the individual *6H*, *4H*, and *15R* spectra occurs which can be spectrally resolved already at 77 K. This feature can be exploited for quick, and spatially resolved, morphology determination of SiC polytype crystals.

Apart from the zero phonon lines α, β, and γ, pronounced phonon sidebands are apparent in the photoluminescence spectra (see Fig.5). Part of them can be assigned to intrinsic lattice phonons of *6H*-SiC and *4H*-SiC [5]. In addition, some sharp phonon features, marked by arrows, are apparent which can be assigned to local vibrational modes (LVMs) of the vanadium impurities. For *6H*-SiC they occur at $E_\alpha - 89$ meV, $E_\beta - 88.3$ meV, and $E_\gamma - 88.6$ meV. These LVM energies are very close to those reported for the isoelectronic Ti^{4+} impurity in *6H*-SiC, where the corresponding values are 90.1, 89.8, and 89.7 meV [6]. We finally note that the origin of the broad lines marked with an asterisk in Fig.5 is not clear at present; they may possibly arise from a low-lying vibronic level of V^{4+} (^2E), since these lines are not observed in the absorption spectrum, as seen below.

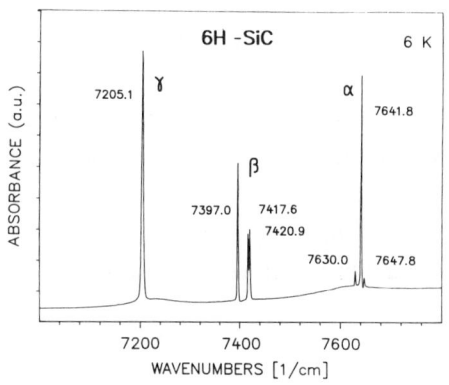

Fig.6. *FTIR absorption spectrum of $V^{4+}(3d^1)$ in 6H-SiC. Wavenumbers quoted refer to the vacuum. Sample thickness: 0.5 mm; T=6 K. After ref. [3].*

The FTIR absorption spectrum of $V^{4+}(3d^1)$ in *6H*-SiC, preferentially observed in compensated material, is shown in Fig.6. Exactly the same zero phonon lines α, β, and γ, which already dominated the luminescence spectra, are also observed in absorption. The doublet at 7417.6 and 7420.9 cm^{-1} appears only at an elevated temperature ("hot" line) in the luminescence spectrum. Both in absorption and emission, the α line can split into four components. This is a manifestation of the stronger axial crystalline electric field acting at the hexagonal substitutional lattice site in *6H*-SiC. We thus conclude that the four components of the α line in *6H*-SiC , and also in *4H*-SiC, arise from transitions between $^2T_2(\Gamma_8)$ and $^2E(\Gamma_8)$ quartet states which are further split by the axial (C_{3v}) crystalline field. For *6H*-SiC this splitting amounts to 17.8 cm^{-1} in the ^2E ground state and to 6.0 cm^{-1} in the $^2T_2$ excited state, as illustrated in Fig.3. In line with this assignment are pronounced polarization effects observed for the four α lines in *6H*-SiC.

Dörnen et al. [7] have very recently performed a detailed Zeeman study of the infrared α lines of $V^{4+}(3d^1)$ in *6H*-SiC; the g-factors of the ^2E ground state Kramers doublet, $g_{//}=1.8\pm01$, $g_\perp=0$ were found to be in full agreement with the ESR-values quoted above. Furthermore, from photo-neutralisation experiments [7] the position of the vanadium donor level for the quasi-cubic β, γ sites in *6H*-SiC was located at $E_V+1.46$ eV, close to our value $E_V+1.6$ eV inferred from photo-ESR on the same crystal.

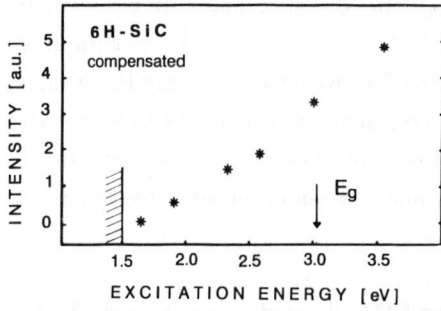

Fig.7. *Photoluminescence excitation spectral dependence of the vanadium emission for a compensated 6H-SiC crystal. The asterisks correspond to various lines of the Kr^+-laser at constant power. After ref. [8].*

It was found that the characteristic intra 3d-shell luminescence of vanadium in SiC can also be excited at below bandgap photon-energies [8]. This is illustrated in Fig.7 for a compensated 6H-SiC sample; a low energy threshold of ~1.5 eV is apparent. It presumably results from photo-ionisation of the neutral charge state $V^{4+}(3d^1)$ from its midgap level (see above) and subsequent carrier recapture into the luminescent $V^{4+}(^2T_2)$ state.

The intra 3d-shell luminescence of neutral vanadium is also observed in *n*-type and *p*-type material. In these cases, photo-neutralisation of the $A^-(3d^2)$ and $D^+(3d^0)$ - states of vanadium must be enforced by the exciting light, before luminescence of the neutral $3d^1$-state can occur.

$V^{3+}_{Si}(3d^2)$, S=1

The electronic ground state of substitutional $V^{3+}(3d^2)$ in SiC is the spin-only triplet S=1 state $^3A_2$. Weak coupling to the lattice phonons, resulting in long spin-lattice relaxation times, is typical in this case, and ESR can be observed already at temperatures far above 4 K. In *6H*-SiC, three sets of spin triplet ESR spectra to be labeled again α, β, and γ are observed, but in *4H*-SiC only two sets, α and β, are observed. The ESR spectrum of $V^{3+}(3d^2)$ on the quasi-cubic sites β and γ in an *n*-type *6H*-SiC sample, $N_D-N_A \approx 2 \times 10^{18} cm^{-3}$, is shown in Fig.8. The strong line in the center at g=2.004 arises from nitrogen donors. Under high resolution this line exhibits the characteristic ^{14}N hyperfine splitting, as shown in the inset. Omitting the nuclear hyperfine interaction $I \cdot \tilde{A} \cdot S$, the S=1 ESR spectra are analyzed by the spin Hamiltonian

$$H = \mu_B g_{//} H_z S_z + \mu_B g_\perp (H_x S_x + H_y S_y) + D \{ S_z^2 - \frac{1}{3} S(S+1) \},$$

where z denotes the c axis of the crystal. The DS_z^2 term is a measure of the axial crystalline electric field gradient acting on the V^{3+} ion. Table I summarizes g factors and D parameters determined for 6H-SiC and 4H-SiC at 77 K. The ESR linewidths were found to vary from sample to sample because of different random strains; these became already apparent under an infrared polarization microscope. However, ESR lines as narrow as ~0.5 G could be observed for the $\Delta m_s = 2$ quasi-forbidden transitions, which are not sensitive to strain, see Fig.9. Furthermore, ^{29}Si ligand hyperfine structure could be resolved in this case which further corroborates our assumption that both infrared and ESR spectra described above arise from vanadium substituting for silicon in SiC. The position of the vanadium acceptor level $V^{4+}(3d^1)/V^{3+}(3d^2) \triangleq A^-/A^0$ could sofar not be determined. We presume it to be located in the upper third of the bandgap.

V^{3+} (3d²)	$g_{//}$	$g_\perp$	|D| (GHz)
6H-α	1.976	1.961	10.70
6H-β	1.961	1.960	2.97
6H-γ	1.963	1.980	0.73
4H-α	1.962	1.958	10.37
4H-β	1.963	1.967	2.65

Table I. g factors and axial field parameters D for $V^{3+}(3d^2)$ on hexagonal (α) and quasi-cubic (β, γ) lattice sites in 6H-SiC and 4H-SiC determined at 77 K.

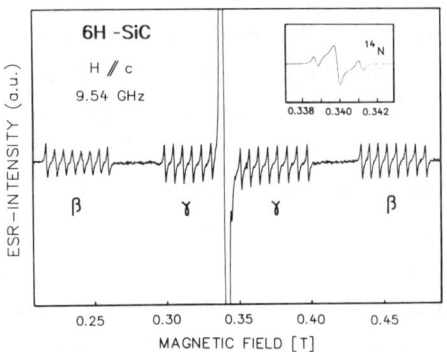

Fig.8. ESR spectrum of the ionized acceptor state $V^{3+}(3d^2)$ in n-type 6H-SiC, under H//c and at 77 K. The two ^{51}V hyperfine octets arising from the hexagonal site α are not shown; they are centered at 0.042 and 0.832 T. The strong line at g=2.004 arises from neutral ^{14}N donors, as shown in the inset.

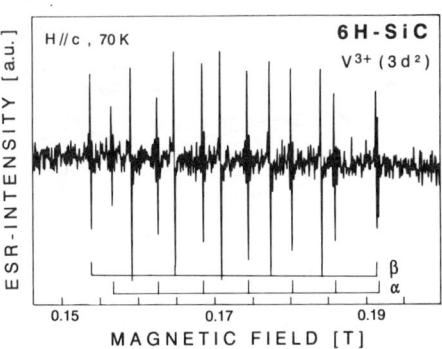

Fig.9. "Forbidden" $\Delta m_S = 2$ transitions of $V^{3+}(3d^2)$ on α and β sites in 6H-SiC. Under H//c only seven ^{51}V hyperfine lines are observed, resulting from the "flip-flop" transitions $\Delta m_S = \pm 2$, $\Delta m_I = \mp 1$.

2. Titanium

The identification of *titanium* as luminescent center responsible for the intense green below bandgap emission of silicon carbide (SiC) crystals has an interesting and controversial history. In the model of Patrick and Choyke [9] it is now agreed that silicon-substitutional isoelectronic neutral Ti($3d^0$) impurities bind excitons. Before recombination, the electron is trapped in the $3d_e$-orbital, thus forming the localized A^--state Ti($3d^1$). In contrast, the holes wavefunction is more delocalized, being bound to the A^--core only by Coulombic attraction. More detailed studies [10] by optically detected magnetic resonance (ODMR) have further revealed that the luminescent state of the green emission in *4H*-SiC and *6H*-SiC is a spin-triplet, S=1, whereas the ground state is diamagnetic, S=0. However, up to now, no direct experimental evidence was given that the presumed ionized acceptor state of titanium, Ti($3d^1$), S=1/2, really lies in the bandgap of silicon carbide.

Recent ESR-work has revealed that the A^--state Ti^{3+}($3d^1$) is stable in *4H*-SiC, E_{gx}=3.265 eV, but not in *6H*-SiC, E_{gx}=3.023 eV, and other lower bandgap SiC-polytypes [11]. The deep acceptor level (-/0) $\hat{=}$ Ti^{3+}($3d^1$)/Ti^{4+}($3d^0$) is estimated [11] to be located only a few tenths of an eV below the conduction band edge of *4H*-SiC. The ESR spectrum assigned to Ti^{3+}($3d^1$) on the hexagonal (α) and quasi-cubic (β) lattice site in *4H*-SiC is shown in Fig.10. The anisotropic g-factor of line α follows the law $g=g_{//} \cos\vartheta$ with $g_{//}$=1.706, similar to that reported above for V^{4+}($3d^1$) on the hexagonal α-site in *6H*-SiC. In contrast, line β exhibits a much weaker angular dependence. The presence of titanium in the α-center is unambiguously evidenced by the characteristic hyperfine structure satellites expected for the odd isotopes ^{47}Ti, 7.3%, I=5/2 and ^{49}Ti, 5.5%, I=7/2, as shown in Fig.11. Surprisingly, it is found that the $g_{//}$-factor of Ti^{3+}($3d^1$) depends strongly on the mass of the individual titanium isotopes, as illustrated in Fig.12. Such effect, resulting from extremely strong electron-phonon coupling has, to our knowledge, not been observed before.

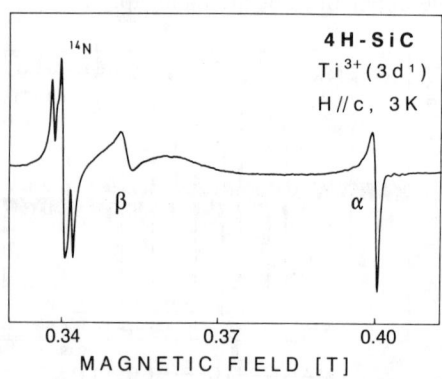

Fig.10. ESR-spectrum of Ti^{3+}($3d^1$) on the hexagonal (α) and quasi-cubic (β) lattice in n-type 4H-SiC. The triplet marked ^{14}N arises from isolated nitrogen donors.

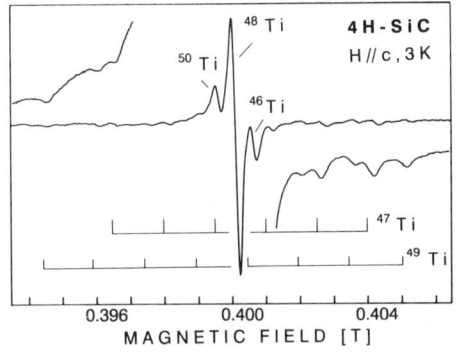
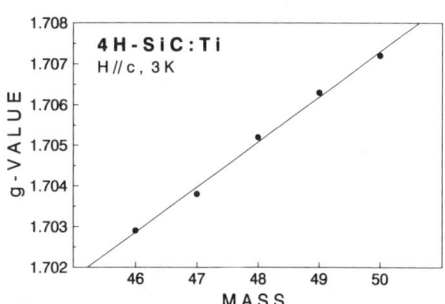

Fig.11. ^{47}Ti and ^{49}Ti hyperfine structure satellites of the α line in Fig.10. Note that the g-value depends on isotope mass.

Fig.12. Dependence of the g-value of $Ti^{3+}(3d^1)$ on mass for the five natural titanium isotopes.

3. Titanium-Nitrogen Donor Pairs

The dominant ESR-signal observed in n-type SiC polytypes arises from the nitrogen donors. Furthermore, the ESR-spectra of ionized vanadium and titanium (*4H*-SiC) acceptor impurities are observed, as discussed above. In addition, some further ESR-lines appear, which can be assigned to neutral titanium-nitrogen donor complexes, $Ti^{3+}(3d^1)$-N_C; their existence was first reported by Vainer et al. [12] for *6H*-SiC. The corresponding ESR-spectrum of such (TiN)° pairs in the *4H*-polytype has recently also been observed [4]. Both in *6H*-SiC and *4H*-SiC the Ti-N pair is to be oriented along the three basal bond directions, its symmetry thus being lowered from C_{3v} to C_{1h}. This can be inferred from the angular dependence of the anisotropic ESR-signal of the Ti-N pair, which is shown in Fig. 13 for the *4H*-polytype. In *6H*-SiC, basal Ti-N pairs are observed on all three lattice sites. However, in the *4H*-polytype, where two lattice sites exist, only one type of Ti-N pairs could be detected by ESR. The characteristic ^{47,49}Ti and ^{14}N hyperfine splittings of the Ti-N pair in *4H*-SiC are shown in Figs. 14 and 15, respectively. The electrical activity of the neutral (TiN)° donor complex arises from the excess 3d-electron

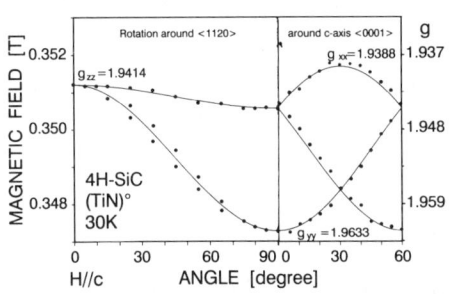

Fig.13. Angular dependences of the (TiN)° pair in 4H-SiC.

in the Ti^{3+} ($3d^1$) configuration. This is created after the nitrogen donor electron has collapsed from its delocalized orbit into the localized $3d^1$-state of the titanium ligand. Photo-ESR experiments [4] performed on a Ti-N pair in *6H*-SiC indicate that the resulting deep donor level is located at $E_C - 0.6$ eV. This value greatly exceeds those of shallow nitrogen donors in *6H*-SiC, which are in the 0.10 - 0.15 eV range.

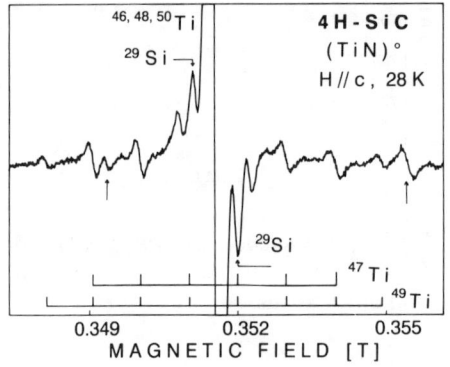

Fig.14. ^{47}Ti and ^{49}Ti hyperfine satellites of the $(TiN)°$ pair in 4H-SiC. The two lines marked by an arrow have not been identified so far.

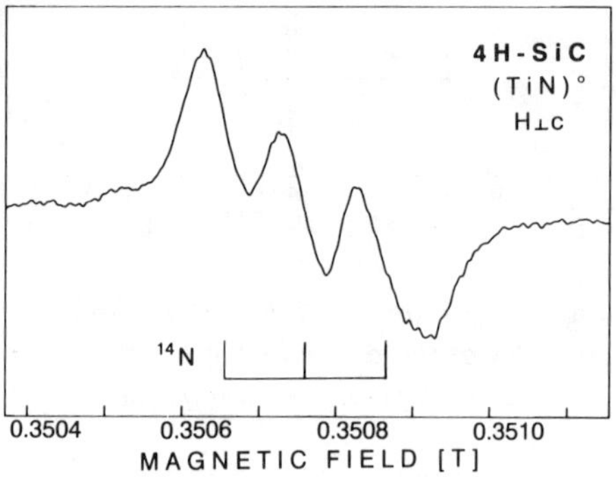

Fig.15. ^{14}N ligand hyperfine splitting of the $(TiN)°$ pair in 4H-SiC.

References

1. P.A.Glasow, in *Amorphous and Crystalline Silicon Carbide*, Springer proceedings in Physics, edited by G.L.Harris and C.Y.-W.Yang (Springer, NewYork,1989), Vol.34, pp.13-33.

2. For a review on recent results in physics and technology of SiC see: G.Pensl and R.Helbig in *Festkörperprobleme / Advances in Solid State Physics*, edited by U.Rössler (Vieweg, Braunschweig, 1990), Vol.30, pp.133-156.

3. J.Schneider, H.D.Müller, K.Maier, W.Wilkening, F.Fuchs, A.Dörnen, S.Leibenzeder and R.Stein, Appl.Phys.Lett. 56, 1184 (1990).

4. K.Maier, J.Schneider, W.Wilkening, S.Leibenzeder and R.Stein, Materials Science and Engineering (Elsevier Sequoia), 1991, in press.

5. D.W.Feldman, J.H.Parker, Jr., W.J.Choyke, and L.Patrick, Phys.Rev. 170, 698 (1968); 173, 787 (1968).

6. A.W.C. van Kemenade and S.H.Hagen, Solid State Commun. 14, 1331 (1974); and references therein.

7. A.Dörnen, Y.Latushko, W.Suttrop, G.Pensl, S.Leibenzeder and R.Stein, this conference.

8. H.D.Müller, PhD-thesis, Freiburg 1991; and to be published.

9. L.Patrick and W.J.Choyke, Phys.Rev.B 10, 5091 (1974).

10. K.M.Lee, Le Si Dang, G.D.Watkins and W.J.Choyke, Phys.Rev.B 32, 2273 (1985).

11. K.Maier, J.Schneider and G.D.Watkins, to be published.

12. V.S.Vainer, V.A.Il'in, V.A.Karachinov, and Yu.M.Tairov, Sov.Phys.Solid State 28, 201 (1986).

PHOTOLUMINESCENCE EXCITATION SPECTROSCOPY OF CUBIC SiC GROWN BY CHEMICAL VAPOR DEPOSITION ON Si SUBSTRATES

J.A. FREITAS, Jr.[1], P.B. KLEIN[2], AND S.G. BISHOP[3]
[1]Sachs-Freeman Associates, Landover, MD 20785, USA
[2]Naval Research Laboratory, Washington, DC 20375, USA
[3]Center for Compound Semiconductor Microelectronics, University of Illinois, Urbana, IL 61801, USA

ABSTRACT

Photoluminescence excitation (PLE) spectroscopy has been applied to the characterization of optical absorption processes in thin film samples of cubic (3C-) SiC grown by chemical vapor deposition (CVD) on Si substrates. Low temperature (6K) PLE spectra have been obtained using a double grating monochromator and a xenon lamp as a tunable source of excitation. For undoped, n-type films of cubic SiC, plots of the integrated PL intensity, in the range 2.4-1.5 eV, as a function of the wavelength of the exciting light are in excellent agreement with optical absorption spectra reported for bulk 3C-SiC. The observed shape of the absorption edge is characteristic of phonon assisted indirect transitions, and spectral features attributable to LA and TA phonons are discernible. No below-gap extrinsic absorption features are observed in the PLE spectra of the undoped films. These results demonstrate the use of the PLE technique for bandedge absorption measurements in thin semiconductor films for which transmission measurements may not be practical. The intense N-Al donor-acceptor pair (DAP) PL bands (2.2-1.5 eV) observed in Al-doped films provide much improved signal-to-noise ratios for the PLE spectra in these samples compared to those obtained in the undoped films. In addition to the characteristic above-gap indirect absorption edge spectrum, the PLE spectra for the Al-doped samples exhibit below-gap, extrinsic absorption features at photon energies corresponding to the nitrogen bound exciton peaks observed in the PL spectra. The extrinsic absorption process which contributes to the excitation of the N-Al DAP PL is apparently a photoneutralization of the compensated (positively charged) nitrogen donors.

1. Introduction

The successful epitaxial growth [1-4] of thin films of cubic SiC by chemical vapor deposition (CVD) on Si substrates has revived interest in this promising wide band-gap semiconductor. Photoluminescence (PL) spectroscopy has played an important role in the characterization [5-10] of both undoped and aluminum-doped thin film samples of CVD cubic SiC. The low temperature PL spectra of most films exhibit a rich variety of characteristic spectral features. However, only a few of these PL bands have been associated with identified impurities or with acceptors and donors of known binding energy. Specific observations include PL spectra attributable to excitons bound to 54 meV neutral nitrogen donors [5-9,11], free-to-bound PL transitions involving aluminum acceptors [8, 12-14], nitrogen-aluminum donor-acceptor pair recombination (N-Al DAP bands) [8,12-15] and a deep DAP band which reveals an unidentified 470 meV acceptor [10].

Photoluminescence excitation (PLE) spectroscopy provides valuable information concerning the optical absorption processes which lead to the excitation of recombination radiation. In PLE spectroscopy the intensity of a selected PL band is recorded as a function of the wavelength of the exciting light. Under some conditions the PLE spectrum constitutes an accurate representation of the absorption spectrum of the sample. (This is an important consideration when analyzing thin film samples for which absorption measurements by conventional optical transmission techniques are problematic.) However, the PLE spectrum incorporates an additional degree of specificity in that it provides a measure of the effectiveness of various absorption processes in the excitation of a *particular* PL band. This is particularly useful when extrinsic (below band gap) absorption processes excite the PL.

Both of these characteristics of PLE spectroscopy are pertinent to the investigation of CVD films of cubic SiC reported here. The PLE spectra obtained from undoped n-type samples provide the first detailed representation of the indirect optical absorption edge of thin film CVD SiC, which is found to be in excellent agreement with the optical absorption spectra reported for bulk (Lely-grown) cubic SiC. In the case of Al-doped films, the PLE spectra exhibit extrinsic absorption features which are attributed to the photoneutralization of compensated donors.

2. Experimental

The thin film samples of cubic SiC, which were obtained from the Naval Research Laboratory, North Carolina State University, and the NASA Lewis Research Center, were all grown by CVD techniques [1,2,4] on Si substrates. Both undoped n-type and Al-doped films have been studied. In all cases the films were removed from their substrates to relieve the strain which results from the ~20% lattice mismatch between the SiC films and the Si substrates.

The PL and PLE spectra described here were obtained with the samples contained in a liquid helium cryostat which provided temperatures ranging from 2 to 300 K. Spectra were acquired in CW mode with excitation provided by an argon (476.5 nm) or krypton (476.2 nm) ion laser or light (460 nm) from a 150W xenon lamp dispersed through a double grating monochromator. The excited luminescence was analyzed by a grating monochromator and detected by a GaAs photomultiplier tube (PMT) which is sensitive to wavelengths shorter than about 900 nm. Appropriate glass filters were used to exclude exciting light from the analyzing monochromator.

The xenon lamp-double grating monochromator combination provided exciting light with wavelength tunable from about 400 to 1000 nm. For the PLE experiments, the luminescence excited by this system was focused onto the entrance slit of a 0.75 m focal length, single grating monochromator. PLE spectra were obtained with the detection monochromator serving as a band pass filter tuned to a particular PL wavelength, or with a mirror replacing the grating in the monochromator so that the integrated PL intensity was recorded. In the latter case, the wavelength limits of the detected PL band were determined by a longpass optical filter placed at the entrance slit and by the long wavelength limit of the GaAs PMT response (~900 nm). The PLE spectra were corrected for the wavelength dependence of the exciting light intensity.

The intensity of the exciting light for the PLE spectra was about 100 μW/cm^2 at the peak of the spectral output for the lamp-monochromator system. PL spectra obtained under these excitation conditions first revealed the importance of recording PL spectra over a broad range excitation intensities. For example, nearly all PL spectra reported in the literature [5-10, 16] for n-type CVD cubic SiC are excited by relatively high power light (>1 W/cm^2). They are dominated by the nitrogen bound exciton (NBE) spectrum (especially the phonon replicas) with its relatively short donor bound exciton radiative life time. In addition to the NBE spectrum, other characteristic features of these high power PL spectra include the 1.972 eV zero phonon line (ZPL) and phonon replicas of the D1 defect band (which has been studied in detail in ion implanted Lely crystals [17] and CVD films [7,8] of cubic SiC), and a broad underlying PL band peaking near 1.8 eV. In contrast, the low power PL spectrum for such undoped samples, excited by light from the lamp and double monochromator system, exhibit only a weak vestige of the NBE spectrum. These spectra are dominated instead by a distant DAP band with peak at about 1.91 eV, the so-called G-band [9,10,18] which is attributed [10] to the pairing of the 54 meV nitrogen donors with an unidentified 470 meV acceptor.

Similarly, the two PL spectra shown in Fig.1 contrast the high- and low-power excitation conditions for an Al-doped sample of cubic SiC. The high power spectrum exhibits sharp line spectra at high energies due to close pair recombination. The low power spectrum is dominated by the long life time distant pair band peaking at about 2.12 eV. Note that the close pair spectra are not observed at low power and that the phonon replicas below the distant pair band exhibit sharper spectral details. It is important to remember that the PLE spectra presented here are obtained under conditions which produce the low power PL spectra of Fig.1. That is, they are dominated by deep PL bands with long lifetime recombination processes.

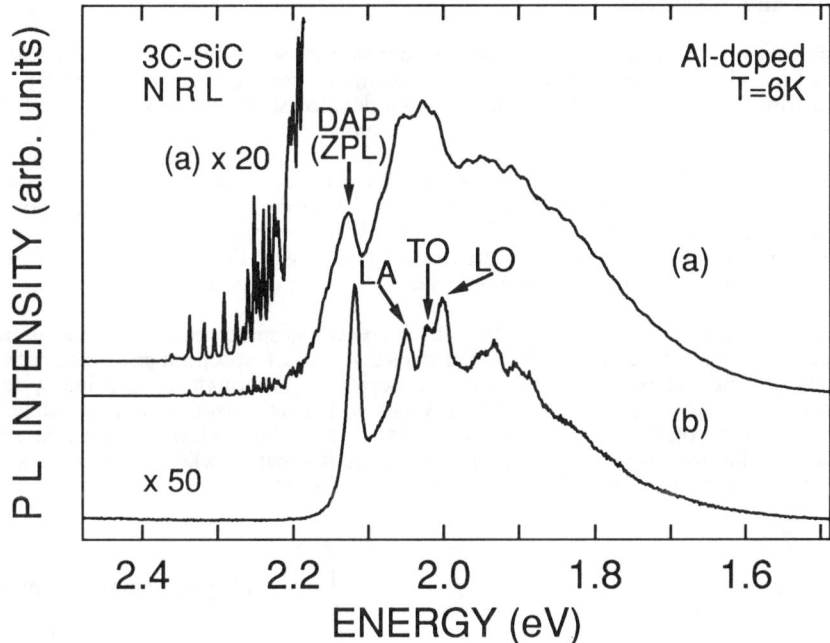

Fig. 1. Low temperature (6K) photoluminescence spectra from an Al-doped SiC film: (a) high power spectrum excited by an argon ion laser (1W/cm^2, 476.5 nm); X20 spectrum shows close donor-acceptor pair spectra; (b) low power spectrum excited by xenon lamp and 0.22 m monochromator (0.1mW/cm^2, 460 nm).

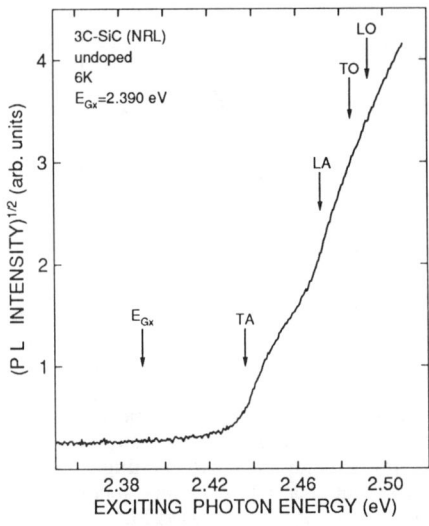

Fig. 2. Photoluminescence excitation spectrum for an un-doped CVD 3C-SiC film deposited with C/Si source gas ration of 2.4. The square root of the integrated photoluminescence intensity (in the spectral range 1.4 - 2.3 eV) is plotted as a function of the photon energy of the exciting light. The exciton energy gap, E_{GX}, and the emitted phonon energies are indicated in the figure.

3. Results and Discussion

Figure 2 shows the 2.35 - 2.50 eV PLE spectrum from an undoped, n-type film of CVD cubic SiC, obtained at 5K under the integrated PL intensity conditions described above. The line shape or intensity distribution of this PLE spectrum exhibits a detailed similarity to the optical absorption spectrum reported by Choyke et al. [11] for bulk, Lely-grown cubic SiC. Note that PL intensity has been plotted on a square root scale in order to illustrate the indirect character of the absorption edge. Choyke et al. described the shape of the edge as characteristic of indirect transitions in which excitons are created. The onsets of optical absorption transitions assisted by the emission of TA, LA, TO, and LO phonons are indicated in Fig. 2. The phonon energies have been determined from the NBE PL spectra [11]. It follows that the exciton energy gap, E_{Gx} = 2.390 eV, is derived by subtracting the TA phonon energy from the observed onset of absorption.

The fact that the PLE spectrum scales closely with the absorption spectrum of Choyke at al. from the band edge up to energies in excess of 2.5 eV, indicates that the PLE spectrum provides a remarkably faithful representation of the interband optical absorption. Although the previously reported absorption measurements were performed on crystals with a light path of approximately 2 mm, Choyke et al. [11] noted that even larger crystals would be preferred for more accurate measurements. Thus the sensitivity of the PLE technique is apparent when one considers the fact that the light path for the thin film SiC samples is only about 10-15 μm.

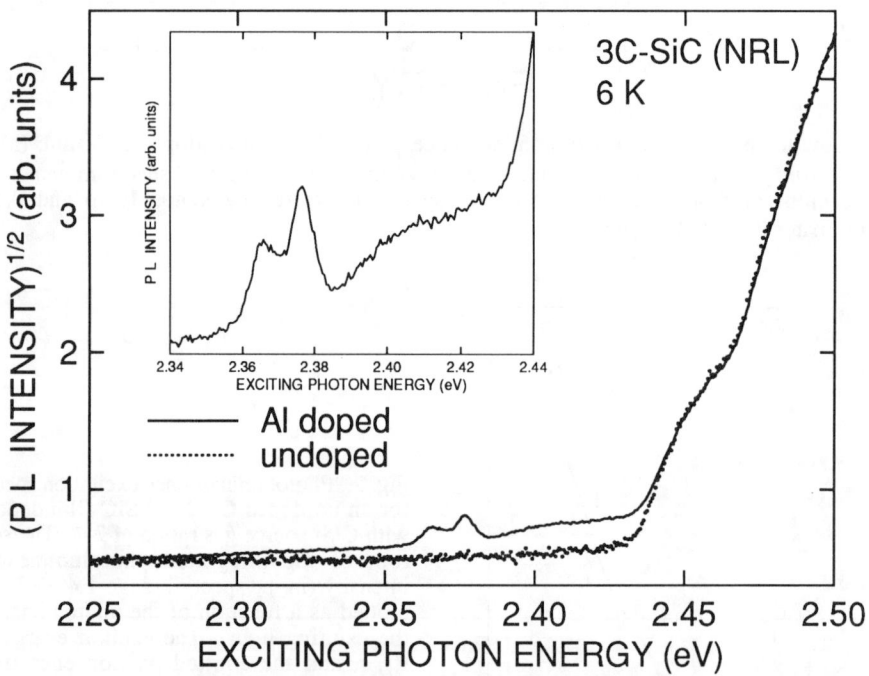

Fig. 3. Photoluminescence excitation spectra for an Al-doped (continuous line) and an undoped (dotted line) CVD 3C-SiC film. The square root of the integrated photoluminescence intensity (in the spectral range 1.4 - 2.3 eV) is plotted as a function of the photon energy of the exciting light. The undoped spectrum has been normalized to the Al-doped spectrum for the best spectral coincidence between 2.45 and 2.50 eV. The insert shows the extrinsic portion of the Al-doped PLE specrum at high gain.

In Fig. 3 the PLE spectrum obtained from an Al-doped sample of cubic SiC is compared to that of the undoped sample presented in Fig. 2. The much greater integrated intensity of the N-Al DAP PL bands and phonon replicas in the Al-doped sample is apparent in the greater signal to noise ratio of its PLE spectrum relative to that of the undoped sample. Furthermore, the PLE spectrum of the Al-doped sample exhibits extrinsic absorption features in the spectral range extending from the onset of interband absorption at about 2.436 eV down to about 2.349 eV.

The extrinsic portion of the Al-doped PLE spectrum is presented at higher gain and resolution, and on an expanded scale in the inset to Fig. 3. It is characterized by an onset at about 2.349 eV, *although this is difficult to determine exactly,* peaks at 2.367 and 2.377 eV, a dip or trough with minimum at about 2.385 eV, and a shoulder which rises to higher energy until it intersects the steeply rising interband PLE at about 2.436 eV.

There are two primary factors which enter into the interpretation of these extrinsic features in the PLE spectrum. First, they are observed only in relatively heavily Al-doped films, and second, the onset, peaks, and shoulder all occur in a spectral range which corresponds to the binding energies and PL bands associated with the known donors in cubic SiC. The first factor indicates that the absorption processes which give rise to these extrinsic PLE features require the presence of large concentrations of compensated (charged) donors. Undoped films are invariably n-type; they exhibit the NBE PL spectra which reveal the presence of the 54 meV donor identified as nitrogen [8,11,13,14,19], and temperature dependent Hall effect measurements which are interpreted in terms of a highly compensated 15-20 meV donor[13,20-23]. In heavily Al-doped samples these donors are compensated and extrinsic optical absorption transitions can photoneutralize the charged donors. The electrons on the resulting neutral donors can then undergo DAP radiative recombination with the holes on the neutral Al acceptors. Although no optical (PL) signature of the shallow (15-20 meV) donor has been identified to date, the N-Al DAP PL spectrum is well documented [8,13,14].

In considering the second factor, the most obvious correlation is between the 2.377 and 2.367 eV PLE peaks and the energies of the ZPL of the NBE bands and the low energy shoulder on the NBE band (believed to be a deeper donor bound exciton [5,8]), respectively, of the near-band edge PL spectrum observed in undoped n-type films [5-9]. Note that these energies are characteristic of the NBE PL spectra in the strained CVD films grown on Si (and subsequently removed from their substrates) which are slightly red-shifted (~1 meV) relative to the NBE spectra reported for bulk Lely crystals of cubic SiC. The obvious suggestion is that this portion of the extrinsic PLE involves the excitation of the nitrogen and unidentified deeper donor bound excitons. Far more speculative is the possible correlation between the approximate 2.349 eV low energy onset of the extrinsic PLE spectrum and the energy 2.349 meV = E_{Gap}- 54 meV (the nitrogen donor binding energy), where the low temperature band gap of cubic SiC is taken to be 2.403 eV. This correlation suggests the hypothesis that 2.349 eV is the onset of extrinsic absorption transitions which photoneutralize the 54 meV nitrogen donors. However, the onset, whose position is highly uncertain, could also be explained as a low energy tail on the apparent donor bound exciton absorption band.

Even more interesting and speculative is the possiblility that the absorption shoulder above the 2.385 eV trough might be associated with photoneutralization of compensated 15-20 meV donors. (The trough is positioned about 18 meV below the band gap.) If this hypothesis were correct, it would represent the first optical manifestation of the 15-20 meV shallow donor which appears to dominate the electrical properties of undoped CVD films of cubic SiC. However, it must be emphasized strongly that this is simply a conjecture and there are equally plausible alternative interpretations of this absorption shoulder.

4. Summary

Photoluminescence excitation spectroscopy has been used to investigate the above gap and extrinsic optical absorption processes which excite the PL bands that characterize CVD films of cubic SiC grown on Si substrates. In undoped films the PLE spectra provide a faithful representation of the indirect optical absorption edge which is consistent with the optical absorption spectrum reported previously for Lely-grown bulk crystals of cubic SiC. The undoped PLE spectra show no evidence of extrinsic (below gap) optical absorption. The PLE spectra of the N-Al DAP bands which dominate the PL spectra of Al-doped films of cubic SiC exhibit extrinsic absorption which is attributed to photoneutralization of compensated shallow donors. The extrinsic PLE spectra of the Al-doped samples contain peaks which correspond to the ZPLs of the donor bound exciton PL bands observed in the undoped films, as well as onsets which, it is speculated, could correspond to the thresholds for photoneutralization of the 54 meV N donor and the unidentified 15-20 meV donor which are pervasive in CVD cubic SiC.

5. Acknowlegements

The authors wish to thank P.E.R.Nordquist, Jr. and M.L. Gipe of the Naval Research Laboratory, H.S. Kong and R.F. Davis of North Carolina State University, and J.A. Powell of the NASA Lewis Research Center for providing the samples which we have studied. This work was supported in part by the Office of Naval Research.

References

1. S. Nishino, J.A. Powell, and H. A. Will, Appl. Phys. Lett. 42 460 (1983).
2. A. Addamiano and P.H. Klein, J. Cryst. Growth 70, 291 (1984).
3. K. Sasaki, E. Sukuma, S. Misawa, S. Yoshida, and S. Gonda, Appl. Phys. Lett. 45, 72 (1984).
4. H.P. Liaw and R.F. Davis, J. Electrochem. Soc. 131, 3014 (1984).
5. S.G. Bishop and J.A. Freitas, Jr., J. Cryst. Growth 106, 38 (1990).
6. J.A. Freitas, Jr., S.G. Bishop, A. Addamiano, P.H. Klein, H.J. Kim, and R.F. Davis, Mats. Res. Soc. Symp. Proc., 46, 581, (1985).
7. J.A. Freitas, Jr., S.G. Bishop, J.A. Edmond, J. Ryu, and R.F. Davis, J. Appl. Phys. 61, 2011 (1987).
8. J.A. Freitas, Jr., S.G. Bishop, P.E. R. Nordquist, Jr., and M.L. Gipe: Appl. Phys. Lett. 52, 1695 (1988).
9. W.J. Choyke, Z.C. Feng, and J.A. Powell, J. Appl. Phys. 64, 3163 (1988).
10. J.A. Freitas, Jr. and S.G. Bishop, Appl. Phys. Lett. 55, 2757 (1989).
11. W.J. Choyke, D.R. Hamilton, and L. Patrick, Phys. Rev. 133, A1163 (1964).
12. G. Zanmarchi, J. Phys. Chem. Solids 29 (1968) 1727.
13. W.E. Carlos, W.J. Moore, P.G. Siebenmann, J.A. Freitas, Jr., R.Kaplan, S.G. Bishop, P.E.R. Nordquist, Jr., M. Kong, and R.F. Davis, in Proc. Mats. Res. Soc. Symp. 97, 253 (1987).
14. S.G. Bishop, J.A. Freitas, Jr., T.A. Kennedy, W.E. Carlos, W.J. Moore, P.E.R. Nordquist, Jr., and M.L. Gipe, in: Amorphous and Crystalline Silicon Carbide, Springer Proc. in Physics, Vol. 34, ed. G.L. Harris and C.Y.-W. Yang (Springer-Verlag, Berlin, Heidelberg, 1987), p. 90.
15. W.J. Choyke and L. Patrick, Phys. Rev. B2, 4959 (1970).
16. H. Okumura, M. Shinohara, E. Muneyama, H. Daimon, M. Yamanaka, E. Sakuma, S. Misawa, K. Endo, and S. Yoshida, Japan. J. Appl. Phys. 27, L116 (1988).
17. W.J. Choyke and L. Patrick, Phys. Rev. B4, 6 (1971).
18. J.A. Freitas, Jr., S.G. Bishop, P.B. Klein, P.E. R. Nordquist, Jr., and M.L. Gipe, in: Amorphous and Crystalline SiC II, Springer Proc. in Physics, Vol. 43, eds. M.M. Rahman, C.Y.-W. Yang, and G.L.Harris (Springer-Verlag, Berlin, Heidelberg, 1988), p. 106.
19. J.A. Freitas, Jr., W.E. Carlos, and S.G. Bishop, in Amorphous and Crystalline SiC III, Howard Universiry, Washington, DC, Apr 11-12, 1990, in press.
20. A. Suzuki, A. Uemoto, M. Shigeta, K. Furukawa, and S. Nakajima, Appl. Phys. Lett. 49, 450 (1986).
21. B. Segall, S.A. Alterovitz, E.J. Haugland, and L.G. Matus, Appl. Phys. Lett 49, 584 (1986).
22. B. Segall, S.A. Alterovitz, E.J. Haugland, and L.G. Matus, Appl. Phys. Lett. 50, 1533 (1987).
23. M. Shinohara, M. Yamanaka, H. Daimon, E. Sakuma, H. Okumura, S. Misawa, K. Endo, and S. Yoshida, Japan J. of Appl. Phys. 127, L434 (1988).

PARAMAGNETIC DEFECTS IN SiC BASED MATERIALS

O. CHAUVET[1], L. ZUPPIROLI[2], J. ARDONCEAU[1], I. SOLOMON[3], Y.C. WANG[4] and R.F. DAVIS[4]

[1] Ecole Polytechnique, L.S.I., 91128 PALAISEAU cedex FRANCE Present adress: (2)
[2] EPFL, IGA/DP, CH 1015 LAUSANNE SWITZERLAND
[3] Ecole Polytechnique, P.M.C., 91128 PALAISEAU cedex FRANCE
[4] North Carolina State University, Dpt Mat. Sci. Eng., RALEIGH NC 27695-7907, U.S.A.

ABSTRACT:

Point defects and impurities play a major role in the electronic properties of silicon carbides. Electron Spin Resonance was used to investigate the magnetic properties of both native and irradiation induced defects. Our experiments covered a wide variety of materials: glow discharge a-$Si_{1-x}C_x$:H with low carbon content (x<15%), electron irradiated ß-SiC single crystals, nanocrystalline SiC fibers and SiC industrial powders. We were able to identify the silicon vacancy, the carbon vacancy and carbon complexes with sp^2 hybridization. A common point in all the materials studied (except the single crystals before irradiation) is the presence of sp^2 carbon, even in low carbon content amorphous SiC. Oxygen impurities play a peculiar role at low temperatures. Evidence is given that the silicon dangling bonds have an almost zero or negative effective Hubbard energy.

INTRODUCTION:

Carbon has the ability to hybridize sp^2 or sp^3. When alloyed with silicon, because of the mismatch between carbon π or σ bonds and silicon σ bonds, in most cases, defects are created: antiphase domains and point defects...Some of them are electronically active. The purpose of this paper is to study by electron spin resonance measurements those which are paramagnetic. This paper is organized as follows: the first part briefly presents the materials which support this study. In a second part, we focus on the carbon incorporation in an amorphous silicon rich compound. In the third part, the paramagnetic defects are identified and the role of oxygen impurities discussed.

PRESENTATION OF MATERIALS

The present study of point defects in disordered SiC structure has required several types of materials: model systems to control part of the parameters, real systems to test the results. The amorphous methylated silicon is an amorphous silicon like system, prepared by glow discharge of silane and methane in a sufficiently low power regime such that carbon is incorporated as methyl group in the silicon host[1]. The electronic properties of these materials have been widely investigated[2]. For our purpose, this system is interesting because the carbon is perfectly sp^3 hybridized due to its incorporation as methyl group. A second model of interest is β-SiC in the form of single crystals. They have been synthesized by chemical vapor deposition (CVD)[3]. In an

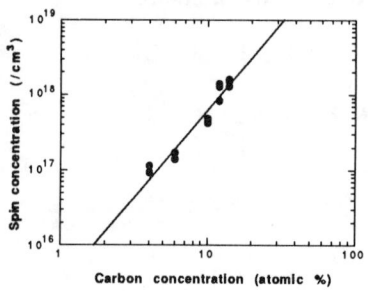 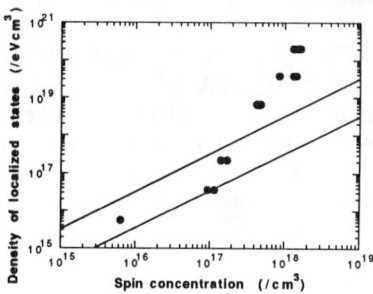

Fig 1: Creation of paramagnetic defects by carbon incorporation in methylated silicon

Fig 2: Density of localized states versus the spin concentration in methylated silicon

ideal single crystal, carbon is also perfectly sp³ hybridized and there are no defects. Electron irradiations have been used to create defects in a controlled way in these crystals. Two energies were used: 2.5 MeV, using the Van der Graaf accelerator of Laboratoire des Solides Irradiés (LSI), where both silicon and carbon atoms were displaced by the incident electrons, 100 keV using a special low energy electron accelerator built in the LSI, where only the carbon atoms were displaced. A great variety of disordered SiC materials were also investigated: ceramic β-SiC powders, amorphous SiC powders and SiC fibers elaborated by polymer pyrolysis.

ESR X-band experiments were performed on a standard ER200D Brucker spectrometer with a helium gas-flow cryostat. Great care was taken to avoid over-modulation or saturation, and to insure the thermalization of the samples. The spin concentration is evaluated with reference to a copper sulfate single crystal, which is also a temperature reference, in contact with the sample.

THE CARBON INCORPORATION:

Let us start with the amorphous hydrogenated silicon. It contains about 10^{15} spins/cm³. These spins arise from the silicon sp³ dangling bonds (DB) with a g value of 2.0055 and a peak-to-peak linewidth of 7 G. When incorporating carbons as methyl groups, the spin concentration ({spin}) is enhanced, as shown on figure 1. Up to 12% carbon content, the created paramagnetic defects are still silicon DB, without any modification of the ESR spectrum. This increase varies as the square of the carbon content (straight line on the plot): and suggests that the incorporation of two methyl groups on two adjacent silicon atoms is required in order to create a DB. Assuming that the carbon is incorporated randomly, we found that $\{spin\} = p \frac{2}{3} \pi R_c^3 \rho^2$ where p is the probability that a pair of methyl radicals incorporated within a distance R_c generates one paramagnetic defect, ρ is the volumic carbon concentration. Assuming that $R_c \sim 2$ Å, the comparison with the experiment gives p~1/1000: even if two methyl groups are incorporated within an atomic distance, only one pair over 1000 creates a paramagnetic defect: the silicon network is thus able to distort sufficiently to allow a "good " incorporation. However, some created defects are not paramagnetic. The density of localized states (DOS) near the Fermi level was measured by space charge limited conduction[4]. The result is plotted in figure 2 and compared to the spin density. The band on the

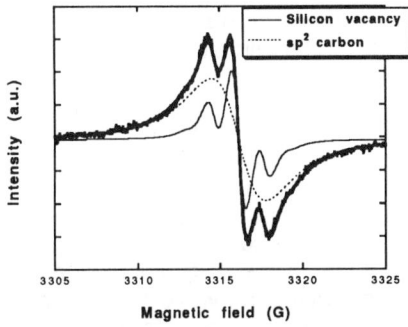
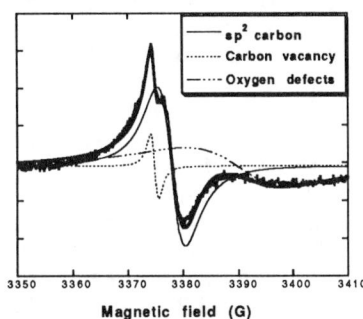

Fig 3: ESR spectrum of 2.5 MeV irradiated β-SiC single crystal

Fig 4: ESR spectrum of 100 keV irradiatied β-SiC single crystal

plot defines a region were these concentrations are equal within a factor 2 (experimental incertitude). Up to 6% of carbon, most of the localized states are paramagnetic: they are silicon DB, the excited states as well as the charged states of a dangling bond (D+, D-) are far from the Fermi level. Above 6% of carbon content, there are much more diamagnetic than paramagnetic states: since the spins concentration is high anyway, the DB are close enough to interact and give a "weak bond" similar to those established around di-vacancies in silicon[5]. Those "weak bonds" are diamagnetic and more extended than the DB, thus contributing to the DOS also by their excited states. Later on, we will propose a structural model for this defect.

Above 15% of carbon content, the ESR spectrum changes: the g factor decreases and reaches the g value of free electrons, while the linewidth decreases too. This is correlated to the appearance of sp^2 carbon, as can be seen in vibrational spectroscopy[2]. Carbon is no more incorporated as methyl group, it becomes reactive and carbon DB or unsaturated C=C bonds appear, that modify the spectrum. These kinds of carbon related bonds are also observed in disorderd SiC, it is thus important to investigate the SiC paramagnetic defects, beginning with the irradiated single-crystals.

THE PARAMAGNETIC DEFECTS OF SiC

Before irradiation, the single crystals contain less than 10^{14} spins/cm^3: no ESR signal was detectable, not even at low temperature, contrary to other works published[6,7]. Figure 3 shows the spectrum after 2.5 MeV irradiation. It is isotropic and composed of a five lines spectrum and a lorentzian contribution. The five lines spectrum is attributed to a spin centered on a silicon site, in hyperfine interaction with one or two ^{29}Si among the twelve silicon first neighbors. As in Itoh's work[8], we attribute this defect to an odd charged state of a silicon vacancy (because silicon is tetravalent), that is to say a carbon sp^3 dangling bond. Before discussing the second contribution, we present the spectrum of the 100 keV irradiated single crystal on the figure 4, observed at 4 K. This spectrum is also isotropic and does not change with temperature. It is composed of 3 lorenztian contributions. The interpretation is easy since only carbon atoms are displaced by the low energy electrons. The position of the weakest contribution, g~2.0048, is the signature of a silicon DB: it is related to an odd charged state of a carbon vacancy. The slight difference with the usual g~2.0055 is either the signature of a carbon rich environment of the DB[9] or of some kind of Jahn-Teller effect. The large contribution at g~2.0000 is characteristic of oxygen related defects in a siliconated material. One possible candidate is a defect similar to the E' center of the silica (a hole bound to an oxygen vacancy)[10]. The presence of oxygen impurities in this material is probable since the walls of the CVD reactor are made of quartz. Secondly, the single crystals are deposited on silicon wafers that are known to be often oxidized on surface: it is then possible that these oxygen related defects are located at the interface between SiC and the substrate.

We should emphasize that the remaining contribution at g~2.003 is the same as in the 2.5 MeV irradiated crystal, observed at the same temperature: same g (2.0028 and 2.0029) same linewidth (4.3 G), same longitudinal relaxation time ($2 \cdot 10^{-5}$ and $3 \cdot 10^{-5}$ s) as deduced from continuous wave saturation of a homogeneous line. So, it is a carbon related defect. We believe that it is a carbon interstitial or di-interstitial. The carbon interstitial has been observed in carbon doped crystalline silicon or in diamond[11-14]. At room temperature, it is less stable than a di-interstitial. The essential point for us is the change in hybridization of this carbon. As a radical, a methyl group does not keep its sp^3 hybridization and relaxes toward a sp^2 configuration (it has been undirectly observed by Watkins in a carbon doped silicon, where the carbon interstitial is not located on a tetraedric site but in a plane configuration[13]). Furthermore, there are topological reasons that 2 carbon interstitials link with a C=C bond rather than C-C bonds because of the size of a tetraedric site (0.98 Å). Thus, this defect is related to an unsaturated sp^2 carbon and is very similar to the defects observed in pyrolitic carbons[12].

It is worth noticing that the silicon DB are very few in comparison with the oxygen or sp^2 carbon related defects. We will return to the stability of these defects, further on.

In summary, we have been able to identify, after irradiation, the silicon or carbon sp^3 dangling bonds, a sp^2 carbon related defect and an oxygen related one. Are these defects always presents in disordered SiC? The sp^3 DB and sp^2 carbon defects are present in every disordered SiC investigated. In crystalline materials, the ESR spectrum is inhomogeneous with two different contributions. As in irradiated single crystals, one arises from the sp^3 DB while the second from sp^2 defects. In amorphous materials, the two contributions are intimately mixed. This behavior is attributed to the fact that the sp^2 carbon is in diluted form in amorphous SiC[15,16], constituting a kind of polymeric network where some C=C appears ponctually. The great exchange ability of sp^2 C bonds is sufficient to homogenize the entire spin system. In crystalline samples, sp^2 carbon is present as free carbon[17]: the 2 spin families are separated in space, giving two different paramagnetic contributions. This condensation of sp^2 carbon occurs because of the lack of a degree of freedom of the surrounding network. It is probable that the sp^2 carbon is mainly located near the grain boundaries.

It is important to note that the ESR line corresponding to the sp^3 DB (or the whole spectrum in the case of amorphous materials) is always centered at g=2.003, very closed to the g value of the carbon DB. It means that the carbon DB are more stable than the silicon DB, as often postulated[9,18].

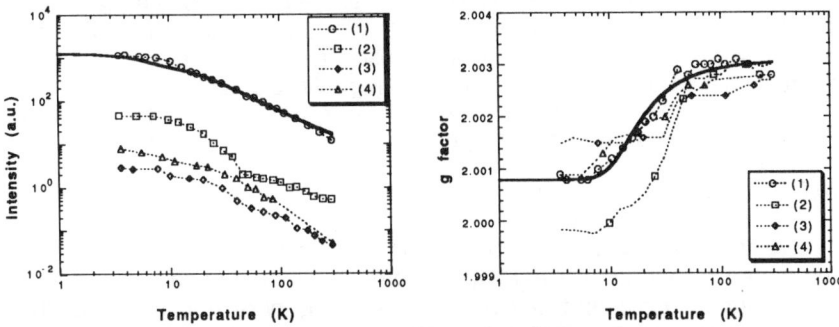

Fig 5: Evolution of the intensity of the line (a) and of the g factor (b) versus the temperature for two crystalline samples (1), (2) and two amorphous ones (3) (4). The bold lines fit this parameters with a two bands model for the sample1.

The oxygen defects are also present in those materials, either with a crystalline or amorphous structure. However, they appear only at low temperatures, with a modification of the spectrum (decrease of the g factor, broadening of the line, saturation of the line intensity), as illustrated on figure 5. We were able to describe these evolutions within a two narrow band model, separated by a gap of 2 meV. The result is shown for sample 1 of fig 5 (NBS's reference β-SiC powder). The lowest band is attributed to oxygen related defects (with a g factor of 2.000) while the second is characteristic of sp^3 DB. When increasing the temperature from 4 K to 300 K, the oxygen related spins displace towards the DB. It is not clear whether this effect is related to silica rich heterogeneities or to a subtle spin dynamic effect around oxygen impurities in the bulk of the material.

Let us turn back to the low paramagnetic stability of the silicon DB, observed whenever there is enough carbon in the material. From an energetic point of view, this stability is usually described in terms of an effective correlation energy: U_{eff} defines the exothermicity of the reaction $2 D^0 \rightarrow D^+ + D^-$. When $U_{eff} > 0$, the dangling bonds are paramagnetic, when $U_{eff} < 0$, the diamagnetic charged states are more stable than the paramagnetic neutral DB. The fact that all the localized states are paramagnetic in low carbon content methylated silicon shows that the previous reaction is exothermic with $U_{eff} > 0$ around some tenth of eV[19-21], as in amorphous silicon. When increasing the carbon content, this is no more the case: it seems that U_{eff} decreases, eventually down to negative values. It has already been proposed in a-Si:H[22,23] or in a-Si$_{1-x}$C$_x$:H[24]. We believe that it is due to the nearby presence of carbons, as illustrated on the following figure:

When interstitial sp^2 carbons come into the silicon host, they distort locally the network, giving to the neighbouring silicon dangling bond a polaronic character. It modifies intimately the stability of the two isolated DB, leading to the formation of an "alternative" bond. Whether we should consider this defect as a D$^+$+D$^-$ pair (or we could talk about a negative U$_{eff}$ of the DB) is not certain. Nevertheless, we think that a microscopic polaronic model of this kind is at the origin of the observed low paramagnetic stability. Furthermore, we have evidence of polaronic carriers in a great variety of disordered carbonated materials (irradiated polymers, SiC fibers...).

CONCLUSION:

We have investigated the paramagnetic defects in a large variety of SiC based materials. Three different kinds of defects have been identified: the sp^3 dangling bonds, unsaturated sp^2 carbons, oxygen related defects. These defects are nearly always present in the studied materials. Noteworthy is the constant presence of sp^2 carbon, even when the carbon content is low. It is present in diluted form in the amorphous materials with polymeric precursors (constituting a kind of polymeric network), while it appears as free carbon in crystalline materials (constituting a composite-like network). This foreign hybridization leads to a local distortion of the surrounding network, which generates silicon dangling bonds with polaronic behavior. It seems that these silicon DB have a very small or even negative effective Hubbard energy. This relaxation of the network occurs as soon as the network has some degrees of freedom: only single crystals or methylated silicon avoid the appearance of sp^2 carbon. Oxygen plays also a major role in the electronic properties at low temperature. A two band model was used to take it into account.

REFERENCE:
1. I. Solomon, M.P. Schmidt, H. Tran Quoc, Phys. Rev.B, 38, 9895 (1988)
2. I. Solomon et al., Phys. Rev. B, 38, 13263 (1988)
3. H.J. Kim, R.F. Davis, J. Appl. Phys., 60, 2897 (1986)
4. I. Solomon, R. Benferhat, H. Tran Quoc, Phys. Rev. B, 30, 3422 (1984)
5. P.A. Lee, J.W. Corbett, Phys Rev. B, 8, 2810 (1973)
6. H.H. Woodbury, G.W. Ludwig, Phys. Rev., 124, 1083 (1961)
7. H. Okamura et al., Jpn. J. Appl. Phys., 27, 1712 (1988)
8. H Itoh. et al., J. Appl. Phys., 66, 4529 (1989)
9. N. Ishii, M. Kumeda, T. Shimizu, Sol. State Com., 41, 143 (1982)
10. A. Stesmans, Y. Wu, J. Non Cryst. Sol., 99, 190 (1988)
11. K.L. Brower, Phys. Rev. B, 9, 2607 (1974)
12. P.R. Brosious, J.W. Corbett, J.C. Bourgoin,Phys Stat Sol A, 21, 677 (1974)
13. G.D. Watkins, K.L. Brower, Phys. Rev. Lett., 36, 1329 (1976)
14. J.N. Lomer, C.M. Welbourn, Phil. Mag. A, 37, 639 (1978)
15. M.A. Petrich, K.K. Gleason, J.A. Reimer, Phys. Rev. B, 36, 9722 (1987)
16. J. Sotiropoulos, G. Weiser, J. Non Cryst. Sol., 92, 95 (1987)
17. T.N. Taylor, J. Mat. Res., 4, 189 (1989)
18. A. Morimoto et al, J. Appl. Phys, 53, 7299 (1982)
19. R.A. Street, D.K. Biegelsen, Sol. State Com., 35, 1159 (1980)
20. J.E. Northrup, Phys. Rev. B, 40, 5875 (1989)
21. J. Koçka, M. Vanecek, F. Schauer, J. Non Cryst. Sol., 97&98, 715 (1987)
22. K. Winer, Phys. Rev. Lett., 63, 1487 (1989)
23. S. Yamasaki et al, Phys Rev. Lett, 65, 756 (1990)
24. Y.A. Zarif'yants et al, Sov. Phys. Semicond., 22, 459 (1988)

ACCEPTORS IN SILICON CARBIDE: ODMR DATA

Pavel G.BARANOV and Nikolai G.ROMANOV

A.F.Ioffe Physico-Technical Institute, Leningrad 194021, USSR

ABSTRACT

The paper presents the results of a study of the acceptors Ga, Al, B and Sc in n- and p-type SiC and spin-dependent recombination processes using optically detected magnetic resonance.

1. Results

We studied epitaxial layers (some ten micron thick) of 6H and 4H polytypes of SiC both of n- and p-type doped with Ga, Al, B and Sc during growing. They were grown by the sublimation sandwich method [1] and kindly supplied by E.N.Mokhov. Bulk materials doped with diffusion and neutron-irradiated and annealed crystals were also studied. The concentration of nitrogen was $10^{17} - 5*10^{18}$ cm^{-3}.

By monitoring the intensity of luminescence which was excited with band-to-band light from the deuterium arc lamp the ODMR spectra of recombining donors and acceptors were recorded at 35 GHz and 1.6 K in 6H- and 4H-SiC doped with Ga, B, Sc and Al [2-6]. First ODMR measurements of SiC:Al were reported in Ref.7.

In gallium and aluminium doped SiC strongly anisotropic signals of shallow acceptors were found. The ODMR spectra recorded in 6H-SiC:Ga epilayer grown on the 6H-SiC:Al substrate are shown in Fig. 1. The ODMR lines correspond to an increase of the luminescence intensity and are identified as donor and acceptor signals. A nearly isotropic ODMR signal with g about 2 which has a partly resolved hyperfine structure belongs to nitrogen. This was proved by observation of this structure in 6H and 4H polytypes of SiC.

The Ga signals have a resolved structure which is due to the hyperfine interaction with the gallium nucleus. Two Ga acceptor signals and three Al acceptor signals seem to belong to three different positions of the impurity in the 6H-SiC lattice. These positions can be classified as either cubic or hexagonal local symmetry sites when considering the first- and second-nearest neighbours. For Ga the signals of two positions are not resolved. The Al and Ga acceptors can be considered as effective-mass-like acceptors. The angular dependence for one of the acceptor signals in 6H- SiC:Al and 6H-SiC:Ga is shown in Fig.2.

A decrease of anisotropy for the gallium acceptors ($g_{//}$ =2.27 and 2.21, $g_\perp$ = 0.6) as compared to the aluminium acceptors ($g_{//}$ =2.41, 2.40, and 2.32, $g_\perp$ = 0) may be due to a larger depth of the Ga acceptor levels. The observed hyperfine structure of Ga acceptors gave the direct measure of the unpaired electron spin density at the gallium nuclei and allowed to estimate the degree of localization of the Ga and Al

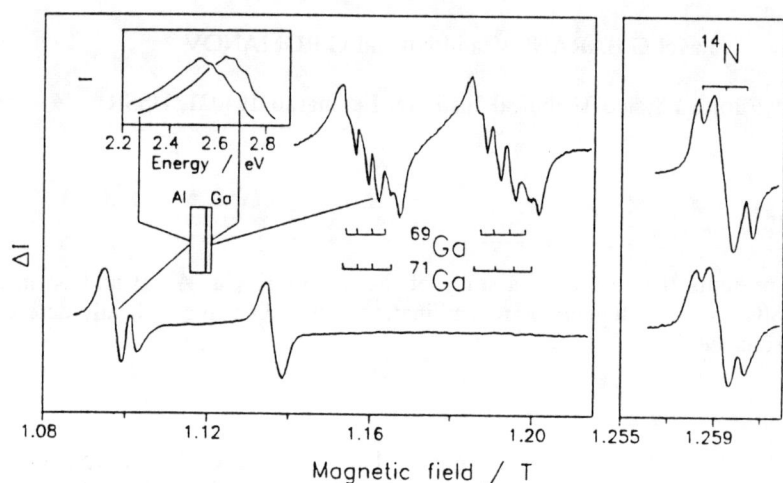

Fig.1 Luminescence and ODMR at 1.6 K and 35.2 GHz in 6H-SiC:Ga (epitaxial layer) and 6H-SiC:Al (substrate) with the magnetic field in the (11$\bar{2}$0) plane directed at 18o to the C_6 axis.

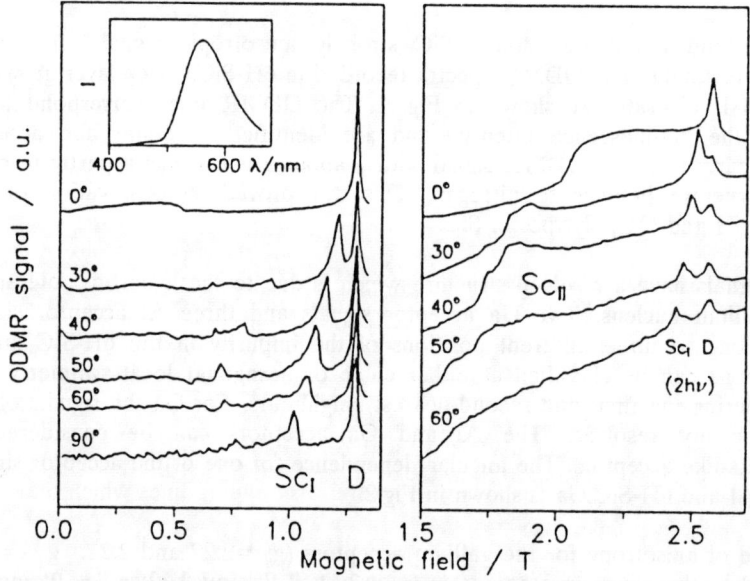

Fig.2 Luminescence spectrum of 6H-SiC doped with scandium and ODMR spectra recorded at 35.3 GHz and 1.6 K with the magnetic field in (11$\bar{2}$0) plane directed at different angles to the C_6 axis.

acceptors at different position in the SiC lattice.

It follows from the experimental values of the hyperfine interaction constants that the density of the wave function of the unpaired electron at the gallium nuclei is 0.56×10^{23} cm^{-3} and 0.68×10^{23} cm^{-3} for the low-field and high-field acceptor signals. A comparison of these values with the density of the wave function of the hybrid sp^3 orbital composed of the 4s and 4p functions of free gallium atoms shows that the localization of an unpaired electron is 5% and 6% for the two Ga signals which can be attributed to the gallium acceptors at the hexagonal and cubic positions respectively. Similar calculations for the aluminium acceptors show that the localization of acceptors is about 2.5% and 3% as compared to the wave function density of sp3 orbital.

Since the emission from a thin layer near the crystal surface could be excited in SiC with the appropriate light the magnetic resonance of the epilayer and the substrate could be selectively studied. This is an important advantage of the optical detection technique. It is also possible to obtain information about the spatial distribution of impurities in the samples.

Two types of luminescence spectra (yellow and red luminescence in 6H-SiC) can be observed in boron doped silicon carbide. ODMR was recorded on both luminescence bands in 6H and 4H polytypes of SiC. The nitrogen donor ODMR and a number of anisotropic signals ascribed to the acceptors were found on the yellow luminescence in 6H-SiC and their analogue in 4H-SiC. An anisotropic ODMR spectrum seems to belong to several types of centres since the relative intensities of the signals vary with the emission wavelength. The observed ODMR spectra do not correspond to any known EPR spectrum in SiC:B and are very different from Al and Ga ODMR spectra as can be seen from angular dependence plotted in Fig 2.

On the red luminescence in boron doped 6H-SiC and its analogue in 4H-SiC the nitrogen donor ODMR was observed together with some ODMR signals overlapping with the hf structure triplet of nitrogen [5].

A very complex ODMR spectrum was found in 6H and 4H polytypes of SiC doped with scandium [6]. Luminescence and ODMR spectra of a 6H-SiC:Sc epitaxial layer are shown in Fig 3. The ODMR spectrum consists of nearly isotropic donor signal and a number of anisotropic lines which seem to belong to different charge states of scandium. In the ODMR spectra shown in Fig 2 one can distinguish different Sc signals which are marked as Sc$_I$ and Sc$_{II}$. They are shown for different angles in the (1120) plane between the C-axis and magnetic field. The spectral dependencies of donor, Sc$_I$ and Sc$_{II}$ ODMR coincide with the emission band. The angular variation of the Sc$_I$ signals can be fitted to S=1/2 spin Hamiltonian with $g_{//} = 1.97$ and $g_{\perp} = 2.49$. The Sc$_I$ signal seems to consist of several overlapping lines which may belong to different lattice sites.

A group of anisotropic lines in the region of g < 2 is marked as Sc$_{II}$. It seems to be possible to describe these signals by S=1/2 spin Hamiltonian for three different sets of parameters. Measurements of ODMR at 24 and 35 GHz confirmed this conclusion. g-values of these three lines are estimated as $g_{//} = 1.21, 1.18, 1.17$ and $g_{\perp}$

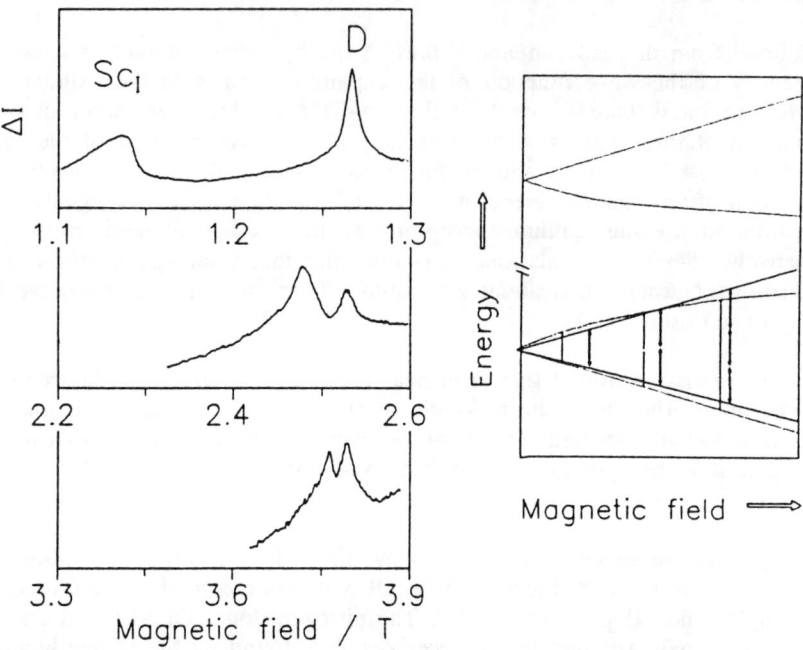

Fig.3 a) "Multiquanta" ODMR in 6H-SiC:Sc recorded at 35.3 GHz and microwave power 500 mW. ; b) a possible model of Sc_I energy levels in SiC:Sc (b).

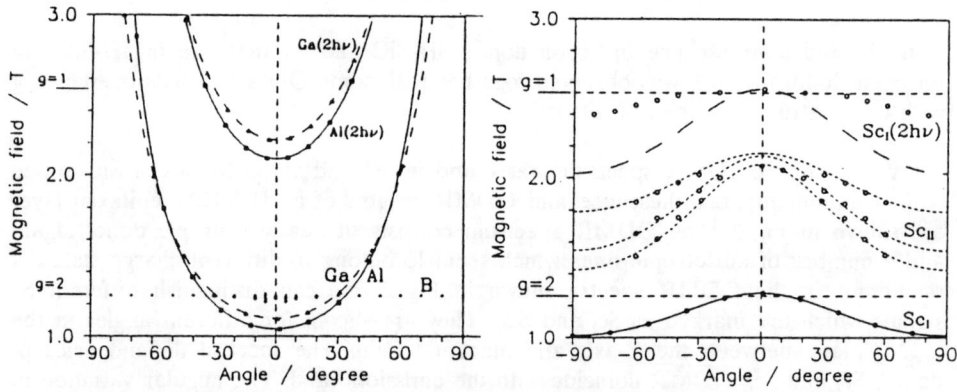

Fig.4 Angular dependences in the $(11\bar{2}0)$ plane of the acceptor ODMR at 35.3 GHz in 6H-SiC doped with boron, aluminium and gallium. The angle 0 corresponds to $B//C_6$.

Fig.5 Angular dependences in the $(11\bar{2}0)$ plane of the acceptor ODMR at 35.3 GHz in 6H-SiC doped with scandium. The angle 0 corresponds to $B//C_6$.

= 1.77, 1.63, 1.41 , respectively. Since the same structure was observed in 4H- SiC:Sc these signals can not be simply related to different lattice sites. The angular dependence of ODMR in 6H-SC:Sc is shown in Fig 4 where circles mark experimental point and curves are the result of a fit with the given parameters.

In ODMR spectra recorded on the intensity of donor-acceptor emission resonance signals which are usually ascribed to multiquanta transitions can be observed [5,6,8]. Normally these "multiquanta resonances" appear at magnetic fields which are exactly twice and three times larger than the "classical resonance" field of donors and acceptors. A different behaviour of the "multiquanta resonances" was found in 6H- and 4H-SiC crystals doped with scandium. In this case the positions of the "multiquanta resonances" of donors were usual but those of the Sc_I acceptors were different from the double and triple field of the "classical" resonance when the magnetic field Bo was not parallel to the hexagonal axis of the crystal (see Fig.4). The positions of the multiquanta resonances of the aluminium acceptors which were recorded in the same sample when exciting the substrate corresponded exactly to the double and triple quanta transition for the $S=1/2$ acceptors and were as shown in Fig 2.

The observed behaviour of the Sc "multiquanta resonances" implies that their energy levels are not linear. This non-linearity may be connected with an influence of other closely lying energy levels. ODMR spectra and a possible model of the energy levels are shown in Fig. 5.

The angular dependence of ODMR of Al and Ga acceptors is in agreement with their effective-mass-like character. The configuration of the valence electron shell of boron is similar to that of aluminium and gallium but the ODMR spectra which are observed in SiC:B are considerably different. This difference seems to be due to the fact that the structure of boron centres is different from that of Al and Ga acceptors which substitute silicon in the crystal lattice. It is probable that boron forms complex centres including vacancies.

A striking difference in the behaviour of Sc as compared to Al and Ga seems to be related with a different valence shell structure of scandium ($3d4s^2$) as compared to that of aluminium ($3s^23p$) and gallium ($4s^24p$). Due to Sc d-electrons quasi d-states of the acceptor in the band gap may exist. These states appear because of interaction of d-levels with the valence band of SiC. We believe that the most probable model of Sc_I might be a neutral acceptor A^0. Sc_{II} may be doubly charged acceptor A^{--}(3d). In addition to the Sc_I and Sc_{II} ODMR a number of lines can be observed in SiC:Sc which may be due to high spin states of Sc.

In 6H and 4H SiC subjected to fast neutron irradiation and annealing a well known green luminescence (D_1-spectrum) is observed. The intensity of this luminescence was found to have a sharp minimum, which is independent on microwaves, at zero magnetic field and a decrease of the emission intensity at EPR of nitrogen [4,5]. The spectral dependence of both the zero field signal and the ODMR signal coincided with the emission spectrum. The signals in some samples were as large as 10% of the luminescence intensity and could be detected in a wide range of temperature (1.6 - 20 K). The observed effects can be explained by the existence of a spin-dependent non-radiative process which is effective at low magnetic fields and at EPR of nitrogen.

Similar results were obtained in SiC with thermal defects, in ion-implanted samples and in SiC crystals grown with the excess of silicon, in which the defect luminescence was observed. In some as-grown epilayers new ODMR signals were obtained. Two isotropic lines with an unusually large splitting (27 mT) which is apparently due to the hyperfine interaction with the ^{29}Si nucleus (4.7%, I=1/2) may belong to centres including Si_C antisite defects in SiC. According to Ref. 9 these defects require the lowest formation energies among other native defects in SiC.

2. Summary

In conclusion, ODMR study of hexagonal polytypes of silicon carbide doped with Ga, Al, B and Sc has proved the donor- acceptor nature of emission. Spatial selectivity of ODMR allowed separate studies of epitaxial layers and substrates. Nitrogen was identified as a donor in all samples. The resolved hyperfine structure of effective-mass-like Ga acceptors gave a direct measure of the localization of the acceptor wavefunction. Different behaviour of ODMR in B doped SiC is connected with different structure of B centres which apparently form complexes. Two charge states of Sc were proposed to explain the ODMR data. Different anisotropy of Sc acceptors as compared to Al and Ga seems to be due to d-electron in Sc valence shell. Observation of multiquanta transitions in ODMR has shown that the energy levels of Sc are not linear probably due to the interaction with highly lying energy levels. A spin-dependent non-radiative process in SiC was found using ODMR.

3. References

1. Yu.A. Vodakov, E.N. Mokhov, Patent USA No.4147572.
2. P.G. Baranov, V.A. Vetrov, N.G. Romanov, and V.I. Sokolov, Soviet Phys. Solid State, 27, 2085 (1985).
3. N.G. Romanov, V.A. Vetrov, P.G. Baranov, E.N. Mokhov, and V.G. Oding, Sov.Phys.-Techn.Phys.Lett. 11, 483 (1985).
4. N.G. Romanov, V.A.Vetrov, and P.G.Baranov, Soviet Phys.-Semiconductors 20, 96 (1986).
5. P.G. Baranov and N.G. Romanov, in:Magnetic Resonance and Related Phenomena (Proc.24th Congress AMPERE),p.85 Poznan 1988.
6. P.G. Baranov, N.G. Romanov, V.A. Vetrov, and V.G. Oding, in: The Physics of Semiconductors (Proc. 20th Int. Conf., Thessaloniki 1990), ed: E.M.Anastassakis and J.D.Jaannopoulos, v.3, p.1855 Singapore-New Jersey-London-Hong Kong: World Scientific 1990.
7. Le Si Dang, K.M. Lee, and G. Watkins, Phys. Rev. Lett. 45, 390 (1980).
8. P.G. Baranov, M.F. Bulanyi, V.A. Vetrov, and N.G. Romanov, Sov.Phys.-Solid State, 26, 2218 (1984).
9. G. Pensl and R. Helbig, Festkörperprobleme, 30, 133 (1990).

LUMINESCENCE AND ABSORPTION OF VANADIUM (V^{4+}): IN 6H-SILICON CARBIDE

A. DÖRNEN[1], Y. LATUSHKO[1,‡], W. SUTTROP[2], G. PENSL[2] S. LEIBENZEDER[3], R. STEIN[3]
[1] 4. Physikalisches Institut, Universität Stuttgart, PO-Box 80 11 40, 7000 Stuttgart 80, F.R.G.
[‡] Laboratory of Spectroscopy of Semiconductors, University of Minsk, 220080 Minsk, USSR
[2] Institut für Angewandte Physik, Universität Erlangen, Staudtstr. 7, 8520 Erlangen, F.R.G.
[3] Corporate Research Laboratory, Siemens AG, 8520 Erlangen, F.R.G.

ABSTRACT

We present an optical study on the lines α (7642 cm^{-1}), β (7397 cm^{-1}), and γ (7205 cm^{-1}) in silicon carbide (6H-polytype), recently attributed to V^{4+}. The line α is analyzed by Zeeman-photoluminescence spectroscopy and the findings are compared to results from electron-paramagnetic resonance (EPR). Our data give strong evidence that the optical defect is identical to the hexagonally coordinated EPR-active center V^{4+}. By absorption measurements we demonstrate that a V^{5+} charge state is present in p-type material. The energetic position of the $V^{4+/5+}$ level is found to be 1.31 eV (species α) and 1.46 eV (species β and γ) above the valence band. Additionally, in p-type material, we detect a new vanadium-related defect ($V^{4+}X$).

Introduction

Silicon carbide is a semiconductor material of high band gap with considerable technological potential [1]. Superior thermal stability and heat conductivity determine this material for devices to be used for high temperature and high power applications. A peculiarity of this material system is the enormous variety of polytypes. The 6H-species is used commercially to fabricate blue light emitting diodes and FETs [1].

Transition metal ions are unavoidable contaminations of silicon carbide with impact to technological properties [2]. Recently, vanadium V^{4+} was investigated by a work combining both, electron-paramagnetic resonance (EPR) and optical spectroscopy [3]. The defect identification follows from a characteristic hyperfine structure found in the EPR spectrum. Optical transitions observed in the same samples were attributed to V^{4+}, too, because of the fine structure found in the zero-phonon transitions and because of a characteristic vibrational side band. Generally, polytypism leads to more complex spectra, since various inequivalent lattice sites appear to have different crystal fields. In this paper we will focus on vanadium in 6H-polytype SiC. For this specific polytype three inequivalent lattice sites are possible (this concerns silicon and carbon sites as well) [4]. While each of these sites have a tetrahedral coordination of the next neighbors, different arrangements exist for the second-nearest neighbors. One site shows a hexagonal symmetry (site α) while two other coordinations have different cubic symmetry (site β and γ).

For each inequivalent site a zero-phonon line and a corresponding vibrational mode spectrum was assigned to V^{4+}. Especially, a fine structure in the line α, consisting of four components, led to identify the corresponding site to be of hexagonal symmetry [3]. The EPR data motivate Zeeman measurements on transition α, which will be reported in the first section.

Experimental

Photoluminescence spectra were measured with conventional grating monochromators, by using a Kr$^+$ laser (647 nm line) or an Ar$^+$ laser (514 nm line) for excitation. Samples were cooled down to 4.2 K and 1.8 K, respectively. We recorded spectra in a spectral region between 1.0 eV and 0.7 eV using a liquid N$_2$ cooled germanium detector. For Zeeman-photoluminescence measurements we used a superconducting split-coil magnet in Voigt and Faraday configuration, respectively;

magnetic fields up to 7.5 Tesla are possible. The absorption measurements were performed with a Bomem DA3.01 Fourier-transform spectrometer, alternatively equipped with an InSb or an InGaAs detector. In several cases we spectroscopically reduced the absorption light source (halogen lamp) by a silicon filter, blocking all light of photon energy higher than 1.2 eV. For photoionization studies prior to the absorption measurements the sample was illuminated with light from a halogen lamp dispersed by a 3/4m-Spex monochromator of spectral resolution of 3 nm (2 meV, approximately).
By Hall measurements we determined the net dopant concentration of the samples.

I. Zeeman measurements (line α)

The line α shows a four-fold fine structure, which is attributed to the symmetry of the lattice site (C_{3v}). From thermalization found in the fine structure, the level scheme in Fig. 1 (right) can be drawn. Schneider [2, 5] explains the results from EPR and from optical experiments by a strong tetrahedral crystal-field splitting of the ^2D ground state of the V^{4+} ion ($3d^1$) into a low energetic ^2E state and a high energetic $^2T_2$ state. The ^2E state couples via spin-orbit interaction to the spin $S_{1/2}$ and finally is split by the strong trigonal field into a Γ_4 and a $\Gamma_{5,6}$ level. The strong trigonal field is believed to split the $^2T_2$ into a high energetic $^2T_0$ and a low energetic $^2T_\pm$. The latter state is split by spin-orbit interaction into Γ_4 and $\Gamma_{5,6}$. Additionally, the states originating from the $^2T_2$ and ^2E level should be subject to a dynamic Jahn-Teller effect. The optical transitions between $^2E(\Gamma_4, \Gamma_{5,6})$ and $^2T_\pm(\Gamma_4, \Gamma_{5,6})$ are identified with the fine structure of line α [2], as indicated in the level scheme of Fig. 1 (right).

For angular dependent Zeeman measurements the sample was rotated around the $\vec{b}$ direction [010], so that the field could be moved by a turn of 90° from the $\vec{c}$ axis [001] to the $\vec{a}$ axis [100]. In Fig. 1 (left) the angular-dependent Zeeman splitting measured at a constant field of 6.9 Tesla is depicted as a function of the angle δ, (δ includes the crystal axis $\vec{c}$ and the magnetic field $\vec{B}$). When the magnetic field is perpendicular to the $\vec{c}$-axis ($\delta = 90°$) no Zeeman splitting occurs at all, indicating that all $g_\perp$ components vanish within the experimental uncertainty of $\Delta g = \pm 0.1$. For any angle δ less than 90° each of the four lines splits symmetrically around the center of gravity. For line 2 and 3 the center of gravity is independent of angle δ, while for line 1 and 4 a quadratic shift is caused by second order effects. We note that no mixing of Zeeman

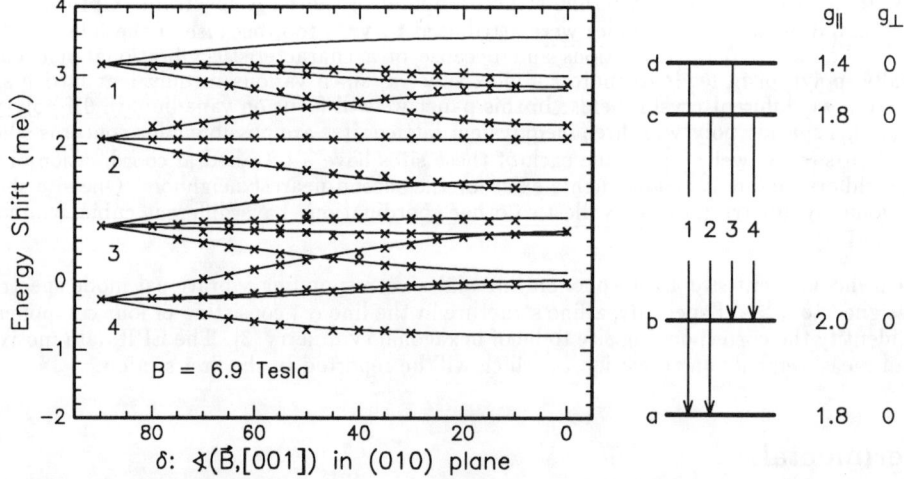

Figure 1: Left: Angular-dependent Zeeman splitting of the four fine-structure components of line α, labelled 1...4. Angle δ includes magnetic field $\vec{B}$ and crystal axis $\vec{c}$. No splitting occurs for $\delta = 90°$. Right: level scheme of line α fine structure and g-factors obtained from Zeeman pattern.

components occur, which would result in a nonlinear splitting being asymmetric to the center of gravity. Besides the quadratic shift the whole Zeeman effect can be explained by a $g_\| H_z$ term for each of the four Kramers doublets. Generally, each of the four fine-structure components of the line α is expected to split into four Zeeman components, corresponding to the two allowed transitions $\Delta m_j = 0$ and the two forbidden transitions $\Delta m_j = \pm 1$. The complete set of four Zeeman components is resolvable for line 3 (Fig. 1), while for line 1, which is rather weak, the high energy component is not observable because of small intensity and thermalization effects. Line 2 and 4 on the other hand show a three-fold splitting. Both allowed transitions coincide since the $g_\|$ values of the two states are identical or almost identical, respectively. From the progress of the forbidden transitions as a function of angle δ, the $g_\|$ values can be determined within an experimental uncertainty of $\Delta g = \pm\, 0.1$. A consistent set of $g_\|$ values, as given in the level scheme of Fig. 1 can be determined when the whole Zeeman pattern is taken into account. The lines drawn in the Zeeman pattern are cosine functions, basing on these g-factors. For line 1 and 4 the additional shift of the center of gravity is added.

Comparing the EPR data ($g_\| = 1.749$, $g_\perp = 0$) [3] with our results ($g_{a\|} = 1.8$, $g_{a\perp} = 0$), we find an excellent agreement within the experimental uncertainty of $\Delta g = \pm\, 0.1$. The optical Zeeman data confirm and consolidate the identification of the EPR center with the optically active defect. Moreover, for the optically active states **a** through **d** of the center α, we have determined the complete set of g-tensors, see Fig. 1 (right). The strong axial character is expressed by the vanishing $g_\perp$ value for all of the g-tensors. We mention briefly that the theory outlined above (Ref. 2), correctly describes the g-factors of ground states **a** and **b** – and hence fits to the EPR data – but fails to explain the g-factors of excited states **c** and **d** [6]. A complete discussion of the data together with a Zeeman study on the lines β and γ will be subject to a following paper.

II. Position of the $V^{4+/5+}$ donor level

For vanadium, replacing a group IV host atom, the quasi-neutral charge state is V^{4+}. In EPR measurements $V^{4+}(3d^1)$ is observable in compensated SiC, while the $V^{3+}(3d^2)$ appears in slightly n-type material [3]. Further on, it was suggested that vanadium is amphoteric. A $V^{5+}(3d^0)$ state (not paramagnetic) was supposed for p-type SiC since no vanadium correlated center was obtained in EPR [3]. In the following, we show by absorption measurements that V^{5+} or V^{4+} can be present in 6H-SiC, depending on the experimental conditions.

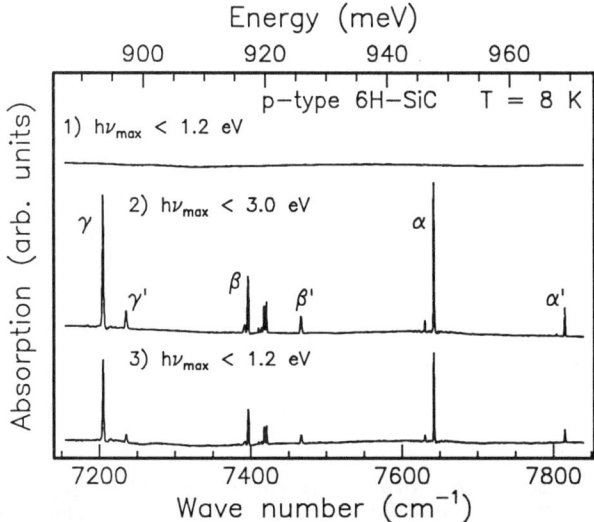

Figure 2: A series of absorption spectra successively measured of an aluminum doped, p-type sample (N_A-N_D = $3\cdot 10^{18}$ cm^{-3}) under various conditions; see text.

Figure 2 summarizes a series of absorption spectra successively taken from an aluminum doped, p-type sample (N_A-N_D = $3 \cdot 10^{18}$ cm^{-3}). The temperature was kept below 10 K. For spectrum 1 the sample was cooled down in darkness and absorption was measured using a silicon filter to block all light with photon energy higher than 1.2 eV [7]. None of the lines α, β, and γ shows up. For spectrum 2 the silicon filter was removed and the complete spectrum of the absorption light source (halogen lamp) is used. The V^{4+} lines appear as they are found in n-type material (the primed lines being characteristic of p-type 6H-SiC belong to a different center and will be discussed in the next section). The transitions still show up, when the silicon filter is inserted back again (spectrum 3), so that the same conditions are present as for spectrum 1. The absorption coefficients of all lines appear to be reduced compared to spectrum 2 but stay constant as long as the sample is kept below 200 K. The initial spectrum 1 can only be reached by heating up the sample to temperatures of approximately 250 K.

The photo-induced absorption is due to an optical charge transfer from vanadium to compensated shallow aluminum acceptor. The photoionization is induced by the high energy spectral components of the absorption light and can be explained as follows. When the p-type sample is cooled down in darkness all vanadium ions are in the V^{5+} charge state, completely compensated by the aluminum acceptors. At least an equal amount of Al$^-$ is present. Filtering the light by a silicon filter removes all spectral components of the light that could ionize the V^{5+} to V^{4+}. Since V^{5+} is optically inactive and no V^{4+} is present, the lines α, β, and γ are not observable (spectrum 1). When the sample is exposed to light of sufficient high photon energy (conditions for spectrum 2) a hole is ionized from V^{5+} into the valence band. The remaining V^{4+} is indicated by the characteristic absorption lines. The observed persistent conversion of the charge state (spectrum 3) shows that the photo-generated hole is trapped by another impurity. Most probably this impurity is the compensated aluminum acceptor (Al$^-$). The Al$^-$ traps a hole by about 300 meV [8]. To restore the initial state, as shown in spectrum 1, the temperature has to be sufficiently high to overcome this barrier. To summarize these results: the optical activation moves the hole, which compensates vanadium to V^{5+}, back to the compensated aluminum acceptor (Al$^-$) and leaves vanadium in the optically active V^{4+} charge state. We note that in n-type material the V^{4+} lines are already present by conditions similar to those for spectrum 1.

When luminescence is measured in p-type material, the optical active V^{4+} charge state is induced by the laser excitation. Hence in luminescence the V^{4+} transitions show up in n- and p-type material.

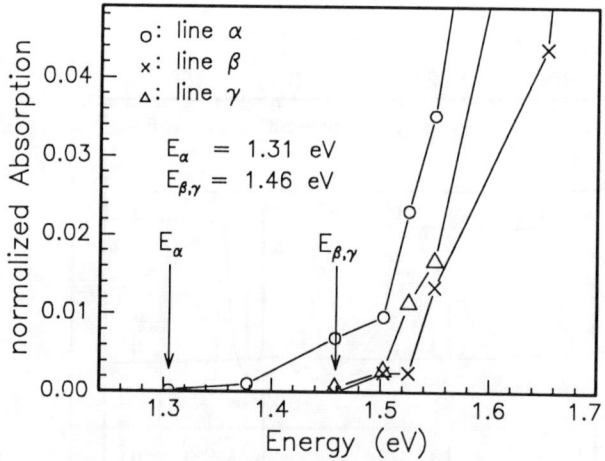

Figure 3: Spectral response of the optical charge conversion. Absorption of line α, β, and γ is plotted via energy E of the light used for photoionization, see text. The absorption of each line is normalized to the value found for E = 2eV (detection limit of normalized absorption is about 0.001), photon flux density: $5 \cdot 10^{14}$ 1/(cm^{2}s) for 10 min.

The peculiar behavior of vanadium as discussed above allows to determine the position of the $V^{4+/5+}$ donor level by absorption measurements carried out in the following manner. The sample is cooled down to T = 6 K in darkness and then illuminated with light of a definite wavelength for a given period of time. After the illumination step we measure the absorption by using a silicon filter, to assure that no photoionization takes place during the measurement. Finally, the sample is heated up to about 250 K so that the original state is prepared again. The whole procedure is repeated for various wavelengths so that the ionization process can be resolved spectroscopically for each of the three species α, β, and γ. Then the absorption coefficients of each line are normalized to the value obtained at an excitation energy of E = 2 eV.

In Fig. 3 the photo-ionization curves obtained from the normalized absorption of the lines α, β, and γ are plotted via the photon energy of the exciting light. The detection limit of normalized absorption is about 0.001, so that the lines can be followed over a range of three orders of magnitude. By comparing the photo-ionization curves of transition α, β, and γ distinct differences show up. For deep donor level $V^{4+/5+}$ we find a trend, which is similar to shallow nitrogen donor and shallow aluminum acceptor [8]: the hexagonally coordinated center α shows an ionization energy ($E_\alpha = 1.31 \pm 0.02$ eV) significantly smaller than the ionization energy found for the cubic-like centers β and γ ($E_{\beta,\gamma} = 1.46 \pm 0.02$ eV).

III. The $V^{4+}X$ center in p-type 6H-SiC

Photoluminescence and absorption spectra in the energetic region between 1.2 eV and 0.7 eV were recorded for several n and p-type samples. In Fig. 4 (lower spectrum) the characteristic V^{4+} lines are depicted as they appear in n-type 6H-SiC, consisting of the lines α, β, and γ and the corresponding local vibrational modes LM_α, LM_β, and LM_γ, indicated by arrows.

As shown in the previous section in p-type material, the V^{5+} charge state is the thermodynamically stable configuration. But by excitation with light of $h\nu > 1.5$ eV V^{4+} is generated and the three characteristic zero phonon lines (α, β, and γ) show up in photoluminescence. Moreover for several p-type samples an additional series of transitions is observed. In Fig. 4

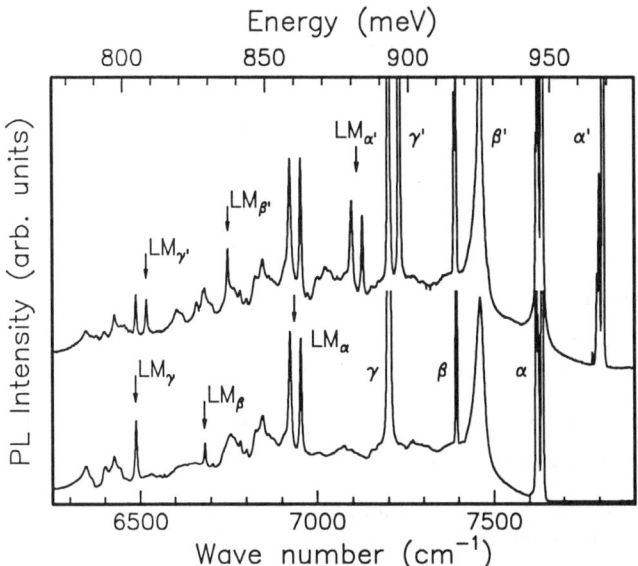

Figure 4: Luminescence of a n-type sample (lower spectrum) compared to a p-type sample (upper spectrum). In p-type material together with the characteristic line spectrum α, β, and γ, another set of lines labelled α', β', and γ' shows up.

(upper spectrum) a typical spectrum of a p-type sample is shown. Obviously the new lines originate from a defect ($V^{4+}X$), which is closely related to the V^{4+} center. To each line of the V^{4+} spectrum a corresponding line of $V^{4+}X$ can be identified. In accordance the additional series is labelled α' (7815.4 cm^{-1}), β' (7466.4 cm^{-1}) and γ' (7235.5 cm^{-1}). The fine structure in transition α' is of the same type as in α. We find 17.2 cm^{-1} (17.6 cm^{-1}) for the ground state splitting and 6.1 cm^{-1} (5.6 cm^{-1}) for the spacings in the excited states of the $V^{4+}X$ (V^{4+}) center. Also the energies found for the vibrational modes are identical to those of the V^{4+} defect: we find for $V^{4+}X$ (V^{4+}), α: 84.9 meV and 88.6 meV (84.9 meV and 88.6 meV) β: 88.4 meV (88.4 meV) and γ: 88.4 meV (88.4 meV).

Since the spectral properties are very similar between both line systems, presumably the $V^{4+}X$ is basically V^{4+} but slightly modified by another impurity in its vicinity. Further work is needed to identify the microscopic structure of this center.

IV. Summary

By Zeeman spectroscopy we find strong evidence that the optical center α is identical to the hexagonally coordinated V^{4+}, recently identified in EPR. We determine the g-tensors for both, the excited states and the ground states. The g-tensors are highly axial, since the hexagonal crystal field completely quenches the $g_\perp$ part. By low temperature absorption measurements, we find the V^{5+} charge state to be present in p-type 6H-SiC. But a persistent optical charge conversion from V^{5+} to V^{4+} can be induced optically, when the sample temperature does not exceed 200 K. Most presumably, compensated aluminum Al$^-$ traps the photo-generated hole. We obtain the position of the $V^{4+/5+}$ donor level by spectroscopically resolving the photoionization. Significantly different ionization energies show up for the hexagonal-like site (α) on one hand and for the cubic-like site (β and γ) on the other hand. In p-type SiC we find a new vanadium related defect ($V^{4+}X$), which appears to be closely related to V^{4+} concerning electronic and vibronic properties.

Acknowledgement

We gratefully acknowledge many helpful and stimulating discussions with J. Schneider and U. Kaufmann.

References

[1] G. Pensl and R. Helbig in *Festkörperprobleme / Advances in Solid State Physics XXX*, ed. by U. Rössler, (F. Vieweg & Sohn, Braunschweig 1990), p. 133.

[2] J. Schneider, this conference.

[3] J. Schneider, H.D. Müller, K. Maier, W. Wilkening, F. Fuchs, A. Dörnen, S. Leibenzeder, and R. Stein, Appl. Phys. Lett. **56**, 1184 (1990).

[4] K.M. Lee, L.S. Dang, G.D. Watkins, and W.J. Choyke, Phys. Rev. B **32**, 2273 (1985).

[5] J. Schneider, private communication.

[6] We performed a complete numerical diagonalization of the $D_2 \otimes S_{1/2}$ state by including crystal field, spin-orbit interaction, Jahn-Teller reduction, and Zeeman operator, according to Ref. 2.

[7] We recall that in FTIR spectroscopy the sample normally is illuminated by a spectroscopically broad light source.

[8] M. Ikeda, H. Matsunami, and T. Tanaka, J. of Lumin. **20**, 111 (1979).

IMPURITY-DEFECT REACTIONS IN ION-IMPLANTED DIAMOND

Aleksey A. GIPPIUS

P.N. Lebedev Physical Institute of the Academy of Sciences of the USSR,
Leninsky prospect 53, Moscow 117924, USSR

ABSTRACT

Reactions involving implanted impurities, radiation defects and background impurities in ion-implanted natural diamonds are discussed with the emphasis on high local density of defects produced by displacement cascades. The results are directly related to problems of implantation doping and hydrogenation of diamond.

1. Introduction

Diamond is sometimes considered as one of two or three major materials at the frontiers of material research [1] and even as the material for electronics of XXI century. As such it would undoubtedly be used in the form of synthetic crystals or films. At present, however, natural diamonds remain important both for applications (heat sinks, semiconducting devices, particle detectors) and for studies of those properties which are common to all forms of diamond and determined by its structure and bonding. An essential feature of natural diamond is high concentration of background impurities of which the most important is nitrogen. Nitrogen, either as isolated substitutional impurity or in various aggregate forms can produce a variety of point defects: combinations of nitrogen atoms, lattice defects and other impurities.

Doping (essential for device applications) is still a serious problem for diamond in spite of definite success in the growth of doped CVD films, so ion implantation remains to be an important alternative. As for any technological process the final properties of ion-implanted solid state structures are the outcome of a succession of reactions on the surface, interfaces and in the crystal volume between defects and impurities, both background and introduced by a given operation. The major factor determining the type and parameters of impurity-defect reactions in ion-implanted solids is high local (within disordered regions) and high average (at higher doses) density of radiation defects.

The present paper deals with luminescent control of reactions involving implanted impurities, radiation defects and background impurities in ion-implanted and annealed natural diamond. It is based on our recent results and, in retrospect, on some of our previous data.

2. Experimental

Luminescent control of defects and impurities in solids is based upon the accumulated data on the nature and properties of appropriate centres. Within the last decade the importance of transition elements as luminescence centres in diamond was recognized first for Ni [2,3] and recently for Ti and Nb [4]. The strongly degenerate levels of these elements are characteristically split according to the symmetry of a given site. High sensitivity of luminescence spectra of transition elements "probes" to various crystal fields makes them a useful tool in studies of lattice disorder, impurity-defect interaction etc. [5]. In our experiments samples of natural diamonds with nitrogen content less than 10^{18} cm^{-3} (type IIa) and more than 10^{19} cm^{-3} (type Ia) were implanted at room temperature by various ions with energies up to 350 keV and isochronally (1 h) annealed up to 1650°C in a vacuum of 10^{-6} Torr. Cathodoluminescence (2-10 keV, 5 μA) was studied in the range 0.4-1.1 μm at 80 K.

3. Results and discussion
3.1 Disordered regions

Considerable fraction of defects created by ion implantation is concentrated within disorder regions produced by displacement cascades [6]. Due to the dynamics of the cascade formation the inner part of a disordered region is vacancy-rich while a large fraction of interstitials is created predominantly at the periphery of the cascade and owing to their high mobility quickly migrate into the surrounding lattice [7]. Our luminescence data [8] confirmed the high local density of intrinsic defects in disordered regions and the increase of the size of their nuclei with the increase of ion mass.

The trend towards the increase of the role of nuclei of disordered regions is evident in our data on Ni implantation. Nickel is known to produce luminescent centres emitting at 884 nm [2,3]. This doublet line (Figure 1a) is rather sensitive to perturbations created by background (mostly nitrogen) impurities and defects, so the implanted Ni ions can be used as "probes" to study their own disordered regions. As can be seen in Figure 1b the widths of the Ni doublet components decrease considerably in the temperature range 1300-1600°C. These temperatures are considerably higher than those characteristic of interstitial atoms (400-600°C), isolated vacancies (800-900°C) and the known multivacancy complexes (100-1200°C) [9]. It means that the broadening of Ni lines in the range 1300-1650°C should be understood not in terms of point defects but rather in terms of gross lattice disorder which survives the annealing temperatures close to the limit of stability of diamond. This disorder is of essentially local character and related to disorder regions at the end of ion tracks. This is borne out by the following data. If the sample implanted by Ni and annealed at 1650°C is again implanted by, for example, Ti, then the lines of Ni doublet are slightly broadened by defects produced by Ti implantation. This broadening disappears

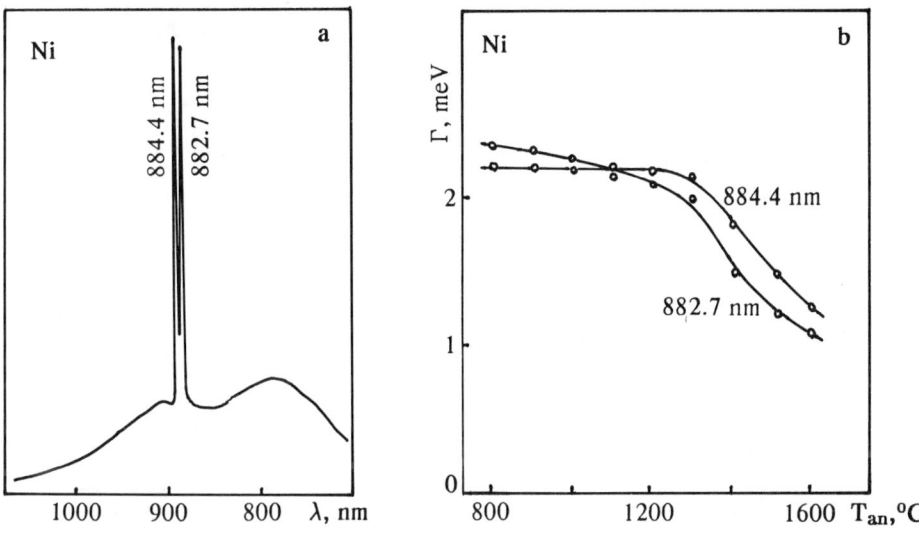

Figure 1. Spectrum (a) and line width Γ versus annealing temperature dependence (b) of Ni related luminescence.

after annealing at 800°C and the line width does not change in the course of further annealing up to 1650°C. It means that Ni centres 'feel' only the mobile component (interstitials and vacancies) of the radiation damage produced by Ti implantation and not the disordered regions surrounding the Ti atoms.

These results are directly related to the problems of ion implantation doping of diamond. Really it is hard to expect that relatively heavy implanted atoms, such as As or Sb, which, similarly to Ni, reside (even after annealing) within disordered regions, would display donor properties which are expected of them if they are in the substitutional position. This pessimistic view is supported by the data of Kalish et al. [10] who found that implanted In atoms are surrounded by strongly damaged regions up to annealing temperature 1800°C. Clearly, to realize the electrical activity of heavy implanted dopants one has to employ high annealing temperature ($\geq$2000°C) with the use of high stabilizing pressure which makes this version of implantation technology rather complicated.

3.2 Nitrogen association and dissociation reactions

Nitrogen as dominant background impurity in natural diamonds is known to exist in various forms ranging from single substitutional atoms to macroscopic "platelets" [14]. The ratio of concentration of various forms determines optical properties of diamond crystals and forms a basis of their classification. The change of the form of nitrogen in diamond was observed as reactions of either association of single nitrogen atoms into aggregates of two or three atoms (at 1600-2000°C and the stabilizing pressure $\approx 6 \cdot 10^9$ Pa [11]) or dissociation of

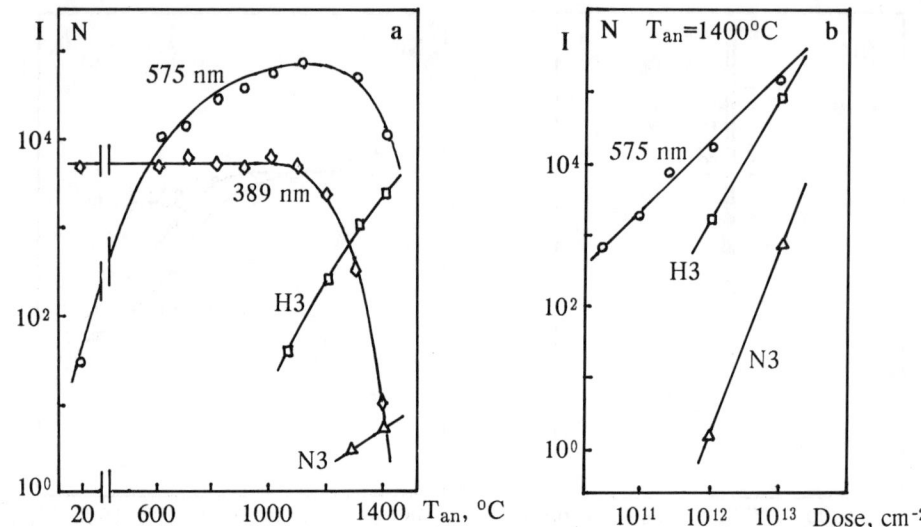

Figure 2. Luminescence intensity I of centres containing one (389 nm, 575 nm), two (H3), and three (N3) nitrogen atoms versus annealing temperature (a) and implantation dose (b).

these aggregates into isolated nitrogen atoms [12]. The process association-dissociation is reversible, its direction depending on temperature and on the ratio of concentrations of isolated and aggregated nitrogen. As practically any reaction in solid state this process should depend upon the density of lattice defects. This was first observed by Collins [13] who found that the association of single nitrogen atoms into pairs (the so called A-form) occurs already at 1500°C in diamonds irradiated by fast electrons. A mechanism of transport of nitrogen atoms via mobile complexes N-V was suggested.

In our "nitrogen free" type IIa samples we also observed the association of implanted (isolated) nitrogen atoms into aggregates. Summing up the behavior of nitrogen atoms implanted in "nitrogen-free" type IIa crystals we see that immediately after the implantation or after annealing at T<1000°C these atoms form interstitial (e.g. 389 nm) and single-atom (575 nm) centres. The well known luminescent centres H3 and N3 (containing vacancies plus 2 or 3 nitrogen atoms respectively) are formed already at the annealing temperatures T≈1000°C (Figure 2a) which are considerably (500-1000°C) lower than those for crystals not irradiated or irradiated by fast electrons. This drastic radiation enhancement of aggregation is due to the high local density of radiation defects. Locally each implanted atom is surrounded by a disorder region, rich of vacancies and ready to provide a "carrier" for nitrogen transport. This situation is to be compared to electron irradiated samples where both vacancies and single nitrogen atoms are randomly distributed.

The importance of high local density of defects for this kind of association reactions is confirmed by the first, second and third power low of "intensity versus dose" dependencies for

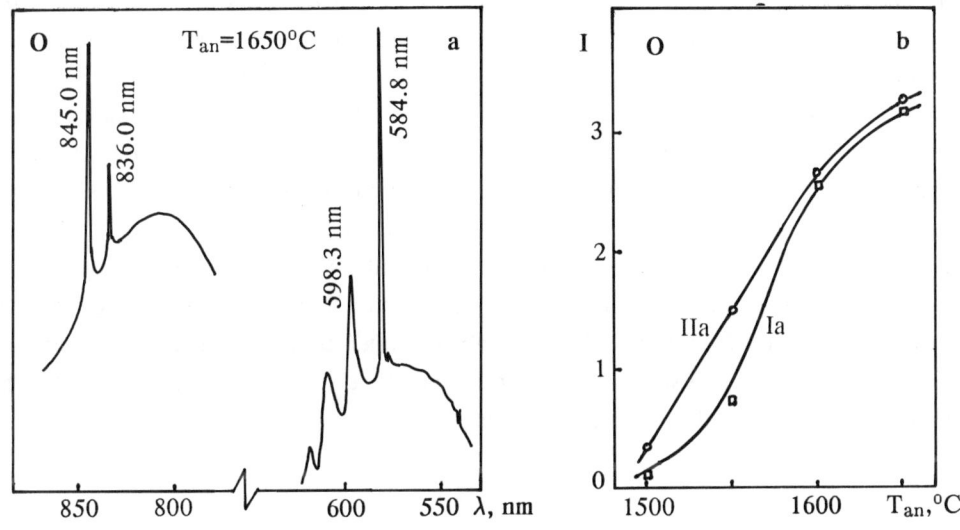

Figure 3. Spectra (a) and annealing behaviour (b) of oxygen related luminescence.

centres containing one, two and three nitrogen atoms respectively (Figure 2b). This simple relation of the power low to the number of nitrogen atoms in a given centre would not hold for randomly distributed vacancies and nitrogen atoms. The kinetics of formation of, for example, H3 centres, containing three (N-V-N) [14] or four (V-N-N-V) [15] components would not follow simple second power low observed in our experiments and understood in terms of two component (NV+NV) reaction.

3.3 Trapping and release

Any process of complex formation can be considered as "trapping" of its components. The terms "trapping" and "release" are generally used when the nature of complexes is not known and the emphasis is on the very fact of disappearance or appearance of some effect due to a given impurity or defect.

In oxygen-implanted diamond we observed several lines (584.8, 598.3, 836, 845 nm, Figure 3a) definitely related to centres containing oxygen (the detailed data will be published elsewhere). The peculiar feature of these centres is that they appear (contrary to practically all optical centres in implanted diamond) only after high temperature (>1500°C) annealing (Figure 3b). This can be understood in terms of release of the implanted oxygen atoms from some configurations where they are kept in optically inactive form. The intensity of these lines is increased if the samples implanted with oxygen and annealed at 1650°C are irradiated by 4 MeV electrons (or bombarded by some other ions) and again annealed at 1650°C. It means that the release of oxygen is enhanced by radiation damage.

Hydrogen was recently shown to passivate electrically active defects in diamond [16]. In any process of hydrogenation or dehydrogenation of diamond it is important to take into account the contribution of hydrogen present in natural diamonds as one of the main background impurities [17]. In this respect it is pertinent to recall our data on luminescent control of hydrogen in diamond [2]. In hydrogen (or deuterium) implanted diamonds we found specific, isotopicly shifted luminescence lines at 545.8 nm (H) and 545.2 (D) respectively. The nonmonotonic annealing curves suggested complicated kinetics of formation and destruction of the centres. An important feature was an additional peak at about 950°C on the annealing curve for hydrogen centres. This was understood as a manifestation of a process which overcompensated the destruction of the centres. This process is believed to be the release of hydrogen from some sources, probably water containing magma droplets [17] where it is "stored". The supply of hydrogen from these sources can essentially affect the properties of heat treated (and irradiated) samples.

REFERENCES

1. R.Roy, Proc. of the Second Int. Conf. on New Diamond Science,1990, Washington,DC (to be published).
2. A.A.Gippius, Vavilov V.S., Zaitsev A.M. and B.S.Jakupbekov, Physica, 116B, 187 (1983).
3. A.T.Collins, J. Phys.: Condens. Matter, 1, 439, (1989).
4. S.A.Kazarian and A.A.Gippius, Kratkie Soobshcheniya po Fizike (FIAN), at press.
5. A.A.Gippius, Ushakov V.V., Yakimkin V.N. and V.S.Vavilov, Nucl. Instr. Meth., B39, 492 (1989).
6. L.C.Kimerling and J.M.Poate, Inst. Phys. Conf. Ser., 23, 126 (1975).
7. R.S.Nelson, Inst. Phys. Conf. Ser., 16, 140 (1973).
8. A.A.Gippius, Zaitsev A.M. and V.S.Vavilov, Fiz. Techn. Poluprov., 16, 404 (1982).
9. J.N.Lomer and C.M.Welbourn, Inst. Phys. Conf. Ser., 31, 339 (1977).
10. R.Kalish, Deicher M., Recknagel E. and Th.Wichert, J. Appl. Phys., 50, 6870 (1979).
11. R.M.Chrenko, Tuft R.E. and H.M.Strong, Nature, 270, 141 (1977).
12. M.R.Brozel, Evans T. and R.F.Stephenson, Proc. Roy. Soc., A361, 109 (1978).
13. A.T.Collins, Inst. Phys. Conf. Ser., 59, 247 (1981).
14. G.Davies, Chem. Phys. Carbon, 13, 1, (1977).
15. A.M.Zaitsev, Gippius A.A. and V.S.Vavilov, Fiz. Techn. Poluprov., 16, 397 (1982).
16. M.I.Landstrass and K.V.Ravi, Appl. Phys. Let., 55, 1391 (1989).
17. J.P.F.Sellschop, Madiba C.C.P. and H.J.Annegarn, Nucl. Instr. Meth., 168, 529 (1980).

NATIVE DEFECT COMPENSATION IN WIDE-BAND-GAP SEMICONDUCTORS

D.B. Laks,[1,2,*] C.G. Van de Walle[3,†], G.F. Neumark[1], and S.T. Pantelides[2]

1. Metallurgy Division, Columbia University, New York, NY 10027 USA
2. IBM T. J. Watson Research Center, Yorktown Heights, NY 10598 USA
3. Philips Laboratories, Briarcliff Manor, NY 10510 USA
* Present address: Solar Energy Research Institute, Golden, CO 80401
† Present address: Xerox Palo Alto Research Center, Palo Alto, CA 94304

ABSTRACT

Wide-band-gap semiconductors are of great technological interest, but are plagued by one major problem: they can easily be doped either n-type or p-type, but not the other. For example, it is comparatively easy to make n-type ZnSe, but difficult to make p-type ZnSe. Compensation by native point defects is the most popular explanation of doping problems in wide-band-gap semiconductors. We determine the concentrations of all native point defects in ZnSe, using accurate *ab initio* total energy calculations. We find that native defect concentrations are too low to cause compensation in stoichiometric ZnSe. Small deviations from stoichiometry can produce enough native defects to compensate doping. However, we find that for either Zn-rich or Se-rich material, native defects will compensate **both** n-type and p-type doping; thus deviations from stoichiometry cannot explain why ZnSe prefers to be n-type. For Li-doped ZnSe we show that the true cause of doping difficulties is twofold: (1) the propensity of Li to become an interstitial donor and (2) the limited solubility of the dopant.

1. Introduction

Wide-band-gap semiconductors have important optoelectronic applications. ZnSe ($E_g = 2.7$ eV) for example, can be used to make a blue semiconductor laser. Greater use of wide-band-gap materials has been hampered by doping difficulties: very few wide-band-gap semiconductors can be doped both n-type and p-type.[1, 2, 3] For instance, ZnTe and diamond can be doped p-type but not n-type, while other wide-band-gap materials can only be made n-type. In spite of recent reports of well-conducting p-type ZnSe,[4, 5] the causes of the doping problems remain unclear.

The most popular explanation of why it is hard to dope wide-band-gap semiconductors is the native defect compensation mechanism.[6, 7] According to the native defect mechanism, p-type ZnSe is compensated by native point defects that are donors. The energy cost to form the native defects would be offset by the energy gained when electrons are transferred from the donor levels of the native defects to the Fermi level (which is near the valence band in p-type material). If the native donor defect levels are near the conduction band edge, then the energy gained by electron transfer would almost equal the width of the band gap (Fig. 1). Consequently, native defect compensation would become more likely as the band gap is increased. n-type doping of

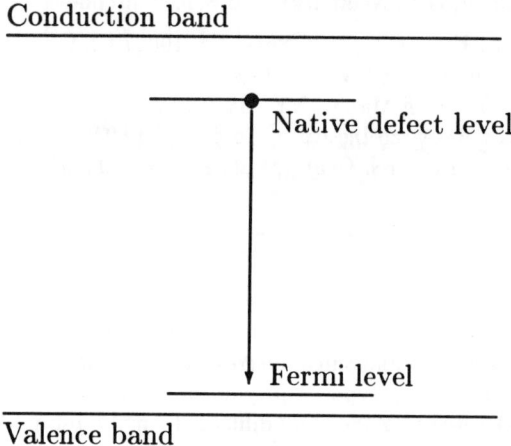

Figure 1: Native defect compensation in a p-type semiconductor. According to the native defect mechanism, native donor defects are formed in wide-band-gap semiconductors. These defects gain back much of their energy of formation by transferring electrons from defect levels near the conduction band to the Fermi level near the valence band. The energy gain due to electron transfer can be of the order of the band-gap energy.

ZnTe and diamond would be compensated by native acceptor defects that transfer their holes to the donor dopants. The appeal of native defect compensation is its universality—it applies to all wide-band-gap materials, all growth methods, and all dopants. Although native defect compensation was proposed some 30 years ago, there is still no convincing evidence either for or against it.

We investigate the native defect compensation method using first-principles total energy calculations. We calculate the total energies of all native point defects in ZnSe, and use these energies to determine the concentrations of these defects. The energies were calculated in a pseudopotential formalism. To get a good description of ZnSe, we include the Zn 3d electrons as valence states, using a mixed basis set of localized functions and plane waves. These are the first defect calculations in a II-VI semiconductor to fully include the effects of the metal d electrons. We find that the concentrations of native defects are far too low to compensate p-type doping of stoichiometric ZnSe. Small deviations from stoichiometry can accommodate large concentrations of native defects. We find, however, that any deviations from stoichiometry compensate n-type as well as p-type ZnSe. Thus, native defects caused by deviations from stoichiometry *cannot* explain why it is so much harder to make p-type ZnSe than n-type ZnSe. We have also determined the native defect concentrations in diamond, and again find that the defect concentrations are too low to compensate doping. Having shown that native defects are not responsible for doping problems in wide-band-gap semiconductors, we will examine doping problems on a case-by-case basis. For the Li_{Zn} acceptor in ZnSe we show that there are two causes of doping difficulties: (1) the presence of interstitial Li donors and (2) the limited solubility of Li in ZnSe.

2. Methods

In this section we describe the theoretical methods used to calculate defect energies. Our calculations use density-functional theory in the local-density approximation (LDA) and norm-conserving pseudopotentials.[8, 9] Supercells (corresponding to 32 atoms of pure ZnSe) are used to represent the defects. These methods, combined with a plane-wave basis set, have been used very successfully in the past for studies of defects in Si and other semiconductors. The II-VI materials, however, present a problem for these methods: zinc contains a tightly-bound set of 3d electrons that cannot be easily represented by a plane-wave basis set. These d electrons may be treated as core states of the pseudopotential, but this results in a very poor description of ZnSe. For example, the lattice constant calculated with a "d-in-the-core" pseudopotential is 5.19 Å compared with the experimental lattice constant of 5.67 Å (Fig. 2).

In order to treat the d electrons as valence states, we use an all-new mixed-basis program, which combines localized functions with the plane waves. The program was carefully optimized, allowing defect calculations in large supercells. We use the Zn 3d pseudo-wave functions as the localized basis functions, and include all plane waves up to a kinetic energy cutoff of 9 Ry. (The eigenvalue problem was solved using an iterative diagonalization scheme.[10]) These methods provide a good description of the lattice constant, bulk modulus and TO phonon frequency of ZnSe (and other materials). Convergence tests were performed for supercell size, the basis set and other calculational parameters, assuring an overall accuracy of better that 0.5 eV.

Density-functional theory used with the LDA consistently predicts band-gaps that are too small compared with experiment. For ZnSe, our calculated band-gap is about 1 eV. This band-gap error will affect formation energies of defects that have occupied electron states in the band gap. The band-gap error will not affect our results for p-type material (where the Fermi level is near the valence band edge) since defect states in the gap will be empty and will not contribute to the total energy of the defect. In n-type material, we can derive a minimum value for the defect energies by assuming that the calculated electron levels in the gap are correctly positioned with respect to the valence band edge. The maximum value is found by assuming that the calculated levels are correctly positioned with respect to the conduction band edge. In our work we will use the worst-case assumptions about defect energies in n-type materials to guarantee that our conclusions are not affected of the LDA band-gap error.

3. Defect energies and concentrations

We use these methods to calculate the formation energies of all native point defects in ZnSe: two types of interstitials (Zn_i and Se_i), two types of vacancies (V_{Zn} and V_{Se}), and two types of antisites (Zn_{Se} and Se_{Zn}). The total energy of each defect was calculated in each relevant charge state, for a total of 29 defect calculations.

The formation energy of a defect may be lowered by the relaxation of nearby host atoms from their perfect crystal sites. We calculate relaxation energies explicitly for the dominant native defects in p-type ZnSe. For interstitial defects we relax the first and second nearest neighbor atoms and also the fourth nearest neighbors that are bonded directly to the first nearest neighbors. For substitutional defects we relax the first nearest neighbors. All of these

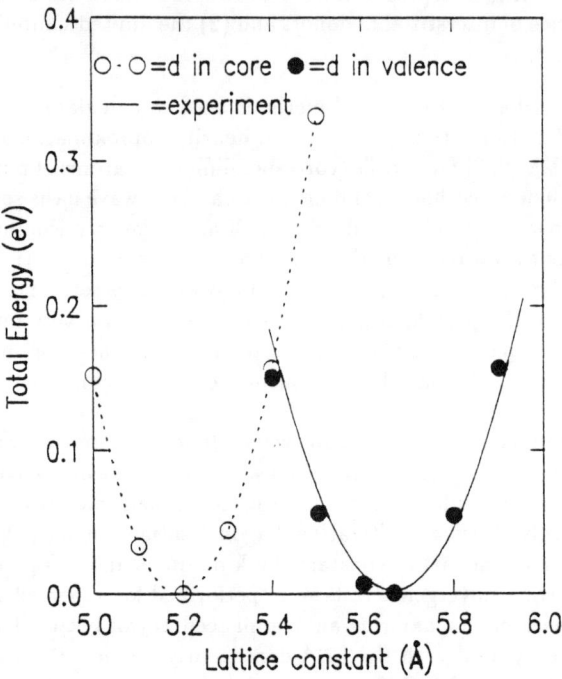

Figure 2: Total energy of pure ZnSe as a function of lattice constant calculated with the Zn 3d electrons treated as core states and as valence states. The experimental values (solid line) are based on the experimental lattice constant and bulk modulus. For comparison, the minimum energy of each curve is set to zero.

relaxations are small: the largest relaxation that we found was 0.2 Å with a relaxation energy of 0.6 eV. For the remaining defects, we assume a relaxation energy of 1 eV; our conclusions remain unchanged even if we allow a 2 eV relaxation energy.

Defect concentrations will also depend on the formation entropy of the defect. In our work we allow a range of 0–10 k_B for the defect formation entropies. By comparison, a recent accurate calculation[11, 12] for the Si self-interstitial found a formation entropy of 5-6 k_B for the ground state. The Si self-interstitial represents an extreme case in that the ground-state configuration has low symmetry, which accounts for half of the formation entropy. It is therefore highly unlikely that the entropies for native defects in ZnSe could be larger than 10 k_B.

In a compound semiconductor like ZnSe, individual native defect energies are not unique numbers: they are a function of the chemical potentials of the Zn and Se atoms, μ_{Zn} and μ_{Se}. The two chemical potentials are constrained by the condition that (in equilibrium) their sum must equal the total energy of a two-atom unit of perfect ZnSe ($\mu_{ZnSe} = \mu_{Zn} + \mu_{Se}$). (We use the total energy of a perfect ZnSe cell at T = 0 K for μ_{ZnSe}.) Given the Zn and Se chemical

	p-type		n-type	
Defect	Concentration (cm^{-3})	Defect	Concentration (cm^{-3})
Zn_i^{++} (T_{Se})	2.5×10^9	V_{Zn}^{--}	1.4×10^{13}
V_{Zn}^0	2.1×10^9	Zn_{Se}^-	5.2×10^{12}
Se_{Zn}^{++}	1.5×10^8	Zn_{Se}^0	1.3×10^{12}
Se_{Zn}^+	1.7×10^7	Zn_{Se}^-	3.5×10^{11}
V_{Zn}^-	8.7×10^6	V_{Se}^0	3.7×10^9
Zn_i^{++} (T_{Zn})	2.2×10^6	Zn_i^0 (T_{Zn})	5.1×10^8
V_{Zn}^{--}	1.2×10^6	Zn_{Se}^+	6.2×10^7
V_{Se}^{++}	8.6×10^5	V_{Se}^+	1.7×10^5

Table 1: Native defect concentrations in stoichiometric ZnSe. Concentrations of defects greater that 10^5 cm^{-3} are shown for ZnSe doped with 10^{18} cm^{-3} acceptors or donors. A formation entropy of 5 k_B per defect is assumed.

potentials, the formation energy of each native defect is well defined and can be derived from a supercell calculation as follows. The total energy of a supercell for the ith defect containing N Zn atoms and M Se atoms, $E(i)$, is calculated. The defect formation energy, $E_{form}(i)$, is then

$$E_{form}(i) = E(i) - N\mu_{Zn} - M\mu_{Se} = E(i) - (N-M)\mu_1 - (N+M)\mu_2 = \epsilon(i) - n(i)\mu_1 \quad (1)$$

where $\mu_1 = (\mu_{Zn} - \mu_{Se})/2$, $\mu_2 = (\mu_{Zn} + \mu_{Se})/2$ (a constant), $n(i) = N - M$, and $\epsilon(i) = E(i) - (N+M)\mu_2$. $n(i)$ is the number of extra Zn atoms that must be added to form the defect (+1 for V_{Se}, -2 for Se_{Zn}, etc.) and is independent of the size of the supercell. Using this prescription all of the defect formation energies, and hence their concentrations, $C(i)$, are unique functions of μ_1. The concentrations, in turn, determine the stoichiometry. Here, we fix the stoichiometry first and then determine $C(i)$. To do this we write $C(i)$ in terms of the total energies and entropies, $S(i)$, of formation:

$$C(i) = e^{S(i)/k_B} e^{-[\epsilon(i) - n(i)\mu_1]/k_B T} = e^{[S(i)/k_B - \epsilon(i)/k_B T]} y^{n(i)} = a(i) y^{n(i)} \quad (2)$$

where $y = \exp(\mu_1/k_B T)$. The stoichiometry parameter is

$$X = -\frac{1}{2}\sum_i n(i)C(i) = -\frac{1}{2}\sum_i n(i)a(i)y^{n(i)} \quad (3)$$

($X = 0$ for perfect stoichiometry, and $X > 0$ for Se-rich). To find defect concentrations as a function of stoichiometry, one simply chooses values of X and the temperature, and solves for y. (The problem is essentially finding a root of a polynomial, which can be done quickly and easily using standard algorithms.)

4. Results

Figure 3 shows the concentrations of minority carriers produced by all native defects for n-type and p-type stoichiometric ZnSe. Individual defect concentrations are presented in Table 1. The results shown are for material with 10^{18} cm^{-3} dopants. The dominant native defects are Zn_i^{++}, V_{Zn}^0, and Se_{Zn}^{++} for p-type, and V_{Zn}^{--} and Zn_{Se}^- for n-type. At MBE growth temperatures (T=600 K) the concentration of minority carriers produced is less than 10^{12} cm^{-3}. For material

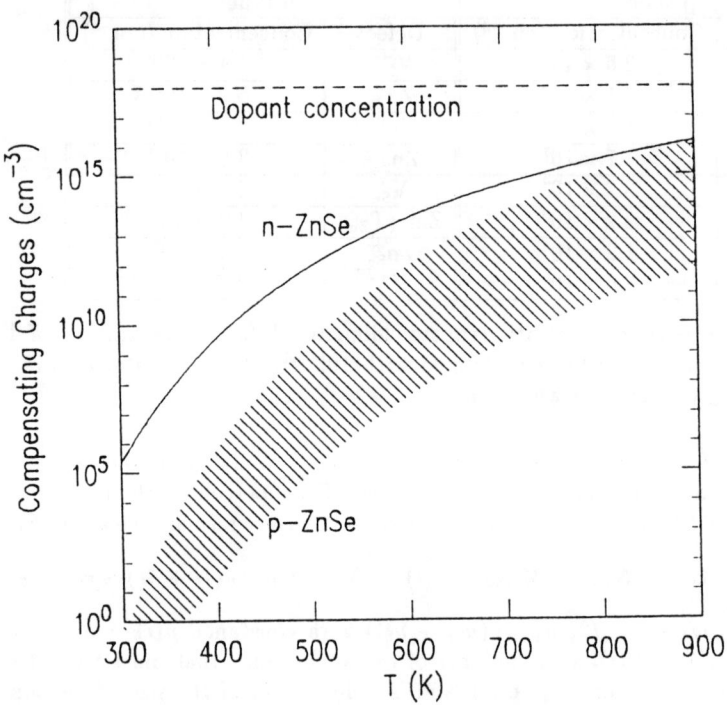

Figure 3: Concentration of minority carriers produced by all native point defects in stoichiometric p-type and n-type ZnSe. (The range of values shown for p-type ZnSe corresponds to defect formation entropies in the range of 0–10 k_B.) For n-type ZnSe, the minimum defect concentrations are shown (derived from the maximum formation energies due to the LDA band-gap uncertainty). This shows that the concentration of native defects in n-type ZnSe is at least as great as it is in p-type ZnSe.

grown at higher temperatures, excess native defects will recombine during cooling, unless the sample is rapidly quenched. (Native defects in ZnSe remain mobile even at temperatures of 400 K,[13] so that kinetic barriers do not prevent the attainment of thermal equilibrium.)

To further support our conclusions, we have derived native defect concentrations for diamond from the first-principles defect energies of Bernholc et al..[14] The doping level is again 10^{18} cm^{-3}. At a CVD-growth temperature of 1100 K, the number of holes produced in n-type diamond by native defects is at most 2×10^{13} cm^{-3} (Fig. 4). Clearly, the concentrations of native defects in both stoichiometric ZnSe and diamond are far too low to produce significant compensation. Furthermore, the native defects are no more likely to compensate p-type ZnSe than the other way around.

Our conclusion that the concentrations of native defects in stoichiometric ZnSe are very low does not mean that native defect compensation in ZnSe never occurs. If the sample is grown

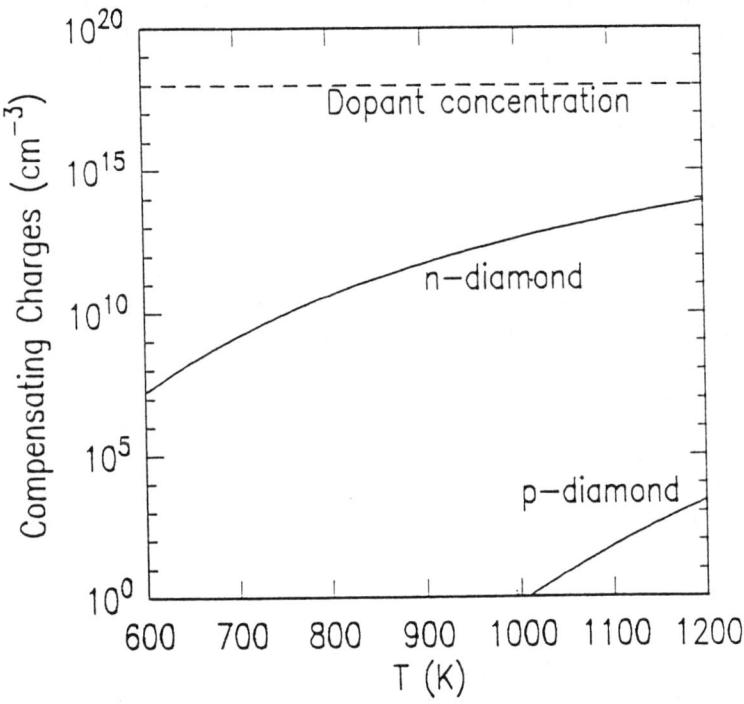

Figure 4: Concentration of minority carriers produced by all native point defects in p-type and n-type diamond. For n-type diamond, the maximum defect concentrations are shown (derived from the minimum formation energies due to the LDA band-gap uncertainty). This shows that the concentration of native defects in n-type diamond is too low to compensate doping.

with even a slight deviation from perfect stoichiometry the concentration of native defects will necessarily be very large, even at $T = 0$ K. (We refer here only to deviations from stoichiometry caused by native defects. In actual crystals, precipitates, dopants and higher dimensional defects may also contribute to the stoichiometry.) Because the density of atomic sites in ZnSe is 4×10^{22} cm^{-3} a deviation from stoichiometry as small as 10^{-4} implies a defect concentration of about 10^{18} cm^{-3}. We find that the native defects that accommodate deviations from stoichiometry are always those that compensate the majority carriers. For p-type ZnSe, the dominant defect is Zn_i in Zn-rich material, and Se_{Zn} in Se-rich material; we find that both are double donors. For n-type ZnSe the dominant (acceptor) defects are Zn_{Se} and V_{Zn} for Zn and Se rich materials, respectively. Similar results were found by Jansen and Sankey.[7] This defect structure is much richer than that used in many previous analyses of native defects in II-VI semiconductors.[15] The difficulty in producing p-type ZnSe cannot be explained by deviations from stoichiometry because any deviation that compensates p-type doping would compensate n-type doping equally well.

The deviations from stoichiometry that we are discussing are too small to measure experimentally, which precludes a *direct* confirmation of our predictions. There is, however, indirect evidence to verify one of our predictions, namely that the zinc vacancy is the dominant native defect in n-type Se-rich ZnSe. As-grown bulk ZnSe samples are highly compensated, and must be annealed in a Zn-rich atmosphere to be made well-conducting. One known cause of this compensation is large numbers of "self-activated" (acceptor) centers, which are donor-V_{Zn} pairs.[16] This shows that zinc vacancies are a prominent defect in as-grown n-type ZnSe. Furthermore, analysis of the Zn-Se phase diagram suggests that ZnSe grown under equilibrium conditions from a melt is Se-rich. Thus, our results for Se-rich n-type ZnSe provide a natural explanation of the occurrence of self-activated centers in ZnSe.

Having settled the native defect compensation issue quantitatively, we now reexamine the notion that native defect compensation increases with the width of the band-gap. Let us restate the standard argument for this trend: for p-type material, imagine a prototypal compensating native donor defect that, when neutral, introduces one electron into a level in the gap (E_L); the formation energy for this defect, E^0, is assumed not to depend on the width of the band gap. The energy gained by transferring the electron from the level in the gap to the Fermi level ($E_L - E_F$, where E_F is the Fermi level) should, in contrast, increase with the width of the gap. Thus, the net energy needed to form compensating defects, $E^0 - (E_L - E_F)$, should decrease as the band gap increases. The flaw in this argument is that it assumes that the level in the gap E_L and E^0 are independent of one another. Actually, the level in the gap is *defined* by $E_L = E^0 - E^+$,[17] where $E^+ + E_F$ is the (Fermi-level dependent) energy of formation of the positive charge state defect (Fig. 5). Using this definition, we find that the net energy required to create a compensating defect is $E^0 - (E_L - E_F) = E^+ + E_F$, independent of the energy of formation of the neutral defect. We see that native-defect compensation will increase with the width of the band gap if and only if $E^+ + E_F$ *decreases* with increasing band gap. The existence of such a trend has not been convincingly established.

5. Dopants in ZnSe

Having eliminated native defects as a generic source of compensation in wide band-gap materials, it is fruitful to identify problems associated with specific dopants. At present, we are examining the technologically important cases of Li_{Zn}, Na_{Zn} and N_{Se} acceptors in ZnSe.[18] Our results for Li doping indicate that two factors inhibit doping. 1) Interstitial Li is a donor. For lightly Li-doped samples, where the Fermi level is far from the valence band edge, substitutional Li_{Zn} has a lower formation energy than Li_i, so that all of the Li will be substitutional acceptors. As the Li content increases the Fermi level moves down, causing the formation energy of the Li_i donor to decrease, and that of the Li_{Zn} acceptor to increase. For heavily Li-doped samples, the Fermi level will be pinned at the value for which the formation energies of the interstitial and substitutional defects are equal. The pinning Fermi level position depends on the zinc chemical potential μ_{Zn}: for low μ_{Zn} (Zn deficient), Li_{Zn} is favored over Li_i, and the pinning value of the Fermi level will be lower. Thus compensation by Li_i can be avoided by growing the crystal in a Zn-deficient environment. 2) A second factor that inhibits Li doping is that the solubility of Li in ZnSe is limited. We calculate the solubility limit for Li in ZnSe to be a few times 10^{18} cm^{-3} at a growth temperature of 600 K. These results explain the experimentally observed saturation of the hole concentration with increasing Li content, and the limited total Li concentration in Li-doped ZnSe.[19, 20]

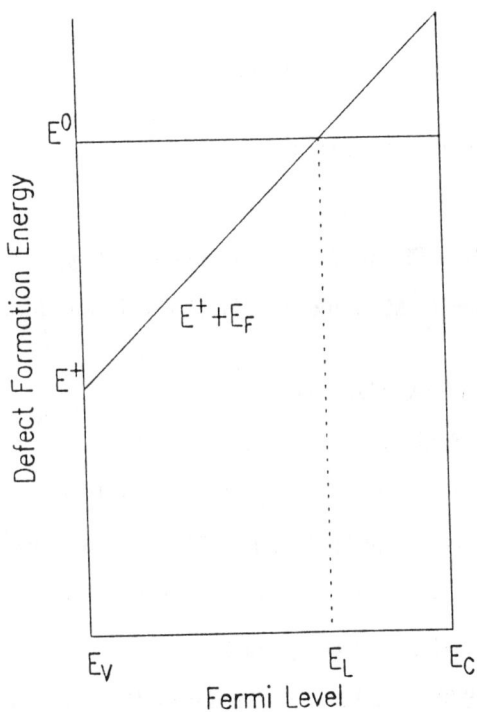

Figure 5: Formation energy of the neutral (E^0) and positively charged (E^+) states of a donor defect, as a function of the Fermi level (E_F). The level in the gap, E_L, is defined as the value of E_F for which $E^0 = E^+$.

6. Summary

In conclusion, we have shown that native defects alone *cannot* be responsible for difficulties in doping the wide band-gap semiconductors ZnSe and diamond. Native defect concentrations in MBE-grown stoichiometric ZnSe are too low to compensate. Deviations from stoichiometry in ZnSe *can* produce large numbers of native defects which, however, compensate n-type as well as p-type material. Therefore native defects produced by deviations from stoichiometry cannot explain why it is much harder to dope ZnSe n-type than p-type. Having eliminated native defects as the source of doping problems in ZnSe, we are examining specific dopant systems. Results indicate that doping problems for Li-doped ZnSe are caused by 1) competition between substitutional and interstitial Li and 2) limited solubility of Li in ZnSe.

We are very grateful to P. Blöchl for many fruitful suggestions, and for making his unpublished work available to us. We are indebted to D. Vanderbilt for his iterative diagonalization program. We acknowledge helpful conversations with R. Bhargava, J.M. DePuydt, T. Marshall, J. Tersoff, and G.D. Watkins. D.B. Laks acknowledges support from an IBM Graduate Fellowship. This work was supported in part by NSF grant ECS-89-21159 and ONR contract

N00014-84-0396.

References

[1] Y. S. Park and B. K. Shin. *Topics in Applied Physics*, volume 17, page 133. Springer, 1977.

[2] R. Bhargava. *J. Crystal Growth*, 59:15, 1982.

[3] G. F. Neumark. *Phys. Rev. Lett.*, 62:1800, 1989.

[4] H. Cheng, J.M. Depuydt, J.E. Potts, and M.A. Haase. *J. Crystal Growth*, 95:512, 1989.

[5] R. M. Park, M. B. Troffer, C. M. Rouleau, J. M. DePuydt, and M. A. Haase. *Appl. Phys. Lett.*, 57:2127, 1990.

[6] G. Mandel. *Phys. Rev.*, 134:A1073, 1964.

[7] R. W. Jansen and O. F. Sankey. *Phys. Rev. B*, 39:3192, 1989.

[8] J. Ihm, A. Zunger, and M.L. Cohen. *J. Phys. C*, 12:4409, 1979.

[9] G.B. Bachelet, D.R. Hamann, and M. Schluter. *Phys. Rev. B*, 26:4199, 1982.

[10] R. Natarajan and D. Vanderbilt. *J. Comput. Phys.*, 82:218, 1989.

[11] P.E. Blöchl and S.T. Pantelides. (To be published).

[12] P.E. Blöchl, D.B. Laks, S.T. Pantelides, E. Smargiassi, R. Car, W. Andreoni, and M. Parrinello. *Proceedings of Int. Conf. on Physics of Semiconductors*. Thessaloniki, 1990.

[13] G.D. Watkins. *Proceedings of International Conference on Science and Technology of Defect Control in Semiconductors*. Yokohama, 1989.

[14] J. Bernholc, A. Antonelli, T. M. Del Sol, Y. Bar-Yam, and S. T. Pantelides. *Phys. Rev. Lett.*, 61:2689, 1988.

[15] A. K. Ray and F. A. Kröger. *J. Electrochem. Soc.*, 125:1348, 1978.

[16] R. K. Watts. *Point Defects in Crystals*. Wiley, 1977.

[17] G. A. Baraff, E.O. Kane, and M. Schlüter. *Phys. Rev. B*, 21:5662, 1980.

[18] C.G. Van de Walle, D.B. Laks, G.F. Neumark, and S.T. Pantelides. (To be published).

[19] J.M. Depuydt, M.A. Haase, H. Cheng, and J.E. Potts. *Appl. Phys. Lett.*, 55:1103, 1989.

[20] T. Marshall and D.A. Cammack. *J. Appl. Phys.*, 69, 1991.

ODMR INVESTIGATIONS OF A - CENTRES IN CdTe

D. M. Hofmann[1,3], P. Omling[1], H. G. Grimmeiss[1], D. Sinerius[2], K. W. Benz[2], and B. K. Meyer[3]

[1]Solid State Department, University of Lund, BOX 118, S - 22100 Lund, Sweden
[2]Kristallographisches Institut, University of Freiburg, Hebelstr. 25, D - 7800 Feiburg, F. R. G.
[3]Physikdepartment E16, Technical University of Munich, James Franck Str.,
D - 8046 Garching, F.R.G.

ABSTRACT

The structure of A - centres in CdTe:Cl has been determined by optically detected magnetic resonance (ODMR) investigations in the luminescence band at 1.42 eV. The centres have trigonal symmetry and the g - factors are $g_{\parallel} = 2.2$ and $g_{\perp} = 0.4$ assuming an effective spin $S = 1/2$. In addition hyperfine interaction with the nearest Tellurium neighbors has been observed. The properties of the A - centres in CdTe are compared to the data of those defects in other II - VI compounds.

1. Introduction

Cation vacancy - donor complexes (A - centres) play an important role in the compensation behaviour of II - VI compounds. In ZnS and ZnSe the properties of those defects have been studied in great detail and it has been demonstrated that they are responsible for the " self - activated " luminescence bands [1,2]. However, there is only limited information about the luminescence and electronic properties of A - centers in the other II - VI compounds, especially in CdTe.
In ZnTe, being an exceptional case in the II - VI family in that sense that inspite of n - type doping only p - type conductivity results, A - centres have possibly been identified [3]. Their optical and structural properties show, however, striking differences compared to ZnS and ZnSe.
In CdTe which can be doped n- and p- type A - centres were claimed to be responsible for a luminescence band located at 1.42 eV, but sofar no experimental proof has been given [4]. Unfortunately the situation in CdTe is complicated by the fact that in this spectral range also donor - acceptor recombinations due to residual impurities Cu and Ag occur [5]. This has led to speculations that A - centers may not be present in the material and that the compensation mechanism in CdTe is governed by extrinsic contaminations [6].
In this study we have been able to identify A - centers in CdTe by ODMR experiments and to correlate the luminescence band. Their optical and magnetic properties show features of intermediate deep defects in this host. Our results support the A - center identification in Cl doped ZnTe.

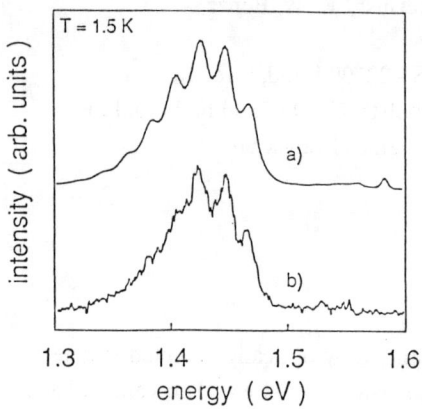

Fig. 1 :

a) Photoluminescence spectrum of the A - centres in CdTe:Cl

b) excitation spectrum of the A - center ODMR signals

2. Experimental Setup

Optical and magnetooptical experiments were performed at 1.7 K with the sample in contact with superfluid helium or at 4.2 K.

The ODMR spectrometer operates at 24 GHz and static magnetic fields up to 2.5 Tesla were available. As excitation light the 514 nm line of an Ar^+ laser was used, typical excitation powers were 200 mW in an unfocused beam. The emission was dispersed through a double monochromator and detected by a Ge detector (North Coast). For the measurements of the magnetic circular polarisation of the emisson (MCPE) a quartz stress modulator and a linear polarizer was set in the emission beam. The samples used were grown either by the Bridgman or the traveling heater technique. Dopants were added into the melt prior growth in concentrations up to 10^{19} cm^{-3}.

3. Results and Discussion

The luminescence spectrum of CdTe:Cl (Fig. 1 a) shows a band centered at 1.42 eV. The zero - phonon line (ZPL) is at 1.478 eV followed by 6 LO-phonon replicas which show an intensity distribution according to the Poisson distribution $I(n) = e^{-S} (S^n/n!)$ with a Huang - Rhy's factor of S = 2.2 ± 0.1. Almost no luminescence due to excitonic recombinations close to the band gap of CdTe at 1.6 eV is observed.

Cu gives also rise to emission in this spectral region (E_{ZPL} = 1.45 eV) and is an omnipresent impurity in CdTe, other defects and impurities causing luminescene at 1.42 eV have been reported in ref. [5,7,8]. The luminescence spectra of those defects or impurities show slightly shifted ZPL's and different Huang - Rhys factors compared to Fig. 1a. However, by the ODMR experiments described in the following it was possible to show that A - centres definitly have the above described photoluminescence in this spectral region.

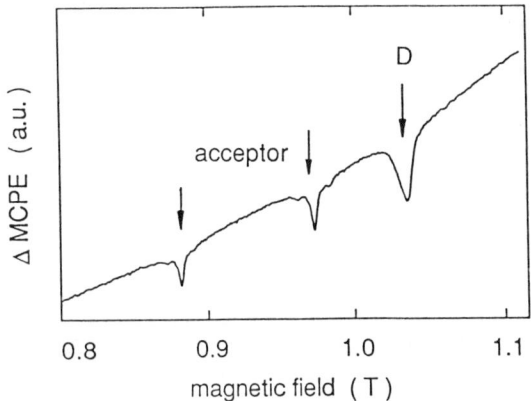

Fig. 2:

MCPE spectrum of CdTe:Cl showing the Cl donor resonance (D) and the resonances identified as originating in the A-center (acceptor)

The ODMR signals obtained in the luminescence of Fig. 1a are shown in Fig. 2. They are measured as decrease of the MCPE which increases with increasing magnetic field (increasing background in Fig.2).

The ODMR excitation spectrum presented in Fig. 1b shows that the resonances originate in the luminescence band of Fig. 1a.

The high field ODMR resonance in Fig.2 is isotropic and has a g-value of 1.69 in agreement with the value obtained for donor centers in electon-spin-resonance experiments [9].

The other ODMR signals in Fig. 2 are anisotropic resolving an acceptor with trigonal symmetry. The corresponding angular dependence is shown in Fig. 3 by full circles, rotating the crystal in a (110) plane. Assuming an effective spin $S = 1/2$ the angular dependence is fitted by the spin Hamiltonian

$$H = \mu_B \cdot B \cdot g_{eff} \cdot S$$

with $g_{eff} = (g_{\parallel,eff}^2 \cos^2\theta + g_{\perp,eff}^2 \sin^2\theta)^{1/2}$

and effective g-values $g_{\parallel,eff} = 2.2$ and $g_{\perp,eff} = 0.4$ (drawn lines in Fig. 3). θ denotes the angle between the magnetic field B and a trigonal <111> symmetry axis.

In contrast A-centres in ZnS and ZnSe have g-values close to the free electron g-value, reflecting the quenched orbital moment of these deep level defects [1].

A-centres act as single acceptors ($V_{Cd}^{--} - Cl_{Te}^{+}$)$^-$, the neutral charge state, when a hole is trapped, is paramagnetic. The level position in CdTe is approx. 120 meV above valence band which is "shallow" compared to ZnS and ZnSe (approx. 1 eV and 0.6 eV respectively) [10]. Shallow acceptors can be described as being formed by a $J = 3/2$ hole from the top of the valence band, which is split by the trigonal pertubation D into $J_z = \pm 1/2$ and $J_z = \pm 3/2$ states. The spin Hamilton operator can thus be written:

$$H = D[J_z^2 - 1/3 J(J+1)] + g_{\parallel}\mu_B B_z J_z + g_{\perp}\mu_B(B_x J_x + B_y J_y)$$

where $g_{\parallel}$ and $g_{\perp}$ are given by $g_{\parallel} = 1/3\, g_{\parallel,eff}$ and $g_{\perp} \cong (2D g_{\perp,eff})/h\nu$, with $h\nu$ the microwave energy [11,12].

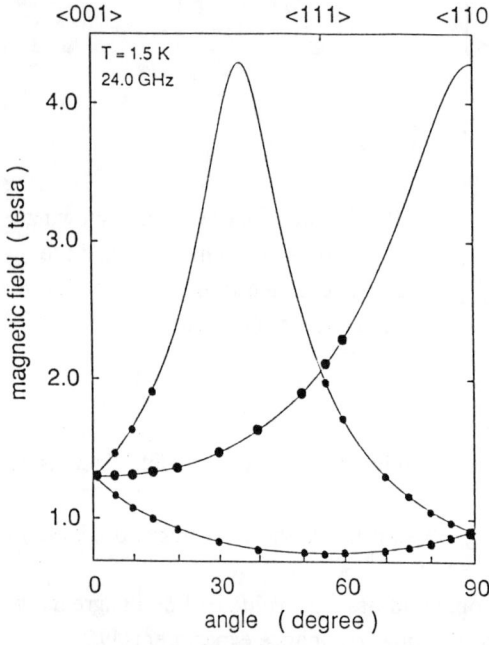

Fig. 3:

Angular dependence of the ODMR lines of the A - center in CdTe:Cl. The magnetic field is rotated in a {110} plane. The experimental data are plotted as filled circles. The calculated angular dependence (trigonal symmetry) is plotted as solid lines

With our measured value of $g_{||,eff}=2.2$ we obtain $g_{||}=0.73$ close to the g - values for shallow effective mass like acceptors in CdTe of g = 0.71 [13].
The splitting of the $J_z=\pm 3/2$ and $J_z=\pm 1/2$ states caused by the trigonal distortion can be estimated from $g_{||,eff}=0.4$ to be D ≃ 0.09 meV [12].
Each of the acceptor lines shown in Fig.2 is accompanied by 4 satelites of lower intensity (Fig. 4). The intensities of the two stronger and the two weaker lines are about 12%, and 1% of the central line respectively. The splittings are, within the experimental accuracy, constant when rotating the crystal with respect to the static magnetic field.
Considering the intensities and positions of the peaks the spectrum cannot explained in terms of a hyperfine interaction with a central nucleus being part of the defect. However, the set of 5 lines can be explained well by the hyperfine interacion with three equivalent Te ligands.
About 92% of the Te isotopes have nuclear spin I = 0 and 8 % have I= 1/2. These are ^{125}Te (I = 1/2, 7% abundance) and ^{123}Te (I = 1/2, 0.9% abundance), the difference in their magnetic moment $\mu=0.8\mu_n$ and $0.73\mu_n$ respectively is to small to be resolved in our experiments.
The nearest neighbor sites can be occupied either by an isotope with I = 0 or I = 1/2. Taking into account the statistical weights the defects with one I = 1/2 Te neighbor should account for 13%, the defects surrounded by two Te atoms with I = 1/2 should account for 0.5% and those with three Te I = 1/2 atoms to 0.001% of the intensity. The latter are difficult to observe, but the experimental data are otherwise in excellent agreement with such a model.
The observation of the Te ligand hyperfine interaction excludes that the defect is located on the Te sublattice. This together with the symmetry rules out any extrinsic impurity to be responsible for the defect, single acceptors such as the group I elements and, e.g. Cu and Ag, have cubic symmetry.

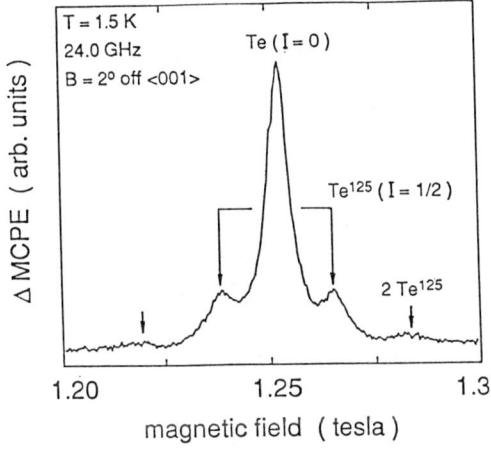

Fig. 4:

The Te ligand interactions of the A - center in CdTe:Cl. The different peaks correspond to 0, 1 and 2 of the 3 Te neighbors having an isotope with nuclear spin $I = 1/2$

We therefore conclude that a Cd - vacancy is at center of the defect surrounded by three Te ligands with a clorine donor atom located in one of the <111> nearest neighbor positions causing the trigonal distortion (Fig. 5). It compensates one of the two negative charges of the Cd - vacancy.

A hyperfine interaction with the clorine donor nucleus ($I = 3/2$) could not be resolved in our experiments. Anyway, it is expected to be small and buried in the linewidth of the Te - hyperfine lines because the hole trapped at the defect is repelled electrostatically by the positive Cl - donor. Experiments like optically detected electron nuclear double resonance (ODENDOR) could possibly resolve these interaction and are planed for the future.

Conclusions

The analysis shows that A - centers in Clorine doped CdTe although having the same atomistic structure as those in the widegap II - VI compounds ZnSe and ZnS their electronic, optical and magnetooptical properties differ remarkably. For A - centers in ZnTe:Cl results have been obtained which are comparable to this investigation [3]. In the Tellurium compounds A - centers behave as "intermediate" deep centres whereas in ZnSe and ZnS they are true deep level defects.

Acknowledgements

One of us (D. M. H.) thanks all collegues at the S. S. D. of Lund Univ. for their kind hospitality during his stay.
This work has been supported by the Deutsche Forschungsgemeinschaft under contract Ho1233/1-1 and Me898/3-1 and by the Swedisch Natural Sciences Research Council and the Swedisch Board for Technical Developments.

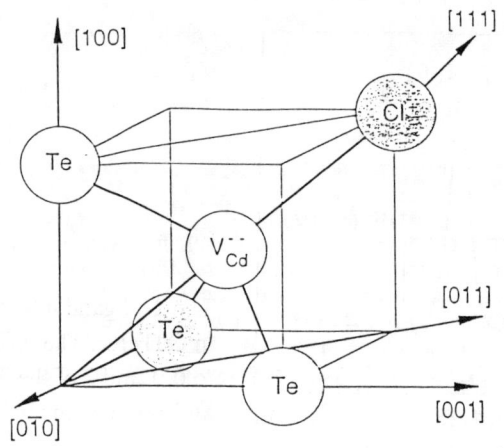

Fig. 5: Structure model of the A - centers in CdTe:Cl

References

[1] J. Schneider, II - VI Semiconducting Compounds, ed. D. G. Thomas (W. A. Benjamin, New York, 1967) p.160
[2] P.H. Kasai and Y. Otomo, J. Chem. Phys. **37**, 1263 (1962)
[3] J. Bittebierre and R. T. Cox, Phys. Rev. B **34**, 2360 (1986)
[4] N. V. Agrinskaya, E. N. Arkad'eva and O. A. Matveev, Sov. Phys. Semicond. **5**, 762 (1971)
[5] J. P. Chamonal, E. Molva, and J. L. Pautrat, Solid State Comm. **43**, 801 (1982)
[6] J.L. Pautrat, J.M. Francou, N. Magmea, E. Molva, and K. Saminadayar, J. of Cryst. Growth **72**, 194 (1984)
[7] C. B. Norris and C. E. Barnes, Rev. Phys. Appl. **12**, 219 (1977)
[8] C. Onodera and T. Taguchi, J. Cryst. Growth **101**, 502 (1990)
[9] K. Saminadayar, D. Galland, E. Molva, Solid State Comm. **49**, 627 (1984)
[10] W. Stadler, B. K. Meyer, D. M. Hofmann, D. Sinerius and K. W. Benz, Material Research Society Symp. Proc. **209**, 445 (1990)
[11] G. E. Pake and T. L. Estle, The Physical Principles of Electron Paramagnetic Resonance (Reading, Massachusetts: Benjamin 1973) Ch. 5
[12] J. L. Patel, J. E. Nicholls, and J. J. Davies, J. Phys. C **14**, 1339 (1981)
[13] E. Molva, J. L. Pautrat, K. Saminadayar, G. Milchberg, and N. Magnea, Phys. Rev. B **30**, 3344 (1984)

PICOSECOND ENERGY TRANSFER BETWEEN EXCITONS AND DEFECTS IN II-IV SEMICONDUCTORS

R.Heitz, C.Fricke, A.Hoffmann and I.Broser
Institut für Festkörperphysik, Technische Universität Berlin, Germany

ABSTRACT

The luminescences of bound excitons in various II-IV semiconductors are studied by means of time resolved spectroscopy at liquid helium temperatures. A general trend of increasing lifetimes with increasing binding energies is in reasonable agreement with the theory of Rashba and Gurgenishvili indicating a predominant radiative decay. It is shown that the investigation of the dynamics of weakly bound excitons provides the possibility to determine parameters of the free exciton. An effective exciton mass of $1.06 m_0$ for ZnO and a free exciton oscillator strength of 0.0014 for ZnS are determined. The limits of the model in case of deeply bound excitons as well as the nonradiative decay channels are discussed. The observed luminescence risetimes due to the formation of bound exciton complexes after generation of free excitons are investigated.

Introduction

In II-IV compounds nearby the free exciton resonances a lot of sharp lines dominate the emission spectra[1]. These lines are due to the radiative decay of excitons weakly bound to different defects always present in these compounds. Due to the high formation probability of such complexes they are the dominant decay channel for intrinsic excitations (e.g. free excitons) in these semiconductors. Even low impurity concentrations suppress the free exciton luminescence due to a drastic shortening of their lifetimes[2]. The giant oscillator strengths connected with bound excitons correspond to extremly short radiative lifetimes, which strongly reactivated the interest in recent years[3,4,5], since the bound excitons become good candidates for fast nonlinear optical devices, suggested for optical or opto-electronic data processing.

In this paper we report on time resolved spectra of free and bound excitons in ZnO, ZnS, CdS and ZnSe at low temperatures. Low excitation densities are employed, since otherwise biexcitons and other nonlinearities superimpose the dynamical behaviour of the bound excitons. The obtained data are compared to recent results for ZnO[5], ZnSe[3,6] and CdS[7] and discussed in the framework of the theory developed by Rashba and Gurgenishvili[8,9] in order to determine the parameters controlling the relevant decay mechanisms. Thereby, it is demonstrated that the investigation of the decay dynamics of weakly bound excitons gives a valuable feedback about basic parameters of free excitons, e.g. their oscillator strength or effective mass. The results are compared to those of excitons deeply bound to Ni-centres and to calometric absorption results.

Experimental

The experimental arrangement consists of an actively modelocked Nd-YAG laser which is either frequency-tripled to pump synchronously a dye laser operating with stilbene 3 for band-band excitation of ZnSe and CdS, or frequency-doubled to pump R6G-dye, subsequently again frequency-doubled to excite ZnS and ZnO above the bandgap. A cavity dumper is used to lower the pulse repetition rate to 3.8MHz and to increase the peak power. The system delivers pulses of about 3ps duration and an average power of up to 300mW. During the luminescence experiments average excitation densities of $0.5 mW/mm^2$ or less are used. The transient luminescence is detected through a 0.75m monochromator by means of time-correlated single photon counting using a micro-channel plate photomultiplier tube (Hamamatsu R2809U). The instrumental time resolution is about 50ps, allowing the determination of lifetimes down to 15ps by deconvolution technics. The results of the subsequent convolutions are given as full lines in fig.1-4. The origin of the luminescence transients is checked by means of time delayed luminescence spectroscopy.

In the analysis of the observed luminescence transients I(t) the formation time τ_r and the decaytime τ_d of bound exciton complexes have to be taken into acount:

$$I(t) \sim (\tau_r - \tau_d)^{-1} [\exp(-t/\tau_r) - \exp(-t/\tau_d)]$$

The rising part of the luminescence intensity depends on the smaller one between τ_r and τ_d and the decaying part on the larger one. To determine the decaytime we performed pump-and-probe measurements under resonant excitation conditions. Average pump powers of about 20mW and probe powers of less than 0.5mW are used.

Results

The time dependence of the bound exciton complexes in high-quality ZnO rods have been investigated. The insert of fig.1 shows the time integrated luminescence spectrum of a not intentionally doped crystal recorded under the same conditions as used for the time resolved measurements. Emission lines due to different localized bound exciton complexes occur. The chemical nature of the participating defects is still not clear but lines near the resonances I_3 at 3.3662eV and I_4 at 3.3629eV have been assigned to (D^0,X)-complexes[10] and the lines I_6 (3.3606eV), I_7 (3.3600eV) and I_9 (3.3567eV) are identified as neutral acceptor bound excitons[11]. The transients of the different bound exciton emission lines, fig.1, reveal a trend of increasing lifetime with increasing binding energy. Only the I_4-complex shows a pecularity, its lifetime is longer than those of the stronger localized I_6- and I_7-complexes. The obtained decaytimes are summarized in tab.I and in qualitative agreement with results reported for I_7 and I_{10}[5]. The observed risetimes of about 60ps to 80ps in the bound exciton luminescences are correlated to the dynamics of the free A-exciton. Under the employed excitation conditions the A-exciton decays almost exponential with a time constant of (60 ± 10)ps during the first 200ps and afterwards becomes slower, probably due to a saturation of the bound exciton recombination channels. This indicates the formation of bound exciton complexes as dominant process lowering the free exciton density in the investigated sample. It has to be noted that for the lines I_4 to I_9 at longer times an additional weak (about 1%) component with a time constant of 1.9ns is observed. Time delayed luminescence spectra show that this slow component is connected with the respective bound exciton emissions, too, indicating a unique slow bound exciton formation channel whose origin is still unclear. In weakly doped crystals the bound exciton emission lines become broader, but at least the decaytimes of the (A^0,X) complexes remains unaltered. Additionally, the free exciton lifetime as well as the luminescence risetimes decrease simultaneously.

The transients of different exciton emission lines of a CdS crystal containing 0.1ppm Ni are shown in fig.2. Not intentionally as well as In- or Ni-doped samples show again that the free exciton decaytime and the bound exciton risetimes decrease simultaneously down to 25ps in doped crystals whereas the decay of the bound exciton complexes remains unaltered. The (A^0,X)-complex (I_1, 2.5356eV) has a lifetime of 650 ± 20ps and the (D^0,X)-complex (I_2, 2.5467eV) has a lifetime of 100 ± 20ps, see tab.I. This values are somewhat shorter than results reported by Henry and Nassau[7], possibly indicating concentration quenching in our crystals or too high excitation densities. We have to exclude both possibilities: the

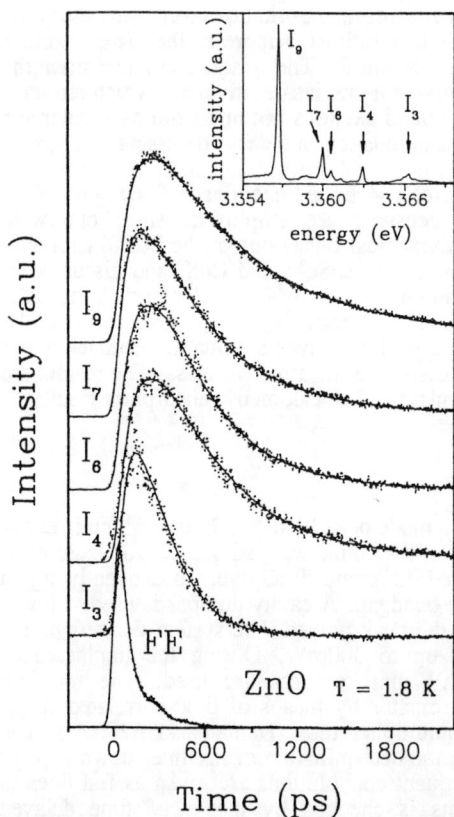

fig.1: *Luminescence transients of exciton emissions in ZnO excited at 4.12eV. The insert gives the time integrated luminescence spectrum.*

experimental results show no resolveable influence of the different used dopand's concentrations ($\leq 10^{16} cm^{-3}$) on the bound exciton lifetimes and luminescence spectra recorded under the employed experimental conditions show no biexciton luminescence, the so called M-band which occur near I_2 and is slower anyhow[12]. In order to investigate the bound exciton dynamics under resonant excitation and to exclude effects of the free exciton density we performed pump-and-probe measurements. Fig.3 shows the results obtained for the I_2-absorption of a thin CdS sample. The insert gives the absorption spectrum ($E \perp c$) with the I_2-resonance at 2.5467eV as observed with the pump beam. No influence of the excitation density is resolved but the much more sensitive pump-and-probe technics reveal small absorption changes induced by the pump-beam. The absorption of the pump-pulse leads to a saturation of the I_2-absorption, whereby the temporal increase of the saturation follows the integral of the pump-pulse. Subsequently, the saturation decays with 110 ± 20ps corresponding well to the 100ps determined in time resolved

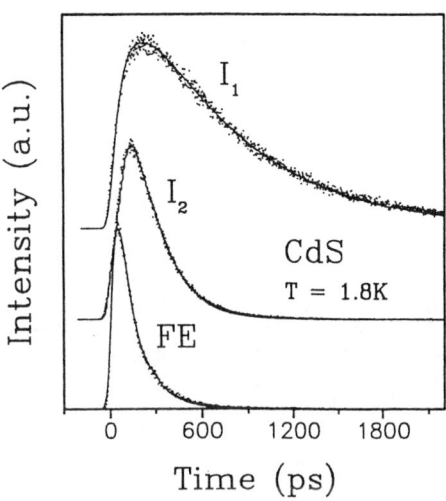

fig.2: Luminescence transients of exciton emissions in CdS excited at 2.78eV.

luminescence experiments. (The decay time of the (A^0,X)-complex has been confirmed, too.) During the temporal coincidence of pump- and probe-pulses fast changes of the pump-and-probe signal are observed, see fig.3. The amplitude ratios of the induced absorption and transmission as well as of the I_2-saturation are sensitive to the experimental conditions indicating three almost independent processes. The induced absorption is connected with the pulse duration wheras the induced fast transmission is connected with the coherence time of the pulses. Therefore, we propose that the fast absorptive change is due to two-photon absorption not connected with the (D^0,X)-complex (it is also seen besides the resonance) and the fast transmittive change to be due to the coherence peak often observed in pump-and-probe signals.

Recently, the I_3- and the I_5-emission line sets occuring in high quality ZnS crystals have been attributed to two different (A^0,X)-complexes[13]. The insert in fig.4 gives the time integrated luminescence spectrum of a typical sample as observed under time resolved conditions. The respective threefold finestructure of the two lines is not resolved due to the experimental resolution. The A-exciton at 3.8000eV is clearly resolved in time delayed spectra. The luminescence transients in fig.4 show exponential decays with 285ps and 150ps for the I_3- and the I_5-complexes, respectively. The free exciton decays almost exponential with (15 ± 10)ps over one decade followed by a weak second component with 150ps corresponding to the I_5-lifetime. Obviously, the impurity correlated processes dominate the lowering of the free exciton concentration but saturate at longer times. Possibly, the reported 'hot'-lines[13] indicate a

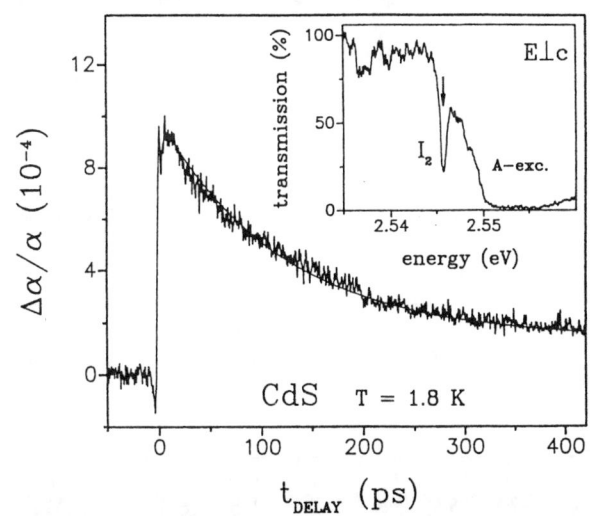

fig.3: Pump-and-probe signal of the I_2 absorption in CdS. The absorption is shown in the insert (E_{pump}, $E_{probe} \perp c$).

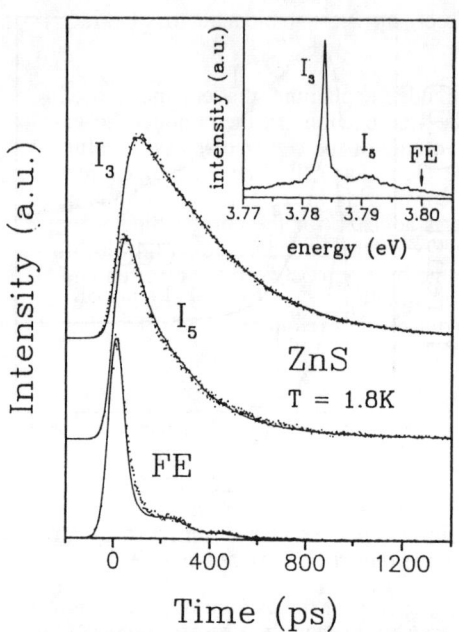

fig.4: Luminescence transients of exciton emissions in ZnS excited at 4.12eV. The insert gives the time integrated luminescence spectrum.

slow thermalization in the ground states of the (A^0,X)-complexes compared to their lifetimes.

On weakly Na- and Li-doped ZnSe samples the free exciton (2.8024eV), the donor bound excitons (I_2) (2.798eV) and three different acceptor bound excitons (I_1) at 2.7931eV (Na), 2.7923eV (Li) and 2.7829eV (Cu) are observed. The decaytimes are in good agreement with those reported[3,6] and are summarized in tab.I. It is clearly resolved that with increasing binding energy the decay becomes slower. The lifetimes of the (A^0,X)-complexes are independent of the dopand's-concentration within the experimental error of about 20ps, but the free exciton lifetime varys from 15ps to 120ps in correlation with the risetimes of the bound exciton complexes. The risetime of the Cu-related (A^0,X)-complex partly exceeds the free exciton lifetime showing efficient energy transfer between weakly localized bound excitons and the stronger localized Cu-related (A^0,X)-complex.

Discussion

The experimental results show that in the investigated materials even for low impurity concentrations the lifetimes of the free A-exciton are strongly reduced due to impurity correlated recombination channels. Thereby, the formation of bound exciton complexes plays the dominant role. Free exciton lifetimes down to a few ten ps are observed. The subsequent decay of the bound exciton complexes is ultrafast, decaytimes down to some ten ps occur in the widegap II-IV semiconductors, tab.I. The bound exciton decaytimes are constant in the investigated doping range ($\leq 10^{16}$cm^{-3}). Recently, a decrease of the lifetime of the In-related (D^0,X)-complex for In-concentrations around 10^{17}cm^{-3} has been reported[14]. In ZnSe an energy transfer between the most delocalized shallow (D^0,X)-complexes and the deep Cu-related acceptor complex is observed.

Two possible recombination processes control the dynamical behaviour of excitons bound to neutral donors or acceptors: the radiative decay and the nonradiative Auger-recombination. The radiative transition probability decreases with increasing localization, whereas the Auger-

tab.I: Experimental (τ_d) and calculated (τ_c) lifetimes and binding energies E_B of bound exciton complexes in ZnO, ZnS, CdS and ZnSe. For the calculation the theory of R&G[9] is used.

	ZnO				ZnS			CdS				ZnSe		
	τ_d (ps)	E_B (meV)	τ_c (ps)		τ_d (ps)	E_B (meV)		τ_d (ps)	E_B (ps)	τ_c (meV)		τ_d (ps)	E_B (meV)	τ_c (ps)
I_3	150	9.7	146				I_2	100	6.1	149	I_2	40	4.5	139
I_4	440	13.0	243											
I_6	350	15.3	309	I_5	150	8.3	I_1	650	17.2	700	I_1^P	100	5.0	164
I_7	420	15.9	326	I_3	285	16.3					I_1^{Na}	280	9.4	422
I_9	600	19.2	433								I_1^{Li}	350	10.2	473
											I_1^{Cu}	1020	19.6	1270

recombination rate increases with increasing localization. As one can see from tab.I our results clearly indicate radiative decay. Obviously, in the direct gap II-IV semiconductors the radiative decay channel dominates.

Rashba and Gurgenishvili[8,9] (R&G) introduced a model explaining the enhanced oscillator strengths of bound excitons compared to those of the free exciton. In their model the gain in oscillator strength is proportional to the ratio of the volume of the center of mass wavefunction of the bound exciton and the volume of a primitive elementar cell. The wavefunction of the bound exciton is derived from effective mass theory considering the interaction with the neutral impurity as delta-function potential whose amplitude is adjusted for the correct binding energy of the complex. Thereby, the giant oscillator strengths of the bound excitons are due to the breakdown of k-conservation and are proportional to the overlapp between the electron and hole wavefunctions. A decrease of overlap with increasing binding energy is caused by the short range binding potential which attracts one charge carrier of the exciton but repels the other one. The lifetimes τ_d are estimated from the following formula:

$$\tau_d = (4.5\ \lambda^2 / (n\ f_{ex}))\ (V / (8\ \pi\ a_{BE}^3)) \qquad (1)$$

where λ is the emitted light wavelength in cm, n is the refractive index, f_{ex} is the oscillator strength of the free exciton and V is the volume of the primitive cell. a_{BE} is the Bohr radius of the bound exciton in the potential of the neutral impurity given by

$$a_{BE} = h / (2\ m_{ex}\ E_B)^{1/2} \qquad (2)$$

m_{ex} is the effective exciton mass and E_B is the binding energy of the exciton-impurity complex. For small binding energies E_B the radiative lifetimes are shorter than the free exciton one.

Taking equation 1 and 2 the lifetimes of the bound exciton complexes can be calculated using known constants of the host material and the observed binding energies. Tab.I gives the calculated lifetimes τ_c together with the experimental values τ_d. The used parameters can be found in[5,3,7], whereby for CdS the effective mass has been replaced[15] by $0.9 m_0$. In general, the experimental and the theoretical lifetimes are in good agreement indicating dominating radiative decay of the bound exciton complexes. Remarkable is the excellent agreement obtained for CdS confirming our experimental results in contradiction to earlier ones[7]. Recently, the nonradiative recombination of excitons in CdS has been investigated by means of calometric absorption spectroscopy[16] (CAS). The results demonstrate nonradiative decay channels for the I_1 and the I_2 lines. Unfortunaly, the measurements have not been quantitative, no quantum efficiencies have been calculated. But if one keep in mind that CAS is extremly sensitive and that the I_1+TA absorption leads to a stronger heating than the I_1 absorption (see fig.4 in[16]), the quantum efficiency of the bound excitons in CdS should be quite close to 1.

The investigation of the dynamics of weakly bound excitons can be used to check or even to determine the parameters used by R&G if they are uncertain or even unknown. In fig.5 the lifetimes of the bound exciton complexes in ZnO are presented as function of the binding energy E_B: $\tau \sim (E_B^{3/2})$. As predicted by the theory of R&G for radiative processes the decaytimes are well proportional to $E_B^{3/2}$ prooving its applicability. But, the absolut values predicted are to small, probably due to an uncertain effective exciton mass. From the experimental lifetimes an effective exciton mass of $1.06 m_0$ is determined instead of the

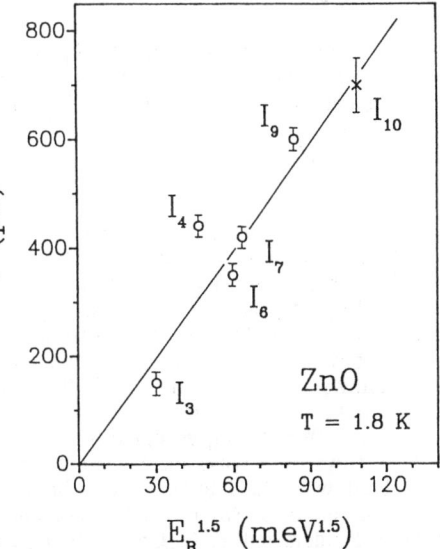

fig.5: Lifetimes τ_d of bound exciton complexes in ZnO in dependence of $E_B^{3/2}$. The data point I_{10} is from ref.[5].

$0.87m_0$ used in the calculation. In ZnS the oscilator strength of the free exciton is still unknown. Using $V=0.0395nm^3$, n~3 and[17] $m_{ex}=0.71m_0$ we obtain $f_{ex}=0.0014$.

Although the theory of R&G gives the general tendency correct, are the deviations sometimes larger than the experimental error, especially for the I_4-line, see fig.5. Despite the fact that the theory of R&G is estonishing powerfull it is simple and neglects a lot of contributions, e.g. correlation effects. Sanders and Chang[18] tried to overcome these shortcomings and also pointed out that for excitons bound to neutral donors the delta-function potential used by R&G should be insufficient. This can be the reason for the deviations observed for the I_4-emission confirming the interpretation as (D^0,X)-complex. Nevertheless, the R&G theory gives good results as far as the exciton binding energy is larger than those of the bound exciton complex. This situation is clearly given in ZnO, ZnS and CdS but not for ZnTe. In ZnTe the Auger-recombination is dominating for larger binding energies[19]. The situauation is completly changed for deep bound excitons. The exciton bound to an isoelectronic Ni^{2+}-centre in CdS has a binding energy of 362meV much larger than the exciton binding energy. In contrast to the calculated 92ns the complex relaxes within a few picoseconds but exclusively nonradiative[20]. The Ni^{2+}-centre with its electronic d^8-configuration provides extreme efficient nonradiative relaxation channels for the bound exciton due to its multilevel structure.

Summary

Time resolved measurements of not intentionally and weakly doped ZnO, CdS, ZnS and ZnSe crystals are presented. A strong decrease of the free exciton lifetimes down to a few ten ps in connection with an increase of the dopand's concentration is observed. Thereby, a saturation of the impurity correlated recombination channels occur after a delay which depends on the excitation density. Lifetimes of excitons weakly bound to neutral donors or acceptors in the range from 40ps to 1ns are observed in these materials, which are independend of the dopand's concentration. Especially, the rich bound exciton spectrum of ZnO allows a detailed comparison with model calculations revealing a reasonable agreement for radiative decay. We suppose that detailed time resolved investigations in direct gap II-VI semiconductors gives a good possibility to determine or check the effective mass or the oscillator strength of the free exciton. In the investigated semiconductors the radiative decay dominates the Auger-recombination.

References

[1] D.G. Thomas and J.J. Hopfield, Phys. Rev. 128, 2135 (1962).
[2] P. Wiesner and U. Heim, Phys. Rev. B 11, 3071 (1975).
[3] T. Steiner, M.L.W. Thewalt and R.N. Bhargava, Solid State Commun. 56, 933 (1985).
[4] F. Minami and K. Era, Solid State Commun. 53, 187 (1985).
[5] V.V. Travnikov, A. Freiberg and S.F. Savikhin, J. Lumin. 47, 107 (1990).
[6] G. Kudlek, A. Hoffmann, R. Heitz, C. Fricke, J.Gutowski, G.F. Neumark and R.N. Bhargava, J. Lumin.48&49, 138 (1991).
[7] C.H. Henry and K. Nassau, Phys. Rev. B 1, 1628 (1970).
[8] E.I. Rashba and G.E. Gurgenishvili, Sov. Phys.-Solid State 4, 759 (1962).
[9] E.I. Rashba, Sov. Phys. Semicond. 8, 807 (1975).
[10] D.C. Reynolds and T.C. Collins, Phys. Rev. 185, 1099 (1969).
[11] J. Gutowski, N. Presser and I. Broser, Phys. Rev. B 38, 9746 (1988).
[12] T. Kobayashi, Y. Segawa and S. Namba, Solid State Commun. 31, 253 (1979).
[13] J. Gutowski, I. Broser and G. Kudlek, Phys. Rev. B 39, 3670 (1989).
[14] C. Fricke, U. Neukirch, R. Heitz, A. Hoffmann and I. Broser, 5th II-VI conference in Tamano 1991, Japan, accepted for publication.
[15] Y. Segawa, Y Aoyagi and S. Namba, J. Phys. Soc. Japan 52, 3664 (1983).
[16] L. Podlowski, J. Gutowski and I. Broser, Appl. Phys. Lett. 57, 455 (1990).
[17] H. Kukimoto, S. Shionoya, T. Koda and R. Hioki, J. Phys. Chem. Solids 29, 935 (1968).
[18] G.D. Sanders and Y.-C. Chang, Phys. Rev. B 28, 5887 (1983).
[19] W. Schmid and P.J. Dean, phys. stat. sol. (b) 110, 591 (1982).
[20] R. Heitz, A. Hoffmann and I. Broser, DPC Leiden 1991, accepted for publication.

LUMINESCENCE OF A 5d-CENTRE IN ZNS

R.Heitz, P. Thurian, A. Hoffmann and I.Broser
Institut für Festkörperphysik, Technischen Universität Berlin, Germany

ABSTRACT

A strong near infrared luminescence band structured by over 70 lines in cubic ZnS has been attributed to a M-centre, recently. Now, we identify the luminescence as an internal W^{2+}-transition on the basis of high resolution spectroscopy and doping experiments. Tungsten occupies a cation place and the crystal-field is sufficiently strong for a low-spin configuration. The rich line pattern is mainly due to the interaction with one resonant phonon mode at 15.5meV with A_1-symmetry and two gap modes with 29.8meV and 32.9meV. W^{2+} forms a deep acceptor 2.548eV above the valance band. A transient shallow acceptor state (W^+,h_{VB}) with a binding energy of 215meV is identified. The influences of stacking faults and of the electron-phonon interaction on the luminescence structure are discussed.

1. Introduction

ZnS crystals exhibit a well known luminescence structure in the near infrared spectral region[1]. Due to the remarkable high number of sharp lines as well as results from Zeeman and excitation measurements the luminescence has been attributed to a M-centre[1,2], which is composed of two neighbouring sulphur vacancies. The difficulties with the interpretation given in[1,2] arose mainly from the lack of an unambigious identification of the zero phonon line(s) (ZPL) and from the superposed Ni^{2+} $^3T_1(P)$-$^3T_1(F)$ luminescence[3]. Recently, the ZPLs have been identified by means of excitation spectroscopy[4]. In the present work it is shown that an internal transition within substitutional W^{2+}-ions explains well all observations. Detailed investigations of the luminescence structure including the ZPL-region in dependence of temperature, magnetic fields, the doping and the excitation conditions are presented. The influences of the polytypic crystal structure and of the electron-phonon interaction are investigated.

2. The luminescence centre

The investigated ZnS crystals exhibit a strong richly structured luminescence in the near infrared spectral region at low temperatures (fig.1). Recently, the weak line group around 1.511eV has been identified as ZPL-region of the luminescence[4], whereas the two weak lines at 1.520eV are due to internal transitions of Ni^{2+}-centres. The ZPLs are followed by acoustical phonon replicas superposed by over 70 sharp lines. New magneto-optical studies reveal for all lines (including the ZPLs) a nonmagnetic ground state and a magnetic triplet with $g=1.82\pm0.08$ for the excited state as indicated in the insert of fig.1. This is in good agreement with reported results[1] for the strong lines between 1.46eV and 1.485eV. The uniform Zeeman behaviour indicates that all emission lines are due to the same electronic transition, the whole luminescence pattern represents phonon replica of the weak zero phonon line at 1.511507eV. The energy positions are summarized in tab.I of ref.[1], but the interpretations given there have to be revised as it will be discussed in the following.

The high energy part of the luminescence spectrum depends strongly on the crystal structure as shown in fig.2. Most of the investigated crystals have a polytypic but preferably cubic crystal structure. The content of hexagonal parts increases from crystal A to crystal C: while crystal A is nearly perfect cubic, crystal C has about 20% polytypic parts. This has been demonstrated by the Ni^{2+}-transitions[3,5] in the same crystals. Most of the ZPLs, fig.2b, become stronger with

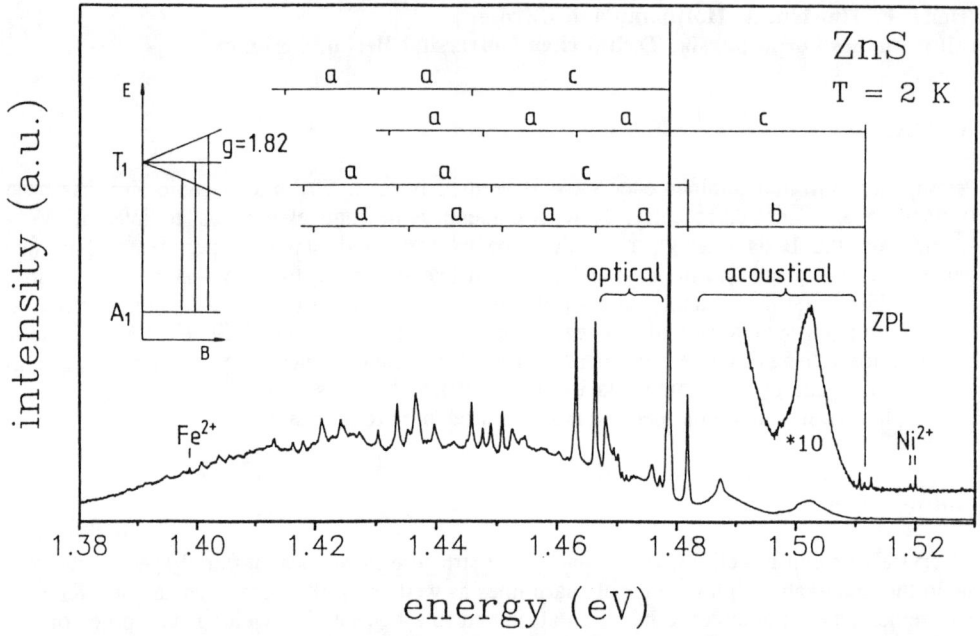

fig.1: Luminescence spectrum of a W-doped cubic ZnS-crystal excited at 2.60eV. Some of the line pairs due to the local phonons with energies of 15.5meV (a), 29.8meV (b) and 32.9meV (c) are marked. The insert shows shematically the Zeeman behaviour of all lines except the marked Ni^{2+}-lines.

increasing polytypic parts. Only the unpolarized cubic emission at 1.51151eV holds its intensity (~0.2%) in relation to the strong phonon replica at 1.47859eV, whereas the axial lines become soon much stronger than the cubic line and are polarized preferably perpendicular to the c-axis of the polytypes (the $<111>_{growth}$-axis of the cubic phase). We attribute these lines to centres on different axial sites as indicated in fig.2b and summarized in tab.I. The energy shifts of about 1meV of the axial lines against the cubic one are very small compared to those known for 3d-elements[5]. But the intensity changes are drastic, indicating a dipole forbidden zero phonon transition in T_d-symmetry (cubic-centre), which becomes allowed in c_{3v}-symmetry. Obviously, the symmetry selection rules are lifted for most of the phonon replica, allmost the whole luminescence band in fig.1 is due to the prevailing cubic centre. However, the situation is different in the one phonon region shown in fig.2a. Shifted by 15.48meV to lower energies a detailed replica of the ZPL-region arises, obviously corresponding to a phonon mode with A_1-symmetry which preserves the selection rules. At least for crystal C also at lower energies phonon replica of the axial lines are seen, the noncubic lines are marked by arrows.

The strong influence of the polytypic crystal structure on the selection rules demonstrates a point-defect as luminescence centre excluding the proposed M-centre. In addition, annealing of the crystals in a sulphur atmosphere supports the luminescence[4], therefore the presence of S-vacancies and especially of M-centres is unprobable. The Zeeman results indicate that the ground state should be a spin and orbital singulet. Among possible defect centres either an electronic d^4-configuration for a substitutional or an electronic d^6-configuration for an interstitial defect meet these demands in the low-spin case with their $^1A_1$-ground states. Since the 3d-elements are well known to have high-spin configurations[6], we doped our crystals with appropriate 4d- and 5d-elements. The principal doping procedure is given in Ref.[7]. Whereas for Nb and Mo the luminescence is not observed, W-doping in the sub-ppm-region leads to a strong

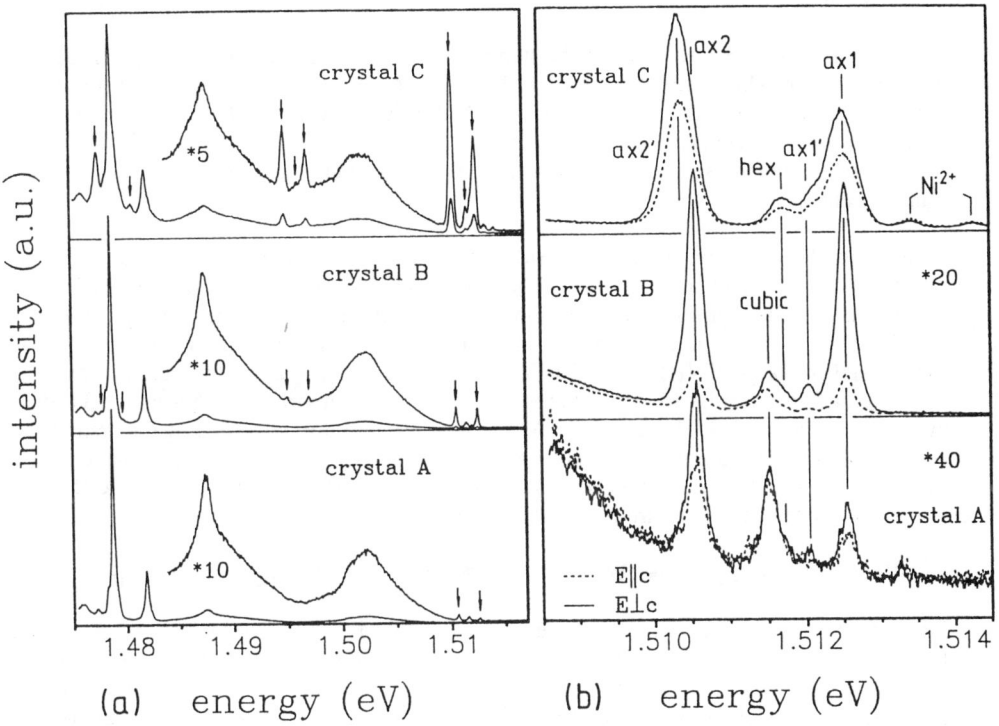

fig.2: Luminescence spectra of the one-phonon (a) and the ZPL (b) region of three different crystals excited at 1.89eV at T=1.8K. Crystal A is allmost perfect cubic and crystal C has strong polytypic parts. Emission lines due to axial centres are marked by arrows.

emission, see fig.1. Therefore, we conclude that at least W can be incorporated in ZnS by indiffusion, occupies an cation-place (substitutional) as W^{2+} and the crystal-field is sufficiently strong for the low-spin configuration with its $^1A_1(e^4)$-ground state. It is known from III-V semiconductors that the ratio Dq/B increases if one goes from 3d- to 5d-transition metals[8] and theoretical calculations[9] show that at least in these compounds for W^{2+}-ions the low-spin situation should hold.

Recently, substitutional W^{2+}-centres have been reported as near infrared luminescence centres in various II-VI-semiconductors[10] but not for ZnS. High-spin configurations and a weaker electron-phonon interaction as for 3d-elements have been stated. The tentative assignment given by Ushakov et al has to be questionized in spite of our results. Due to the doping procedure by ion implantation and the high doping level in the 1000ppm range it is not sure, that the luminescence centres are realy isolated substitutional W^{2+}-ions. Our results indicate a similar electron-phonon interaction for 5d- as for 3d-elements. Additional, the larger mass of 5d-ions leads to stronger localized phonons and thus to a lot of sharp phonon replicas.

tab.I: Energy positions of the ZPLs of the different W^{2+}-centres occuring in polytypic ZnS-crystals.

centre	E (eV)	ΔE (meV)
ax2'	1.51035	-1.16
ax2	1.51055	-0.96
cubic	1.51151	-
hex	1.51168	+0.17
ax1'	1.51205	+0.54
ax1	1.51254	+1.03

3. The term structure

The identification of the excited state of the luminescence is not trivial. A hint is given by the luminescence lifetime which we determined to be $7\mu s$. The lifetime indicates a spin-forbidden transition but is to short for a quintett-singulett transition. Due to the strong spin-orbit interaction typical for 5d-elements it is not any more reasonable to employ the crystal-field states as it is usual for 3d-elements, but we propose the $^3T_1(e^3t_2^1)$-state to give the main contribution to the luminescent state.

As quoted above the ZPLs are observed in excitation spectra[4], too, demonstrating the presence of W^{2+}-centres in the unexcited crystals. The excitation spectrum of the luminescence, see fig.3, reveal further absorption processes with their maxima at 1.80eV, 2.44eV and 2.70eV. The 1.80eV band is mainly due to the spin-allowed $^1A_1$-$^1T_1$ transition. Probably, the lack of finestructure and the large FWHM of 155meV are due to a strong mixing with the nearby $^3T_1$, $^3T_2$ and $^5T_2$ states by the spin-orbit interaction, which is more important for 5d- as for 3d-ions. The two bands around 2.45eV and 2.70eV are connected with charge transfer transitions between the W-centre and the valenceband as has been shown by excitation measurements under additional irradiation[1]. The blue band with its maximum at 2.70eV is due to a charge transfer process and the subsequent recapture of the generated free hole:

$$W^{2+}(^1A_1) + h\nu_{excit.} \rightarrow W^+ + h_{VB} \rightarrow (W^{2+})^* \rightarrow W^{2+}(^1A_1) + h\nu_{IR}$$

A $I^{2/3}$-fit of the low energy onset of the excitation band yields an ionization energy of (2.548 ± 0.010)eV for the deep W^{2+}-acceptor in ZnS. The green band with its maximum at 2.44eV starts with two sharp resonances at 2.33325eV and 2.33551eV followed by a series of phonon replicas, insert of fig.3. The first prominent phonon replica correspond to a phonon of 15.2meV in good agreement with the local A_1-mode mentioned above. The lines vanish above T=25K and do not shift in their spectral position in this temperature region. The band is due to the formation of a shallow acceptor state (W^+,h_{VB}) as commonly known for 3d-ions in ZnS[11,12]:

$$W^{2+}(^1A_1) + h\nu_{excit.} \rightarrow (W^+,h_{VB}) \rightarrow (W^{2+})^* \rightarrow W^{2+}(^1A_1) + h\nu_{IR}$$

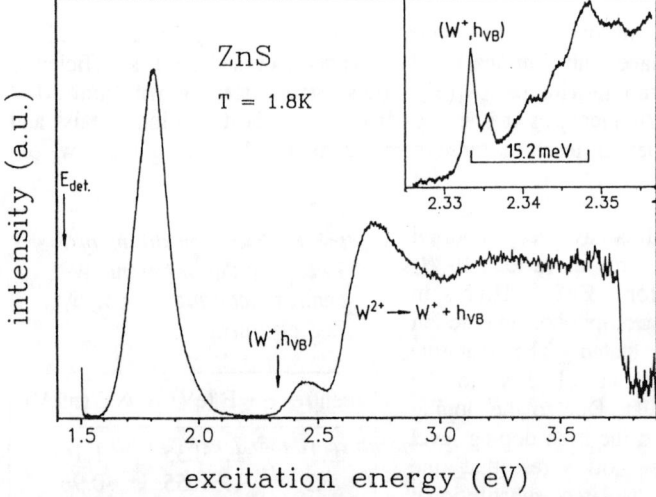

The binding energy of the shallow acceptor complex (W^+,h_{VB}) is (215 ± 10)meV. This binding energy is larger than expected for a shallow acceptor, which can be due to central cell corrections or mixing with high lying 5d-states.

fig.3: Excitation spectrum of the luminescence band given in fig.1. The insert shows enlarged the region of the transient shallow acceptor state (W^+,h_{VB}).

4. The electron-phonon interaction

The luminescence structure consisting of over 70 sharp lines (fig.1) is somewhat strange for an internal transition of a transition metal. A detailed analysis of the energy distances show that all prominent resonances in the high energy part as well as nearly all resonances in the low energy part are due to multiphonon processes from the cubic ZPL, whereby only three phonon modes are involved: One resonant mode at 15.5meV and two gap modes at 29.8meV and 32.9meV. The first replica of the resonant mode is not seen for a cubic crystal (fig.1) but occur for the polytypic centres (fig.2a) due to its A_1-symmetry. The two strong lines around 1.48eV are the one phonon replica involving the two gap modes. The phonon modes of the axial centres differ slightly from the cubic one and at higher quantum numbers the phonon energies decrease slightly by up to 100μeV. Some of the phonon spacings are indicated in fig.1. Further, a weak coupling to acoustical as well as optical phonons of the host crystal is observed indicated in fig.1. The given interpretation of the lines as well as the involved localized phonon modes differ from those in earlier works[1,2]. Two aspects should be remarked: At first the observed line pattern demonstrates the absence of any Jahn-Teller coupling in the ground state as expected for a $^1A_1(e^4)$-state but indicates a strong rearrangement of the surrounding in the excited $^3T_1$-state. Secondly, the dominating interaction with distinct local phonons is due to the large mass difference between the W-ions and the replaced Zn-ions.

A strong emission line seperated by 3.38meV from the cubic ZPL occurs at higher temperatures, shown in fig.4 for the nearly perfect cubic crystal A. Crystal C shows a somewhat distorted lineshape but still the cubic hot line dominates. The hot line originates from a thermally populated excited state. The experimental derived term sheme is given as insert in fig.4. From symmetry selection rules and the observed Zeeman behaviour the ZPL seen at helium temperatures is due to a forbidden T_1-A_1 transition

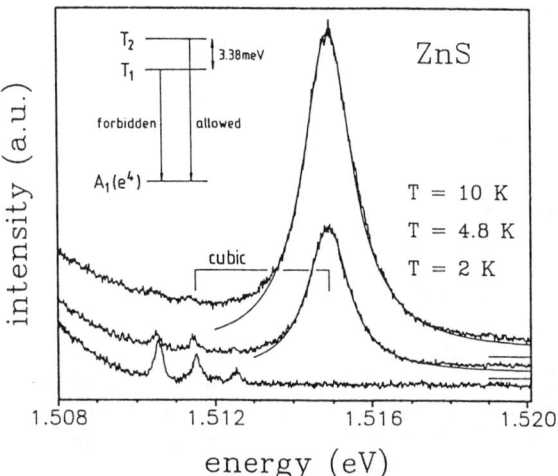

fig.4: Temperature dependence of the ZPL region of crystal A excited at 1.88eV. The hot line is fitted with an Lorenzian lineshape.

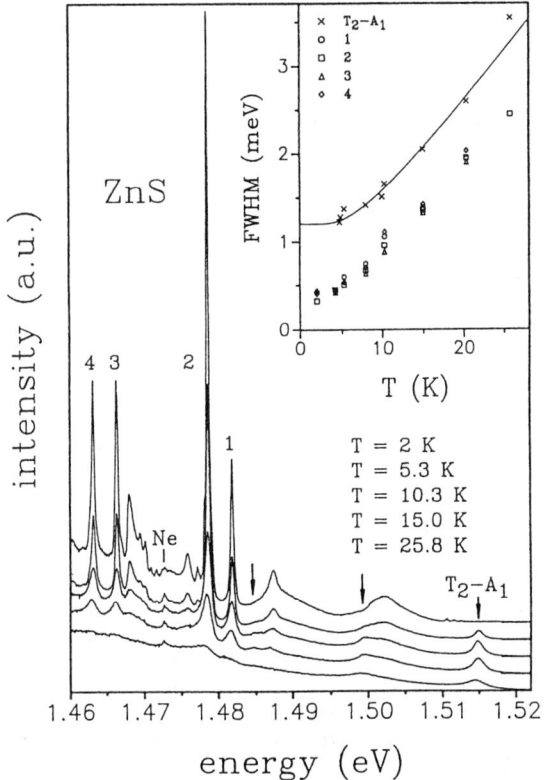

fig.5: Temperature dependence of the one phonon region of crystal A excited at 1.88eV. The lines occuring at higher temperatures are marked by arrows. The insert shows the development of the FWHMs. The FWHM of the hot ZPL is fitted by a direct phonon decay.

and the high temperature line is due to an allowed T_2-A_1 transition. The estimated ratio of the oscillator strengths is over 1000:1. Further finestructure in the excited state can not be excluded, but the Zeeman results give no reason for a further state near the T_1-state. W-ions on axial sites experience a symmetry reduction to c_{3v} connected with a splitting of the T_1-state in one A_2- and one E-component. Now, the A_1-E transition is allowed for the polarization perpendicular to the c-axis as observed for the polytypes, see fig.2b. In the framework of a Tanabe-Sugano diagram[10] the origin of these states is not clear, since the strong spin-orbit interaction leads to a mixing of the crystal-field states. Theoretical calculation including the spin-orbit interaction on the same stage as the crystal-field are nessesary.

Fig.5 show the temperature dependence of the one phonon region of the luminescence. The lines become broader and weaker, above 30K the lines vanish. Above 4.2K the hot ZPL and their phonon replica occur as indicated by arrows in fig.5. The development of the FWHMs of the hot T_2-A_1 ZPL and some prominent phonon replica are given in the insert of fig.5. The FHWMs increase allmost linear with temperature above 10K. Whereas the low temperature lines show an inhomogenious broadening of a few hundret μeV at 2K the T_2-A_1 transition is homogenious broadened, as is shown by the Lorenzian fits in fig.4. Their FWHM is fitted with the assumption of a direct decay to the T_1-state under the emission of low frequency phonons. An active phonon of 1.70meV and a spontaneous emission rate of 1.8THz have been used, indicating a strong coupling to low frequency acoustical phonons in the excited state.

5. Conclusion

In conclusion we have shown that the incorporation of the 5d-transition metal W leads to a strong richly structured luminescence band ZnS. W is incorporated on cation sites as W^{2+} and forms a deep acceptor 2.548eV above the valence band. Excitation resonances due to the formation of a shallow acceptor state (W^+,h_{VB}) with a binding energy of 215meV are recognized. Mainly three local phonon modes are involved in the luminescence process. The absence of any Jahn-Teller interaction in the ground state gives the oppurtunity to study the local vibrational modes of the defect centre.

6. Acknowledgement

We wish to thank Dr R.Broser for supplying the crystals and E.Birkicht for the doping. The work has been partly supported by the Deutsche Forschungsgemeinschaft.

7. References

[1] I. Broser, R. Germer, F. Seliger and H.-J. Schulz, J. Phys. Chem. 41, 101 (1980).
[2] R. Germer, Phys. Rev. B 27, 2412 (1983).
[3] I. Broser, R. Broser, E. Birkicht and A. Hoffmann, J. Lumin. 31/32, 424 (1984).
[4] A. Hoffmann, I. Broser and P. Thomsen-Schmidt, J. Lumin. 40/41, 321 (1988).
[5] T. Buch, A. Geoffroy and B. Lambert, J. Physique C 3, 159 (1974).
[6] A. Zunger, Solid State Phys. 39, 275 (1986).
[7] I. Broser, A. Hoffmann, R. Germer, R. Broser and E. Birkicht, Phys. Rev. B 33, 8196 (1986).
[8] V.S. Vavilov, V.V. Ushakov and A.A. Gippius, Physica 177B&118B, 191 (1983).
[9] N. Makiuchi, T.C. Macedo, M.J. Caldas and A. Fazio, Defect and Diffusion Forum 61/62, 145 (1989).
[10] V.V. Ushakov and A.A. Gippius, J. Cryst. Growth 101, 458 (1990).
[11] D.J. Robbins, J. Lumin. 24/25, 137 (1981).
[12] R. Heitz, P. Thurian, A. Hoffmann and I. Broser, send to J. Phys. C.

INTERSTITIAL DEFECTS IN II-VI SEMICONDUCTORS: ROLE OF THE CATION d STATES†

J. T. Schick,[1] C. G. Morgan-Pond,[2] and J. I. Landman[2]

[1] Deparment of Physics, Villanova University, Villanova, PA 19085, USA
[2] Deparment of Physics and Astronomy, Wayne State University, Detroit, MI 48202, USA

ABSTRACT

Calculations of the localized electronic defect states due to tetrahedral and hexagonal interstitials and vacancies in CdTe have been carried out both with an sp^3 and with an sp^3d^5 valence band basis set including full accounting of orbital overlap. Both cation and anion defects are considered. We discuss the effects of the Cd d orbitals on the formation of the localized defect states and compare the results of the full spd calculations to experiment.

1. Introduction

Since point defects, including interstitials, play a major role in determining the electronic properties of II-VI semiconductors, a full understanding of these properties requires an accurate description of the important point defects in these materials. With the exception of some recent work which considers tetrahedral interstitials and some non-interstitial defects in ZnSe [1], most calculations of interstitial properties in II-VI semiconductors neglect the effects of the cation d states on the bonding, though calculations of bulk properties have shown [2] that these effects are often important in II-VI semiconductors containing Zn, Cd, or Hg.

The interaction of the anion p orbitals with the d orbitals on neighboring cations, which have energies close to the valence band in these II-VI materials, is not forbidden by symmetry at the Γ point in the zincblende lattice. Both d and p orbitals have a T_2 (or Γ_{15}) symmetry representation, and these two T_2 representations can interact. As expected in second order perturbation theory, the p- and d-derived T_2 levels repel each other, so that the d-derived T_2 level is lowered in energy, and the p-derived T_2 level which forms the valence band maximum is pushed up. Although tight-binding, pseudopotential, and other calculations which use adjustable parameters can be adjusted to fit chosen gaps in the band structure even when they do not explicitly include the d orbitals in the valence band, the ability of such calculations to reproduce an accurate picture of all the effects of this p-d repulsion has been questioned [2].

In this paper, we examine the differences in the electronic structure of the defect states obtained in a tight-binding calculation with parameters chosen to fit the bulk band structure using an sp basis only, and a similar calculation using an spd basis, for Cd and Te interstitials and vacancies in CdTe. We consider the two tetrahedral positions and the hexagonal position for the interstitial. The two tetrahedral positions are a natural starting point when searching for the preferred interstitial position, and the hexagonal point is the point half-way along a line between two adjacent tetrahedral sites, where the open channel for diffusion between tetrahedral sites is narrowest. In Section 2, the method and parameters of the calculation are described in greater detail and in Section 3, the results are discussed for each defect.

2. Setup

For the defect calculation, we employ a 64-atom, cubic supercell which has a side length of twice the lattice constant of the bulk material, and we add one atom for interstitials, or remove one atom for vacancies. Interstitials are placed in one of the two tetrahedral positions: one on the alternate Cd sublattice, with Te nearest neighbors, and the other on the alternate Te sublattice, with Cd nearest neighbors, or in the hexagonal position. In a tetrahedral position, the interstitial has four nearest neighbors arranged at the vertices of a tetrahedron, six second nearest neighbors, and twelve third nearest neighbors. The hexagonal interstitial has six nearest neighbors, three Cd and three Te, which form a puckered hexagonal ring, eight second nearest neighbors, and six third nearest neighbors.

In order to maximize the transferability of parameters between the bulk and the various defect environments described above we use 5s, 5p, and 4d Slater orbitals as basis states for the tight-binding calculation. The Slater orbitals are of the form

$$\Phi_{nlm}(r,\theta,\phi) = Nr^{n-1}e^{-\mu r/a_o} Y_{lm}(\theta,\phi), \qquad (1)$$

where a_o is the Bohr radius, N is the normalization factor, and Y_{lm} is a spherical harmonic. For Cd 5s, 5p, and 4d orbitals $\mu = 1.638$, 1.412, and 3.969, respectively, and for Te 5s, 5p, and 4d orbitals $\mu = 2.506$, 2.158, and 4.9906, respectively, as determined in Hartree-Fock calculations by Clementi and Roetti [3]. We interpolated the value for the Cd p state from the available Cd data.

Since the tight-binding basis states constructed in this manner are not orthogonal, the overlap matrix is not the identity matrix. In this case the eigenvalues are determined by the relation

$$H\psi = ES\psi, \qquad (2)$$

where S is the overlap matrix of the tight-binding states and H is the Hamiltonian of the system. The overlap integrals of the atomic states are calculated analytically, in order to reduce computational intensity, using a method outlined by Mulliken et al.[4].

We assume that the distance dependence of the interactions between pairs of orbitals is the same as that of the overlap between those two orbitals. This means that there are a maximum of sixteen free parameters in our band structure, six for the on-site energies and the remaining ten for the hopping interactions. The hopping interaction h between orbital ℓ on one atom and orbital ℓ' on a neighboring atom participating in a bond of symmetry m is given by the relation $h_{\ell\ell'm} = M_{\ell\ell'm} S_{\ell\ell'm}$, where $S_{\ell\ell'm}$ is the overlap and $M_{\ell\ell'm}$ is a multiplier which we adjust to produce the best band structure. Interactions are included to a distance of 85% of a lattice spacing, which means that up to third-nearest neighbors are included. The sixteen parameters are determined through fitting to spin-orbit averaged band structures of Chelikowsky and Cohen [5]. We fit the sp^3 and sp^3d^5 bases separately and the results of these fits are presented in Table 1. The zero of energy for each set of parameters is arbitrary. Note that the anion d on-site energy has been arbitraily set to -500, to remove these orbitals from the calculation.

Table 1.   On-site energies and hopping integral multipliers in eV.

	parameter	sp-only	spd
on-site energies	anion-s	-11.000	-9.842
	anion-p	-0.975	-1.026
	anion-d	-----	-500.000
	cation-s	-5.800	-5.670
	cation-p	-1.825	-1.875
	cation-d	-----	-10.000
hopping multipliers	ss-sigma	-14.754	-14.000
	sp-sigma	-9.886	-9.916
	pp-sigma	-3.390	-3.712
	pp-pi	-1.599	-1.802
	sd-sigma	-----	-32.533
	pd-sigma	-----	45.315
	pd-pi	-----	34.032
	dd-sigma	-----	-2.031
	dd-pi	-----	-6.860
	dd-delta	-----	-0.013

In Fig. 1 we plot the band structure for the sp basis and in Fig. 2 we plot the same for the spd fit. In general, the results look quite similar, except for the addition of a d band at approximately the experimental energy, 10.5 eV below the valence band maximum [6]. However, a close examination of the top of the valence band reveals that the band is too flat for the sp case in comparison to the spd basis because of the requirement that the conduction bands had to be reproduced reasonably well. When the d orbitals are included, we obtain a better overall fit of the bands because p-d repulsion causes the valence band maximum (VBM) to be pushed up in energy. Since the VBM is not pushed up by p-d repulsion in the sp fit, the total width of the valence band at the Γ point is adjusted by lowering the anion s on-site energy, which lowers the energy of the bottom of the valence band.

Charge transfer in the bulk is taken into account through the fitting. For the defect, we have previously seen that there are small deviations (< 0.2 electrons on the interstitial itself) from the electronic occupancy in the bulk for various interstititial defects [7], so additional corrections due to charge transfer have not been included in these defect calculations. In the following, we reference all energies to the bulk VBM.

3. Results

For defects with tetrahedral symmetry, as well as for bulk states at the Γ point, inclusion of d orbitals in the basis for the valence band adds states of T_2 and E symmetry to the A_1 and T_2 states derived from the s and p orbitals. For hexagonal interstitials, inclusion of the d orbitals adds states of A and E symmetry. The new d-derived states can interact with s- and p-derived states of the same symmetry representation, in the same way at the T_2 (or Γ_{15}) bulk states derived from p and d orbitals at the Γ point interact.

In general, we find that the tight-binding calculations using either an sp^3 basis or an sp^3d^5 basis to fit the band structure produce similar deep defect levels and electrical behavior for the defects considered. The largest differences between the sp and spd descriptions of defects with tetrahedral symmetry occur for defects involving addition or removal of Cd atoms, which have a d shell close in energy to the valence band.

Tetrahedral tellurium interstitials:
The defect states for the tetradedral Te interstitial between Te involve the interaction of the Te interstitial orbitals with the bonding hybrids on its Te neighbors. The effects of the Cd d orbitals on the defect states are due to smaller interactions with more distant neighbors. Since neither the interstitial nor its nearest neighbors have a d shell close in energy to the VBM, we might expect the energy of the localized states for this defect to be unaffected by the inclusion of the d orbitals. When the d orbitals are included, pushing up the energy of the VBM, we might then expect an overall downward shift of the defect levels, measured relative to the valence band. In fact, we do find that both the A_1 level and the T_2 level in the gap are about 0.5 eV lower in the spd description than in the sp description. In the spd calculation the A_1 level is nearly resonant with the VBM and the T_2 level, which has six states occupied with 4 electrons for a neutral defect, occurs at 1.0 eV above the VBM.

Theoretical calculations [8, 9] predict that the preferred tetrahedral position for Te interstitials is between Cd. The Te interstitial between Cd has a T_2 level with six states occupied by four electrons at about 1.1 eV in both the sp and spd calculations. This interstitial also has an A_1 level at 0.3 eV above the VBM in the sp description and an A_1 resonance at about 0.3 eV below the VBM in the spd description. We may interpret the apparent downward shift of the A_1 level, measured relative to the VBM, as resulting from the p-d repulsion pushing up both the VBM and the T_2 defect level when d-orbitals are included.

Hexagonal interstitials:
A more complicated situation occurs for hexagonal interstitials, due to their lower symmetry. For these interstitials, A levels derived from the s orbitals on the interstitial can mix with A levels derived from the p and d orbitals, whereas for defects with tetrahedral symmetry, as for bulk states at the Γ

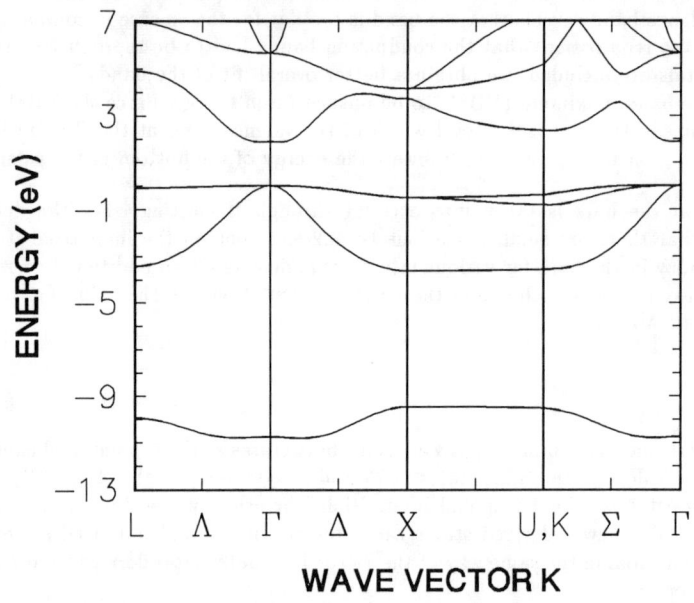

FIG. 1 CdTe Band structure using s and p orbitals.

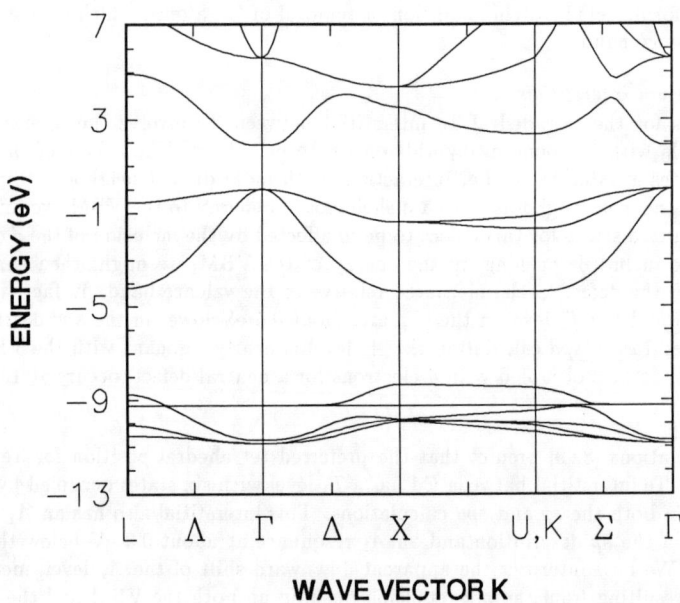

FIG. 2 CdTe Band structure using s, p, and d orbitals

point, only interactions between p- and d-derived states can occur.

For the hexagonal Te interstitial, we observe an E level, with four states occupied by two electrons, at 1.4 eV above the VBM in the sp description and at 1.5 eV in the spd description. Two levels of A symmetry appear at 0.5 eV and 0.9 eV in the sp calculation, while in the spd calculation one of the A levels has moved below the VBM and the other level appears at 0.5 eV. As for the tetrahedral Te interstitials, the level shifts resulting from the contribution of the d orbitals to the bonding around the hexagonal Te interstitial do not change its basic electrical behavior. The partially occupied levels observed in the midgap region for neutral Te interstitials in the hexagonal and both tetrahedral positions suggest that simple migration (without a change in charge state) of the neutral Te interstitial along the tetrahedral-hexagonal-tetrahedral path may be energetically favorable. This migration would be consistent with the interpretation of experimentally observed Te self-diffusion behavior as resulting from a rapidly diffusing neutral Te interstitial [10, 11].

For the Cd interstitial in the hexagonal position, we see unoccupied A and E levels and a doubly occupied A level at 1.7 eV, 1.3 eV, and 0.4 eV, respectively, in the sp calculation. When d orbitals are included, the E level shifts upward by 0.4 eV, both A levels remain unshifted, and a new E level or resonance appears close to the VBM. Due to the finite size of the supercell, this new E level appears to interact with the bulk E state at the VBM, and the resulting level repulsion leads to one state at 0.2 eV and the second at -0.3 eV. The properties of the hexagonal Cd interstitial may not be as interesting from a practical standpoint as the properties of the hexagonal Te interstitial, since both theoretical calculations [8] and experimental work [12] indicate that simple interstitial diffusion down the tetrahedral-hexagonal-tetrahedral channels in the lattice is not favored for Cd.

Tetrahedral cadmium interstitials:
High temperature conductivity in Cd-rich CdTe is dominated by double donor native defects with donor levels about 0.2 eV or less below the conduction band, which are postulated to be Cd interstitials [13, 14]. In the sp description, the tetrahedral Cd interstitial between Te has a T_2 deep level in the gap at 1.6 eV, with two electrons occupying its six states, and an A_1 level at 1.8 eV. The T_2 level is pushed up by 0.2 eV when the d states are included in the calculation; leaving the last two electrons in the A_1 level which remains at 1.8 eV (lower than the T_2 level by 0.02 eV). The spd description of this interstitial is therefore consistent with the experimental characterization of the shallow donor native defects seen in high temperature electrical measurements on Cd-rich CdTe. Earlier tight-binding calculations suggest that the preferred tetrahedral position for the Cd interstitial is between Te [8], although subsequent linear muffin tin orbital calculations using the atomic spheres approximation (LMTO-ASA) suggest that the preferred tetrahedral position for Cd interstitals is between Cd [9].

The tetrahedral Cd interstitial between Cd has a T_2 level, occupied by two electrons for the neutral interstitial, at 1.2 eV and an A_1 level at 1.5 eV in the sp description. When d orbitals are included, the T_2 level is pushed up by 0.6 eV, so that the last two electrons fall to occupy the A_1 level, which has remained relatively constant, at 1.4 eV. The large magnitude of the T_2 level shift results from the strong p-d repulsion which raises the energy of the localized defect state formed from the p orbitals on the Cd interstitial and the bonding hybrids on its Cd nearest neighbors, since all of these atoms have excess Cd nearest neighbors, each of which has interacting d orbitals.

Vacancies:
The Te vacancy has a single level of T_2 symmetry in the gap which is completely unoccupied for the neutral vacancy, above a completely filled valence band. In the spd description, this level is roughly resonant with the VBM, however, due to the finite size of the supercell, it mixes with the VBM bulk state, and is pushed up to 0.2eV, while the VBM state is pushed down to -0.2 eV. In the sp description, the T_2 defect level is at 0.5 eV. Since the localized defect states of the Te vacancy are created from dangling bonds on the neighboring Cd atoms, while the Cd d orbitals mainly affect the energies of the bonding hybrids on the Te atoms in the lattice, the energy of these localized states is largely unaffected by the addition of Cd d orbitals in the valence band. Therefore, we may interpret the downward shift of the localized vacancy level, measured from the VBM, when d orbitals are included, as being due to the upward shift of the VBM.

The localized states of the Cd vacancy are created from dangling bonds on the neighboring Te atoms. When the Cd has been removed, the p-d repulsion felt by the bonding hybrids on the Te neighbors is reduced, allowing the localized T_2 defect levels formed from these hybrids to fall in energy relative to the VBM in the spd description. Therefore, while the Cd vacancy has a T_2 level at 0.5 eV in the sp description, with four electrons occupying its six states, in the spd description this partially occupied level has fallen below the VBM, and two additional T_2 levels have moved out of the conduction band, to sit just above and just below 1.5 eV. Although experimental identification of isolated vacancy levels is difficult, the shallow double acceptor levels of the Cd interstitial resulting from the full spd calculation are consistent with several experimental identifications of Cd vacancy levels [13, 15].

REFERENCES

†This work was supported in part by the Night Vision and Electro-Optics Laboratories and the ARO under ARO Grant No. DAAL03-87-0061, by the Institute for Manufacturing Research, and by the NSF under Grant No. DMR900044P, and utilized the Cray Y-MP at the Pittsburgh Supercomputing Center.

[1] D. B. Laks et al., Phys. Rev. Lett. **66**, 648 (1991).
[2] S.-H. Wei and A. Zunger, Phys. Rev. B **37**, 8958 (1988).
[3] E. Clementi and C. Roetti, At. Dat. Nucl. Dat. Tables **14**, 177 (1974).
[4] R.S. Mulliken, C.A. Rieke, D. Orloff, and H. Orloff, J. Chem. Phys. **17**, 1248 (1949).
[5] J.R. Chelikowsky and M.L. Cohen, Phys. Rev. B **14**, 556 (1976).
[6] L. Ley, R. A. Pollak, F. F. McFeely, S. P. Kowalczyk, and D. A. Shirley, Phys. Rev. b **9**, 600 (1974).
[7] J.T. Schick and C.M. Morgan-Pond, J. Vac. Sci. Technol. A **8**, 1108 (1990).
[8] C. G. Morgan-Pond, J. T. Schick, and S. Goettig, J. Vac. Sci. Technol. A **7**, 354 (1989).
[9] M. A. Berding, A. Sher, A.-B. Chen, and R. Patrice, J. Vac. Sci. Technol. A **8**, 1103 (1990).
[10] H. H. Woodbury and R. B. Hall, Phys. Rev. **157**, 641 (1967).
[11] P. M. Borsenberger and D. A. Stevenson, J. Chem. Phys. Sol. **29**, 1277 (1968).
[12] M.-F. Sung Tang and D. A. Stevenson, J. Vac. Sci. Technol. A **7**, 544 (1989).
[13] S. S. Chern, H. R. Vydyanath, and F. A. Kroger, J. Solid State Chem. **14**, 33 (1978).
[14] Other references reviewed in K. Zanio, *Cadmium Telluride, Semiconductors and Semimetals*, Vol. 13, ed. by R. K. Willardson and A. C. Beer (New York: Academic Press, 1978), pp. 116-129.
[15] Other references given on p. 148 of K. Zanio, *Cadmium Telluride, Semiconductors and Semimetals*, Vol. 13, ed. by R. K. Willardson and A. C. Beer (New York: Academic Press, 1978), pp. 116-129.

PAC Study of the Acceptor Li in II-VI Semiconductors

H. Wolf, Th. Krings, U. Ott, U. Hornauer, and Th. Wichert

Technische Physik, Universität des Saarlandes, 6600 Saarbrücken, FRG

Abstract

For the II-VI semiconductors CdS, ZnS, and CdTe the state of Li atoms was investigated using the donor ^{111}In as radioactive probe atom for perturbed angular correlation (PAC) experiments. From the absence of In_M^+-Li_M^- pairing and the simultaneous formation of In_M^+-V_M^{--} pairs which was observed in all these semiconductors it is concluded that Li is incorporated rather interstitially as Li_i^+ accompanied by a vacancy V_M^{--} than substitutionally on the group II site as an acceptor Li_M^-.

1. Introduction

Detailed information on the atomic configuration of dopant atoms is much less available for II-VI semiconductors than for elemental or III-V semiconductors. But, in II-VI semiconductors doping, in particular *p*-doping, poses a crucial problem because of the strong interaction of the dopant atoms with intrinsic defects leading to self-compensation phenomena[1]. Since Li introduced on the group-II site is regarded as a promising candidate for *p*-doping of II-VI semiconductors we have begun an investigation of the incorporation of Li into II-VI compounds on an atomic scale. Li atoms that occupy the site of the group II element and act as an acceptor should form coulombic bound pairs with the donors that are also present in the crystal. Such a pairing is easily detectable by the perturbed angular correlation technique (PAC) using the radioactive donor ^{111}In as a probe atom as has been successfully proven for Si, Ge and several III-V semiconductors[2,3]. In contrast, PAC studies of defects in II-VI semiconductors are just at their beginning, e.g. in ^{111}In doped ZnO[4,5], CdTe[6,7], and CdS[8-10]. In this paper, first results on a comparative study of the behavior of Li in CdS, ZnS, and CdTe will be presented.

2. Experimental Details

The PAC technique employed here measures the defect specific electric field gradient tensor (EFG) which at a particular lattice site is determined by the local electric charge distribution. Working with the radioactive probe atom ^{111}In, which decays with a half-life of 2.8 days to its daughter isotope ^{111}Cd (see Fig. 1), the EFG will generate three frequencies ω_n in the PAC time spectrum R(t) which can be described by

(1) $$R(t) = S_0 + \sum_{n=1}^{3} S_n \cos(\omega_n t)$$

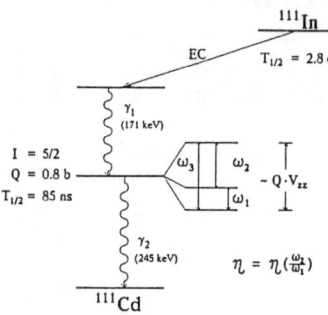

Figure 1. Nuclear decay schema of the PAC probe atom ^{111}In showing the β-decay (electron capture) and the subsequent emission of the two γ rays used for the measurement of the γγ angular correlation.

This spectrum is obtained by measuring the coincidence rate of the two γ-rays emitted by the excited ^{111}Cd nucleus as a function of the time which has passed during the detection of both γ-rays. The different components V_{ii} of the traceless tensor are obtained from the frequencies ω_n. The largest component V_{zz} is usually expressed by the coupling constant $\nu_Q = eQV_{zz}/h$ where Q is the nuclear quadrupole moment of the isomeric ^{111}Cd state ($T_{1/2}$ = 85 ns) used for the detection of the EFG. As a second parameter the asymmetry parameter $\eta = (V_{xx} - V_{yy})/V_{zz}$ is used ($0 \leq \eta \leq 1$), which is deduced from the measured frequency ratio ω_2/ω_1 (Fig. 1). The coefficients S_o and S_n depend on the orientation of the EFG tensor with respect to the host lattice, i.e. on the orientation of the formed ^{111}In-defect complex; in addition, the coefficients depend on the fraction of ^{111}In atoms that are involved in this complex. A detailed description of PAC is found elsewhere[2].

For the PAC experiments single crystals were cut into pieces of typically 5 to 10 mm^3 size and were diffused inside of an evacuated quartz ampoule with about 10 to 20 μCi ^{111}In from a carrier free ^{111}InCl$_3$ solution. Into CdS (Eagle Picher) and ZnS (Eagle Picher) this diffusion happened under S$_2$ overpressure at 1073 K for 90 min.; into CdTe (R. Triboulet, CNRS, Meudon) the diffusion happened under Cd overpressure at 1023 K for 90 min. The PAC spectra were recorded at 295 K using a 4 detector set-up.

3. Experimental Results

For the II-VI semiconductors CdS, ZnS, and CdTe Fig. 2 shows the PAC time spectra along with their Fourier transforms that were obtained after diffusion of the respective semiconductor with Li. In all three cases, the absence of a Li specific EFG shows that ^{111}In$^+$-Li$^-$ pairs were not formed in those compounds but rather In$_M^+$-V$_M^{--}$ complexes consisting of the radioactive donor ^{111}In$_M^+$ and the acceptor vacancy V$_M^{--}$, both being situated in the

Table I. Site specific EFGs observed at the probe ^{111}In/^{111}Cd at 295 K after Li diffusion. Listed are the strength V_{zz} and the asymmetry parameter η of the EFG tensor.

CRYSTAL	$\nu_Q = eQV_{zz}/h$	$(V_{xx}-V_{yy})/V_{zz}$	COMPLEX
CdS	72.4(2) MHz	0.35(2)	In$_{Cd}$-V$_{Cd}$ [1]
	78.7(2) MHz	0.21(2)	In$_{Cd}$-V$_{Cd}$ [2]
ZnS	81.5(4) MHz	0.16(4)	In$_{Zn}$-V$_{Zn}$
CdTe	60.0(4) MHz	0.10(4)	In$_{Cd}$-V$_{Cd}$

[1] V$_{Cd}$ above or below the basal plane (Fig. 4, left)
[2] V$_{Cd}$ in the basal plane (Fig. 4, right)

sublattice of the metallic group II element Cd or Zn. The identical complexes, as recognized by the identical EFGs, are formed - only less pronounced - during diffusion of these II-VI compounds with the radioactive ^{111}In probe atoms.

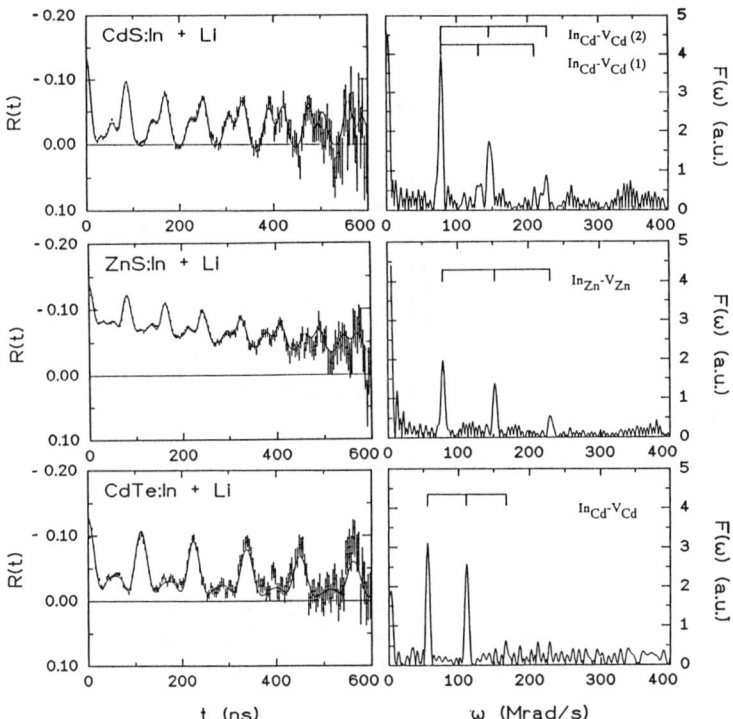

Figure 2. PAC time spectra and their associated Fourier transforms measured after Li diffusion. The corresponding values of the different EFGs are listed in Table I.

<u>CdS</u> : The ^{111}In doped crystals were diffused with Li at 793 K (30 min.). A least-squares fit to the measured PAC time spectrum (Fig. 2, top) yields that two different EFGs what is also visible by the two frequencies triplets in the associated Fourier transform. Their coupling constants ν_Q along with the respective asymmetry parameters η are listed in Table I. Both EFGs are already known to be caused by the trapping of Cd vacancies[9,10].

<u>ZnS</u>: The ^{111}In doped crystals were diffused with Li at 793 K (30 min.). The least-squares fit to the PAC spectrum (Fig. 2, middle) yields a single EFG that is characterized by ν_Q = 81.5 MHz and η = 0.16. The identical EFG is also found directly after the diffusion of ^{111}In into the ZnS crystal; only the fraction of the ^{111}In-defect complexes was again increased by the presence of Li.

The influence of the Li-diffusion temperature on the fraction of the formed In-defect complexes shows Fig. 3 (left panel); here, the diffusion time was always 30 min.. In order to separate the effects caused by the presence of Li from possible modification of the crystal solely caused by the respective diffusion temperature, a second ZnS crystals was identically treated, however, without Li. The first data point results from the doping of the crystals with ^{111}In at 1073 K. It is evident that there is no influence of Li on the In-defect fraction below 580 K. Above this temperature, however, Li seems to diffuse into ZnS thereby stabilizing and enhancing the In-defect fraction considerably as compared to the ZnS crystal that was heated to the same temperatures without Li.

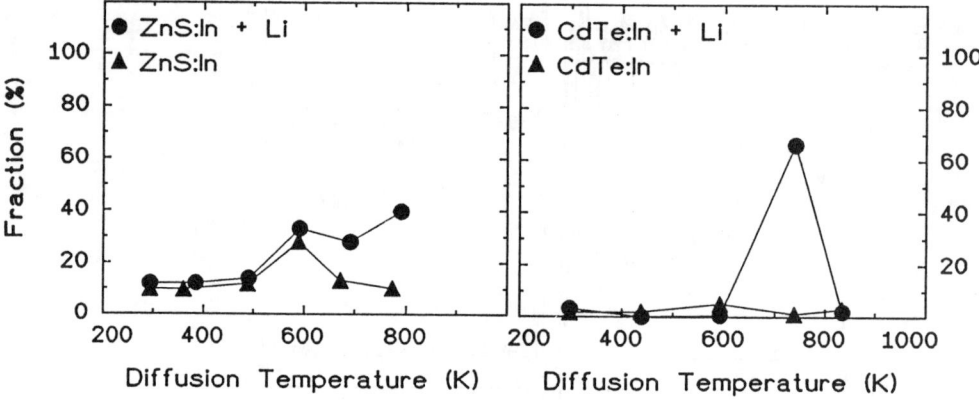

Figure 3. Influence of the diffusion temperature of Li on the fraction of In_M^+-V_M^{--} pairs observed at 295 K.

CdTe: After diffusion of an ^{111}In doped crystal with Li at 720 K (30 min.) the PAC spectrum shown in Fig. 2 (bottom) again reveals no Li specific EFG. For the observed EFG that is characterized by ν_Q = 60.0 MHz and η = 0.10 is well-known from previous PAC experiments and has been identified as the In_{Cd}^+-V_{Cd}^{--} pair[6,7].

As for ZnS, the influence of the Li diffusion temperature on the formation of the In_{Cd}^+-V_{Cd}^{--} pairs was investigated (Fig. 3, right panel). Whereas temperatures up to 820 K are not able to produce any In_{Cd}^+-V_{Cd}^{--} pairs in CdTe the presence of Li gives rise to a very pronounced increase in the pair fraction at 720 K. The strong decrease of the formed complexes above this temperature can be explained by a too small binding energy of the pairs.

4. Discussion

For CdS, ZnS and CdTe the PAC results show:
i) Because no donor-acceptor pairing of the type In_M^+-Li_M^- is observable Li doping does not lead to a measurable population of the Li_M acceptor states at the respective diffusion temperatures.
ii) Li doping enhances the concentration of acceptor vacancies V_M^{--} what becomes visible through the enhanced formation of In_M^+-V_M^{--} pairs. A detailed identification of this complex has been performed for CdS[9,10] as well as in CdTe[6,7]. The similar EFGs observed for the In-defect complexes in the three II-VI compounds (see Table I) as well as the similar formation conditions represent a strong hint that also the complex in ZnS is caused by a trapped V_M^{--} defect and, therefore, corresponds to an In_{Zn}^+-V_{Zn}^{--} pair.

From these results it is concluded that the incorporation of Li is governed by the reaction

(2) $\quad Li_M^- \rightleftarrows Li_i^+ + V_M^{--}$

which, at least at the time of Li diffusion, is strongly shifted to the right hand site resulting in the formation of interstitial Li_i^+ and a vacancy V_M^{--}; this complex still acts as a single acceptor and, therefore, can produce p-conductivity in a II-VI crystal. Since both elements, In and Li, form donor states it is easily understandable that no In-Li pairing occurs, and since the introduction of Li is accompanied by the formation of a vacancy an enhancement of In_M^+-V_M^{--} pairs has to occur, as well.

For Li in ZnSe also Sasaki and Oguchi based on density-functional calculations arrived at the conclusion that the acceptor Li_{Zn} is unstable against disintegration into a vacancy V_{Zn} and interstitial Li_i[11]. On the other hand, the formation of Li_i donors in CdS contradicts photoluminescence experiments by Henry et al. which claim the formation of Li_{Cd} states in the CdS lattice[12,13]. It should be noted that the behavior of Li observed by PAC is common to all, here investigated II-VI compounds in spite of their different band gap energies and bonding characters, and, second, that the states of both the donor In and the acceptor Li are affected by the same intrinsic defect, the metallic vacancy V_M.

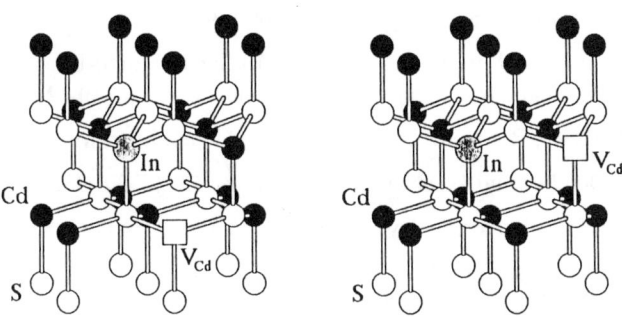

Figure 4. Atomistic model of CdS showing the two different In_{Cd}^+-V_{Cd}^{--} complexes associated with the two different EFGs (see Table I).

The fact that for CdS two different EFGs are observed (see Table I) which both belong to the In_{Cd}^+-V_{Cd}^{--} pair is easily explainable on the basis of the hexagonal lattice structure of CdS: The distances between In_{Cd} and V_{Cd} are slightly different if the trapped vacancy resides above/below or inside of the basal plane of the hexagonal lattice (see Fig. 4). This effect is strongly reduced for the ZnS lattice because its c/a ratio of 1.6368 is closer to the ideal ratio of c/a = 1.6330, at which both distances become equal, than the ratio c/a = 1.6238 of CdS; in the cubic lattice of CdTe the respective distances are obviously identical. Consequently, for ZnS and CdTe only a single EFG characterizes the In_M^+-V_M^{--} pairs.

The data in Fig. 3 clearly illustrate that a minimum temperature is required for the observation of the Li induced increase in the fraction of In_{Cd}^+-V_{Cd}^{--} pairs. This temperature is about 580 K and 720 K for ZnS and CdTe, respectively, and might be indicative for the onset of the diffusivity of Li atoms in these compounds on a μm scale. Further studies, however, are necessary in order to understand this correlation more quantitatively, as well as the influence of the higher diffusion temperatures on the observed pair fractions.

Information on complexes formed between a group III donor and a metallic vacancy can also be obtained from electron spin resonance b(ESR) experiments. Up to now, this information is only available for ZnS where the formation of D_{Zn}-V_{Zn} pairs (D = Al, Ga, In), so-called A-centers, has been reported[14], whereas for CdS and CdTe similar information is not available. Like the EFG tensor in the PAC experiment also the corresponding g tensor, characterizing these pairs in the ESR experiment, is not axially symmetric.

References

1. Y. Marfaing, Prog. Crystal Growth Charact. **4**, 317 (1981).
2. Th. Wichert, M. Deicher, G. Grübel, R. Keller, N. Schulz und H. Skudlik, Appl. Phys. A **48**, 59 (1989).
3. N. Achtziger, A. Baurichter, S. Deubler, D. Forkel, H. Plank, M. Puschmann, H. Wolf und W. Witthuhn, Mater. Sci. Engineer. **B4**, 169 (1989).
4. H. Wolf, S. Deubler, D.Forkel, H. Foettinger, M. Iwatschenko-Borho, F. Meyer, M. Renn, W. Witthuhn and R. Helbig, Materials Science Forum **10-12**, 863 (1986).
5. S. Deubler, J. Meier, R. Schütz und W. Witthuhn, Nucl. Instr. Meth. B, in print (1991).
6. R. Kalish, M. Deicher und G. Schatz, J. Appl. Phys. **53**, 4793 (1982).
7. D. Wegner und E.A. Meyer, J. Phys.: Condens. Matter **1**, 5403 (1989).
8. E. Bertholdt, M. Frank, F. Gubitz, W. Kreische, Ch. Ott, B. Röseler, F. Schwab, K. Stammler und G. Weeske, Appl. Phys. Lett. **58**, 461 (1991).
9. R. Magerle, M. Deicher, U. Desnica, R. Keller, W. Pfeiffer, F. Pleiter, H. Skudlik und Th. Wichert, Appl. Surface Science, in print (1991).
10. H. Wolf, Th. Krings und Th. Wichert, Nucl. Instr. Meth. B, in print (1991).
11. T. Sasaki, T. Oguchi und H. Katayama-Yoshida, Phys. Rev. B **43**, 9362 (1991).
12. C.H. Henry, K. Nassau und J.W. Shiever, Physical Review B **4**, 2453 (1971).
13. R.N. Bhargava, Journal of Crystal Growth **59**, 15-26 (1982).
14. J. Schneider, Mat.Res.Soc.Symp.Proc. **46**, 13 (1985).

GENERATION OF METASTABLE SHALLOW DONORS UNDER COOLING IN HEXAGONAL II-VI SEMICONDUCTORS

N.E.KORSUNSKAYA, I.V.MARKEVICH, E.P.SHULGA, I.A.DROZDOVA and M.K.SHEINKMAN
Institute of Semiconductors, Ukrainian Academy of Sciences, Kiev 252650, USSR

ABSTRACT

A specific structure which contains metastable centers is found on (0001) plane of CdS, CdSe, CdSSe crystals. The structure does not take place on freshly cleaved surface and arises gradually when crystal is kept after cleavage in the air or inert gas. Metastable centers are shown to be in thin (~100Å) layer near (0001) plane surface. In the metastable state they are shallow donors. The height of the barrier separating the metastable state from the ground one is 0.05eV. The distinctive feature of these centers is that their transition into metastable state is induced by temperature decrease, which results in sharp rise of (0001) plane conductivity under cooling in 300-77K temperature range. The density of shallow donors at 77K reaches $\sim 10^{18} cm^{-3}$. It is supposed that lattice rearrangement which results in creation of shallow donors is stimulated by pyroelectric field and/or mechanical stresses arising under cooling.

I. Introduction

Metastable centers excite great interest and are now investigated intensively (see, for example, [1-3]). Defects which can transfer from the ground state into the metastable one under illumination, carrier injection or external pressure have been observed in a number of semiconductors (Si [4], InSb [5], ZnCdTe:Cl [6], CdF_2 [7], AlGaAs [8], etc.). In many cases such defects act in metastable state as shallow donors, so their transition in this state results in considerable rise of conductivity [5-8]. Recently we have found similar defects in hexagonal II-VI semiconductors. The distinctive feature of these defects, however, is that their transition into the metastable state is induced by the temperature decrease.

Earlier a bizarre effect was observed by us on (0001) plane of CdS, CdSe and CdSSe single crystals, namely, a sharp increase of conductivity under cooling [9]. Detailed investigations described below have led us to the conclusion that it is creation of shallow donors in thin layer near (0001) plane surface that is responsible for the anomalous temperature dependence of conductivity. It have been shown also that metastable centers in near surface layer of (0001) plane are absent initially and arise gradually after crystal cleavage.

II. Experimental

Undoped highly resistive ($\rho_{300K} \geq 10^8 \Omega cm$) CdS single crystals obtained through zone sublimation [10] were used. Samples were cut off from the boule and then cleft as shown in fig.1a. Cadmium (0001) and sulfur (000$\bar{1}$) planes were identified by means of chemical etching [11]. Ohmic indium contacts were melted either after cleavage on basal planes or before cleavage on prismatic planes (fig.1b). The latter fashion was used when characteristics of freshly cleaved surface were investigated.

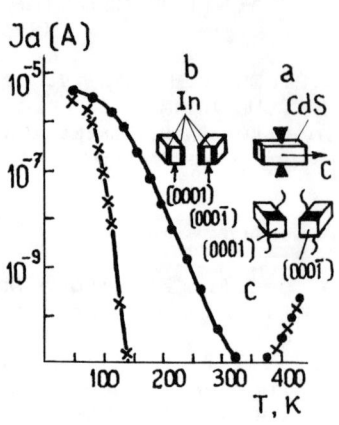

Fig.1.a. Scheme of crystal cleavage.
b. Appearance of mounted samples.
c. Temperature dependence of dark current on (Cd) plane under cooling (.) and under heating (x).

III. Results

It have been found that when highly resistive CdS crystal is cooled from 400 to 77K in dark the dark current through (Cd) plane at first decreases with activation enegy 0.5-0.7 eV, but at 320-300K begins to rise rapidly (fig.1c). At 77K this "anomalous" current J_a proves to be 10^4-10^6 times grater than at 300K, the more fast the cooling the grater J_a. The rise of J_a is accompanied with photosensitivity increase and appearance of persistent photoconductivity (PPC). Heating results in sharp drop of J_a, so at any temperature in the 100-300K range the current measured under heating is much lower than that measured under cooling (fig.1c). When the sample is heated to ~100K and then cooled again, J_a proves to be considerably lower than after direct cooling from 400K. Initial J_a value can be restored by heating to 300-400K and subsequent cooling. The diminution and following restoration of J_a can be made many times, the results being well reproducible. As was mentioned above, J_a could be also regulated by the change of cooling rate. Thus, we can control "anomalous" conductivity and investigate (Cd) plane characteristics at various J_a values. The maximum value of $J_a = J_a^{max}$ has been obtained by quick cooling (~0.5degree/s) from 400K. Photosensitivity and PPC values correlate with J_a one. When after preliminary cooling from 400K the sample is kept at the constant temperature T_r, the relaxation of J_a is observed. The higher T_r the greater relaxation rate V (fig.2a). The value of activation energy of this relaxation process obtained from lgV(1/T_r) dependence was found to be 0.05eV

(fig.2b). Relaxed current can be restored again by cooling from 300-400K.
Hall effect measurements show that J_a rise under cooling and its drop under heating or relaxation are due to the change of free electron density. The mobility of "anomalous" electrons at 77K proves to be about $10 cm^2/Vs$ and surface electron density at $J_a=J_a^{max}$ reaches $10^{11}-10^{12} cm^{-2}$.
Photo Hall effect under illumination with intrinsic light ($\lambda=490nm$) which is absorbed in $10^{-5} cm$ near surface layer have been also investigated on both basal as well as prismatic planes. The photoelectron mobility has been found to be almost the same and reaches about $300 cm^2/Vs$ for all planes. Thus, the thickness of conductive layer is less than $10^{-5} cm$.
In 77-4.2K temperature range J_a reduces under cooling and restores after subsequent heating (fig.1c). The plot of $lgJ_a(1/T)$ function is the straight line which slope depends on "anomalous" conductivity value at 77K. At $J_a \ll J_a^{max}$ the activation energy obtained from this slope $E_a=0.035eV$. The increase of J_a results in the decrease of E_a and at $J_a=J_a^{max}$ $E_a \leq 0.005eV$. By the other words, at T<77K "anomalous" conductivity behaves as "normal" dark conductivity does [13,14]. It is known that when density of noncompensated shallow donors N_d reaches $10^{18} cm^{-3}$ degeneration of conductivity takes place [14]. It is reasonably to think that J_a decrease at T<77K is due to

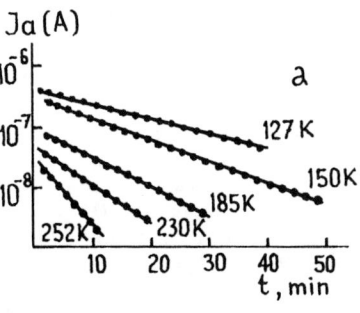

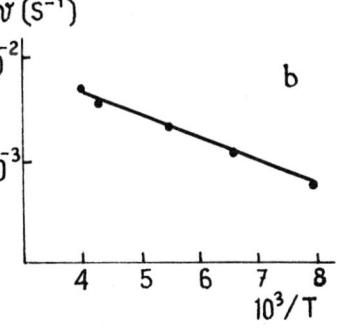

Fig.2. a. Isothermal relaxation of J_a at various temperatures after cooling of the sample from 400K. b.Temperature dependence of J_a relaxation rate.

freezing of "anomalous" electrons at hydrogenlike donors. Then bulk "anomalous" electron density at $J_a=J_a^{max}$, when E_d is negligible, is about $10^{16} cm^{-3}$. Therefore, the thickness of conductive layer is about 100A.
In exciton luminescence spectra of (Cd)and (S) planes I_1 and I_2 lines are observed which are known to be due to recombination of excitons bound to neutral shallow acceptors and neutral hydrogenlike donors correspondingly [15] (fig.3). The ratio of

these lines intensities $W_{I_2}/W_{I_1} = \alpha$ is considered to give the information about ratio of donor to acceptor densities in the crystal subsurface layer where exciting light is absorbed (10^{-5} cm) [16,17]. It have been found that α on (Cd) plane of the certain sample depends on J_a value. When J_a is reduced by heating of the sample to T=100K, α is decreased too (fig.3b). It is necessary to note that α value on (Cd) plane always proves to be essentially greater than that on (S) plane of the same crystal (fig.3a). It has been found that all effects described above do not take place on freshly cleaved surface. When the sample is quickly cooled to 77K at once after cleavage, (0001) plane has immeasurably small dark conductivity and low photosensitivity, PPC being absent. Holding of the sample after cleavage at 300K for a time interval $\Delta t=20-30$ s results in appearance of noticeable J_a. The more Δt the greater J_a, the latter reaching saturation value J_a^S at $\Delta t=20-30$ hours. The cleavage of the crystal and its holding at 300K was carried out in the air as well as in helium gas, the results in both cases being practically identical (fig.4). The rise of the temperature hastens J_a saturation process: at 400K the latter comes to end at $\Delta t \sim 10^2$ s. The rise of J_a with Δt on freshly cleaved surface is accompanied with the increase of α value in (Cd) plane exciton luminescence. At the same time α on (S) plane remains unchanged.

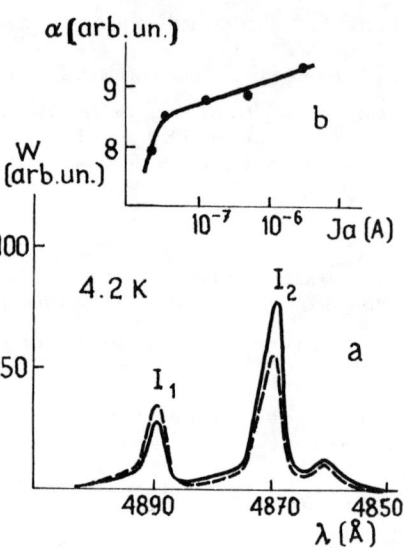

Fig.3. a.Exciton luminescence spectra on (Cd) plane (solid curve) and (S) plane (dashed curve) of the same crystal. b.Dependence of $\alpha = W_{I_2}/W_{I_1}$ on J_a value on (Cd) plane.

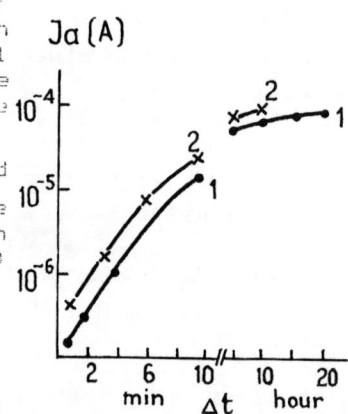

Fig.4. Dependence of J_a on time after crystal cleavage: 1-in the air; 2-in the helium gas.

IV. Discussion

It is well known that II-VI hexagonal compounds are pyroelectrics [18]. Under cooling (0001) plane acquires positive charge [19], and so creation of conductive layer near (0001) plane surface may first of all be thought to cause by compensation of pyroelectric field with crystal electrons. Some experimental data, however, contradict this simple electronic model.

i. If the conductive layer is due to compensation of pyroelectric field, the same effect would be observed on (S) plane under heating. In reality, however, increase of (S) plane conductivity under heating does not take place.

ii. The major part of "anomalous" electrons arises in 170-80K temperature range (fig.2), where more than $10^{18} cm^{-2} degree^{-1}$ free electrons are added under cooling. At the same time CdS pyroelectric constant in 300-77K temperature range is equal to $\sim 2 \cdot 10^{-10}$ Coulomb/cm^2 degree [19], and so only $1.3 \cdot 10^9 cm^{-2} degree^{-1}$ electrons are required to compensate pyroelectric field.

iii. Since equilibrium free electron density in investigated crystals is very low ($<10^7 cm^{-3}$ at 300K), electrons required for pyroelectric field compensation have to be supplied from deep centers by either thermal generation or pyroelectric field ionization. At T<170K, however, thermal ionization of deep centers is negligible. On the other hand, pyroelectric field arising in CdS crystal after cooling from 300 to 77K $E_p \sim 6 \cdot 10^4 V/cm^2$ is too week to ionize deep centers.

iv. Electronic model is not be able to explain the rise of α with the J_a increase. One has to conclude, therefore, that either creation of hydrogenlike donors or destruction of acceptors takes place on (Cd) plane.

The former seems to be more probable, because it is difficult to believe that initial density of donors in undoped highly resistive crystals is so large as $\sim 10^{18} cm^{-3}$. Another argument for donor creation is that mobility of electrons behind conductive layer is much greater than that of "anomalous" electrons. Destruction of acceptors would lead to the increase of electron mobility μ_n in subsurface layer. At the same time creation of donors have to result in μ_n reduction.

So, we can to state out that the increase of conductivity of (0001) plane occur due to not electronic, but ionic process, namely creation of hydrogenlike donors under cooling.

Highly conductive state of (0001) plane is nonequilibrium (metastable) one: when cooling is stopped the relaxation into the ground state occurs. Activation energy of this process 0.05eV is the height of the barrier separating the metastable state from the ground one. Lattice rearrangement which results in the creation of shallow donors may be supposed to be induced by pyroelectric field and/or mechanical stresses which arise under cooling.

Thus, investigations described above show that (0001) plane subsurface layer has a specific structure which contains metastable centers.

This structure, however, does not take place on freshly cleaved surface. The increase of J_a and α with Δt leads to the conclusion that it emerges gradually after crystal cleavage. The rough estimation of emergence process activation energy gives the value of 0.6-0.7eV.

The nature of observed metastable centers, mechanism of their transition into metastable state, as well as the mechanism of creation of specific structure on (0001) plane is the matter of further investigations.

REFERENCES

1. J.M.Langer, "Lecture Notes in Physics", Springer-Verlag, Berlin, 122, p.123 (1980).
2. G.A.Baraff, Proc. 14th Int. Conf. on Defects in Semiconductors, (Paris, 1986) part 1, p.372.
3. M. Scheffler, Festkorperprobleme 29, 231 (1989).
4. M.T.Asom, J.L.Benton, R.Sauer, L.C.Kimerling, Appl. Phys. Lett., 51, 256 (1987).
5. S.Porowski, M.Konczykowski, J.Chroboczek, Phys. Stat. Sol. (b), 63, 291 (1974).
6. B.C.Burkey, R.P.Khosla, J.R.Fisher, D.L.Loux, J. Appl. Phys., 47, 1095 (1976).
7. U.Piekara, J.M.Langer, B.Krukowska-Flude, Solid St. Com., 23, 583 (1977).
8. T.N.Theis, Proc. 14th Int. Conf. on Defects in Semiconductors, (Paris, 1986) part 1, p.393.
9. N.E.Korsunskaya, I.V.Markevich, E.P.Shulga, Ukr. Fiz. Zh., 33, 1673 (1988).
10. E.V.Markov, A.A.Davydov, Izvestia Acad. Nauk SSSR, Ser. Neorg. Mater., 7, 575 (1971).
11. S.J.Czyzak, J. Appl. Phys., 31, 94 (1960).
12. "Physics and Chemistry of II-VI Compounds" edited by M.Aven and J.S.Prener (Amsterdam, 1967) Chap.7.
13. Ibid, Chap.11.
14. M.Itacura, H.Toyoda, J. Phys. Soc. Jap., 18, 150 (1963).
15. "Physics and Chemistry of II-VI Compounds" edited by M.Aven and J.S.Prener (Amsterdam, 1967) Chap.8.
16. M.A.Subhan, M.N.Islam, J.Woods, J. Phys. Chem. Sol., 33, 229 (1972).
17. A.P.Ahojan, N.E.Korsunskaya, I.V.Markevich, Zh. Prikl. Spektr., 49, 859 (1988).
18. H.A.Klasens, Nature, 158, 306 (1946).
19. W.J.Minkus, Phys. Rev. 136, A1277 (1965).

Strain Relief in Thin Films: Can We Control It?

F.K. LeGoues

IBM Research Division, T.J. Watson Research Center, Yorktown Heights, New York 10598, USA.

ABSTRACT

In this paper, we review several experiments which tackle the questions of the importance of growth conditions on the introduction, and microstructure, of strain relieving defects, and show that the nucleation of dislocations is the critical step in determining and controlling the formation of defects.

In the first set of experiments, the growth of Ge on Si(001) and Si(111) was changed from islands to layer-by layer by using a surfactant. This has a dramatic effect on the nucleation of defects. Indeed, in the Ge/Si(001) system, the nucleation of dislocations is suppressed and novel V-shaped defects are formed. For the Ge/Si(111) system, the same dislocations are formed both during islanding and layer-by-layer growth. But layer-by-layer growth forces dislocations to nucleate at the surface as partials, which glide to the interface leaving a stacking fault threading through the thin film. These faults are later annihilated by the glide of a second partial dislocation, resulting in a perfect, relaxed Ge film.

Secondly, we review the anomalous strain relaxation observed in compositionally graded layers of SiGe/Si(001). In this case, Frank-Read type sources of dislocations are observed, which results in dislocations being injected deep into the Si substrate, leaving the top of the film itself defect free. We discuss the mechanisms involved and show that this method can be used to grow relaxed, defect free SiGe alloy films of arbitrary thickness and composition.

I) INTRODUCTION

Strain and strain relieving defects play a considerable role in determining the electronic properties as well as the microstructure of epitaxial layers. Strain can be used in technology to manufacture materials with specific properties — e.g. change the band gap in very thin Si/Ge superlattices[1], but must be understood and controlled in order to obtain the desired properties reproducibly. Considerable progress has been achieved recently in the field of SiGe heterojunction type devices[2]. This was done by using the band offset provided by alloying Si with Ge, while keeping the SiGe film strained, i.e, at the lattice parameter of the Si substrate. This completely avoids the issue of defect formation and of defects threading through the thin film, and is is relatively straightforward, as it only requires that, given a certain Ge composition (and thus a certain strain), the SiGe thin film be kept below a given critical thickness. While there are serious discrepancies between theories[3,4] used to calculate this critical thickness and experimental results[5], it is obvious that, for a given misfit between thin film and substrate, there is a thickness under which the introduction of dislocations actually raises the total energy of the system, so that there is no driving force for dislocation formation. Thus, there is a critical thickness under which the thin film will be stable throughout the processing steps involved in the fabrication of the device. This thickness is better know empirically than understood theoretically, but, for the purpose of making devices, this is quite appropriate.

The problems that have been encountered when trying to correlate the measured critical thickness to theory are two fold: Firstly, the theoretical critical thickness relates only to the formation of "the first dislocation", i.e, it actually cannot be detected by any practical means: by the time usual techniques detect dislocations, they are generally a very high number of them. Thus, the experiments are actually more a measure of the limitation of the instrument being used than a real measure of the elusive critical thickness. Secondly, the problem is too complicated to be completely modeled so that some simplifying assumptions have to be made, which tend to shed some doubts on the theoretical projections. For example, the most often used theoretical critical thickness is the one that was calculated by Matthews and Blakeslee[4]. Two major assumptions were made. First, it was assumed that dislocations were present in the substrate, so that it was only necessary to bend the dislocations along the interface, not to nucleate them. This can be extended to cases where nucleation is so easy that the limiting step during relaxation is the motion of the dislocations. We will show in this paper that, given today's almost perfect substrates and cleaning procedures, neither of these applies. The second major assumption was that the interaction between dislocations could be neglected. It has been shown by Hull[6] and by Freund[7] that interaction between dislocations is critical in determining the final microstructure, and in particular, the density of threading dislocations at the end of the relaxation process. We will show here that this interaction can be overcome to obtain perfect, relaxed thin films.

In this paper, we describe two sets of experiments which address these issues. In the first set of experiments, we study the effect of growth morphology upon introduction of strain relieving defects. We demonstrate that the nucleation of dislocations can be drastically altered by changing the growth mode from a island type of growth to a layer-by-layer type of growth. In effect, layer-by-layer growth suppresses low energy nucleation sites generally provided by the edges of islands. This work demonstrates the critical role played by nucleation sites for dislocation, and the fact that the notion of critical thickness is meaningless unless nucleation energy of dislocations is taken into account. In the second set of experiments, we show that the final microstructure of a relaxed film can be controlled by the simple "trick" of grading the film compositionally, from the composition of the substrate, to that of the desired overlayer. In this case again, we limit the number of nucleation sites, and force the dislocation to move on a "graded" set of planes, thus greatly reducing interaction between dislocations. We will show that this gives us enough control of the dislocations to achieve virtually perfect films of arbitrary composition and lattice parameter.

II) Surfactant aided epitaxy.

In a recent series of papers, Copel et al[8,9], and Horn-von Hoegen et al[10] have shown that the growth mode of Ge on Si can be drastically altered by using a surfactant, and LeGoues et al[11,12,13] have studied the effect of that change upon the final microstructure of the Ge thin film. Ge tends to grow layer-by-layer both on Si(100) and on Si(111) because it has a lower surface energy than Si. This layer-by-layer growth is limited though, because of the 4% size difference between the Si and Ge lattices, causing strain in the Ge film. Indeed, after only three monolayers (ML) of Ge, the strain in the Ge film is high enough to destabilize the layer-by-layer growth and, from this point on, islanding occurs. By using a surfactant whose surface energy is lower than that of both the Ge-terminated surface and the Si-terminated surface, Copel et al[8,9] and Horn-von Hoegen et al[10] have shown that the Ge surface mobility can be decreased enough to prevent islanding altogether so that layer-by-layer growth can be maintained for arbitrary thicknesses.

II-a) Experiment.

Ge films were grown by molecular beam epitaxy (MBE) on Si(001) and Si(111), both with and without a surfactant. For the Si(100) case, arsenic was used as a surfactant, and the growth was done at a temperature of about 500°C. For the case of Si(111), antimony was used, and the films were grown at temperature of about 610°C. In both cases, the temperature is rather critical, since it it important to achieve a full monolayer coverage of the surfactant, while preventing desorbtion during growth and interdiffusion of Si in the Ge film. Prior to growth, the substrates were cleaned by mild sputtering and heating to about 1050°C. The samples were prepared for TEM observation by mechanical thinning to about 20μm followed by ion-milling at liquid nitrogen temperature to electron transparency. They were then observed in a Phillips 430 operating at 300kV and in a Jeol 4000 microscope operating at 400kV.

II-b) Growth on Si(001): Formation of V-shaped defects.

Figure 1 shows cross-sectional micrographs of 6 ML of Ge grown on Si(001), with and without a surfactant. Clearly, the surfactant has prevented islanding, and layer-by-layer growth has been achieved. For such thin films, both cases result in the Ge being strained to match the Si substrate (note that this is true even for the islands: there are no dislocations in any of the islands). In the island case though, it is expected that the side of the islands will act as a nucleation site for dislocation half-loops. This will results in a very defective microstructure, with numerous threading dislocations being left in the Ge film at the end of the relaxation process. It has been argued that, during layer-by-layer growth, dislocations would have to be formed as half-loops from the surface. This is unlikely though because calculations of the energy necessary to homogeneously form such half-loops would require strain far in excess of that generated by the 4% misfit between Si and Ge[14]. Here, we can finally do a controlled test of this idea: Indeed, we are growing the same Ge film, at the same temperature, but one has grown layer-by-layer, while the other one has islanded. Figure 2 shows a 12 ML Ge film grown layer-by-layer. Strain relieving defects have been formed, but they are not the expected dislocations. These so called "V-shaped" defects are similar to deformation twinning and relieve the strain by tilting a few planes of the Ge film so as to "pack more atoms in the same space", i.e, planes are tilted so that the closed packed planes are perpendicular to the direction of maximum strain. A complete description of the V-shaped defects is given in refs. 11 and 12. Fig. 3 shows another defect, after further Ge growth. Here, we note two important features.

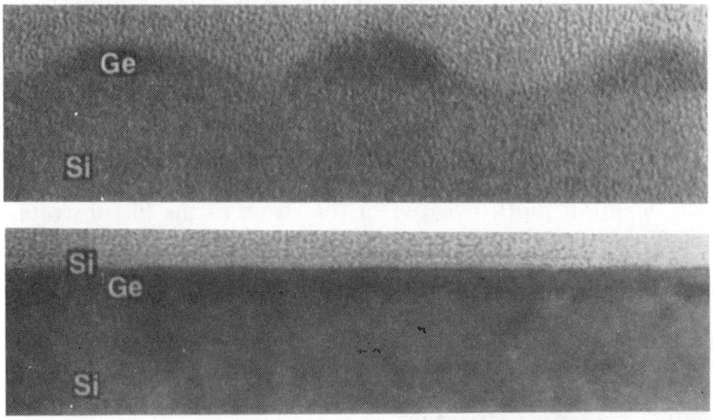

Figure 1: Cross-sectional micrographs of 6 ML of Ge on Si(001).
a) grown without a surfactant.
b) grown with an As surfactant.

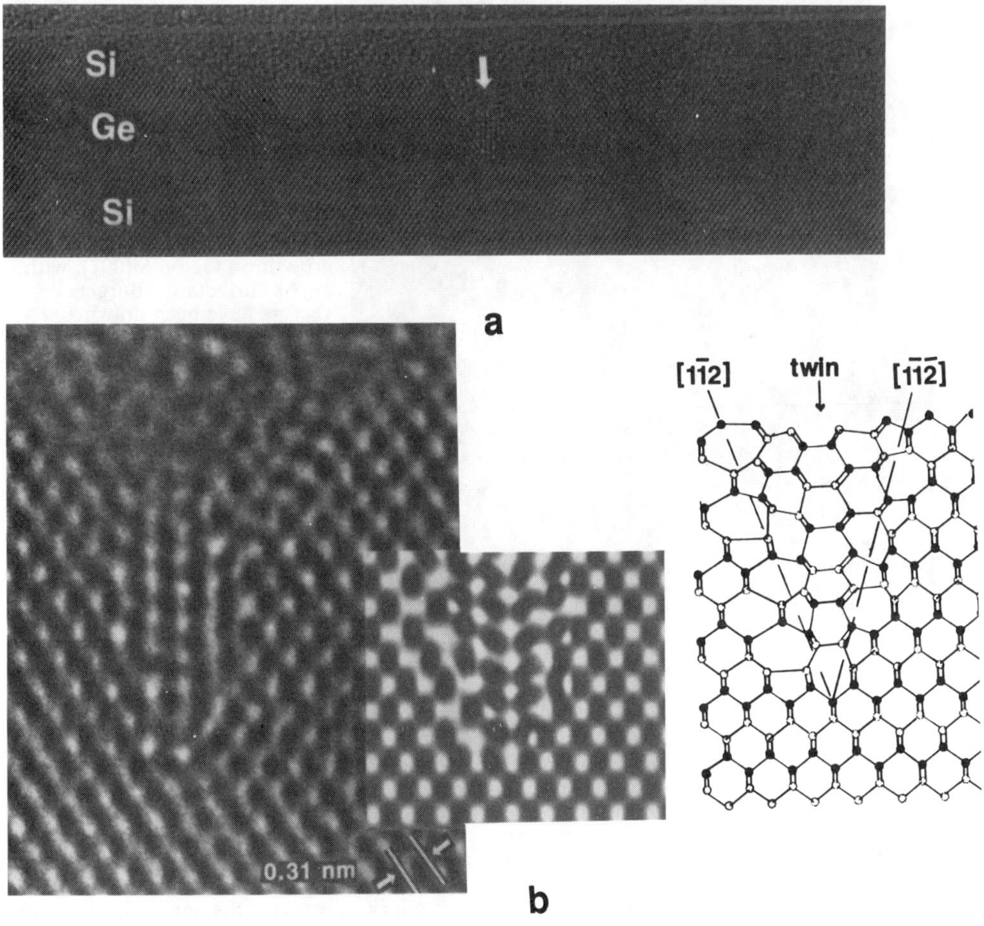

Figure 2: Cross-sectional micrograph of 12Ml Ge on Si(001), grown with a surfactant.
a) general view.
b) detail of the V-shaped defect. An atomic model, and simulation are also included.

First, the defect itself has acted as a nucleation site for a 90° dislocation, which has been injected in the substrate itself. Secondly, since the defect has been overgrown, we can now draw a complete Burger's circuit around it and demonstrate that it indeed acts as a 90° dislocation itself. Thus, the total strain is relieved partly by the V-shaped defects, and partly by dislocations in the substrate. In ref. 12, we demonstrated that the dislocations were injected in the substrate because of the repulsive forces between the dislocation and the defect. This results in the Si substrate itself being severely strained for a few monolayers, in order to meet the Ge lattice "halfway". Thus this system, even after complete relaxation ,i.e, when the top of the Ge film has its natural lattice parameter, will still be under considerable strain both because of the strain in the Si substrate, and because the bottom of the Ge layer has a lattice parameter varying continually as the V-shaped defect grows.

Contrary to what appears to be the case in Fig. 3, in most cases the overgrown Ge on top of the defect is not perfect. Indeed, numerous twins and stacking faults are generated during this process, so that these films are not directly usable for electronic applications.

Figure 3: Detail of a V-shaped defect after further growth on Ge on Si(001), with an As surfactant. Burgers vectors have been drawn around the defect, and around the dislocation that has been injected in the Si substrate.

II-c) Growth on Si(111): Formation of perfect, relaxed Ge thin films.

When Ge is grown on an antimony terminated Si(111) substrate, again, layer-by-layer growth is achieved. Unlike the case of Ge/As/Si(001) though, the type of defects that are formed is not changed by the growth morphology. The same dissociated dislocations are formed in both cases, but the location at which the nucleation events occur is changed, which results in very different final microstructures. Figure 4 shows high resolution micrographs of about 60 ML of Ge deposited on Si(111) with and without a surfactant. When island growth occurs, the edges of the islands provide numerous nucleation sites for dislocations. The dislocations then glide as partials along the (111) plane. Each partial dislocation generates a stacking fault as it glides toward the center of the island. If a complementary partial then glides on exactly the same plane, it will annihilate the stacking fault, but this is statistically unlikely because of the plethora of nucleation sites provided by the edges of the islands. Thus, the microstructure shown on Fig. 4a results, where the island is full of stacking fault and twins. When the islands finally coalesce, these twins and stacking faults will thread to the surface, resulting in a very defected Ge layer. On the other hand, when Ge is grown layer-by-layer on Si(111), the perfect microstructure shown on Fig. 4b results. Here, the strain is relieved by a series of partial dislocations located on one single plane at the interface. The dislocations are the same as in the island case, but, unlike the island case, they are restricted to the plane of the interface, defining a patchwork of alternatively perfect and faulted areas in this single plane, leaving the film itself defect free and relaxed.

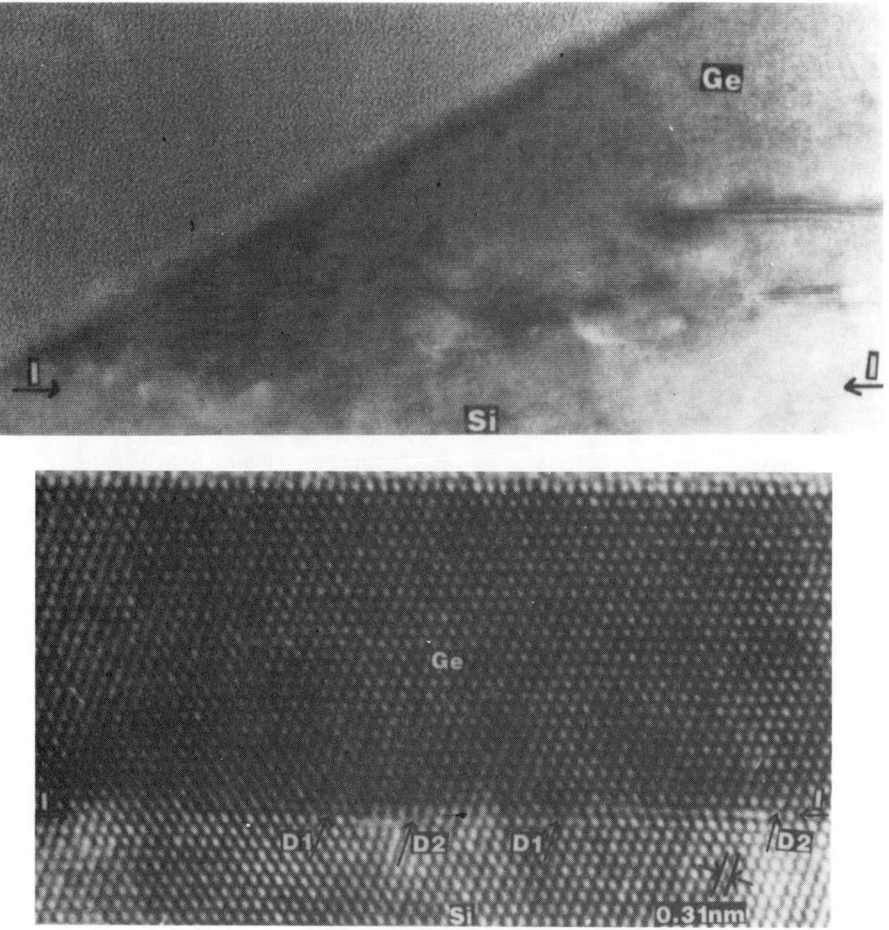

Figure 4: High resolution micrographs of about 60ML of Ge grown on Si(111). a) without a surfactant (side of an island). b) with antimony as a surfactant.

Since the film is grown layer-by-layer, the only possible nucleation site for dislocations is the surface. Surprisingly, the dislocations that are observed on Fig. 4b cannot have formed in this configuration at the surface. In order to understand why this has to be the case, let us consider the case of the set of dislocations imaged in Fig. 4b. Since this is a $[1\bar{1}0]$ cross-section, the dislocation marked D_1, which is directly imaged here as an extra lattice fringe, has to have a Burger's vector equal to $1/6[11\bar{2}]$. D_1 thus has a Burger's vector perpendicular to the plane of the cross-section, which is why it Burger's vector can be directly imaged. D_2 corresponds to the complementary dislocation, i.e, it restores the perfect lattice at the end of the stacking fault and has thus a Burgers vector equal to $1/6[2\bar{1}\bar{1}]$. Thus, the "total" dislocation ($D_1 + D_2$) has a Burger's vector equal to $1/2[10\bar{1}]$. The partials as well as the full dislocation can all glide on the (111) plane of the interface, and ONLY on this plane. Thus it is impossible that they had formed as half-loops from the surface, since this would require glide along another {111} plane, i.e, $(\bar{1}11)$, $(1\bar{1}1)$, or $(11\bar{1})$. This final microstructure thus must have evolved from the formation and reaction of dislocations that have formed at the surface and glided on one the these planes. This is illustrated schematically in Fig. 5. As shown in Fig. 5a, the $1/2[10\bar{1}]$ Burgers vector can be dissociated into two other Burgers vectors. $1/2[10\bar{1}] = 1/2[0\bar{1}\bar{1}] + 1/2[110]$. $1/2[0\bar{1}\bar{1}]$ can glide toward the interface on the $(1\bar{1}1)$ plane as a half-loop from the surface. When the loop reaches the interface, it can cross-slip onto the plane of the interface, forming the dislocation observed experimentally in Fig. 4b, plus a full edge dislocation with a Burgers vector equal to 1/2[110]. This last dislocation can then climb out of the film, which incidentally takes care of the "extra" atoms in the film.

But dislocations in this system tend to form as partials, so that this reaction does not happen as one single step. Fig. 5b describes the sets of partials involved. First, a partial dislocation with Burgers vector equal to 1/6[$\bar{1}\bar{1}$2] loops from the surface and glides toward the interface. the first partial to be nucleated will be the one with the highest driving force for nucleation, i.e, the one whose effective Burgers vector in the plane of the interface is largest. The glide of this partial dislocation generates a stacking fault in the film, leaving it defective. This stage of the film growth in depicted in Fig. 6. Here, the first partial has glided to the interface and cross-slipped unto the plane of the interface, leaving a stacking fault along the interface, and threading through the film. Because they are no other favorable nucleation sites for dislocations, the intersection between the stacking fault and the surface acts as a prefered nucleation site for the second partial, 1/6[1$\bar{2}$1], which, upon gliding to the interface, annihilates the previously formed stacking faults. We have thus a "self-annihilating" defect. At the end of the growth all of the stacking faults formed at intermediate growth thicknesses have been annihilated, leaving the film relaxed and defect free. A more detailed description of this mechanism can be found in ref. 13.

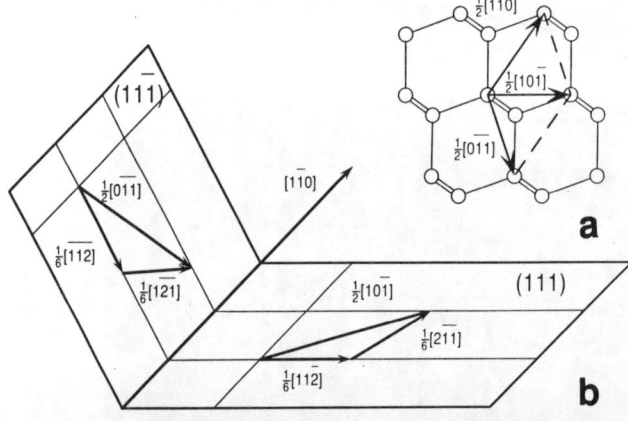

Figure 5: Schematic representation of the dislocations involved in the relaxation process.
a) Burger's vectors of the full dislocations.
b) Burger's vector, line and glide plane of the partial dislocations participating in the relaxation.

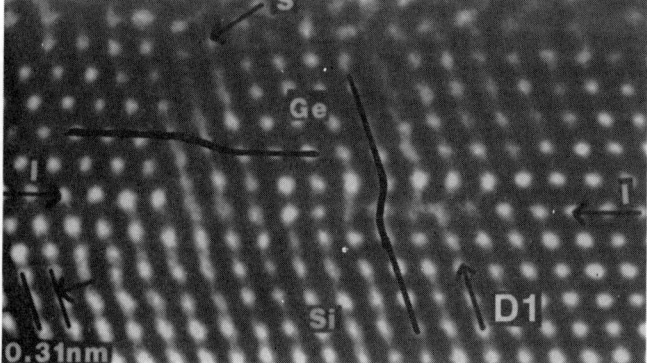

Figure 6: Cross-sectional micrograph of 15ML Ge grown on Si(111) with an Sb surfactant. The dislocation D_1 has been highlighted, as well as the stacking fault along the interface, and the stacking fault threading through the Ge film.

II-d) Conclusions.

This study of the effect of growth morphology on the introduction of strain relieving defects highlights the critical importance of the nucleation energy and site on the final microstructure of thin films. In the case of Ge/Si(001), layer-by-layer growth renders dislocation formation unfavorable. Instead, V-shaped defects are formed. These actually play the role of a dislocation with extended core, since a Burgers vector can be drawn around them, once they have been overgrown. In this case, it is obvious that the calculations, à la Matthews and Blakeslee, which involve classical elasticity cannot be of any value: Here the "dislocation core" is larger that the thickness of the film in which it forms. It also shows that the nucleation energy is a critical factor in calculating critical thickness. This renders the problem significantly more complicated since we now need to address where, and how, are dislocations formed.

The case of Ge/Si(111) is simpler since the same dislocations are formed both during layer-by-layer growth and islanding. Nonetheless it demonstrates the importance of the location of the dislocation formation: Here, by limiting the sites at which dislocations can be formed, we have in effect complete control on the film microstructure, and can indeed achieve films that are relaxed and defect free.

III) Compositionally graded films.

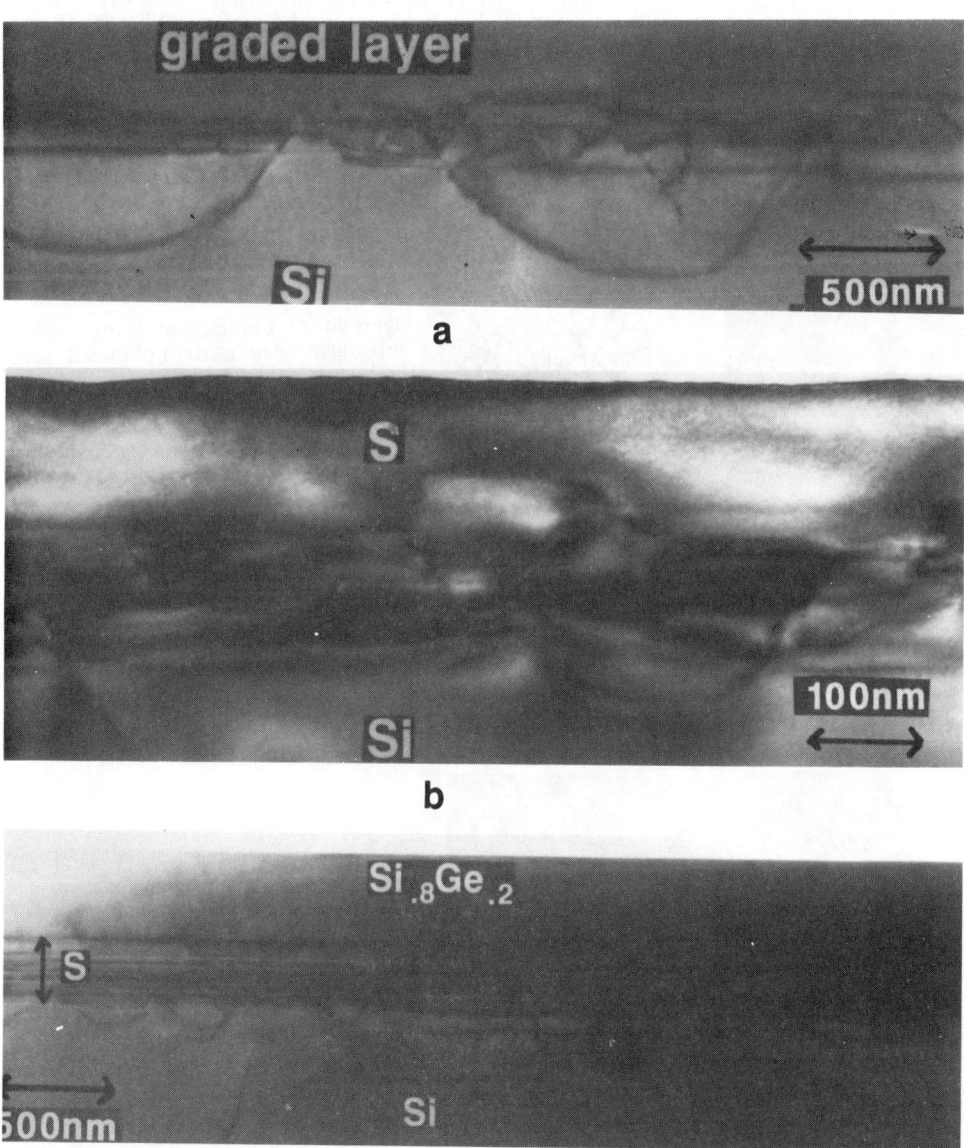

Figure 7: Relaxed SiGe alloys films grown by grading the composition, and showing the anomalous relaxation process. a) The Ge composition increases linearly. b) The Ge composition increases in a step-wise manner. c) The graded buffer consist of a superlattice (marked "S"), corresponding to: 200Å $Si_{95}Ge_5$/50Å Si/200Å $Si_{90}Ge_{10}$/50Å Si/ ($Si_{85}Ge_{15}$/50Å Si) three times/ $Si_{82}Ge_{18}$/50Å Si.

Here we review recently published data[15], relating to the control of threading dislocations in a relaxed film by grading the film compositionally. In this case, we not only control the nucleation of dislocations, but also their velocity to such an extent that relaxed SiGe films of arbitrary thickness and composition can be grown. We review the conditions necessary to achieve this and describe possible extension of this technique to other materials.

III-a) Experiment.

SiGe thin films were grown both by UHV-chemical vapor deposition (CVD) and MBE at a temperature of about 500°C. The cleaning procedures used before the growth are described in refs. 16 and 17 respectively. TEM samples were prepared in the same fashion as described in the previous section. The observations were done at 300kV, except for the convergent beam patterns, which were obtained at 100kV.

III-b) Results.
III-b-1) Perfect, relaxed films of arbitrary composition.

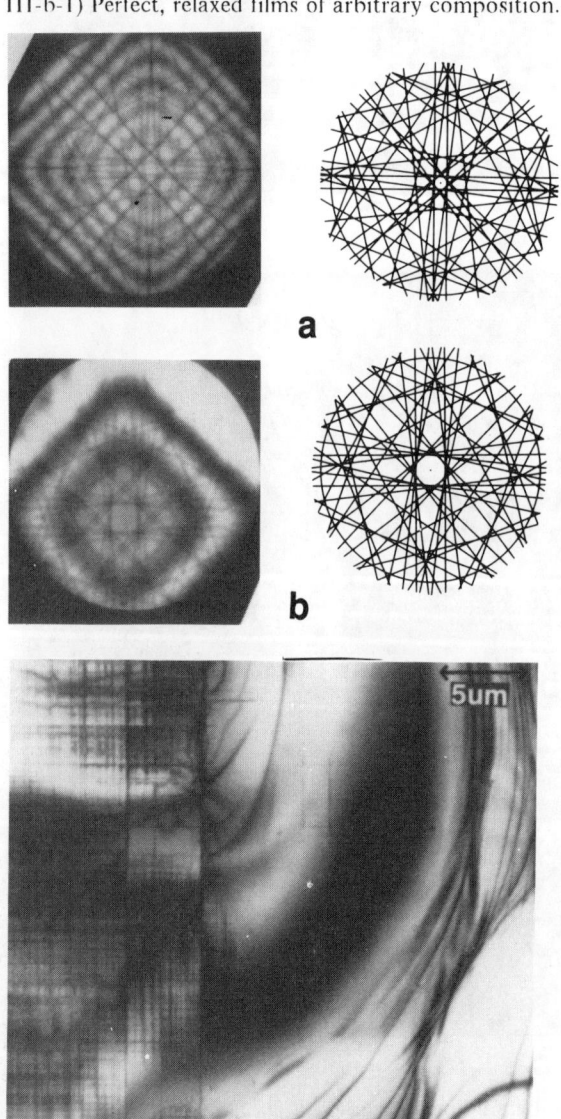

Figure 8: a) Center spot of the convergent beam pattern obtained from the Si substrate, and correspnding simulation.
b) Center spot of the convergent beam pattern obtained form the top layer of the sample shown in Fig. 6c, and corresponding simulation.
c) Planar view of the sample shown in Fig. 6c: the thickness of the TEM sample increases in the direction of the arrow.

Fig. 7 shows several samples grown by UHV-CVD, where the top layer is relaxed, and defect free. In fig. 7a, the Ge composition varies linearly from 0 at the substrate to about 25% at the top of the film. Surprisingly, no threading dislocations are present at the top of the film, while numerous dislocations seem to loop deep inside the Si substrate. Fig. 7b shows a sample were the grading has been done in a step-wise manner, i.e, the Ge composition increases from one of the superlattice layer to the next. Again, the top layer, containing 20%Ge is relaxed and dislocation free. Fig. 7c shows a sample where the graded layer consists of a superlattice, where Ge-rich layers of increasing Ge concentration are separated by Si-rich layers. The same phenomenon occurs here too. This same sample was also used to ascertain the complete relaxation of the top layer: Cross-sectional views were prepared perpendicular to the (100) direction (instead of the more usual (110) direction), and cooled to about -140°C in the TEM in order to obtain clear convergent beam patterns. Convergent beam diffraction is extremely sensitive to very small distortions of the lattice, so that tetragonal distortion due to strain, as well as changes in the lattice parameter, can be readily detected. Fig. 8 shows the center spots from the convergent beam pattern obtained from the Si substrate, and from the top-most layer of the sample shown in 7c. Fig. 8 also includes simulated patterns, which show that, using the Si lattice as a reference, the pattern obtained from the top layer can be reproduced if we assumed a completely relaxed SiGe lattice containing 20% Ge. The complete relaxation can actually be deduced simply by noting that the square symmetry is retained from 8a to 8b, demonstrating the lack of tetragonal distortion in the top layer. In order to quantify the quality of this relaxed top layer, planar view TEM was done, Fig. 8c. Here, we are looking at a wedge, i.e, toward the right of the picture, the TEM sample is very thin, so that we are only imaging the very top portion of the sample. As the TEM sample gets thicker (toward the left of the picture), we start probing deeper into the sample, and intersecting the dislocations that are buried in the superlattice and the substrate.

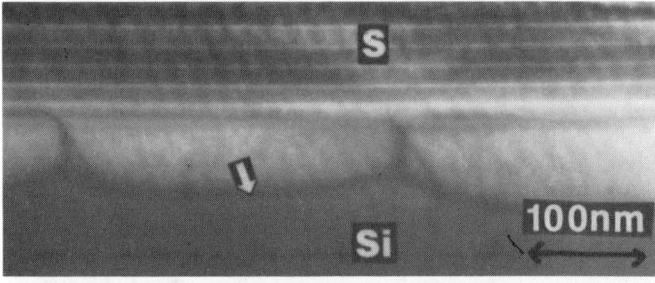

a

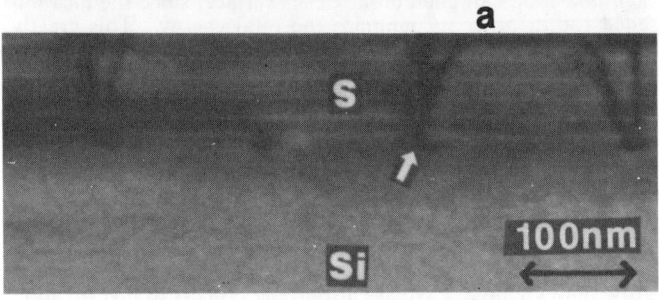

b

c

Figure 9: Graded samples demonstrating conditions for the anomalous relaxation to occur.
a) Sample grown by MBE on a clean starting surface.
b) Sample grown by MBE on a surface where particulates are present.
c) Sample grown by CVD, in the same conditions as the sample shown in Fig. 6a, but the Ge concentration in the thin film is kept constant at 25%.

By working at very low magnification, and probing several of the thin areas, we obtain an upper limit for the number of threading dislocations of $10^4/cm^2$, which correspond to a reduction of about 7 orders of magnitude compared to similar layers grown without the graded buffer. It is worth noting that this means that such materials are now device grade materials, that can be, and have been[18,19], used for novel device structures. The defect density was corroborated independently by etch pit counts, and the same number of $10^4/cm^2$ was obtained.

III-b-2) Conditions and Mechanism.

We have shown in Fig. 7 that, in order for this new mechanism to operate, the buffer layer can either be linearly, or step graded. Indeed, the step grading can be done in a rather complicated manner, i.e, as a superlattice, were the layers of increasing Ge concentrations are separated by pure Si, or low Ge concentration layers. Fig. 9 illustrates the other conditions necessary for this phenomenon to occur: First, as shown in fig. 9a, the sample can be grown by MBE instead of CVD. Secondly, the condition of the initial growth interface is critical. If particulates are left at this interface, they will pin-down threading dislocations and results in a very defective final microstructure, Fig. 9b. Fig. 9c illustrates the fact that a clean interface is not enough for this mechanism to occurs: here, the sample was grown in exactly the same conditions as the one shown on Fig. 7a, but the Ge content was kept constant. Thus, the grading itself plays a determining role.

The mechanism proposed for this anomalous strain relief was described in in ref. 15. This mechanism has to be able to explain the two striking experimental facts, first that the top layer is relaxed and defect free, and, secondly, that dislocations are injected deep into the Si substrate itself. Fig. 10 demonstrates that the dislocations that are very deep inside the substrate are in fact part of a pile-up originating near the interface. It also shows, through a $\vec{g}\cdot\vec{b}$ analysis, that all of the dislocations in one pile-up have the same Burgers vector, equal to $1/2[10\bar{1}]$, i.e glissile on the (111) plane. This strongly suggest the presence of a Frank-Read type of dislocation source at, or near, the substrate/overlayer interface. We propose that the grading, as well as the very clean growth surface, make it very difficult to nucleate dislocations. An initial network probably forms at defects (possibly particulates or "diamond-defects" similar to those described by Eaglesham et al[20]), significantly past the critical thickness (as defined by Matthews and Blakeslee), and thus are under enough strain so that the threading part can move all the way to the edges of the wafer. After this initial network has formed, the intersections between dislocations start acting as Frank-Read sources, as shown in Fig. 11. Each new loop formed pushes the previous one further into the substrate, explaining the presence of very deep dislocation pile-ups in the Si substrate. Further, because of the grading, each new loop can glide on a "fresh" surface, since the location of the dislocation needs to be graded too in order to minimize the total energy. This greatly minimizes the interaction between dislocations, and thus the possibility of pinning by intersecting dislocations[6,7]. This in turn explains the lack on threading dislocations in the top layer: all of the threading parts have been able to move unimpaired to the edges of the wafer.

III-b-3) Other materials - Outlook for future uses of this technique.

We have presented only results concerning SiGe films grown onto Si(100), with a maximum composition of 25%Ge. We have successfully grown layers containing up to 60%Ge, fully relaxed and defect free. These can be used for devices that require straining the Si lattice, e.g, tunnelling junction, or as a substrate for strain symmetrized superlattices. This technique could be extended straightforwardly to grow pure Ge on Si. We are also in the process of investigating this technique for different materials, e.g. GaInAs/GaAs. The process described here is really quite general, and, given that in these systems, the same type of dislocation as in SiGe are formed, the same phenomenon should occur. Similarly, a number of II-VI systems should display the same behavior. Eventually, this technique could also be used to manufacture perfect substrates of arbitrary composition/lattice parameters, on which to grow devices or novel materials.

IV) General Conclusions.

These experiments have demonstrated that it is time forgo the notion of critical thickness, at least the one that have been defined by Matthews and Blakeslee (and refined, but never completely changed, ever since). The idea was certainly valid at the time it was developed, first because the substrates themselves were so defective that the nucleation of dislocation was never an issue, and, secondly, because the cleaning procedures had not reached the level of today, thus probably leaving numerous particulates that could also act as nucleation sites. Todays substrates and cleaning procedures make it impossible to omit the nucleation of dislocations in any calculation.

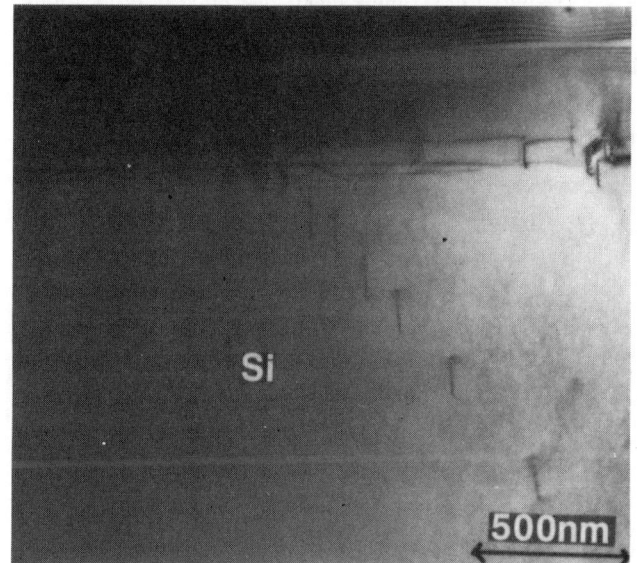

Figure 10: a) Cross-sectional sample, tilted about 30° around (110), so that dislocations that were running perpendicular to the electron beam in Fig. 7 and 9 are now imaged as lines. A pile-up, with its origin in the bottom part of the graded layer is clearly seen.

b-f) $\vec{g}\cdot\vec{b}$ analysis of a series of dislocation loops, showing that all of the loops have the same Burgers vector, equal to $1/2[10\bar{1}]$.

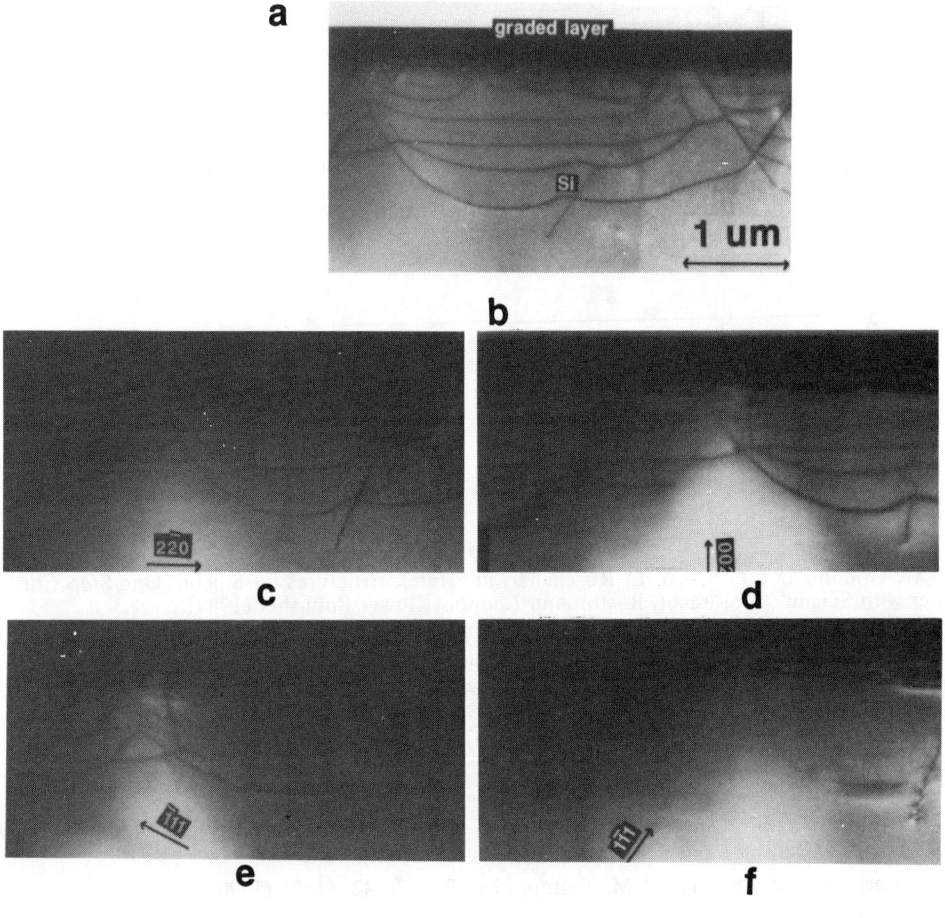

Indeed, comparison between experimental results and a Matthews-Blakeslee type of critical thickness are probably only useful as a check of how clean the system is: the dirtier the growth, the closer to the theory the data will be since dirt will certainly make nucleation of dislocations easier. In theory, it would be possible to develop a new critical thickness taking into account the barrier to nucleation. Unfortunately, this critical thickness would be critically dependant upon exactly how dislocations are formed: In the case on Ge/As/Si(100), we have actually demonstrated that it was possible to completely prevent the nucleation of dislocation because other defects have a lower nucleation energy barrier. In this case, the critical thickness would be the thickness necessary to form the V-shaped defects, not dislocations. At this point in time, the problem of including nucleation energy is thus impractical, mostly because, in most cases, we don't known where the dislocations are formed.

We have nonetheless reached a point where, through careful growth techniques, we can indeed control the formation and migration of defects/dislocations to an extend where electronic grade materials can be grown relaxed and defect free. We have shown two techniques that achieve this goal: In the first one, we have controlled the growth mode through control of the surface energetics. This results in the formation of defect free, relaxed Ge on Si(111). The idea of using surfactants could clearly be used with other systems, and at other composition of SiGe alloys. The second technique seems almost simplistic and consists of using a buffer layer of graded composition to reach the lattice parameter and composition desired. Although the idea here is very simple, its workability critically depends on how good the growth is, e.g, the initial condition of the growth interface determines the final microstructure. This is probably why we are only now discovering this new phenomenon. It certainly opens the door to a wide array of novel structures, materials, and devices.

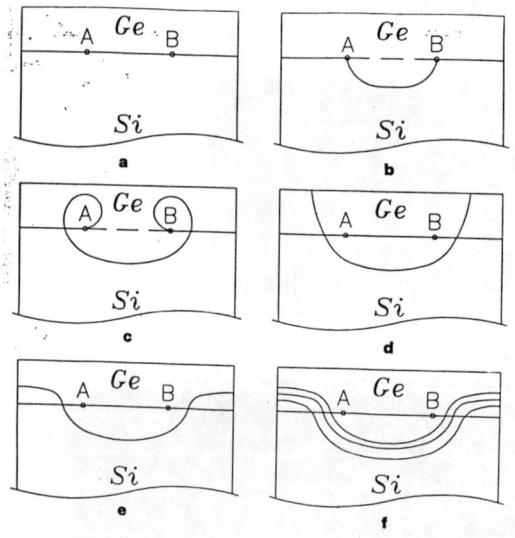

Figure 11: Schematic representation of the Frank-Read mechanism for dislocation formation, applied to this case.

REFERENCES

1. D.W. Goodman, Y.I. Nissin, E. Rosencher, in "Heterostructures on Silicon: One Step Further with Silicon", Dordrecht, Boston and London Kluwer Publisher (1989).
2. G.L. Patton, J.H. Comfort, B.S. Meyerson, E.F. Crabbe, G.J. Scilla, E. DeFresart, J.M.C. Stork, J.Y.-C. Sun, D.L. Harame, and J. Burghartz, Electron. Dev. Lett., 11, 171 (1990).
3. J.H. van der Merwe, J. Appl. Phys. 34, 117 (1963)
4. J.W. Matthews and A.E. Blakeslee, J. Crystal. Growth, 29, 273 (1975)
5. R. People and J.C. Bean, Appl. Phys. Lett., 47, 322 (1985)
6. R. Hull, J.C. Bean, C. Buescher, J. Appl. Phys., 66, 5837, (1989).
7. L.B. Freund, J. Appl. Phys., 68, 2073, (1990).
8. M.Copel, M.C. Reuter and R.M. Tromp, Phys. Rev. Lett., 62, 632 (1989).
9. M. Copel, M.C. Reuter, and R.M. Tromp, Phys Rev. B, 42, 11679 (1990)

10. M. Horn-von Hoegen, F.K. LeGoues, M. Copel and R.M. Tromp, submitted for publication in Phys. Rev. Lett.
11. F.K. LeGoues, M. Copel and R.M. Tromp, Phys. Rev. Lett., **63**, 1826, (1989).
12. F.K. LeGoues, M. Copel and R. Tromp, Phys. Rev. B, **42**, 11690 (1990)
13. F.K. LeGoues, M. Horn-von Hoegen, M. Copel and R. Tromp, submitted for publication in Phys. Rev. B.
14. S.V. Kamat and J.P. Hirth, J. Appl. Phys., **67**, 6844 (1990).
15. F.K. LeGoues, B.S. Meyerson and J.M. Morar, "Anomalous Strain Relaxation in SiGe thin Films and Supelattices", Phys. Rev. Letters, **22**, 2903 (1991).
16. B.S. Meyerson, F. Himpsel and K.J. Uram, Appl. Phys. Lett., **57**, 1034 (1990).
17. (100) surface were prepared similarly to those described in: J.F. Morar and M. Wittmer, Phys. Rev. B, **37**, 2618 (1988)
18. K. Ismail, B.S. Meyerson and P.J. Wang, Appl. Phys. Lett., **58**, 2117, (1991).
19. K. Ismail, B.S. Meyerson, and P.J. Wang, Appl. Phys. Lett., in press
20. D.J. Eaglesham, E.P. Kvam, D.M. Maher, C.J. Humphreys and J.C. Bean, Phil. Mag., **59**, 1059 (1989)
21. J.P. Hirthe and J. Lothe, in "Theory of Dislocations", second edition, John Wiley & Sons, (1982).
22. R. Hull. J.C. Bean, D.J. Werder, and R.E. Leibenguth, Phys. Rev.B., ,**40**, 1681, (1989).

COMPOSITION MODULATION EFFECTS ON THE GENERATION OF DEFECTS IN $In_{0.54}Ga_{0.46}As$ STRAINED LAYERS

F.Peiró, A.Cornet, A.Herms, J.R.Morante, S.Clark*, R.H.Williams*
LCMM. Dept. Física Aplicada i Electrònica. Universitat de Barcelona. Diagonal,645. Barcelona 08028.
* Dept. Physics and Astronomy. University of Wales. PO BOX 913. Cardiff. UK.

ABSTRACT

The influence of the presence of a composition modulation on the generation of defects in $In_{0.54}Ga_{0.46}As$ strained layers grown by MBE on InP substrates is reported. The most important feature found in all the samples is the existence of a tweed-like quasiperiodic structure with a strong dark contrast roughly along both the [001] and [010] directions. The wavelength Λ of this quasiperiodic structure has been found to be dependent on the layer thickness. In a first step, the strain relaxation is absorbed by the presence of a modulation of composition. For layer thickness bigger than 0.5 μm, defects start to appear because their nucleation is more favourable than increasing elastic energy in the layer. The interface dislocation network, expected in the growth of mismatched layers, has not nucleated in any of the samples, even for the thickest one. The elastic energy associated with the modulation induced strain has been taken into account to explain this behaviour.

INTRODUCTION

Lattice mismatched $In_xGa_{1-x}As$ layers grown on InP substrates are now of interest as they can be used to fabricate optoelectronic devices working in the wavelength range >2 μm [1]. The performance and reliability of these semiconductor devices are known to depend initially upon the perfection of the materials, especially for injection devices such as lasers and light-emitting diodes where high current densities flow through the active regions during operation. It is essential for the active region to be free from structural defects because they act as nonradiative recombination centers [2,3]. The defects usually discussed are those more directly related to the lattice mismatch between the overgrowth layer and the substrate, namely misfit dislocations [4] which appear when the layer thickness exceeds the critical value, t_c. However, misfit dislocations are not the only way available for the accomodation of the misfit strain. Stacking faults inside the epilayer can also greatly relax the misfit strain [5]. Despite many theoretical models [6,7] have been developed to describe the transition between the strained and relaxed systems, experimental understanding of the strain relaxation process is still very limited.

In this work we studied the morphological defects in $In_{0.54}Ga_{0.46}As$ strained layers grown by MBE on InP substrates, when the growth conditions used favour composition modulations to be present. In order to explain the relaxation mechanism we have taken into account the elastic energy associated with the modulation induced strain.

EXPERIMENTAL PROCEDURES

All the samples were grown using a VG Semicon V80H Molecular Beam Epitaxy (MBE) system. Growth was carried out at 515°C on InP (100) semi-insulating Fe-doped substrates. The substrates were etched prior to growth, at 50° C in $H_2SO_4 : H_2O : H_2O_2$ prepared in the ratio 7:1:1. Five epilayers were grown at a fixed alloy composition of $x = 54.3\% + 0.2\%$ with different thicknesses as shown in Table I. On initiation of growth a step increase in temperature of the group III sources was employed to limit In and Ga flux transients.

SAMPLE	THICKNESS (μm)	Λ (nm)
A	0.29	405
B	0.49	365
C	0.74	320
D	0.98	240
E	1.96	235

Table I. *Sample characteristics. The measurement error of the coarse structure wavelength (Λ) has been estimated to be $\pm 10\%$.*

TEM studies were performed on plan view and cross section samples. The cross-sectional specimens were thinned by I^+ bombardment. Planar view specimens were prepared by mechanical and ion beam milling. As it is well known, the use of Ar^+ to etch the samples is not very advisable in samples grown on InP substrates because the preferential etch of P produce In islands, which difficult the observation of interfaces. However, in this study, the controlled use of Ar^+ etching allows us to know, by the presence of these islands, which zone of the sample we are on. First of all, samples were etched from the substrate side and after from the layer surface. By changing the etching times, we were able to obtain samples with their thinnest regions at different distances from the interface. The observations have been performed using an Hitachi H-800 NA microscope operating at 200 keV.

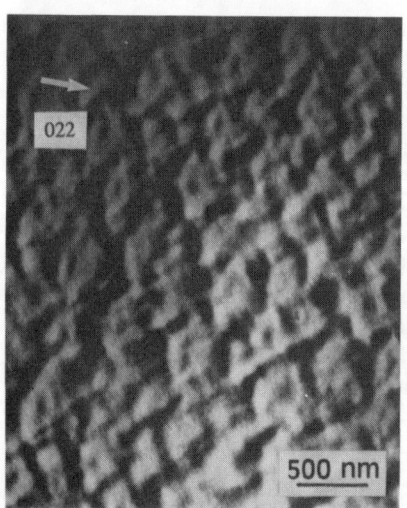

Figure 1. *Plan view [100] micrograph of the sample B. The arrow marks $g = 0\bar{2}2$. The coarse modulation lies along both [010] and [001] directions.*

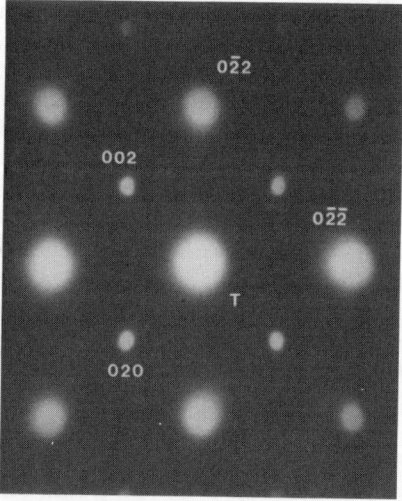

Figure 2. *Diffraction pattern of sample B*

RESULTS

The most important feature that has been found in all the samples, is the existence of a tweedlike structure with strong dark contrast in the [001] and [010] directions. The micrograph reproduced in figure 1 has been obtained with the **g**=0$\bar{2}$2 reflection, using the bright field two-beam diffraction method. Quasiperiodic contrast modulation lies in the growth plane. The 022 reflection reveals in esence the same features, and the contrast is reversed by inverting the vector **g**. When the sample is imaged in **g**=004, only the set of bands perpendicular to g remains visible. A similar behaviour occurs for **g**=040, which make the bands lying on [010] to disappear. Furthermore, in the diffraction pattern (fig.2), <002> spots exhibit a weaker intensity than the <004> ones, as expected also from the Treacy calculations of diffraction patterns of a composition modulated structure [8].

Henoc et al [9] suggested, from Energy Dispersive X-Ray Analysis (EDX), that the presence of a composition modulation at the layer-substrate interface is responsible for the coarse pattern observed. Although the mechanism of this modulation is not well established [10,11], the role played by the associated elastic energy must be taken into account to explain the stability of III-V alloys grown on InP [12].

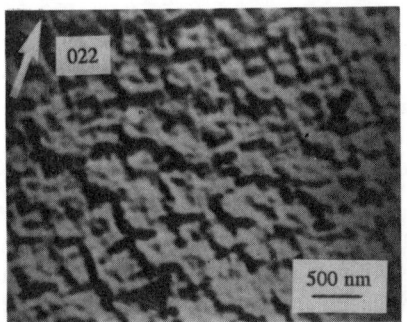 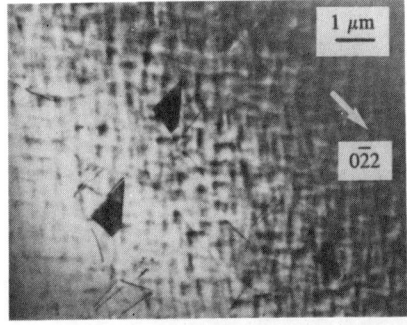

Figure 3. *Bright field two beam condition image of sample C with g=0$\bar{2}$2. Stacking faults and threading dislocations start to appear.*

Figure 4. *Bright field two beam condition image of sample E, with g = 0$\bar{2}$2.*

Epilayers with thicknesses beyond the critical value, t_c have been analyzed to study the influence of the composition modulation on strain relaxation. Figures 2,3 and 4 are an example of the coarse pattern variation as layer thickness increases. The contrast modulation is evident for all the cases. In Table I, we summarize the thicknesses as well as the wavelength Λ of the tweedlike structure present in each sample. The modulation wavelength has been found to decrease as layer thickness increases. It should be noted that samples D and E have similar values of Λ, despite the large difference on their thickness values. The measurements of the wavelengths shown in table I, have been performed by averaging the results from different regions in the specimen. We have not found any difference in the peridiocity between the two orthogonal sets.

It is worthwhile to point out that the interface dislocation network, expected in growth of mismatched layers, has not nucleated in any of the samples, even for the thickest one. However, misfit dislocations are not the only way available for acommodation of the misfit strain [5]. Among other mechanisms, stacking faults inside the epilayer can also greatly relax the misfit strain. In our case, the presence of stacking faults is more significant in thicker samples (figure 3). Defects start to appear in sample C and their densities rise as layer thickness increases, reaching a value of about 10^7 cm^{-2} for the thickest sample.

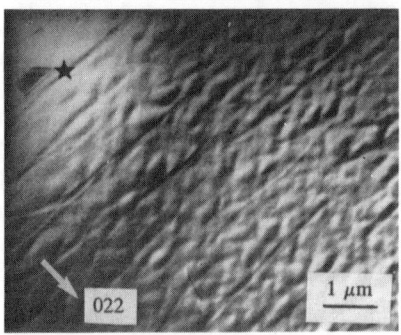

Figure 5. *Micrograph of sample E imaged with $g=022$. The asterisc marks the same region that in figure 6. Dislocation lines, stacking faults and coarse structure are in contrast.*

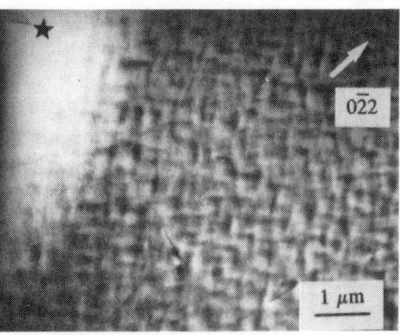

Figure 6. *Micrograph of sample E, $g=0\bar{2}2$, showing the dark bands lying along 010 and 001. $R=<2\bar{1}1>$ type stacking fault are still present.*

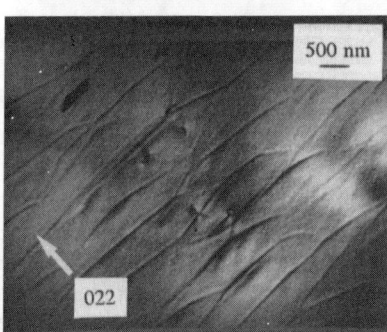

Figure 7. *Dislocation lines and stacking faults upward from the interface in sample E.*

In order to study the effects of the composition modulation on the generation of defects we have carried out plan view TEM observations at different distances from the interface. Here we present the results for the specimen with the thickest layer (1.96µm). Figure 4 is a bright field, $g=0\bar{2}2$, two beam condition image of this sample in the region near the interface. Figures 5 and 6 correspond to two different reflections, $g=022$ and $g=0\bar{2}2$ respectively, of a region farer from the interface that the one shown in figure 4. The star in the figures refers to the same area of the sample. Comparing figures 4 and 6, we can observe that the contrast of the coarse structure decreases as the separation from the interface increases. At the same time, dislocations lines start to appear. The general feature is that most of the dislocations appear under 022 (Fig. 5), 040 and 004 reflections. Under $02\bar{2}$ reflection, the dislocations disappear

(Fig. 6) and the stacking faults are still in contrast. Based on the **g.b** criterion for the visibility of dislocations, where **g** is the diffraction vector and **b** the Burgers vector of the dislocations, we can conclude that most of the dislocations are misfit lying on $<02\bar{2}>$. Finally, when the upper region of the layer is imaged (i.e. region near interface has been etched from the substrate) only stacking faults and misfit dislocations are observed (Fig. 7). Remark that the coarse structure has disappear.

DISCUSSION

It is well known that in heteroepitaxial growth the epilayer-substrate strain is initially accomodated elastically. However, at sufficient large epilayer thicknesses, $>t_c$, the accomodation of strain by nucleation and propagation of defects (misfit dislocations, stacking faults) is more favourable than the increase of the elastic energy of the system. On the other hand, Glas [12] showed that the stress relaxation of the system can also be related to a modulation of the lattice parameter. So, in our case of low mismatched InGaAs layers, a balance of the elastic energy in the system must include both contributions: defects and composition modulation. The relaxation of the stress tends to an stabilisation of the system leading to a determinated composition modulation depending on the total energy, which is obviously related to the layer thickness.

In this framework, our results could be explained by the following considerations: In the thinnest samples, the energetic balance of the system leads to a presence of a modulation of composition. As the layer thickness increases, there is a diminution of the wavelength in order to absorb the increment of elastic energy introduced by the greater thickness. However, for the thickest samples this diminution of the wavelength is likely to be not enough to absorb the excess energy. So, like in mismatched homogeneous layers, when layer thickness exceeds a critical value, the relaxation of the energy by means of defect nucleation is more favourable than increasing elastic energy in the strained layer. In this way, note that despite the large difference in thickness between E and D samples, their wavelengths are very similar, being the density of stacking faults higher in the thickest one.

To know the critical value at which defects begin to appear, a balance of the elastic energy in the system including the nucleation of defects and the energy associated to the modulation of composition must be done. So, this critical value will depend on the growth parameters (temperature and composition) influencing the thermodinamical stability of the system and on the strain present in the sample because of the composition and layer thickness. In our samples, grown at 515°C and with a low mismatch, we have found 0.5 μm for the critical value.

The existence of a composition modulation at the interface suggests that regions of localized strain are present, limiting the propagation and interaction of dislocations to form misfit segments generally observed in mismatched layers. On the contrary, when a region above the interface is imaged (Fig.6) a set of misfit dislocations lying on $<022>$ can be observed. In this case the structure is not affected by the modulation induced strain because, as suggested by Strunk [13], the propagation mechanism of such dislocations is a climbing process.

CONCLUSION

We have studied the influence of the presence of a composition modulation on the generation of defects in $In_{0.54}Ga_{0.46}As$ layers grown by MBE on InP substrates. Despite all samples studied have thicknesses larger than the theoretical critical thickness, the expected interface dislocation network has not nucleated in any of the samples, even for the thickest ones. The presence of regions of localized strain associated with this modulation has been taken into account to explain the observed behaviour.

ACKNOWLEDGEMENTS

This work has been partially funded by a ESPRIT Research Project (Reference 3086).

REFERENCES

[1] P. Kightley, R.I. Taylor, A.J. Moseley, P.D. Augustus, A.C. Marshall and R.J.M. Griffiths, Inst. Phys. Conf. Ser. 100, 187 (1989).
[2] A.K. Chin, C.L. Zipfel, S. Mahajan, F. Ermanis and M.A. DiGiuseppe, Appl. Phys. Lett. 41, 555 (1982).
[3] P.M. Petroff, R.A. Logan and A. Savage, Phys. Rev. Lett. 44, 287 (1980).
[4] J.W. Matthews, in "Epitaxial Growth", J.W. Matthews Ed., Academic Press, New York (1975).
[5] S.N.G. Chu, A.T. Macrander, K.E. Strege and W.D. Johnston, Jr., J. Appl. Phys. 57, 249 (1985).
[6] J.W. Matthews, J. Vac. Sc. Technol., 12, 126 (1975).
[7] R. People and J.C. Bean, Appl. Phys. Lett. 49, 229 (1986)
[8] M.M.J. Treacy, J.M. Gibson and A. Howie, Philos. Mag. A 51, 389 (1985).
[9] P. Henoc P, A. Izrael, M. Quillec and H. Launois Appl. Phys. Let. 40, 963 (1982).
[10] S. Mahajan, M.A. Shadid and D.E. Laughlin Inst. Phys. Conf. Ser. 100, 143 (1989).
[11] F. Glas NATO ASI series B203 (Plenum Press, New York) pp. 217-233 (1989).
[12] F. Glas J. Appl. Phys. 62, 3201 (1987).
[13] H.P. Strung NATO ASI series B203 (Plenum Press, New York) pp. 217-233 (1989).

ELECTRON-TRAPPING DEFECTS IN MBE-GROWN RELAXED n-$In_{0.05}Ga_{0.95}As$ ON GALLIUM ARSENIDE

A C Irvine[†], L K Howard[*], and D W Palmer[†]

[†]Physics and Astronomy Division, University of Sussex, Brighton BN1-9QH, UK.
[*]Physics Department, University of Surrey, Guildford GU2-5XH, UK.

ABSTRACT

We have studied, by capacitance-voltage measurements and deep-level transient spectroscopy, the electron traps present in 1.5μm-thick relaxed n-$In_{0.05}Ga_{0.95}As$ grown by molecular beam epitaxy (MBE) on an n-GaAs substrate. We find that the capacitance-voltage data indicate a significant decrease in the effective space-charge concentration in the InGaAs close to the hetero-interface, suggesting the presence of a high concentration of electron traps in that region. Our DLTS measurements indicate two electron-trapping levels in the $In_{0.05}Ga_{0.95}As$ layer, one at E_c-0.56eV and the other at E_c-0.80eV. The shallower level is assigned, on the basis of its concentration profile and unusual electron-capture characteristics, to extended defects such as dislocations, and we have elsewhere reported strong evidence that the E_c-0.80eV level in this MBE-grown InGaAs corresponds to the EL2 defect commonly found in bulk-grown GaAs. We find however that the E_c-0.56eV and E_c-0.80eV trap concentration profiles cannot account for the observed carrier depletion near the $In_{0.05}Ga_{0.95}As$-GaAs hetero-structure interface, and propose that it is due to electron trapping on other dislocation-related acceptor levels in the lower half of the bandgap.

1. Introduction

The lattice mismatched $In_xGa_{1-x}As$/GaAs system has considerable potential for the fabrication of electron devices such as high electron mobility transistors (HEMTs) [1] and hetero-junction bipolar transistors (HBTs) [2]. However, the difference in the respective lattice constants means that the growth of $In_xGa_{1-x}As$ on GaAs can proceed pseudomorphically (with the mismatch taken up by elastic strain in the $In_xGa_{1-x}As$) only until the layer reaches a certain thickness, the critical thickness. Thereafter, the $In_xGa_{1-x}As$ relaxes towards its bulk lattice constant by plastic deformation via dislocation generation with a consequent degradation of material quality [3]. Clearly the understanding of electrical degradation of $In_xGa_{1-x}As$ is of great importance for the future growth and optimisation of practical devices in the $In_xGa_{1-x}As$/GaAs system.

Transmission electron microscopy studies of relaxed $In_xGa_{1-x}As$ on GaAs [4-7] for indium compositions x of less than 0.25 have shown that the relaxation of strain is accompanied by the formation of misfit dislocations close to the junction of the two mismatched layers. These misfit dislocations may be produced by the glide of threading dislocations inherited from the substrate or, if the density of threaders is low, the misfits may arise from the nucleation and subsequent expansion of half-loops at the surface [8]. In either case, the process of dislocation glide may result in the interaction of the dislocations with other defects, either point defects or extended defects, and the generation of new defect structures. Similarly, dislocation climb may occur non-conservatively, with the dislocation acting as a sink or source of point defects [9].

In the present work we investigate the presence of electrically active defects in relaxed $In_{0.05}Ga_{0.95}As$ layers containing dislocations.

2. Experimental Procedures

The $In_{0.05}Ga_{0.95}As$ layer used in this study was grown by molecular beam epitaxy (MBE) at RSRE, Malvern, Worcs, UK, in a Vacuum Generators VG V80H system. Onto a (100)-oriented n^+-GaAs substrate (n = (5-9)x10^{17} cm^{-3}) were deposited in succession a 1.17μm GaAs buffer layer, a 1.5μm thickness of $In_{0.05}Ga_{0.95}As$ and, finally, a thin (200Å) GaAs cap layer for better Schottky contacting. All epitaxial layers were silicon doped to $n = 1 \times 10^{16}$ cm^{-3}. The buffer layer was grown at a substrate temperature of 580°C, with the subsequent layers grown at 520°C in order to prevent significant indium desorption. Growth proceeded at 1 monolayer/sec under a group V to group III flux-ratio of 5:1.

Ohmic back contacts were made to the sample by alloying tin at 290°C in an atmosphere of HCl. Prior to this, the samples were cleaned by a succesion of rinses in an ultrasonic bath; firstly in methanol, then in isopropyl alcohol, and finally in distilled water. A similar rinse followed the ohmic contacting and, after a brief deoxidising rinse in 10% HCl solution, 1mm gold Schottky front contacts were deposited by vacuum evaporation at 10^{-5} torr through a stainless steel mask. The reverse breakdown obtained from these diodes at room temperature was often greater than 20V, allowing the layer to be depleted as far as the substrate. Capacitance-voltage profiling and deep-level transient spectroscopy (DLTS) [10] measurements were made using a computerized data collection system employing a Boonton 72B capacitance meter, box-car signal analyser and digital transient recorder.

3. Results and Discussion

Figure 1 shows the effective space-charge concentration N$^+$ (as obtained from the C-V data in the usual way) in the epilayers as a function of depth at two temperatures, 40K and 375K. The most striking feature is the broad dip in effective space-charge concentration at the In$_{0.05}$Ga$_{0.95}$As-GaAs interface, and we interpret this in terms of a high density of electron-trapping defects in this region. It is to be noted that the general form of the N$^+$ profile changes little between the two temperatures; this will be discussed below.

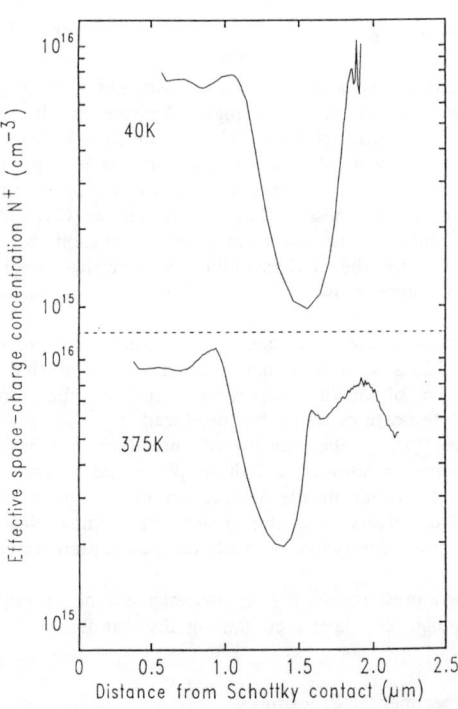

Fig.1 Effective space-charge concentration N$^+$ as a function of distance from the Schottky contact, evaluated from 1 MHz C-V measurements at the temperatures indicated. The In$_{0.05}$Ga$_{0.95}$As-GaAs interface was at 1.5µm from the Schottky contact.

DLTS measurements at 40-380K revealed the presence of two electron trapping levels in the In$_{0.05}$Ga$_{0.95}$As layer. By varying the bias conditions (and therefore the measurement depths) for different DLTS scans, it was apparent that the concentration of each of the two traps was strongly position-dependent. Figure 2, showing DLTS spectra for the various bias conditions indicated, illustrates this. It was noted that the 290K peak appeared at slightly lower temperatures under conditions of lower bias at constant pulse height; because of the way in which the depletion depth depends on applied reverse bias, this implies an enhanced electron emission under higher field conditions.

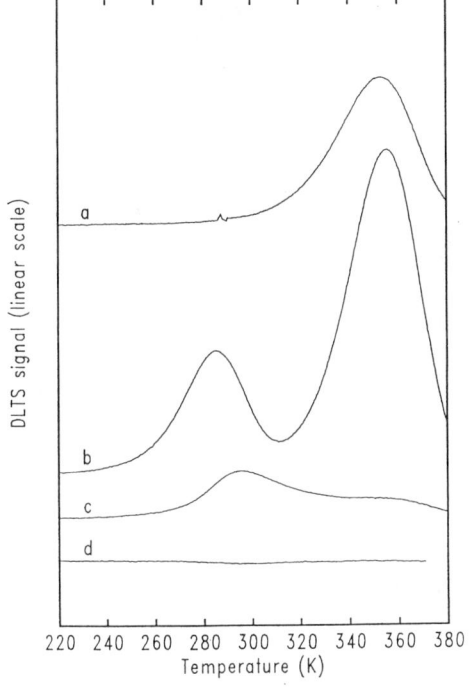
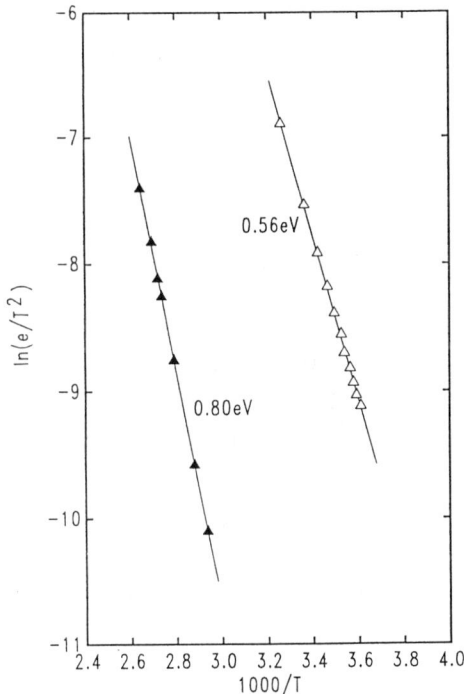

Fig.2 DLTS spectra for different values of steady reverse bias V_r and filling pulse bias V_p:
(a) $V_r=-4V$, $V_p=-2V$;
(b) $V_r=-8V$, $V_p=-6V$;
(c) $V_r=-10V$, $V_p=-8V$;
(d) $V_r=-12V$, $V_p=-10V$.
The emission rate window was set at $20.2s^{-1}$ and the filling pulse duration was 2ms.

Fig.3 Arrhenius plots for the two DLTS peaks shown in figure 2. Data points for the $E_c-0.56eV$ peak are shown for $V_r=-8V$, $V_p=-6V$. In the case of the $E_c-0.80eV$ peak, the condition $V_r=-2V$, $V_p=0V$ was used.

Arrhenius plots for the two traps under the field conditions marked are given in figure 3, yielding trap energies of $(0.560\pm0.006)eV$ and $(0.796\pm0.006)eV$ for the 290K peak and the 360K peak respectively. On the basis of the emission behaviour of the $E_c-0.80eV$ trap over the $In_xGa_{1-x}As$ composition range $0.05 \leq x \leq 0.20$ in this and other samples, and because of the distinctive photo-capacitance quenching that we observe, we have argued [11] that the $E_c-0.80eV$ trap is due to the well-known EL2 level, not previously observed in MBE-grown $In_xGa_{1-x}As$.

We have studied also the dependences of the DLTS amplitudes of the two peaks on the duration of the electron-filling pulse, and we have found that a pulse of at least 100ms was required to fill the $E_c-0.56eV$ trap, with the $E_c-0.80eV$ (EL2) level filling about an order of magnitude faster. The DLTS peak height for each trap is plotted as a function of the filling pulse duration in figure 4.

It is seen in figure 4 that the height of the $E_c-0.56$eV peak is approximately proportional to the logarithm of the filling duration over two decades. As considered by Wosinski [12], such behaviour is characteristic of rows of defects such as may occur at dislocations; the mechanism of the effect is that the filling rate of individual traps is progressively reduced as the filling of defects on the row changes the row's electric potential. In the same study, Wosinski described the observation by DLTS of a trapping level, ED1, produced by plastic deformation of bulk GaAs, which showed just such a characteristic. Although the activation energy, 0.68eV, obtained for ED1 in that study, differs from our value of 0.56eV for the 290K peak, the similarity of the electron-capture behaviour suggests that Wosinski's ED1 and the $E_c-0.56$ level of the present study may involve similar defect arrangements.

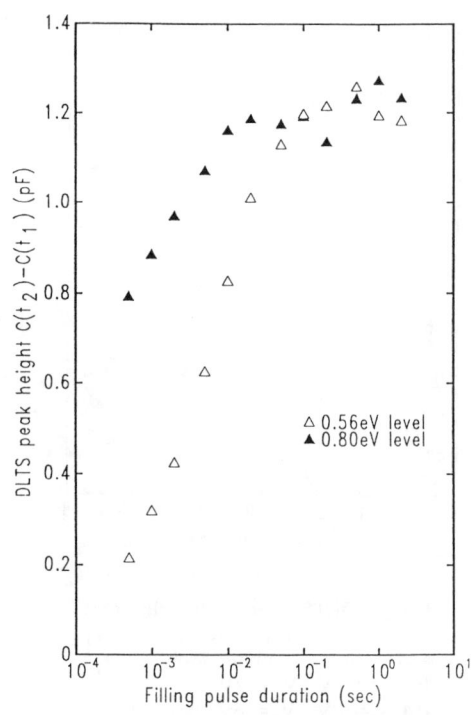

Fig.4 DLTS peak height as a function of filling pulse duration for the $E_c-0.56$eV level (△) and the $E_c-0.80$eV level (▲), measured at the temperatures, 290K and 360K respectively, of the centres of the DLTS peaks shown in figure 2. For each set of data, the DLTS biasing conditions were $V_r=-8V$, $V_p=-6V$.

Concerning the filling of the $E_c-0.80$eV level, which we identify with EL2 defects, we note that measurements that we have made on VPE n-GaAs using the same equipment and experimental arrangement have shown essentially complete filling of EL2 centres in that material for all the filling-pulse durations for which data are shown in figure 4, and this is in agreement with data for EL2 in VPE n-GaAs given by Mircea et al [13]. However, although that work [13] reported that the filling of EL2 traps in VPE n-$In_xGa_{1-x}As$ was slower than that of such centres in VPE n-GaAs, our data of figure 4 show that many of the EL2 traps in our MBE $In_{0.05}Ga_{0.95}As$ were filled even more slowly. We suggest that some or all of the EL2 centres in our $In_{0.05}Ga_{0.95}As$ were present near other electron traps (perhaps related to dislocations), such that, like the $E_c-0.56$eV trap, their rate of filling by electrons was reduced by the effect of nearby, already filled traps.

In order to investigate the spatial dependences of the trap concentrations, further DLTS data were taken with the reverse bias during the emission period held constant at 12V and the pulse height changed for successive DLTS measurements. For 12V reverse bias, the inner boundary of the depletion region was in the GaAs buffer layer for which the carrier concentration was well known and fairly constant with temperature. This experimental procedure therefore enabled a reasonably accurate study of the trap concentration to be carried out. Because of the strongly non-uniform effective space-charge concentration in our sample, care had to be taken in the analysis of trap concentration data. The method we chose for determining the trap distribution was that of Lefèvre and Schulz [14], employing the double integration of the measured space-charge concentration from the depletion region boundary towards the Schottky contact in order to determine the points at which the trapping levels crossed the Fermi level. For each of the two traps, the peak height was scaled up on the basis of figure 4 in order to compensate for insufficient filling time. The results of this analysis are presented in figure 5. The markedly non-uniform nature of the calculated profiles suggests that neither trap is impurity-related; a formation mechanism based on dislocation production and propagation seems the most likely origin of the two traps. While the E_c–0.56eV level seems, on the evidence of its strongly logarithmic filling characteristics (figure 4), to be related to dislocation core states, the E_c–0.80eV level may be produced by a dislocation climb mechanism such as that proposed by Figielski [15].

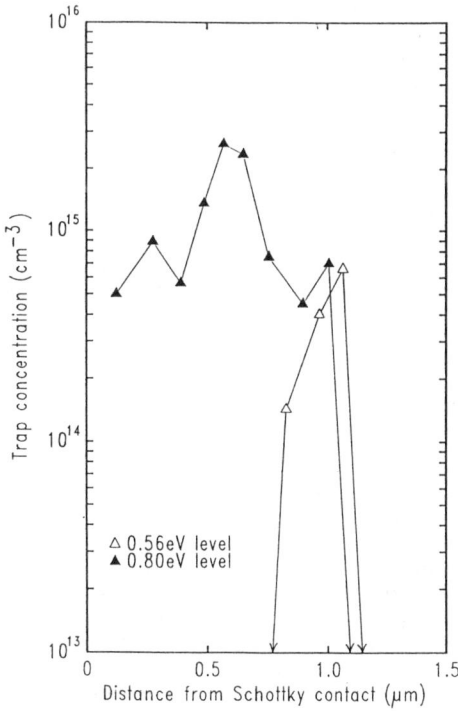

Fig.5 Trap concentration for the E_c–0.56eV level ($\triangle$) and the E_c–0.80eV level ($\blacktriangle$) as a function of distance from the Schottky contact. The concentrations were determined from the DLTS data by the method outlined in the text. The $In_{0.05}Ga_{0.95}As$–GaAs interface was at 1.5μm from the Schottky contact.

In order to test whether band bending at the InGaAs/GaAs conduction band discontinuity caused any serious analytical problems, we have developed a numerical computer program similar to that which Jeong et al [16] used for C–V simulation, but with provision for a DLTS bias pulse. It was found that, for traps having energy levels similar to those which we observe, the difference in peak heights due to the expected conduction band discontinuity of 0.05eV was within the uncertainty limits of our experimental determination of trap concentrations.

Clearly the evidence of figure 5 suggests that the two levels that we detect by DLTS are not, on their own, sufficient in concentration to produce the large dip in N^+ shown in figure 1. Considering also the fact that the N^+ plot of figure 1 changes little in form between the two measurement temperatures, we can reasonably conclude that the N^+ dip is likely to be largely due to compensation by acceptor levels in the lower half of the band-gap. Such electron traps would not be observable by the standard DLTS technique because of the high temperature that would be needed for emission of the trapped electrons to the conduction band. Evidence for the creation of electron traps below mid-gap has been reported in studies of plastic deformation in GaAs [17–19].

As is shown in figure 5, our analysis shows no evidence that the two traps are present in the 0.4μm region of InGaAs nearest to the hetero-interface. Ioannou et al [20] have reported a similar observation on MBE-grown indium-doped (0.6%) GaAs for the grown-in traps M1, M3 and M4; the apparent absence of those traps from the dislocation zone was suggested to be a consequence of point defect absorption during the non-conservative climb of dislocations in that region. In the case of the present work, this would seem an unlikely, though not impossible, mechanism if we are to argue that the defects are created by dislocation-related processes in the first place. Fermi level pinning on deep traps of concentration greater than or of the order of 10^{16} cm^{-3} at or near the interface would seem to be a more likely explanation for the apparent non-appearance of our two levels in that region. This process, similar to that observed by Woodall et al [21], would mean that any E_c-0.56eV and E_c-0.80eV levels near the InGaAs-GaAs interface would be held above the Fermi level for any given reverse bias and therefore would not be filled by the conventional DLTS technique. It must be stressed, though, that no rectification attributable to the hetero-interface has been observed for our samples. In addition, the good agreement of the N^+ data of figure 1 with the interface position indicates that there can be no large series capacitance effect arising from the InGaAs-GaAs interface.

In conclusion, we observe by DLTS the presence of two electron-trapping levels, at E_c-0.56eV and E_c-0.80eV, in $In_{0.05}Ga_{0.95}As$ MBE-grown on GaAs, and suggest that these levels arise from dislocation-related defect processes. However, the measured total concentration of the defects is insufficient to produce the observed strong dip in the effective space-charge concentration N^+ near the InGaAs-GaAs interface, and we propose that that reduction in N^+ is due mainly to additional acceptor traps, arising from misfit dislocations and having energy levels in the lower half of the band gap.

This work was carried out in a research collaboration between the Physics and Astronomy Division of the University of Sussex and the Strained Layers Structures Group of the University of Surrey, and we acknowledge the financial support of the UK Science and Engineering Research Council.

REFERENCES

1. J.J.Rosenberg, M.Benlamri, P.D.Kirchner, J.M.Woodall and G.D.Pettit,
    IEEE Electron Dev.Lett. EDL-6, 491-493 (1985)
2. L.P.Ramberg, P.M.Enquist, Y.-K.Chen, F.E.Najjar, L.F.Eastman, E.A.Fitzgerald and
    K.L.Kavanagh, J.Appl.Phys. 61, 1234-1236 (1987)
3. D.J.Dunstan and A.R.Adams, Semicond.Sci.Technol. 5, 1202-1208 (1990)
4. E.A.Fitzgerald, D.G.Ast, P.D.Kirchner, G.D.Pettit and J.M.Woodall,
    J.Appl.Phys. 63, 693-703 (1988)
5. K.L.Kavanagh, M.A.Capano, L.W.Hobbs, J.C.Barbour, P.M.J.Marée, W.Schaff, J.W.Mayer,
    D.Pettit, J.M.Woodall, J.A.Stroscio and R.M.Feenstra,
    J.Appl.Phys. 64, 4843-4852 (1988)
6. K.H.Chang, P.K.Bhattacharya and R.Gibala, J.Appl.Phys. 66, 2993-2998 (1989)
7. V.Krishnamoorthy, P.Ribas and R.M.Park, Appl.Phys.Lett. 58, 2000-2002 (1991)
8. P.M.J.Marée, J.C.Barbour, J.F.van der Veen, K.L.Kavanagh, C.W.T.Bulle-Lieuwma and
    M.P.A.Viegers, J.Appl.Phys. 62, 4413-4420 (1987)
9. E.R.Weber, H.Ennen, U.Kaufmann, J.Windscheif, J.Schneider and T.Wosinski,
    J.Appl.Phys. 53, 6140-6143 (1982)
10. D.V.Lang, J.Appl.Phys. 45, 3023-3032 (1974)
11. A.C.Irvine and D.W.Palmer, submitted for publication
12. T.Wosinski, J.Appl.Phys. 65, 1566-1570 (1989)
13. A.Mircea, A.Mitonneau, J.Hallais and M.Jaros, Phys.Rev.B 16, 3665-3675 (1977)
14. H.Lefèvre and M.Schulz, Appl.Phys. 12, 45-53 (1977)
15. T.Figielski, Appl.Phys. A 36 217-219 (1985)
16. J.Jeong, T.Schlesinger and A.G.Milnes, IEEE Trans.Electron Dev. ED-34, 1911-1918 (1987)
17. T.Ishida, K.Maeda and S.Takeuchi, Appl.Phys. 21, 257-261 (1980)
18. M.Suezawa and K.Sumino, Jap.J.Appl.Phys. 25, 533-537 (1986)
19. M.Skowronski, J.Lagowski, M.Milshtein, C.H.Kang, F.P.Dabkowski, A.Hennel and H.C.Gatos,
    J.Appl.Phys. 62, 3791-3798 (1987)
20. D.E.Ioannou, Y.J.Huang and A.A.Iliadis, Appl.Phys.Lett. 52, 2258-2260 (1988)
21. J.M.Woodall, G.D.Pettit, T.N.Jackson, C.Lanza, K.L.Kavanagh and J.W.Mayer,
    Phys.Rev.Lett. 51, 1783-1786 (1983)

ATOMIC ORDERING IN (110)InGaAs AND ITS INFLUENCE ON ELECTRON MOBILITY

O. UEDA AND Y. NAKATA
Fujitsu Laboratories Ltd., 10-1 Morinosato-Wakamiya, Atsugi 243-01, Japan

ABSTRACT

Atomic ordering in InGaAs grown on (110) InP substrates by molecular beam epitaxy, has been studied by transmission electron microscopy. In the electron diffraction pattern from the InGaAs, superstructure spots associated with CuAu-I type ordered structure are found. When the tilting angle of the substrates increases, the ordering becomes stronger. The ordering is also stronger in crystals grown on substrates tilted toward the <001> or the <00$\bar{1}$> direction than those on substrates tilted toward the <1$\bar{1}$0> direction. From these results, one can conclude that atomic steps on the growth surface play an important role in the formation of ordered structures. The ordering becomes stronger when the growth temperature increases in the range 360-485°C. In high resolution images of the crystal, doubling in periodicity of 220 and 200 lattice fringes is found, which is associated with CuAu-I type ordered structure. Moreover, anti-phase boundaries are very often observed in the ordered regions. It is also found that ordering is not perfect, and that ordered regions are plate-like microdomains lying on planes slightly tilted from the (110) plane. We have fabricated InGaAs/N-InAlAs heterostructures with a strongly ordered InGaAs channel layer. The measured two-dimensional electron gas mobilities from these structures are found to be 100,000 cm^2/Vs in the <001> direction and 161,000 cm^2/Vs in the <$\bar{1}$10> direction with a sheet electron concentration (N_s) of 9.5×10^{11} cm^{-2} at 6 K. The latter mobility is much higher than both calculated alloy scattering limited mobility and the best experimental results for lattice-matched InGaAs/N-InAlAs systems. The mobility enhancement in the <110> direction is considered to be achieved by the suppression of alloy scattering due to the occurrence of ordering in the InGaAs channel layer.

1. Introduction

In most III-V alloy semiconductors, phase separation due to bulk and/or surface spinodal decomposition [1-3] and atomic ordering on the growth surface [4-14], are major issues for thermal stability of the crystals. In particular, atomic ordering in alloy semiconductors grown on (001) substrates has been extensively studied [4-14]. In most cases, the observed ordered structure is of CuPt type in which sublattice ordering occurs on the (111) planes with doubling in periodicity [4-12]. It has also been confirmed that ordered (001) InGaP has band gap energies 50-80 meV smaller than those in non-ordered crystals [6-8], being very consistent with theoretical calculations [15]. On the other hand, as previously reported by Kuan et al. [13,14], a CuAu-I type ordered phase is present in AlGaAs [13] and InGaAs [14] crystals grown on (110) substrates. However, they have only shown transmission electron diffraction (TED) patterns corresponding to CuAu-I structure. In this paper, we describe a more detailed transmission electron microscopic (TEM) study on the microstructure of ordered InGaAs grown on (110) InP substrates by molecular beam epitaxy (MBE). We also report a strong enhancement of electron mobility at low temperature due to the ordering.

2. Experimental procedures

Undoped-InGaAs crystals grown on (110) InP substrates by MBE, are examined in this study. Crystals are grown on (110) just InP substrates, or InP substrates tilted toward the <001>, <00$\bar{1}$> or <1$\bar{1}$0> direction, and at 360-450°C. During growth, substrates are rotated so as to achieve good uniformity in composition, thickness and carrier concentration. V/III partial pressure ratios are kept at 400 or 60 (only crystals grown on substrates tilted toward the <00$\bar{1}$> direction). The epi-layers are approximately 1.0 μm thick. The lattice-mismatch between the InGaAs layer and the InP substrate is within $\pm 1 \times 10^{-3}$. Thin specimens for TEM are prepared by chemical etching for plan-view observation and by ion milling for cross-sectional observation. TEM observation is carried out in an ABT (or ISI) ultra-high resolution analytical electron microscope EM-002B operated at 200 kV.

3. Results and discussion

3.1 Identification of CuAu-I type structures by TED and substrate orientation dependence of the degree of ordering

Figures 1(a) and 1(b) show TED patterns from a (110) plan-view and a (1$\bar{1}$0) cross-section of InGaAs crystal grown on a (110) InP substrate 3° tilted toward the <1$\bar{1}$0> direction. In both figures, superstructure spots are observed at positions indexed as 001, 110, 1$\bar{1}$0, 112, 1$\bar{1}$2, ...etc. This particular set of superstructure spots are associated with CuAu-I type ordered structure [1,10]. In this structure, Ga atoms preferentially occupy the (0, 0, 0) and (1/2, 1/2, 0) sites and In atoms preferentially occupy the (1/2, 0, 1/2) and (0, 1/2, 1/2) sites in each unit cell. Therefore, a perfectly ordered InGaAs crystal consists of alternating InAs and GaAs monolayers, i.e., $(InAs)_1/(GaAs)_1$ monolayer superlattices, when viewed along either the <1$\bar{1}$0> growth direction or the <001> direction . It should be noted that the superstructure spots shown in Fig. 1(b) are extremely streaky, and that the streaks are "S"-shaped, slightly tilting from the [110] direction, which is also observed in the TED patterns from (1$\bar{1}$0) cross-sections of ordered InGaP (CuPt-type) crystals grown on (001) GaAs tilted substrates. This may be due to either (1) presence of anti-phase boundaries in the ordered region or (2) the fact that ordered regions are plate-like microdomains (see Fig. 5), although computer simulations are required to resolve this issue.

Figures 2(a)-2(c) show TED patterns from plan-views of InGaAs crystals grown on (110) InP substrates with different tilting angles, 0°, 3° and 5°, toward the <001> direction. In crystals grown on a (110) just InP substrate (Fig. 2(a)), the intensity of superstructure spots is very weak, indicating very weak ordering. However, when a substrate tilted 3° toward the <001> direction is used, the ordering becomes stronger as shown in Fig. 2(b). Furthermore, the ordering becomes much stronger in a crystal grown on a substrate tilted 5° toward the <001> direction (Fig. 2(c)). Since tilting of the substrate orientation toward the <001> or the <00$\bar{1}$> direction introduces periodic arrays of steps on the substrate surface, one can conclude that atomic steps on the growth surface play an important role in the formation of the ordered structures. It is also established that the ordering is stronger in crystals grown on substrates tilted toward the <001> or the <00$\bar{1}$> direction than those on substrates tilted toward the <1$\bar{1}$0> direction. This may be explained as follows: when the (110) surface is tilted toward the <001>, the steps are very straight, giving rise to strong ordering, but wavy steps with many kinks are formed when the (110) surface is tilted toward the <1$\bar{1}$0>, rather eliminating the ordering (in the latter case, only kinks can enhance the ordering).

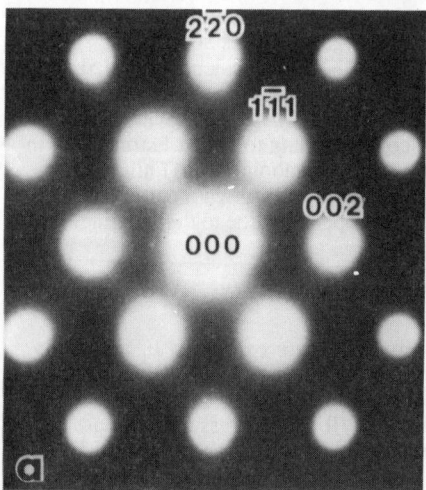 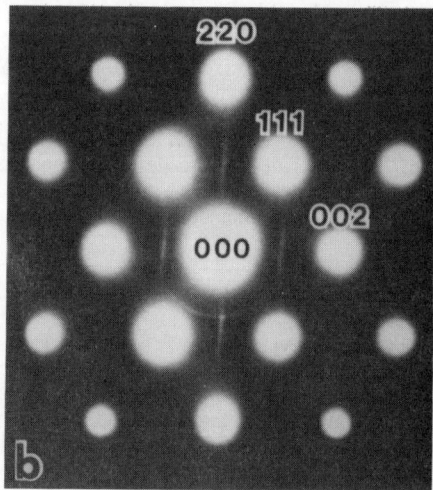

Fig. 1 A TED pattern from a plan-view and a cross-section of InGaAs crystal grown on a (110) InP substrate 3° tilted toward the <1$\bar{1}$0> direction (T_g=485°C).
(a) (110) planview; (b) (1$\bar{1}$0) cross-section.

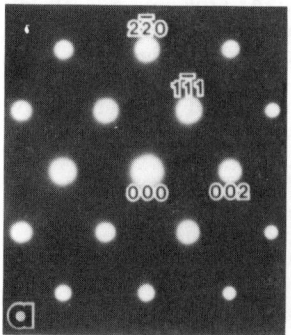 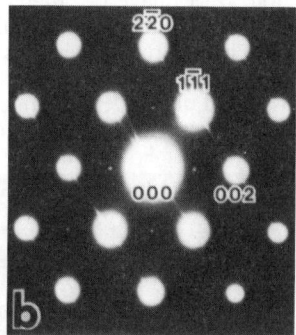 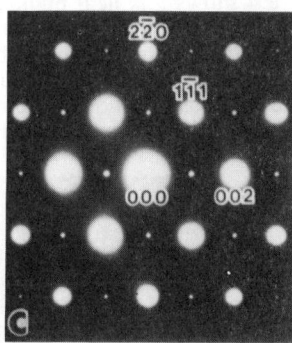

Fig. 2 TED patterns from plan-view of InGaAs layers grown on substrates with different tilting angles (T_g=480°C).
(a) (110) just InP substrate; (b) (110) InP substrate tilted 3° toward the <001> direction; (c) (110) InP substrate tilted 5° toward the <001> direction.

3.2 Growth temperature dependence of the degree of ordering

In order to clarify the influence of growth temperature on the generation of ordered structure, we grew InGaAs crystals on (110) InP substrates 3° tilted toward the <110> direction, at various temperatures in a range 435-485°C. From TED analysis, it was found that in this temperature range, the degree of ordering increases with growth temperature. Figures 3(a) and 3(b) show TED patterns from (110) cross-sections of InGaAs crystals grown at 435°C and 485°C, respectively. Another TED analysis has been carried out on crystals grown on substrates tilted towards the <001> direction. It has also been found that degree of ordering increases with an increase in growth temperature in the range of 360-450°C. These findings can be explained by the fact that the mobility of deposited atoms on the growth surface also increases with temperature and that the generation of ordered structures is thought to be strongly related to the migration and reconstruction of deposited atoms [4,8].

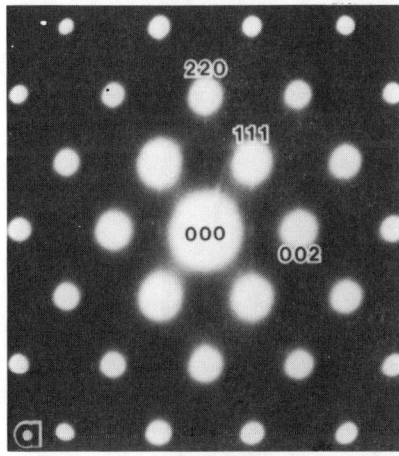 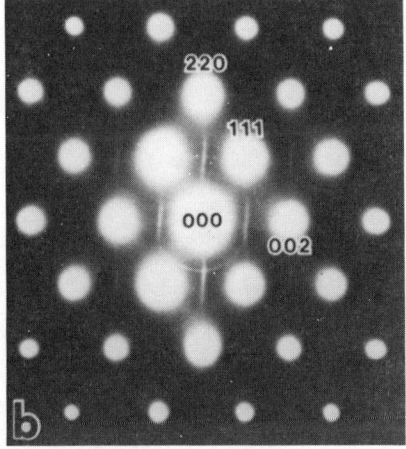

Fig. 3 TED patterns from ($\bar{1}$10) cross-sections of InGaAs crystals grown on (110) InP tilted 3° toward the <110> direction at 435 and 485°C.
(a) T_g=435°C; (b) T_g=485°C.

3.3 Microstructural characterization of CuAu-I type structure

High resolution TEM analysis was carried out to evaluate the ordered structure on an atomic scale. Figure 4 shows a typical high resolution TEM image of ordered InGaAs grown on a (110) InP substrate 3° tilted toward the <001> direction. Doubling of the 002 (see arrows denoted by X_1-X_3) and the 220 (see arrows denoted by Y_1-Y_4) lattice fringes, which is associated with the CuAu-I type structure, is observed locally, suggesting that the crystal consists of both well ordered and non-ordered regions. It should be emphasized that planar defects are often observed in high resolution images, e.g., the region denoted by APB, where a phase-shift of the 002 lattice fringes is

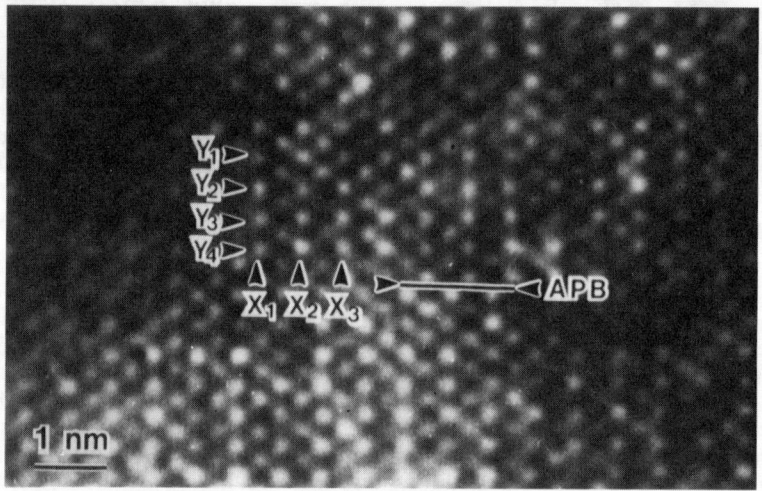

Fig. 4 A high resolution TEM image of ordered InGaAs crystal.

observed. These are expected to be anti-phase boundaries as suggested by Kuan et al. [16]. In a (001) cross-section high resolution TEM image, doubling of both the 220 and the 220 lattice fringes is observed. From computer simulation of CuAu-I type structure in InGaAs [16], it has been found that the position of In-As pairs exhibits a brighter spot than that of Ga-As pairs at an optimum focus condition (Δf=-38 nm, thickness=8-25 nm).

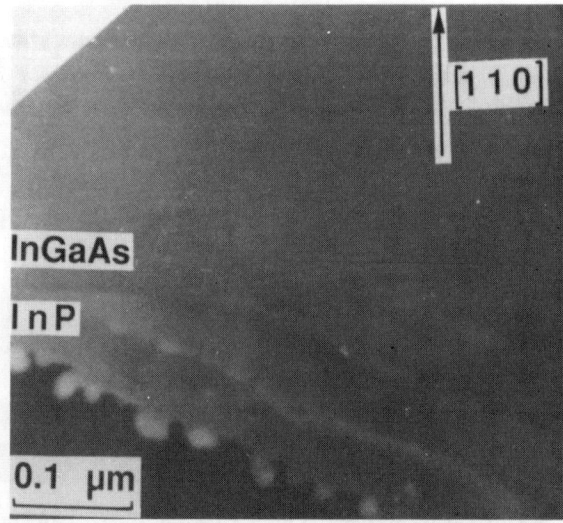

Fig. 5 A cross-sectional dark field image of an ordered InGaAs crystal from one of the one of the superstructure spots.

In order to clarify the cross-sectional shape of the ordered region, a dark field image was taken directly from one of the superstructure spots, as shown in Fig. 5. In this image brighter regions are ordered since the intensity is associated with only the superstructure spot. It is found that the ordered regions are plate-like micodomains 2-4 nm thick, lying on a plane slightly tilted from the (110) plane. These results are very consistent with the "streaky" superstruture spots and tilting of the streaks in the TED pattern. On the basis of these results, it is considered that the generation of CuAu-I type structure takes place by i) migration of deposited atoms on the growth surface, and ii) rearrangement of migrating atoms from the step edges over two atomic layers.

3. 4 Enhancement of two-dimensional electron gas mobility in InGaAs/N-InAlAs heterostructures

In order to investigate the electrical properties of ordered (110) InGaAs, we studied two-dimensional electron gas (2DEG) mobility in InGaAs/N-InAlAs heterostructures with a strongly ordered InGaAs layer. We grew selectively doped InGaAs/N-InAlAs heterostructures on (110) InP substrates 3° and 5° tilted toward the $<00\bar{1}>$ direction. These heterostructures consist of a 600 nm-thick undoped InAlAs buffer layer, a 600 nm-thick undoped InGaAs channel layer, a 10 nm-thick undoped InAlAs spacer layer, a 90 nm-thick Si-doped InGaAs cap layer with a carrier concentration of 3×10^{17} cm^{-3}, and a 10 nm-thick Si-doped InGaAs cap layer with a carrier concentration of 3×10^{17} cm^{-3}. The 2DEG mobility and sheet electron concentration (N_s) are plotted as a function of temperature in Fig. 6. The 2DEG mobility is highly anisotropic: in the $<001>$ direction, the mobility (denoted by closed squares) saturates at 100,000 cm^2/Vs and is comparable to that in the sample grown on a (001) substrate (see broken line), while in the $<110>$ direction, the mobility (denoted by closed triangles and closed circles) is much higher than the alloy scattering limited mobility calculated at N_s of 1×10^{12} cm^{-2} by Takeda [17] (denoted by an arrow), at lower temperatures. The 2DEG mobility in the $<\bar{1}10>$ direction in a sample grown on 3° tilted substrate reached 161,000 cm^2/Vs with N_s of 9.5×10^{11} cm^{-2} at 6 K (101,000 cm^2/Vs with N_s of 9.6×10^{11} cm^{-2} at 77 K). This is the highest mobility, to our knowledge, ever reported for lattice-matched InGaAs/N-InAlAs heterostructure systems. This mobility enhancement in the $<\bar{1}10>$ direction is considered to be caused by the suppression of alloy scattering due to the formation of CuAu-I ordering in the InGaAs channel layer. On the other hand, the 2DEG mobility in the $<001>$ direction compares with that of the non-ordered sample. The difference in the 2DEG mobility in different directions could be explained as follows: Since the plate-like ordered and disordered regions are stacked on nearly the (110) plane on the tilted substrates, stripes of ordered and disordered regions are formed along the $<\bar{1}10>$ direction at the

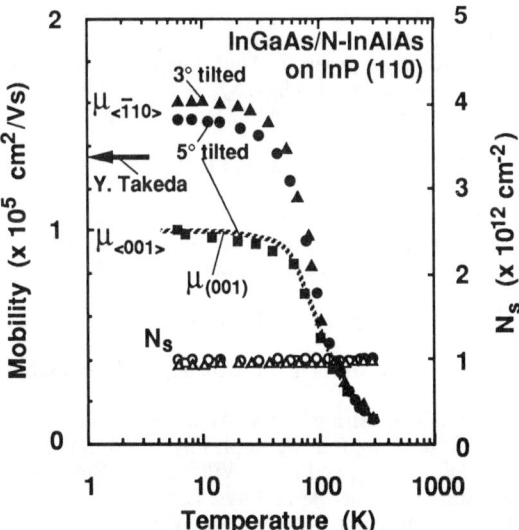

Fig. 6 Temperature dependence of the 2DEG mobility and the sheet electron concentration in the $<\bar{1}10>$ and $<001>$ directions in selectively doped InGaAs/N-InAlAs heterostructures grown on (110) InP substrate 3° and 5° tilted toward the $<00\bar{1}>$ direction.

InGaAs side of the hetero-interface where the 2DEG accumulates. Thus, along the <001> direction, i.e., normal to the stripes, electrons travel across the disordered regions. Along the <1̄10> direction, i.e., parallel to the stripes, the 2DEG formed in both the ordered and disordered regions move along the stripes. The electrons in the ordered regions should be unaffected by alloy scattering in the disordered regions.

The highest obtained 2DEG mobility in the <1̄10> direction, however, is lower than the theoretical prediction in an alloy-scatter-free $(InAs)_1(GaAs)_1$ MSL [17]. One of the reasons for this is that the InGaAs layers are not completely ordered even in the <1̄10> direction. Moreover, the ordered regions include many anti-phase boundaries. The 2DEG mobility in the <1̄10> direction could be improved by reducing the number of such boundaries.

4. Conclusions

We have studied in detail atomic ordering in MBE-grown InGaAs crystals on (110) InP substrates by TEM. First, the existence of CuAu-I type structure in the crystal was confirmed from TED analyses. We have found that when the tilting angle of the substrates increases, the ordering becomes stronger. It is also shown that the ordering is stronger in crystals grown on substrates tilted toward the <001> or <001̄> direction than those on substrates tilted toward the <11̄0> direction. From these results, it is concluded that atomic steps on the growth surface play an important role in the formation of ordered structures. The ordering also becomes stronger when the growth temperature increases in the range 360-485°C. In high resolution images of the crystal, doubling in periodicity of 220 and 200 lattice fringes is found, which is associated with CuAu-I type structure. Moreover, APB's are very often observed in the ordered regions. It is also found that ordering is not perfect, and that ordered regions are plate-like microdomains lying on planes slightly tilted from the (110) plane. We have fabricated InGaAs/N-InAlAs heterostructures with a strongly ordered InGaAs channel layer. The measured 2DEG mobilities from these structures are found to be 100,000 cm^2/Vs in the <001> direction and 161,000 cm^2/Vs in the <1̄10> direction with a sheet electron concentration (N_s) of 9.5×10^{11} cm^{-2} at 6 K. The <1̄10> mobility is much higher than both calculated alloy scattering limited mobility and the best experimental results for lattice-matched InGaAs/N-InAlAs systems. The mobility enhancement in the <1̄10> direction is considered to be achieved by the suppression of alloy scattering due to the occurrence in the InGaAs channel layer.

References

1. P. Henoc, A. Izrael, M. Quillec and H. Launois, Appl. Phys. Lett. 40, 963 (1982).
2. O. Ueda, S. Isozumi and S. Komiya, Japan. J. Appl. Phys. 23, 241 (1984).
3. A. G. Norman and G. R. Booker, J. Appl. Phys. 57, 4715 (1985).
4. O. Ueda, M. Takikawa, J. Komeno and I. Umebu, Japan. J. Appl. Phys. 26, L1824 (1987).
5. O. Ueda, M. Takikawa, T. Takechi, J. Komeno and I. Umebu, J. Crystal Growth 93, 418 (1988).
6. M. Kondow, H. Kakibayashi and S. Minagawa, J. Crystal Growth 88, 291 (1988).
7. A. Gomyo, T. Suzuki and S. Iijima, Phys. Rev. Lett. 60, 2645 (1988).
8. O. Ueda, M. Hoshino, M. Takechi, M. Ozeki, T. Kato and T. Matsumoto, J. Appl. Phys. 68, 4268 (1990).
9. M. A. Shahid, S. Mahajan, D. E. Laughlin and H. M. Cox, Phys. Rev. Lett. 58, 2567 (1987).
10. O. Ueda, T. Fujii, Y. Nakata, H. Yamada and I. Umebu, J. Crystal Growth 95, 38 (1989).
11. H. R. Jen, M. J. Cherng and G. B. Stringfellow, Appl. Phys. Lett. 48, 1603 (1986).
12. H. R. Jen, K. Y. Ma and G. B. Stringfellow, Appl. Phys. Lett. 54, 1154 (1989).
13. T. S. Kuan, T. F. Kuech, W. I. Wang and E. L. Wilkie, Phys Rev. Lett. 54, 210 (1985).
14. T. S. Kuan, W. I. Wang and L. Wilkie, Appl. Phys. Lett. 51, 51 (1987).
15. S. -H. Wei and A. Zunger, Appl. Phys. Lett. 56, 662 (1990).
16. O. Ueda and T. Nakamura (unpublished).
17. Y. Takeda and A. Sasaki, Japan. J. Appl. Phys. 24, 1307 (1985).

DOPANT DIFFUSION IN $Si_{0.7}Ge_{0.3}$

DANIEL MATHIOT[1] AND JEAN CLAUDE DUPUY[2]
[1]MCNC, PO Box 12889, Research Triangle Park, NC 27709-2889 (USA) - on leave of absence from CNET-CNS, BP 98, 38243 Meylan Cedex (France)
[2]Laboratoire de Physique de la Matière (CNRS-URA 358), INSA Lyon, 20 Avenue Albert Einstein, 69621 Villeurbanne Cedex (France)

ABSTRACT

We report on the results of an experimental study of the diffusion of boron and phosphorus in $Si_{0.7}Ge_{0.3}$ alloys for temperatures between 800 and 1050°C. When the dopant concentration is lower than the intrinsic carrier concentration at the diffusion temperature, the diffusion has a simple Fickian behavior. The corresponding intrinsic diffusivities have an activation energy of 1.62 eV for P, and 1.79 eV for B. On the other hand, at high concentration, the diffusion profiles of P are similar to what is observed in pure Si, with "kink and tail" shapes. The high concentration B profiles are unsymmetrical, with a pronounced shoulder on the near surface side. These observations suggest that the diffusion of these dopant atoms in SiGe alloys involves rather complex couplings with the lattice point defects, as in the case of diffusion in pure Si.

1. Introduction

In the last years a lot of efforts have been made to develop new devices taking advantages of the heterojunction formed between silicon and silicon-germanium alloys. This has lead to a strong renewal in studies of the properties of these alloys (see, e.g., [1] for a recent overview). However most of these studies deal with the electrical or optical properties of the SiGe alloys. Although the fabrication of performing devices would probably involve doping of the material, there is to our knowledge no data concerning the diffusivity of impurities in SiGe materials.

In addition to the technological interest reported above, the study of dopant diffusion in SiGe alloys may also have a more fundamental aspect. Indeed, the basic mechanism of diffusion of dopant impurities in group IV semiconductors is still controversial. Whereas it is now widely accepted that diffusion in Ge is mediated via single vacancies, there is no consensus yet on the majority point defect responsible for diffusion in Si. Thus it is possible that the study of dopant diffusion in SiGe alloys could give some useful informations, and then help to solve the remaining question.

The aim of this work was thus to perform an experimental study of the diffusion of phosphorus and boron in $Si_{0.7}Ge_{0.3}$, in the 800-1050°C temperature range.

2. Experiments

For this study we used commercial, thick (10μm) epitaxial layers of SiGe alloys grown on (100) Si substrates by chemical vapor deposition. The SiGe layers were p-type doped, with a boron concentration of the order of 10^{15} cm^{-3}. The Ge content was measured by Rutherford Back Scattering (RBS), and was found to be 0.30±0.01.

For thick epitaxial layers, the strain arising from the difference in the lattice parameters is accommodated by the generation of an interfacial dislocation array. These misfit dislocations can also induce some threading dislocations which could extend across the whole epitaxial layer. These dislocations, acting as alternative paths, could possibly affect the dopant diffusion behavior in our samples. However DLTS measurements performed on the surface of our epitaxial layers failed to detect any signal which could be attributed to extended defects, whereas the same measurements made after chemically etching the top 5 μm reveal the presence of such defects. This indicates that any existing dislocations are confined in the first micrometers near the epitaxial layer/substrate interface, and do not extend up to the surface of the samples. We believe thus that our results are representative of the diffusion behavior in bulk $Si_{0.7}Ge_{0.3}$.

Dopants were introduced into our samples by ion implantation at an energy of 300 keV, with doses of 3×10^{13} and 3×10^{15} cm^{-2}. In order to minimize channeling effects, the implantations were performed in a random-like direction, the wafers being tilted from the (100) normal direction. The wafers were then cut in individual samples which were isothermally annealed in a conventional furnace under pure argon flow. The anneals were performed between 800 and 1050 °C, for times varying between 30 minutes and 4 hours. The temperature was kept constant at ± 1°C during the anneals, the absolute temperature accuracy being of the order of 3°C.

The dopant profiles were analyzed by Secondary Ion Mass Spectroscopy (SIMS), with a CAMECA IMS 4F apparatus. The depth calibration was obtained by measuring the crater depth at the end of the analysis, and assuming a constant erosion rate. The absolute concentrations were calculated by comparison between the measured integrated doses and the implanted ones.

3. Results

We have found that the as-implanted profiles are well described by simple Gaussian curves. The corresponding parameters (mean projected range and standard deviations) are given in Table I. The given values are the average between several experimental profiles.

TABLE I : Distribution parameters for ion implanted B and P in $Si_{0.7}Ge_{0.3}$

impurity	R_p (μm)	σ (μm)
Boron	0.730	0.106
Phosphorus	0.378	0.101

In these conditions, the profiles obtained after diffusion for a time t should be also Gaussian curves, if the diffusivity D is constant. The standard deviation σ' measured after diffusion is linked to the initial value σ through the relationship :

$$\sigma'^2 = \sigma^2 + 2 D.t \qquad (1)$$

As illustrated in Figure 1 in the case of phosphorus, this simple behavior is obtained for

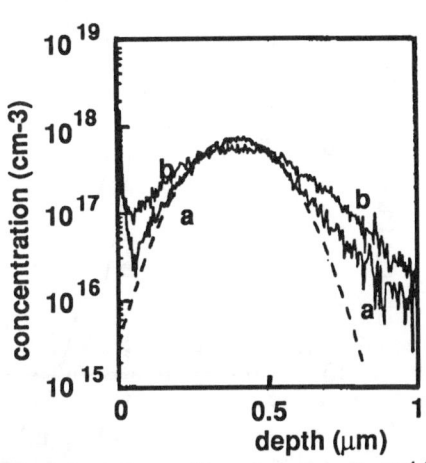

Figure 1 : P profiles for the 3x10^13 implants. (a) as-implanted, (b) after 30 min at 950°C

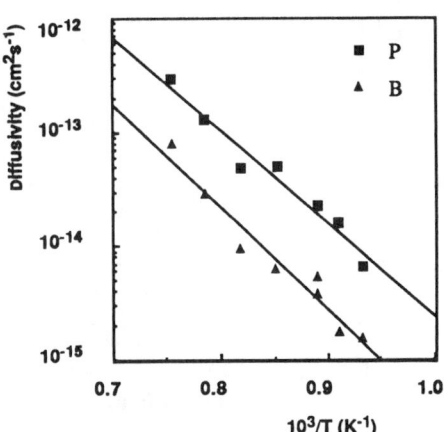

Figure 2 : Intrinsic diffusivities of P and B in $Si_{0.7}Ge_{0.3}$

the redistribution profiles corresponding to the lowest implantation dose. For these low dose implants, the maximum dopant concentrations are lower than the intrinsic carrier concentration in the whole temperature range [2], and then Equ.(1) allows the determination of the intrinsic dopant diffusivity in $Si_{0.7}Ge_{0.3}$, simply from the broadening of the Gaussian profile. It is known that ion mixing during SIMS analysis induces a broadening of the measured profiles. However, this "instrumental" broadening can be described, in first order, as a Gaussian broadening of standard deviation σ_i, such that the measured standard deviation σ_m is given by :

$$\sigma_m^2 = \sigma^2 + \sigma_i^2 \quad (2)$$

Thus, by making the difference between the standard deviations measured on the as-implanted and diffused profiles, the σ_i^2 term cancels. The measured profile broadening is then the actual value of 2Dt.

The corresponding values of the intrinsic P and B diffusivities are plotted in Figure 2, and the temperature variations are well described by the Arrhenius laws :

P : $D_i = 3.7 \times 10^{-7} \exp(-1.62/kT)$ cm²s⁻¹ (3a)

B : $D_i = 3.6 \times 10^{-7} \exp(-1.79/kT)$ cm²s⁻¹ (3b)

Let us note that, due to the rather limited temperature range investigated, the error on the activation energy could be rather large, probably of the order of a few tenths of eV.

On the other hand the profiles measured for the highest dose are more complicated, as shown on Figures 3 and 4. In the case of phosphorus, for not too high temperatures, the

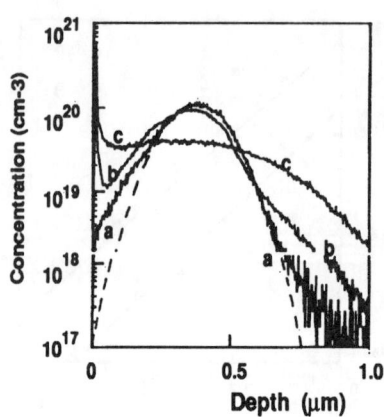

Figure 3 : P profiles for the 3×10^{15} implants. (a) as-implanted, (b) after 150 min at 828°C, and (c) after 30 min at 1000°C

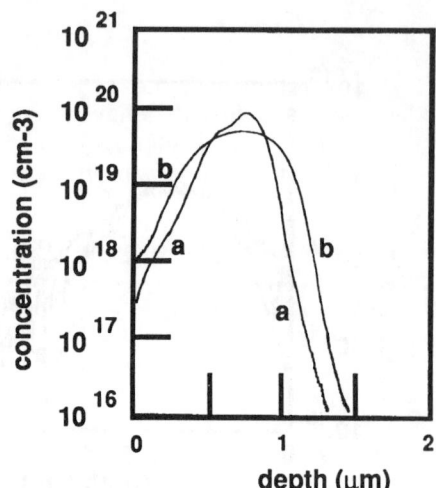

Figure 4 : B profiles for the 3×10^{15} implants. (a) after 45 min at 903°C, and (b) after 20 min at 1050°C

profiles exhibit a "kink and tail" shape analogous to what is observed in pure silicon (Fig.3b). In the case of boron, highly unsymmetrical profiles are observed, with the formation of a pronounced shoulder in the near surface side of the profiles (Fig.4a). At higher temperatures (Figs.3c, 4b), the larger dopant redistributions lead to more regular profiles, although these profiles can no longer be described by simple Gaussian curves.

4. Discussion

First of all we would like to emphasize that the diffusion anomalies observed at high dopant concentration are probably not linked to problems caused by residual ion implantation related defects. Although such defects can play a role for short time diffusions, we believe that they are negligible in our case. Indeed, the melting point of the $Si_{0.7}Ge_{0.3}$ alloy is about 1485 K, to be compared to a value of 1685 K for pure Si. Then, a scaling with respect with these melting points indicates that the 850-900°C temperature range where we observe the highest profile anomalies in $Si_{0.7}Ge_{0.3}$ corresponds roughly to a 950-1000°C range in pure Si. At these temperatures it is well established in the case of Si that the diffusion enhancements associated with the annealing of the residual implant damage are a transient phenomena which has no effect for long time annealings (see, e.g., Ref.3). It is quite reasonable to think that the same thing is true in the present experiments.

As shown on Figs.3 and 4, the deviations from a simple diffusion behavior are effective mainly at the lowest temperatures, where the difference between the maximum dopant concentration and the intrinsic carrier density is the highest. In fact, these observations are quite analogous to what is observed in pure silicon, where the dopant diffusion mechanisms involve rather complex couplings with the lattice point defects [4]. This remark leads us to the conclusion that dopant diffusion in the $Si_{0.7}Ge_{0.3}$ alloys takes place by the same basic mechanisms as in Si, i.e. via the diffusion of (dopant,vacancy) and /or (dopant,interstitial) pairs [4]. Since nothing is known on the properties of these pairs in the SiGe alloys, it was not possible to perform simulations of the high concentration diffusion profiles.

However a rough estimate of the binding energy of the (dopant,defect) pair which dominates the diffusion behavior can be obtained from the activation energy of D_i. Indeed, assuming that the dopant atoms diffuse primarily by coupling with one kind of defect, the activation energy of D_i is, in first order, the difference between the self-diffusion activation energy and the binding energy of the pair. An estimate of the self-diffusion activation energy in $Si_{0.7}Ge_{0.3}$ is given by the results of McVay and DuCharme, who measured Ge diffusivities in SiGe alloys as a function of the Ge contents [5]. These authors found that for Ge contents from 30 to 100 percent, the activation energy is nearly constant, with a value of about 3.0 eV. Using this value, and the values found in the present study for D_i of P and B (Equ.3), we found that the (P, defect) pair has a binding energy of about 1.38 eV, while the (B,defect) pair binding energy is about 1.21 eV. The same calculation in pure Si, with a self-diffusion activation energy of about 5 eV [6,7], and an intrinsic activation energy of 3.66 eV for P diffusion, and of 3.60 eV for B [4], leads respectively to a value of about 1.34 eV and 1.40 for the (P,defect) and (B,defect) binding energies in Si. The fact that these values are very similar to the corresponding values found in the SiGe alloy strengthens the idea that dopant diffusion takes place by analogous mechanisms in the two materials.

Before to conclude, we would like to point out another surprising point concerning P and B diffusion in $Si_{0.7}Ge_{0.3}$, which is the extremely low value of the pre-exponential factor of the intrinsic diffusivities (Equ.3). Indeed this value corresponds to a negative entropy of diffusion of about -9 k. Although the error on the pre-exponential term is probably very high (and thus the finding of a unique value for B and P may be fortuitous), we think that the negative sign of the entropy is significant. Indeed a positive value would require a pre-exponential factor above about 5×10^{-3} cm^2s^{-1}, i.e. more than four orders of magnitude higher than our experimental estimation. Such a highly negative entropy of diffusion is in contrast with what is observed in pure Si [4], and remains to be explained.

5. Conclusion

We have experimentally studied the diffusion of phosphorus and boron in $Si_{0.7}Ge_{0.3}$ for temperatures between 800 and 1050°C. The diffusion profiles show that, when the dopant concentration is below the intrinsic carrier concentration at the diffusion temperature, the diffusion has a simple Fickian behavior. The corresponding intrinsic diffusivities have activation energies of 1.62 eV and 1.79 eV for P and B respectively. The measured pre-exponential factors are surprisingly low, corresponding to diffusion entropies of about -9 k.

On the other hand, at high concentration, the dopant diffusion profiles exhibits strong anomalies, more or less similar to what is observed in pure Si. This suggests that the diffusion mechanisms are analogous in the two materials, involving strong couplings with the lattice point defects.

Acknowledgments

The authors wish to thank J.C.Oberlin who performed the RBS analysis, and A.Brun and A.Grouillet for the ion implantations.

References

1. T.P.Pearsall : CRC Critical Rev. in Solid State and Mater.Sci. **15**, 551 (1989)
2. D.Mathiot and J.C.Dupuy : Appl. Phys. Lett. **59**, 93 (1991)
3. R.B.Fair, J.J.Wortman, and J.Liu : J. Electrochem. Soc. **131**, 2387 (1984)
4. D.Mathiot and J.C.Pfister : J. Appl. Phys. **55**, 3518 (1984)
5. G.L.McVay and A.R.DuCharme : Phys. Rev. B **9**, 627 (1974)
6. H.J.Mayer, H.Mehrer, and K.Maier : Inst. Phys. Conf. Ser. **31**, 186 (1977)
7. L.Kalinowski and R.Seguin : Appl. Phys. Lett. **35**, 211 (1979)

CHARACTERISATION OF DISLOCATIONS IN THE PRESENCE OF TRANSITION METAL CONTAMINATION

V. Higgs[1], E. C. Lightowlers[1], C. E. Norman[2] and P. Kightley[3]

[1] Department of Physics, King's College London, Strand, London WC2R 2LS, UK.
[2] Department of Materials, Imperial College, London SW7 2BP, UK.
   (Present address: MASPEC Institute, Via Chiavari 18/A, 43100, Parma, Italy.)
[3] Department of Materials Science and Engineering, University of Liverpool, Liverpool, L69 3BX, UK.

ABSTRACT

Photoluminescence (PL) spectroscopy, electron beam induced current (EBIC) measurements, transmission electron microscopy (TEM) and defect etching have been used to characterize dislocations in plastically deformed high purity float zone (FZ) Si, epitaxial stacking faults (ESF) in Si epilayers grown by low pressure chemical vapour deposition (LPCVD), and oxidation induced stacking faults (OISF) in high purity FZ Si. For each case, with very low levels of transition metal contamination, dislocation related D-band luminescence could not be detected or was very weak, and no EBIC contrast could be observed. Deliberate contamination with very low levels of Cu, Fe, Ni, Ag or Au gave rise to both intense D-band luminescence and strong EBIC contrast. Higher levels of contamination eventually quenched the luminescence but did not affect the EBIC contrast, and at very high levels precipitation could be observed by TEM. Some preliminary cathodoluminescence (CL) imaging measurements have revealed patterns of slip lines in plastically deformed lightly contaminated FZ Si similar to those seen in EBIC measurements. We show that the D3 and D4 originate mainly at the slip lines, whereas the D1 and D2 bands are dominant between the slip lines.

1. INTRODUCTION

The origin of the optical and electrical properties of dislocations in Si has been the subject of many detailed investigations[1,2]. The optical studies have concentrated on correlating the observed spectral features with specific dislocation types and then trying to understand the electronic transitions involved. Numerous other techniques have also been employed to correlate the observed properties with deep levels. However, the origin of the electronic states involved is still unclear.

The role of impurities must be seriously considered, especially the transition metals. These metals are fast diffusers and are known to form deep levels which can readily affect the electrical and optical properties of Si. They are easily incorporated accidentally during sample preparation or can be already present in the starting material. This makes it difficult to generate defects under clean conditions unless extreme care is used during sample preparation in conjunction with sensitive analytical techniques to check the cleanliness. Also the purity of the starting material must be known to ensure that the bulk impurity levels are as low as possible. In addition, in order to investigate the effects of the transition metals it is important to be able to control their deliberate introduction both at low and high levels.

We have recently demonstrated[3] that dislocations in material free from transition metal contamination ($< 10^{11} cm^{-3}$) give no observable D-band luminescence and no EBIC contrast, whilst after deliberate Cu contamination ($\approx 10^{13} cm^{-3}$) both appear strongly. This effect is illustrated in Figure 1. The PL spectrum from a sample plastically deformed at 750°C containing a dislocation density of $10^7 cm^{-2}$ shows very weak D-band luminescence (Figure 1a). After deliberate contamination with 4×10^{12} atoms.cm^{-2} and annealing at 900°C for

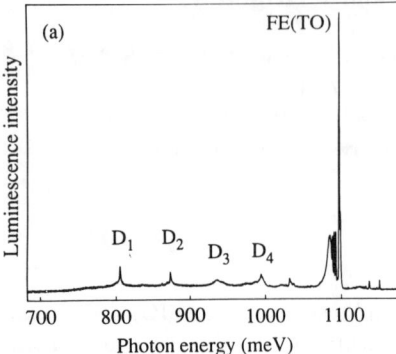 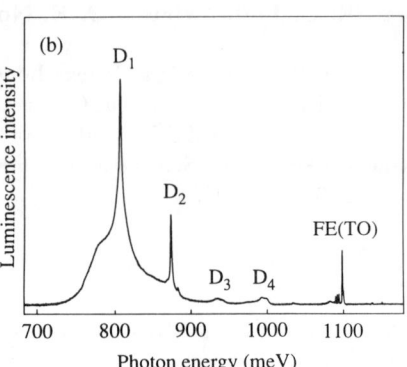

Figure 1. Photoluminescence spectra from FZ Si plastically deformed at 750°C, dislocation density $\approx 10^7 \text{cm}^{-2}$, a) before contamination, b) after deliberate Cu contamination ($\approx 4 \times 10^{12} \text{atoms.cm}^{-2}$) and annealing at 900°C for 1 hour.

1 hour the D-bands are now the dominant features in the spectrum (Figure 1b). In this paper, we report an extension of the initial investigation. Dislocations produced in high purity FZ Si, epitaxial stacking faults (ESF) and oxidation induced stacking faults (OISF) have been intentionally contaminated with a range of transition metals (Cu, Ni, Fe, Ag or Au). The effect of the level of contamination in terms of surface concentration ($10^{11} - 10^{16} \text{atoms.cm}^{-2}$) has also been investigated for Cu and Ni.

Further insight into understanding the origins of radiative recombination at dislocations may be obtained using cathodoluminescence (CL) imaging. The high spatial resolution combined with the ability to record CL spectra and obtain both panchromatic and monochromatic images has made it an extremely useful technique for defect studies of compound semiconductors. However, in the past, the application of this technique to Si has been limited due to the very low luminescence efficiency of silicon and the relatively low sensitivity of detectors. A preliminary investigation of CL imaging of dislocations in plastically deformed and Cu contaminated FZ Si is also reported in this paper.

2. EXPERIMENTAL

Deformation-induced dislocations were produced in samples cut from an ingot of ultra high purity FZ Si, which had a total transition metal content $< 10^{11} \text{atoms.cm}^{-3}$. The samples were stressed uniaxially and heated in a stress cell made totally from high purity quartz. A wide range of epitaxial layers (T=850-1150°C, P=40-760 torr and thicknesses $\approx 1 - 15\mu m$) were grown by low pressure chemical vapour deposition (LPCVD) using a commercial Applied Materials reactor (AMC7811) at GEC Marconi. For comparison, epitaxial wafers from two different commercial manufacturers were also analysed. In addition, high purity FZ Si wafers were given a special surface treatment and then oxidised under clean conditions to produce OISF's.

For the PL measurements, the samples were immersed in liquid helium at 4.2K and the luminescence was excited using an Ar^+ laser tuned to either 457.9nm or 514.5nm. The luminescence was analysed using a Nicolet 60SX Fourier transform spectrometer fitted with a Ge diode detector (North Coast). Preferential defect etching was used to determine the nature of defects present and their density. The defects were also analysed in more detail using TEM and EBIC. Atomic absorption (AA) spectroscopy was used to determine the transition metal impurity levels in the epitaxial layers; the epitaxial layers were selectively removed using ultra pure chemical etching solutions and then analysed using a Perkin Elmer PE3030. Similar experiments were carried out on whole wafers to determine suitable starting material for producing OISF's. The detection limit for transition metal impurities in the epitaxial layers was $\approx 2 \times 10^{13}$ atoms.cm^{-3} and $\approx 10^{11}$ atoms.cm^{-3} in the bulk.

The controlled introduction of the transition metal contaminants onto the sample surface was carried out by backplating from an appropriate transition metal salt solution. The concentration of the metal ion in solution was measured by AA and the surface contamination level was checked using total X-ray reflection fluorescence (TXRF). Additional TXRF measurements were carried out to determine the uniformity and reproducibility of the plating process and checks were also made for cross contamination (detection limit $< 10^{11}$ atoms.cm^{-2}). The samples were RCA cleaned prior to contamination and then annealed in RCA cleaned quartz furnace tubes in flowing nitrogen gas immediately after contamination. Adjacent sections were RCA cleaned without deliberate contamination and then annealed. Control experiments were also carried out using blank plating solutions.

CL measurements were made at ≈ 10 K using the CL mode of a scanning electron microscope (SEM) fitted with a specially designed CL monochromator (MonoCL, Oxford Instruments) directly attached to the SEM. This system uses a retractable parabolic mirror to collect the light which can then be reflected into a 0.3m grating monochromator or directly onto a Ge diode detector (North Coast). Spectra were measured using an optical chopper and standard lock-in techniques. Both panchromatic and monochromatic images were obtained using band pass filters centered on the D-bands. An integral framestore was used to control the electron beam and for signal averaging to obtain good quality images. The spatial resolution was estimated to be $\approx 10 \mu$m for our typical operating conditions[4].

3. RESULTS AND DISCUSSION

The plastically deformed high purity FZ Si samples had either very weak or no observable D-band luminescence and no detectable EBIC contrast. Following light surface contamination ($\approx 4 \times 10^{12}$ atoms.cm^{-2}) by various metals, the D-band features (D1-D4) were the dominant luminescence features observed and the dislocations could be detected by EBIC. The positions of the D-bands remained constant for the different metals.

No D-band luminescence was observed from either the OISF samples or from ESF's in epilayers grown at low temperature (T=850°C) and low pressures (< 120 torr). However, at both higher temperature (> 850°C) and higher growth pressures (> 120 torr) weak D-band luminescence was detected. All the commercially grown epilayers that contained ESF's showed much stronger D-band (D1 and D2 only) features. The metal content of all these layers was determined using AA. On inspection of the AA analysis it was clear when comparing layers of both similar thickness and ESF density that D-band luminescence was only observed when there was a relatively high level of transition metals present ($> 2 \times 10^{13}$ atoms.cm^{-3}). The total metal content of the commercially grown epilayers was usually at least 2-3 orders of magnitude higher than that in material grown at GEC Marconi. TEM examination revealed no unusual features associated with the structure of the ESF's. Also, no evidence was found for metal precipitation.

A selection of epilayer samples that contained a range ESF densities ($10^4 - 10^7$cm^{-2}) and showed no D-bands was deliberately contaminated and then annealed. Following deliberate contamination with either Cu, Fe, Ni, Ag or Au ($\approx 4 \times 10^{12}$ atoms.cm^{-2}) the D-bands (D1 and D2 only) could be clearly observed. Similar results were observed for the OISF's. In addition, deliberate contamination of the epilayer samples that already contained D-band luminescence produced a dramatic increase in the D-band luminescence intensity. As with the deformed FZ material, the positions of the D-band features were approximately constant for the different metals.

These results are consistent with those we reported previously[3], and show that transition metal contamination plays an essential role in both the D-band luminescence and EBIC contrast.

For both types of stacking fault it was found that as the level of contamination (Cu or Ni) was further increased ($> 4 \times 10^{12}$ cm^{-2}) the D-band features decreased in intensity until eventually they could no longer be observed. Figure 2 shows the variation of the intensities of the D1 and D2 bands as a function of Cu contamination (measured by TXRF) for an epilayer sample containing 5×10^7 cm^{-2} ESF's. The D-band peak intensities were normalized to the phosphorus bound exciton peak arising from intentional doping of the epilayers. These results show that the level of contamination is also an important parameter in considering the mechanisms involved in dislocation related luminescence. These results also explain previous PL investigations on impurity decoration of dislocations[5,6], where transition metal contamination was found to have no effect; in those investigations much higher levels of contamination sufficient to quench the D-band luminescence were being used.

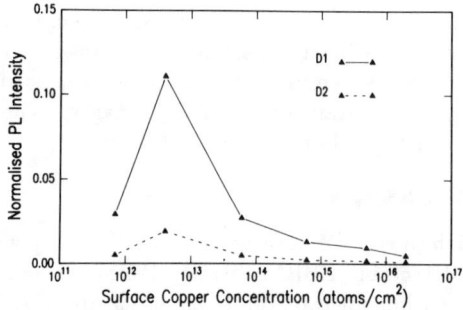

Figure 2. D-band intensity (D1 and D2) as a function of surface Cu contamination (measured by TXRF) for an epilayer sample containing 5×10^7 cm^{-2} ESF's.

Subsequent TEM analysis has revealed that there are two different regimes of decoration around the stacking fault's. At low levels ($< 10^{15}$ atoms.cm^{-2}) of contamination there is no evidence of precipitation found on the stacking faults. As the level of contamination is increased further metal related precipitates can be observed on the bounding partials in TEM.

EBIC investigations have been carried out on these epitaxial layers and the bulk FZ samples containing OISF's, before and after contamination[7]. It was found that EBIC contrast was observed from all the defects following contamination, whether contaminated on the atomic scale or by precipitates, whereas the dislocation related luminescence features decreased in intensity at high contamination levels until they can no longer be observed. However, at this stage of contamination donor bound exciton luminescence can still be observed from the epilayer and not all the luminescence has been quenched. We suggest that as the level of contamination is increased and microprecipitates are formed on the partial dislocations they absorb the centers responsible for dislocation related radiative recombination. This is consistent with the conclusions reached in recently reported DLTS measurements on OISF's deliberately contaminated with Au, Pt, Cu and Ag[8].

In order to investigate further the relationship of the D-bands with the dislocation structure we have carried out some preliminary CL imaging and CL spectroscopy measurements. Figure 3a shows an panchromatic CL image of a plastically deformed FZ Si sample containing $10^8 cm^{-2}$ dislocations, after Cu contamination and annealing. Two intersecting [110] slip planes can be seen quite clearly and Figure 3b shows the corresponding EBIC image of the same sample. CL spectra were recorded on the slip lines ($\approx 30 \mu m$ apart) and in between the slip lines.

Figure 3. a) Panchromatic CL image, and b) EBIC image of FZ Si plastically deformed at 800°C and lightly contaminated.

Figure 4a shows the CL spectrum on the slip band, where D3 and D4 are the dominant features. As one moves away from the slip band D1 and D2 become the dominant features (Figure 4b). This periodic variation in the spectra was found all over the sample surface and the same behaviour was observed on other samples containing single slip. These results suggest that D3 and D4 may be more strongly related with the dislocation core structure than D1 and D2. However, monochromatic images of the dislocation slip lines could be obtained using filters to transmit radiation only from D1 or D2. The PL spectrum of this sample is similar to the spectrum shown in Figure 1b in which D1 and D2 are the stronger features. Further measurements are planned to explore the differences in PL and CL spectra.

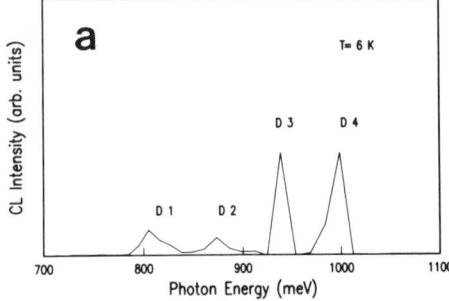
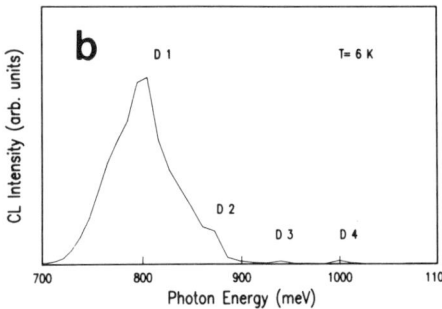

Figure 4. CL spectra from FZ Si plastically deformed at 800°C and lightly Cu contaminated a) on a slip line, b) in between the slip lines.

These investigations have shown that transition metal contamination plays an important role in the production of D-band luminescence from plastically deformed Si and from samples containing ESF's or OISF's. Further experiments will be carried out combining EBIC with cathodoluminescence imaging to investigate the relationship between specific defect types, the metal contaminant and the luminescence.

4. ACKNOWLEDGEMENTS

The authors would like to thank M. R. Goulding from G.E.C. Marconi for providing the epitaxial samples, A. Brinklow from G.E.C. Marconi for carrying out the AA measurements, E. Schemmel from Atomika Technische Physik GmbH for making the TXRF facilities available for our use, and F. Tothill and P. Wright from Oxford Instruments for allowing us to carry out the cathodoluminescence experiments. We would also like to thank D. B. Holt from Imperial College for advice on CL imaging and S. Tajbakhsh from King's College London for assistance in image processing. This work was supported by the Science and Engineering Research Council.

5. REFERENCES

1. R. Sauer, J. Weber, J. Stolz, E. R. Weber, K. H. Küsters and H. Alexander, Appl. Phys. **A36**, 1 (1985).
2. P. Omling, E. R. Weber, L. Montelius, H. Alexander and J. Michel, Phys. Rev. **B32**, 6571 (1985).
3. V. Higgs, C. E. Norman, E. C. Lightowlers and P. Kightley Proc. 20th Int. Conf. on the Physics of Semiconductors, edited by E. M. Anastassakis and J. D. Joannopoulos, Vol.1, p. 706-709, 1990.
4. B. G. Yacobi and D. B. Holt, Cathodoluminescence Microscopy of Inorganic Solids, Plenum Press, NY 1990, p. 58-59.
5. M. Suezawa and K. Sumino, Phys. Stat. Sol. (a) **78**, 639, (1983).
6. K. Weronek and J. Weber, Proc. of POLYSE 90, Springer-Verlag, to be published.
7. V. Higgs, C.E. Norman, E. C. Lightowlers and P. Kightley, Proc. of the 7th International conference on the Microscopy of Semiconductor Materials, Oxford, March 1991, to be published.
8. A. R. Peaker, M. Kaniewska, J. Kaniewski, J. H. Evans, B. Hamilton and G. Lorimer, Proc. Electrochem. Soc. 179th Meeting, Defects in Silicon II, May 1991, Washington, USA, to be published.

CORRELATION OF THE D-BAND PHOTOLUMINESCENCE WITH SPATIAL PROPERTIES OF DISLOCATIONS IN SILICON

K. WERONEK[1], J. WEBER[1], A. HÖPNER[1], F. ERNST[2], R. BUCHNER[3], M. STEFANIAK[4], AND H. ALEXANDER[4]

[1]Max-Planck-Institut für Festkörperforschung, Heisenbergstr. 1,
7000 Stuttgart 80, Federal Republic of Germany
[2]Max-Planck-Institut für Metallforschung, Seestr. 92,
7000 Stuttgart 1, Federal Republic of Germany
[3]Fraunhofer-Institut für Festkörpertechnologie, Paul-Gerhardt-Allee 42,
8000 München 60, Federal Republic of Germany
[4]Universität zu Köln, II. Physikalisches Institut, Zülpicherstr. 77,
5000 Köln 41, Federal Republic of Germany

ABSTRACT

The D-band recombination in silicon is found to be independent of impurities trapped at dislocations. Deliberate contamination of high purity silicon samples, containing dislocations, with copper or iron results in a drastic decrease of the D-band photoluminescence. After a reduction of the copper and iron concentration, the D-bands reappear and increase in intensity by two orders of magnitude. The intensities and polarizations of the D-band photoluminescence depend strongly on the direction of detection and on the structure of the dislocation. We correlate the D_1/D_2-recombination with the stacking fault between two Shockley partial dislocations.

1. Introduction

Silicon samples containing dislocations exhibit a characteristic photoluminescence (PL) spectrum at low temperatures. Drozdov et al. labeled the PL-bands D_1-D_4 [1]. In samples which were plastically deformed, bent, or twisted at high temperatures, or in which precipitates are formed, the D-band PL always shows up at the same energy position: D_1: 0.812 eV, D_2: 0.875 eV, D_3: 0.934 eV, D_4: 1.00 eV. The relative and total intensities, however, of the D-band spectrum depend strongly on the conditions during the generation process of the dislocations. Extensive studies were performed to determine the nature of the D-band PL. The bands can be grouped into pairs D_1/D_2 and D_3/D_4, according to their similar optical properties. For details of the early work on the D-band PL see Ref. [2].

The role of impurities in the generation of the D_1-D_4 bands is not understood. Whereas the dislocation density is not modified by low-temperature annealing after long times, the PL-bands broaden and decrease in intensity [3,4]. No direct influence of specific impurities was found in the D-band spectrum. Doping with oxygen or transition metals (TM), such as Au, produced a broadening of the D-bands. Higgs et al. [5], however, observed the D-band PL only after additional TM doping (Cu doping in particular) in their deformed crystals. It was suggested that the D-band PL is only associated with those dislocations that are decorated with TMs.

In this paper, we examine the origin of the D-band PL: are the D-bands due to recombination at the dislocation core or due to recombination at impurities bound to dislocations? We use a clean method to generate the dislocations, and by deliberately doping with different TMs, we study their influence on the D-band PL. A careful transmission electron microscopy (TEM) study of the investigated samples reveals the different types of dislocations. We correlate the well defined spatial properties of the dislocations with the PL intensities and polarizations of the D-bands.

2. Experimental

In order to generate dislocations in high-purity float-zone (FZ) silicon wafers, a method was adopted that is otherwise used for annealing of ion implantation damage or recrystallization of amorphous materials. Heating with a focussed cw Ar-laser beam (20 W, TEM00, multi line mode) melts the surface layer of the silicon wafer. The melted zone was scanned across the wafer, leading to parallel stripes (separation ~ 40 μm) of recrystallized Si with a high dislocation density ($\sim 5 \times 10^7$ cm^{-2}). The scanning speed was 10 cm/s. Two different wafer temperatures [room temperature (RT) and 500°C] were used to study diffusion processes during the generation of the dislocations. Before the laser treatment, the wafers were cleaned by a standard RCA cleaning procedure [6] to ensure a low TM concentration on the surface [7]. A wafer holder made out of quartz was used to avoid direct TM contamination.

After laser recrystallization, the samples were cut and scratched with a rod of high-purity TM on the back side. Heat treatment was performed in a quartz tube ($T = 850$°C). A flow of high purity nitrogen or argon gas was blown across the sample during the heat treatment. After the diffusion process, the samples were quenched in ethylene glycol at RT. With this diffusion method no unintentional contamination with TMs occurs.

Photoluminescence was excited using the 514 nm line of an Ar laser. The samples were mounted in a He-bath cryostat. PL signals were analyzed by a 1-m grating monochromator and detected with a liquid nitrogen cooled Ge-detector. The signals were processed in standard lock-in technique. The whole setup was controlled by a desktop computer.

The polarization of the PL light relative to the crystal directions was determined by an analyzer in front of the entrance slit of the monochromator. A careful calibration of the setup was necessary to detect small polarization differences.

The cleaning process applied to the laser melted samples is a standard procedure in device production [8]. Annealing at temperatures between 800°C and 1100°C in HCl containing atmospheres results in the formation of TM chlorides which are transported away by the stream of the HCl atmosphere. To avoid oxidation of the samples, the HCl gas has to be dry and oxygen free.

Another set of samples was plastically deformed at 800°C along the [213]-axis to create a well defined dislocation geometry. Only FZ Si (B, P, undoped) was used for deformation to avoid the creation of dislocations in the strain field around oxygen precipitates. The special stress geometry is shown in Fig. 2(a). In these samples, most of the dislocations are in the $(1\bar{1}1)$-glide plane ($\vec{p}$). Fig. 2(e) gives a TEM micrograph of the (011) plane of these samples which shows the preferential $(1\bar{1}1)$-glide planes.

3. Influence of transition metals in the D-band recombination

All samples exhibit intense D-band PL after laser melting and subsequent recrystallization. For direct comparison with the TM diffused samples, other samples were subjected to the same thermal treatment, but without the TM surface contamination. After Cu diffusion for 5 min at 850°C, the D-band PL intensity is reduced and disappears for longer diffusion times. The presence of Cu in the samples is confirmed by the known Cu-related PL at 1.22 μm (1.014 eV) [9]. Iron doping results in the same quenching of the D-band PL. To verify our result, Cu diffused samples were subjected to annealing in a HCl atmosphere at 800°C. After 90 min,

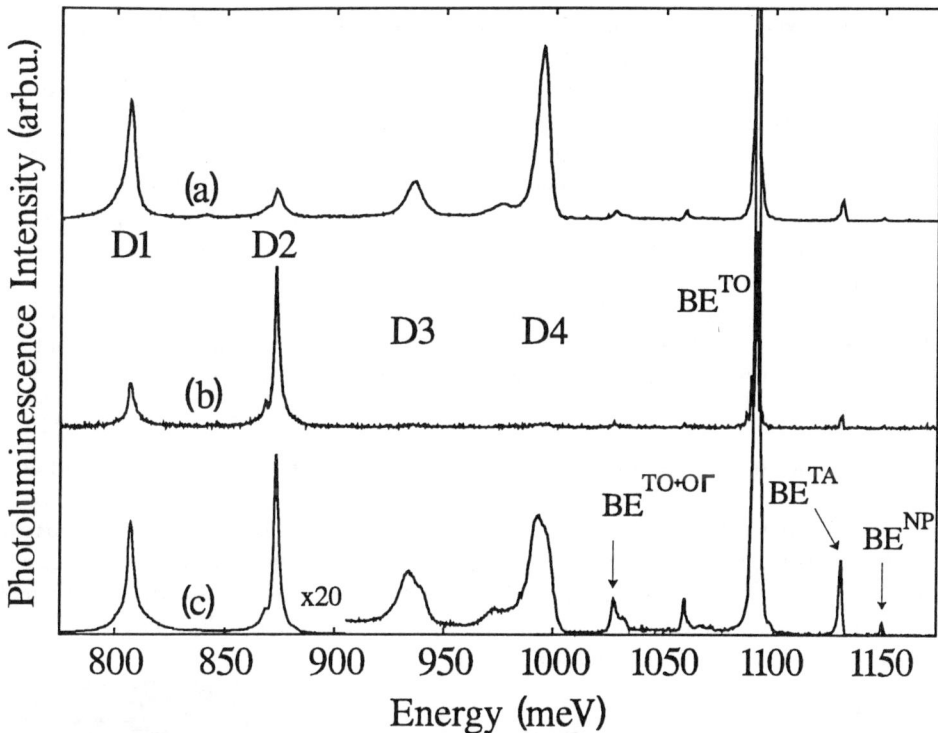

FIGURE 1: (a) Photoluminescence spectrum of a laser melted sample of FZ-Si:B (5×10^{14} cm^3). (b) The same sample as in (a), but additionally annealed at 800°C for 6 h in Argon. (c) Same sample as in (a), but additionally annealed at 800°C for 6 h in HCl.

the D_2/D_1 bands reappeared, whereas the Cu-PL decreased. After 6 h, the D_4/D_3 bands reappeared, and the Cu-PL had almost vanished. The reduction in the Cu concentration supports the efficiency of the employed cleaning process. We applied the heat treatment in a HCl atmosphere to the laser melted and recrystallized samples in order to generate high purity samples with dislocations. The PL spectrum of an original laser melted sample is given in Fig. 1(a). The intensities of the D-bands of these samples are approximately 5 times larger compared to the samples which contained a much higher dislocation density after deformation. For reference a piece of the sample was annealed at 800°C for 6 h in an Ar atmosphere. The result of the PL measurements is shown in Fig. 1(b). An increase of D_2 and a decrease of D_1 by a factor of 5 is detected, whereas D_4/D_3 almost vanish. We explain this behavior of the D_3/D_4 with the passivation of their corresponding centers due to residual Cu. Etch pit analysis and TEM on this sample gave no evidence of a reduction in the dislocation density after the thermal treatment. After heat treatment of the laser melted sample in HCl for 6h at 800°C, a drastic increase in the total PL-intensity occurs [Fig. 1(c)]. This behavior is indicative of a much lower concentration of nonradiative recombination centers. The energy separation of the D_4 and D_3 bands is identical to the separation of the BE^{NP} and BE^{TO} lines. This suggests that D_3 is the TO-phonon replica of D_4.

The D_1/D_2 bands show an increase in intensity by a factor of 20-100 in the samples annealed in HCl gas. The lines have a smaller half width (e.g. 0.9 meV for D_2) at low excitation density compared to the original laser melted sample. From the energy separation of D_1 and D_2 we assign D_1 to the O^Γ phonon replica of D_2.

In our doping experiments we find no enhancement of the D-band PL after TM incorporation into the dislocations; removal of residual TMs from our samples by a heat treatment in a HCl atmosphere results in a drastic increase of the D_1 and D_2 bands. These results indicate that Cu and Fe suppress the D-band PL by forming precipitates in the vicinity of dislocations. These precipitates, mostly detected at kinks and jogs by TEM, are smallest in HCl-treated samples (< 5 nm). From our experiments, however, we cannot rule out the possibility that single TM atoms in the dislocation core (e.g. at the kinks) are the origin of the D-band PL. The residual tiny TM contamination, remaining in the wafer even after the HCl treatment, would be enough to account for such a decoration.

We exclude the possibility that TM-precipitates in the vicinity of dislocations are responsible for the D-band PL. Our PL experiments are in agreement with a recombination at intrinsic states of the dislocations, which might be disturbed by individual TM atoms.

4. Orientational dependence of the intensity and polarization of the D-band PL

Samples with a well defined topology of dislocations allow us to correlate the D-bands with distinct features of the dislocations. Stress along the [213]-axis of a Si-crystal favors single slip within the $(1\bar{1}1)$ plane (with Burgers vector $\vec{b} = \pm[011]$). The geometry of the deformed samples is given in Fig. 2(a). Fig. 2(b) shows the view on top of the primary glide plane. Practically all dislocations are in this plane [Fig. 2(e)].

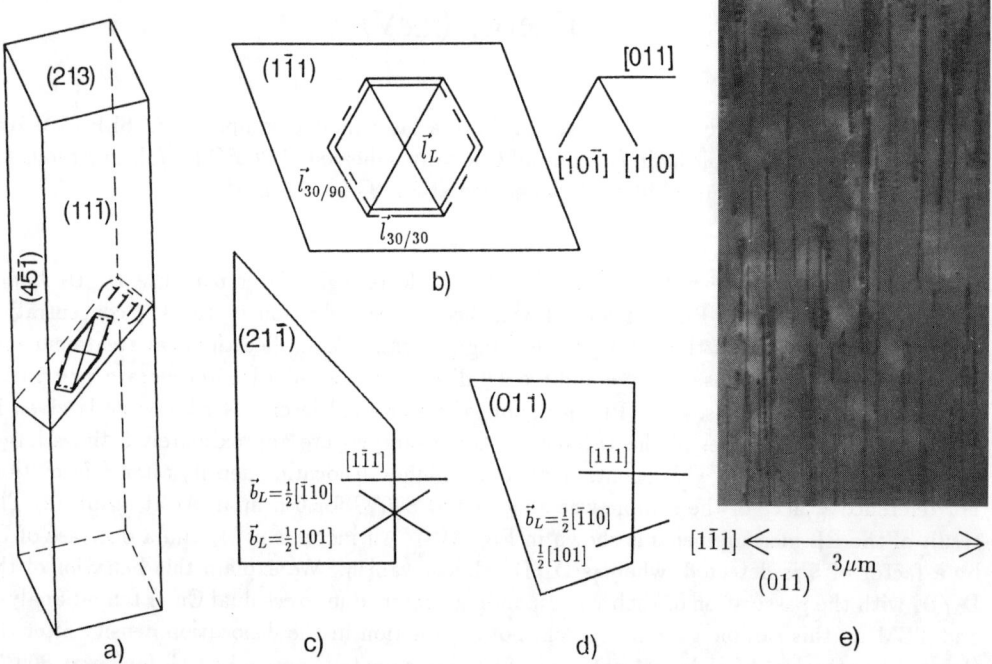

FIGURE 2: (a) Geometry of plastically deformed samples. (b) View onto the primary glide plane $\vec{p} = (1\bar{1}1)$. (c) Projection of the Burgers vectors $\vec{b}_L$ of the Lomer dislocations in the $(21\bar{1})$ plane. (d) Projection of the Burgers vectors $\vec{b}_L$ of the Lomer dislocations in the (011) plane. (e) TEM of the (011) plane, showing the primary glide system $(1\bar{1}1)$.

The PL intensity of the different D-bands was detected for propagation of the PL emission along the 3 orthogonal directions ($[011]$, $[21\bar{1}]$, $[1\bar{1}1]$), which are correlated with the primary glide system [Fig. 2(b)]. All D-bands have similar intensities for detection along the $[011]$ and $[21\bar{1}]$ direction. But the bands have substantially weaker intensities when detected along the $[1\bar{1}1]$ direction. The D_3/D_4 bands show roughly twice the intensity for observation along $[21\bar{1}]$ and $[011]$ compared to the $[1\bar{1}1]$ direction. No PL-intensity could be detected for D_1 and D_2 along the $[1\bar{1}1]$ direction.

The linear polarization of the D-band PL was also measured for the 3 different propagation directions. Our results confirm in principle the results of Izotov and Shteĭnman [10] but are at variance with data reported by Suezawa et al. [11].

Due to lack of space, we will concentrate here only on the D_1/D_2 bands; a full account of our results will be published elsewhere. The D_1/D_2 bands show a pronounced linear polarization with the electric field vector $\vec{E} \parallel [1\bar{1}1]$ if detected along the $[011]$ and $[21\bar{1}]$ propagation direction of the light. There is almost no intensity emitted along the $[1\bar{1}1]$ crystal direction. The PL of the D_1/D_2 bands is simply explained by a dipole transition with $\vec{E}$ oriented perpendicular to the principle glide system.

By a careful TEM study, we have analyzed the type and geometry of dislocations in our deformed samples. Table 1 summarizes the results and gives the geometrical properties of the dislocations. Our results are identical to other studies on samples deformed under the same conditions [12]. From the table, two possible dislocation candidates can be correlated with the polarization behavior of the D_1/D_2 bands.

Type (Origin)	Symbol	$\vec{l}$	$\vec{b}$	Remarks
90°/30°(60°)	$d_{90/30}$	$[110],[10\bar{1}]$	$\frac{1}{6}[121],\frac{1}{6}[\bar{1}12](\frac{1}{2}[011])$	highest density
30°/30°(screw)	$d_{30/30}$	$[011]$	$\frac{1}{6}[121],\frac{1}{6}[\bar{1}12](\frac{1}{2}[011])$	neglectable density
Lomer(60°+60°)	d_L	$[110]$	$\frac{1}{2}[\bar{1}10]$	$\frac{1}{2}[\bar{1}10]=\frac{1}{2}[011]+\frac{1}{2}[\bar{1}0\bar{1}]$
Lomer(60°+60°)	d_L	$[10\bar{1}]$	$\frac{1}{2}[101]$	$\frac{1}{2}[101]=\frac{1}{2}[110]+\frac{1}{2}[0\bar{1}1]$
Stacking Fault	SF	$\vec{p}=(1\bar{1}1)$	$-\frac{1}{3}[1\bar{1}1]$(intrinsic)	between $d_{90/30}$

Table 1: Types of dislocations in samples, found in plastically deformed samples ($T=800°C$, deformation along the $[213]$-axis)

The polarization properties of the D_1/D_2 bands could be correlated with the geometry of the Lomer dislocations which we find in significant concentrations in our samples. Under the stress geometry, two different Lomer dislocations are possible, which form by the combination of two 60° dislocations out of different glide systems. If we assume a polarization of the D_1/D_2 bands parallel to the Burgers vector of the two Lomer dislocations, a linear polarization $\vec{E} \parallel [1\bar{1}1]$ of the D_1/D_2 bands is possible under observation in the $[21\bar{1}]$ direction [Fig. 2(c)] and was actually detected. However, along the $[011]$ direction we would expect a small tilt (21°) from the $[1\bar{1}1]$ direction [Fig. 2(d)]. We tried to determine the tilt angle very accurately with a resolution of 0.5° in several different samples. However, we could only detect an angle smaller than 4°, which is due to the inaccuracy of the mounting of the sample in the cryostat. We therefore have to exclude the Lomer dislocations as possible origin of the D_1/D_2 band PL.

The only feature in Table 1 which has a $[1\bar{1}1]$ symmetry is the stacking fault. The Shockley partial dislocations move at higher temperatures with different velocities, generating in between them intrinsic stacking faults. In equilibrium the splitting width of the Shockley partials is

5.8 nm [13]. Under the assumption that the dipole transition is perpendicular to the plane of the stacking fault, we can explain the intensity and polarization behavior of the D_1/D_2 bands.

5. Summary

Dislocations were generated by laser melting and recrystallization of high purity Si. The PL intensity of the D-bands is correlated with the Cu contamination of the samples. High Cu contaminations passivate the D-band PL, whereas in clean samples the D-band PL is drastically increased.

The intensities and linear polarizations of the D-bands are detected from different directions of plastically deformed samples. We could correlate the polarization properties with geometric features of the dislocations, which we determined by TEM. We propose that the D_1/D_2 recombination originates from the intrinsic stacking fault between two Shockley partial dislocations. The D_3/D_4 bands show a more complex polarization behavior and cannot be directly correlated with the geometry of dislocations in our samples. Further studies have to clarify which electronic levels of the dislocations are responsible for the D-band PL, and which dislocation feature gives rise to the D_3/D_4 bands.

We thank H.-J. Queisser for his steady interest in this work and H. Gottschalk for helpful discussions. We acknowledge the technical assistance of G. Neumann, A. Heidenreich, W. Heinz and W. Krause. This work was supported by the Bundesministerium für Forschung und Technologie under contract NT 2786.

References

[1] N.A. Drozdov, A.A. Patrin, V.D. Tkachev, Pis'ma Zh. Eksp. Teor. Fiz. **23**, 651 (1976) [Sov. Phys. JETP Lett. **23**, 597 (1976)].

[2] R. Sauer, J. Weber, J. Stolz, E.R. Weber, K.-H. Küsters, and H. Alexander, Appl. Phys. A **36**, 1 (1985).

[3] Yu.A. Osip'yan, A.M. Rtishchev, and E.A Shteĭnman, Fiz. Tverd. Tela **26**, 1772 (1986) [Sov. Phys.-Solid State **26**, 1072 (1986)].

[4] A.N. Izotov, Yu.A. Osip'yan, and E.A Shteĭnman, Fiz. Tverd. Tela **28**, 1172 (1986) [Sov. Phys.-Solid State **28**, 655 (1986)].

[5] V. Higgs, E.C. Lightowlers, G. Davies, F. Schäffler and E. Kasper, Semicond. Sci. Techn. **4**, 593-598 (1989).

[6] W. Kern and D. Puotinen, RCA Rev. **31**, 187 (1970).

[7] K. Graff, Mat. Sci. Eng. **B4**, 63 (1989).

[8] Microelectronic Materials and Processes, ed. by R.A. Levy (Kluwer Academic Publishers, Dordrecht/Boston/London, 1989), p. 31.

[9] J. Weber, H. Bauch, and R. Sauer, Phys. Rev. B **25**, 7688 (1982).

[10] A.N. Izotov and E.A Shteĭnman, Sov. Phys. Solid State **28**(4),1986.

[11] M. Suezawa and K. Sumino, Phys. Status Solidi A **78**, 639 (1983).

[12] H. Alexander, J. de Physique **40**, C6-1 (1979).

[13] H. Gottschalk, J. de Physique **40**, C6-127 (1979).

PHOTOLUMINESCENCE AND ELECTRONIC STRUCTURE OF DISLOCATIONS IN Si CRYSTALS

Yu. LELIKOV, Yu. REBANE, S. RUVIMOV, D. TARHIN,
A. SITNIKOVA AND Yu. SHRETER
A.F. Ioffe Physico-Technical Institute, Academy of Sciences,
194021, Leningrad, USSR

ABSTRACT

Photoluminescence (PL) spectra of various extended defects in Si have been studied. A scheme of classification of the PL lines is proposed according to the squared edge components of the Burgers vectors. A possibility of passivation of dislocation core by transition metal impurities is discussed.

1. Introduction

As shown earlier, all the main features of the dislocation-related PL can be accounted for by assuming it to originate from the radiative recombination of the electrons and holes bound in the 1d-dislocation bands split off from the bulk band edges by the strain field. Within this approach, a classification of the dislocation-related PL lines according to the square modulus of the Burgers vector has been proposed [1].

In the present work this concept is further developed and used for the classification of PL spectra, obtained for the first time, of the main structural defects in silicon such as A, B, C, D microdefects, various stacking faults (SF) and rod-like defects (RLD).

The studies of the effect of quenching and annealing on the spectra suggest a possibility of passivation of dislocation core states by low concentrations of fast transition metal diffusers (Cu, Ni, etc.)

2. Electronic structure of dislocations and classification of spectral lines

Bound electronic states at dislocations can originate from the three main factors:
- long-range strain fields of dislocations;
- dangling bonds, impurities or defects in dislocation cores;
- impurity clouds and intrinsic point defects in the vicinity of dislocations.

Long-range strain field is a stable inherent property of the dislocation which does not depend on the structure of its core and the degree of decoration by impurities, and is determined by its topological charge, i.e. the Burgers vector. Therefore the basis of the dislocation electronic structure consist of the one-dimensional (1d-) energy bands split off from the edges of the corresponding bulk bands by the strain field. The dislocation proper can be considered as an 1d-semiconductor with a smaller band gap which is in contact with the bulk semiconductor.

Dislocation core defects, such as kinks, jogs, dangling bonds and impurities in cores result in an effective doping of the dislocation 1d-semiconductor, i.e. determine in it a Fermi level position and create carrier concentrations in 1d-dislocation bands, (mobility, lifetime etc.).

Impurities and defects around a dislocation (Cottrell clouds) are responsible for the contact-type phenomena between the 1d-dislocation and bulk semiconductors and affect such parameters as the Reed's cylinder radius (width of the space charge region), cross section of carrier trapping by a dislocation etc.

In principle, different experiments may reveal dislocation states of all three types. Indeed, ESR permits studying states with unpaired electron spins localized primarily at dangling bonds, core defects and impurities, whereas the EDSR technique [2] detects mobile states with unpaired spins in 1d-bands.

The Hall effect permits one to determine the Fermi level in a system made up of a 1d-dislocation and a bulk semiconductors, and DLTS - the defect and impurity ionization energies in the dislocation core and in its vicinity.

PL reveals the states in 1d-bands, since they possess the largest trapping cross sections for non-equilibrium carriers.

Thus the PL and EDSR techniques permit one to detect primarily the intrinsic properties of a dislocation determined by its Burgers vector, while the ESR, DLTS and Hall effect allow investigation of its extrinsic properties depending on core defects and doping level.

By scaling Schrödinger's equation, one can readily show that the carrier binding energy in the strain field of a dislocation falling off as 1/r should be proportional to the square of the Burgers vector modulus for a given direction. Calculations [3] demonstrate that the contribution of the screw component b_s of the Burgers vector to the carrier binding energy is about an order of magnitude smaller than that of the edge component b_e. Therefore in the first approximation the carrier binding energy for the 1d-bands is proportional to b_e^2 and the position of the dislocation-related PL lines can be roughly evaluated by the expression:

$$\hbar\omega = E_g - A \cdot (b_e/a)^2 \qquad (1)$$

where a is the lattice constant, E_g is the band gap width, and the fitting constant A can be estimated as $A = (\Xi b_e)^2 m / 4\pi\hbar^2$ [1] where Ξ is an averaged constant of the deformation potential, and m is the mean effective carrier mass.

Since many structural defects are bounded by dislocations of different types, eq.(1) permits one also to classify the PL lines of these defects.

In Table 1 are listed estimated sums of binding energies E of carriers at dislocations with [1$\bar{1}$0] axes and different Burgers vectors, as well as the positions of the PL lines found by (1) for the parameters E_g = 1170 meV, A = 800 meV. The last column contains the experimental positions of some lines.

Equation (1) does not take into account the dissociation of dislocations. It is well known, however, that most of the 60°-dislocations in Si dissociate into 90° and 30°-partials. Since, as follows from (1) the bound carrier states at a 90°-partial dislocation are deeper, it provides the main contribution to the PL. As for the effect of the strain field of a 30°-partial, it may be considered as a perturbation [4]. Therefore in a crystal containing dislocations with different widths of SF one observes line series [5] corresponding to different separations between partials, which varies discretely with a step equal to the lattice constant of the SF [4]. Thus according to (1) PL-spectrum of a 60°-dislocations with different degrees of dissociation should extend from the line at

Table 1

Dislocation type	b/a	$(b_e/a)^2$	E, meV	$\hbar\omega$, meV	Experiment
Screw non-split	1/2[110]	0	0	-	-
30°-partial glide	1/6[211]	1/24	30	1140	-
Stair-rod partial	1/6[110]	1/18	45	1125	1030
[113]-defect	1/3[001]	1/9	90	1080	903
90°-partial glide	1/6[112]	1/6	130	1040	1025
Frank partial	1/3[111]	1/3	270	900	870(D2)
60°-non-split	1/2[101]	3/8	300	870	-
Lomer-Cottrell	1/2[110]	1/2	400	770	807(D1)

line at 1040 meV corresponding to an isolated 90°-partial. It is an intermediate line D4 ($\hbar\omega$ = 1000 meV), that corresponds to a 60°-dislocation with an equilibrium SF width of 50 Å.

3. Photoluminescence spectra of structural defects in silicon

Reliable attribution of the PL lines to the structural defects in Si is complicated by difficulties in preparing crystals containing only one particular defect type. Therefore the compiling of an atlas of structural defect spectral lines requires a systematic study of the PL spectra of crystals with different defects prepared by different techniques. The present work is an attempt at such a study.

The PL-spectra and TEM micrographs obtained are given below in Figs.1,a-k, where the corresponding Burgers vector increases from a to k. All the spectra were measured using an Ar laser with λ = 0.5145 μm, W = 300 mW, at T = 4.2 K.

Fig.1,c,f shows PL spectra and TEM-micrograph of silicon crystals prepared by uniaxial two-stage deformation along the [213] and [111] axes (the first stage τ = 1.2 MPa, T = 800°C, ε = 0.3%, second stage: τ = 250 MPa for [213], and τ = 170 MPa for [111], T = 420°C, t = 30 min, $\varepsilon \approx$ 0%).

In the first case crystal contains perfect Shockley loops and PL-spectrum consists of series lines corresponding to non-equilibrium 60°-dislocations [5] and weak lines $\hbar\omega$ = 807 meV (D1) and $\hbar\omega$ = 870 meV (D2) (Fig.1,c). The latter had different identifications as lines linked with point defects near a dislocation [5], with oxidation-induced SF (OSF) [6] and with Lomer-Cottrell dislocations (L-C) [1].

In the second case three glide systems are excited and TEM picture shows a lot of L-C dislocation (Fig.1,f). The intensity of D1 line has grown which supports the interpretation of D1 as due to the L-C dislocations. As for D2, it can be preliminary attributed to sessile Frank partials which may form in the climb of dissociated 60°-dislocations [7], or in condensation of non-equilibrium interstitial atoms or vacancies into Frank dislocation loops during deformation.

Fig.1,e presents spectra and TEM micrographs of Si crystals prepared by oxidation of plates with "softly" damaged (100) surface in an oxygen ambient at T = 1050°C for 10 min. Under these conditions, the formation of defects of two types was observed in the crystals, namely, OSFs, lying in (111) planes and bounded by Frank dislocations with a Burgers vector 1/3 [111], and half loops extended along the surface and made up of perfect dislocations with bounded primarily by L-C dislocations. As seen from the figure, in this case the spectrum exhibits the principal line D1 and the weak line D2. Since D1 has already been identified as due to the Lomer-Cottrell dislocation emission, the line D2 can be attributed to the emission taking place due to Frank dislocations.

Fig.1,g presents a spectrum and TEM micrograph of an FZ crystal prepared by the zone stopping technique. This crystal contains only A-type microdefects which, as seen from TEM contrast, consist mainly of L-C dislocations. The spectrum has one principal line D1 in agreement with the above interpretation.

Shown in Figs.1,h-k are PL spectra of crystals with A + B, B,C,D microdefects respectively. All the spectra reveal primarily a broad emission band in the region of D1 and D2 which agrees with the concept of these microdefects as being loops lying in the (111) and (110) planes and bounded by Frank or L-C dislocations [8]. The width of these bands depends apparently on the spread in the dimensions of these loops ($50 \div 10^4$ Å).

Figs.1,a,b present spectra and TEM micrographs of two crystals with intrinsic epitaxial SF (ESF) obtained by epitaxial growth of the damaged surface. The samples differ by impurity concentrations. As seen from the micrographs, the samples contain mainly complete or incomplete tetrahedrals. The edges of the complete tetrahedral are formed by stair-rod dislocations, but the incomplete ones terminate in sessile Frank partials or Shockley partials [9]. Therefore the broad line in the Fig.1,a at $\hbar\omega = 1030$ meV can apparently be attributed to a stair-rod dislocation decorated by shallow impurities, and that in the Fig.1,b at $\hbar\omega = 1025$ meV, to an isolated $90°$ Shockley partial.

Fig.1,d shows spectra and TEM micrographs of a crystal containing rod-like defects (RLD) and dislocation dipoles. Since the dislocation dipoles consist of non-dissociated $60°$ dislocations or L-C dislocations, their emission lines should be close to D1 and the line $\hbar\omega = 903$ meV observed at shorter wavelengths can be reasonably attributed to RLDs, which, by Bender [10], are bounded by dislocations with a Burgers vector 1/3[001].

Thus, all these data provide an atlas of the main extended structural defects in silicon. As can be seen from Table 1, there is an approximate correspondence of the line positions to formula (1), which proves the key role of dislocation strain field in the luminescence formation.

In conclusion we consider the effect of quenching and annealing of Si crystals with growth-in dislocations on the spectra of dislocation-related PL.

Recently Higgs et al [11] revealed that in Si crystals with undecorated dislocations there is no dislocation-related PL. Luminescence appeared, however, after light contamination (0.003 - 0.1 monolayer) of crystals by transition metal impurities (Cu,Ni,Fe). The positions of the lines are independent of the impurity type. At the same time larger amounts of transition metal impurities were found to suppress the dislocation-related PL.

These results can be explained by assuming that small concentration of transition metal impurities passivate deep recombination centers in the dislocation core and increase the carrier lifetime in the 1d-bands. When introduced in high concentrations, however, transition metal impurities result in the formation of precipitates on dislocations and in a decrease of the lifetime.

To vary the transition metal concentration in dislocation cores, we used quenching and annealing of FZ Si crystals ($N_i = 10^{12}$ cm^{-3}) with a low concentration of growth-in dislocations

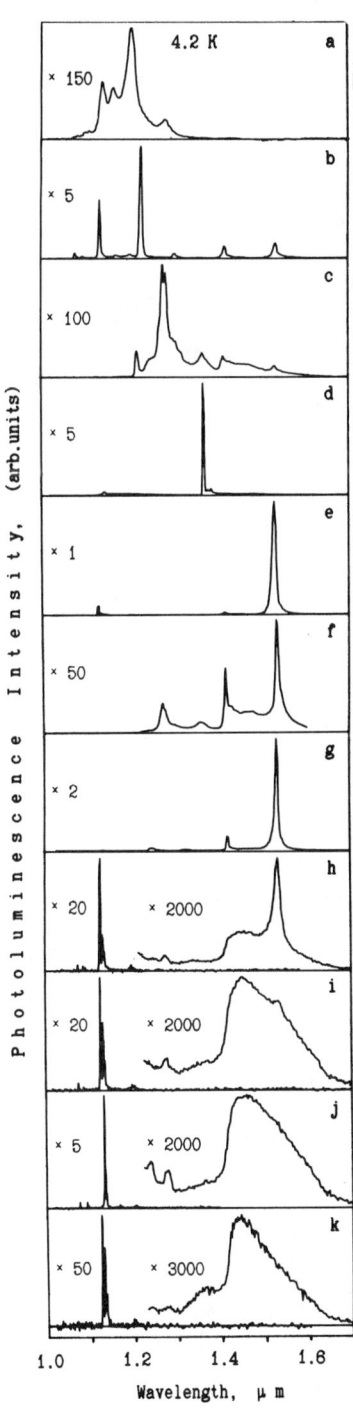
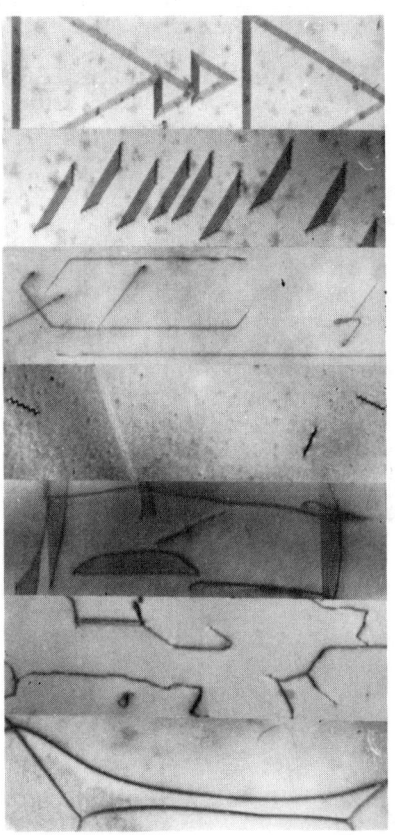

Fig.1 PL-spectra and TEM-micrograhps of various defects in Si crystals
a - perfect epilayer SF
d - imperfect epilayer SF
c - Shockley loops
d - rod-like defects
e - oxidation-induced defects
f - L-C dislocations
g,h,i,j,k - A,A+B,B,C,D-microdefects respectively. The scale used in the TEM-micrographs is 1.5, 1.0, 0.7, 0.7, 0.5, 0.4, 0.7 μm in 1 cm from a to g, respectively.

D4 and D1, as well as the weaker D3 and D2.

The quenching was performed by dropping the sample into silicone oil with $T = 20°C$, and annealing, by maintaining it for 10 min in argon ambient. The line D4 disappeared at quenching temperatures $T > 400°C$, and D1, at $T > 500°C$. This can be explained by the glide of the dissociated 60°-dislocations (D4) and the climb of L-C dislocations (D1) under thermoelastic stress and their escape from the impurity atmosphere [12].

Under annealing, the intensity of D4 recovered at $T_0 = 450°C$, while that of D1 recovered partially at $500°C$, with a dramatic increase at $T = 900°C$. The recovery of the luminescence of D4 and, partially, of D1 at $T = 450 - 500°C$ can be attributed to the passivation of core dislocation states by diffusing transition metal impurities which are activated at these temperatures [13]. The dramatic increase of D1 intensity when annealing at $T = 900°C$ can be accounted for by the growth of sessile L-C dislocation loops [10].

It was also found that the PL did not disappear when the same crystals contaminated intentionally with copper (10^{13} cm^{-3}) were quenched. Neither did the quenching affect the PL in crystals with a high dislocation density, $N_d > 10^7$ cm^{-2}. These facts can be explained by the impossibility for dislocations to escape from copper impurities because in the first case the impurities are spread throughout the crystal and in the second case dislocations become less mobile due to their interaction.

All these data are consistent with our hypothesis of the passivation of core dislocation states by transition metal impurities, but of course a more direct evidence, based on the lifetime measurements is still needed.

The authors are grateful to Prof.H.Alexander and his colleagues for providing us with samples of deformed Si, as well as to Drs.O.Aleksandrov and R.Vitman for samples with ESFs and microdefects.

REFERENCES

1. Yu.Rebane and Yu.Shreter, in *Polycrystalline Semiconductors II*, eds.J.H.Werner and H.P.Strunk (Springer Verlag, 1990) in press.
2. V.Kveder, T.Mchedlidte, Yu.Osip'yan and A.Shalynin Sov.Phys JETP, 66, 838 (1987).
3. S.Winter, Phys.Stat.Sol.(b) 90, 289 (1978).
4. Yu.Lelikov, Yu.Rebane and Yu.Shreter, Inst.Phys.Cong.Ser. N 104, 113 (1989).
5. R.Sauer, Ch.Kisielowski-Kemmerich and H.Alexander, Phys.Rev.Lett. 57, 1472 (1986).
6. A.Peaker, B.Hamilton, G.Lahiji, I.Ture and G.Lorimer, Europ.Mater.Research Ser.Sump.Strasburg, Paper B, 2, 1 (1989).
7. J.Thibault-Dessaux and J.L.Putaux, Inst.Phys. Conf.Ser, N 104, 1 (1989).
8. K.Ravi, *Imperfections and impurities in semiconductor silicon* (A.Wiley-Interscience Publication, N-Y,, 1981) p.95.
9. G.Booker, Disc.Farad.Soc. N 38, 298 (1964).
10. H.Bender, Phys.Stat.Sol., (a) 86, 245 (1984).
11. V.Higgs, E.Lightowers, G.Davies, F.Schaffler and E.Kasper, Semicond. Sci. Technol. 4, 593 (1989)
12. L.Milevskii, Fiz.Tverd.Tela, 2, 2158 (1960).
13. E.Weber and D.Gilles in *Defect Control in Semiconductors*, ed.K.Sumino (North-Holland, 1990) v.1, p.89.

16th International Conference on Defects in Semiconductors (ICDS-16), Bethlehem, July 22-26, 1991.

CHARACTERIZATION OF POINT DEFECTS IN SI CRYSTALS BY HIGHLY SPATIALLY RESOLVED PHOTOLUMINESCENCE

MICHIO TAJIMA[1], HIROSHI TAKENO[2] AND TAKAO ABE[2]
[1]Institute of Space and Astronautical Science, Yoshinodai, Sagamihara 229, Japan
[2]Shin-Etsu Handotai, Annaka, Gunma 379-01, Japan

ABSTRACT

Microscopic distributions around dislocations of both band-edge emission at 1.09 eV and deep-level emission at 0.77 eV at room temperature have been observed for the first time in a Czochralski-grown Si crystal, cooled rapidly by detaching it from the melt and subsequently annealed at 1000°C. The intensity variation reflects the distribution of vacancies which are frozen-in during rapid cooling and are absorbed by dislocations. The 0.77 eV band is considered to originate in small oxygen precipitates in the embryonic stage.

1. Introduction

Vacancies and interstitials play a key role in the oxygen precipitation process in Czochralski(CZ)-grown Si crystals. Despite many efforts to measure them directly, none has yet been successful. Edge-type dislocations are known to act as both sinks and sources for the point defects. As a result, anomalous oxygen precipitation occurs in the vicinity of the dislocations after annealing.[1] The purpose of this study was to analyze the behavior of the point defects by photoluminescence (PL) mapping technique. We have observed for the first time the microscopic distribution around dislocations of both band-edge and deep-level emissions at room temperature. The association of the deep-level emission with the point defects is discussed.

2. Experimental

A <100> Si ingot with a diameter of 75 mm was grown by a conventional puller. The growing ingot was detached from the melt in order to freeze-in point defects. In this particular ingot, slip dislocations were introduced into the crystal accidentally. Wafers were sliced from the ingot parallel to the growth direction, and then subjected to isothermal annealing at 1000°C for 15 min - 16 h. For comparison, a float-zoned (FZ) crystal with dislocations the same order of magnitude as the CZ sample was prepared. The FZ sample was subjected to the identical isothermal annealing.

The PL spectroscopic analysis and PL intensity mapping were performed by two apparatus. One was used for spectroscopy and macroscopic mapping with a spatial resolution down to 100 μm.[2] The sample was mounted on an XY stage and irradiated with the 488 nm line of an Ar ion laser or the 647 nm line of a Kr ion laser. The beam diameter and incident power of the lasers was about 100 μm and 1 - 10 mW on the sample surface, respectively. The PL from the sample was dispersed with a grating monochromator and detected by a Ge *pin* diode. The spectral response of the measurement system was calibrated with blackbody radiation. For low temperature measurements, the sample was mounted in a temperature variable cryostat.

The other apparatus has been newly developed for microscopic mapping and has a spatial resolution down to 1 μm. The sample was mounted on a rotation, tilt, and

XYZ-translation stage. The 488 nm line of an Ar ion laser was focused onto the sample using a near infrared (NIR) objective of magnifications ranging from 10x to 100x. The beam diameter was varied from 1 to 200 μm with an incident power of 1 - 10 mW. The PL from the sample was collected from the front surface using the same objective. Specific PL bands for intensity mapping were extracted by a combination of filters. A Ge *pin* diode or a photomultiplier was used for the PL detection. The wavelength coverage of the system ranges from 500 to 1800 nm.

The sample was first investigated by the former apparatus, and the area of interest then studied further by the latter system. The concentration of interstitial oxygen with its variation across a sample was determined by infrared (IR) absorption spectroscopy. The distributions of dislocations and oxygen precipitates were visualized by X-ray topography.

3. Results

Dislocations introduced into the sample were revealed by X-ray topography. Figure 1 (a) shows the X-ray topograph of the sample after annealing at 1000°C for 4 h. The topograph is just the same as the one taken before the annealing. Straight white lines are slip dislocations. The contrast changes depending on the measurement conditions. Detailed PL analysis was made in the area outlined by the square.

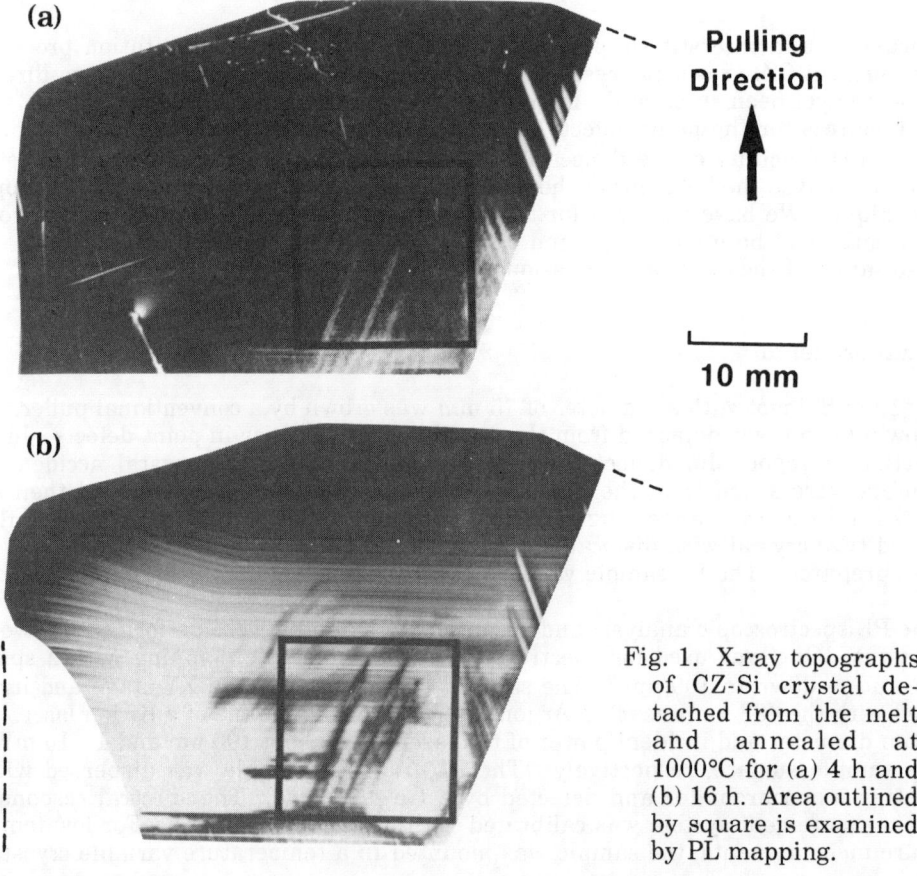

Fig. 1. X-ray topographs of CZ-Si crystal detached from the melt and annealed at 1000°C for (a) 4 h and (b) 16 h. Area outlined by square is examined by PL mapping.

When the annealing time exceeds 8 h, white contrast with a striation pattern is vaguely visible in the X-ray topograph. This white contrast becomes intense with increase in the annealing time and is due to the oxygen precipitates. This is supported by the fact that the concentration of the interstitial oxygen starts to decrease after 8 h annealing in the white contrast region.[1] The X-ray topograph after 16 h annealing is shown in Fig. 1 (b). The sample is not identical with the one in Fig. 1 (a), but is sliced from an adjacent part of the ingot. The area outlined by the square corresponds with that outlined in Fig. 1 (a). It should be noted that the white contrast disappears in the vicinity of dislocations. Correspondingly, the interstitial oxygen concentration is higher in this region, indicating the retardation of oxygen precipitation around dislocations.

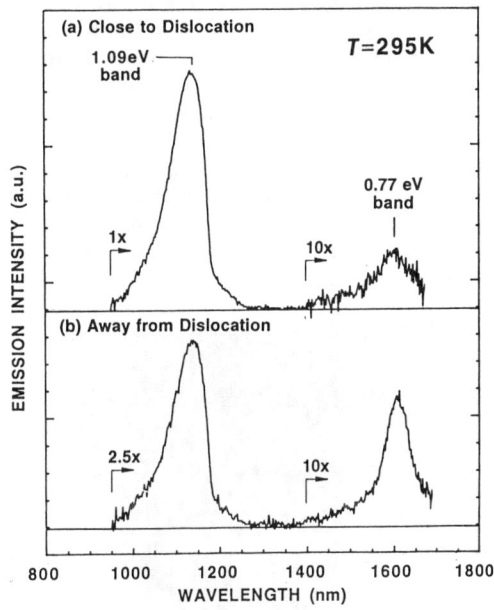

Fig. 2. PL spectra of CZ-Si sample annealed at 1000°C for 4 h: (a) close to dislocation, (b) away from dislocation.

PL spectra of the samples were measured at room temperature. In the as-grown sample only band-edge emission with a peak at 1.09 eV is observed. A short time anneal at 1000°C for less than 1 h does not induce any substantial change in the PL spectra. When the annealing time exceeds 4 h, deep-level emission with a peak at 0.77 eV appears besides the band-edge emission, as shown in Fig. 2. The PL intensities of the two bands vary depending on the measurement position; close to dislocations or away from dislocations.

In order to examine the intensity variation in more detail the PL intensities of the two bands were mapped in the outlined area in Fig. 1 (a), as shown in Fig. 3. The intensity of the 0.77 eV band is reduced along dislocation lines, while that of the 1.09 eV band is raised along these lines. The two PL intensity patterns make a complementary contrast. A periodic intensity fluctuation is recognized perpendicular to the growth direction, possibly corresponding to the growth striation. The spatial resolution of 100 μm in Fig. 3 was not high enough to investigate the intensity variation in the vicinity of dislocations. We measured microscopic intensity variations with a resolution of 10 μm in the area outlined by the square in Fig. 3.

Microscopic mappings in Fig. 4 reveal characteristic intensity variations. The intensity of the 0.77 eV band is raised along the dislocation line with a width of about 30 - 40 μm. This core region is surrounded by a low-intensity region about 100 μm wide (denuded zone). The outer background region shows the highest intensity. A complementary intensity profile is observed for the 1.09 eV band, as shown in Fig. 4 (b). These intensity patterns correspond more to Fig. 1 (b) (after 16 h anneal) than to Fig. 1 (a) (after 4 h anneal), although the PL mappings are on the same sample as in Fig. 1 (a).

The intensity patterns after 16 h annealing look similar to Figs. 3 and 4. The widths of the core region and the denuded zone are extended to about 50 - 80 μm and 200 μm, respectively. In the as-grown sample, the intensity of the 1.09 eV band is decreased along dislocation lines with a width of about 20 - 40 μm. The denuded zone is not formed around dislocations.

For the dislocated FZ sample, only the 1.09 eV band is observed before and after annealing. The intensity variation of this band is the same as for the as-grown CZ sample, and the intensity is decreased along the dislocation core region. The annealing at 1000°C neither changes the intensity pattern of the 1.09 eV band nor induces the 0.77 eV band.

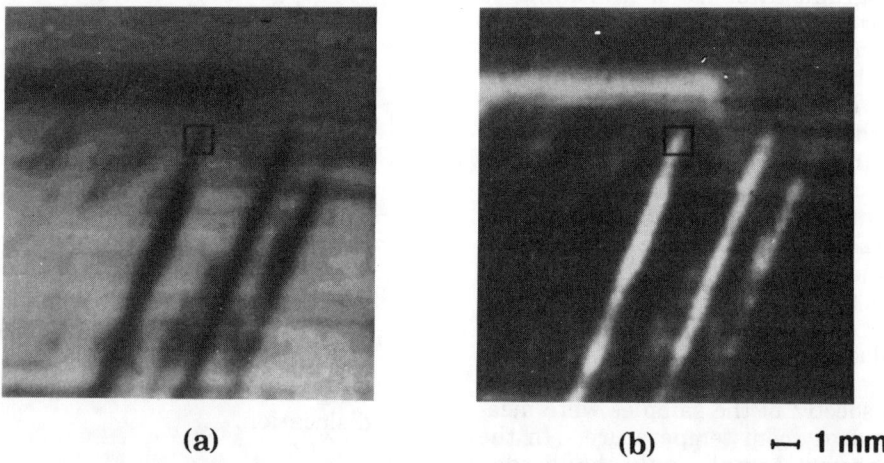

Fig. 3. PL mapping of (a) 0.77 eV band and (b) 1.09 eV band in area outlined by square in Fig. 1 (a) (CZ-Si sample annealed at 1000°C for 4 h) with a spatial resolution of 100μm. Whiter contrast indicates higher intensity. Area outlined by small square is examined by highly resolved PL mapping in Fig. 4.

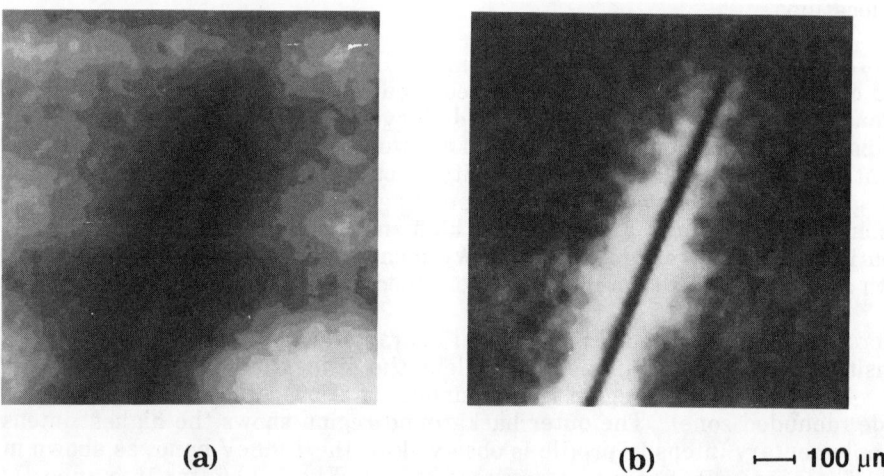

Fig. 4. Microscopic PL mapping of (a) 0.77 eV band and (b) 1.09 eV band around dislocation outlined by square in Fig. 3 (CZ-Si sample annealed at 1000°C for 4 h) with a spatial resolution of 10 μm. Whiter contrast indicates higher intensity.

Temperature dependence of the 0.77 eV band was investigated as shown in Fig. 5. The 0.77 eV band is primarily traceable to the line at 0.813 eV and secondarily to the line at 0.874 eV at liquid helium temperature. The two lines look similar to the so-called D1 and D2 lines, respectively, which are reported to be due to dislocations.[3-6] However, it should be pointed out that the peak position of the 0.813 eV line is substantially different from the D1 line (0.808 eV),[7] which appears in plastically deformed FZ crystals[4] and CZ crystals involving dislocation loops punched out from oxygen precipitates.[6]

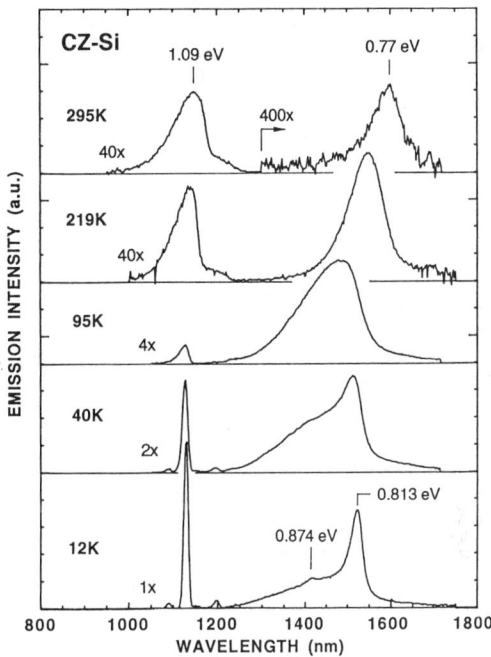

Fig. 5. Temperature dependence of PL spectra of CZ-Si sample annealed at 1000°C for 4 h.

4. Discussion

An annealing at 1000°C for 16 h barely induces oxygen precipitation in conventional CZ Si wafers. The oxygen precipitation in the present sample is due to the excess vacancies frozen-in by rapid cooling.[1] The presence of vacancies enhances the oxygen precipitation by the reaction:

$$2Si + 2O_i + vacancy \rightarrow SiO_2$$

Because vacancies are absorbed by dislocations, the oxygen precipitation is retarded in the vicinity of the dislocations.

The 0.77 eV PL band appears after 4 h anneal, and the oxygen precipitates are not detected by either IR spectroscopy or X-ray topography. The intensity variation of the 0.77 eV band corresponds to the X-ray topograph after the oxygen precipitation occurs: the intensity of the band is raised where the oxygen precipitates will arise after prolonged annealing. The 0.77 eV band does not appear in the as-grown sample. These results lead us to suggest that the origin of the 0.77 eV band is the initial stage of oxygen precipitates or the extended form of the nucleation center for the oxygen precipitation. Possible candidates are small SiO_2 precipitates and small agglomerates involving vacancy and oxygen. The participation of oxygen in the 0.77 eV band is decisive, because this band does not appear in the dislocated FZ sample. The 0.77 eV band is different from the deep level emissions associated with the thermal donors and the new donors,[8] which are also believed to be the early stage of oxygen precipitates.

The small precipitates or agglomerates are gettered by dislocations through their strain field, which pushes up the intensity of the 0.77 eV band along the core region of the dislocations. These small precipitates or agglomerates form deep levels which act as radiative recombination centers for the 0.77 eV band. The complementary intensity contrast for the 1.09 eV band is explainable as being due to the consumption of excited carriers at the deep levels.

In the remaining paragraph we will discuss the correlation between the present 0.77 eV band and the dislocation-related D1 and D2 lines. The 0.77 eV band is traceable to the lines at 0.813 eV and 0.874 eV at liquid helium temperature. It is clear that the 0.77 eV band is not directly related to dislocations, because its intensity becomes highest away from dislocations (Fig. 3), and because there is a substantial difference in the peak position between the 0.813 eV line and the D1 line (0.808 eV). The closeness of the 0.813 eV line to the D1 line and the pairing[4] with the 0.874 eV line or the D2 line, however, suggests some common features between the 0.77 eV line and the D1 and D2 lines.

It is worthwhile to point out that the association of point defects with the D1 and D2 lines has been suggested previously. Sauer et al. interpreted that the D1 and D2 lines are related to point defects in the strain field of dislocations.[4] Higgs et al. reported that the D1 through D4 lines are due to metal atoms or point defect complexes trapped in the strain fields of dislocations.[5] We can rule out the association of metal atoms in our case: if metal contamination during the annealing had caused the 0.77 eV band, the band would also have appeared in the dislocated FZ sample.

5. Conclusion

The microscopic intensity variation of the deep level emission at 0.77 eV at room temperature has been observed for the first time in a CZ Si crystal, cooled rapidly by detaching it from the melt and subsequently annealed at 1000°C. The intensity variation in the vicinity of dislocations is correlated with the distribution of vacancies which are frozen-in during the rapid cooling and are absorbed by the dislocations. Although the 0.77 eV band is traceable to the lines close to the D1 and D2 lines at liquid helium temperature, the band is concluded not to originate in dislocations themselves. The deep levels responsible for the 0.77 eV band are suggested to be due to the embryonic stage of oxygen precipitates, possibly small SiO_2 precipitates or small agglomerates involving oxygen and vacancy.

Acknowledgements

The authors would like to thank R. Shimizu and T. Ishikawa of Atago Bussan for their helpful cooperation in setting up the microscopic PL mapping apparatus, and M. Warashina for his help in the PL measurement.

References

1. H. Harada, T. Abe and J. Chikawa in *Semiconductor Silicon 1986*, eds. H. R. Huff, T. Abe and B. O. Kolbesen (Electrochem. Soc., Pennington, 1986) p.76.
2. M. Tajima, J. Cryst. Growth 103, 1 (1990).
3. N. A. Drozdov, A. A. Partin and V. D. Tkachev, Sov. Phys.-JETP Lett. 23, 597 (1976).
4. R. Sauer, J. Weber and J. Stolz, Appl. Phys. A 36, 1 (1985).
5. V. Higgs, E. C. Lightowlers and P. Kightley, Mat. Res. Soc. Symp. Proc. 163, 57 (1990).
6. M. Tajima and Y. Matsushita, Jpn. J. Appl. Phys. 22, L589 (1983).
7. Although the peak position of the D1 line was reported to be 0.812 eV in Ref. 3, that position was 0.808 eV in our results on several deformed Si samples, which agrees with the results in Ref. 4.
8. M. Tajima, T. Masui and T. Abe in *Semiconductor Silicon 1990*, eds. H. R. Huff, K. G. Barraclough and J. Chikawa (Electrochem. Soc., Pennington, 1990) p.994.

THEORETICAL STUDY ON THE STRUCTURE AND PROPERTIES OF DISLOCATIONS IN SEMICONDUCTORS

Kinichi Masuda-Jindo

Department of Materials Science and Engineering, Tokyo Institute of Technology, Nagatsuta, Midoriku, Yokohama 227, Japan

The atomic configulations and electronic states of dislocations in covalent semiconductors are studied using an LCAO (linear combination of atomic orbitals) recursion electronic theory. Partiqular attention will be focused on the determination of band gap states associated with the dislocation line and point like singularities, "solitons", in the dislocation core region. Using the calculated electronic states of the dislocations, we discuss the effects of impurity doping and non-radiative recombination of the injected carriers on the dislocation motion in the semiconductors.

1. INTRODUCTION

It has been well established that the dislocation mobility in semiconductors is affected quite significantly by doping of electrically active impurities [1]. The effect of n-doping is quite large for Si and Ge, and it increases the dislocation velocity by reducing the apparent activation energy of dislocation motion. The behavior of p-doping is anomalous, but for high concentrations of acceptors, the velocity also increases (decreases) when compared with intrinsic Si (Ge). On the other hand, dislocation motion in covalent semiconductors (e.g., GaAs, InP, GaP and Si) is strongly enhanced by irradiation of electron beam or laser light [2,3]. The observed excitation enhancement of the dislocation motion can be interpreted in terms of the reduction in activation energy of non-radiative recombination of injected carriers at the dislocation core [4]. In the present study, we focus our attention to the electronic states associated with dislocations in covalent semiconductors. We calculate the atomic configurations and local electronic states of dislocations in Si crystals using the LCAO (linear combination atomic orbitals) recursion electronic theory [5].

Using the calculated electronic states of the dislocations, we also discuss the effects of impurity doping and non-radiative recombination of the injected carriers on the dislocation motion in the semiconductors. We will show that the point like singularities "solitons" exsisting in the reconstructed core are responsible for the deep levels of the dislocation cores. This conclusion is identical to that of the earlier work by Heggie and Jones [6] in the sense that point like irregularities play an important role in the elemental process of the dislocation motion. However, the present calculations are in distinction with the previous ones in the following points: We have found that "solitons" in the reconstructed dislocation core with very small atomic displacements Δ (for details, see Fig.1) can produce the prominent deep levels in the band gap. This is in accordance with the recent theoretical speculation by Maeda and Takeuchi [4] on the dislocation mobility and experimental result on the dislocation core using the high resolution electron microscopy observations (no evidence of strong core reconstruction) [7].

2. PRINCIPLE OF CALCULATIONS

To calculate the atomic configuration of the dislocation core, we use the LCAO recursion theory and the quenched molecular dynamics method [8]. We assume that the total energy of the system can be given by a sum of the band structure energy E_b and the pairwise repulsive energy E_r contributions. The band structure energy E_b can be calculated from the electronic Green's functions of the continued fraction form:

$$G_{ii}(E) = 1/[E - a_1 - b_1/(E - a_2 - \ldots - b_n/(E - a_{n+1} - b_{n+1}/E - \ldots)], \quad (1)$$

where a_i and b_i are the recursion coefficents and obtained by usual recursion technique [5]. The local density of staes (DOS) $\rho_i(E)$ on atomic site i can then be calculated from the Green's function $G_{ii}(E)$ as

$$\rho_i(E) = -(1/\pi) \lim_{s \to 0} \mathrm{Im}\, G_{ii}(E + is). \quad (2)$$

The atomic energy levels and two center hopping integrals are taken from Ref.9 for the atomic configuration calculation of the dislocations. This set of transferable TB parameters reproduces well the equilibrium volumes of close packed structures of Si and is suitable for extensive molecular dynamics simulations. On the other hand, the minimal basis sp^3s* basis functions proposed by Vogl et al. [10] are used for the electronic structure calculations of the dislocations. In this model, the addition of an excited s-like states, s*, to the usual sp^3 minimal basis set has the effect of reducing the energy of the indirect conduction band minimum by coupling to the anti-bonding p-like conduction band state. This model has been applied to interpret successfully data on point defects, bulk and surface core excitons and semiconductor surface states [11]. The atomic energy levels E_s, E_p and E_s* are shifted rigidly so as to ensure the local charge neutrality in the crystal with dislocations.

To terminate the recursion coefficients, we also use two different termination schemes: (1) For the atomic configuration calculation of the dislocation, we use the simple termination scheme proposed by Beer and Pettifor [12], with the exact recursion coefficients up to the fourth level. We use the more elaborate average termination procedure of Ref.13 for the electronic structure calculation of the dislocations: The recursion coefficients are calculated up to 34th level for the clusters of about 32800 atoms.

Table 1. TB parameters used in the present calculation (eV).

material	E_s	E_p	E_s*	$ss\sigma$	$sp\sigma$	$s*p\sigma$	$pp\sigma$	$pp\pi$
Si	-4.2	1.715	6.69	-2.075	2.481	2.327	2.716	-0.715
Ge	-5.88	1.61	6.39	-1.695	2.366	2.260	2.853	-0.823

The atomic relaxation calculation is preformed by using the quenched molecular dynamics method, i.e., by integrating the Newtonian equation of motion with the central difference algolithm [8]. We use the following explicit expressions:

$$r_i(t+\Delta t) = 2r_i(t) - r_i(t-\Delta t) + [F_i(t)/m]\Delta t^2 + 0(\Delta t^4) , \quad (3)$$

$$v_i(t) = (1/2)[r_i(t+\Delta t)-r_i(t-\Delta t)/\Delta t] + 0(\Delta t^3) , \quad (4)$$

where $r_i(t)$, $v_i(t)$ are the position and velocity of the atom i at time t and $F_i(t)$ is the force acting on the atom i at this time. The minimum energy atomic configuration can be determined by using the quenching procedure, i.e., the velocity of an atom i is cancelled when the product $F_i(t)v_i(t)$ is negative.

3. RESULTS AND DISCUSSIONS

In a diamond cubic crystal, the important dislocations are the 60°, screw and 90°(edge) perfect dislocations. The first one dissociates into a 30° and 90° partial dislocations while the others split into a pair of 30° and 60° partial dislocations, respectively. All the partials are separated by intrinsic stacking faults. The plastic flow occurs, primarily, through the motion of 30° and 90° glide partials lying on {111} planes. These partials, which have line

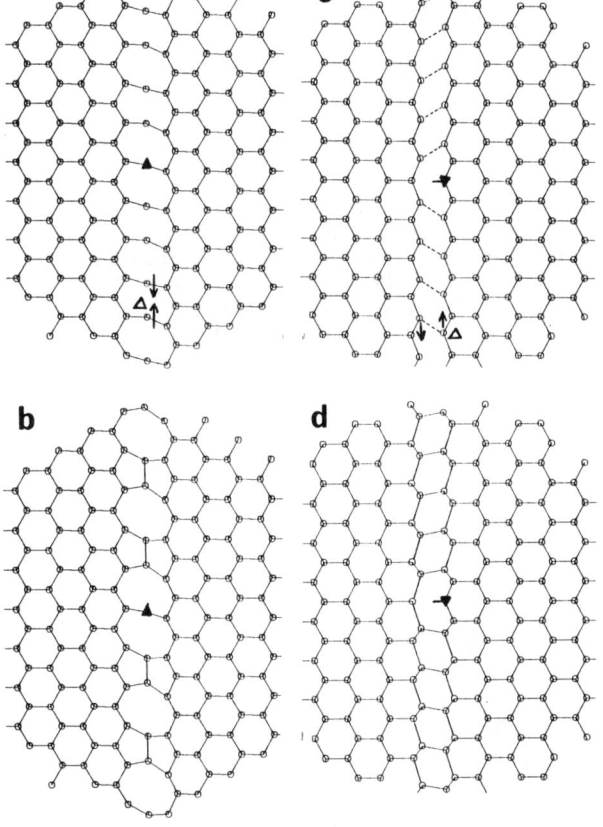

Fig.1 Atomic configurations of reconstructed 30° partial (a and b) and 90° partial (c and d) dislocations in a diamond cubic crystal. a) and c) [b) and d)] show the reconstructed structures with small atomic displacements of $\Delta=0.1d_0$ [large atomic displacements of $\Delta=0.32d_0$]: Δ denotes the magnitude of the atomic displacement in the direction of the arrow and is given periodically along the dislocation line to produce a reconstructed structure.

directions along <1$\bar{1}$0> are believed to be reconstructed into a structure with no dangling bonds. In view of this, we have performed the elecronic structure calculations for 30° and 90° partial dislocations. We have found, from the atomistic simulation by the quenched molecular dynamics method, that the atomic displacements for the reconstructed cores (with no dangling bond states) Δ are rather small and exsistence of fully reconstructed cores are not realized easily in the diamond cubic crystals; slightly larger atomic displacements $\Delta=0.15d_0$ (d_0 being the nearest-neighbour distance of the perfect Si crystal), are obtained for 30°partial dislocations in Si. This is due to the fact that the energy reduction due to the reconstruction of the dangling bond states is not so effectve compared to the increase of the elastic energies due to the distorted bonds.

Before discussing the electronic states of the dislocations, we briefly summarize the recent theories of enhanced dislocation motion.

3a. Elemental Process of Dislocation Motion

The dislocation motion has been discussed in detail by Hirth and Lothe [14] on the basis of the abrupt kink model. The dislocation velocity is generally written in two different forms according to whether the kink-kink collision occurs or not

$$v = \begin{cases} 2d\sqrt{Jv_k} & (X<<L), \quad (5-a) \\ dJL & (X>>L). \quad (5-b) \end{cases}$$

Here J is the frequency of kink-pair formation per unit length of the dislocation, v_k is the lateral velocity of a kink along the Peierls valley, d is the interval of Peierls valleys, X is the mean free path of migration of a kink, and L is the segment length of a dislocation. The condition X<<L and X>>L, respectively, correspond to the dislocation motion with and without kink-kink collisions. Maeda and Takeuchi [4] have derived J directly from a set of rate equations which represent the equilibrium jump frequency of kinks at each kink site (0,1,2,... in Fig.2). The result of J is written as

$$J = v_s(bd\tau/kT)\exp[-\frac{1}{kT}(\sum_{i=0}^{p-1}(\Delta E_{i+}-\Delta E_{(i+1)-}) + E_{diff})], \quad (6)$$

where v_s is the trial jump frequency of straight-dislocation sites,

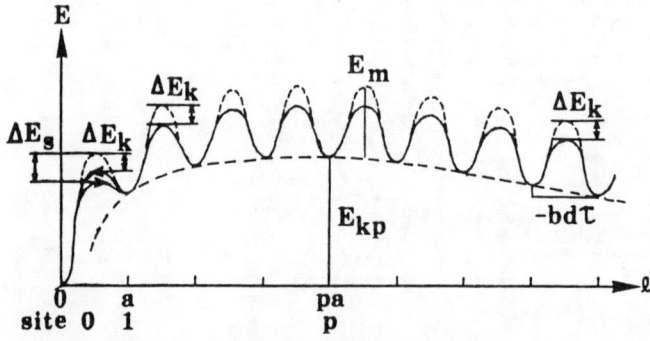

Fig.2 Potential barriers for dislocation motion.

b is the Burgers vector, τ is the shear stress component in the direction of b, and E_{diff} is the effective activation energy barrier to diffusive kink migration. In the summation of the above eq.(6), p is the site number at which the energy of a kink pair assumes the maximum value, ΔE_{i+} and $\Delta E_{(i+1)-}$ denote the potential barriers to the processes of site i → site (i+1) and site i ← site (i+1), respectively. On the other hand, the lateral kink velocity v_k is obtained by means of the Einstein relation in diffusive phenomenon and given in a form of the Ahrenius type, with activation energy of E_{diff}. One can then discuss the elementary process of the dislocation motion by estimating J and v_k, i.e., the summation term and E_{diff} in eq.(6) taking into account the experimental conditions.

3b. Electronic States and Dislocation Motion

In Fig.3, we present the calculated local electronic DOS on the atom (soliton site, marked by Δ in Fig.1c) in the core of 90°partial dislocation in Si, together with s- (3a) and p- (3b) partial DOS.

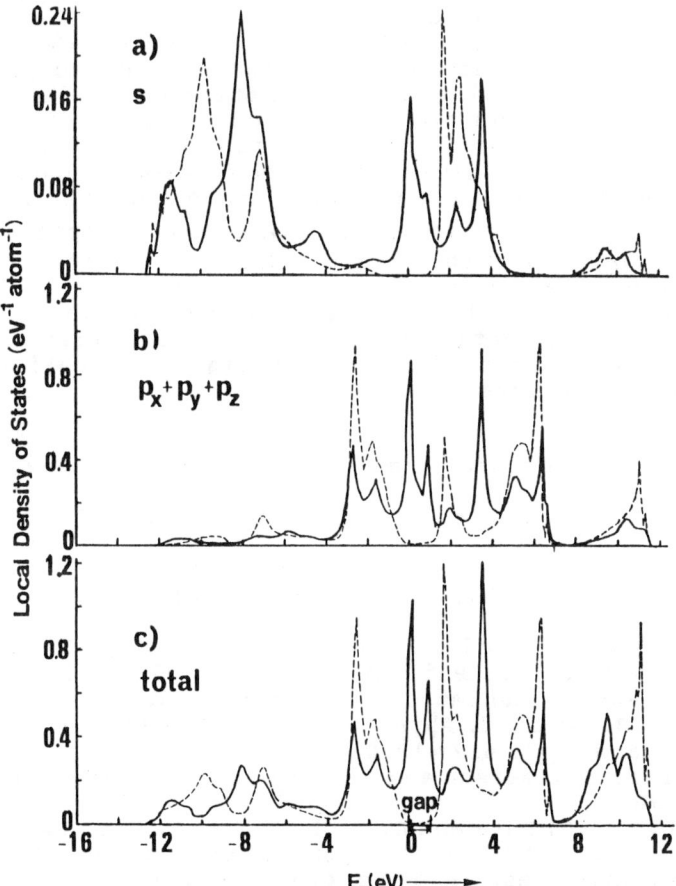

Fig.3 Electronic DOS of atom ("soliton" site) in the core of 90° partial dislocation in Si (dashed curves; perfect lattice).

One can see in Fig.3 that s- and p- (and s*) partial DOS are strongly deformed due to the variation of the atomic configuration of the dislocation core. In these calculations, it is interesting that the prominent deep levels of "solitons" (located near the center of the band gap) appear even for the reconstructed core with small atomic displacements of $\Delta \cong 0.1 d_0$. This means that there can be deep centers along the straight dislocation line in much smaller density than that of the geometrical dangling bonds [15].(We have obtained a broadened DOS structure in the band gap for the unreconstructed core, indicating the dislocation states extend over the entire gap region.) In particular, the present electronic structure calculation is important in conjunction with the discussion of Sec.3a: By using eqs.5 and 6 and the experimental facts that (i) X>>L is realized under usual experimental conditions and (ii) the pre-exponential factor of the dislocation velocity expression is propotional to the excitation intensity I, one can show that the reduction in activation energy ΔE is simply equal to ΔE_s (when X<<L, $\Delta E=(\Delta E_s+\Delta E_k)/2$), where ΔE_s and ΔE_k are the energies released upon non-radiative capture of excited carriers at the straight dislocation site and kink site, respectively. This implies that even if the enhancement of kink migration occurs, it does not contribute to the dislocation mobility enhancement. Therefore, the observed reduction in the activation energy (0.68~0.82 eV for intrinsic Si) of the enhanced dislocation motion corresponds to the deep energy levels associated with the straight dislocation site. Furthermore, it must be noted that clear evidence of core reconstruction of the partial dislocations has not been given yet [7]. Summarizing these theoretical speculations and experimental observations, we come to the conclusion that the point-like singularities such as solitons (with small atomic displacement Δ) or segregated impurities in the straight dislocation core can produce the prominent deep levels in the energy band gap.

Finally, we note that the doping effects of the electrically active impurities on the dislocation motion can also be understood in terms of the deep levels associated with the straight dislocation sites relative to the Fermi level.

REFERENCES
1. J.R. Patel, L.R. Testardi And P.E. Freeland, Phys. Rev. B13, 3548 (1976).
2. K. Maeda, M. Sato, A. Kubo and S. Takeuchi, J. Appl. Phys. 54, 161 (1983).
3. K.H. Kusters and H. Alexander, Physica 116B, 594 (1983).
4. N. Maeda and S. Takeuchi, Inst. Phys. Conf. Ser. No.104: Chapter 3, 303 (1989).
5. R. Haydock, V. Heine and M.J. Kelly, J. Phys. C8, 2591 (1975).
6. M. Heggie and R. Jones, Phil. Mag. 48, 365, 379 (1983).
7. S. Takeuchi and M. Hashimoto, to be published.
8. J.S. Luo and B. Legrand, Phys. Rev. B38, 1728 (1988).
9. S. Sawada, Vacuum 41, 612 (1990).
10. P. Vogl, H.P. Hjalmarson and J.D. Dow, J. Phys. Chem. Solids, 44, 365 (1983).
11. C. Mailhiot, C.B. Duke and D.J. Chadi, Surf. Sci. 149, 366 (1985)
12. N. Beer and D.G. Pettifor, "Electronic Structure of Complex Systems", Plenum Press, (New York) P.769 (1984).
13. K. Masuda-Jindo, Phys. Rev. B41, 8407 (1990).
14. J.P. Hirth and L. Lothe, "Theory of Dislocations" (McGraw Hill, New York) P.531 (1968).
15. M. Suezawa and K. Sumino, phys. stat. sol., 78a, 639 (1983).

SOLID STATE PROCESSES AT THE ATOMIC LEVEL

A. OURMAZD, F.H. BAUMANN, M. BODE, Y. KIM, AND J.A. RENTSCHLER

AT&T Bell Labs, Holmdel, NJ 07733, USA

ABSTRACT

Recent developments in high resolution transmission electron microscopy allow the composition of materials to be mapped at near-atomic sensitivity and resolution. We describe how such quantitative chemical mapping techniques can be used to study important solid state processes. Examples include chemical stability of multilayers, diffusion of native point defects, and interaction of individual energetic ions with solids.

1. INTRODUCTION

Scientifically, solid state processes are of fundamental interest, because they involve a rich variety of defect reactions. Technologically, controlling solid-state prcoesses is essential for the fabrication of new materials and devices. Many solid state processes involve the substitution of certain atoms on the lattice with other, chemically different species, leaving the structure essentially unaltered. Microscopic understanding of such phenomena is needed to make contact with theory, and to achieve adequate process control for ultra-large scale integration of semiconductor devices.

The transmission electron microscope is now routinely used to reveal the atomic-scale structure of materials. But until recently, it has been ineffective in revealing the atomic details of the large variety of solid state reactions that involve only compositional changes. This, and the qualitative nature of TEM analsyis have restricted the role of electron microscopy to the examination of extended defects, leaving point defects and their reactions inaccessible to direct microscopic examination.

Chemical lattice imaging [1-3], however, is a TEM based technique capable of revealing changes in the sample composition with atom-column resolution. The application of recently developed vector pattern recognition algorithms [4] to chemical lattice images allows the composition of individual atomic columns to be determined with near-atomic sensitivity; single- and double-atom substitutions in individual atomic columns of typical semiconductors can be detected at ~1σ (~60% confidence) and 2σ (90% confidence) levels, respectively. "Chemical Mapping" is thus a quantitative means for studying microscopic changes in the composition of materials at the atomic level [3].

In this paper, we outline the principle of quantitative chemical mapping, and briefly describe some of its applications. Examples will include the chemical stability of interfaces, the intrinsic thermodynamic properties of native point defects, and the fundamentals of ion-solid interactions. Details can be found in the references provided in the text.

2. CHEMICAL LATTICE IMAGING

Compositional changes in a material that involve changes in the atomic occupancy of a subset of lattice sites necessarily cause substantial changes in a set of reflections, that we name chemical. An example is the (200) reflection in the zinc blende system; when the primary electron beam enters the sample along a <100> direction, the (200) reflection occurs only because of chemical differences in the occupants of the two fcc sublattices. Similar reflections exist in all materials whose compositional changes involve changes in the occupancy of a subset of lattice sites. At the simplest level, chemical lattice imaging seeks first to use dynamical interactions to maximize the intensity of such chemical reflections, and then to use the bandpass characteristics of the objective lens to enhance their contribution to the image [1-3].

Fig. 1(a) is a chemical lattice image of a GaAs layer between its two AlGaAs neighbors. Note that although the structure of the sample remains zinc-blende throughout, the image changes strongly on crossing the interface. This is because in a chemical lattice image, the compositional information in the sample is encoded into the details of the patterns that combine in a mosaic to form the image. We have developed a pattern recognition approach, which examines each unit cell of the image, deduces its composition by comparing it with a model (template) image, and yields a confidence level for this determination [4]. In semiconductors, it is thus possible to detect single- and double-atom substitutions in individual atomic columns of materials with ~60% and ~90% confidence, respectively. Fig. 1(b) is a quantitative chemical map, obtained by pattern recognition analysis of the chemical lattice image shown in Fig. 1(a). In Fig. 1(b) the height represents the composition, and the colors provide statistical information, with color changes corresponding to compositional changes with an error probability of less than 3 parts in 10^3.

3. CHEMICAL STABILITY OF INTERFACES

The mechanical stability of strained interfaces has received extensive attention. However, even interfaces between materials with no lattice parameter difference are far from equilibrium. On crossing a modern GaAs/AlGaAs interface, for example, the Al concentration changes by several orders of magnitude in a few lattice spacings. As originally pointed out by Cahn [5], such systems relax by interdiffusion, which can take novel pathways. It is thus scientifically worthwhile, and technologically important to investigate the stability of interfaces against interdiffusion. In most semiconductors, the modest diffusivities of point defects limit substantial relaxation at room temperature. However, an interface can relax during thermal annealing, in-diffusion of dopants, or ion-implantation.

Given chemical maps of the type shown in Fig. 1(b), it is straightforward to make accurate measurements of the interdiffusion coefficient at single interfaces as follows [6,7]. The composition profile across a given interface is measured in two pieces of the same sample, one of which has been annealed in bulk form. An example is shown in Fig. 2, where composition

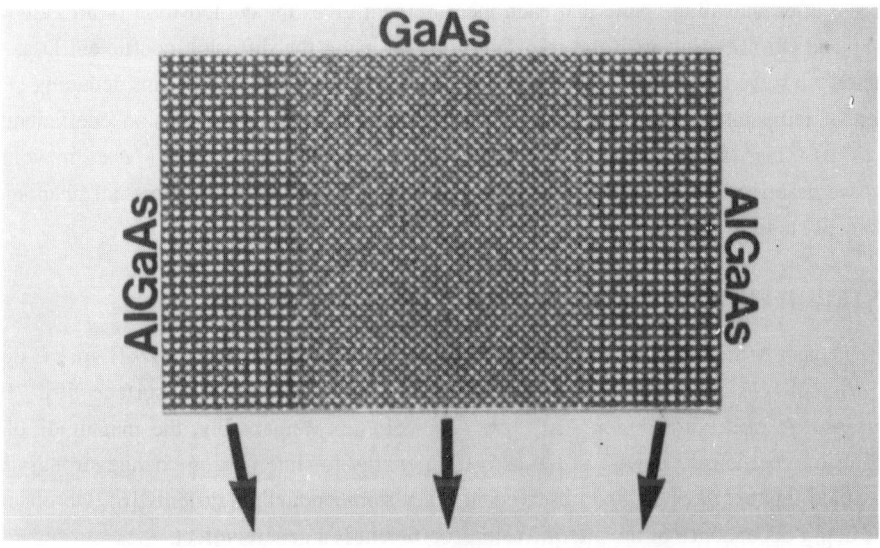
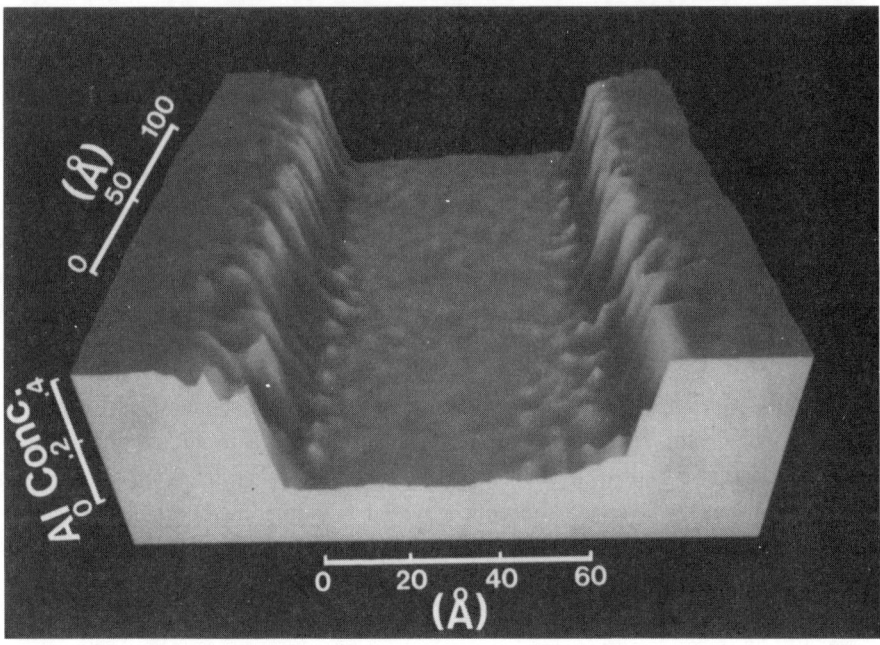

Fig. 1   (a) Chemical lattice image of a GaAs quantum well between two $Al_{0.4}Ga_{0.6}As$ barriers. (b) three-dimensional representation of a quantitative chemical map obtained by vector pattern recognition analysis of the chemical lattice image shown in (a). Height represents the local composition, and color changes represent three standard deviation changes in the signal, i.e. changes in composition with an error probability of less than 3 parts in 10^3.

profiles for two $In_xAl_{1-x}As$ strained layers, imbedded in an $In_{0.52}Al_{0.48}As$ matrix are shown before and after annealing. Note that each measurement gives the composition of an individual atomic plane [8]. Starting with the initial profile and using the diffusion coefficient D as free parameter, we solve the diffusion equation to fit the final (annealed) profile, thus deducing D as a function of temperature and interface depth [6-9]. In this way, interdiffusion coefficients as small as $10^{-21} cm^2/s$ can be measured, sampling volumes as small as $10^{-19} cm^3$ in volume. Below, we describe two examples, where this capability reveals new fundamental phenomena, even in well-studied systems.

3.1. INTERDIFFUSION DUE TO THERMAL ANNEALING

Fig. 3 is an Arrhenius plot of the interdiffusion coefficient D vs $1/kT$ for C-doped $GaAs/Al_{0.4}Ga_{0.6}As$ interfaces at three different depths beneath the surface [9]. Each measurement is made in a region $\sim 10^{-18} cm^{-3}$ in volume. Remarkably, the magnitude of the interdiffusion coefficient, as well as the activation energy for intermixing change strongly with depth. Since this behavior is also observed in other semiconducting systems [6], we conclude that the depth-dependence of the interdiffusion coefficient is a general effect.

We have established that the depth dependence of the interdiffusion coefficient is related to the injection of point defects from the sample surface. In particular, interdiffusion in semiconducting systems is assisted by the presence of native point defects (interstitials and vacancies), whose concentration is often negligible in as-grown samples. For interdiffusion to occur, such native defects must be injected from the sample surface during the anneal. The interdiffusion coefficient is a sensitive function of the concentration of these defects at the particular interface studied, and thus can be used to investigate the microscopics of native point defect diffusion in multilayered systems. Indeed, it is thus possible to measure the formation energy and migration energy of a given native defect (interstitial or vacancy) as a function of its charge state [10].

Returning to interdiffusion, two important points emerge. First that the interdiffusion coefficient varies strongly with depth. Thus a measurement of this parameter is meaningful only if it refers to a single interface at a known depth. Second, it follows that the interface stability is also depth-dependent. Thus the layer depth must be regarded as an important design parameter in the fabrication of modern devices. This effect assumes additional importance when interdiffusion is also concentration dependent, leading to strong intermixing at very low temperatures [11].

3.2. ION IMPLANTATION

Intermixing at an interface can be caused by the passage of (low energy) native point defects, or (high energy) ions implanted into the sample. This suggests using interfacial intermixing as a means of study the interaction of high energy ions with a solid [12,13]. We now describe experiments that reveal the microscopics of the way individual 320 keV Ga^+ ions interact with a GaAs/AlAs multilayer held at 77 K during implantation.

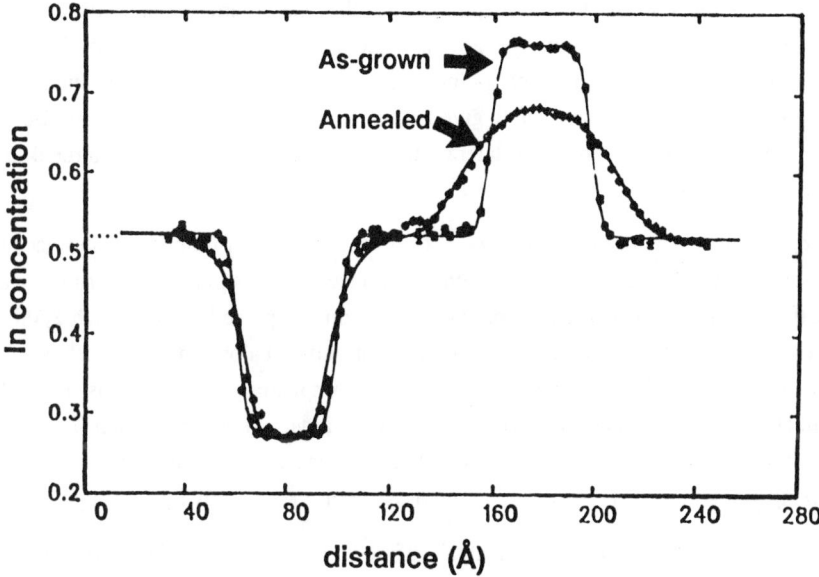

Fig. 2 Indium concentration profile across two $In_xAl_{1-x}As$ layers embedded in an $In_{.52}Al_{.48}As$ matrix, before and after annealing. Each point gives the composition of a 0.15 μm segment of an individual atomic plane. Error bars are plotted, but are often too small to be seen. Note highly nonlinear nature of the interdiffusion [8].

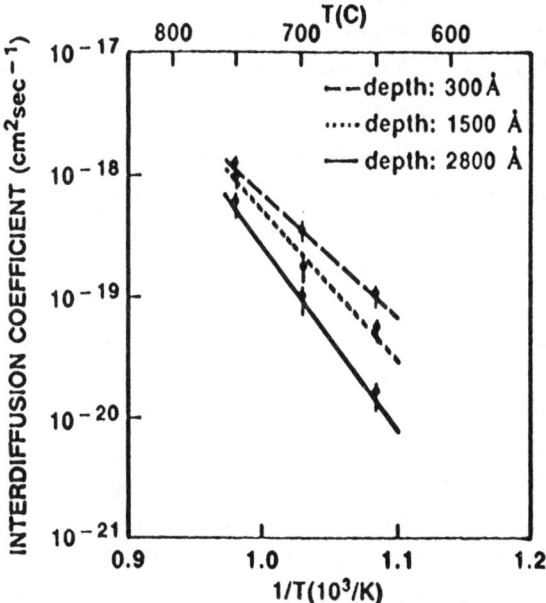

Fig. 3 Arrhenius plot of the interdiffusion coefficient at C:GaAs/AlGaAs interfaces at three different depths.

Fig. 4 is a chemical lattice image of one period of the as-grown AlAs/GaAs multilayer, together with composition profiles for each interface. The GaAs layer is situated 1400 Å beneath the surface and is thus close to the depth where the maximum damage during subsequent implantation is expected to occur. The growth direction is from bottom to top, the implantation direction from top to bottom. Each point on the profiles of Fig. 4 represents the average composition of a 1μm segment of a given atomic plane before implantation. Both top and bottom interfaces display excellent lateral uniformity, and can be characterized by similar characteristic widths.

After implantation to a dose of $5\times10^{12} cm^{-2}$, chemical analysis of individual interfaces located at depths between 1000 Å and 1700 Å beneath the surface reveals significant intermixing across the top interfaces, although (on average) only one Ga^+ ion has passed through each 2000 Å^2 of the specimen. As shown in Fig. 5, the intermixing is not uniform along the top interfaces, but shows large fluctuations on the 50 lateral scale. Kinematic implantation simulations using the TRIM program [14] show, that under our experimental conditions, a single implanted Ga^+ ion creates a damage track ~50 wide. This suggests that the observed fluctuations in the degree of intermixing along the interface are due to the passage of one or a few ions through each segment.

Using a statistical approach to the analysis of such data, we have deduced the microscopic damage signature created by a single ion as it passes through an interface [13]. At a dose of 5×10^{12} ions per cm^2, the damage signature consists of a core ~ 15 in diameter, within which the interfacial composition width is increased from its initial value of 2.7 to ~20. This is substantially larger than expected from kinematic (TRIM) simulations (~ 5 Å). Our approach also allows us to determine the damage signature caused by the successive passage of n ions ($n=1,2,3,...$), and thus to identify how many ions have pierced a given segment of the interface. We have thus obtained images which are microscopic records of the passage of ions through an interface [13].

Remarkably, there are major differences between the behavior of adjacent interfaces [12,13]. Fig. 6 shows neighboring top (AlAs on GaAs) and bottom (GaAs on AlAs) interfaces after implantation to a dose of 5×10^{12} ions $/cm^2$. The composition profiles (Fig. 6) and interface width histograms [13] extracted from such images clearly establish that the top interface is substantially broadened by implantation, but the bottom interface is left practically unaltered. Implanting 320 keV Ga^+ ions into our 2000 Å thick multilayer is roughly analogous to firing a bullet through a telephone book. We find that only every other sheet has developed a hole! This remarkable effect, totally unexpected on kinematic grounds, is due to, and can be controlled by means of an electric field. During implantation (at 77K) of the sample we have so far considered, the Fermi level is most likely pinned at midgap at the surface, and close to the valence band at 3000 Å from the surface, placing the multilayer in an electric field. We find that the large intermixing asymmetry observed at the top and bottom interfaces is due to the drift of the implantation damage in the electric field to the (AlAs on GaAs) interfaces, where it is trapped [13]. As described by Tersoff [15], this trapping is due to the discontinuities in the

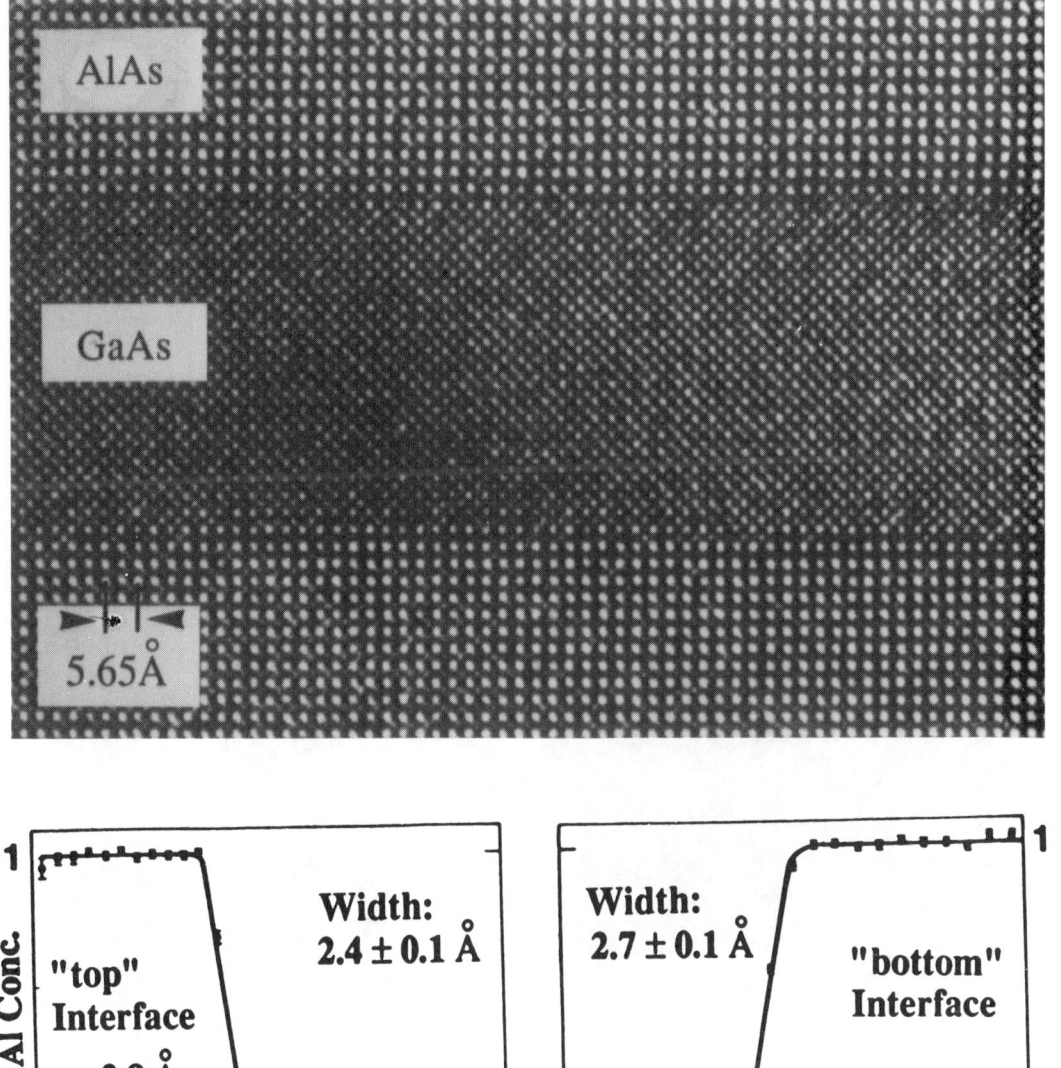

Fig. 4 Chemical lattice image of an AlAs/GaAs/AlAs period, together with composition profiles for the two AlAs/GaAs interfaces. Each data point represents the composition of a 1μm segment of an atomic plane parallel to the interface.

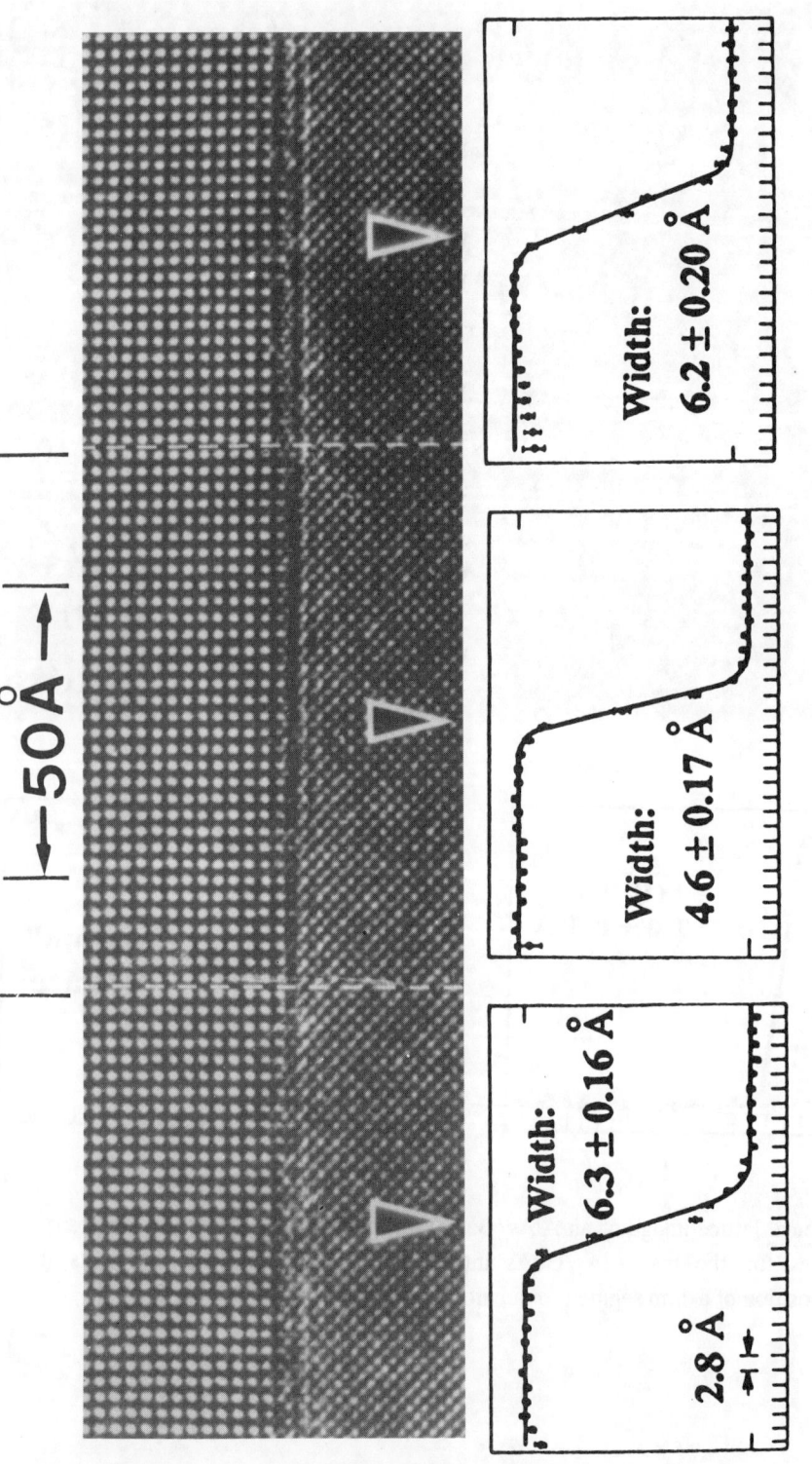

Fig. 5 Chemical lattice image of an (AlAs on GaAs) interface, implanted with 5×10^{12} ions/cm^2. The chemical profiles, obtained by averaging over 50 Å segments of planes parallel to the interface, show substantial variations in the *GaAs→AlAs* transition width.

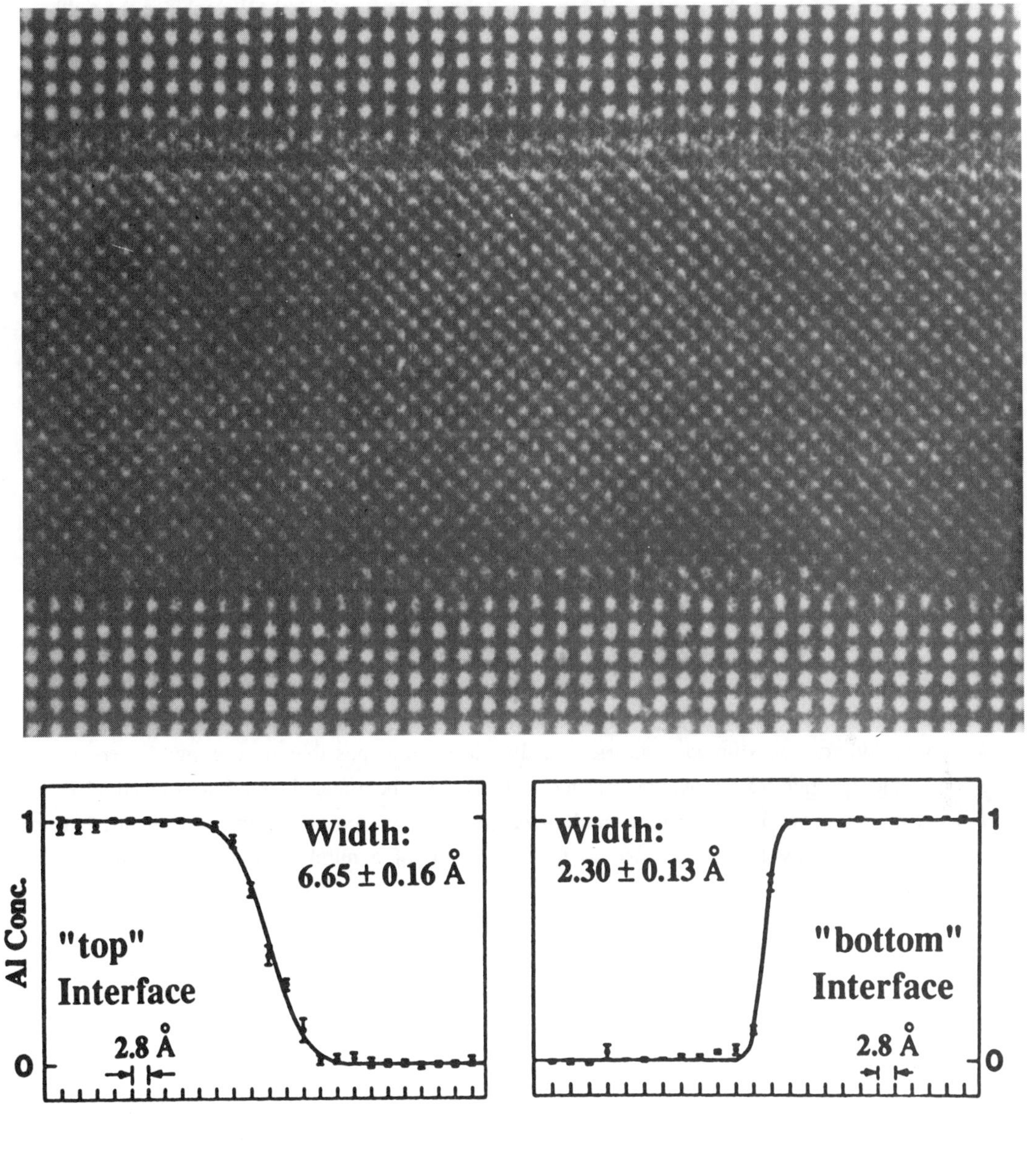

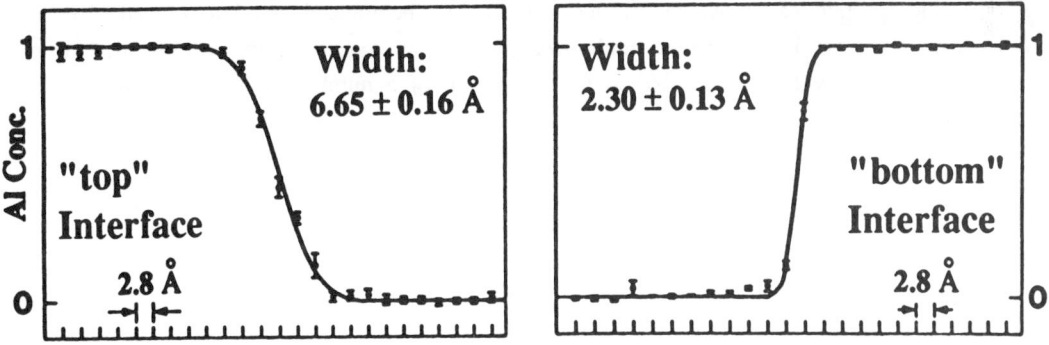

Fig. 6 Chemical lattice image of an AlAs/GaAs/AlAs period, implanted with 5×10^{12} ions/cm^2. The "top" interface (AlAs on GaAs) is strongly intermixed, while the "bottom" interface (GaAs on AlAs) is not.

bandstructure and the formation enthalpies of the defects involved. Experimental proof of influence of the electric field rests on our ability to reverse the asymmetry of the intermixing between the top and bottom interfaces by reversing the electric field [13]. We achieve the field reversal by embedding "intrinsic" multilayers in *p-i-n* and *n-i-p* structures, and observe a strong reversal in the intermixing asymmetry between the top and bottom interfaces. This firmly establishes the strong influence of the electric field in determining the location of the defect agglomerates responsible for the intermixing, and offers the tantalizing prospect of steering defects in solids by means of electric fields.

4. SUMMARY AND CONCLUSIONS

It is now possible to map the composition of materials at near-atomic sensitivity and resolution. This provides immediate access to a wide range of solid-state processes, that involve atomic substitutions on the lattice, but leave the structure essentially unaltered. The ability to measure interdiffusion coefficients as small at $10^{-22} cm^2/s$ in regions as small as $10^{-19} cm^3$ in volume, reveals a host of unexpected phenomena. Examples include highly nonlinear interdiffusion [6,8], the dominant effect of the surface in determining layer and device stability [6,11], and exotic forms of chemical relaxation in strained solids [8]. Equally important, chemical mapping techniques allow one to repeat the early experiments of high energy physics in the solid state. Just as a stack of photographic emulsion layers can be used to track the passage of cosmic radiation, the intermixing at a series of chemical interfaces (such as GaAs/AlAs) can be used to record the arrival and passage of point defects, be they high energy implanted ions [13], or low energy native point defects injected during an anneal [6,7,10]. The sensitivity of chemical mapping allows one to study such processes long before steady-state has been reached, providing access to hitherto unexplored regimes. Finally, it is now possible to use multilayers as microscopic laboratories, in which selected defects may be trapped and interrogated [10]. Indeed, the combination of "designer multilayers" grown by modern epitaxy and quantitative chemical microscopy has already led to the realization of a number of hitherto Gedanken experiments.

REFERENCES

[1] A. Ourmazd, W.T. Tsang, J.A. Rentschler, and D.W. Taylor, Appl. Phys. Lett. **20**, 1417 (1987).

[2] A. Ourmazd, D.W. Taylor, J. Cunningham, and C.W.Tu, Phys. Rev. Lett. **62**, 933 (1989).

[3] A. Ourmazd, F. H. Baumann, M. Bode, and Y. Kim, Ultramicroscopy **34**, 237 (1990).

[4] A. Ourmazd, D.W. Taylor, M. Bode, and Y. Kim, Science **246**, 1571 (1989).

[5] J.W. Cahn, Acta Metall. **9**, 525 (1961).

[6] Y. Kim, A. Ourmazd, M. Bode and R.D. Feldman, Phys. Rev. Lett. **63**, 636 (1989).

[7] A. Ourmazd, Y. Kim, and M. Bode, Mat. Res. Soc. Proc. **163**, 639 (1990).

[8] F.H. Baumann, A. Ourmazd, and T.Y. Chang, Bull. Am. Phys. Soc. **36**, 474 (1991).

[9] Y. Kim, A. Ourmazd, R.J. Malik, and J.A. Rentschler, Mat. Res. Soc. Proc. **159**, 351 (1990).

[10] J-L. Rouviere, Y. Kim, A. Ourmazd, J. Cunningham, R.J. Malik, and J.A. Rentschler, to be published.

[11] Y. Kim, A. Ourmazd, and R.D. Feldman, J. Vac. Sci. Technol. **A8**, 1116 (1990).

[12] M.Bode, A. Ourmazd, J.A. Rentschler, M. Hong, L.C. Feldman, J.P. Mannaerts, Mat. Res. Soc. Symp. Proc. Vol. **157**, 197 (1990).

[13] M.Bode, A. Ourmazd, J. Cunningham, and M. Hong, Phys. Rev. Lett. **67**, 843 (1991).

[14] J.P. Biersack, Nucl. Instrum. Method B **19**, 32 (1987).

[15] J. Tersoff, Phys. Rev. Lett. **65**, 887 (1990).

THEORY OF Zn-ENHANCED DISORDERING IN GaAs/AlAs SUPERLATTICES

Q.-M. ZHANG, C. WANG, and J. BERNHOLC
Department of Physics, North Carolina State University
Raleigh, NC 27695-8202

Abstract

We use the Car-Parrinello method to study the Zn-enhanced interdiffusion problem in GaAs/AlAs superlattices. The energetics of several mechanisms for the diffusion of Zn impurity have been examined. It is found that a pair consisting of a substitutional Zn acceptor and an interstitial group III atom has a substantially lower formation energy than an isolated interstitial. The low formation energy of this pair results in the interstitial kick-out mechanism having a much lower activation energy than the ones involving vacancies or the dissociative (Frank-Turnbull or Longini) mechanism. The lowest energy path for the interchange of group III atoms involves a kick-out of Zn by a group III interstitial, followed by a fast Zn interstitial diffusion and a subsequent ejection of another group III atom into the interstitial channel. The activation energies for these processes, determined by following the kick-out trajectories and including a full relaxation of all the atoms, are in good agreement with the experimental data.

Introduction

Experimental studies on Zn diffusion in a GaAs/AlAs superlattice revealed a remarkable fact: that the layer structure becomes disordered during Zn diffusion at a much lower annealing temperature (~500-600 °C) than without Zn (~900 °C) [1]. The interdiffusion between the group III elements is thus enhanced by several orders of magnitude by Zn diffusion. Furthermore, disordering occurs only in those regions of the material where Zn is present. This phenomenon has potential applications in opto-electronic devices such as solid state lasers and optical waveguides. Although similar impurity-induced layer disordering has later been found in other III-V superlattices or through doping with other elements [2-7], Zn-enhanced interdiffusion in GaAs/AlAs has been most extensively studied.

Several models have been proposed to explain the Zn-enhanced interdiffusion in GaAs/AlAs superlattices. Since the intermixing is caused by Zn diffusion, the discussion focused on this aspect. In the dissociative (also called Frank-Turnbull [8] or Longini [9]) mechanism, an interstitial Zn-vacancy pair is formed and Zn diffuses rapidly as an interstitial but slowly as a substitutional atom. The left-behind vacancy can accommodate a neighboring group III atom and hence contributes to the movement of Ga and Al atoms [1]. Other vacancy-based mechanisms assume that a substitutional Zn and a nearest neighbor vacancy form a pair which migrates through a series of nearest or second nearest neighbor hops [10, 11]. Another mechanism, which was called a "kick-out" mechanism, was proposed by Gösele and Morehead [12]. In this mechanism, an interstitial Zn joins the group III sublattice by pushing the host atom away and creating a group III interstitial. Tan and Gösele further suggested that Fermi-level lowering due to p-type doping enhances the interdiffusion [13].

The recent progress in the methods of computational physics [14] allows us to calculate total energies of large systems and to simulate the essential aspects of these processes from the first principles. In the present work, we use the Car-Parrinello method to study the energetics of interdiffusion. We focus mainly on GaAs because the diffusion of Zn in GaAs is substantially slower than in $Al_xGa_{1-x}As$ [15]. By comparing the activation energies for the various mechanisms, we search for the lowest energy diffusion path for Zn and the exchange path for Ga and Al atoms.

Calculations

The calculations are performed using the Car-Parrinello (CP) method [16]. The electrons are described by the density-functional theory [17], the local-density approximation for exchange and correlation [18, 19], and norm-conserving pseudopotentials [20]. The supercell size corresponds to 64 host atoms. The electronic wavefunctions are expanded in plane waves with the kinetic energy cutoff of 14 Ry. The pseudopotentials used for all the atoms except Zn were given by Bachelet, Hamann and Schlüter [21] and refitted by Gonze, Käckell and Scheffler [22] to eliminate a Ga 'ghost' state. For Zn, a soft core pseudopotential, which includes the 3d electrons in the core, was constructed using Hamann's program [23]. Before being applied to the diffusion problem, the potentials were tested in perfect bulk GaAs, ZnTe, and ZnSe calculations. For GaAs, the experimental lattice constant and the bulk modulus was reproduced within 1.3 and 12% respectively. As usual in the local density theory, the computed cohesive energy was 9% greater than the experimental value. For ZnSe (ZnTe) the lattice constant, the bulk modulus, and the cohesive energy differed by 1.6, -48, and -4% (0.9, -42, and -5%) from the experimental values. The discrepancies here are due to the freezing of Zn 3d electrons (see [24]). They probably lead to slight underestimates in the calculated diffusion barriers for Zn.

A remarkable advantage of the CP method is that it can treat the motion of atoms and electrons simultaneously. The Newton equations of motion for atoms are solved using forces derived from the local density equations. All the atoms in the supercell are fully relaxed in the presence of a defect. An ab-initio Molecular Dynamics (MD) simulation is possible for some dynamical processes. However, due to the large activation energies for Zn diffusion and cation exchange, a direct ab-initio MD simulation is unaffordable at present. Hence we investigate individual mechanisms through total energy calculations and search for saddle points using adiabatic trajectories (see below).

Under equilibrium conditions, the diffusion coefficient of an atom, D, can be written as
$$D = D_0 \exp(-E_a/k_B T) \qquad (1)$$
where D_0 is the pre-exponential factor including the entropy contribution, k_B is Boltzmann constant, and E_a is the activation energy for the diffusion. The activation energy is the sum of the formation and migration energies of the defect. The formation energies are derived from total energy calculations for the supercell. For impurities, we

Table I. The formation energies for neutral T_d symmetry native point defects for perfectly stoichiometric GaAs.

	V_{Ga}	V_{As}	As_{Ga}	Ga_{As}	As_{TGa}	As_{TAs}	Ga_{TGa}	Ga_{TAs}
[eV]	4.0	3.9	1.7	2.0	3.6	4.1	1.5	1.3

quote formation energies with respect to a crystal in contact with a bulk impurity reservoir. The migration energies are extracted from total energy differences between the saddle points and the initial states.

A saddle point can often be located by examining the symmetry of a Potential Energy (PE) surface. However, when the PE is not symmetric, such as in the kick-out processes we investigate, the location of the saddle point is unknown. A point by point calculation of PE along the trajectory becomes costly. As an alternative, we propose a new, much more efficient procedure to determine the migration barrier, which we call an "adiabatic trajectory" simulation. The main idea is that a constant small speed is assigned to the diffusing atom while the remaining atoms continuously relax in response to its motion. As in a real CP simulation, the system moves along the lowest energy Born-Oppenheimer surface [16]. In cases where level crossing occurs, we use the finite temperature CP formalism [25] for stability. During the simulation, the velocities of the remaining atoms are decomposed according to the direction of the forces acting on them. The perpendicular components are set to zero and the parallel component is reduced by a constant factor if it is antiparallel to the force. This procedure removes the excess energy introduced by the constant speed motion of the diffusing atom and leads to a faster relaxation of the whole system. We estimate that it is four times faster than a point by point calculation. In the cases studied below, we move the diffusing atom with a speed corresponding to the mean speed at room temperature.

Results and Discussion

At first, we calculated the formation energies for all eight possible native point defects with T_d symmetry in GaAs. They are: two vacancies V_{Ga} and V_{As}, two antisites As_{Ga} and Ga_{As}, and four tetrahedral interstitials As_{TGa}, As_{TAs}, Ga_{TGa}, and Ga_{TAs}, where the subscripts indicate the nearest neighbor atoms. The energy gain upon relaxation of all the atoms ranges from 0.05 eV (V_{Ga}) to 0.6 eV (As_{TAs}). The nearest-neighbor relaxation is not always dominant. For example, the relaxation distances from Ga_{TGa} to its first and second neighbors are 0.09 Å and 0.01 Å, respectively, but 0.03 Å and 0.07 Å for Ga_{TAs}. The formation energies are given in Table I. The Ga interstitial (Ga_I) has the lowest formation energy and is thus the preferred point defect in stoichiometric GaAs. In p-type GaAs it exists as Ga_I^+. The table shows also that in As-rich GaAs the lowest energy stoichiometry-compensating defect is As_{Ga}, which is the main component of the EL2 defect [26, 27].

Turning to Zn, it is a well-known Ga-site substitutional acceptor. We obtain 0.86 for the formation energy for Zn^0 in GaAs. The Zn diffusion mechanism is still disputed although a substitutional-interstitial path is generally accepted. Two mechanisms were proposed based on this path [9, 12]. Both of them claimed a good fit to the experimental diffusion profiles [28, 29]. The dissociative model [9] suggests that when GaAs is heavily doped with Zn, a small portion of Zn could become interstitial and rapidly diffuse through the process

$$Zn_{Ga} \leftrightarrow Zn_I + V_{Ga}. \quad (2)$$

The kick-out mechanism [12] involves a kick-out by an interstitial group III atom of a substitutional Zn to an interstitial site

$$Zn_{Ga} + Ga_I \leftrightarrow Zn_I. \quad (3)$$

We calculated the total energies of the system along the path of eq. (2) and obtained an energy barrier of 3.6 eV. For the Zn_{Ga}-Ga_{TAs} pair, we found that if GaAs is in contact with a Ga reservoir, its formation energy is only 0.2 eV greater than that of an isolated Zn_{Ga}. The reason for the low formation energy is Coulombic attraction, since it can be thought of as Zn^-_{Ga}-Ga^+_{TAs}. This pair introduced no defect states in the forbidden gap in our calculations.

An adiabatic trajectory simulation was carried out to determine the PE along the <100> Ga_{TAs} kicking trajectory, since the saddle point can not be determined by symmetry in this case. The PE along the trajectory is shown in Fig. 1. A barrier of 1.8 eV is obtained. When Zn becomes an interstitial, it can move very fast along the low electron density channel with a migration energy of 0.2 eV and then kick-in into another substitutional site. Assuming that the Zn_{Ga}-Ga_{TAs} pair migrates in the neutral charge state, the activation energy for this process is 2.0 eV.

Another less obvious kick-out process is also possible [30]. In this <111> kick-out, a Ga_{TGa} interstitial pushes a nearest neighbor Zn_{Ga} along the <111> direction onto an As site. The As atom moves into the interstitial channel and then comes around and pushes Zn along the <11-1> direction onto another Ga site. The last push regenerates the Ga interstitial. The process can then repeat itself. The resulting picture is Zn diffusing

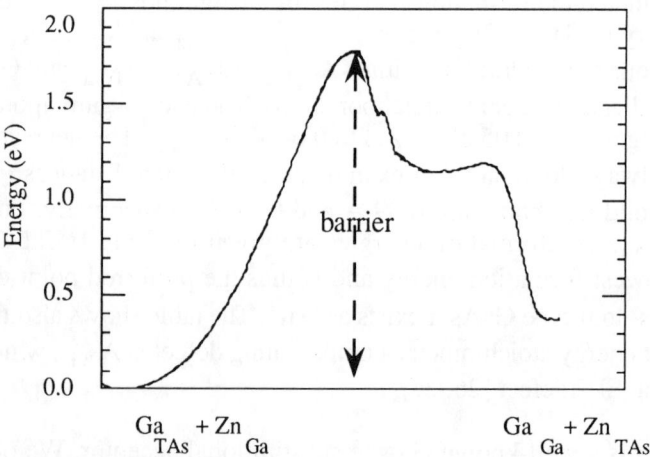

Fig.1 The potential energy surface along the <100> kick-out trajectory.

along the bonding chain in the (110) plane. The migration barrier for this process is 1.9 eV, with the saddle point occurring when As_I is near the symmetric hexagonal site. The activation energy, which includes the formation energy of the pair, is 2.1 eV.

We have also examined two vacancy mechanisms [10, 11]. Both of them start with a Zn_{Ga}-V_{As} pair and involve nearest-neighbor or second-nearest-neighbor jumps. We omit the detailed description of the processes here and just mention that the activation energies are 3.6 eV for Shaw's model [10] and 5.5 eV for Van Vechten's model [11].

After comparing all the activation energies, we conclude that the kick-out mechanism is the lowest energy one for Zn diffusion in GaAs. In the <100> kick-out, Zn interstitial can move in the interstitial channel with a barrier of only 0.2 eV, while in the case of the <111> kick-out each migration step has to overcome a barrier of 1.9 eV. Therefore, the dominant process is the <100> kick-out. Its activation energy of ~ 2.0 eV is in good agreement with the earlier experimental results of 2.5 eV [31] and 2.1-3.1 eV [32]. Although the Zn diffusion in GaAs is likely to be the rate-determining step, similar calculations were also done in AlAs. The formation energy of the Zn_{Al} - Al_{TAs} pair in AlAs is 0.4 eV. For the <100> kick-out the migration energy is 1.2 eV, resulting in an activation energy of 1.6 eV. In the <111> kick-out the migration energy is 2.0 eV, leading to an activation energy of 2.4 eV. The <100> kick-out dominates thus in AlAs as well. The lower barrier for Zn diffusion in AlAs is in agreement with experimental data, which show a faster diffusion of Zn in $Al_xGa_{1-x}As$ than in pure GaAs [15].

After determining the lowest energy path for Zn diffusion, we can discuss the role of Zn in enhancing the interdiffusion. It appears that the principal role of Zn is to supply cation interstitials through the formation of Zn_{III} – III_I pairs. The breakage of the pair provides positively charged cation interstitials which disorder the lattice. If Zn atoms are injected at the GaAs surface as interstitials, they diffuse fast with an activation energy of only 0.2 eV and thus penetrate deep into the sample. Along the diffusion path Zn atoms become substitutional and form either Zn_{Ga}-Ga_{TAs} or Zn_{Al} - Al_{TAs} pairs with formation energies in the pure materials of 0.2 or 0.4 eV, respectively. Our computed pair breakage energies in the 64-atom supercell are ~ 0.5 eV. We also carried out adiabatic trajectory simulations for the <100> kick-out of Ga and and Al by the positively charged cations:

$$Al_I^+ + Ga_{Ga} \leftrightarrow Ga_I^+ + Al_{Ga} \qquad \text{in GaAs} \qquad (4)$$
$$Ga_I^+ + Al_{Al} \leftrightarrow Al_I^+ + Ga_{Al}. \qquad \text{in AlAs} \qquad (5)$$

The migration barriers for these processes are 1.6 and 1.1 eV, respectively. Assuming that Zn has already diffused in and the pairs have formed, these are also the activation energies for interdiffusion in the pure materials. They will vary somewhat during interdiffusion, since they should depend on the composition of the alloy (cf. eq. 4-5). Experimentally, Lee and Laidig [32] observe an average activation energy for interdiffusion of ~ 1 eV in GaAs/AlAs superlattices *after* the indiffusion of Zn.

References

1. W. D. Laidig, N. Holonyak, Jr., M. D. Camras, K. Hess, J. J. Coleman, P. D.

Dapkus, and J. Bardeen, Appl. Phys. Lett. **38**, 776 (1981).
2. M. D. Camras, N. Holonyak, Jr., K. Hess, M. J. Ludowise, and C. R. Lewis, Appl. Phys. Lett. **42**, 185 (1983).
3. W. D. Laidig, J. W. Lee, P. K. Chiang, L. W. Simpson, and S. M. Bedair, J. Appl. Phys. **54**, 6382 (1983).
4. K. Meehan, N. Holonyak Jr., J. M. Brown, M. A. Nixon, P. Gavrilovic, and R. D. Burnham, Appl. Phys. Lett. **45**, 549 (1984).
5. M. Kawabe, N. Matsuura, N. Shimizu, F. Hasegawa, and Y. Nannichi, Jpn. J. Appl. Phys. **23**, L623 (1984).
6. N. Kamata, K. Kobayashi, K. Endo, T. Suzuki, and A. Misu, Jpn. J. Appl. Phys. **26**, 1092 (1987).
7. R. W. Kaliski, D. W. Nam, D. G. Deppe, N. Holonyak, Jr., K. C. Hsieh, and R. D. Burnham, J. Appl. Phys. **62**, 998 (1987).
8. F. C. Frank and D. Turnbull, Phys. Rev. **104**, 617 1956).
9. R. L. Longini, Solid State Electron. **5**, 127 (1962).
10. D. Shaw, Phys. Stat. Solidi (a) **86**, 629 (1984).
11. J. A. Van Vechten, J. Phys. C: Solid State Phys. **17**, L933 (1984).
12. U. Gösele, and F. Morehead, J. Appl. Phys. **52**, 4617 (1981).
13. P. Y. Tan and U. Gösele, J. Appl. Phys. **61**, 1841 (1987).
14. See, e.g., the paper by R. Car in these Proceedings.
15. C. P. Lee, S. Margalit, and A. Yariv, Solid State Electronics **21**, 905 (1978).
16. R. Car, and M. Parrinello, Phys. Rev. Lett. **55**, 2471 (1985).
17. P. Hohenberg and W. Kohn, Phys. Rev. **136**, B864 (1964).
18. W. Kohn, and L. J. Sham, Phys. Rev. **140**, A1133 (1965).
19. D. M. Ceperley, and B. J. Alder, Phys. Rev. Lett. **45**, 566 (1980), J. Perdew and A. Zunger, Phys. Rev. B **23**, 5048 (1981).
20. D. R. Hamann, M. Schlüter, and C. Chiang, Phys. Rev. Lett. **43**, 1494 (1979).
21. G. B. Bachelet, D. R. Hamann, and M. Schlüter, Phys. Rev. B **26**, 4199 (1982).
22. X. Gonze, P. Käckell, and M. Scheffler, Phys. Rev. B **41**, 12264 (1990).
23. D. R. Hamann, Phys. Rev. B **40**, 10391 (1989).
24. S.-H. Wei, and Alex Zunger, Phys. Rev. B **37**, 8958 (1988).
25. M.P. Grumbach, D. Hohl, R. M. Martin, and R. Car, Bull. Am. Phys. Soc. **36**, 836 (1991).
26. J. Dabrowski and M. Scheffler, Phys. Rev. Lett. **60**, 2183 (1988).
27. D. J. Chadi and K. J. Chang, Phys. Rev. Lett. **60**, 2187 (1988).
28. A. H. van Ommen, J. Appl. Phys. **54**, 5055 (1983).
29. K. B. Kahen, Appl. Phys. Lett. **55**, 2117 (1989).
30. C. S. Nichols, C. G. Van de Walle, and S. T. Pantelides, Phys. Rev. **40**, 5484 (1989).
31. B. Goldstein, Phys. Rev. **118**, 1024 (1960).
32. J. W. Lee and W. D.Laidig, J. Electron. Mat. **13**, 147 (1984).

SPATIAL PARTITION OF PHOTOCARRIERS TRAPPED AT DEEP DEFECTS IN MULTIPLE QUANTUM WELLS

D. D. NOLTE[1], R. M. BRUBAKER[1], Q. N. WANG[1] AND M. R. MELLOCH[2]
[1]Dept. of Physics, Purdue University, W. Lafayette, IN 47907
[2]School of Electrical Engineering, Purdue University, W. Lafayette, IN 47907

ABSTRACT

We have observed photocarrier trapping at defects within the quantum barriers of implant-damaged GaAs/AlGaAs multiple quantum well samples. Internal space-charge electric fields are generated by illuminating the samples with two coherent pump laser beams that generate photorefractive gratings. The gratings diffract a probe laser, giving a sensitive means to monitor trapped space-charge. Photocarriers generated in multiple quantum well structures experience spatially different dynamics, depending on whether they are in the well regions or the barrier regions. Selective choice of bandgaps and optical excitation wavelengths control where and how photocarriers trap at defects. Using this method, we are able to isolate effects from defects within quantum barriers.

1. Introduction

Deep level defects have strongly localized wavefunctions that extend over only several ångstroms. These point defects therefore are not strongly perturbed by band-edge discontinuities in quantum-well and multilayer structures. A given defect is relatively invariant to spatial changes in bandstructure, having the same properties within quantum well regions as in quantum barriers. On the other hand, little information is currently available about the *identities* of defects in one layer relative to another. Furthermore, the roles that defects play in carrier dynamics, such as transport or recombination, vary markedly depending on where the defects are situated.

We are able to directly observe the space-charge fields generated by photocarriers trapped at defects in quantum barriers. The effects of the defects in the barriers can be separated from the effects of defects in the wells. This is accomplished by partitioning photocarriers between barriers and wells by judicious choice of bandgaps and laser excitation energies. The presence of space-charge fields within the quantum wells is detected by the photorefractive effect. In this effect, intersecting coherent laser beams write periodic space-charge gratings that alter the optical properties through the Franz-Keldysh effect on quantum-confined excitons. The resulting index and absorption gratings diffract a probe laser. The photorefractive effect is extremely sensitive to small changes in electric fields and index, and provides a sensitive measure of the role of defects in trapping photocarriers. In our experiments, we are able to separate out the effects of defects in the barriers by a careful balance of photorefractive grating formation by space-charge in the barriers against screening of the gratings by space-charge in the wells.

2. Semi-Insulating Quantum Wells

Our multiple quantum wells consist of 60 periods of $Al_{0.3}Ga_{0.7}As/GaAs$ with 75 Å wells and 100 Å barriers. The layers are made semi-insulating by hydrogen implantation at 160 keV. The implant causes radiation damage, generating defects nearly midgap that pin the Fermi level. A dose of 10^{12} cm^{-2} is sufficient to make the layers semi-insulating, without adversely affecting the width of the excitons. On the other hand, a dose of 10^{13} cm^{-2} broadens the exciton by approximately 3 meV. This broadening of the exciton corresponds approximately with one defect per exciton volume, or $N_D \geq 10^{17}$ cm^{-3}. This concentration of defects includes both electrically active defects, as well as neutral defects. During implantation, a large fraction of generated defects and complexes can have energy levels that lie outside of the bandgap, producing neutral defects. These defects can broaden the exciton lines through strain fields. However, these defects do not participate in compensation, and cannot trap space-charge.

The exciton width plays a central role in our investigation of trapped space-charge. Space-charge generates electric fields that alter the absorption of the excitons. Furthermore, absorption changes are accompanied by changes in the refractive index. In our samples, the electric field is applied in the plane of the quantum wells, generating the Franz-Keldysh electro-optic effect for quantum-confined excitons. The field causes lifetime broadening of the exciton absorption. Sharper excitons generate larger electro-optic effects. In our study, therefore, we rely on the relatively sharp absorption lines of the room-temperature excitons to provide a measure of space-charge fields in the samples.

3. Photo-Induced Space-Charge Electric Fields

To generate space-charge in the quantum well structures, we illuminate the sample with two coherent laser beams. The coherent interference between the beams generates interference fringes. When a voltage bias is applied across the sample, this spatially inhomogeneous illumination causes transport that generates trapped space-charge which screens the field in the bright fringes. The spatially modulated space charge field causes the Franz-Keldysh effect, converting the electric field grating into an absorption and index grating. The periodic modulation of these gratings act as diffraction gratings that diffract a probe beam. This process of grating generation and diffraction is called the photorefractive effect[1]. The photorefractive effect is extremely sensitive to small electric

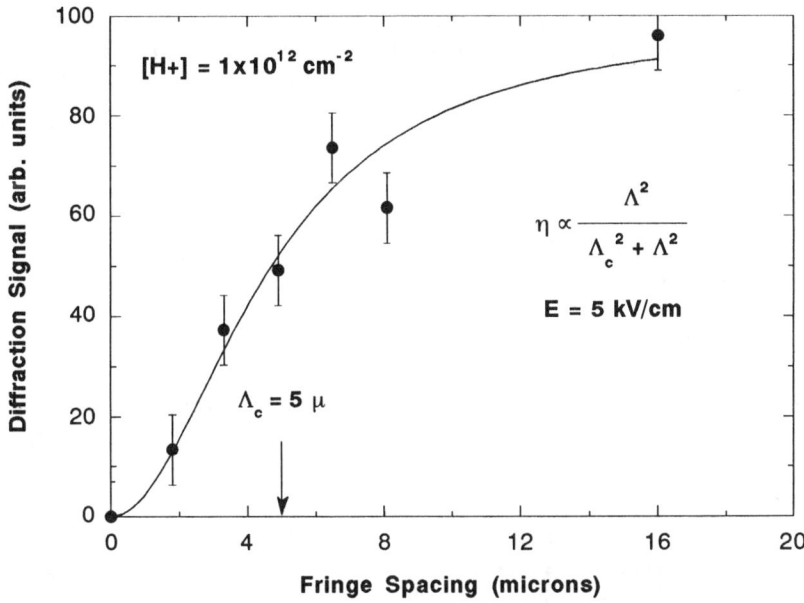

Fig. 1   Diffraction signal η as a function of holographic fringe spacing Λ. The cutoff spacing $\Lambda_c = 5~\mu$ corresponds to an effective deep level trap concentration of 4×10^{14} cm^{-3}. The sample was implanted with [H$^+$] = 1×10^{12} cm^{-2} at 160 keV.

fields, and small concentrations of space-charge. The effect is also sensitive to the defect density. For a given applied electric field, the screening is limited by the trap density. The diffraction signal as a function of interference fringe spacing is shown in Fig. 1. The characteristic screening length Λ_c for our sample is 5 microns. The screening length corresponds to the trap density through

$$N_{sc}^{eff} = f(1-f) N_T = \frac{2\pi\varepsilon\varepsilon_0}{e\Lambda_c} E_0$$

where E_0 is the applied electric field, and f is the occupancy of the trap. A 5 micron screening length yields an effective trap density of 4×10^{14} cm^{-3}. No information is available from this measurement of the defect occupancy f, so the effective number of traps can be as much as an order of magnitude smaller than N_T, the total number of electrically active traps. For hydrogen implantation at 160 keV, the ratio of neutral defects (obtained from the exciton broadening) relative to the number of electrically active traps, is

$$N_{neut}/N_{elect} \geq 10^2$$

Therefore, many more neutral defects are generated by the proton implant than electrically active defects.

4. Spatial Partition of Photocarriers

One of the goals of our study is to isolate the effects from the subset of deep defects that are in the AlGaAs barriers. This is possible by choosing appropriate excitation energies for the lasers. In particular, we write the interference fringes with a HeNe laser with a wavelength of 633 nm. The bright fringes generate carriers both in the wells and in the barriers. The carriers transport to screen the applied field, generating spatially modulated space-charge fields and refractive index gratings. The gratings are probed with a laser tuned close to the band-edge of the GaAs wells. This probe laser generates carriers only in the quantum wells. When the probe laser intensity is comparable to or larger than the HeNe pump intensity, then the photoconductivity of the probe laser erases the space-charge that is stored in the wells. The space-charge gratings in the barriers, on the other hand, are inaccessible to the erasure caused by the probe. Therefore, the refractive index gratings persist up to relatively large probe intensities[2]. Under sufficiently large probe laser intensities, the isolated space-charge is erased by transport from the wells into the barriers. This transport can occur by quantum tunneling, phonon-assisted tunneling or thermionic emission. These processes are shown in Fig. 2, and have been modeled to predict the experimental behavior. The relaxation of the isolated charge trapped at defects in the barriers therefore can give

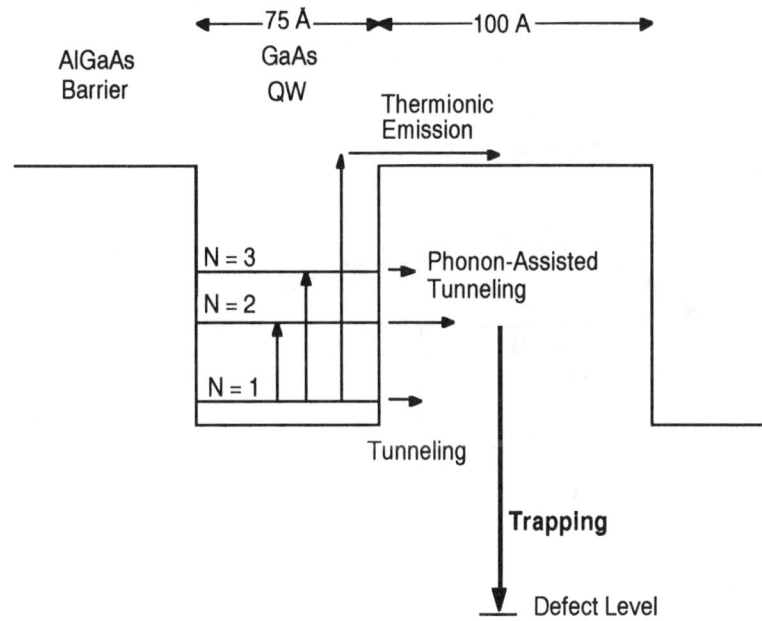

Fig. 2 Trapping of photocarriers into deep level defects through phonon-assisted tunneling and thermionic emission.

a direct measure of these transport processes as well as characteristic properties of the defects in the barriers. The persistence of the space-charge trapped in the 100 Å AlGaAs quantum barriers is shown in Fig. 3, related to the expected results for a bulk sample with no isolation of the space-charge, and to the expected results for a 200 Å barrier. The excess diffraction signal at large probe-to-pump ratios is caused by the subset of deep defects in the barriers.

5. Discussion

In this paper, we have demonstrated that a subset of deep defects in a multilayer structure can be isolated for separate study. Specifically, defects in the quantum barriers can support trapped space-charge and contribute to the photorefractive effect, while defects in the wells make no contribution. This spatial segregation of photocarriers can be combined with spectroscopic techniques[3] to provide a new technique for studying defects in quantum barriers.

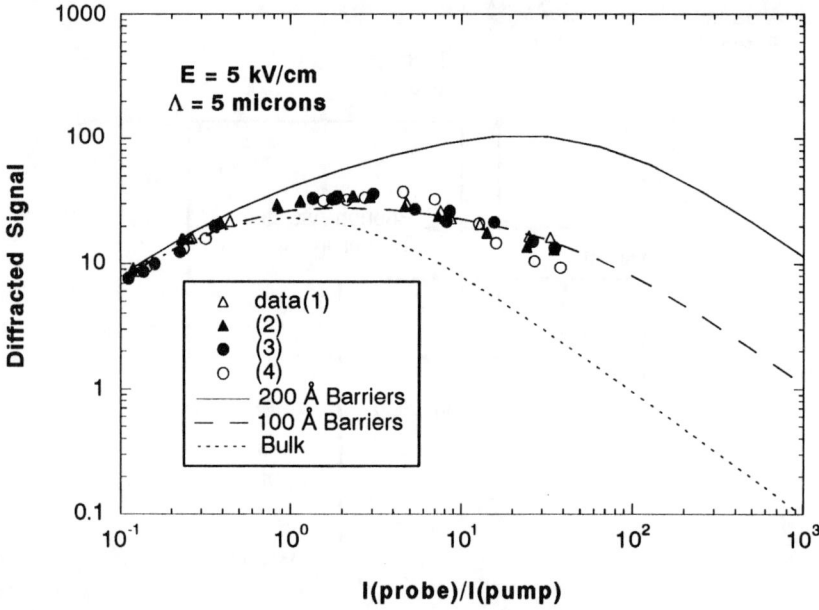

Fig. 3   Diffraction signal as a function of probe/pump intensity ratio for data from 100 Å barriers, compared with simulation of a 200 Å wide barrier sample, and simulation of a bulk sample. The persistence of the diffraction up to large probe intensities is a consequence of the space-charge isolated in the AlGaAs barriers.

6. References

1. P. Günter and J. P Huignard, *Photorefractive Materials and Their Applications* (Springer-Verlag, 1988)

2. D. D. Nolte, Q. N. Wang and M. R. Melloch, Appl. Phys. Lett. **58**, 2067(1991)

3. D. D. Nolte, D. H. Olson, A. M. Glass, Appl. Phys. Lett. **56**, 163(1990)

PICOSECOND DYNAMICS OF EXCITON CAPTURE, EMISSION AND RECOMBINATION AT SHALLOW IMPURITIES IN CENTER-DOPED ALGAAS/GAAS QUANTUM WELLS.

C.I. HARRIS AND H.KALT
Max-Planck-Institut für Festkörperforschung, 7000 Stuttgart 80, Germany.

B.MONEMAR, P.O.HOLTZ AND J.P.BERGMAN
Department of Physics and Measurement Technology, Linköping University, S-15183 Linköping, Sweden

M.SUNDARAM, J.L. MERZ AND A.C. GOSSARD.
Center for studies of Quantized Electronic Structures (QUEST), University of California at Santa Barbara, CA 93016, USA

ABSTRACT

The interaction of free and bound excitons in intentionally doped AlGaAs/GaAs quantum wells has been studied using a time resolved photoluminescence technique. The dependence of the dynamic processes on doping concentration, well width and temperature is discussed. In particular the behaviour of excitons bound to acceptors and donors are compared and the importance of thermalization in determining the observed lifetime highlighted.

1. Introduction

Studies to date on the dynamics of excitons in quantum wells (QW's) have concentrated primarily on the properties of free excitons (FEs) in undoped samples [1] [2]. Localization due to potential fluctuations caused by interface roughness is found to be important to the understanding of the dynamic properties of these excitons [3] [4]. Detailed spectroscopic work on bound excitons (BEs) in quantum wells has recently been reported [5] [6]. However the dynamic interaction between FEs and BEs over short time scales ($\tau < 200$ps) has not been treated in detail. A study restricted to acceptor doped samples and time scales ($\tau > 200$ps) was however recently published [7]. In this work we have investigated the full temporal development with ps time-resolution of both FE and BE emission in doped QWs as a function of doping and temperature.

2. Samples and Experimental Procedure.

A large number of multiple quantum well samples with different well widths and doping densities have been studied. The samples consisting of 50 periods were grown using molecular beam epitaxy (MBE) at 680°C with non-interrupted growth. The well width was varied between 5nm and 15nm for different samples, while the undoped AlGaAs barriers were kept at 15nm thickness. The GaAs layers were doped in the central 20% of the well to a volume concentration varying from $3 \cdot 10^{16}$ up to $5 \cdot 10^{17}$ cm^{-3} for different samples. Be was used for acceptor doping, while Si was employed for donor doping. The transient photoluminescence (PL) data were obtained with a synchroscan streak camera with a temporal resolution of approximately 10ps. The sample temperature could be varied down to 5K. A tuneable synchronously pumped dye laser was used for excitation, with a typical pulse length of 5ps.

3 Results and Discussion.

3.1 Acceptor doped samples.

Fig.1. illustrates time-resolved PL spectra for a set of Be doped 15nm samples, with the doping varied between $3 \cdot 10^{16}$ and $5 \cdot 10^{17}$ cm^{-3}, i.e. a 2D density varying between $9 \cdot 10^9$ and $2 \cdot 10^{11}$ cm^{-2} per QW. The spectra are taken with excitation resonant with the FE, and are delayed 36ps from the center of the laser pulse. These data illustrate, that narrow well-resolved BE peaks are only observed at doping levels below 10^{17} cm^{-3} (approx. $2 \cdot 10^{10}$ cm^{-2}). At higher doping levels the acceptors interact considerably, leading to severe broadening of the BE spectra and spectral diffusion in the broad BE band. This behaviour is analogous to observations for acceptor BEs in bulk semiconductors [8]. The data to be discussed below are therefore mainly selected from the samples with lower doping levels (approx. 10^{10} cm^{-2}). The data in Fig.1 also suggest that at lower doping levels and short delay times a transition which we assign to the biexciton is present in the spectra, and may influence the analysis of the dynamics of the FE-BE interaction. Since we however restrict the discussion here to data above 5K, the biexciton is effectively thermalized with the FE, and therefore does not significantly influence the analysis of the FE and BE kinetics. It is interesting to note that the presence of the biexciton is gradually lost at higher doping levels, indicating that the formation of biexcitons is limited by the preferential capture to the bound site.

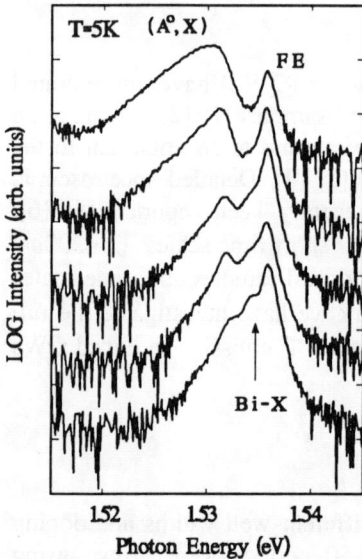

Figure 1: Short time (t=36ps) PL spectra for acceptor doped samples with doping density from $3 \cdot 10^{16}$ (bottom curve) up to $5 \cdot 10^{17}$ cm^{-3} (top curve)

The binding energies for BEs associated with neutral acceptors in AlGaAs/GaAs QWs are dependent on QW width Lz, varying from a value of approximately 6.5meV at Lz=5nm to approximately 3.5meV for Lz=15nm [5]. This implies that at low temperatures such as the 5K used in the present work, the thermal emission of BEs up to FEs is weak. It is therefore possible to study the kinetics of BE recombination independently, if excitation resonant with the BE is employed. On the other hand, if the excitation is resonant with the FE, capture to the lower energy BE state will be efficient in the early time response. These processes are illustrated in Fig.2 for the same set of Be doped samples as in Fig.1. Resonant excitation in the FE (Fig. 2(a)) instantaneously creates a FE population which then decays via two channels: (1) Capture to the BE state, and (2) recombination (mainly radiative). The

PL transients in Fig. 2(a) are dominated by capture to BE states during the first ~100ps, this capture process is increasingly important at higher doping levels (the capture rate is proportional to the number of neutral acceptors. From the behaviour of the PL decay at longer times the FE recombination lifetime can be extracted, via a computer simulation of the transient with the proper rate equations [9]. For excitation resonant with the BE (Fig. 2(b)), at 5K (or below) very little thermal excitation to the FE state is observed in the streak camera data. Therefore these transients merely reflect the recombination of the BEs, and can be used to directly calculate the true BE lifetime. The BE lifetime has been found to vary between 350ps and 500ps, when the QW width varies between 5nm and 15nm [7] [10], while we find a typical value of 400ps for the FE lifetime at 5K for Lz=10nm.

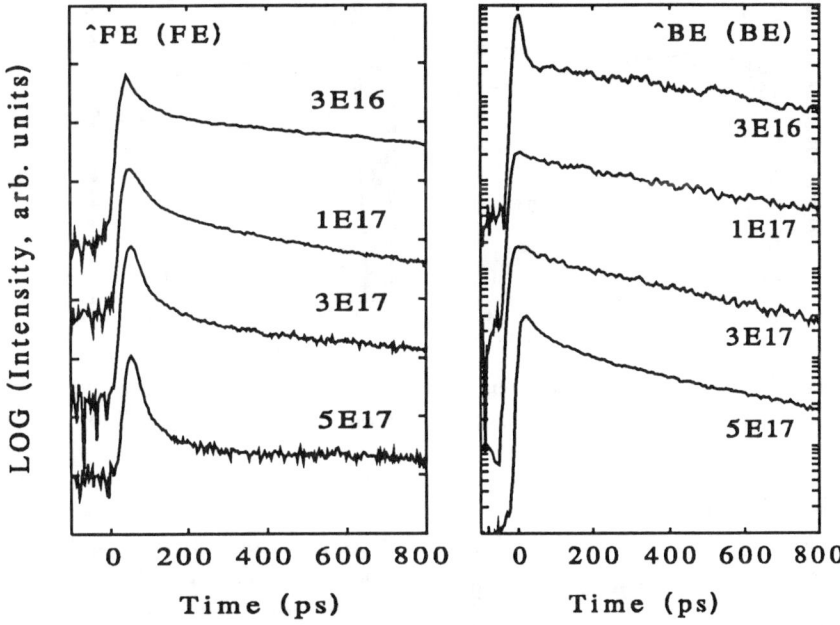

Figure 2: Doping density dependence of time resolved PL transient for excitation resonant with the FE (a) and BE (b).

Fig 2(b) also shows that the BE lifetime is not significantly dependent on the doping level. A slight decrease in τ_{BE} at the highest doping concentration shown in Fig. 2(b) is explained by spectral diffusion of BEs to lower energies than those detected in the integration window. This relaxation process is associated with the localization of BEs to regions of increasing local potential. Such processes can be directly compared and understood from equivalent behaviour in highly doped bulk semiconductors [11]. The temperature dependence of the exciton kinetics in acceptor doped QWs is illustrated in Fig. 3. Transients for both the FE (3(a)) and BE (3(b)) with resonant excitation are shown. At 15K and 25K the resonantly excited FE and BE transients are very similar, indicating that the populations are effectively thermalized already at 15K. As mentioned above this thermalization is not effective at 5K, where the FE and BE decays are only weakly coupled. The decay time at 25K which is common to both BE and FE is found to be larger than the 5K value. This is in part explained by the increase in the FE recombination time with temperature [1] [7].

The initial fast decay observed for the FE in Fig. 3(a) due to capture at the BE occurs over a shorter time scale at higher T due to the increased thermal emission. This initial decay also contains an intrinsic effect due to the heating of the initially resonantly created excitons (with K~0, i.e. corresponding to a temperature of ~1K) to the actual lattice temperature, whereby most of these excitons are scattered up to sufficiently large K-values that they are no longer radiative [12]. In fact a similar but weaker behaviour is also observed for the BE with excitation resonant with the BE (see Fig, 3(b)). This effect is not well understood at present, although a BE Auger mechanism has been suggested [9].

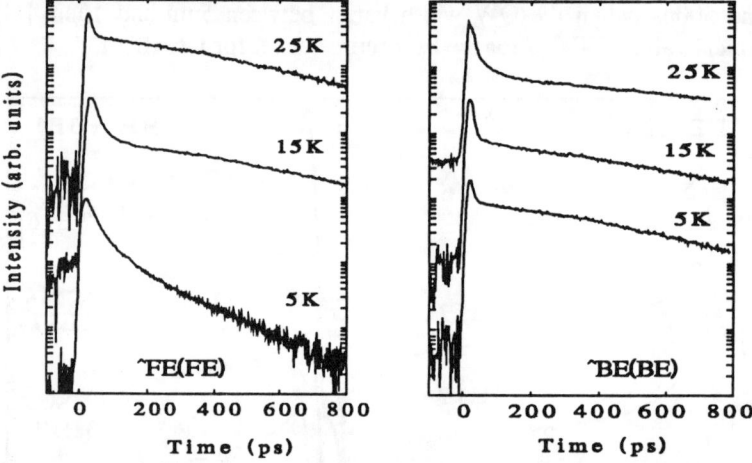

Figure 3: Variation of the PL transients with temperature for a 10nm wide QW with low doping. The behaviour for excitation resonant with FE (a) and BE (b).

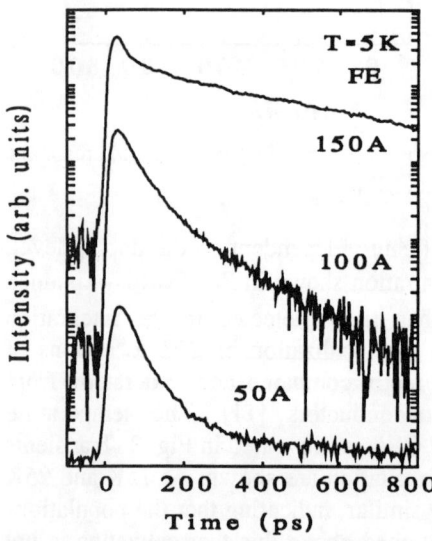

Figure 4: Well width dependence of exciton capture from FE to BE for acceptor doped samples (doping density approx. $3 \cdot 10^{16}$ cm^{-3})

The dependence on well width of the excitonic PL transients for these acceptor doped samples is also quite strong as is shown in Fig. 4, where the FE PL is detected at 5K with excitation resonant with the FE. The thermalization between BE and FE is more efficient for Lz=15nm, due to the significantly lower BE binding energy in this case. For Lz=50A the capture to the BE is very strong, and the BE is the dominant recombination channel once this capture has occurred [13].

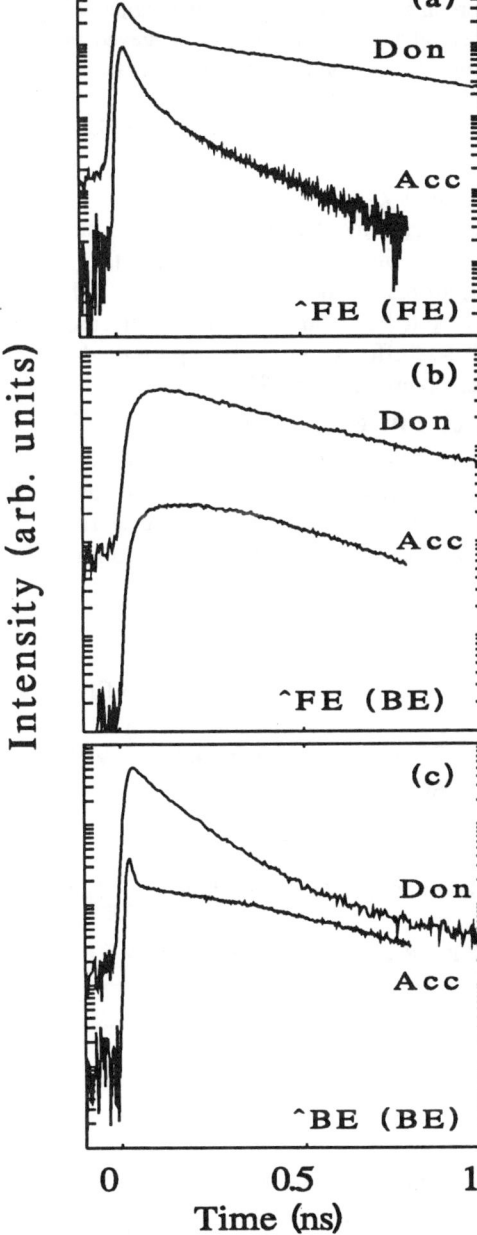

3.2. Donor Doped samples.

The binding energy of the donor BE in AlGaAs/GaAs is much smaller than for acceptors, typically about 2.2meV at Lz=10nm, compared to ~4.5meV for the acceptor BE. This means that BEs bound at neutral shallow donors will be close to being thermalized even at 5K. As a result the kinetics will be quite different from the acceptor case. These differences are illustrated at 5K in Fig 5. The excitation is resonant with the FE in figures (a) and (b), and resonant with the BE for (c). In (a) the FE PL is detected while in (b) and (c) the BE is detected. Fig 5(a) illustrates the fast thermalization of the FE in the donor doped sample. In contrast the acceptor case is dominated by capture to the BE, and essentially no thermalization between FE and BE is evident. Fig. 5(b) also indicates the faster thermalization for the donor BE, with excitation in the FE, which occurs in the donor case. In Fig 5(c) the thermal emission from BE to FE dominates the shape of the BE decay for the donor system. Whereas essentially no thermal emission is observed for the accetors ar 5K. These strong thermal emission effects for the BE make it difficult to evaluate any accurate values for donor BE lifetimes from these studies, temperatures lower than 5K are required for this purpose.

Fig. 6 illustrates the temperature dependence of the BE transient upon excitation resonant with the the donor BE (Lz=10nm). These data can be contrasted with Fig. 3(b), for the case of the acceptor BEs of the same well width. The thermalization process is faster at higher T,

Figure 5: Comparison of the exciton dynamics in acceptor and donor doped 10nm quantum wells at 5K under resonant excitation conditions.

as expected and the PL decay time evaluated at the long decay times coincides with the corresponding FE decay at all temperatures for the donor case. This is very different for the acceptor BE 3(b), where the effective decay time at 5K is significantly different from the FE decay time.

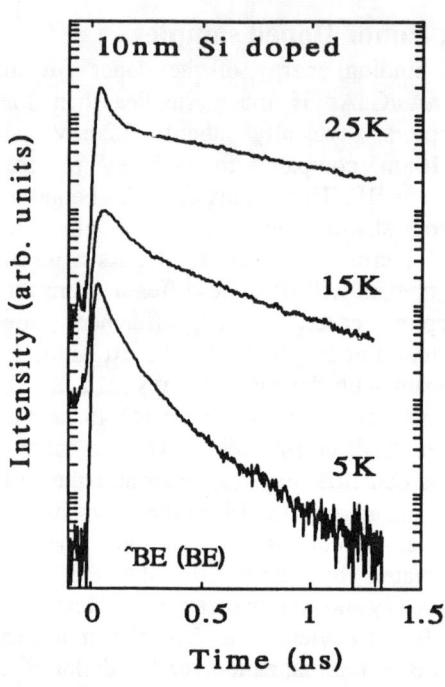

Figure 6: Temperature dependence of the thermalization for the donor doped sample. Thermalized equilibrium is achieved at lwer temperatures than for the acceptor case.

The discussion above has been limited to the case of rather low excitation densities (typically $< 10^9$ cm^{-2}). At higher densities additional mechanisms come in to play such as exciton scattering effects, which significantly alter the short time behaviour of both FEs and BEs. A full discussion of these effects will be presented in a forthcoming paper [9].

Acknowledgements

The authors would like to thank K. Rother for technical assistance. One of us (CIH) acknowledges the financial support of the Commission of the European Communities under project SC1000494.

References.

[1] J. Feldmann, G.Peter, E.O.Göbel, P.Dawson, K.Moore, C.T. Foxon and R.J.Elliot, Phys. Rev. Lett. **59**, 2337 (1987).
[2] J.Kuhl, A.Honold, L.Schultheis and C.W.Tu, Festkörperprobleme **29**, 157 (1989).
[3] J.Hegarty and M.D.Sturge, Surface Science **196**, 555 (1988).
[4] H. Stolz, D.Schwarz, W. van der Osten and G. Weimann, Superlattices and Microstructures **6**, 271 (1989).
[5] P.O.Holtz, M.Sundaram, K.Doughty, J.L.Merz and A.C.Gossard, Phys. Rev. B **40**, 12398 (1989).
[6] D.C.Reynolds, K.K.Bajaj, C.E.Leak, C.E.Stutz, R.L.Jones, K.R.Evans, R.W.Yu and W.M-Theis, Phys. Rev. B **40**, 6210 (1989).
[7] J.P.Bergman, P.O.Holtz, B.Monemar, M.Sundaram, J.L.Merz and A.C. Gossard, Phys. Rev. B **43**, 4765 (1991).
[8] P.J.Dean and A.M.White, Solid State Elec. **21**, 1351 (1978).
[9] C.I.Harris, H.Kalt, B.Monemar, J.P.Bergman, P.O.Holtz, M.Sundaram, J.L.Merz and A.C.Gossard, unpublished.
[10] B. Monemar, H.Kalt, C.I.Harris, J.P.Bergman, P.O.Holtz, M.Sundaram, J.L.Merz and A.C.Gossard, Superlattices and Microstructures **9**, 281 (1990).
[11] see for example B.Monemar, N. Magnea and P.O.Holtz, Phys. Rev. B **33**, 7375, (1985).
[12] L.C.Andreani, F.Tassone and F.Basanni, Solid State Commun. **77**, 641 (1991).
[13] B.Monemar, H.Kalt, C.I.Harris, J.P.Bergman, P.O.Holtz, M. Sundaram, J.L.Merz, A.C.Gossard and K.Köhler, proceedings ICPS 20 1990 **2**, 1549 Thessalonika.

SPECTROSCOPY OF SHALLOW DONOR IMPURITIES IN GaAs/GaAlAs MULTI-QUANTUM WELLS

Janette L DUNN, Elaine PEARL and Colin A BATES
Physics Department, The University, Nottingham, NG7 2RD, UK

ABSTRACT

The spectra observed from silicon-doped GaAs/GaAlAs multi-quantum well structures in experiments such as far infrared (FIR) photo-conductivity contain peaks which have been identified with transitions from the ground state to several excited states of a confined Si impurity in magnetic fields of up to 10 T. Such systems are of particular interest as some of the observed features of the spectra can be explained in terms of a low-field 'hydrogenic' model, whilst other features are consistent with strong-field descriptions, where Landau-like behaviour dominates.

Details of a theoretical model are given which predicts transitions from the 1s-like ground state of a hydrogenic impurity to various excited states. These results are then used to determine the energies of so-called metastable states, which are not bound in the low-field limit. Transitions to both bound and metastable states are identified in the observed experimental data.

1. Introduction

There has been much discussion in the literature concerning hydrogen-like shallow-donor impurities in semiconductors, both in the bulk and in multi-quantum wells (MQW's). Transitions to $2p_{+1}$- and $2p_{-1}$-like excited states have clearly been identified in FIR photoconductivity experiments. However, there are several contradictory pictures for the assignments of the remaining transitions, particularly in MQW's. The underlying cause of the problem is that for the magnetic fields of up to about 10T used experimentally, the behaviour of the impurities is intermediate between that obtained using a low-field hydrogenic picture and that of a high-field Landau type formalism.

In this paper, a theoretical model will be developed using states which are principally hydrogen-like, but which incorporate some of the required Landau-like behaviour appropriate to higher fields. An analysis of the number of nodes of the predicted eigenstates in the z-direction will be used as an additional aid to identification of the states. In addition, transitions to various metastable states will be identified using knowledge of the bound hydrogenic states.

The model used for the hydrogenic states is based on that of Greene and Lane[1] and is an extension of our previous work[2,3]. The model of Greene and Lane[1] has been extended to include the effects of higher excited states and of states with angular momentum m = ±2. It has also been modified to allow for a non-isotropic effective mass, and to include the effects of odd-parity states which are important for impurities which are not located at the centre of a quantum well.

2. The Theoretical Model

In the model used, an anisotropic effective mass is introduced by defining a parallel effective mass $m_{\parallel}$ for motion in the x-y plane and a perpendicular effective mass $m_{\perp}$ for motion in the z-direction. Relative effective masses $m_{r}^{\perp} = m_{\perp} / m^{*}$ etc. are also defined, where $m^{*} = 0.067\, m_{e}$ is the usual GaAs effective mass. The effective masses in the barriers may be different to those in the wells.

The Hamiltonian for a shallow donor impurity in a MQW with an applied perpendicular magnetic field can then be written in the dimensionless form

$$\mathcal{H} = -\frac{1}{m_r}\nabla^2 + \frac{1}{m_r'}(\gamma l_z + \frac{1}{4}\gamma^2\rho^2) - \frac{2}{r} + V_B(z) \qquad (2.1)$$

using cylindrical coordinates with origin at the centre of a quantum well and with the z-axis defined to be the MQW growth axis, and where

$$\frac{1}{m_r}\nabla^2 = \frac{1}{m_r'}(\frac{1}{\rho}\frac{\partial}{\partial\rho}(\rho\frac{\partial}{\partial\rho}) + \frac{1}{\rho^2}\frac{\partial^2}{\partial\phi^2}) + \frac{1}{m_r^\perp}\frac{\partial^2}{\partial z^2} \qquad (2.2)$$

The unit of length is the GaAs effective Bohr radius a_o = 98.7 Å, the unit of energy is the GaAs effective Rydberg R = 5.83 meV, and the quantity γ is a dimensionless measure of magnetic field, related to the field B in Tesla by the relation γ = 0.15 B. $V_B(z)$ is an energy operator for the square well potential, defined to have the value zero in the wells and a constant value V_o in the barriers. For $Ga_{1-x}Al_xAs$ MQW's, V_o is taken to be 85% of the total band gap difference ΔE_g between GaAs and GaAlAs. The position of the impurity electron is $r = [\rho^2 + (z - z_I)^2]^{1/2}$, where z_I specifies the z-coordinate of the impurity nucleus.

Many attempts have been made to solve the Hamiltonian (2.1) for MQW's and the analogous Hamiltonian for the bulk. Although exact analytical solutions can not be found for the intermediate magnetic fields of interest. The 1s to $2p_{\pm 1}$ energy splittings were predicted to a reasonable degree of accuracy using a variational approach in which hydrogenic states were expanded in terms of Gaussian-type basis sets[1,4]. This has the advantage that, although the states are constructed using a weak-field formalism, the Gaussian behaviour of Landau states is partially incorporated in the wavefunctions, making them more appropriate for larger fields. This basic approach will be developed here, although without resort to a variational procedure.

To a first approximation, the eigenstates ψ of a hydrogenic impurity in a quantum well can be expected to be a product of bulk hydrogen states ψ_{nlm} and states $f(z)$ which are solutions of the standard square well problem (ie. cos(kz) in the barriers and $[Ae^{\kappa z} + Be^{-\kappa z}]$ in the wells). The parameters k, κ, A and B are fixed by ensuring that ψ and $(1/m_r^\perp)\partial\psi/\partial z$ are continuous across the well boundaries.

The angular momentum operator l_z commutes with the Hamiltonian (2.1), so its eigenvalue m must be a good quantum number in this system. For impurities located at the centre of a quantum well, the z-type parity π_z [= $(-1)^{l+m}$] is also a good quantum number. However, n and l will not be good quantum numbers. Hence, approximate eigenstates of the on-centre MQW problem can be found by taking linear combinations of hydrogenic states with different n and l values, but with the same value of m and of π_z. For off-centre impurities, it is necessary to allow mixing between states of different parity.

The Slater-type exponential factors $e^{-r/n}$ in the standard hydrogen states for the bulk can be expanded in terms of basis sets of Gaussians[5]

$$e^{-\frac{r}{n}} = \sum_i C_i e^{-a_i r^2} \qquad (2.3)$$

where the C_i and a_i are parameters whose values are determined numerically. A further field-dependent parameter δ was included as an additive factor to the a_i in the Gaussians in ρ to allow for the expected reductions in orbit sizes in the x-y plane as the magnetic field strength increases. Their final choice, which is followed here, was δ = 0.1 γ, although the results are not very sensitive to the precise value of δ used.

A basis set of Gaussian-type functions for the MQW problem is constructed by choosing a set of numbers A_k for each (m, π_z) set from the sets of a_i in the Gaussian expansions for each of the states in that set. Here, it is the range of values chosen which most strongly influences the final results rather than the precise values used.

The hydrogenic states in a quantum well are thus assumed to be a linear combination of states Ψ_i given by

$$\Psi_i = f(z)(z-z_I)^{q_i} \rho^{|m|} e^{im\phi} e^{-\alpha_i(z-z_I)^2} e^{-(\beta_i+\delta)\rho^2} \tag{2.4}$$

where $q_i = (\pi_z + |\pi_z|)/2$. Following Greene and Lane[1], the parameters α_i and β_i are taken from the sets A_k, with the restriction that, if the A_k are listed in order of magnitude, $\alpha_i = A_k \to \beta_i = A_k$ or $A_{k\pm 1}$, ie. that the arguments of the Gaussians in ρ and $(z - z_I)$ are 'similar' in magnitude. The number of terms k is chosen to be large enough for the resultant wavefunctions to be good, but small enough for calculations with the states to be manageable. A value for k of 5 or 6 for each (m, π_z) set was taken as a reasonable compromise, which gives a total number of basis states i for each set of $(3k-2) = 13$ or 16.

The final values for A_k used here for all sets of states are the same as those used by Greene and Lane[1] for the m = 0 and m = ±1 even parity states, namely 13.4, 2.01, 0.454, 0.123, 0.0324 and 0.00717. Many other sets of A_k covering a similar range of values have also been investigated, and found to give very similar results, at least for the lower-lying states.

Approximate eigenvalues and eigenstates of the Hamiltonian $\mathcal{H}$ are found by numerically solving the generalised eigenvalue problem $\mathcal{H}\psi = EU\psi$. In the original papers of Greene and co-workers, eigenstates were obtained by carrying out a variational procedure. Here, in order to save computing time, a strict variational procedure is not employed. Instead, the optimal parameters of Greene and Lane[1] are used directly in calculations, although the parameters are changed slightly 'by hand' in order to observe the effect on the final results.

According to simple electric dipole selection rules, transitions from the 1s-like ground state are only allowed to states with m = ±1 for the Faraday configuration used experimentally. For on-centre impurities, transitions must also conserve π_z. However, there is evidence that in the bulk, transitions are weakly observed to both states with m = 0 and m = ±2, as well as to states with odd z-parity, due to various weak perturbation effects. Hence, the possibility of transitions to all states are investigated.

The states of a hydrogenic impurity can be labelled in the alternative high-field notation (N, m, ν), where N is the principle Landau quantum number, m the usual magnetic quantum number and ν gives the number of nodes of the wavefunction in the z-direction. The z-parity of these states is thus carried entirely by the quantum number ν, ie. $\pi_z = (-1)^\nu$. In this notation, the states (N, m, ν) exist for all integers $N \geq 0$ with $m \leq N$. However, only the states with N = 0 and with N = m for $N \geq 1$ are truly bound, and extrapolate back to the hydrogen-like states in weak fields. It is these states which are predicted by our model above. The remaining states are called metastable states, and are considered in more detail below.

3. Results and Comparison with Experiment

FIR photoconductivity measurements on $GaAs/Ga_{0.67}Al_{0.33}As$ MQW's have recently been made in fields of up to 8T.[6] The wells and barriers were approximately 150 Å wide, and the central 50 Å of each well were doped with silicon to a concentration of approximately 1×10^{16} cm^{-3}. Some data is also available on wells of other widths[7-10]. These have been combined to determine, although somewhat approximately, those results which would have occurred with a 150 Å -wide sample.

The $2p_{+1}$ and $2p_{-1}$ transitions were fitted to a reasonable degree of accuracy by Greene and Lane[1]. However, the $2p_{+1}$ - $2p_{-1}$ splitting is observed to be slightly less than the value of precisely 2γ which must occur if an isotropic effective mass is used in the calculations. An improved fit to the data can be found by choosing a parallel effective mass $m_{\prime}$ of around $0.069m_e$ and perpendicular mass $m_{\perp}$ of $0.068m_e$ for the 150Å wide wells considered here. The effective masses in the barriers were taken to be larger than those in the wells by the usual additive factor $0.083m_e x$.

The final results for transitions to bound excited states of on-centre impurities are shown as solid lines in figure 1. Also shown is the data of Grimes et al[6] averaged over all samples (Δ) and the average of the data for the $2p_{+1}$ and $2p_{-1}$ transitions as observed by other workers (o)[8-11]. Denoting the states by labels which seem most appropriate by analogy to pure hydrogen, the transitions presented are to final states (from bottom right to top left) $2p_{-1} \equiv (0,-1,0)$, $3p_{-1} \equiv (0,-1,2)$, $2p_{+1} \equiv (1,1,0)$, $3d_{+1} \equiv (1,1,1)$, $3p_{+1} \equiv (1,1,2)$ and $3d_{+2} \equiv (2,2,0)$. A possible transition to a further m = 2 line is shown as a dotted line.

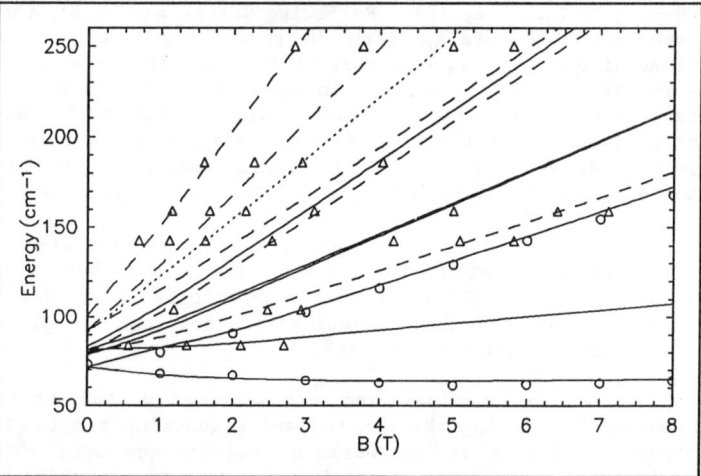

Figure 1: Transitions from 1s

The diagonalisation procedure of the model in section 2 produces $(3k-2)$ states for each (m,π_z) set. However, not all of these states have a physical meaning. According to the high-field picture, only one truly bound hydrogenic state is allowed with a given value of m and given number of nodes ν. Thus the number of nodes of each of the predicted eigenstates can be examined and only the lowest-energy state with a given value of ν for each m selected. The states chosen all have the correct number of nodes to satisfy this condition.

According to the electric dipole selection rules, transitions from 1s to states of odd-parity such as $3d_{+1}$ should be forbidden for on-centre impurities. However, they will be allowed for off-centre impurities, where states of different parity are mixed together, and may be weakly allowed for on-centre impurities due to various perturbation effects. The theoretical prediction for $3p_{+1}$ is almost coincident with that of $3d_{+1}$, making the identification of these lines in the experimental data unclear. However, the experimental points in this region are observed to behave somewhat differently to those of the other points (having a different line shape and being sensitive to different experimental conditions), which is consistent with the data being a sum of two different transitions. The identification of the line labelled $3d_{+2} \equiv (2,2,0)$ can not be made conclusively because the data which fits this line also fits the (2,1,0) metastable state (see below), to which transitions are expected to be much stronger.

Similar results to those above have been obtained for finite values of z_I. Generally speaking, the pattern of energy levels predicted is very similar to the on-centre case but with a shift to lower energy. Most of the lower energy transitions are ~ 3 to 5 cm^{-1} lower than the corresponding on-centre cases for z_I = 20 Å and 20 to 30 cm^{-1} lower for z_I = 50 Å.

Further experimental points have been observed which can not be fitted to states derived from a hydrogenic model. It is well-known in bulk GaAs that transitions are seen to the metastable states[11]. Hence, it is likely that transitions to metastable states will also occur in MQW's. One characteristic feature of transitions to metastable states is that lines occur in pairs separated by the cyclotron resonance energy $\hbar\omega_c$ (= 2γ) or multiples thereof. In particular,

$$E(N,m,\nu) - E(N-m,-m,\nu) = m\hbar\omega_c \qquad (3.1)$$

where $E(N,m,\nu)$ is the energy of the state (N,m,ν). The observed data does indeed have several pairs of states with $\hbar\omega_c$ separations. On a hydrogenic picture, there is no fundamental reason why any pairs of states other that nl_{+1} and nl_{-1} should be separated by $\hbar\omega_c$, and it is unlikely that all of the higher-energy states observed which obey this pattern are associated with m = -1, as these levels tend to be much lower in energy.

Unfortunately, expressions for the metastable states are not generally known. However, a semi-qualitative attempt can be made to identify some of the metastable states. In the bulk, the (1,-1,0) metastable state has been identified just above and almost parallel to the 2p$_{+1}$ level[11-13]. A similarly placed level occurs in the quantum well case, so this is identified with the (1,-1,0) metastable state. The (2,1,0) metastable state can then be identified $\hbar\omega_c$ above this. As mentioned above, this latter line coincides with the position of the (2,2,0) hydrogenic line. The experimental observations near this point could be a composite of both lines, although the metastable contribution can be expected to dominate due to the electric dipole selection rules.

As the Coulomb potential is rotationally symmetric, it can not mix together Landau states with different values of m. The number of nodes ν must also be conserved. However, states with different N's will be mixed together. The effect is to increase the separation of the states (N,m,ν) and $(N+1,m,\nu)$ to a value slightly larger than $\hbar\omega_c$ which is approximately constant (decreasing slightly with increasing N and increasing field). As the positions of the (1,1,0) and (2,1,0) lines have already been identified, this allows the positions of the (3,1,0) and (4,1,0) metastable states to be determined. The position of the (2,-1,0) metastable state can then be fixed by further use of (3.1).

The results for all of the metastable states identified are shown as dashed lines on figure 1, where it can be seen that the agreement between experiment and theory is good. From bottom right to top left, they are (1,-1,0), (2,1,0), (2,-1,0), (3,1,0) and (4,1,0).

Physical pictures for the behaviour of the impurity electron in various of its states have been obtained by plotting the probability density as a function of position. Plots of Ψ^2 integrated over ρ and ϕ have been made in order to show the probability of being at a given z position. Some typical plots for zero field are shown in Figure 2.

Probability density plots show that the 1s ground state is almost entirely localised in the central well at all field strengths for 150 Å-wide wells. The 2p$_{+1}$ state is also strongly localised in the central well, but does extend into next-nearest wells in small fields. In larger fields, the localisation in the central well is found to become almost total. In contrast, the odd-parity 3d$_{+1}$ wavefunction

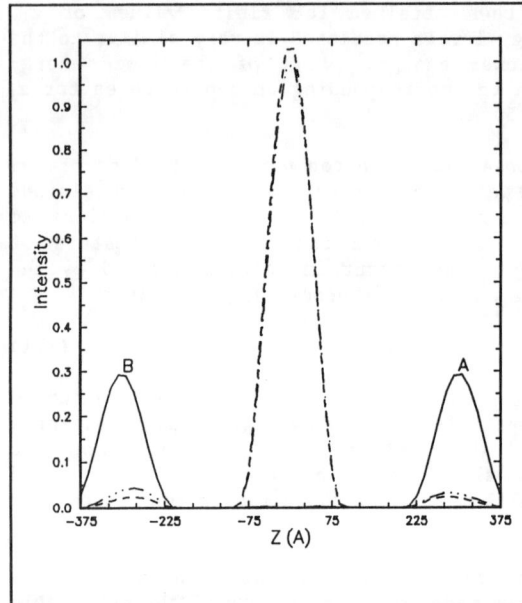

Figure 2: Probability Densities for the $2p_{+1}$ (_ _ _) and $3d_{+1}$ (___) states of on-centre impurities and the $2p_{+1}$ state of impurities with $z_I = 50$ Å (_..._).

is very small in the central well and spreads out over several wells in small fields. Contour plots of the probability density as a function of z and ρ (integrated over ϕ) show that in zero field, the electron is confined to approximately circular orbits in the x-y plane of radius 50Å for the 1s state and 200 Å for the $2p_{+1}$ state.

5. Discussion

Using the results of a theoretical model for hydrogenic impurities and knowledge of metastable states in the bulk, it has been possible to explain almost all of the recent FIR photoconductivity results[6] on GaAs/ GaAlAs MQW's in terms of transitions from the 1s-like ground state of a Hydrogenic impurity. There is some evidence that transitions to states with m = 2 and to odd-parity states may be weakly observed due to weak perturbations and the effects of off-centre impurities. However, some of the higher states can not be conclusively identified at this stage, and further work is necessary.

Acknowledgements

We are grateful to our colleagues Dr J M Chamberlain, Dr R Grimes and Dr M Stanaway at Nottingham for providing results of their experiments prior to publication. One of us (EP) wishes to thank the SERC for a Research Studentship.

References

1. R.L. Greene and P. Lane, Phys. Rev. B **34**, 8639 (1986)
2. J.L. Dunn, E. Pearl, R.T. Grimes, M.B. Stanaway and J.M. Chamberlain, Mat. Sci. Forum **65-66**, 117 (1990)
3. J.L. Dunn and E. Pearl, J. Phys.: Condens. Matter, Submitted for publication.
4. R.L. Greene and K.K. Bajaj, Phys. Rev. B **31**, 913 (1985)
5. S. Huzinaga, J. Chem. Phys. **42**, 1293 (1965)
6. R.T. Grimes, M.B. Stanaway, J.M. Chamberlain, J.L. Dunn, M. Henini, O.H. Hughes and G. Hill, Semicond. Sci. Technol **5**, 305 (1990)
7. Y-H Chang, B.D. McCombe, J-M Mercy, A.A. Reeder, J. Ralston, and G.A. Wicks, Phys. Rev. Lett **61**, 1408 (1988)
8. N.C. Jarosik, B.D. McCombe, B.V. Shanabrook, J. Comas, J. Ralston, and G. Wicks, Phys. Rev. Lett. **54**, 1283 (1985)
9. J-P Cheng and B.D. McCombe, Phys. Rev. B **42**, 7626 (1990)
10. J-P Cheng, W.J. Li and B.D. McCombe, Mat. Sci. Forum **65-66**, 99 (1990)
11. A.v. Klarenbosch, T.O. Klaassen, W.Th. Wenckebach and C.T. Foxon, J. Appl. Phys. **67**, 6323 (1990)
12. S. Narita and M. Miyao, Solid State Commun. **9**, 2161 (1971)
13. H.P. Wagner and W. Prettl, Solid State Commun. **66**, 367 (1988)

EXCITONS BOUND AT SHALLOW IMPURITIES IN GaAs/AlGaAs QUANTUM WELLS WITH VARYING DOPING LEVEL

P.O. Holtz[1], B. Monemar[1], M. Sundaram[2], J.L. Merz[2], A.C. Gossard[2], C.I. Harris[3] and H. Kalt[3]

[1] Department of Physics and Measurement Technology, Linköping University, S-581 83 Linköping, Sweden
[2] QUEST Center, University of California at Santa Barbara, Santa Barbara, CA 93106, USA
[3] Max-Planck-Institut für Festkörperphysik, D-7000 Stuttgart 80, Germany

ABSTRACT

A spectroscopic study of the properties of excitons bound (BE) at shallow impurities, both acceptors and donors, in narrow GaAs/AlGaAs quantum wells (QWs) is presented. The doping conditions in the QWs have been varied in a systematic way up to the degenerate limit. Several novel aspects of the excitonic properties are demonstrated: The first observation of the two electron transition related to the donor BE in selective photoluminescence, yielding information about the electronic structure of the donor. For the case of acceptors confined in narrow QWs, we demonstrate that the capture process into the acceptor BE occurs predominantly from the localized exciton (LE) state, i.e. a free exciton localized in the locally lowest potential due to interface roughness. Accordingly, exceptionally small BE linewidths are achieved upon excitation resonant with the LE also for narrow QWs. For higher acceptor concentrations, the normal BE peak is replaced by a broader feature appearing with a larger binding energy and higher intensity level than the BE. A plausible candidate for this novel band is the exciton bound at interacting acceptors.

1. INTRODUCTION

While the bound excitons (BEs) have been extensively studied in bulk material and we have by now a fairly detailed knowledge about the BE properties, the corresponding level of knowledge is considerably lower for the BEs in the two-dimensional (2D) case. The first observation of a BE in a quantum well (QW) in photoluminescence (PL) was reported less than ten years ago by R.C. Miller et al [1]. A binding energy of about 6.5 meV was found for the shallow acceptor BE in their 46 Å wide GaAs/AlGaAs QW to be compared with 2.9 meV in bulk GaAs [2]. The acceptor BE binding energy has been shown to increase with decreasing QW width down to $L_z = 50$ Å [3]. However, the BE binding energy is expected to reach a maximum for a certain L_z to be back on the bulk value for $L_z=0$ in accordance with the impurity binding energy, although this has not been experimentally demonstrated so far for the BE. Accurate determinations of the BE binding energy from conventional PL measurements are difficult to perform for small L_z, since the exciton linewidths are increasing with decreasing L_z and the BE is barely resolved from the free exciton (FE). The accuracy is significantly improved by the observation of the BE in selective PL (SPL) experiments upon excitation resonant with the FE, as described in more detail below [4]. Also, the observation of two hole transitions (THTs) of the BE [3], has allowed accurate determinations of the BE energies. By detecting such a THT peak in PLE measurements, it is possible to observe the BE also in PLE spectra, usually exhibiting a significantly smaller linewidth than in PL.

For the case of the donor BE in GaAs/AlGaAs QWs, there has been even more uncertainty about the the interpretation of the recombinations close to the bandgap. This uncertainty is mainly due to the fact that the donors are more shallow and the confined donor BE appears at an energy position very close to the free-to-bound (FB) recombination involving the confined donor [5]. We present in this study the first observation of two electron transitions (TETs) of the donor BE in a QW by using SPL. The observation of TETs provides a verification of the interpretation of the donor BE, but yields also information about the excited states of the confined donor and a way to accurately determine the binding energies of the donor. In addition, we report briefly on the differences in the BE spectra that occur, when the doping level in the QW is increased up to the degenerate limit.

2. SAMPLES AND EXPERIMENTAL

The samples were grown by MBE on a semiinsulating substrate. On top of the GaAs buffer, 50 periods of GaAs QWs separated by 150Å wide $Al_{0.3}Ga_{0.7}As$ barriers and finally a 50Å wide GaAs cap layer were grown. The central 20% of the QWs were doped with Be in the p-type QWs and with Si in the n-type wells. The doping level has been varied in a wide range: From $2 \cdot 10^9$ up to $1.5 \cdot 10^{12}$ cm^{-2} for both the p-type and n-type doping. Also the QW widths have been varied in a wide range, 50 - 150Å, but most of the results reported here for the acceptor doped structures originate from measurements on 150Å wide QWs, while most of the studies on donor doped structures were performed on 100Å wide QWs.

The 5145Å line from a cw Ar^+ ion laser has been used for PL spectra with above bandgap excitation. The same Ar^+ laser pumping a solid state Titanium Sapphire tunable laser was employed for the SPL and PLE measurements. All measurements reported here were performed at temperatures below 2K.

3. ACCEPTOR DOPED QWs

3.1. LOW DOPING

In striking contrast to bulk material, the intrinsic FE normally dominates the recombination processes up to fairly high impurity levels due to the effect of confinement on both the electron and hole [6]. The acceptor BE does not appear in the PL spectrum until the doping level is above $\approx 1.5 \cdot 10^9$ cm^{-2} in a 100 Å wide QW. Also at moderate doping levels (say $p = 1.5 \cdot 10^{10}$ cm^{-2}), only the FE and acceptor BE are observed in a typical PL spectrum of a similar QW with above bandgap excitation. However, with selective excitation close to or resonant with the excitons, new features, e.g. the FB emission involving the neutral Be acceptor, the THT and resonant Raman scattering (RRS) satellites, appear in the SPL spectra [3]. Such studies have provided important information about the excited states of the confined impurities, and their dependence on the QW width and impurity position in the QW.

The intensity ratio BE/FE is further reduced and the exciton lines broaden with decreasing QW width and increasing localization effects. Due to these facts, it is usually not possible to resolve the FE-BE peaks for such narrow QWs and the BE binding energies can not be accurately determined. However, similar SPL experiments as described above also provide information on the exciton localization and exciton binding energies. Due to interface roughness, also the FE gets localized at the locally lowest potential. Thus the "free excitons" can be more properly treated as

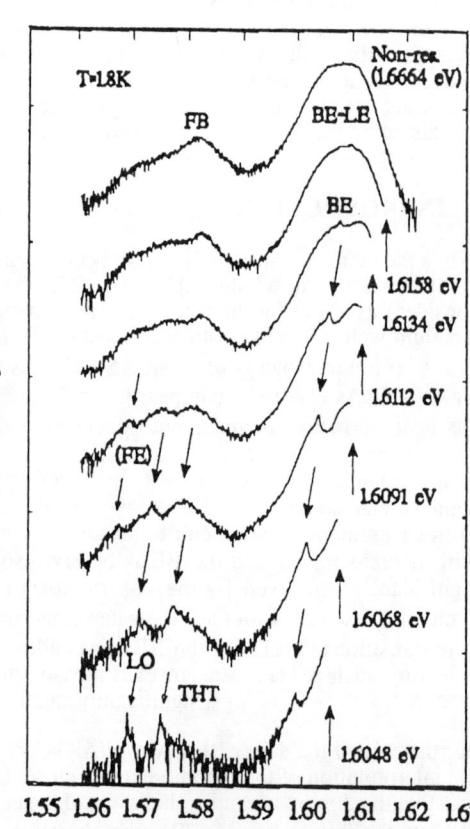

Fig. 1: A synopsis of SPL spectra for a 50 Å QW doped with Be to level of $1.5 \cdot 10^{10}$ cm^{-2}

localized excitons (LEs), trapped at interface fluctuation potentials. This localization process of excitons has been further investigated by picosecond transient measurements, which directly demonstrate the spectral diffusion of excitons. Furthermore, by exciting resonantly with the LE, a sharp and well-defined BE peak appears in the SPL spectrum also for narrow QWs, as illustrated in Fig. 1 for a 50Å wide QW. This is in contrast to the case with non-resonant excitation, when it is usually not possible to resolve the FE-BE peaks for such narrow QWs. Thus, this kind of measurements allow an accurate determination of the BE binding energies.

As mentioned above, the FE normally dominates the PL spectrum in narrow QWs. This effect is even more striking in PLE spectra, as is illustrated in Fig. 2 (the lower spectrum) for a 71Å wide QW with the FB band detected. In fact, the first observation of the BE in PLE was not reported until recently [7], when the THT peak was detected (the upper spectrum in Fig. 2). In bulk material, on the other hand, this method, to look for the enhancement of the BE in PLE, when the THT is detected, is a common method to verify the interpretation of THTs.

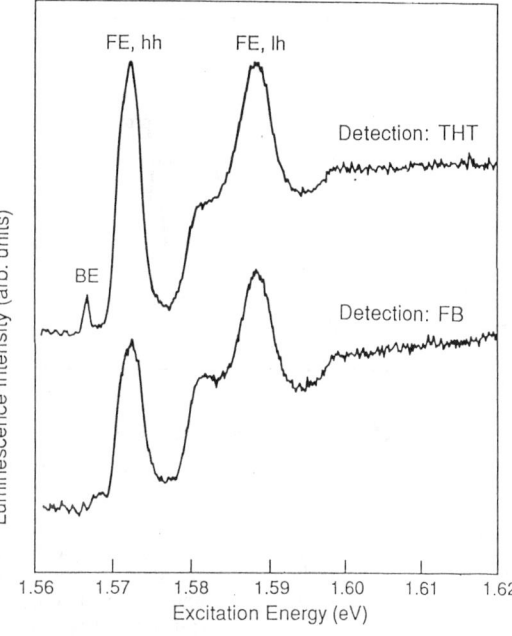

Fig. 2: PLE spectra for a 71Å wide acceptor doped QW with two different detection energies

3.2. HIGH DOPING

When the acceptor concentration is successively increased, a wing on the low energy side of the BE is rapidly increasing in intensity to be dominating already at an acceptor concentration of $5 \cdot 10^{10}$ cm^{-2} for the case of a 150Å wide QW, while the FE intensity is decreasing. Fig. 3 shows such a PL spectrum for a 150Å QW, which is acceptor doped to a level of $1.5 \cdot 10^{11}$ cm^{-2} (corresponding to a spatial separation between the acceptors of about 260Å on average). As can be seen, the FE timeintegrated intensity is more than two orders of magnitude lower than the new acceptor related band. The energy separation between the FE and this new band is about 7.4 meV, to be compared with the corresponding FE - BE energy separation of 4.2 meV in a low doped QW of the same width.

The PLE spectrum of the same sample is shown in Fig. 4, when detecting the low energy wing of the novel band. In addition to the hh- and lh-states of the FE, a new band appears also in this PLE spectrum. This band is relatively strong compared with e.g. the weak BE peak as observed in PLE spectra in a moderately doped QW of the same width [5]. This new feature appears 5.0 meV below the FE hh peak, i.e. with an energy separation smaller than in the PL spectrum, but still with a larger separation from the FE peak than the normal BE. The PLE spectrum reflects mainly the density of states, while a relaxation to states at a lower potential corresponding to larger L_z occurs before the recombination process observed in PL. This is a well-known mechanism caused by interface roughness, which gives rise to e.g. the Stokes' shift. However, the normal Stokes' shift for the FE in the same structure is 3.6 meV, while the corresponding downshift in PL for this novel band is significantly larger, 6.0 meV. Thus, the additional downshift in energy has to be due to a different relaxation process.

One plausible candidate for this new band is excitons bound at interacting acceptors. The interaction between two or more acceptors should increase the potential binding the exciton compared to a single

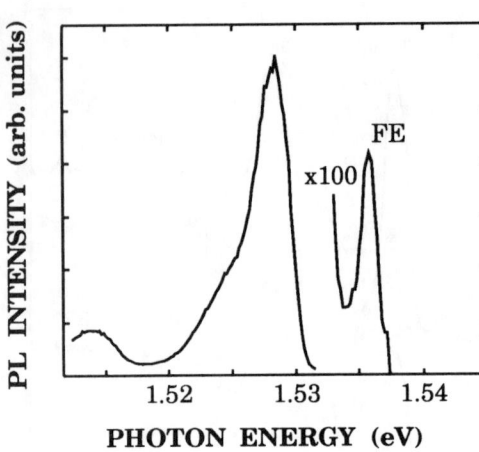
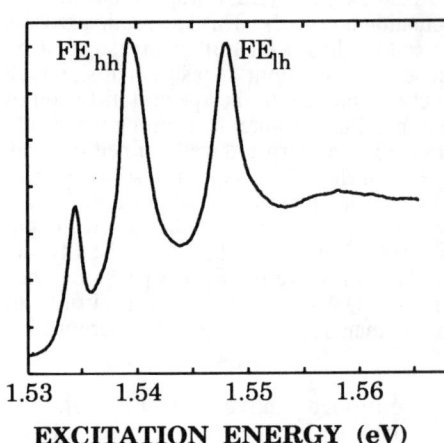

Fig. 3: PL spectrum of a 150Å wide QW, center doped with Be to a level of $1.5 \cdot 10^{11}$ cm^{-2}

Fig. 4: PLE spectrum of the same QW as shown in Fig. 3

acceptor. A distribution of binding energies due to the varying spatial separation between the acceptors and possibly also the number of interacting acceptors is expected. Then excitons could be transferred from one site to another site with a higher binding energy. Such a transfer process should give rise to an additional downshift in energy in the timeintegrated PL spectrum compared to the PLE spectrum.

Similar transformations from narrow BE lines to broader and asymmetric bands at lower energies have earlier been observed for bulk material, such as GaP and ZnTe, at higher doping levels [8], but have not been reported for the 2D case. This behavior has been explained in terms of pairing and cluster formation and simple calculations of the exciton binding energy as a function of the acceptor concentration for a H_2-like acceptor pair have been presented for the bulk case [8].

4. DONOR DOPED QWs

4.1. LOW DOPING

Similarly to the acceptor doped QW, the FE and donor BE dominate the PL spectrum for moderately donor doped QWs. However, a broadening on the low energy side of the donor BE appears already at doping levels at about $2 \cdot 10^9$ cm^{-2} (for a 100Å wide, Si doped QW). This broadening is likely due to the FB transition involving the recombination between an electron from the Si donor and a hole from the valence band. With the objective to resolve the TET peak, we have tried to minimize the intensity of the FB band, since it easily can hide the weak TET features. However, it appeared to be significantly more difficult to monitor the TET peaks of the donor BE than the corresponding THTs in acceptor doped QWs. This can possibly be explained by the tail of the FB transition described above, which partly overlaps with the TET. Also, the fact that the donor states are considerably more shallow than the acceptor states means that the extent of the donor wave function is comparable with or larger than the QW width. In particular, the final 2s state of the donor is widely spread and could be significantly perturbed by the relatively narrow QW. However, when the experimental conditions were opitimized (corresponding to a doping level of $1 \cdot 10^{10}$ cm^{-2} in the central 20% of the QW), a weak TET feature appeared in the SPL spectrum, when the excitation was close to or resonant with the donor BE, as shown in Fig. 5. The energy position of the observed TET satellite yields a value on the 1s - 2s energy separation of 10.6 meV, which is in excellent agreement with the theoretical prediction, 10.6 meV [9], obtained if appropriate values on the effective masses and dielectric constants are used, i.e. with different values for the well and barrier materials. This point seems to be important, since a significant discrepancy from the experimentally determined value is derived (9.9

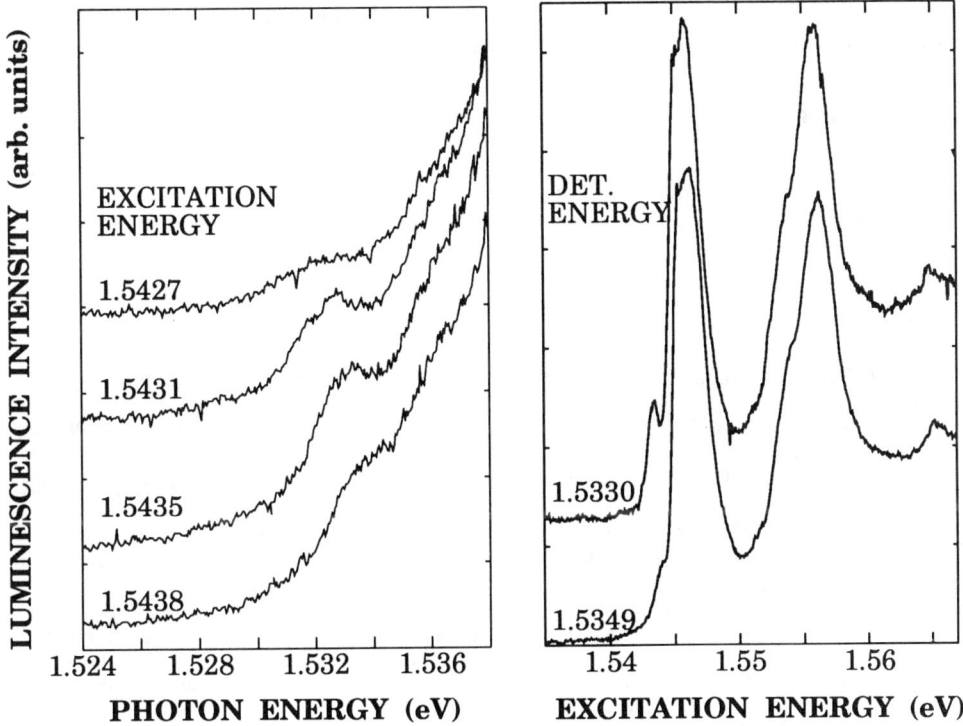

Fig. 5 SPL spectra for a 100Å wide Si doped QW with excitation close to or resonant with the BE (at 1.5435 eV)

Fig. 6 PLE spectra for the same Si doped QW as used in Fig. 5

meV), if the same values on the effective masses and dielectric constants are used for the two media 10.

Similarly to the acceptor case described above, the interpretation of the TET satellites has been verified by PLE measurements on the same sample. When the TET peak is resonantly detected (the upper spectrum in Fig. 6), the donor BE appears in the PLE spectrum in addition to the usual FE related peaks. For comparison, a reference PLE spectrum with a non-resonant detection is also shown (the lower spectrum in Fig. 6), in which the BE has almost disappeared.

4.2. HIGH DOPING

With increasing donor concentration in the QW (at about $2 \cdot 10^{17}$ cm^{-3} corresponding to a sheet density of $1 \cdot 10^{11}$ cm^{-2}), the BE peak broadens further and shifts towards lower energies [11]. However, in the case of donor doped QWs, the interpretation of the spectra is not straight forward and it is difficult to say whether this downshift has the same origin as proposed for the acceptor case described above, i.e. the exciton bound at interacting neutral donors. Several donor related recombinations are expected to appear at approximately the same energy position: The FB emission involving a free hole recombining with a bound Si donor electron, the exciton bound at the ionized Si donor, and the D$^-$ related emission.

When the donor concentration is further increased, the intensity of the exciton peaks decrease first in PL and at at higher doping levels also in PLE. Above a doping level of about $1 \cdot 10^{12}$ cm^{-2}, all exciton peaks are completely quenched in both PL and PLE. Instead an emission band originating from a free to free recombination is observed in addition to the FB emission at lower energies. These bands are upshifted in energy both in PL and PLE, when the doping level is further increased, due to bandfilling effects (E_F is raised in the first electron subband). A more detailed report of this case of degenerately Si-doped samples will be published separately [11].

ACKNOWLEDGEMENTS

The work performed at University of California at Santa Barbara was partially supported by the NSF Science and Technology Center for Quantized Electronic Structures (QUEST). We also acknowledge the Viktor and Erna Hasselblad Foundation for the financial support of the solid state Titanium Sapphire tunable laser.

REFERENCES

1. R.C. Miller, A.C. Gossard, W.T. Tsang, and O. Munteanu, Solid State Commun. 43, 519 (1982)
2. A.M. White, P.J. Dean, L.L. Taylor, R. C. Clarke, D.J. Ashen, and J.B. Mullin, J. Physics C5, 1727 (1972)
3. P.O. Holtz, M. Sundaram, K. Doughty, J.L. Merz, and A.C. Gossard, Phys. Rev. B40, 12338 (1989)
4. B. Monemar, P.O. Holtz, J.P. Bergman, C. I. Harris, H. Kalt, M. Sundaram, J.L. Merz, and A.C. Gossard, presented at the Int. Conference on Electronic Properties of Two-Dimensional Systems, Nara, Japan, July 1991
5. X. Liu, A. Petrou, B.D. McCombe, J. Ralston, and G. Wicks, Phys. Rev. B38, 8522 (1988)
6. W.M. Chen, P.O. Holtz, B. Monemar, M. Sundaram, J.L. Merz, and A.C. Gossard, Layered Structures - Heteroepitaxy, Superlattices, Strain and Metastability, ed. by L.J. Schowalter, F.H. Pollak, B.W. Dodson, and J.E. Cunningham, Materials Research Society Symposium Proceedings, Boston, MA, USA, 1989
7. P.O. Holtz, M. Sundaram, J.L. Merz, and A.C. Gossard, Phys. Rev. B40, 10021 (1989)
8. See e.g. E. Molva and N. Magnea, Phys. Stat. Sol (b) 102, 475 (1980)
9. S. Fraizzoli, F. Bassani, and R. Buczko, Phys. Rev. B41, 5096 (1990)
10. M. Stopa and S. DasSarma, Phys. Rev. B40, 8466 (1989)
11. C. I. Harris, B. Monemar, H. Kalt, and K. Köhler, submitted Phys. Rev. B

SCANNING TUNNELING MICROSCOPY STUDIES OF SEMICONDUCTOR SURFACE DEFECTS

J. E. Demuth
IBM Research Division, Thomas J. Watson Research Center
P.O. Box 218, Yorktown Heights, NY 10598

ABSTRACT

A perspective of recent scanning tunneling microscopy (STM) studies is presented that reveals the wealth of intrinsic and extrinsic defects arising on semiconductor surfaces. A brief review is given of the principles of STM, its limitations and related proximity probe methods. Specific applications considered include atom-resolved imaging, spectroscopy, carrier dynamics and ESR with the STM. Intrinsic defects on cleaved and annealed surfaces of Si and Ge, cleaved GaAs(110) and InSb(110), MBE grown GaAs(100) and epitaxial Ge on Si(111) are reviewed. Transition metal impurities found for both Si(100) and (111) surfaces as well as electrically active defects observed by STM on oxidized Si surfaces are discussed.

1. Introduction

In order to understand the relation of surface defects to the well-established field of bulk semiconductor defects, one must consider the current measurement capabilities required, the trends in semiconductor technology and the development of new methods to probe surfaces. An outsider can characterize the study of defects in semiconductors as a field which has focused primarily on bulk defects which can be well-characterized using a variety of solid state measurement techniques. This field has also considered non-bulk defects, such as arise at the important Si/SiO_2 interface. Such interface defects have been more complicated to study due to their lower density associated with reduced dimensionality. For most bulk defects, and particularly for interface defects, the overall low defect densities requires highly sensitive measurement methods together with methods or ways to produce sufficient defects for investigation.

Given the trend in semiconductor device technology to smaller structures, thinner active channels, heterostructures and band gap engineered materials, defects are likely to continue to play an important role /1/. In particular, interfacial defects and related defects in thin layers are likely to occur which are associated with stresses or the surface process involved in the fabrication of these materials. In addition to the possible technological need to identify and understand defects in these new materials and layered structures, new classes of defects may prevail. Here surfaces and interfaces may enhance the generation of specific bulk defects which may have altered properties associated with the different environment, local bonding and kinetics at surface and interfaces. Given the richness of nature, it is likely that many interesting defects at surfaces and interfaces will arise. Whether we can detect, characterize or relate these to important properties of these systems remains an open question.

New opportunities to examine and study defects on surfaces have recently arisen with the development of the scanning tunneling microscope (STM) /2/. This technique allows the direct imaging and interrogation of surface defects with atomic scale resolution. It is fair to say that the richness of surface phenomena unveiled by the STM has already exceeded early expectations and has markedly altered the way surface scientists envision surfaces /3/. With the refinement of the STM together with the extension of these atom-resolved techniques, further opportunities will arise to understand surface defects. Thus, both the challenge and opportunity to study surface defects exists, even though it is unclear whether surface defects will play a significant role in future solid state devices. The ability to observe and characterize surface defects for the first time is intriguing and likely holds many surprises.

In this paper I will briefly summarize some of the recent STM studies of semiconductor surfaces that reveal the richness of surface defects and provide references for interested readers. Where possible I have attempted to mention surface phenomena which can be related to bulk defects and their properties. A central theme of these discussions concerns the use of STM to investigate and understanding surface defects together with promising new STM capabilities such as ballistic electron emission microscopy /4,5/, spatially resolved carrier dynamics /6,7/ and ESR with the STM /8/. This paper is organized as follows: section 2 introduces the STM, its capabilities and limitations; section 3, intrinsic surface defects; section 4, transition metal impurities, section 5, compound semiconductors and section 6, oxidized silicon surfaces and spin detection with the STM.

2. STM and related proximity probes

Scanning tunneling microscopy (STM) was developed by Binnig and Rohrer /2/ and now has numerous related methods that probe surface electronic structure, forces, electrical potentials and chemical potentials with varying degrees of lateral resolution. In STM a fine tip is raster scanned along the surface by piezoelectric transducers with a few volts or less bias between the tip and sample. When the tip is of the order of 5 Å from the surfaces, tunneling occurs, and the measured tunnel current can be stabilized to a constant value during scanning using a feedback circuit. The feedback error signal extends or contracts the piezo to maintain a constant overlap of the tip-sample wavefunctions. This constant current topograph obtained reflects a contour of constant charge density of the surfaces /9/. The sp bonding of semiconductors, the localized nature of Si dangling bonds and their extension into the vacuum, provide rich topographic features in atomic scale STM images of semiconductors.

For semiconductors bias voltage are typically a few eV with tunnel currents set to $\sim$ 1/nA for Si and a factor 10 less for GaAs and other compound semiconductors which degrade at higher currents /10,11/. Since the bias voltage determines which occupied (or empty) electronic states of the semiconductors are sampled when tunneling out of (or into) the semiconductor, these topographs also contain spatial information about the density of states (DOS) at the surface /12/. Measurement of I/V curves and reduction to normalized conductance (dI/dv / I/V) provides spectra which closely resembles the surface DOS near the Γ point of the surface Brillouin zone. Such spectra can be obtained by momentarily interrupting the feedback loop during a scan /11/. Differential images at different energies /15/ or more generally atom resolved spectra pixel by pixel provide spectroscopic images of the surface electronic structure /16/. Such maps provide a way to identify the origin of new surface states associated with particular vacancies /17/, substitutional defects /18/ or metallic islands /19/, and can also provide fundamental insight to the bonding at surfaces /20/. The conduction and valence band edges, together with many surface states lie in the $\pm 3eV$ range easily accessed by the tip single bias. Higher tip-sample bias voltage sense image states /21/ and can also lead to tip instabilities.

One of the underlying problems in interpreting STM images is separating the positions of atoms, ie, surface geometry, from the electron density contours. In many cases even general knowledge about the nature of the electronic states, apriori, greatly simplifies interpreting images. For example, tunneling out of the valence band edge or into the conduction band edge of cleaved compound semiconductors allows one to image the anion or cation derived states. The observed change density contours can be related to detailed atom positions through theoretical calculations /10/. Generally after deposition of foreign atoms and/or high temperature annealling, semiconductor surfaces become greatly rearranged which precludes direct theoretical modeling. STM topographies and spatially resolved spectroscopy greatly complement classical techniques by defining overall local periodic structures and delineating local inhomogeneities. The ability of the STM to define growth, coverage or annealing conditions which produce single, well-defined homogeneous phases has proven to be extremely valuable for other quantitative macroscopic measurements and their meaningful analysis.

While the STM is the most developed and widely applied of the proximital probe methods, several others have been developed. These rely on sensing and feeding back on local forces (atomic force microscopy, AFM) /23/, local thermal conduction /24/, capacitance /25/, electrical /26/ and chemical /27/ potentials or current injected thru the surface layer into the substrate (ballistic electron emission microscopy, BEEM) /4,5/. In some cases, such as in STM spectroscopy, the feedback control is momentarily or periodically interrupted so that some other quantities can be measured. AFM's main advantage is for use on insulating samples and to minimize the electronic contributions to tunneling that tend to dominate tunneling when at atomic resolution. Atomic resolution has been achieved in AFM /28/ but does not appear to be as common as with STM. Using magnetic tips or tips with spin polarized states also allow magnetic properties to be measured /29/. BEEM allows the homogeneity and electronic structure of buried interfaces to be determined, but lacks atomic resolution due to defocusing of the injected electron beam /4,5/. Other methods such as capacitance microscopy /25/ or thermal microscopy /24/ require larger tips for sensing and have not yet yielded atomic resolution.

3. Intrinsic semiconductor surface defects

All surfaces, either cleaved or thermally annealed are defective with respect to the bulk. The way the surface is formed determines its overall perfection and the new surface structures that form. Cleavage along a lattice plane results in a sheet of half occupied, broken bonds. Given the higher energy to maintain these dangling bonds, surface lattice distortions arise, called reconstruction which reconfigure bond charges to form pair bonds and minimize the electronic energy. III-IV and II-V compound semiconductors produce flat, idealized cleavage plans along (110) directions. The atoms on these surfaces show small but periodic displacements of anions and cation pairs of the outermost atoms /10/. Si and Ge cleave along <111> directions and produce a markedly different bonding configuration forming π-bonded chains of surface atoms /14/. In these, periodic lattice distortion leads to new double spaced superstructures. The most prevalent defects existing on cleaved surfaces are due to the occurrence of steps and domains. Given the 3 fold symmetry of the (111) surfaces, three possible cleave domains can form. On Si(111) numerous 2x1 domain boundaries and atomic steps are observed. Sample quality and cleavage techniques appear to be the most important in altering surface domain production and step density /10/.

More relevant to most practical semiconductor surfaces are the even lower energy equilibrium reconstructions which arise upon high temperature annealing at elevated temperatures. Si and Ge reconstruct to form 7x7 and 2x8 superstructures at 890° and 300°C, respectively. The 7x7 structure consists of a very stable arrangement of several complex local structures including ad-atoms (Si atoms atop 3 fold coordinated surface silicon atoms which otherwise would have dangling bonds), Si dimer bonds, several missing Si atoms or "corner holes" and stacking faults arranged within the top two Si bilayers of the surface /30/. Since this superstructure can form independently at different locations on the surface, numerous misfit boundaries between different domains arise. These boundaries appear to be sinks for impurity atoms and also serve as sites for subsequent nucleation and growth /31/. Recent studies of the transition from the 2x1 to 7x7 reconstruction reveals defect formation in the surface associated with the removal of some Si atoms to form the higher density 7x7 reconstruction /32/. Ge(111) is remarkably simpler than Si and reconstructs to form ad-atoms atop the surface which order approximately every second lattice position but predominately form a 2x8 superstructure /33/. Surface domain boundaries tend to permeate this surface due to the small energy differences between the different local orderings of the Ge ad-atom. The trade off in local strain energies and the reduction in electronic energy by various local bonding configurations can readily account for these different structures /34/. Interestingly epitaxial Si grown on Si(111) 7x7 at lower temperatures can assume a 5x5 reconstruction analogous to the 7x7 structures due to the different stress at lower growth temperatures /30/. Local stress from the substrate also encourages epitaxial Ge to grow on Si(111) in a 7x7 like superstructure initially, a 5x5 at higher coverages and then revert to the bulk-like 2x8 structure for thick layers /36/. Steps on both Si(111) 7x7 and Ge(111) 2x8 are well-defined and have reproducible structures /35/.

Thermally annealed Si(100) surfaces show a simpler form of reconstruction where Si atoms on each layer pairs together to form simple dimers which form double spaced rows along the surface /37/. The dimerization on each layer rotates by 90° between layers due to the orthogonal projection of the tetrahedial bonds along each (100) layer. This bond rotation produces a variety of step terminations and interesting epitaxial growth habits on this surface. Antiphase boundaries are only observed in low temperature growth of Si in Si(100). They act as nucleation sites for subsequent layer formation and promote multilayer growth instead of layer by layer growth /38/. As subsequent layers grow, the initial defects on the starting surface remain in the underlying layers with improved perfection /37,38/ - the well known regrowth procedure for MBE growth on Si(100).

One of the current complications in understanding surface defects of thermal annealed samples is distinguishing intrinsic surface defects from extrinsic defects. Unfortunately, STM lacks a quantitative experimental approach to identify elements. As sample cleaning procedures and UHV system cleanliness have improved over time, defect densities have become reduced. Some extrinsic defects associated with dopants and metal impurities have been identified and are discussed in section 4. Bulk p-type dopants have also been invoked to explain missing atomic features, such as for example adatoms on Si(111) 7x7 surfaces, since when substituted for Si atoms can reduce valence bond charge and can eliminate expected occupied dangling bonds /39/. Studies of As and Sb terminated surfaces confirm these general ideas of conservation of valence bonding and electron counting /40/. However, details of local strain fields and competing lower energy structures produce alternate, unexpected new structures and arrangements of valence bond charge. Several point defects in elemental semiconductors have shown time dependent changes in atomic positions at room temperature, ie. atoms hopping between two bonding positions /41/. In contrast, atoms near steps always tend to be stable. Such metastable structures may also be induced by the local fields or forces associated the with probe tip. STM studies at temperatures up to 300°C have revealed atomic motion and the dynamic nature on both Si and Ge surfaces, even concerted motions of islands /42/!

Electrical properties of defects on the technologically important Si(100) surface have been studied using tunneling spectroscopy /3/. Certain defects produce asymetric charge densities on Si dimers producing alternating buckled dimers /37/. Pairs of half dimers are frequently observed and produce states near the Fermi level /3/. Carrier recombination centers have been deduced by measuring the induced photovoltage with the STM tip on different defects /6,7/ and for epitaxial structures on the (111) surfaces /43/. On Si(100), certain arrangements of atoms at steps, particular terminations of epitaxial Si islands and the pairs of half dimer defects provide strongly enhanced surface recombination /44/.

4. Transition metal impurities

Transition metal impurities have long plagued studies of thermally annealed elemental semiconductors surfaces with small impurity levels in many areas producing totally new surface reconstructions. Nickel from stainless steel components in vacuum systems has produced impurity stabilized reconstruction on both Si(111) and Si(100) surfaces /45,46/. The presence of these impurities on the order of 0.5 to 2% have been determined with other chemically sensitive surface methods such as Auger spectroscopy. On Si(100) this Ni impurity manifests itself as pairs of dimer vacancies which at higher concentrations align between several neighboring rows of dimers to form a 2x8 periodicity /45/. On Si(111) a new $\sqrt{19}$ x $\sqrt{19}$ arrangement of Si atoms is observed. In neither case are Ni atoms believed to be directly observed in STM since the Ni s-states are diffuse and have a low DOS while the Ni d-states are too spacial contracted to be directly involved in tunneling. In various STM studies of Si(111) samples, small quantities of metal impurities frequently appear as if they were swept out of the local 7x7 structures after annealing to produce donut shaped structures approximately 6 Å in diameter at or near 7x7 domain grain boundaries. Pd Mo, Ni and Co impurities all appear to produce similar donut shaped grain boundary impurities /47,48/. Pd and Mo impurities not only get concentrated or squeezed into the boundaries between different

7x7 domains but also get pushed into the inside lower terraces of stepped Si(111) surfaces /47/.

Recent STM studies of low coverages (0.02 monolayer) of cobalt on Si(111) annealed to 600°C has revealed a single phase of these disordered donut structures all centered over similar lattice sites /48/. Medium energy, ion channeling and blocking experiments on this same system has revealed the location of the cobalt atoms to be in substitutional sites with six silicon atoms atop the cobalt and bonded to adjacent Si atoms /48/. Interestingly these-ring like structures are also generally observed near the perifery of Pd_2Si islands formed after low temperature deposition and appear to be involved in surface diffusion of Pd at lower temperatures /47/.

In addition to this novel ring structure other defects are observed on metalized Si surfaces. Different epitaxial silicides grown on Si(111) show varying degrees of perfection which are associated with local stress arising from lattice mismatch. Top surfaces of some silicides exhibit large numbers of missing local stress arising from atom defects /47/ or silicon ad-atom superstructures depending upon coverage and temperature /51/. Lattice defects have also been observed and characterized for several non transition metals. Missing atoms as well as Si substitutional defects have been spectroscopically characterized for the $\sqrt{3} \times \sqrt{3}$ Al ad-layer structure on Si(111) /18/. These silicon atoms provide an extra electron and produce localized states near the Fermi level.

5. Compound semiconductors

As noted earlier, compound semiconductors produce uniform step free (110) cleavage planes. Local defects indicative of missing atoms arise on both GaAs and In Sb (110) surfaces /11,51/. In particular excess charge and its local redistribution near vacancies have been used as evidence for anion vacancies in InSb(110) while larger vacancies with minimal charge redistribution are taken as Schottky defects /12/. The most dramatic manifestation of charge redistribution and transfer on semiconductors arises for O ad atoms on GaAs /51/. Bulk dislocations have been generated in compound semiconductors by straining the crystal /53/. Examination of the resulting cleaved surfaces shows the expected structure of various dislocations at the surface. Interestingly no evidence of charging at these GaAs dislocations have been observed. Surface dislocations have also been observed on epitaxial Ge on Si(111) /35/. A network of dislocations at the interface of thin Ge layers on Si(111) is visible in the STM topograph of the Ge surface, presumably via the strain arising at the interface /54/.

Recent in-situ STM studies of MBE grown GaAs(100) /55/ have revealed highly imperfect, irregular structures with well defined subunits. These units on average combine to form the long range periodicities seen by diffraction methods.

6. Oxidized silicon surfaces

UHV oxidation of silicon in the monolayer regime results in an atomically structureless undulating surface topography even starting from atomically flat Si(100) surface /56/. The small spacings between Si-O bonds together with the amorphous nature of SiO/SiO_2 are envoked to account for the lack of atomic features. Despite the lack of atomic scale features, two striking electronic properties have been noted in STM work on these surfaces. The first is the temporary trapping of charge within the oxide layer at specific sites /56,57/. Here the coulomb field of a trapped electron blocks tunneling and markedly reduces the tunnel current. After a milli to microsecond the electron is released and the tunnel current returns. The trapping and decay rate of this process is usually sufficiently rapid that the fluctuations are integrated by the feedback loop and do not show up in the topographs. Single and double electron traps are found, and trapping sites observed generally clustered together in different regions of the sample and are distributed at different depths in the oxide layer /56/. The second property is the occurrence of RF noise at special atomic locations on oxidized Si(111) /8/. This property is attributed to the precession of individual paramagnetic centers in the external magnetic field which modulates the tunneling signal. Given the significance of detecting individual spins with STM and the wide use of ESR in semiconductor defect studies we discuss this phenomena in more detail. Such spin resolved STM may also provide additional chemical and structural information than currently possible with STM alone.

A simple classical argument suggests that the local magnetic field ~ 1 Å away from an isolated electron spin should produce a Lorentz force on the tunneling electrons which is comparable to their electrostatic field and thereby deflect the tunneling electrons. The Larmor precession of the magnetic dipole arising in a static magnetic field will then produce a modulation in the tunnel current at the Larmor frequency. The magnitude of this modulation is unknown since the detailed coupling mechanism between the spin and tunneling electron is certainly more complex than the classical argument suggests. For example, indirect magnetic interactions of the dipole with the substrate DOS may arise /58/ or direct spin-orbit interactions may occur /59/. Another important point that pervades all measurements of isolated quantum systems is that the tunneling electrons perturb and change the quantum state of the system. One can consider, the extra RF impedance at the Larmor frequency seen in our tunneling detector to be quantum mechanically derived from the dephasing process or excitations between the tunneling electrons and the local spin.

In our initial experiments the RF system allowed the detection of a 0.25nA RF modulation /8/. One of the difficulties of these experiments is the infrequency of observation of these RF sources on the surface. The overall response of the RF noise spectrum analyzer was relatively slow making scanning both frequency and position tedious. Recent work by Welland et. al. /60/ has improved the detection sensitivity by adding a modulation coil to modulate the magnetic field and has used lock-in detection techniques to increase signal to noise. This work also formed a high density of reproducible signals by depositing large organic free radical molecules on surfaces.

One can also ask why free spins are not observed on other semiconductor surfaces. The absence of STM spin precession from the high density of the dangling bonds on clean Si(111) 7x7 is reasonable given the fact that ultra-sensitive ESR studies conclude that these dangling bonds are too strongly interacting with one another to be observed /61/. Similarly, the dangling bond defects of Si in the $\sqrt{3} \times \sqrt{3}$ Al/Si(111) structure is expected to produce free spins /18/ and were also investigated in these early studies /8/. The absence of RF signals from these local dangling Si bonds can again be attributed by their high density which allow them to interact with one another which presumably broadens the "signal".

Given the magnetic field dependence and reproducibility of the Larmor signals seen on oxidized silicon, it is likely that the signal arises from a paramagnetic center in the oxide. Several possible spin centers in SiO_2 interfaces are known to exist - most notably the P_b center /62/. However, the occurrence of the spin signal at irregular structures in the oxide suggest the possibility of a trapped electron or superoxide (O^2) species in a dislocation. One limitation of these first measurements is that the uncertainties in the magnetic field ($\sim \frac{1}{2}\%$) were sufficient to prevent chemical use of the determined g value. Also, the high local electric fields of the tip near a local magnetic moment is expected to modify the determined g-value. Thus, although there are still many open issues and questions about probing free spins with the STM, the potential exists.

7. Conclusions

Recent advances in surface science, most notably the development and use of the STM, has shown the large variety and degree of defects existing at semiconductor surfaces. Many analogies to bulk defects and properties for vacancies, anti-site defects, charge trapping, metastability, and transition metal impurities at surfaces have been observed. As the STM is gradually developed and improved as a physical probe to measure properties, more detailed and specific information about such defects is likely to become available. Initial STM work to probe individual paramagnetic centers in semiconductors is encouraging but primitive with many improvements still possible.

8. Acknowledgements

The author wishes to thank M. Azzaro for the preparation of this manuscript as well as his numerous collaborators who have participated in much of the work cited in this review.

9. References:

1. D.V. Lang, Materials Science Forum, Vol **38-41**, 13 (1989).
2. G. Binning and H. Rohrer, Helv. Phys Acta **55**, 762 (1982.
3. See for example, J.E. Demuth, in <u>Between Science and Technology</u>, Ed. A. Sarlemijn and P. Kroes (North Holland, 1990) p. 57.
4. W.J. Kaiser and L.D. Bell, Phys. Rev. Lett. **60**, 1406 (1988).
5. M. Prietsch and R. Ludeke, Phys. Rev. Lett, **66**, 2511 (1991).
6. R.J. Hamers and K. Markert, Phys. Rev. Lett. **64**, 1051 (1990).
7. Y. Kuk, R.S. Becker, P.J. Silverman and G.P. Kochanski, Phys. Rev. Lett. **65**, 456 (1990).
8. Y. Manassen, R.J. Hamers, J.E. Demuth and A.J. Castellano, Jr., Phys Rev. Lett. **62**, 2531 (1989).
9. J. Tersoff and D.R. Hamann, Phys. Rev. B **31**, 805 (1985).
10. R. Feenstra, J.S. Stroscio, J. Tersoff and A.D. Fein, Phys, Rev. Lett. **58**, 1192 (1987).

11. L.J. Whitman, J.A. Stroscio, R.A. Dragoset and R.J. Celotta, J. Vac Sci and Technol, **B9**, 770 (1990).
12. R.J. Hamers, in Annual Review of Phys. Chem., Ed. H.L. Strauss, G.T. Babcock and C.B. Moore, (Annual Reviews, Inc; Palo Alto, Ca), V **40**, 531 (1989).
13. N.D. Lang, Phys. Rev. B **34**, 5947 (1986).
14. J. Stroscio, R. Feenstra and A.P. Fein, Phys. Rev. Lett. **57**, 2579 (1986).
15. R.J. Hamers, R.M. Tromp and J.E. Demuth, Phys. Rev. Lett. **56**, 1972 (1986).
16. Ph. Avouris and R. Wolkow, Phys. Rev. B **39**, 5091 (1989).
17. J.E. Demuth, R.J. Hamers and R.M. Tromp, MRS Symposium Proceedings, Vol. **77**, 1 (1987).
18. R.J. Hamers and J.E. Demuth, Phys. Rev. Lett. **60**, 2527 (1988).
19. R. Feenstra and P. Martensson, Phys. Rev. Lett. **61**, 447 (1988).
20. J. Northrop, Phys. Rev. Lett., **57**, 154 (1986).
21. R.S. Becker, J.A. Golovchenko, D.R. Hamann and B.S. Swartzentruber, Phys. Rev. Lett. **55**, 2032 (1985).
22. See for example, H.K. Wickramasinghe, J. Vac. Sci and Technol. **48**, 363 (1990).
23. G. Binnig, C.F. Quate and Ch. Gerber, Phys. Rev. Lett. **56**, 930 (1986).
24. C.L. Williams and H.K. Wickramasinghe, Appl. Phys. Lett. **49**, 1587 (1986).
25. Y. Martin, D.W. Abraham, H.K. Wickramasinghe, Appl. Phys. Lett. **52**, 1103 (1988).
26. P. Muralt, D.W. Pohl and W. Denk, IBM Jour. of Res. and Div. **30**, 443 (1986).
27. C.C. Williams and H.K. Wickramasinghe, J. Vac. Sci and Technol. **B9**, 537 (1991).
28. E. Mayer, H. Heinzelmann, H. Rudin and H.J. Guntherodt, J. Vac. Sci. and Technol. **B9**, 1329 (1991).
29. R. Wiesendanger and H.J. Guntherodt, R.J. Gambino and R. Ruf, Phys. Rev. Lett. **65**, 247 (1990).
30. K. Takayanagi, Y. Tanishiro, M. Takahashi, H. Motoyoshi and K. Yagi, J. Vac. Sci and Technol. **A3**, 1502 (1985).
31. see for example, U. Koehler, J.E. Demuth and R.J. Hamers, J. Vac. Sci. Technol. **A7**, 2860 (1989).
32. R.M. Feenstra and M.A. Lutz, Surface Sci **243**, 243 (1991).
33. R.S. Becker, J.A. Golovchenko and B.S. Swartzentruber, Phys. Rev. Lett. **54**, 2678 (1985).
34. R.D. Meade and D. Vanderbilt, Phys. Rev. B **40**, 3905 (1989).
35. See for example, J.E. Griffith, J.A. Kubby, P.E. Wierenga, R.S. Becker and J.S. Vickers, J. Vac. Sci. Technol. **A6**, 493 (1988).
36. U. Koehler, O. Jusko, G. Pietsch, B. Müller and M. Henzler, Surface Sci. **248**, 321 (1991).
37. R.M. Tromp, R.J. Hamers and J.E. Demuth, Phys. Rev. Lett. **55**, 1303 (1985).
38. R.J. Hamers, U.R. Koehler and J.E. Demuth, J. of Ultra Microscopy **31**, 10 (1989).
39. J.E. Demuth, R.J. Hamers, R.M. Tromp and M.E. Welland, IBM Jour. of Res. and Dev. **30**, 396 (1986).

40. R.D. Bringans, R.I.G. Uhrberg, R.Z. Bachrach and J.E. Northrop, Phys. Rev. Lett. **55**, 533 (1985).
41. J.A. Golovchenko and R.S. Becker, Nature **325**, 417 (1987).
42. R.M. Feenstra, A.J. Slavin, G.A. Held and M.A. Lutz, Phys. Rev. Lett. **66**, 3257 (1991).
43. D.G. Cahill and R.J. Hamers, Phys. Rev. B **44**, 123 (1991).
44. D.G. Cahill and R.J. Hamers, J. Vac. Sci. and Technol. **B9**, 564 (1991).
45. R.J. Wilson and S. Chiang, Phys. Rev. Lett. **58**, 2575 (1987).
46. H. Niehus, U.K. Hoehler, M. Copel and J.E. Demuth, J. of Ultra Microscopy **152**, 735 (1987).
47. U. Koehler, R.J. Hamers and J.E. Demuth, private communcations.
48. P.A. Bennett and M. Copel, to be published.
49. U. Koehler, J.E. Demuth and R.J. Hamers, Phys. Rev. Lett. **60**, 2499 (1988).
50. J. Rowe and R. Becker, private communications.
51. R.M. Feenstra and A.P. Fein, Phys. Rev. B **32**, 1394 (1985).
52. J. Stroscio, R.M. Feenstra and A.P. Fein, Phys. Rev. Lett. **58**, 1668 (1987).
53. G. Cox, D. Szynka, U. Poppe, K.H. Graf and H. Urban, Phys. Rev. Lett. **64**, 2405 (1990).
54. G. Meyer, B. Voigtlander and N.M. Amer, to be publsihed.
55. M.D. Pashley, K.W. Haberern and J.M. Gaines, J. Vac. Sci. and Technol. **B9**, 938 (1991).
56. R.M. Koch and R.J. Hamers, Surface Science **181**, 333 (1987).
57. M.E. Welland and R.H. Koch, Appl. Phys. Lett. **48**, 333 (1987).
58. R. Mezei and A. Zawadowski, Phys. Rev. B**3**, 187 (1971).
59. D. Shackal and Y. Manassen, to be published.
60. A.W. McKinnon and M.E. Welland, Proceedings of 1991 International Conference on STM.
61. K. Baberschke, private communications.
62. E.H. Poindexter, G.J. Gerardi, M.E. Rueckel and P.J. Caplan, J. Appl. Phys. **56**, 2844 (1984).

THE ATOMIC AND ELECTRONIC STRUCTURE OF ORDERED BURIED B(2 × 1) LAYERS IN Si(100)

M. Needels, M. S. Hybertsen, and M. Schluter
AT&T Bell Laboratories, Murray Hill, NJ 07974, USA

ABSTRACT

It has recently been demonstrated that one half monolayer of boron forms a (2 × 1) reconstruction on the Si(100) surface and this periodicity is preserved after subsequent epitaxial Si overgrowth. The atomic geometry and electronic band structures of an ideal half monolayer of substitutional boron are determined by first principles total energy calculations. Two predictions are made. First, the Si epi layer is displaced ~0.35 Å with respect to the substrate. Second, the hole induced by each boron is distributed across three distinct deep impurity bands.

1. Introduction

It has recently been found in semiconductors that the symmetry of certain ordered adatom reconstructions can be preserved when epitaxial overlayers are grown at low temperatures.[1] Consequently, the overgrowth results in a "δ-doped" buried layer of impurity atoms with a high degree of lateral translational order. We study here the system of boron on Si(100) which has recently been prepared in the laboratory. Low-T (~300°C) Si overgrowth over B on a Si(100), in contrast to Si(111),[2] surface preserves not only the original boron order, but also results in a crystalline instead of an amorphous Si overlayer. This system is thus a good candidate for producing electrically-active ordered doping layers in crystalline silicon with the potential for high carrier mobility.[2,3]

2. Calculational Method

We have performed *ab initio* total energy calculations of substitutional boron both in the Si(100)-(2 × 1) surface and in a (100)-(2 × 1) buried layer using methods described elsewhere.[4] In brief, the calculations are done within the Density Functional Theory framework, using the Local Density Approximation (LDA) with the Perdew-Zunger parameterization of the Ceperley-Alder electron gas correlation energy. We use norm-conserving, separable, non-local pseudopotentials, optimized for convergent plane wave expansions. The cutoff energy for the plane wave expansion is 20 Ry. For the Si(100) surface, we use a slab supercell geometry ten layers thick and hydrogen terminated on one side. For the buried boron layer, we use a supercell twenty atomic layers thick in the (100) direction. This amounts to ~ 9000 plane waves with the 20 Ry cutoff. Eight k-points in the irreducible quarter of the (2 × 1) Brillouin zone are used to construct the charge density and 45 k-points are used to compute the Fermi surface. The self-consistent LDA equations are solved using on an iterative modified Car-Parinello scheme.

3. Results and Discussion

We first discuss the results of our calculations on the Si(100) surface, which forms a (2 × 1) "dimer" reconstruction.[5] When boron is deposited on the on this surface, a new reconstruction occurs with a similar (2 × 1) periodicity.[1] This pattern is found to be stable with 1/2 monolayer boron coverage and becomes increasingly disordered at higher coverages. Using this information, we conduct a limited search, in order to understand how an ordered, buried δ-doped layer might be formed, for the equilibrium substitutional configuration of the boron (2 × 1) covered Si(100) surface. We find that dimer formation remains favorable. However, some sites are much lower in energy than others. Figure 1 shows a side view of the Si(100) surface with the sites where we considered a substitutional boron atom labelled A-D. We find the sub-surface position B to be 0.11, 0.65, and 0.40 eV lower in energy than sites, A, C, and D, respectively. We thus believe that boron occupies a subsurface position on Si(100), in analogy to Si(111).[6] No experimental data are available yet for the boron position on the Si(100) surface. Since we find position A only slightly higher in energy, it is likely that an experimentally prepared surface (30 sec anneal at 450°C) contains some disordered mixture of A and B positions. Nevertheless, high quality δ-doped buried layers are more likely to be grown because the lowest energy position for the substituitional boron atom is subsurface in contrast to a site on the surface.

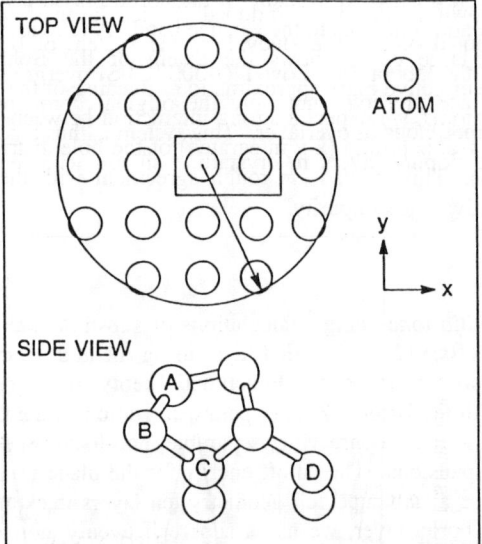

Fig. 1 Sketch of the structural arrangement on Si(100) (2 × 1). The top figure shows a (2 × 1) unit cell with a boron acceptor Bohr-radius superimposed. The bottom figure shows a side view of the surface with individual atomic sites for boron substitution labelled A-D.

We now discuss the buried boron layer, beginning with the geometric structure. For our calculations, we assume a perfect substitutional (2 × 1) pattern for a half monolayer of boron buried in crystalline silicon, but allow for both atomic and volumetric relaxation. Consequently, we have one boron atom in our unit cell. This model of the buried layer is reasonable for the following reasons. It has been found from transmission electron diffraction and from grazing incidence x-ray diffraction that an epitaxial layer of Si can be grown at 300°C on this surface without qualitative change of the (2 × 1) pattern.[3] Segregation studies using Auger spectroscopy indicate that at least 50% of the boron remains in the original ordered layer. Finally, channeling studies suggest substitutional boron sites.[3] Our most important geometric result is a large contraction of the interlayer spacing in the z-axis direction. Because boron has a covalent radius 0.29 angstroms smaller than that of silicon, we expect a sizable reduction in volume/atom near the boron layer. However, the (2 × 1) symmetry constrains the x-axis and y-axis inter-layer spacings to remain at their bulk values but allows the z-axis spacings to change. We calculate a rigid body displacement of the Si layers, upon crossing the boron layer, of $\Delta z = -0.35 \text{Å}$, with the distortion confined to two to three layers on each side. Beating patterns seen in grazing incidence x-ray scattering[7] can be fit with $\Delta z = -0.45 \pm 0.1 \text{Å}$, with the distortion extending over fewer than 4 layers on either side, which is in excellent agreement with the theoretical predictions.

Finally, we discuss the electronic structure of the boron buried layers. Each boron atom in the δ-doped layer adds one hole to the valence band. A top view of this layer is shown in Figure 1. The large circle shows the extent of the Bohr radius of the hole wavefunction in the effective mass approximation. Because of the large overlap between neighboring boron atoms, we expect a strong interaction between their electronic states. Figure 2 shows the z-axis profile (x-y integrated) of the hole distribution, which extends over ~7Å (FWHM). This result is in rough agreement with the (9.4Å) width of an isolated boron effective mass impurity.

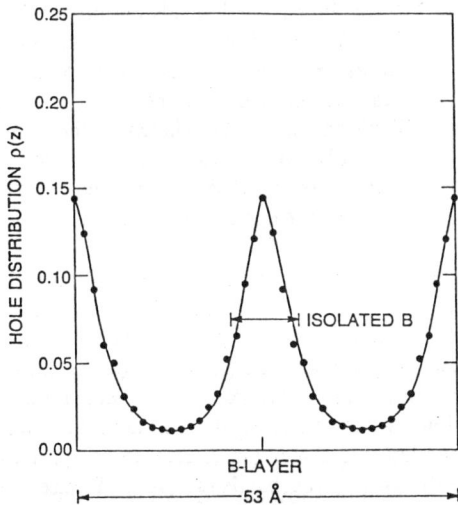

Fig. 2 Calculated hole distribution profile along the (100) direction, integrated over the (2 × 1) unit cell. The width at half maximum is comparable to that of an isolated boron impurity.

Because the wavefunctions of the (2 × 1) ordered boron atoms overlap strongly in the x-y plane, a deep level 2-D impurity bandstructure results instead of a shallow impurity level. Furthermore, the anisotropic crystal field ($x \neq y \neq z$) splits the p-like states into three bands with different masses. The bandstructure is shown in Figure 3, where the Fermi-energy is the zero of energy. As usual, the projected Si bulk bandstructure is superimposed onto the impurity bandstructure. The Fermi-level lies slightly above the valence band maximum, consistent with the experimental finding that 100% of the boron dopants are electrically active.[3]

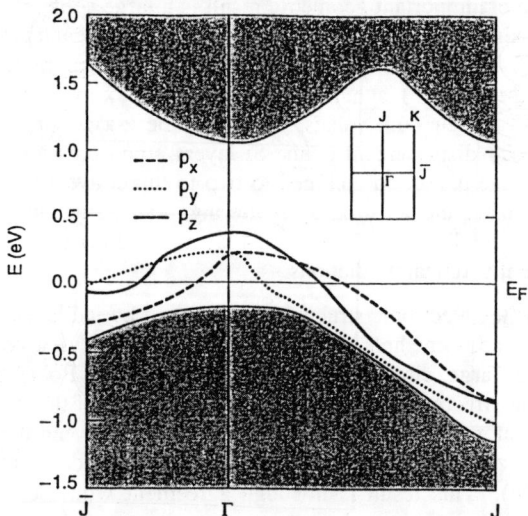

Fig. 3 Two-dimensional bandstructure of Si(100) (2 × 1) B. The boron induced impurity bands are coded according to their p_x (dashed line), p_y (dotted line), or p_z (full line) character. Overlayed is the projected Si bulk band-structure, with the bulk conduction band rigidly shifted so the gap matches the experimental value.

It is instructive to further analyze the impurity bandstructure. Figure 4 shows charge density contour plots of p holes in the individual three impurity bands (at $\mathbf{k} = 0$). The contours indicate p-like functions, strongly localized at the boron atoms. At $\mathbf{k} = 0$ the crystal field places the p_z orbital above the p_x orbital (along the short (2 × 1) direction) and the p_y orbital, which are nearly degenerate. As expected, the dispersion of the planar p orbitals is stronger along the orbital direction (σ) than perpendicular to it (π). Because of the strong orbital overlap this dispersion difference is approximately independent of short or long (2 × 1) direction. The bandstructure in Figure 3 is coded to reflect the orbital character and band crossings. Keeping in mind that this is an LDA bandstructure, there is some uncertainty in the position of the gap states with respect to the band edges.

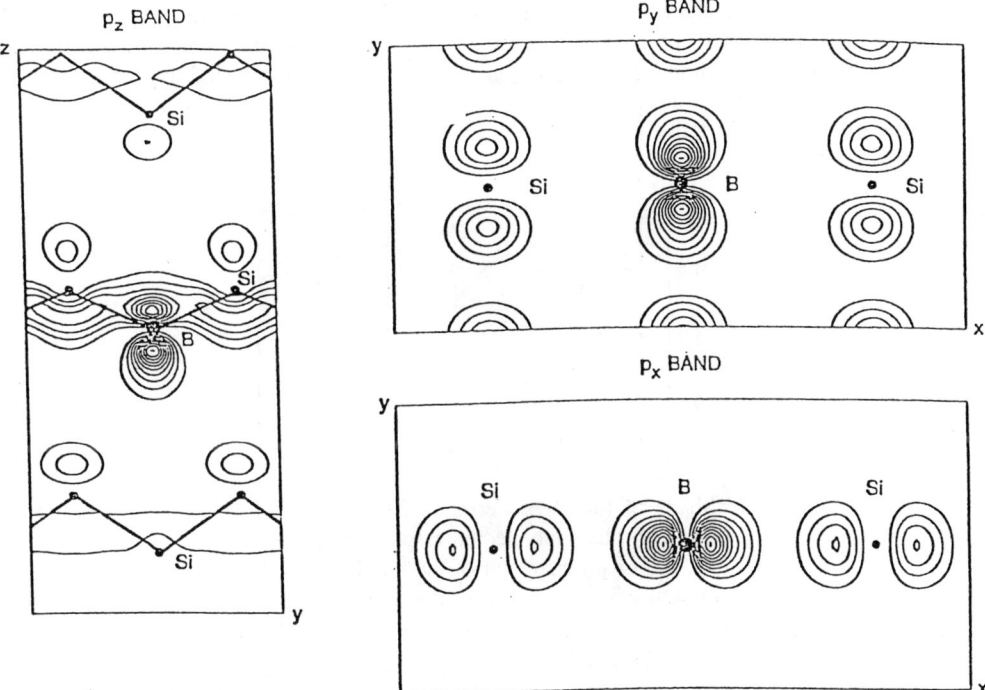

Fig. 4 Charge density contour plots of hole carrier distributions arising from the three impurity bands at k = 0 in Figure 3.

As seen from Figure 3 the Fermi-level crosses all three impurity bands giving rise to three separate sheets of the Fermi surface with varing orbital character (see Figure 5). There exists the possibility for a series of optical interband transitions in the 0.5-1.0 eV range. Carriers are all hole like and strongly confined (within ~4Å) to the 2D dopant layer. Their effective masses in the x, y plane are very anisotropic. Along the dispersive σ-directions the p_x, p_y masses are comparable to the Si light hole mass, while along the π-direction they are very heavy. The p_z mass is more isotropic and similar to the Si heavy hole mass.

We are currently investigating how much residual boron disorder will limit electron mobilities. The confined 2D nature of the hole wavefunction strongly samples this disorder. More knowledge about the disorder will help to understand details of the impurity scattering,[8] and hopefully suggest avenues to increase mobilities that could lead to new improved semiconductor devices.

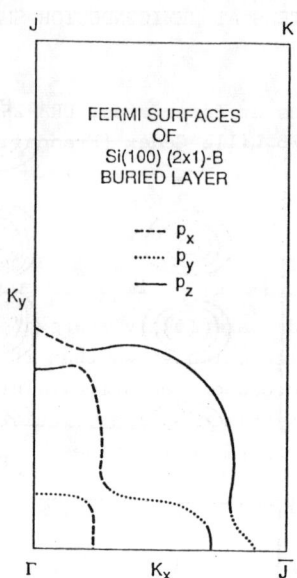

Fig. 5 Shape of the three Fermi-surface sheets in the (x-y) plane labelled corresponding to the p_x, p_y, or p_z character of the bands, shown in Figure 3.

Acknowledgements

We are grateful to R. L. Headrick, B. E. Weir, L. C. Feldman, and J. R. Patel for valuable discussions and making their data available prior to publication. We acknowledge a grant of computer time from the San Diego Supercomputer Center, where some of these calculations were performed.

References

1. R. L. Headrick, L. C. Feldman, and I. K. Robinson, *Appl. Phys. Lett.* **55** , 442 (1989).
2. R. L. Headrick, B. E. Weir, A. F. J. Levi, B. Freer, J. Bevk, and L. C. Feldman, *J. Vac. Sci. Tech. A* **9** , 2269 (1991).
3. R. L. Headrick, B. E. Weir, A. F. J. Levi, D. J. Eaglesham, and L. C. Feldman, *Appl. Phys. Lett.* **57** , 2779 (1990).
4. M. Needels, J. D. Joannopoulous, Y. Bar-Yam, and S. T. Pantelides, *Phys. Rev. B* **43** , 4208 (1991).
5. R. E. Schlier and H. E. Farnsworth, *J. Chem. Phys.* **30** , 917 (1959).
6. R. L. Headrick, I. K. Robinson, E. Vleig, and L. C. Feldman, *Phys. Rev. Lett.* **63** , 1253 (1989); P. Bedrossian, R. D. Meade, K. Mortensen, D. M. Chen, J. A. Golovchenko, and D. Vanderbilt, *ibid.* , 1257 (1989); I. W. Lyo, E. Kaxiras, and Ph. Avouris, *ibid.* , 1261 (1989).
7. B. E. Weir, R. L. Headrick, and L. C. Feldman, to be published.
8. A. F. J. Levi, S. L. McCall, and P. M. Platzman, *Appl. Phys. Lett.* **54** , 940 (1989).

NEGATIVE U SYSTEMS AT SEMICONDUCTOR SURFACES

G. ALLAN AND M. LANNOO
Laboratoire d'Etudes des Surfaces et Interfaces, URA 253 CNRS,
ISEN, 41, Boulevard Vauban, 59046 Lille Cedex (France)

ABSTRACT

General arguments are presented which tend to show that near semiconductor surfaces, adatom-substrate bonds are likely to exhibit negative U behavior. A direct simple tight-binding calculation shows that it would be notably the case of gold and alkali atoms adsorbed on GaAs(110). This explains the pairing of adatoms as observed in STM experiments. The calculated ionization levels also are in good agreement with the experimental values.

1. Introduction

A controversy still exists to explain the formation of the Schottky barrier. The defect model assumes that the metal Fermi level is pinned by deep levels due to defects (for example antisites) located near the metal-semiconductor interface. The other popular model assumes that the Fermi level is pinned by the Metal Induced Gap States close to a so called 'Charge neutrality Level'. From the experimental point of view, scanning tunneling microscopy (STM) experiments give informations not only on the atomic structure of the adsorbate and on the adsorption site but also on its electronic structure [1-6]. The STM current versus voltage exhibits two peaks respectively close to the conduction and the valence bands characteristic of the adsorbate. These peaks would correspond to positive or negative ionization of the adatom depending on the sign of the current between the STM tip and the sample. Photoemission [7-9] measurements show the same picture of the occupied states as STM. High resolution electron energy loss spectroscopy has also been used to get the excitation spectrum [10].

On the other hand, theoretical calculations have tried to determine the stable adsorption site and the electronic density of states close to the semiconductor band gap [11-15]. Due to the odd number of electrons on the adatom, the theory predicts that the Fermi level is pinned by the adatom-substrate bonding state in a high density of states energy region. This has never been observed, the surface being non metallic at low coverages [1,7,9-10]. Pairs, clusters or chains of adatoms are also observed by STM which tends to support the idea of an attractive interaction between the adatoms.

In the first section, starting from the hydrogen molecule we show that a monovalent atom adsorbed on the empty Ga dangling bond on the GaAs(110) surface is a good candidate to get a negative U center. The concept of negative U was first introduced at a general level by Anderson [16] and was observed for the vacancy in Silicon [17-18]. Since that time, it has been commonly accepted for numerous point defects in semiconductors.

Then in the following section, we develop a simple tight-binding model of the adsorption on a Ga dangling bond. It allows to calculate the bond ionization energies which are compared to the STM measurements. Finally we discuss some other cases where the same effect could occur.

2. The negative U center

Let us consider a bond between two atoms like it occurs in molecular hydrogen [19]. The total energy of the bond $E_{tot}(N)$ is a function of the number N of electrons which occupy the bond and which can be equal to 0, 1 or 2. The ionization level $\epsilon(N+1,N)$ is defined by

$$\epsilon(N+1,N) = E_{tot}(N+1) - E_{tot}(N) \tag{1}$$

which obviously is equal to minus the ionization energy from the bond in state N+1 to state N. In our case (N=0, 1 or 2) two levels $\epsilon(2,1)$ and $\epsilon(1,0)$ can be defined and the system effective U^* is given by

$$U^* = \epsilon(2,1) - \epsilon(1,0) \tag{2}$$

The U^* value is fixed by two contributions. The first one is the electron-electron interaction U_e which only exists for N=2 and is positive. This is the only contribution in the case of free atoms. For covalent bonds, there is an other contribution due to the fact that the bond strength increases with N. This adds a negative term U_r to U_e essentially due to atomic relaxation.

To show the cancellation between U_r and U_e, let us consider the simplest case of the hydrogen molecule. The total energy $E_{tot}(N)$ of the molecule as a function of N is

$$E_{tot}(N=2) = 2 E_H - \Delta E(H_2)$$

$$E_{tot}(N=1) = E_H - \Delta E(H_2^+) \tag{3}$$

$$E_{tot}(N=0) = 0$$

where E_H is the total energy of the free hydrogen atom. $\Delta E(H_2)$ and $\Delta E(H_2^+)$ are the binding energies of the molecule H_2 and H_2^+ (respectively equal to 4.72 and 2.78 eV). This leads to an experimental value U^* equal to 0.84 eV much smaller than the electron-electron interaction U_e one can estimate from the size of the molecule to be larger than 18 eV. The change of the molecule binding energy from N=1 to N=2 is also accompanied by a reduction of the atomic distance from 0.106 nm to 0.074 nm showing the increase of the interatomic bond strength.

For defects in semiconductors, both U_e and U_r are reduced from the free bond value: U_e by a factor roughly equal to the material dielectric constant and U_r by the existence of elastic restoring forces between the defect and the crystal. The result is that U_r can be smaller or larger than U_e leading to positive or negative U^* situations.

According to the preceeding arguments, for an atom adsorbed on a semiconductor surface, the electron-electron repulsion U_e would also be reduced but in this case by the semiconductor dielectric constant near the surface ϵ_s which is in the classical limit equal to $(\epsilon+1)/2$. On the other hand, the adatom-substrate bond has one free end and there will be no reduction of the relaxation energy compared to the free bond. Then U_e would be smaller than U_r and the adatom is a good candidate for a negative U^* center.

3. Tight-binding model of the adatom-substrate bond

To check the validity of the preceeding conclusion for the adatom, we have developed a molecular model for the adsorption of gold or alkali atoms on GaAs(110) [20]. We consider that adsorption occurs "on top of the Ga dangling bond" which gives a maximum binding energy [14] in our model and which is also observed by STM[1,6]. We use a nearest-neighbors' tight binding approximation as described in [14]. The interactions between the adatom and the surface are taken from Harrison's law [21]. The Au d electrons are neglected so that Au adatoms are treated like alkali atoms. The Ga atom nearest neighbor of an adatom is taken at its unrelaxed surface position whereas all the others surface atoms are assumed to stay at their free surface relaxed positions. The reason for this comes from the fact that the relaxation of surface atoms is directly related to the broken bond character and to the filling of the dangling bond state which are modified only for the Ga atoms first nearest neighbors of the adatoms. Such a result has been recently demonstrated for the adsorption of Na atoms on GaAs(110) [15]. We also consider that, for the adsorption in the neutral state, the adatom-Ga bond length is the sum of both atomic radii. This generally gives good values for most conventional single bonds.

In a molecular model, we only consider the interaction between the s adatom state and an isolated sp^3 dangling bond. The corresponding effective tight-binding intreaction between these two states is adjusted to give the same levels as a full model in the case N=1. The s state level E_s depends on the orbital occupancy n_s through the Coulomb interaction U_s. E_s is determined from the free atom Hartree-Fock value E_{HF} [22] by adding half the Coulomb term U_{HF} to compensate for the change in the selfexchange between the free and and the adsorbed atoms:

$$E_s = E_{HF} + \frac{U_{HF}}{2} + \left(\frac{U_{HF}}{2} - \frac{e^2}{2(R+\lambda)} \right) (n_s - 1) \qquad (4)$$

$$E_s = E_{s0} + U_s n_s$$

where U_s also takes into account the reduction of the intraatomic Coulomb term by the image charge potential we have taken here in the limit of large substrate dielectric constant.

The dangling bond orbital energy E_d also depends of its electron occupation which is equal to $(N-n_s)$

$$E_d = E_{d0} + U_d (N-n_s) \qquad (5)$$

where U_d is the dangling bond intraatomic Coulomb interaction [23] and N the bonding state occupancy (N=0,1 or 2).

In the molecular model, the bonding state E_B which depends on the electron occupation N is given by

$$E_B(N) = \frac{E_s + E_d}{2} - \sqrt{\delta_N^2 + \beta_N^2} \tag{6}$$

where β is the effective tight-binding interaction between the s adatom state and the sp^3 dangling bond. It depends on the bond occupancy N through the interatomic distance R_N

$$\beta_N = \beta_0 \exp(-qR_N) \tag{7}$$

δ_N is simply equal to $(E_s - E_d)/2$. A repulsive term $C_0 \exp(-qR_N)$ is added to the bond energy to account for the short range interatomic repulsion. Then in order to get the total energy $E_{tot}(N)$ we must substract the Coulomb terms counted twice in the one-electron energy

$$E_{tot}(N) = N \left(\frac{E_s + E_d}{2} - \sqrt{\delta_N^2 + \beta_N^2} \right) - U_s \frac{n_s^2}{2} - U_d \frac{(N-n_s)^2}{2} + C_0 \exp(-pR_N) \tag{8}$$

The selfconsistent solution is obtained by minimization of $E_{tot}(N)$ with respect to n_s keeping N and R_N constant. Then the equilibrium distance is calculated by minimization with respect to R_N. One generally estimate that p is close to 2q [24]. In the case p=2q, the calculation is analytic and from equations (1) and (2), one gets [20]:

$$U^* \sim - \sqrt{\delta_1^2 + \beta_1^2} \frac{\left(1 - \frac{U_d U_s}{4(\delta_1^2 + \beta_1^2)}\right)}{\left(1 + \frac{U_s + U_d}{4\sqrt{\delta_1^2 + \beta_1^2}}\right)} \tag{9}$$

Let us first remark that U^* is always smaller than U_s and U_d and that the experimental value (~1.7 eV) is then incompatible with the estimated value of U_d (~0.6 eV [23]).

To compare with the preceeding section, let us consider the case of a covalent bond where $U_s = U_d = U_H$. We get

$$U^* = \frac{U_H}{2} - \beta_1 \tag{10}$$

The contribution U_e in such a simple model is simply equal to half the free atom value U_H as an extra electron is shared by the two atoms of the bond. One can also see that the reduction due to the bond relaxation U_r is not negligible as it is equal to the hopping integral β_1 between the two atoms for N=1.

In the case of an ionic bond as a gold or alkali adatom on GaAs(110), U^* is negative provided $U_d U_s < 4(\delta_1^2 + \beta_1^2)$. The dangling bong Coulomb repulsion is small (~0.6 eV [23]) as U_s which is of the order of 1 eV. As $\sqrt{\delta_1^2 + \beta_1^2}$ is close to 1.2 eV, we get for this system a negative U^* value (~-0.6 eV), a conclusion which holds true in a broad range of values of the parameters.

4. Comparison with experimental results

In view of equation (2), a negative U^* value means that the levels $\epsilon(2,1)$ and $\epsilon(1,0)$ occur in inverted order. In other words, that also means that one gets

$$E_{tot}(N=0) + E_{tot}(N=2) < 2 E_{tot}(N=1) \tag{11}$$

which means that for a globally neutral interface the stable situation corresponds to an equal number of positively (N=0) and negatively (N=2) charged adatom-Ga bonds. This will clearly favor the formation of pairs of bonds in opposite charge states in view of their electrostatic interaction equal to minus $e^2/\epsilon_s d$ where d is the interbond distance and ϵ_s the surface dielectric constant. The total energy of such a pair containing two electrons (two in a bond, zero in the other) is

$$E_p(N=2) = E_{tot}(N=2) + E_{tot}(N=0) - \frac{e^2}{\epsilon_s d} \tag{12}$$

which represents the basic stable entity. For such a pair the energy levels correspond to ionization of either one electron or one hole. The corresponding total energies $E_p(N=3)$ and $E_p(N=1)$ are respectively equal to $E_{tot}(N=2) + E_{tot}(N=1)$ and $E_{tot}(N=1) + E_{tot}(N=0)$ with no electrostatic contribution since one of the member of the pair (with N=1) is neutral. The corresponding ionization levels of the pair are

$$\epsilon_p(3,2) = \epsilon(1,0) + \frac{e^2}{\epsilon_s d} \tag{13}$$

$$\epsilon_p(2,1) = \epsilon(2,1) - \frac{e^2}{\epsilon_s d}$$

Their energy distance is equal to $\left(\frac{+2e^2}{\epsilon_s d} - U^*\right)$, i.e. in the normal order since U^* is negative. A reasonable value of d can be estimated of the order of 10 a.u. With ϵ_s close to 7, this gives $e^2/\epsilon_s d \sim 0.4$ eV. Thus the distance between the two levels comes out to be 1.4 eV in our model.

This prediction of these ionization levels at $E_c - 0.1$ eV and $E_v - 0.4$ eV corresponds quite well to the peaks observed by STM which are located at $E_c - 0.5$ eV and $E_v - 0.7$ eV for Au adatoms [1]. For Cs adsorbed on GaAs(110), the bond with the Ga atom is more ionic. We find the ionization levels equal to $E_c + 0.3$ eV and $E_v + 0.5$ eV whereas the experimental values are $E_c + 0.2$ eV and $E_v + 0.25$ eV [6-7].

5. Conclusion

We have developped a simple tight-binding model to study the relaxation of the bond between a monovalent atom adsorbed on GaAs(110). In agreement with simple arguments deduced from molecular hydrogen, we have found that such a system is a negative U center. The present results should be applicable to other situations with unfilled bonding states. This would also be the case for example of monovalent adatoms on GaAs(111) Ga terminated surface. Due to Ga vacancies in the surface plane, the electronic structure of this surface is quite similar to the GaAs(110) one with a full As derived dangling bond state and an empty Ga one. If gold or alkali atoms are adsorbed on the Ga dangling bonds, one would also get negative U centers for this system.

References

1. R.M. Feenstra, Phys. Rev. Lett. 63, 1412 (1989).
2. R.M. Feenstra, J. Vac. Sci. Technol. B7, 925 (1989).
3. R.M. Feenstra, P. Martenson, and R. Ludeke, Mat. Res. Symp. Proc. 139, 15 (1989).
4. P.N. First, J.A. Stroscio, R.A. Dragoset, D.T. Pierce, and R.J. Celotta, Phys. Rev. Lett. 63, 1416 (1989).
5. P.N. First, R.A. Dragoset, J.A. Stroscio, R.J. Celotta, and R.M. Feenstra, J. Vac. Sci. Technol. A7, 2868 (1989).
6. L.J. Whitman, J.A. Stroscio, R.A. Dragoset, and R.J. Celotta, Phys. Rev. Lett. 66, 1338 (1991).
7. K.O. Magnusson, and B. Reihl, Phys. Rev. B40, 5864 (1989).
8. K.O. Magnusson, and B. Reihl, Phys. Rev. B40, 7814 (1989).
9. T. Maeda Wong, N.J. Di Nardo, D. Heskett, and E.W. Plummer, Phys. Rev. B41, 12342 (1990).
10. N.J. Di Nardo, T. Maeda Wong, and E.W. Plummer, Phys. Rev. Lett. 65, 2177 (1990).
11. C.Y. Fong, L.H. Yang, and I.P. Batra, Phys. Rev. B40, 6120 (1989).
12. J.E. Klepeis, and W.A. Harrison, Phys. Rev. B40, 5810 (1989).
13. J. Ortega, and F. Flores, Phys. Rev. Lett. 63, 2500 (1989).
14. G. Allan, M. Lannoo, and C. Priester, J. Vac. Sci. Technol. B8, 980 (1990).
15. J. Hebenstreit, M. Heinemann, and M. Scheffler(To be published).
16. P.W. Anderson, Phys. Rev. Lett. 34, 953 (1975).
17. G.A. Baraff, E.O. Kane, And M. Schlüter, Phys. Rev. Lett. 43, 956 (1979); Phys. Rev. B21, 3583 (1980).
18. G.D. Watkins, and J.R. Troxell, Phys. Rev. Lett. 44, 593 (1980).
19. G. Allan, and M. Lannoo, Phys. Rev. Lett. 66, 1209 (1991).
20. G. Allan, and M. Lannoo, J. Vac. Sci. Technol. (To be published).
21. W.A. Harrison, Phys. Rev. B24, 5635 (1981).
22. W.A. Harrison, Phys. Rev. B31, 2121 (1985).
23. M. Lannoo and P. Friedel in *Basic Theory of Crystalline Surfaces* (Springer Verlag, Berlin, 1991).
24. G. Allan, and M. Lannoo, J. Phys. (Paris) 44, 1355 (1983).

AB INITIO CALCULATIONS ON EFFECT OF Ga—S BONDS ON PASSIVATION OF GaAs SURFACE — A PROPOSAL FOR NEW SURFACE TREATMENT

TAKAHISA OHNO AND KAZUMI WADA
NTT LSI Laboratories, 3-1, Morinosato Wakamiya, Atsugi-shi, Kanagawa243-01, Japan

ABSTRACT

In order to elucidate the passivating effects of the sulfide solution treatment on GaAs surfaces, we have investigated the structural and electronic properties of the sulfur-adsorbed GaAs surfaces by using the *ab initio* pseudopotential method. The optimal adsorption site of sulfur atoms is determined, and Ga—S bonds are found to be stably formed on the sulfur-treated GaAs surface. We have shown that the stable Ga—S bonds remarkably reduce the surface state density in the GaAs midgap region and shift the Fermi level toward the valence band maximum of GaAs. The sulfur passivation of GaAs surfaces can be explained quite well in terms of the formation of Ga—S bond on the sulfur-treated surface, without introducing any disorder or defect near the surface. Based on the calculated results, we propose a new sub-surface structure which can passivate GaAs surfaces more effectively than the sulfur treatment alone. In our new structures, atomic layers of other semiconductor with lower valence-band-maximum are deposited on GaAs surface before the sulfur treatment, which can be experimentally achieved by the atomic layer passivation (ALP) technique. It is shown that the proposed structures further improve the surface electronic properties of GaAs because of the band discontinuity at semiconductor interfaces as well as the III—S bond formation.

1. Introduction

The development of GaAs technology has been impeded due to the poor electronic properties of GaAs surfaces characterized by a high density of surface states. Recently, sulfide solution treatments have been reported to effectively improve GaAs surface properties.[1,2] The deposition of a thin Na_2S film onto the GaAs(001) surface can enhance the photoluminescence (PL) intensity relative to the untreated surface.[1] The $(NH_4)_2S_x$ treatment benefits the performance of minority-carrier devices sensitive to surface recombination, such as heterojunction bipolar transistor. Controllable Schottky barrier heights are achieved on the $(NH_4)_2S_x$ treated GaAs(001) surface.[2] The sulfur treatments have been shown to passivate not only the GaAs(001) surface but the (110) and (111) surfaces as well.[3,4] These experiments have revealed that the sulfur treatments drastically reduce the surface state density of GaAs. Conventional explanation of this drastic reduction is based on the assumption that the sulfur treatment reduces the number of midgap As antisite defect states. This assumption, however, has not been justified yet.

In this paper, we present a first-principles study of the structural and electronic properties of both Ga- and As-terminated GaAs(001)-(1x1) surfaces adsorbed with a monolayer of sulfur atom. We have determined the optimal adsorption site of S atoms by minimizing the total energy, and have evaluated the surface electronic structure for the resulting optimal adsorption configurations. The mechanism of sulfur passivation of the GaAs surface is explained in terms of the dangling bond termination by S atoms. Finally, we propose a new method of surface passivation in which we make use of the III—S bond formation on surfaces and the band discontinuity at semiconductor interfaces.

2. Method of calculations

We have performed the first-principles total energy calculations based on the density functional formalism. Semirelativistic norm-conserving nonlocal pseudopotentials and a momentum-space formalism were employed.³ The wave functions are expanded in a plane-wave basis set with a kinetic-energy cutoff of 7.29 Ry. Slater's Xα formalism is adopted for the exchange and correlation energy in the local density approximation. Parameter α is fixed to be 0.7.

The surface is simulated by a slab geometry. The unit supercell for the GaAs(001) surface contains four GaAs layers (i.e., eight atomic layers) plus a vacuum region equivalent to about four GaAs layers in thickness. The slab of four GaAs-layers exposes a different surface on each side, namely the Ga- and As-terminated surfaces. This polar character of the GaAs(001) surface introduces some complications to the slab calculations. One problem is an artificial charge transfer from one surface of the slab to the other. This is because the surface states of cation dangling bonds are located above those of anion dangling bonds. Another problem is the interaction between the two surface states through the slab. To avoid the artificial charge transfer and to decouple the two surfaces of the slab, we terminate the one surface of no interest by fictitious H atoms. When we are interested in the Ga-terminated surface, for example, we place two fictitious H atoms with 0.75 nuclear and 0.75 electronic charges on each surface As atom at the other side of the slab, which has two dangling bonds containing 1.25 electrons. When the length of the As—H bond is properly given, the fictitious H atom forms a completely filled bonding state together with the As dangling bond and prevents the charge transfer from the other surface of the slab. The surface states associated with the As atoms energetically leave the band gap region due to the bond formation with the H atoms. As a result, only the surface states associated with the Ga atoms remain in the band gap region. When the As-terminated surface is investigated, fictitious H atoms with 1.25 electrons are deposited to terminate the surface Ga atoms on the other side of the slab. The pseudo-potentials of the fictitious H atoms with 0.75 (or 1.25) electrons are obtained by multiplying that of a normal H atom by 0.75 (or 1.25). Fig.1 presents the slab model for describing the Ga-terminated GaAs(001) surface adsorbed by a S monolayer. This method is analogous to what was done for the GaAs(001) surface by Qian et al.[6]

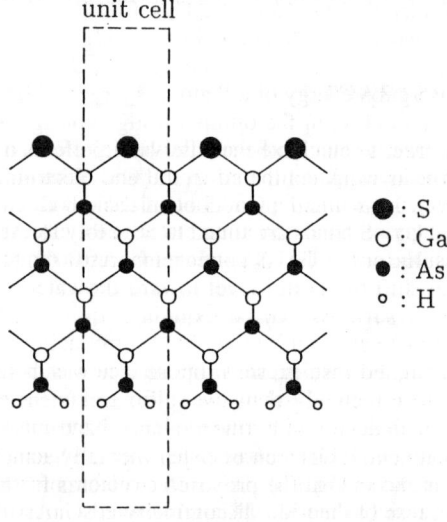

Fig.1. Slab geometry for the Ga-terminated GaAs(001) surface adsorbed with a S monolayer. Fictitious H atoms with 0.75 electrons are deposited on the surface As atoms.

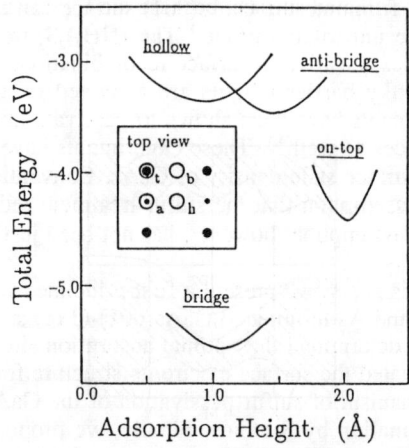

Fig.2. Calculated total energies of the S-adsorbed Ga-terminated GaAs(001)-(1x1) surfaces as a function of the S adsorption height. Four possible adsorption sites are considered. Total energies are measured relative to the energy in case the S atoms are far away from the surface.

3. Adsorption of sulfur atoms

First, we address the optimal adsorption geometries of sulfur atoms on the GaAs(001) surface.[7] We take into account four possible adsorption sites, i.e., the bridge site, on-top site, anti-bridge site and hollow site. The calculated total energies of the S-adsorbed Ga-terminated GaAs(001)-(1x1) surfaces are plotted in Fig.2 as a function of the vertical distance between the S and Ga atomic layers. In these calculations the Ga and As atoms are fixed in bulk lattice positions. It is shown from Fig.2 that the bridge site is the most stable adsorption site and the on-top site is the second stablest position. The anti-bridge and hollow sites are found to be energetically very unfavourable compared with the bridge and on-top sites. The adsorption energy of a S atom at the bridge site is calculated to be 5.6 eV, which is larger by 1.2 eV than at the on-top site. This difference in adsorption energy stems from the fact that a S atom is bonded to two Ga atoms in the bridge site, while it is bonded to only one Ga atom in the on-top site. For the As-terminated surface, S atoms also adsorb most stably on the bridge site, and the adsorption energy of a S atom is larger by 0.8 eV at the bridge position than at the on-top position. Furthermore, in the optimal bridge configurations, the S adsorption energy on the As-terminated surface is found to be smaller by 1.3 eV than that on the Ga-terminated surface. That is, the As—S bond is weak compared to the Ga—S bond. As mentioned below, the As—S bond contains 2.25 electrons while the Ga—S bond has 2.0 electrons. The excess electrons weaken the As—S bond by occupying an antibonding state. The weak As—S bonds are consistent with the recent experimental results for the $(NH_4)_2S_x$-treated GaAs(001), (111)Ga and (111)As surfaces.[3]

3-1. Effect of Ga—S bonds

In Fig.3(a) we present the surface electronic structure of the Ga-terminated GaAs(001)-(1x1) surface with a S monolayer adsorbed in the energy-optimized bridge configuration. There are two surface state bands (the D1 and D2 bands) near the valence-band maximum (VBM) of GaAs. These two bands are associated with the S dangling bonds, as shown from the calculated charge-density contours in Fig.3(b). This surface electronic structure can be explained by using a simple tight-binding picture. On the ideal Ga-terminated GaAs(001) surface, surface states are present at the midgap region, which originate from the Ga dangling bonds. By the bridge-site adsorption of S atoms on this Ga-terminated surface, a bonding and an antibonding state are formed between the Ga dangling bond and the S-sp^3 orbitals. The bonding state, which is the Ga—S covalent bond, lies within the GaAs valence bands and is completely filled. The antibonding state is located within the GaAs conduction bands and empty. The formation of Ga—S bond leaves each surface S atom two dangling bonds containing 1.75 electrons. These S dangling bonds generate the surface state D1 and D2 bands, which is located near the VBM of GaAs, much lower in energy than those of the Ga-terminated surface, because the S potential is much deeper than the Ga potential.

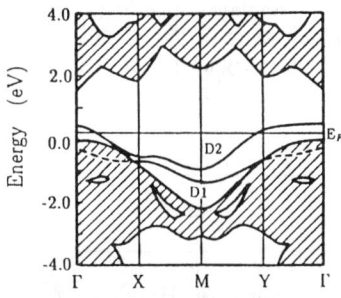 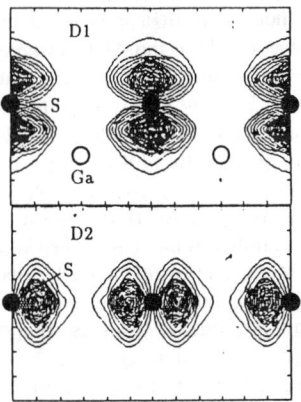

Fig.3(a). Calculated electronic structure of the Ga-terminated GaAs(001)-(1x1) surface with a S monolayer adsorbed on the optimal bridge position. The surface state D1 and D2 bands originate from the S dangling bonds.

Fig.3(b). Charge density contour plots of the D1 and D2 states at the M point for the S-adsorbed Ga-terminated GaAs(001)-(1x1) surface. The plot for the D1 state is in a (110) plane including the Ga-S bonds, and the plot for the D2 state in a (-110) plane including no Ga-S bonds.

In this way, the adsorption of a sulfur monolayer on the Ga-terminated GaAs(001) surface replaces the midgap Ga-related surface states with the S-related surface states near the VBM, resulting in the shift of the Fermi level toward the VBM. The Fermi level shift is consistent with the experiments showing that the band bending for the unpassivated n-type GaAs surface is increased after sulfur treatments.[8] The deep level transient spectroscopy (DLTS) measurement has shown that the midgap level in as-grown GaAs(001) samples is greatly reduced after sulfur treatments, while the level near the VBM remains with a high density.[9] The calculated surface state bands are also in agreement with this experiment. The surface states near the VBM will be inefficient recombination centers due to their high thermal emission probability.[10] Thus, the surface recombination velocity will be drastically reduced after the sulfur adsorption.

The fact that the S dangling bonds are not completely occupied indicates that possibility of a dimer formation of S atoms in the bridge position. Actually, the total energy calculations have shown that the surface S atoms can be dimerized at the S-adsorbed Ga-terminated GaAs(001) surface. The reduction in the S—S bond length, however, is much smaller than that of the Ga—Ga dimer at the Ga-terminated surface. Furthermore, the S—S dimer formation energy of 0.13 eV is one order of magnitude smaller than that of 1.70 eV for the Ga—Ga dimerization. This is because the S dangling bond is not completely but nearly fully occupied. The S—S dimer formation is consistent with the observed (2x1) structure of the S-adsorbed GaAs(001) surface.[3] We have found that the weak dimers of S atoms do not significantly affect the surface electronic structure. In addition, the nearly filled S dangling bonds indicate that the S-treated surface is resistant to contamination.

3-2. Effect of As—S bonds

The surface electronic structure of the As-terminated GaAs(001)-(1x1) surface adsorbed with a S monolayer in the energy-optimized bridge configuration is presented in Fig.4(a). The surface state D1 and D2 bands, which originate from the S-sp^3 orbitals, are both completely filled. That is, the S dangling bonds are fully occupied at this surface. This indicates that no driving force toward dimerization exists on this surface, which is confirmed by the total energy calculations. The most striking feature of this electronic structure is the appearance of a surface A band besides the D1 and D2 bands. The A band has a large dispersion nearly crossing the GaAs energy gap, and is occupied with 0.5 electrons. It is shown that the A band is the As—S antibonding state, from the charge density contour plot in Fig.4(b). For the S-adsorbed Ga-terminated surface, the Ga—S antibonding state is located within the GaAs conduction bands, and can hardly be distinguished from other states due to resonance. This is due to the Ga dangling bond lying above the As dangling bond. Because of the A band as well as the D2 band, the surface state density of the As-terminated GaAs(001) surface is not reduced in the midgap region by the adsorption of a S monolayer. This means that the surface recombination velocity will not be reduced at the As-terminated surface. It should be noted that the S adsorption has quite different effects on the Ga- and As- terminated GaAs surfaces.

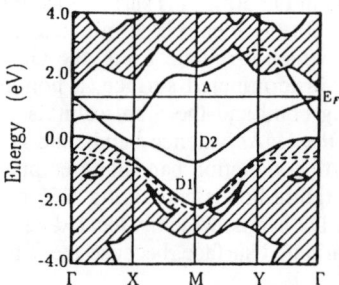

Fig.4(a). Calculated electronic structure of the As-terminated GaAs(001)-(1x1) surface with a S monolayer adsorbed on the optimal bridge position. The surface state D1 and D2 bands originate from the S dangling bonds. The A band is the As-S antibonding state.

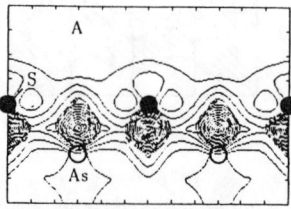

Fig.4(b). Charge density contour plot of the A state at the Γ point for the S-adsorbed As-terminated GaAs(001)-(1x1) surface. The plot is in a (110) plane including the As-S bonds.

We have found that the formation of stable Ga—S bonds remarkably reduces the midgap surface state density on the GaAs(111) surface as well as on the (001) surface. Furthermore, the adsorption of other chalcogen atoms(Se and Te) is found to exhibit the passivating effect similar to the S adsorption.[11] Spindt et al. have discussed the sulfur passivation in terms of the antisite defect model of GaAs interface states.[12] Within their model, it has been postulated that the sulfur treatment reduces the number of the As antisite defects, which results in the improvement in surface electronic properties. The mechanism of the reduction in the number of As antisites, however, has not been clearly presented. We have shown in this paper that the passivating effect of sulfur treatment can be quite well explained in terms of the formation of the stable Ga—S bond at the treated surface and the resultant dangling bond termination by S atoms, without introducing any unclear defects or disorder near the surface. These results also suggest that the dangling bond at the surface is a more reliable candidate for the origin of the GaAs surface states, than the defect state near the surface.

4. New method of surface passivation

Although the midgap surface states disappears, the sulfur treatment can't totally eliminate the surface states of GaAs, that is, there are still surface states near the VBM. Accordingly, we propose a new sub-surface structure which can passivate GaAs surfaces more effectively than the sulfur treatment alone. In our new structures, atomic layers of other semiconductors with lower VBM (such as GaP, AlAs and InP) are deposited on GaAs surface before the sulfur treatment, as shown in Fig.5. These structures can be experimentally achieved by the atomic layer passivation (ALP) technique.[13] we have found that the III—S bonds formed on these semiconductors have the passivating effects similar to the Ga—S bonds on GaAs. Therefore, our ALP treatment is expected to shift the surface states downward against the VBM of GaAs by the valence band discontinuity at the semiconductor interface.

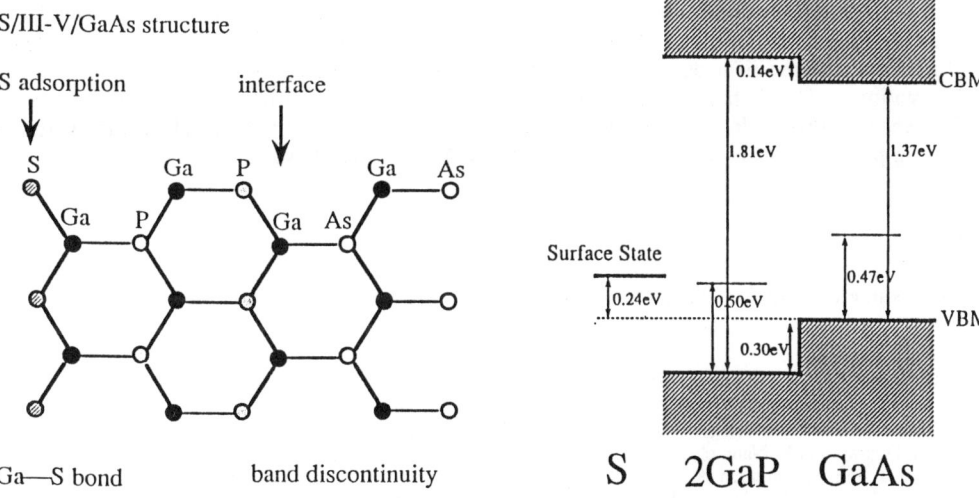

Fig.5. Example of the proposed sub-surface structure. This figure shows the atomic positions of the S/(GaP)$_2$/GaAs(001) structure, where two layers of GaP are deposited on the Ga-terminated GaAs(001) surface and then a S monolayer is adsorbed on the deposited GaP layers.

Fig.6. Electronic structure of the S/(GaP)$_2$/GaAs(001) structure. The horizontal line within the band gap of GaAs (or GaP) represents the position of the surface state maximum on the bulk GaAs (or the strained bulk GaP) adsorbed with a S monolayer, and the horizontal line drawn on the left is that for the S/(GaP)$_2$/GaAs structure.

We confirm that the passivating effect is enhanced by the ALP treatment on the basis of *ab initio* calculations. In Fig.6 presents the electronic structure of the $S/(GaP)_2/GaAs(001)$, where two layers of GaP are deposited on the Ga-terminated GaAs(001) surface and then a S monolayer is adsorbed on the deposited GaP layers. The atomic positions of the GaP layers as well as the S atom are determined by minimizing the total energy. Since the lattice constant of GaP is smaller by 4% than that of GaAs, the adsorbed GaP layers are expanded in the direction parallel to the interface and compressed perpendicular to the interface. In the region two atomic layers away from the GaP/GaAs interface, the averaged self-consistent potential is almost the same as that in the bulk GaAs or the strained bulk GaP.[14] Thus, the band lineup at the GaP/GaAs interface can be determined by combining the potential averages and band structures of the GaAs and the strained GaP. The obtained valence band discontinuity is 0.30 eV, as shown in Fig.6. For the strained GaP adsorbed with a S monolayer, the surface state maximum is found to be located 0.50 eV above the VBM of GaP. For the $S/(GaP)_2/GaAs(001)$, this surface state maximum is expected to shift downward against the VBM of GaAs by the valence band discontinuity (0.30 eV) at the GaP/GaAs interface. In fact, the calculated surface state maximum is located 0.24 eV above the VBM of GaAs. In this way, the proposed sub-surface structures on GaAs can shift the surface states downward against the VBM of GaAs, because of the valence band discontinuity at semiconductor interfaces as well as the III—S bond formation on surfaces. If a species of adsorbate generates surface states near the conduction band minimum (CBM) of semiconductors different from S atoms, we can passivate GaAs surface effectively by depositing atomic layers of other semiconductors with higher CBM and then the adsorbates. In this case, the surface states shift upward against the CBM of GaAs by the conduction band discontinuity. Furthermore, the carriers are confined in the GaAs substrate by the conduction and valence band discontinuities. As a result, the carriers won't see the surface states virtually and the surface electronic properties are expected to be much improved.

5. Conclusions

We have investigated the structural and electronic properties of the sulfur-adsorbed GaAs surfaces by using the *ab initio* pseudopotential method. It is found that the stable Ga—S bonds remarkably reduces the midgap surface state density of GaAs and moves the Fermi level toward the valence band maximum. The sulfur passivation of GaAs surfaces can be explained quite well in terms of the formation of Ga—S bond on the sulfur-treated surface, without introducing any disorder or defect near the surface. In terms of the calculated results, we propose a new sub-surface structure which can passivate GaAs surfaces more effectively than the simple sulfur treatment.

Acknowledgements

We would like to thank Dr. Y. Wada for the valuable discussions. We also gratefully acknowledge the encouragement and support of Dr. K. Hirata and Dr. A. Yoshii.

References

[1] C. J. Sandroff, R. N. Nottenburg, J. C. Bischoff and R. Bhat, Appl. Phys. Lett. 51, 33 (1987).
[2] J. Fan, H. Oigawa and Y. Nannichi, Jpn. J. Appl. Phys. 27, L1331 and L2125 (1988).
[3] H. Oigawa, J. Fan, Y. Nannichi, K. Ando, K. Saiki and A. Koma, Jpn. J. Appl. Phys. 28, L340 (1989).
[4] K. Ueno, T. Shimada, K. Saiki and A. Koma, Appl. Phys. Lett. 56, 327 (1990).
[5] J. Ihm, A. Zunger and M. L. Cohen, J. Phys. C12, 4409 (1979).
[6] G.-X. Qian, R. M. Martin and D. J. Chadi, Phys. Rev. B38, 7649 (1988).
[7] T. Ohno and K. Shiraishi, Phys. Rev. B42 11194 (1990).
[8] C. J. Spindt, D. Liu, K. Miyano, P. L. Meissner, T. T. Chiang, T. Kendelewicz, I. Lindau and W. E. Spicer, Appl. Phys. Lett. 55, 861 (1989).
[9] D. Liu, T. Zhang, R. A. LaRue, J. S. Harris, Jr. and T. W. Sigmon, Appl. Phys. Lett. 53, 1059 (1988)
[10] W. Shockley and W. T. Read, Phys. Rev. 87, 835 (1952).
[11] T. Ohno, to be published in Surf. Sci.
[12] C. J. Spindt and W. E. Spicer, Appl. Phys. Lett. 55, 1653 (1989).
[13] Y. Wada, Y. Mada and K. Wada, to be published in "Extended abstracts of Solid State Devices and Materials '91".
[14] A. Taguchi and T. Ohno, Phys. Rev. B39, 7803 (1989).

TWO-DIMENSIONALLY LOCALIZED VIBRATIONAL MODE DUE TO Al ATOMS SUBSTITUTING FOR Ga ONE-MONOLAYER IN GaAs

HARUHIKO ONO and TOSHIO BABA
Fundamental Research Laboratories, NEC Corporation,
34 Miyukigaoka, Tsukuba, Ibaraki 305, JAPAN

ABSTRACT

We demonstrate evidences of a vibrational mode localized in a single slab of Al atoms substituting for a (001) Ga monatomic layer in GaAs. Infrared absorption spectra of a single slab of one-monolayer (1-ML) AlAs sandwiched by GaAs or AlAs/GaAs superlattices were investigated by Fourier transform spectroscopy. A specific absorption peak was observed at 358 cm^{-1} with a line width of 3 cm^{-1} in a sample with a single slab of 1-ML AlAs. Polarization dependence of the peak strength reveals the anisotropy of E mode vibration, in which Al atoms vibrate parallel to the Al layer. This 358 cm^{-1} peak is concluded to be a phonon mode due to two-dimensionally distributed Al atoms in three dimensional GaAs crystal.

1. Introduction

It is well known that a light impurity atom substituted for a heavy host atom in a crystal has a localized vibrational mode (LVM). For this reason, light impurity atoms substituted for a monatomic layer of host atoms may be expected to have a specific two-dimensionally localized vibrational mode (2D-LVM). Well controlled AlAs/GaAs superlattices can be obtained by using a recent crystal growth technique of molecular beam epitaxy (MBE). Free electrons in the conduction band confine in GaAs layers of superlattices, being the two-dimensional electron gas, since the band gap is narrower in GaAs than in AlAs. The phonon properties in such semiconductor superlattices have been investigated by Raman scattering measurements[1-4]. The results have revealed that the optical phonons also confine in individual layers of a superlattice. Recently, the phonon confinement was observed in ultra-thin-layer superlattices[5-7]. However, these experiments can not demonstrate the existence of 2D-LVM, because the distances between AlAs layers in such superlattices were too close to neglect interactions between the layers. In order to detect the proper 2D-LVM, we should measure for a sample with a single slab of monatomic impurity layer. This paper demonstrates evidences of such a 2D-LVM localized in a single slab of Al atoms substituted for a (001) Ga monatomic layer in GaAs, using Fourier transform infrared spectroscopy (FTIR).

2. Experimental

Samples used were grown on a (001) semi-insulating GaAs substrate by MBE. A single slab of one-monolayer (1-ML) AlAs, which corresponds to a single monatomic Al layer substituted for a Ga layer, or AlAs/GaAs superlattices were sandwiched between 500Å-thick GaAs epitaxial layers. The Al densities per unit area for all samples were controlled to be constant as 1.2×10^{16} cm^{-2}. Infrared absorption spectra of the samples were obtained by FTIR at near liquid helium temperature. The FTIR measurements were performed at a resolution of 0.1 cm^{-1} with a Si bolometer cooled at 4K as an infrared detector. A reference spectrum of as-received substrate was subtracted from the sample spectra, in order to eliminate the background due to lattice phonons in GaAs substrate. The layer structures of samples were confirmed by high resolution transmission electron microscopy[8]. For the sake of comparison, MBE-grown $Al_{0.5}Ga_{0.5}As$ and AlAs, and Al-doped bulk GaAs were also investigated.

3. Results and discussion

Figure 1 shows absorption spectra of Al-related vibrational modes observed in this work. The spectra (a) and (b) in Fig.1 were obtained from a sample with a single slab of 1-ML AlAs, and from a sample with 20 periods of 1-ML AlAs/14-ML GaAs, respectively. They show an distinct peak at 358 cm^{-1}. The latter sample contains 20 slabs of monatomic Al layer separated by 40 Å each other. Comparing the spectra (a) and (b), one can recognize that the spectrum (b) is exactly 20 times the spectrum (a). This fact indicates that the peak is absolutely due to the Al layers in the samples. Since the 358 cm^{-1} peak was observed in a non-periodic sample with a single slab of 1-ML AlAs, it clearly does not originate from the periodicity of superlattices. Therefore, we suppose that the peak is a localized phonon being characteristic of a monatomic Al layer, i.e., 2D-LVM. Spectra (c) and (d) are Al-LVM in a bulk GaAs crystal and TO-phonon mode in an MBE-grown AlAs layer, respectively. The Al-LVM is thought to be a vibrational mode relating to zero-dimensionally distributed Al atoms, and the AlAs TO-phonon relating to three-dimensionally distributed Al atoms. The relationship between the Al distribution and the vibrational frequencies will be discussed below.

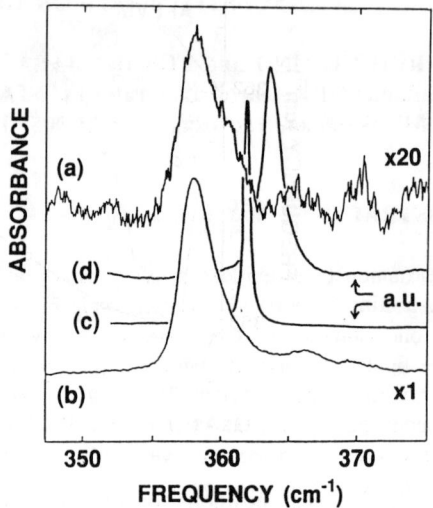

Fig.1 Al-related phonons in,
(a) 1-ML AlAs
(b) 20 periods of (AlAs)$_1$/(GaAs)$_{14}$
(c) Al-doped bulk GaAs
(d) 50Å-thick AlAs

For the purpose of confirmation that the observed 358 cm^{-1} peak is a proper 2D-LVM, we investigated the change of the spectra for superlattices whose layers were modulated systematically, as shown in Fig.2. The superlattice samples are categorized into three series. In the series I, AlAs layers in 1-ML AlAs/14-ML GaAs were replaced by AlGaAs layers. In this case, Al atoms

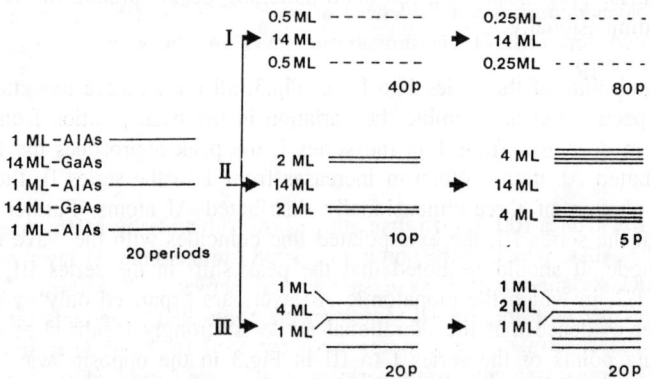

Fig.2 Superlattice samples used for experiments.

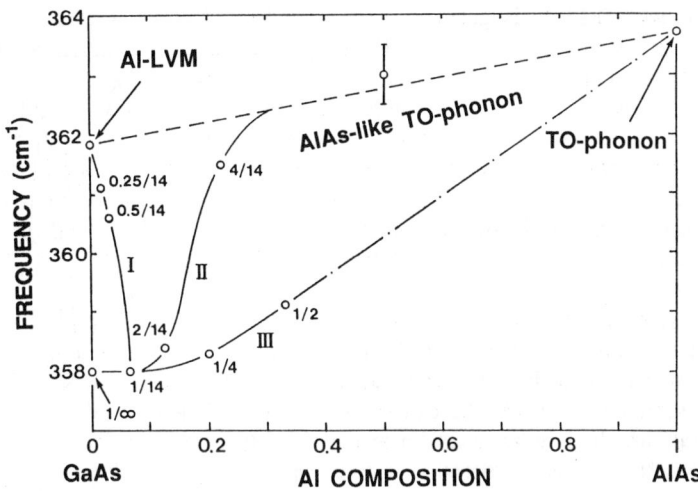

Fig.3 Peak positions for all observed samples plotted against nominal Al composition. The number m/n means m-ML AlAs / n-ML GaAs.

become isolated from each other being distributed zero-dimensionally. The series II increases AlAs layers keeping GaAs layers 14-ML. This cause the Al distribution to be three-dimensional. Moreover, in the series III, GaAs layers are decreased keeping AlAs layers 1-ML. Since the neighboring Al monatomic layers approach each other, in the last series, one can check the interactions between the Al layers. Peak positions obtained for all the samples in the present study are summarized in Fig.3. Figure 3 shows the peak positions plotted in wave numbers against the Al composition, x. In the superlattice samples, the Al compositions are the average values in the superlattice region. The TO-phonon on AlAs was observed at 363.7 cm^{-1}, and the impurity LVM of isolated Al in GaAs was observed at 361.8 cm^{-1}, as shown in Fig.1. The line between them is the AlAs-like TO-phonon modes in alloy AlGaAs[9]. If the Al atoms are distributed at random, the TO-phonon peak must be on this line. Actually the peak in $Al_{0.5}Ga_{0.5}As$ was observed at 363 cm^{-1}, which is just on the TO-phonon line. It is noted that all the peaks for samples having modulated structure were observed at positions lower than the AlAs-like TO-phonon line in AlGaAs. This means that the bond softening occurs around the Al atoms, when they distribute two-dimensionally.

If one traces the data points of the series I to III in Fig.3, all the data are recognized to change systematically as expected. Let us examine the variation in the peak position from the standard sample 1/14. When m decreases from 1 in the series I, the peak approaches the LVM of zero-dimensionally distributed Al atoms. When m increases from 1 in the series II, the peak rapidly approaches the TO-phonon of three-dimensionally distributed Al atoms. Furthermore, when n decreases from 14 in the series III, the extrapolated line coincides with the wave number of the AlAs TO-phonon mode. It should be noted that the peak shift in the series III is quite small even in the sample 1/2, in which the monatomic Al layers are separated only by a 2-ML GaAs layer. This is a clear evidence that the vibrational mode is strongly localized at each Al layer. If one traces the data points of the series I to III in Fig.3 in the opposite way, all the points converge into a specific wave number, 358 cm^{-1}, which was observed for a sample having a single monatomic Al layer embedded in GaAs. Therefore, our argument that the 358 cm^{-1} peak is the 2D-LVM due to two-dimensionally distributed Al atoms, provides a quite natural and consistent explanation for all the data described in Fig.3.

Further evidence for 2D-LVM is given in Fig.4, in which the line widths and the integrated intensities of the peaks in the series II and III are plotted against the nominal Al composition. If one traces the data points in a similar way to the case in Fig.3, all the points vary systematically as a function of x. The sample with separated monatomic Al layers has the smallest line width, and the largest integrated intensity that corresponds to the absorption cross section due to photon-phonon interactions. This fact shows that the vibrational mode due to the monatomic Al layer consists of a single vibrational mode. When the interactions between the neighbor Al layers are not negligible, various vibrational modes occur and then the peak width increases and the absorption cross section decreases, as is seen in Fig.4.

We next investigate the symmetry behavior of the 2D-LVM. The symmetry of the (001) monatomic Al layer being substituted for a Ga layer in GaAs is the point group D_{2d}. Thus, the vibrational modes are divided into singlet B_2 and doublet E mode vibration. The B_2 mode concerns atoms vibrating perpendicular to the layer and the E mode parallel to the layer. When the incident light, which is a transversal wave, propagates perpendicular to the Al layer, the photons interact only with the E mode. Therefore, the 2D-LVM we observed at 358 cm^{-1} must be the E mode. In order to confirm this idea, we further measured the polarization dependence of 2D-LVM intensity in an $(AlAs)_1(GaAs)_{14}$ sample, and Al-LVM intensity in a bulk GaAs sample as a reference. The samples were mounted inclining at an angle of 45° to the incident light, as shown in Fig.5(a). The LVM spectra were observed with the incident light polarized at

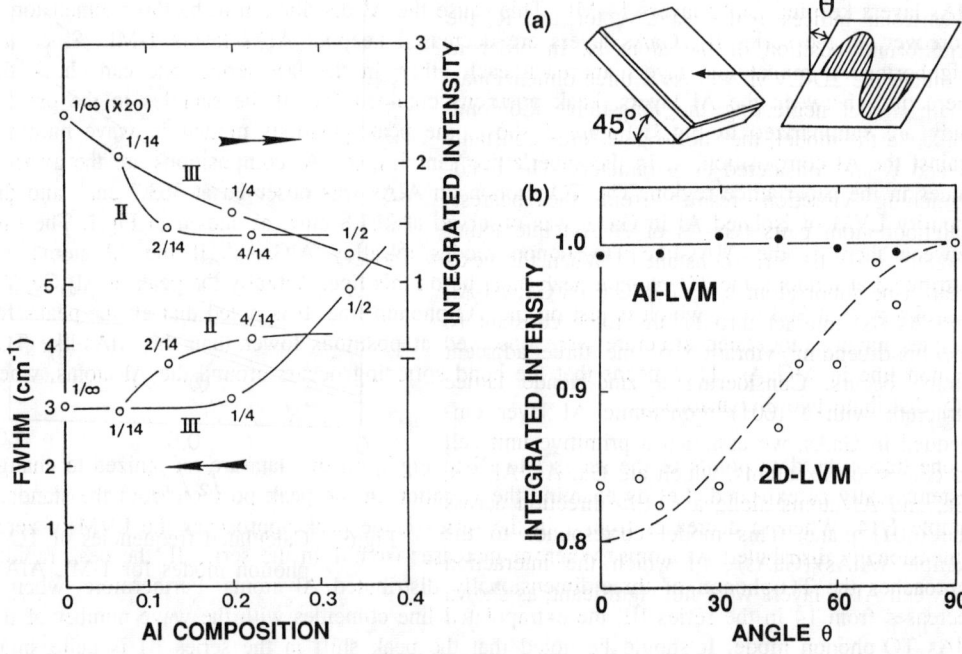

Fig.4 Full width at half maximum (FWHM) and integrated intensity plotted against nominal Al composition.

Fig.5 (a) Schematic diagram for the polarization experiments.
(b) Integrated intensity plotted against the polarization angle.

θ from the plane of incidence. Figure 5(b) shows integrated absorption of the peak plotted against the polarization angle θ. The open circles are for the 358 cm^{-1} peak in the (AlAs)$_1$(GaAs)$_{14}$, and the solid circles for the Al-LVM in the bulk GaAs. The integrated intensities are normalized by that when the incident light was polarized at θ = 90°. As shown in Fig.5(b), the peak intensity of the 358 cm^{-1} takes minimum at θ = 0°, and maximum at θ = 90°. In contrast to this, impurity LVM due to Al in GaAs did never depend on the polarization angle. This is caused by the T_d symmetry of the Al atom giving rise to the triplet-degenerated vibrational mode. The above experiments clearly show the anisotropic behavior of the 358 cm^{-1} peak and indicate that the peak is due to the E mode vibration of 2D-LVM.

Finally, we estimate the 2D-LVM by calculating the normal vibrational modes of three-dimensional lattice in which the atoms vibrate around the s-th site on the l-th unit cell. Under a harmonic approximation the frequency ω of the normal vibration can be obtained by solving the 3n × 3n equation,

$$\det(D_{\alpha\beta}(ss',q) - \omega^2 \delta_{\alpha\beta} \delta_{ss'}) = 0. \qquad (1)$$

The dynamical matrix $D_{\alpha\beta}(ss',q)$ is given by

$$D_{\alpha\beta}(ss',q) = (M_s M_{s'})^{-1/2} \sum \phi_{\alpha\beta}(ls,l's') \exp[iq(R_{l's'}-R_{ls})]. \qquad (2)$$

Where, M_s is the mass of the atom s, $\phi_{\alpha\beta}(ls,l's')$ the force constant between atoms ls and l's', α and β cartesian indices, q the wave vector, and R_{ls} the equilibrium position of the atom ls. In order to estimate the 2D-LVM, we use a three-dimensional-chain model neglecting the long-range Coulomb force. In the model, the short-range force constants f_1 and f_2 are considered as parameters. The bond-stretching vibration results from the nearest-neighbor atom pairs (Ga-As or Al-As) interacting directly with the force constant f_1. When the next-nearest-neighbor atom pairs (Ga-Ga, Al-Ga, Al-Al, and As-As) interact through the force constant f_2, the bond-bending vibration of the three adjacent atoms occurs. Considering a zinc-blende lattice structure with a (001) monatomic Al layer embedded in GaAs, we assumed a primitive unit cell to consist of six atoms, which are Ga, As, Al, As, Ga, and As atoms along a <110> direction across the (001) plane. This model corresponds to the sample (AlAs)$_1$(GaAs)$_2$, in which the interaction between AlAs layers were not significant, as shown in Fig.3.

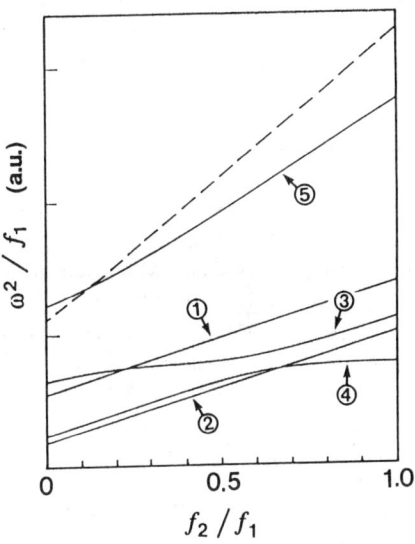

Fig.6 Calculated frequencies of TO-phonon modes for 1-ML AlAs.

The TO-phonon modes at Γ point (q=0) were calculated for the case when the incident light comes along the [001] direction. Figure 6 shows the results of calculated frequency ω as a function of the ratio f_2/f_1. Since six atoms were assumed to be in a unit cell, five vibrational modes appear for the TO-phonon. Two modes (1 and 2 in Fig.6) correspond to the GaAs bulk phonon, because the modes also appear in the independent calculation for GaAs. Other two (3 and 4) correspond to the GaAs-like phonon perturbed by the Al layer. The mode 5 has the highest frequency apart from the GaAs-like phonon band. The broken line in Fig.6 is for the

AlAs bulk phonon mode calculated in the same way. Since the mode 5 is located near the AlAs bulk phonon mode, the mode is considered to be the vibrational mode strongly localized at the Al layer; i.e., 2D-LVM. If the force constant ratio $f_2/f_1 > 0.11$, the 2D-LVM appears at a frequency lower than AlAs TO-phonon, being in agreement with the experimental results, though the ratio f_2/f_1 has not been established in the literature. The above calculation with a simple model shows that a vibrational mode localized in a monatomic Al layer appears around the AlAs TO-phonon frequency, being sensitive to the force constant ratio f_2/f_1.

4. Summary

We presented clear evidences of the 2D-LVM in a single slab of monatomic Al layers embedded in GaAs using infrared absorption spectroscopy. The 2D-LVM observed at 358 cm^{-1} is strongly localized in a monatomic Al layer and is damped abruptly towards the direction perpendicular to the layer, while the impurity LVM is strongly localized at the impurity atom zero-dimensionally. The 2D-LVM peak in Fig.1 seems to be asymmetric and rather broad, compared with the impurity LVM. The LVM due to Al in GaAs has a sharp symmetric peak with a width of 0.5 cm^{-1}. The asymmetry and the broadness of the 2D-LVM might be caused by some imperfections of the Al layer, such as steps. If one made a perfect monatomic Al layer in GaAs, one could obtain a sharp symmetric 2D-LVM peak.

The authors are thankful to N.Ikarashi for his helpful observations using high resolution transmission electron microscopy. They also thank K.Ishida for his valuable discussion during the work, and Y.Matsumoto and H.Watanabe for their encouragement.

References

1. C.Colvard, R.Merlin, M.V.Klein, and A.C.Gossard, Phys.Rev.Lett. 45, 298 (1980).
2. B.Jusserand, D.Paquet, and A.Regreny, Phys.Rev. B30, 6245 (1984).
3. C.Colvard, T.A.Gant, M.V.Klein, R.Merlin, R.Fischer, H.Morkoc, and A.C.Gossard, Phys.Rev. B31, 2080 (1985).
4. A.K.Sood, J.Menendez, M.Cardona, and K.Ploog, Phys.Rev.Lett. 54, 2111 (1985).
5. M.Nakayama, K.Kubota, H.Kato, and N.Sano, J.Appl.Phys. 60, 3289 (1986).
6. A.Ishibashi, M.Itabashi, Y.Mori, K.Kaneko, S.Kawado, and N.Watanabe, Phys.Rev. B33, 2887 (1986).
7. T.Toriyama, N.Kobayashi, and Y.Horikoshi, Jpn.J.Appl.Phys. 25, 1895 (1986).
8. N.Ikarashi, A.Sakai, T.Baba, K.Ishida, J.Motohisa, and H.Sakaki, Mat.Res.Soc.Symp.Proc. Vol.183, edited by R.Sinclair, D.J.Smith, and U.Dahmen (San Francisco, California, 1990) p187.
9. O.K.Kim and W.G.Spitzer, J.Appl.Phys. 50, 4362 (1979).

O SURROUNDING OF P_b DEFECTS AT THE (111)Si/SiO$_2$ INTERFACE

A. STESMANS
Department of Physics, Universiteit Leuven, 3001 Leuven, Belgium

ABSTRACT

First observation of partially resolved ^{17}O hyperfine (HF) structure in electron spin resonance spectra of [111]P_b defects at the interface of ^{17}O-enriched (51.24%) (111)Si/SiO$_2$ structures in combination with conscientious monitoring of defect density and related dipolar interaction has revealed details on the P_b's immediate O surrounding in the silica side. It is found that the unpaired sp^3 hybrid has its strongest HF interaction, of HF splitting constant $a_{\|1}$=2.7±0.15 G, with only one O site in a first shell, a next interaction of $a_{\|2}$=1.1 G with one O site in a second shell, and a third interaction of $a_{\|3}$≈0.2 G with two equivalent O sites in a third shell. While complementing the P_b model, the results conflict with a previously proposed symmetric ditrigonal ring silica cover of P_b defects.

I. Introduction

Thermal oxidation or nitridation of Si at whatever temperature(T) is attended with the generation of intrinsic trivalent-Si defects at the interface.[1-3] The dominant defect at the (111)Si/SiO$_2$ interface —termed P_b center— has been identified by the electron spin resonance technique (ESR) as an unpaired sp^3 orbital on an interfacial Si atom backbonded to 3 Si atoms in the bulk and pointing into a microvoid[2,4] (schematically denoted as °Si≡Si$_3$). It exhibits C_{3v} symmetry and accounts for 50-100% of all electrically-active trapping and recombination fast interface states. Noteworthy is that only the [111]P_b variant with unpaired sp^3 hybrid ⊥ interface (sketched in Fig. 1) is observed in a conventional as-oxidized (1 atm dry O$_2$; 900-950°C) (111)Si/SiO$_2$ structure.[2,5] Such interface, typically[2,6] comprises about 5x10^{12} P_b's cm^{-2}.

The °Si≡Si$_3$ model accounts well for most of the experimental observations on the P_b defect, of which the main properties are primarily set by the Si substrate. Yet, this model is incomplete in the sense that it only reflects the Si (substrate) "side" of the defect. A full model of this prototype interface defect located right at a sharply bordered interface would also incorporate the surrounding structure at the insulator side. That additional insight could provide information on the physical mechanism(s) leading to P_b formation.

The P_b's immediate oxide surrounding is still unknown, except for the fact that O is not incorporated in the immediate bonding structure of P_b. This has been concluded from a K-band ESR study[7] on ^{17}O enriched (51.26%) and ordinary (0.037% ^{17}O natural abundance) (111)Si/SiO$_2$, which, rather than observing any ^{17}O (nuclear spin I=5/2) P_b hyperfine (HF) structure, found that the mere effect of enrichment was a broadening of the peak-to-peak linewidth ΔB_{pp} from about 1.4 to 4.2 G; The observation of ^{17}O P_b HF structure, though, would provide a clue to map the O surrounding. A similar conclusion was reached from a comparative study[3] of Si oxide and nitride showing that the insulator's influence on the P_b ESR features is only of secondary nature.

Recent experiments,[6] however, have led to a more thorough understanding of the P_b signal structure. In particular the dipole-dipole (DD) interaction between P_b's has been identified, mainly as a result of the optimization of a reversible H-passivation method. The natural line shape, that is the shape unaffected by DD

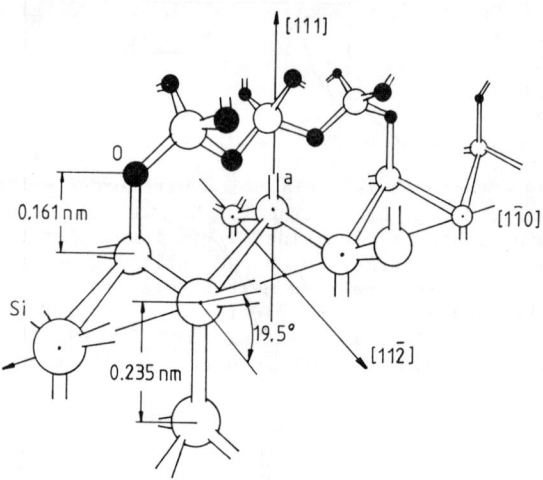

Fig.1: Ball-and-stick model of a [111]P_b defect (entity a) at the (111)Si/SiO$_2$ interface.

broadening, has been well characterised and a unique ΔB_{pp}-[P_b] relationship has been revealed. These results show that the P_b concentration is an important parameter and dictate that any comparison, either experimentally or theoretical-computationally, regarding the effect of ^{17}O enrichment must be carried out between otherwise *strictly identical* samples in order to keep [P_b], and thus DD effects, unaltered. In particular, this means that to simulate the P_b spectrum of ^{17}O enriched Si/SiO$_2$, the calculated ^{17}O HF histogram has to be convoluted with the correct spectrum of unenriched Si/SiO$_2$ of *equal* [P_b].

This renewed insight has opened a new perspective for ^{17}O P_b measurements. Together with a low-T ESR spectrometry optimized to record clear *undistorted* resonances and well controlled *reproducible* sample preparation, this has allowed the observation of ^{17}O P_b HF structure, which is the subject of the present work.

II. Experimental details

Slices of a commercial Czochralski-grown two-side polished (111)Si wafer (p type, 10 Ωcm) were oxidized at 920±10°C either in ordinary oxygen (purity>99.999%) or oxygen enriched to 51.24% ^{17}O at a pressure of 0.2 atm for about 75 min resulting in an oxide thickness of ≈145 Å. Particular attention was paid to cycle both samples fully identically, apart of course from the differing ^{17}O ambient. Post-oxidation thermal treatments in vacuum and hydrogen[6] were applied to optimize the P_b density at the balance of a favorable signal-to-noise (S/N) ratio and an acceptable DD broadening. The final P_b density was (4.9±0.3)x10^{12} and (5.2±0.5)x10^{12} cm^{-2} for the unenriched and enriched sample, respectively —well equal within experimental accuracy as strictly required for the present purpose.

K-band (≈20.1 GHz) ESR absorption-derivative $dP_{\mu a}/dB$ signals ($P_{\mu a}$ representing the absorbed microwave power) were measured at 4.3 K for the applied magnetic induction **B** perpendicular to the (111) interface (**B**$\|$[111]). The high saturability of the P_b signal at 4.3K imposed reduction of the microwave power P_μ incident on the TE$_{011}$ cavity of loaded Q of ≈15000 to ≤ 0.5 nW to record undistorted signals.

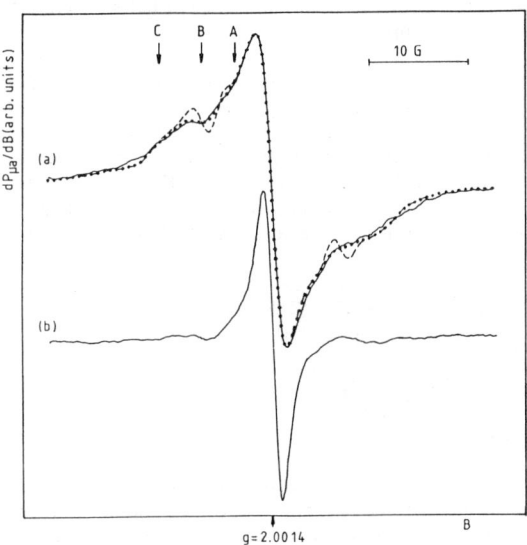

Fig.2: K-band ESR spectra (solid curves) of the [111]P_b defect measured at 4.3 K with P_μ=-65 dBm and $B\perp$(111) interface of (111)Si/SiO$_2$ grown in enriched ^{17}O (51.24%) (a) and ordinary oxygen (b). The dotted curve is a computer simulation based on ^{17}O HF histogram calculation using the parameters summarized in Table I.

III. ESR spectra

The P_b signals of both the unenriched and enriched samples are shown in Fig. 2. The unenriched sample is characterised by ΔB_{pp}=1.69±0.03 G and a line shape factor $k \equiv I/[V_p(\Delta B_{pp})^2]$=5.6±0.3, where I and $2V_p$ represent the signal intensity (area under the integrated derivative spectrum) and and peak-to-peak height of the $dP_{\mu a}/dB$ signal, respectively. The corresponding density $[P_b]$=(4.9±0.3)x10^{12} cm^{-2} refers[9] to a DD broadening of 0.4 G. The distinct effect of ^{17}O enrichment is clear from a comparison of both spectra. While showing a broadening of the P_b response to ΔB_{pp}=2.71±0.06 G, there, interestingly, also appears (partially) resolved ^{17}O HF structure, which has three main characteristics as indicated on Fig.2(a): a kink at position A, a resolved peak at B, and a broadly extending shoulder indicated as region C. The increase of the line shape factor for the enriched sample to 10.1±0.8 just confirms the appearance of additional structure. Simulation of this structure based on the calculation of the ^{17}O HF stick diagram for various O shell surroundings and correct convolution provides information on the immediate O surrounding of the P_b defect.

A clear result is that the observed ^{17}O HF splittings are much smaller than that resulting from HF interaction with a central ^{29}Si nucleus (isotropic HF splitting ≈156 G),[4] showing again that, unlike Si, oxygen is not an immediate part of the P_b structure.

IV. ESR spectra simulation

Simulation of the P_b signal of the enriched sample starts from the calculation of the HF spectral histogram for the configuration of a P_b defect that is surrounded by r shells each containing n_i *equivalent* O sites, where r=4 and n_i=0,..,10. Each O site has a probability p=0.5124 for being occupied by an ^{17}O atom. Within the localized hybrid-orbital (LHO) picture, such histogram is simply obtained by accumulating the shifts in resonance field caused by each surrounding ^{17}O nucleus *individually* and the corresponding statistical amplitude for each 'line' (stick). This histogram is then convoluted with the 'unenriched' experimental spectrum of Fig.2(b) to obtain a simulation.

The simulation task now consists in calculating the convoluted HF histogram for each (physically reasonable) shell configuration, that is each set of n_1, n_2, n_3, n_4 values, and selecting the best fit. Since it is anticipated that O is not an immediate part of the P_b bonding structure, ^{17}O atoms in more distant shells are expected to cause only small, hardly resolvable HF splittings. Hence why the number of shells (r) considered is limited to 4. The histogram H(B) for one n_1, n_2, n_3, n_4 shell configuration is the accumulation of all pairs

$$H(B) = \{ (\prod_{i=1}^{4} 6^{-k_i} P_{k_i}^{n_i}, B_0 + \sum_{i=1}^{4} (\sum_{j=0}^{k_i} M_{I,j} a_{\|i})) \}, \quad (1)$$

where

$$P_{k_i}^{n_i} = \frac{n_i!}{k_i!(n_i-k_i)!} p^{k_i}(1-p)^{n_i-k_i} \quad (2),$$

$M_{I,0}=0$, and k_i and $M_{I,j}(j>0)$ are understood to run through all values $0, 1, \ldots, n_i$ and $-5/2, -3/2, \ldots, 5/2$, respectively, to cycle all possible combinations. B_0 represents the ESR resonance field in absence of HF interaction while $a_{\|i}$ is the HF splitting constant for $B\|[111]$ resulting from interaction of the unpaired P_b sp^3 hybrid with ^{17}O nuclei in shell i.

The best fitting result is shown in Fig.2 by the dashed curve, while the corresponding fitting parameters are summarized in Table I. This interestingly reveals that the P_b unpaired electron has its strongest HF interaction with only one O site (shell 1) characterised by $a_{\|1}=2.7$ G, which has a probability $p=0.5124$ of being occupied by an ^{17}O nucleus. The second strongest HF interaction is again only with one O site (shell 2) of $a_{\|2}=1.1$ G, while the 3th level interaction is with two equivalent O sites in shell 3. The 4th level HF interaction, of $a_{\|4}\approx 0.1$ G, with 10 equivalent O sites is to be seen as indicative rather than a correct physical result. The essential point for this shell is that it contains a *large* number of almost equivalent O sites of *small* HF interaction when occupied by ^{17}O atoms. They, in fact, represent the distant hemi-spherical cloud of small-$a_{\|i}$ O sites of which the mere effect is to slightly blur (broaden) the spectrum. Incorporation of this 'shell' only improves slightly the overall fitting quality, but is not essential.

Though the fit is not yet perfect, it accounts well for all characteristic details (cf. A, B, & C). And it needs to be mentioned that, regarding the first two O shells surrounding, the fit is unique. No reasonable fit can be produced if allowing more than one equivalent O site in the first or second shell.

Table I: ^{17}O HF splitting data and shell distributions of O sites in the immediate oxide neighborhood of $[111]P_b$ centers in $(111)Si/SiO_2$ as obtained by fitting 20.1 GHz ESR spectra measured at 4.3 K for $B\|[111]$.

Shell #	n_i	$a_{\|i}$ (G)	$\Delta a_{\|i} = \sigma_{M_{I,i}}/(2|M_I|)$ (G)
1	1	2.7±0.15	0.60
2	1	1.1±0.1	0.25
3	2	0.2	-
4	10	0.1	-

The final, almost perfect fit, as shown by the dotted curve in Fig.2, is obtained by additionally incorporating a spread in $a_{\|i}$, which fades the peaky structure. Such spread is known to exist[2,4,8] as a result of the interface strain and/or randomness of the overlaying SiO_x film. This causes slight variations in the positions of the surrounding O sites from P_b site to P_b site, even within one shell, with attendant alterations in wave function overlap and HF interaction strenghts. The spread in $a_{\|i}$ has been incorporated by replacing each line (stick) in the histogram shifted over $M_I a_{\|i}$ by a Gaussian distribution of equal intensity and standard deviation $\sigma_{MI,i} = 2|M_I|\Delta a_{\|i}$, that is a relative spread of $2|M_I|\Delta a_{\|i}/(2|M_I|a_{\|i}) = \Delta a_{\|i}/a_{\|i} \approx 22\%$. Note the increasing impact of $\Delta a_{\|i}$ with increasing field shift.

A remarkable result of the analysis is that, like the Si side, the immediate oxide side of the P_b center reproduces very well from P_b site to P_b site, at least what concerns the first three shells surrounding. This opposes a random matching of SiO_x to c-Si and is in favor of a kind of *epitaxial* transition.[9,10]

Another interesting result is that the relative spread $\Delta a_{\|}/a_{\|}$ in HF interactions is significantly larger for ^{17}O nuclei, that is $\approx 22\%$, than for central ^{29}Si atoms,[4] for which $\Delta a_{iso}/a_{iso} \approx \Delta a_{\|}/a_{\|} \approx 9.5\%$, where a_{iso} is the isotropic part of the HF interaction. This is as expected since the c-Si side of the P_b defect is a much more rigid structure than the oxide side: The Si-O-Si bond angle is much more flexible[11] than the rigid tetrahedral Si-Si-Si angle, which makes that the interfacial strain will be largely adapted by the overlaying SiO_2.

V. Discussion

So far, most HF data on defects have been interpreted along the simplified LHO picture (see, e.g., Refs. 4,12). Within this model the present data regarding the immediate O surrounding of P_b indicate that the defect has only one O atom in a nearest position, one O atom at a slightly larger distance, 2 equivalent O atoms in a third neighbor shell, and many more O atoms in more remote shells. The LHO model, however, ignores spin-polarization effects of atomic cores by valence levels[10] which makes the model more interpretative rather than predictive. Hence, it is not necessarily so that the O site leading to the strongest ^{17}O HF interaction is also the O site *nearest* to the core of the P_b defect, etc. Yet, the interpretation *is* conclusive about the 'symmetry' of the O surrounding, that is, strongest HF interaction with *only one* O site, closely followed by a weaker HF interaction again with *only one* O site.

The fact that O is not part of the central bonding structure of P_b points to a microvoid like structure of the SiO_2 cap overlaying the P_b defect. There is so far one such model[10] —an axial microvoid model— that pictures the P_b silica cap as an axially-symmetric puckered ditrigonal ring of six SiO_4 tetrahedra —an epitaxial tridymite-like crystalline transition.[9,10] Symmetry considerations, however, show that our results conflict with this otherwise attractive model. There are various ways to interpret this finding.

In a first one, still accepting the basic correctness of the tridymite-like concept, the results could perhaps refer to a distorted symmetry of the six-membered SiO_4 ring cap; if this ring would be off center, for example, this would indeed result in the removal of the axial symmetry of the ^{17}O P_b HF interaction. Previously equivalent O sites could then lead to (slightly) different ^{17}O P_b HF interaction strenghts, which would be more in agreement with observations. A difficult to meet requirement for such modified tridymite-like model, however, might be that, along the present results, the distortion introduced should reproduce largely identically over the numerous P_b sites.

Another interpretation could conclude the inappropriateness of the tridymite-like ring cap thus bearing out the need for another microvoid model. Along one suggestion, the data could well be in line with a matching zig-zag overlay consisting of 3 fairly linearly arrayed SiO_4 tetrahedra, as pictured in Fig.1. A slight asymmetric relaxation of this array resulting in an asymmetric positioning of two O atoms nearest to the P_b core would well agree with the present first two 'shells' ^{17}O HF interactions. It is realized though that the confirmation of such model will require an in-depth analysis of the *global* c-Si/SiO_2 matching within the framework of the correct oxidation mechanism.

It is clear that the correct evaluation of the present results in terms of deriving the correct oxide side of the P_b cluster will require substantial additional theoretical work, starting from detailed quantummechanical calculations on those novel microvoid cluster models which are deemed appropriate. The comparison of the derived symmetry and ^{17}O P_b HF interaction strengths with experiment will then select the correct terminating Si cap overlaying P_b defects.

VI. Conclusions

Optimized ESR measurements on [111]P_b defects in ^{17}O enriched (111)Si/SiO_2 structures have revealed the shell symmetry of O sites in the immediate silica surrounding, which, apparently, conflicts with the axial microvoid model based on the concept of the ditrigonal ring SiO_2 cap of P_b. The strenght of the ^{17}O HF interactions indicate that O is not incorporated in the central bonding structure of the P_b defect, as expected.

If the P_b's are seen as somehow co-establishing the intrinsic Si/SiO_2 interface structure rather than being loose results of interface adaptation,[13] the present new insight may add to uncover this structure and the closely linked oxidation mechanism. Much is expected from calculations incorporating the new insight.

REFERENCES

1. E. H. Poindexter, P. J. Caplan, B. E. Deal, and R. R. Razouk, J. Appl. Phys. 52, 879 (1981)
2. K. L. Brower, Phys. Rev. 33, 4471 (1986)
3. A. Stesmans and G. Van Gorp, Phys. Rev. 39, 2864 (1989)
4. K. L. Brower, Appl. Phys. Lett. 43, 1111(1983)
5. A. Stesmans, Appl. Phys. Lett. 48, 972 (1986)
6. A. Stesmans and G. Van Gorp, Phys. Rev. 42, 3765 (1990)
7. K. L. Brower, Z. Phys. Chem. Nue Folge 151, 177 (1987)
8. A. Stesmans and J. Braet, in *Insulating Films on Semiconductors*, edited by J. J. Simone and J. Buxo (Amsterdam, North Holland, 1986), p.25
9. A. Ourmazd, D. W. Taylor, J. A. Rentschler, and J. Bevk, Phys. Rev. Lett. 59, 213 (1987)
10. M. Cook and C. T. White, Phys. Rev. B38, 9674 (1988)
11. A. G. Revesz and G. V. Gibbs, in *Proceedings of the Conference on the Physics of MOS Insulators*, edited by G. Lukovsky, S. T. Pantelides, and F. L. Galeener (Pergamon, New York, 1980), p.92
12. G. D. Watkins and J. W. Corbett, Phys. Rev. 134, A1359 (1964); A. H. Edwards, Phys. Rev. B36, 9638 (1987)
13 A. Stesmans and G. Van Gorp, Appl. Phys. Lett. 57, 2663 (1991)

^{17}O HYPERFINE STUDY OF THE P_b CENTER

J.H. STATHIS[a], S. RIGO[b], I. TRIMAILLE[b], and M.S. CROWDER[c]

[a]IBM Research Division, T.J. Watson Research Center, Yorktown Heights, NY USA 10598
[b]Group de Physique des Solides, Université Paris VII, Tour 23, 2 Place Jussieu, 75251 Paris Cedex 05
[c]IBM Storage System Products Division, San Jose, CA USA 95193

ABSTRACT

Using ^{17}O-enriched thermal oxide on silicon, we have measured the hyperfine interaction between dangling bonds at the (111) interface (P_b centers) and oxygen atoms in the SiO_2. Our analysis indicates that each P_b center interacts weakly with only a single oxygen atom.

1. Introduction

The P_b center is a silicon dangling-bond pointing into the SiO_2 from the Si side of a Si/SiO_2 interface[1]. It is one of the most important point defects in terms of its influence on silicon technology, and has been extensively studied using electron paramagnetic resonance (EPR). A combination of ^{29}Si hyperfine measurements[2] and theoretical models[3] have revealed the essential structure of this defect at the {111} interface: a trivalent (3-coordinated) silicon situated exactly at the interface, with the dangling bond pointed in the [111] direction normal to the interface into the oxide overlayer, and back-bonded to three silicon atoms in the Si substrate. Electrical measurements[4] have further revealed its energy levels with respect to the Si band gap.

In spite of such study, many aspects of its structure remain unknown. In particular, we address here the question of the position of oxygen atoms surrounding the Si dangling bond. It can be seen that our knowledge of the structure of this defect is entirely from the *silicon* side of the interface. Yet the dangling bond, if it points into the oxide, must point *at* something. Are the nearest oxygen atoms arrayed in specific, well-defined positions with respect to the central silicon, or are their positions random? How many oxygen atoms are involved, or does this number vary from site to site? Several ideas exist in the literature. First is the idea that the dangling bond juts into a small void in the oxide. The structure of such a void might be random, or it might consist of a well-defined cage structure such as that suggested by Cook and White[3]. On the other hand, the dangling bond might actually point into the SiO_2 network itself, since the much smaller dielectric constant (larger band gap) of the insulator compared to the semiconductor may make the actual interaction small. In this context, one must again ask whether the structure of the oxide over the dangling bond has a well-defined form (such as the idea suggested by Pantelides[5] in which the dangling bond points at a single oxygen directly over the silicon atom) or is simply random. The resolution of these various possibilities will answer the question of whether the oxide itself plays any essential role in the formation or electronic properties of the P_b center. The understanding of the oxide side of the P_b defect is essential to a complete understanding of its structure and origin.

EPR hyperfine measurements represent the best means of addressing this question. Hyperfine structure can be directly related to the number of nuclei that are part of the structure of a defect, and to the overlap of the spin density with these nuclei. Just as the ^{29}Si hyperfine structure enabled us to learn about the Si side of the defect, ^{17}O hyperfine measurements can be used to tell us about the oxide side. To enhance the concentration of ^{17}O (I=5/2, natural abundance .037%) oxides were grown in 55.65% isotopically enriched oxygen. EPR results on these samples are described in this paper. Preliminary results were given previously[6]. By deconvolution of the ^{17}O-broadened spectrum, we are able to ascertain that the unpaired election on the P_b center interacts weakly with only a single oxygen atom.

We have also performed the first S-band (4 GHz) EPR measurements of the P_b center, comparing them to X-band (10 GHz) results on the same samples, for both normal (^{16}O) and ^{17}O-enriched samples. From these EPR measurements at different frequencies we confirm earlier ideas about the the origin of the EPR linewidth.

2. Experimental Details

The samples were made from (111) oriented wafers, resistivity 5-8 kΩ-cm, polished on both sides to a thickness of 5-6 mil. Prior to oxidation, the wafers were cleaved into 2mm×20mm strips, then were cleaned using standard procedures. The purpose of cleaving *prior* to oxidation was to minimize the EPR signal from the cleaved edges. We found that cleaning and oxidizing the cleaved edges greatly reduces this signal, compared to cleaving after oxidation. Unfortunately some residual edge damage signal is still present. An additional step of lightly etching the edges of the samples prior to oxidation may have been able to reduce this signal further, but we found that after oxidation the usual etchant was ineffective.

The samples were initially oxidized in atmospheric-pressure dry oxygen of normal isotopic ratio at 800°C for approximately 5.25 hours, to a thickness of approximately 140 Å. Following this, all samples received futher oxidation at 800°C and 38 mbar for approximately 5 hours in either normal oxygen or in oxygen that was isotopically enriched to 55.65% ^{17}O, adding about 40 Å to the oxide thickness. The interfacial region is formed during the second oxidation step, with the isotopic make-up of the oxide at the interface equal to that of the gas[7]. EPR measurements were performed on Bruker ER-200 systems at X-band (10 GHz) and S-band (4 GHz) at room temperature. The concentrations of P_b centers in the normal and ^{17}O-enriched samples were the same within 10%, about 3.7×10^{12}cm^{-2}, determined by double numerical integration and comparision with a calibrated ruby standard.

3. Results

Most studies of the P_b center have been performed at 10 GHz or 20 GHz. However, as a general rule, weak hyperfine interactions are better resolved the lower the microwave frequency. This is because the hyperfine splitting (in magnetic field units) is independent of frequency, while some other sources of broadening, such as inhomogenous broadening caused by random variations in the defect structure, are proportional to frequency. To determine which broadening mechanism dominates the linewidth of the P_b center, we performed measurements on the same samples at both S-band (4 GHz) and X-band (10 GHz). Although multi-frequency data exist in the literature, in the form of X-band (10 GHz) and K-band (20 GHz) measurements by different groups, this is the first time multi-frequency measurements have been performed on *identical* samples. This is important because the linewidth of the P_b center varies depending on defect concentration[8] and other processing conditions.

Comparisons of 4 GHz and 10 GHz measurements of the P_b center, prepared with either normal or ^{17}O-enriched oxygen, are shown in Figures 1 and 2. These spectra are plotted in a somewhat unconventional fashion: Each curve is the *difference* between spectra taken with the magnetic field parallel to [111] direction and with the field normal to this direction. The spectrum thus appears as a superposition of the resonance with H∥[111] and an *inverted* resonance with H⊥[111]. This allows the resonaces from both orientations to be displayed simultaneously.

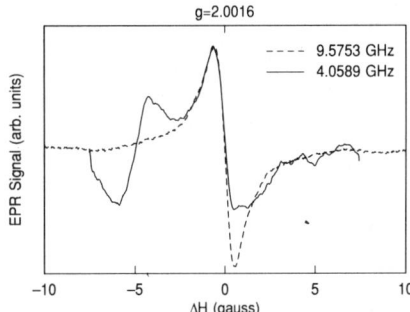

1. Comparison of 10 GHz and 4 GHz spectra of P_b centers in samples grown in normal oxygen. The spectra have been shifted along the field axis to align them at $g_\parallel$=2.0016. The peak-to-peak width of the $g_\parallel$ resonances are the same at the two frequencies, within experimental error. Note that the data are EPR spectra at two different orientations, subtracted to eliminate background signals and to allow display of both orientations on the same curve. The $g_\perp$ resonance (toward lower field) therefore appears inverted. The 4 GHz spectrum is slightly distorted on the high field side by a background signal that did not subtract completely.

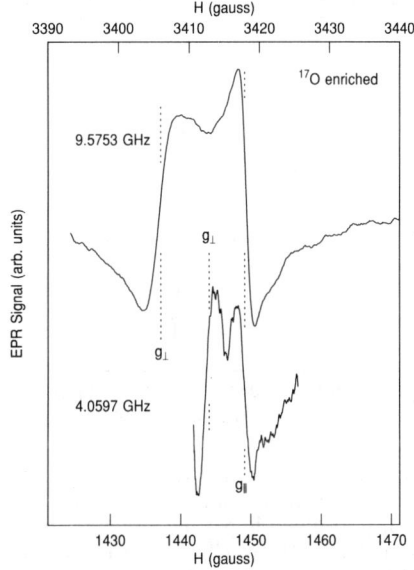

2. Comparison of 10 GHz and 4 GHz spectra of P_b centers in samples grown in ^{17}O-enriched oxygen. The field axes are arranged so that the spectra are aligned at $g_\parallel$. Note that the $g_\perp$ linewidth scales with frequency. See Figure 1 for a description of the line shapes.

For the normal-oxygen samples, we find the linewidth for H∥[111] to be the same, within experimental error, at 4 and 10 GHz (Figure 1). (The shoulder on the high-field side of the S-band spectrum is a background artifact.) This confirms earlier assertions[9] that this linewidth results from either unresolved hyperfine or dipolar interactions. Thus, there appears to be no advantage in going to lower frequency for the present study, since there is no reduction in the intrinsic linewidth and therefore no expected improvement in resolving the individual ^{17}O hyperfine lines.

For ^{17}O-enriched samples, the linewidth for H∥[111] is independent of frequency, as expected for hyperfine broadening. The H⊥[111] resonance, however, is considerably narrower at S-band, confirming the idea[9] that at X-band (and higher frequencies) this linewidth is determined by g variations.

The X-band EPR spectrum of P_b centers from an ^{17}O enriched interface is shown in Figure 3, which compares the spectrum from a sample grown in normal oxygen under identical conditions. The feature on the low field side is the resonance due to the scribed edges of the Si substrate, and does not concern us here.

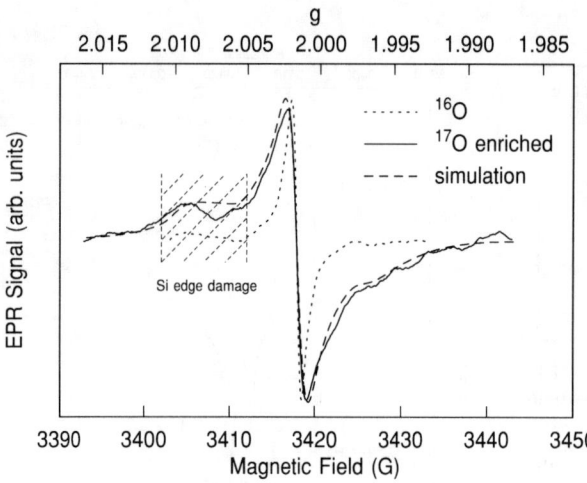

3. EPR spectra of P_b centers in a normal oxide and an ^{17}O-enriched oxide. The dashed curve is a simulation based on the deconvolution discussed in the text.

The broadening of the resonance due to ^{17}O is apparent. The ^{17}O-enriched spectrum in Figure 3 also exhibits faint shoulders visible especially on the high field side. These may be partially resolved hyperfine structure, but this interpretation must await confirmation and further analysis. Brower[10], who was the first to report the ^{17}O hyperfine broadening of the P_b center, did not observe this structure. However, we see from Figure 2 that the S-band spectrum also hints at such structure.

Lacking unambiguously discernible hyperfine structure, we will not make at this point any attempt at a precise determination of the positions of oxygen atoms around the dangling bond. Nonetheless, we are able to address the simpler question of *how many* oxygens are interacting with the dangling bond, without mapping their precise positions.

Our analysis relies on the fact that the ^{17}O enrichment of the interface is only about 50%. From this, it follows that some percentage of the dangling bond sites must involve only ^{16}O (I=0). The percentage will depend exponentially on the *number* of oxygen atoms which are part of the structure of the P_b center. If N oxygen atoms comprise the oxide side of the P_b center, then for an ^{17}O enrichment ε the fraction of sites with *only* ^{16}O nuclei will be $(1-\varepsilon)^N$. This gives the fractional intensity of the expected central (non hyperfine-broadened) line. The spectrum in the ^{17}O-enriched sample should consist of a central line with this intensity, with the remainder of the signal in a a broader background arising from those sites with one or more ^{17}O nuclei. This broad component would consist of a series of resolved lines if the number N of oxygen atoms is small and their positions relative to the defect are the same at every defect. It will be of gaussian shape if a large number of oxygens are involved *or* if their position varies from site to site so that the individual resonances cannot be resolved.

To separate these two components we performed a simple fourier-transform deconvolution and digital filtering, using the unenriched spectrum as the deconvolution function. The result is shown in Figure 4. We obtain a sharply peaked spectrum, with broad wings. The resolution of our deconvolution procedure, given our particular choice of digital filtering, is shown in the figure.

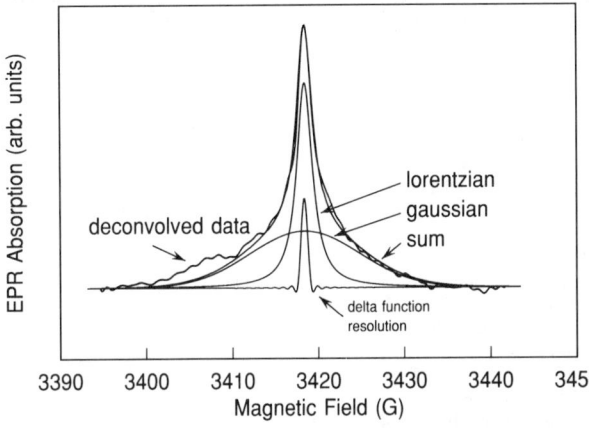

4. Deconvolution of the EPR spectrum of P_b centers in ^{17}O-enriched oxide. The fit to a sum of lorentzian and gaussian components is shown.

We find that the deconvolved line can be very accurately described as a sum of two lines: a broad gaussian and a narrower lorentzian as illustrated in Figure 4.

We interpret the gaussian component as the envelope of the hyperfine lines for those sites with one or more ^{17}O nuclei near the defect. Comparing the integral of this line with that of the central line (the lorentzian component) we find that the hyperfine-broadened component of the spectrum is 56% of the total. This is in precise agreement with the predicted value if each P_b center interacts predominantly with only one oxygen atom. Because the hyperfine structure appears as only a broad gaussian line, not as individual resolved lines, the position of this single oxygen must vary randomly from one defect site to the next. In this sense, the structure of the P_b center is not precisely that suggested by Pantelides[5] in which an oxygen atom sits directly over and is bonded to the silicon atom, the oxygen thus being 3-fold coordinated. Calculations[11] and experiment[12] have indicated that the interaction with the oxygen atom would be too strong in this configuration. According to our results, the mean hyperfine coupling A_{iso} of the dangling bond to the oxygen is only about 2.5 gauss, corresponding to only about 0.15% wave function amplitude on the oxygen 2s orbital.

It is possible that, because of the noise present in the data, our deconvolution procedure is not capable of resolving the hyperfine lines individually; we are presently exploring other deconvolution methods. If such individual hyperfine lines can be resolved it will enable a more precise determination of the position of the oxygen atom and the distribution of these positions. This will not change the main conclusion of this paper which is that the dominant interaction is with a single oxygen.

As explained above, the sites with only ^{16}O are expected to give an unbroadened central line. We interpret the lorentzian component in Figure 2 as this central component. While a naive model predicts that this line should appear as a delta function, the additional broadening could result from a variety of causes, such as hyperfine interactions with more distant ^{17}O in the oxide bulk, or superexchange with nearby defects mediated by these ^{17}O nuclei[13].

Going back to Figure 3, we illustrate (by the dashed line) the *simulation* of the ^{17}O-enriched spectrum based on the above deconvolution. To generate this curve, the fitted deconvolution (sum of lorentzian and gaussian) was convolved with the spectrum of the unenriched sample. The fit to the original data for the enriched sample is quite good, including the shoulder at 3425 G, demonstrating the internal consistency of our deconvolution. Note that the Si edge damage is not included in the model, so the fit is not expected to match the data in this region.

4. Summary

In conclusion, our analysis of the EPR spectrum of the P_b center in a partially ^{17}O-enriched oxide shows that the P_b center dangling bond interacts weakly with a *single* oxygen atom. Within the resolution of our analysis, the position of this oxygen atom relative to the dangling bond is not well-defined, but varies randomly from site to site.

REFERENCES

[1] E. H. Poindexter and P. J. Caplan, Prog. Surface Science **14**, 210 (1983).
[2] K. L. Brower, Appl. Phys. Lett. **43**, 1111 (1983).
[3] M. Cook and C. T. White, Phys. Rev. B. **38**, 9674 (1988).
[4] E. H. Poindexter, G. J. Gerardi, M. E. Rueckel, P. J. Caplan, N. M. Johnson, and D. K. Biegelsen, J. Appl. Phys. **56**, 2844 (1984).
[5] S. T. Pantelides, Phys. Rev. Lett. **57**, 2979 (1986).
[6] J. H. Stathis, S. Rigo, and I. Trimaille, Solid State Commun. **79**, 119 (1991).
[7] F. Rochet, B. Agius, and S. Rigo, J. Electrochem. Soc. **131**, 914 (1984).
[8] A. Stesmans and G. Van Gorp, Phys. Rev. B **42**, 3765 (1990).
[9] K. L. Brower, Phys. Rev. B **33**, 4471 (1986).
[10] K. L. Brower, Z. Phys. Chem. Neue Folge **151**, 177 (1987).
[11] A. H. Edwards, Appl. Surf. Science **39**, 309 (1989).
[12] N. M. Johnson, W. Shan, and G. Yu, Phys. Rev. B **39**, 3431 (1989).
[13] C. P. Poole, *Electron Spin Resonance* (Wiley, New York, 1983).

DEFECTS INDUCED BY HIGH ELECTRIC FIELD STRESS AND THE TRIVALENT SILICON DEFECTS AT THE Si-SiO$_2$ INTERFACE

D. VUILLAUME, A. MIR and D. GOGUENHEIM
Laboratoire d'Etude des Surfaces et Interfaces, URA 253 CNRS,
Institut Supérieur d'Electronique du Nord, 41 Bd. vauban, 59046 Lille, France

ABSTRACT

The interface states created at the <100> Si-SiO$_2$ interface by the injection of electrons from the substrate under the application of a high electric field (8-10 MV/cm) across the oxide are studied by deep level transient spectroscopy (DLTS). The temperature and field dependence of the creation mechanism are investigated. For an oxide field above a critical value of 8.3-8.9 MV/cm and stress at low temperature (100K), defects are only created in the range 0.2-0.3 eV below the conduction band ("P_{b0}-like" defects), while at lower field both "P_{b0}-like" and "P_{b1}-like" (at energy 0.4-0.5 eV) are created together. However, in the former case, a long-term time-dependent formation of "P_{b1}-like" defects during storage at room temperature after the damaging source has been turned-off is evidenced.

1. Introduction

The electronically active defects generated in MOS devices are known to reduce the lifetime of VLSI circuits. For instance, the high electric field applied into the gate oxide of EEPROM, leads to degradation of device performances because the stress generates defects at the Si-SiO$_2$ interface and into the oxide. With the reduction of oxide thicknesses the knowledge of the effects induced by high electric field stress (HEFS) is an important challenge for the reliability of deep-submicronic devices. However, despite the fact that this subject has gained a great interest these last 10 years, the creation mechanism as well as the microscopic nature of these HEFS-induced interface defects are not fully understood. Two main mechanisms involve the role of interfacial trapped-holes [1,2] or the diffusion of hydrogen-related species in the oxide towards the interface [3-6] as the precursors for the interface state creation. Concerning the nature of these HEFS-induced defects, some work have pointed-out the possible relation with the trivalent silicon defects (Si dangling bond, the so-called P_b center) at the Si-SiO$_2$ interface [7-10] but no definitive picture emerges at the moment. It has also been emphasized that a long-term time-dependent evolution of interface states takes place after the source of HEFS or ionizing radiation has been turned-off [11-14].

This paper gets new insights on the stress temperature and oxide field dependences of the creation mechanism of the HEFS-induced interface defects. The possible nature of these defects will be discussed in the light of the P_b center properties at the <100> Si-SiO$_2$ interface.

2. Devices and experiments

The n-type (phosphorous doped to $2.5 \times 10^{15} cm^{-3}$) MOS capacitors were fabricated on <100> oriented Si surfaces by dry oxidation at 1050°C resulting in an oxide thickness of 750Å. Oxydation was followed by a post oxidation anneal (POA) in N_2+H_2 at 400°C during 1 hour. Aluminium gate was evaporated and followed by a post metallization anneal (PMA) in forming gas. For the sake of comparison with the properties of P_b centers, as-oxidized samples have been made with a similar oxidation process, except for the POA and PMA in order to achieve a high density of P_b centers [15].

The HEFS were performed by Fowler-Nordheim tunnelling (FNT) injection of electrons from the Si substrate (positive bias on the gate) under a constant gate voltage mode. The average oxide fields were in the range 8.3-8.9 MV/cm and the time variation of the injected current was measured by a HP4140 picoamperemeter and integrated to give the injected charges. The sample were mounted in a continuous-flow liquid nitrogen cryostat (Biorad-Polaron DL4960) to perform the stress in the temperature range 100 to 450K and the DLTS measurements. High-frequency (1MHz) capacitance-voltage (HFCV) characteristics were systematically performed to test the electrical quality of the devices. The measurement of the energy distribution of the interface state density $D_{it}(E)$ and electron capture cross-section were carried out by deep level transient spectroscopy [16,17]. We have also used the HFCV measurement at low temperature (100K) (Jenq method [18]) to measure the energy-integrated state density after having performed stress at low temperarure (LT) and to avoid the extra-generation of interface states during the warm-up of the stressed samples to room temperature (RT) [1]. Charge trapping in the oxide films was determined from the mid-gap voltage shift measured by the HFCV.

3. Results and discussion

Figure 1 shows the increase of the interface state density ΔD_{it} measured at several energies below the CB for sample having received an electron injection fluence of $Q_{inj}=5.6 \times 10^{15} cm^{-2}$ at an oxide field $E_{ox}=8.3MV/cm$. DLTS measurements have been made after RT annealing of the stressed sample during 1 hour. This time is long enough to obtain a stabilized state density. These results include the generation of defects during RT anneals [1,19]. The generation of interface states by the HEFS alone is measured by the Jenq technique [18] immediately after the stress (fig. 1). Basically, the creation mechanism is thermally activated above 180K, and at lower temperatures the generation of interface states takes place during the RT anneal after the HEFS has been turned-off. Thermally activated creation has been also reported by DiMaria et al. [6] for electron traps creation at higher injected charge fluences. We have recently reported [20-22] that these results at temperatures above 180K are consistent with the model involving the diffusion of hydrogen-related species released at the anode side by hot-electrons which pill-up at the Si-SiO$_2$ interface where they are able to create defects.

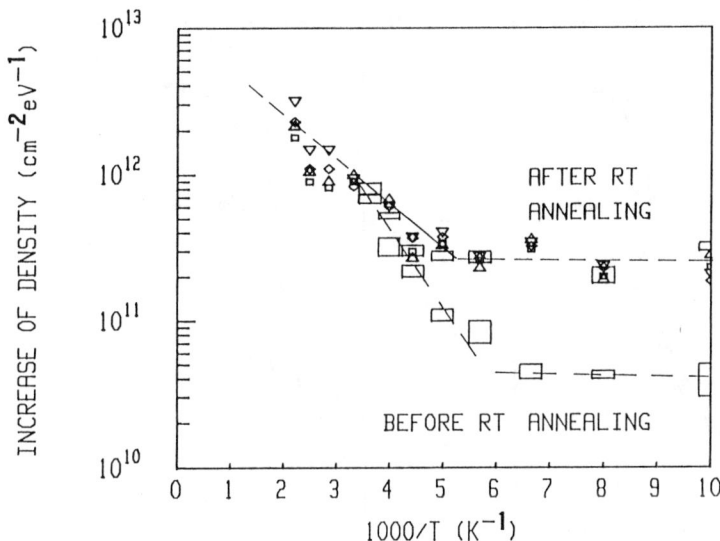

Figure 1 : ΔD_{it} as a function of the inverse of the stress temperature measured immediatly after the HEFS and after RT annealing.

In this paper we focusse on the mechanism responsible for the formation of interface during RT anneal after HEFS performed at low temperature (100K). Figure 2 show the LT-CV obtained at oxide fields E_{ox}=8.3MV/cm and 8.9MV/cm. For HEFS at 8.3MV/cm the following features emerge : we have not detected any significant positive charge generation after stressing. Only a weak positive charge could be deduced from the CV shift corresponding to less than 4.8×10^{10} positive charges/cm². This charge is one decade lower than the interface state density generated by the subsequent RT anneal during 1h as shown by LT-CV and DLTS measurements (fig. 3). We therefore conclude that the trapped hole model [1,2] does not satisfactorily explain our results. In this model the generation of interface states is ascribed to a transformation process from interfacial positive charges (trapped holes) generated during the HEFS which turn into interface states during the RT anneal [1,2]. DLTS measurements (fig. 3) show that interface states are created in the wide energy range in the upper mid-gap. We have recently reported [10] that the energetical distribution and the isochronal anneal properties of theses HEFS-induced interface states are strictly similar to those of the <100> P_b centers measured by DLTS on as-oxidized samples. However, the energetical dependences of their electron capture cross-sections are different. We have suggested that HEFS-induced defect should be a non-isolated P_b-center, i.e. a P_b defect with an unknown species X in its neighboring sites (P_b-X). The presence of this species X, interacting with P_b would be expected to stiffen the interatomic forces, leading to a strong modification of the capture cross-sections. This (P_b-X) model is consistent with the similarity of the state density distributions of the HEFS and as-oxidized samples because the number of defects sites at the <100> Si surfaces are equal in both cases. If the interaction between P_b and X is not too strong, the annealing properties of a (P_b-X) defect would resemble that of the sole P_b center.

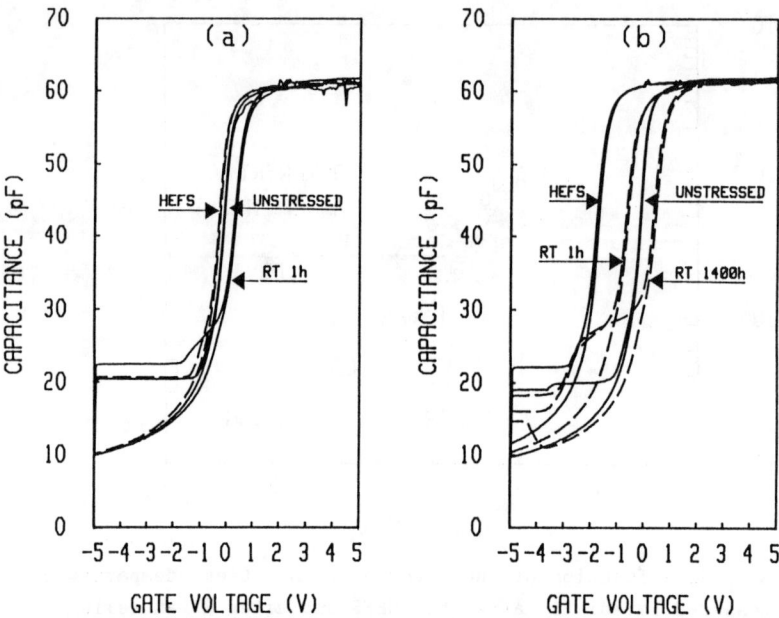

Figure 2 : Low temperature CV curves measured on unstressed samples, immediately after the HEFS and after RT annealing, (a) HEFS at an oxide field of 8.3 MV/cm, (b) 8.9 MV/cm.

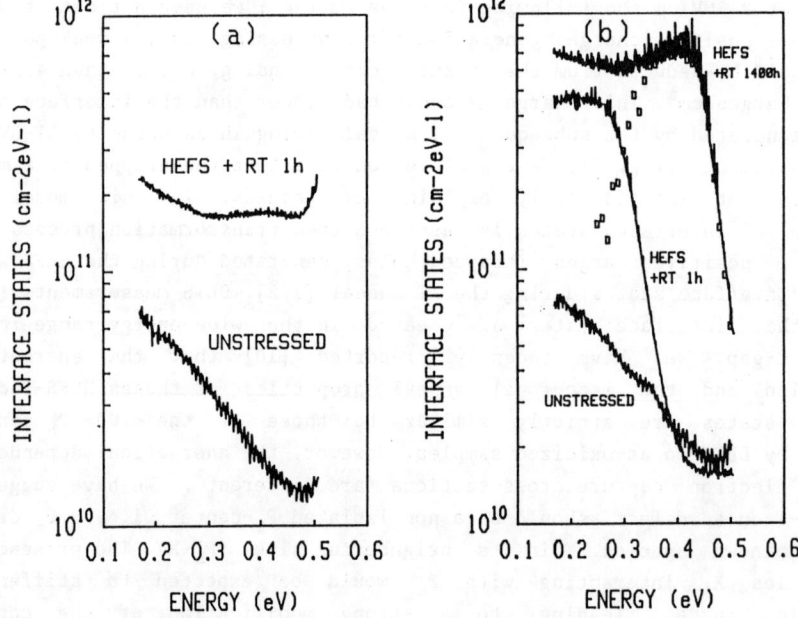

Figure 3 : DLTS spectra measured on unstressed sample and after different times of RT anneal : (a) E_{ox}=8.3 MV/cm, RT anneal 1h (b) 8.9MV/cm, RT anneals 1h and ~1400h. Squares are the states generated after the first 1h RT anneal.

A very different behavior is observed for the HEFS stress at 8.9MV/cm. A large interfacial positive charge ($\sim 4 \times 10^{11} cm^{-2}$) is generated by the low temperature stress (fig. 2). This positive charge decreases after RT anneal during 1h while the density of interface state increases. This behavior is consistent with the earlier experiments in favor of the trapped hole model [1,2]. In our work, DLTS measurements have shown more interesting features (fig. 3). After RT anneal during 1 hour, states are only generated in the range 0.2-0.3 eV below CB. This energy corresponds to the characteristic energy of the P_{b0} centers at the <100> Si surface [23,24] and in the light of our earlier results (recalled above) [10] we call this defect "P_{b0}-like" in the following. After storage at RT during ~1400h, we have observed that additional interface states have appeared in the range 0.4-0.5eV. By substracting the two DLTS curves, we observe that these interface states exhibit a density peaked at $\sim E_c - 0.4eV$ (fig. 3, squares). This energy distribution resemble that of the P_{b1} center measured at $E_c - 0.42$ eV $\pm$ 0.02 eV by DLTS on our as-oxidized samples [24]. We call this HEFS-induced defect a "P_{b1}-like" defect. Moreover, during this long time storage at RT, the positive charge has completely disappeared as shown by the LT-CV measurement (fig. 2).

4. Conclusions

We have found that a critical electric field (between 8.3 and 8.9 MV/cm) separates two behaviors for the creation of defects in MOS structures submitted to HEFS. Below it, no interfacial positive charge is created during the stress at low temperature and both "P_{b0}-like" and "P_{b1}-like" defect are created by RT anneal. This creation is completed after a 1 hour anneal and DLTS measurement performed after one year storage at room temperature have given the same state density. Above it, the generation of positive charge is evidenced and a transformation process of this charge into interface states is observed as predicted by the trapped hole model in agreement with earlier work [1,2]. However, a "P_{b0}-like" defect is rapidely created (after 1 hour anneal at RT), while we have observed a long-term time-dependent (~1400 hours) creation of a "P_{b1}-like" defect. Our results on the long-term time-dependent creation of interface defects are not similar to those reported by Ma and coworkers [11-14]. Over the same time period (~1000-2000 h) they have observed a decrease of interface states peaked at $\sim E_v + 0.75eV$ associated with the appearance of a peak at $\sim E_v + 0.35eV$. Our DLTS measurements are not able to investigate states below mid-gap on n-type MOS capacitors but we have observed an increase of the states at $\sim E_c - 0.4$ eV in contrast with the results of Ma and coworkers. However, quite similar results to those shown in fig. 3 have been observed on irradiated samples [25]. It is intersting to notice that both damage processes yield similar defect creation mechanism. More extensive measurements (detailed time-dependent creation kinetics, gate bias and temperature effects...) are under progress to analyse the mechanism responsible for the behavior evidenced in the present work.

ACKNOWLEDGEMENTS

We are indebted to J.C. Bourgoin for fruitful discussions and to M. Jourdain and G. Vincent for supplying the samples. This work was supported under contract by the "GCIS". One of us (DG) acknowledges financial support from a grant of the French Minister of Research and Technology.

REFERENCES

1. G. Hu and W.C. Johnson, Appl. Phys. Lett. 36, 590 (1980).
2. S.J. Wang, J.M. Sung and S.A. Lyon, Appl. Phys. Lett. 52, 1431 (1988).
3. C.T. Sah, J.Y.C. Sun and J.J. Tsou, J. Appl. Phys. 54, 5864 (1983).
4. C.T. Sah, J.Y.C. Sun and J.J. Tsou, J. Appl. Phys. 55, 1525 (1984).
5. D.J. DiMaria, Appl. Phys. Lett. 51, 1431 (1987).
6. D.J. DiMaria and J.W. Stasiak, J. Appl. Phys. 65, 2342 (1989).
7. W.L. Warren and P.M. Lenahan, Appl. Phys. Lett. 49, 1296 (1986).
8. H. Miki, M. Noguchi, K. Yogokawa, B. Kim, K. Asada and T. Sugano, IEEE Trans. Electron Devices 35, 2245 (1988).
9. G.J. Gerardi, E.H. Poindexter, P.J. Caplan, M. Harmartz, W.R. Buchwald and N.M. Johnson, J. Electrochem. Soc. 136, 2609 (1990).
10. D. Vuillaume, D. Goguenheim and J.C. Bourgoin, Appl. Phys. Lett. 58, 490 (1991).
11. E.F. DaSilva, Y. Nishioka and T.P. Ma, Appl. Phys. Lett. 51, 270 (1987).
12. Y. Nishioka, E.F. DaSilva and T.P. Ma, IEEE Electron Dev. Lett. EDL-8, 566 (1987).
13. Y. Nashioka and T.P. Ma, Appl. Phys. Lett. 53, 1744 (1988).
14. T.P. Ma, Semicond. Sci. Technol. 4, 1061 (1989).
15. E.H. Poindexter and P.J. Caplan, Prog. Surf. Sci. 14, 201 (1983).
16. M. Schulz and N.M. Johnson, Appl. Phys. Lett. 31, 662 (1977).
17. D. Goguenheim, D. Vuillaume, G. Vincent and N.M. Johnson, J. Appl. Phys. 68, 1104 (1990).
18. C.S. Jenq, PhD Dissertation, Princeton Univ., 1977 (unpublished), see a brief description in Ref. [1].
19. J.K. Wu, S.A. Lyon and W.C. Johnson, Appl. Phys. Lett. 42, 585 (1983).
20. D. Vuillaume, R. Bouchakour, M. Jourdain, G. Salave and A. El-Hdiy, in Proc. of INFOS91 (Liverpool, 1991), in press.
21. D. Vuillaume, R. Bouchakour, M. Jourdain, A. El-Hdiy and G. Salace, J. Appl. Phys., submitted.
22. D. Vuillaume, Appl. Phys. Lett., to be published.
23. G.J. Gerardi, E.H. Poindexter, P.J. Caplan and N.M. Johnson, Appl. Phys. Lett. 49, 348 (1986).
24. D. Vuillaume, D. Goguenheim and G. Vincent, Appl. Phys. Lett. 57, 1206 (1990).
25. R.E. Stahlbush, B.J. Mrstik and R.K. Lawrence, IEEE Trans. Nucl. Sci. NS-37, 1641 (1990).

INTERSTITIAL DEFECT REACTIONS IN SILICON PROCESSED BY REACTIVE ION ETCHING

J. L. BENTON and M. A. KENNEDY
AT&T Bell Laboratories, Murray Hill, New Jersey 07974

J. MICHEL and L. C. KIMERLING
Massachusetts Institute of Technology, Cambridge, Massachusetts 02193, and
AT&T Bell Laboratories, Murray Hill, New Jersey 07974

ABSTRACT

Reactive Ion Etching (RIE) of Silicon introduces interstitial defects into two distinct regions of the near surface of the material. The *surface damage region* extends to 1000 Å below the surface, contains a high concentration of interstitial defects (Si_i, C_i and B_i), and exhibits donor characteristics. The *defect reaction region* is created by the diffusion of interstitial defects into the bulk where they are trapped by impurities, forming associates. The extent of these associative defect reactions depends on the concentration of impurities and their interaction distances, and is observed to depths ≥1 µm. Photoluminescence (PL) experiments reveal the recombination enhanced diffusion of C_i into this second region where it is trapped by O_i in Czochralski grown Si, forming C_iO_i defect pairs with a diffusion length of 500 Å. In float-zoned Si, carbon is the dominant impurity, and the formation of C_sC_i pairs is observed with a diffusion length of 5000 Å. Spreading Resistance (SR) measurements allow the investigation of the reaction $B_i + O_i \rightarrow B_iO_i$. Isochronal anneals indicate that these carbon, oxygen and boron interstitial defects recover by 400 °C.

Introduction

Reactive ion etching (RIE) is a key process for pattern transfer in semiconductor device technology. The anisotropic nature of RIE is critical for the realization of silicon submicron design specifications. This work shows that the physical, electrical, and chemical reactions in the near surface region of RIE processed silicon introduce interstitial defects which control the properties of the underlying device material.

Impurity implantation displacement damage at the energies used for RIE (100-400 eV) is estimated to be contained in a surface region less than 50 Å. In addition to pure physical processes, RIE includes chemical interactions of the etching species which result in the weakening and breaking of Si bonds and which add to the extent of the damage. The relevant literature presents conflicting pictures of the type and depth of RIE damage. Measurements of current-voltage characteristics and Schottky Barrier heights show that RIE affects the electrical properties of Si, introducing donors at the etched surface.[1] Other electrical experiments[2][3][4] have described the depths of RIE induced defect penetration from 200 Å to 7µm. Transmission electron spectroscopy (TEM) has shown a modified silicon layer containing lattice damage at depths exceeding 3000 Å.[5] Earlier photoluminescence (PL) results[6] indicate that defects are confined to within 200 Å. The understanding of the interstitial defect reactions which occur during RIE provides the explanation for these perplexing anomalies.

We have previously reported[7][8] that silicon surfaces reactive etched in plasmas consist of two distinct regions, a *surface damage region* extending 1000 Å from the surface and a *point defect reaction region* which can extend to depths >1µm. In this work we survey the interstitial defect reactions with occur during RIE and monitor their recovery.

Experimental Procedure

Samples used in these studies were Czochralski (CZ) and floating zone (FZ) Si doped with either boron or phosphorus at $10^{14} - 10^{16}$ cm.$^{-3}$ Reactive ion etching in either $CF_4 + 8\%O_2$, $SF_6 + 8\%O_2$, or $SiCl_4$ was performed in a commercial parallel plate reactor. The reactor parameters were, a gas flow rate of 10 sccm, chamber pressure of 150 mtorr, rf power of 250 watts, dc bias of 175 volts, electrode temperature of 18°C, pump down time of 1 hour and a base pressure of 2×10^{-6} torr.

Immediately after RIE, the samples were immersed in liquid helium for PL measurements. The spectrometer consisted of a 0.75 meter monochromator with a liquid nitrogen cooled Ge detector and an argon-ion laser (514 nm). After PL, the samples were beveled at an angle less than 0.5°, and the carrier concentration was measured to 4μm depths using a Solid State Measurements 150 spreading resistance system.

Isochronal anneals were performed in flowing argon. To prevent contamination, samples were cleaned using a solvent degreasing step followed by a dilute HF dip, and then placed between between two silicon wafers during heat treatment.

Results

It is important to understand, when characterizing plasma etched silicon, that there are two distinct regions of damage. The shallow *surface damage region* is confined to 1000Å, is highly defective, and is a source of self-interstitials. Rutherford Backscattering data on silicon etched in $CF_4 + 8\%O_2$ indicates the presence of 3% F atoms in the first 1000Å, and transmission electron microscopy shows a high density of extended defects in this same region[8]. Current-voltage traces exhibit high leakage current in the reversed-biased direction until 1000Å have been removed from the RIE treated silicon surface,[7][8] indicating the presence of generation-recombination centers in this region. The enhancement of p-type Schottky barrier heights after RIE has been explained as the formation of compensating donors.[1] Spreading resistance measurements converted to carrier concentrations indicate the presence of donors in the near surface region in n-and p-type silicon reactive etched in CF_4, SF_6 or $SiCl_4$.

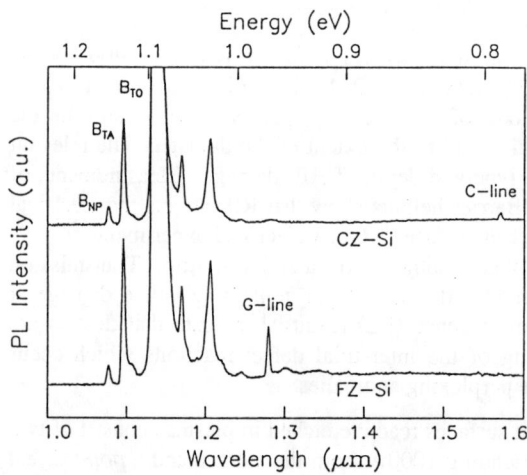

Figure 1. PL spectra of silicon Reactive Ion Etched in $CF_4 + 8\%O_2$. The G-line ($C_i C_s$) is present in FZ material and the C-line ($C_i O_i$) is observed in CZ Si.

An examination of PL spectra taken on silicon after RIE in any of the above mentioned gases gives evidence of the presence of interstitial related defects. The PL spectra of p-type silicon after RIE, shown in Figure 1, are representative of all our samples. The boron bound exciton peak and its phonon replicas are seen in the energy range 1.0 to 1.2 eV and are present in the control samples before RIE. After plasma processing, the Si PL spectrum includes the G-line, 969 meV, which has been identified as the interstitial carbon complex, C_iC_s,[9] and/or the C-line, 790 meV, proven to be the C_iO_i pair.[10] Depth profiles were obtained by anodic oxidation of the sample followed by an HF oxide strip which sequentially removed layers of 250 Å thickness. PL measurements suggest a high concentration of nonradiative centers in the *surface damage region* which dominate the luminescence process. The composite results of the electrical and optical measurements verify that a high concentration of defects are created in the near surface region by RIE and that it acts as a source of interstitial defects.

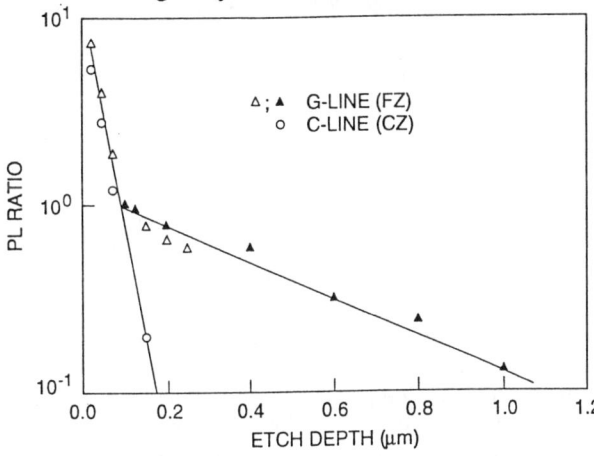

Figure 2. Photoluminescence peak intensity ratio versus depth in Si after RIE in $CF_4+8\%O_2$. In CZ material, the C-line (C_i-O_i) data is presented; in FZ Si, the G-line (C_i-C_s) is plotted.

In addition to the *surface damage region*, a second *defect reaction region* has been detected in previous PL experiments.[7] Figure 2 summarizes the supporting data. Relative defect concentrations of C_iC_s and C_iO_i are plotted as a function of depth from the Si surface after RIE in CF_4. The relative concentrations are determined by computing the ratio of the G-line or the C-line to the B_{TA} phonon replica intensity. The exponential decay of these defect concentrations suggests a trapping mechanism. Interstitial carbon, C_i, created near the surface by the RIE, diffuses into the bulk of the Si where it is trapped by oxygen in CZ material, forming C_iO_i pairs; or trapped by carbon in FZ silicon, making C_iC_s defects. The trapping length for these reactions is estimated using the relationship

$$L_C = (D_{C_i}\tau_{C_i})^{1/2} \qquad (1)$$

where D_{C_i} is the diffusivity of interstitial carbon in Si and τ_{C_i} is the time constant for the diffusion limited capture kinetics which is inversely related to the reaction radius for pair formation, R, and the concentration of the trapping impurity, [I],

$$\tau_{C_i} = (4\pi R D_{C_i}[I])^{-1/2} \qquad (2)$$

Substitution of a 1 Å reaction radius for interstitial trapping by a neutral impurity and an oxygen concentration of 10^{18} cm^{-3} in CZ Si gives a diffusion length for the formation of C_iO_i of $L_D = 500$ Å. Employing the same reasoning to estimate the depth of C_iC_s formation in FZ Si containing 10^{16} cm^{-3} substitutional carbon yields $L_D = 5000$ Å. The predominant defect product depends on the branching ratios which is discussed in following section. We conclude, therefore, that the size of the *defect reaction region* formed by RIE is governed by interstitial pair formation and is dependent on the impurity concentrations present in the starting material.

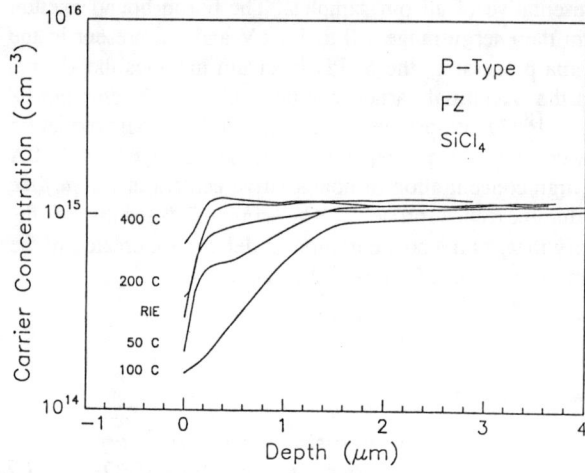

Figure 3. Carrier concentration profiles of boron doped FZ Si after RIE. The depth of the donor concentration varies with isochronal (20 min.) anneal, showing the two separate defect regions.

Evidence for additional interstitial defect reactions is seen in the spreading resistance measurements of boron doped, reactive etched Si. The carrier concentration profiles of p-type, FZ Si etched in $SiCl_4$ are presented in Figure 3. The data are presented for the sample after RIE and after 20 minute isochronal anneals. The presence of two separate regions of dopant compensation becomes obvious after heat treatments at 50°C and 100°C. The *surface damage region* is confined to 1000Å and survives heat treatment at 400°C, while the *defect reaction region* extends >1μm in depth and recovers by 200°C. Comparison of SR profiles of p-type FZ samples, $[O] = 10^{16} cm^{-3}$, with CZ silicon, $[O] = 10^{18} cm^{-3}$, simultaneously reactive ion etched verifies that the depth of the defect reactions depends on the concentration of oxygen.

The defects observed by SR are in the same *defect reaction region* as the defect pairs seen by PL, but they occur only in boron doped Si and therefore, cannot be $C_i C_s$ or $C_i O_i$ pairs. Based on the following observations, we believe that the SR data indicates the presence of $B_i O_i$ pairs.

Figure 4. Isochronal annealing curves of the $B_i O_i$ defect taken from the SR data, and of $C_i C_s$ pair based on PL G-line ratios.

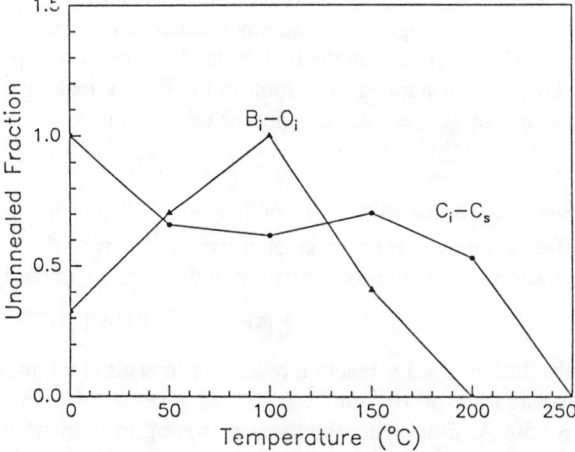

First, the donor defects are only seen in p-type material. Second, the concentration of donors, as measured by SR after annealing at 100°C, increases with the concentrations of both boron and oxygen. Simple diffusion length calculations using equations (1) and (2) predict B_iO_i pair formation at depths corresponding to the observed changes in carrier concentrations. And finally, the isochronal annealing data shown in Figure 4 agrees with the temperatures of growth and recovery published for the B_iO_i pair seen by Deep Level Transient Spectroscopy (DLTS) in electron irradiated silicon.[11]

Discussion and Conclusion

The comprehensive picture of the interstitial defect reactions in reactive ion etched silicon has its basis in the DLTS studies of high energy electron irradiated silicon[8][12]. The process of RIE creates self-interstitials and vacancies in the *surface damage region* of the silicon. The Si_i is then captured by either C_s or B_s to produce the interstitial defects, C_i or B_i, by the replacement mechanism.[13] The pure thermal diffusion of these elemental defects would be less than 100Å for the temperatures and times typical of RIE. However, the interstitial defects undergo recombination enhanced diffusion caused by the presence of ultraviolet light in the plasma. The illumination results in the injection of minority carriers which recombine at C_i or B_i resulting in the observed long range diffusion[8] of these defects in the *defect reaction region*. Subsequently, the interstitial atoms are trapped by the background impurities forming defect pairs which are observed experimentally. The depth of the diffusion-limited trapping and the probability of forming specific pairs depends on the relative concentrations of the reactants.

The reaction hierarcies determine the final defect products produced by RIE. The carbon branch in both n- and p-type silicon will be

$$Si_i + C_s \rightarrow Si_s + C_i \qquad (3)$$

$$C_i + X \rightarrow C_iX \qquad (4)$$

where X is oxygen, carbon or phosphorus. In FZ silicon, RIE produces C_iC_s which anneals at 200°C as shown in Figure 4. The C_iO_i pair is observed in CZ material and anneals at 400°C. The formation of C_iP_s is not expected during RIE because it is known to undergo recombination enhanced dissociation.[14] The B_i display parallel reactions,

$$Si_i + B_s \rightarrow Si_s + B_i \qquad (5)$$

$$B_i + X \rightarrow B_iX \qquad (6)$$

where the reactant, X, is oxygen, carbon or boron. In low resistivity silicon where the concentration of boron is $\geq 5 \times 10^{16}$ cm^{-3}, B_iB_s pairs are expected, but to date, there are no experiments on this system. Spreading resistance measurements give evidence that B_i does react with oxygen creating the B_iO_i pair. The dissociation of this pair at 200°C should feed the creation of the B_iB_s pair, but as yet, no identifying signal has been detected. The B_iB_s defect recovers at 400°C.

Reactive ion etching modifies the electrical and optical properties of silicon. The *surface damage region* contains a high concentration of nonradiative defects which are generation-recombination centers. Interstitial defects created in this region show enhanced diffusion into the *defect reaction region* where they are trapped by bulk impurities. The depth and defect contents of the latter region are determined by known interstitial defect reactions.

References

1. J. Stephen, S. Fonash, S. Ashok and Ranbir Singh, *Appl. Phys. Lett.* **39**(5), 423, (1981).

2. H. Matsumoto and T. Sugano, *J. Electrochem. Soc.,* **129**(12), 2823 (1982).

3. A. Henry, O. O. Awadelkarim, J. L. Lindstrom, G. S. Oehrlein, *J. Appl. Phys.* **66**(11), 5388 (1989).

4. S. W. Pang, D. D. Rathman, D. J. Silversmith, R. W. Mountain and P. D. DeGraff, *J. Appl. Phys.* **54**(6), 3272 (1983).

5. H. P. Strunk, H. Cerva and E. G. Mohr, *J. Electrochem. Soc.* **135**(11), 2876, (1988).

6. J. Weber and M. Singh, *Appl. Phys. Lett* **49**(23), 1617, (1986).

7. J. L. Benton, J. Michel, L. C. Kimerling, B. E. Weir, R. Gottscho, *J. Elect. Mat.* **20**(9), 643 (1991).

8. J. L. Benton, B. E. Weir, D. J. Eaglesham, R. A. Gottscho, J. Michel and L. C. Kimerling, *J. Vac. Sci. Tech. B,* Jan/Feb (1992).

9. K. D. O'Donnell, K. M. Lee and G. D. Watkins, *Physica (Ultrecht) B* **116**, 258 (1983).

10. G. Davis, E. C. Lightowlers, R. Woolley, R. C. Newman and A. S. Oates, *J. Phys. C,* **17,** L499 (1984).

11. P. J. Drevinsky, C. E. Caefer, L. C. Kimerling and J. L. Benton,

12. L. C. Kimerling, M. T. Asom, J. L. Benton, P. J. Drevinsky and C. E. Caefer, *Materials Science Forum***, 38-41,** 141, (1989).

13. G. D. Watkins, in *Radiation Damage in Semiconductors,* edited by N. B. Urli and J. M. Corbett, (Inst. of Phys. Conf. Ser. 31, London, 1977), p.273.

14. M. T. Asom, J. L. Benton, R. Sauer, and L. C. Kimerling, *Appl. Phys. Lett.,***51**,(4), 256 (1987).

ANOMALOUS DAMAGE DEPTHS IN LOW-ENERGY ION BEAM PROCESSED III-V SEMICONDUCTORS

S. J. PEARTON[1], F. REN[1], T. R. FULLOWAN[1], R. F. KOPF[1], W. S. HOBSON[1], C. R. ABERNATHY[1], A. KATZ[1], U. K. CHAKRABARTI[1] and V. SWAMINATHAN[2]

[1] AT&T Bell Laboratories, Murray Hill, NJ 07974, USA.
[2] AT&T Bell Laboratories, Solid State Technology Center, Breinigsville, PA 18031, USA.

ABSTRACT

Dry etching processes such as reactive ion etching (RIE) or ion milling typically introduce deep level defects into the semiconductor which compensate the shallow level doping over many hundreds to several thousand angstroms. This depth of damage introduction cannot be explained without invoking rapid diffusion of point defects into the semiconductor. For example, we show using a selectively doped GaAs/AlGaAs heterostructure that ion bombardment with 400 eV O^+ ions leads to damage depths $>410 \text{Å}$. The projected range (R_p) of such ions in GaAs is only 22 Å with a straggle (ΔR_p) of ~ 11 Å. Assuming a Gaussion distribution for O^+ ions directly implanted into the sample from the plasma, the oxygen distribution would fall to 10^{-3} of its peak value at $R_p + 3.72 \, \Delta R_p$ i.e. 63 Å. Channelling of these low energy ions can increase the incorporation depth by several times, but this alone cannot explain the results. Rapid diffusion of point defects created near the surface is required to understand the depth of carrier compensation. Substantial damage depths can be observed even for very low ion energy bombardment (≤ 150 eV), with implications for plasma processing of III-V device structures.

Introduction

One of the major issues with dry etching of GaAs and related materials is the introduction of surface damage by energetic ion bombardment. This has been investigated for a variety of dry etching methods using both inert and reactive ion bombardment. The degree of damage has been found to be inversely proportional to the ion mass, and directly proportional to the ion energy.[1] Sidewall damage has also been studied.[2,3] The most obvious effect of near-surface damage is a reduction in the carrier concentration between 200-1000 Å from the surface. This is considerably deeper than the projected range of ions crossing the plasma sheath and must be explained either by channelling of these relatively low-energy particles or recombination-enhanced motion of defects, or a combination of both. In at least some cases where H_2 is involved in the etch mixture, there appears also to be passivation of donors and acceptors by association with hydrogen. This is a well-known effect from studies of the role of atomic hydrogen in passivating shallow level dopants in semiconductors.

In this paper we will given several examples of the anomalously large damage depth in dry etched III-V materials and devices, and show that this cannot be accounted for by channelling alone. It is necessary to involve an enhanced diffusion mechanism, possibly recombination-enhanced migration of point defects, in order to explain the results.

Experimental

A variety of samples were used for these experiments. Nominally Si-doped ($n = 10^{17} \text{ cm}^{-3}$) bulk InP wafers were exposed to $1 C_2H_6 : 10 H_2$ or $19 \, CCl_2F_2 : 10_2$, 4 mTorr, $0.85 \text{ W} \cdot \text{cm}^{-2}$ plasmas, with DC biases of ~ 380 V on the sample. The dopant profiles after RIE were obtained from Hg-probe capacitance-voltage (C-V) measurements. GaAs-AlGaAs HEMT structures consisting of a 410Å n^+ GaAs cap layer, 300Å n^+ AlGaAs donor layer, 25Å undoped AlGaAs spacer layer, 3000Å undoped GaAs buffer and 10 period, 40Å AlGaAs/40Å GaAs superlattice on a GaAs substrate were used for measurement of the saturated drain-source current at 2V reverse bias as a function of the DC bias on the sample during exposures to a 30 mTorr O_2 discharge. A multi-quantum well GaAs/AlGaAs consisting of 500Å AlGaAs barriers on either side of GaAs layers 15, 30, 50, 70 and 100Å thick was used to provide a unique spectral signature of damage at specific depths in the sample.[4] Photoluminescence measurements on the change in intensity of the individual luminescence peaks from the wells establishes the depth profile of the induced damage.

Results and Discussion

Figure 1 shows near-surface carrier profiles in etched InP as a function of post-RIE treatments (30 sec anneals). For the case of C_2H_6/H_2 etching, the reduction in the net carrier concentration extends to ~1000Å and the initial profile is recovered by 500°C annealing. This reduction in the doping density is assumed to be due to the creation of deep acceptors in the InP by incident ions from the discharge, which trap the conduction electrons in the material and are not thermally ionized at room temperature. We rule out hydrogen passivation of the shallow dopants in the InP as contributing to the carrier loss, since this appears to be significant only for shallow acceptor dopants. The depth of the carrier reduction is also considerably greater than the projected range of ions crossing the plasma sheath. In this case, axial and planar channelling of the ions might explain the results. For example, the projected range of 380 eV H^+ ions is estimated to be ~38Å with a straggle of 98Å from a Monte Carlo simulation[7] and this is much less than the carrier removal depth. If we take the relation that the final ion distribution has fallen to 10^{-3} of its peak value at a distance $R_p + 3.72 \Delta R_p$, where ΔR_p is the longitudinal straggle then the hydrogen should be present to a depth of only ~365Å. Channelling however can increase this distance by factors up to ~5, and damage created by the channelling ions alone could account for the depth of carrier removal. In the case of CCl_2F_2/O_2 RIE the profiles in Fig. 1 are more complicated because of chemical changes to the surface, but show a similar effect.[8]

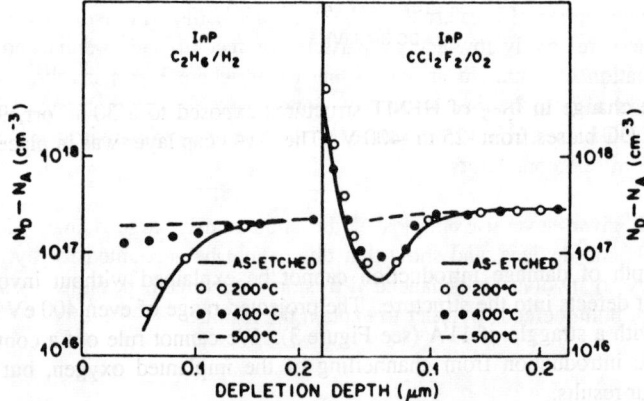

Figure 1. Carrier profiles in uniformly doped n-type InP etched in either a C_2H_6/H_2 or CCl_2F_2/O_2 discharge for 4 min, as a function of post-RIE annealing temperature.

We have also observed that bulk GaAs and InP samples ion milled with 100-500 eV Ar$^+$ ions show the presence of a high concentration ($>10^{10}$ cm^{-2}) of dislocation loops at a depth of 130-460Å below the surface.[9] This again is much greater than the mean range of 500 eV Ar$^+$ ions (24Å, with a straggle of 12Å).

Figure 2 shows the saturated drain-source current from the HEMT structure as a function of the DC bias on the sample during a 5 min O$_2$ plasma exposure. There is no measurable change in I$_{DSS}$ until the ion energy is >150 eV. It is important to remember that I$_{DSS}$ will only begin to decrease when damage has permeated the GaAs contact layer and reached the AlGaAs donor layer, i.e. to a depth of 410Å. The damage is expected to be in the form of point defects which create deep levels in the AlGaAs bandgap, trapping free carriers and removing them from the conduction process in the HEMT. As the DC bias on the sample is increased above -200 V, the I$_{DSS}$ values rapidly decrease as more traps are introduced into the AlGaAs donor layer.

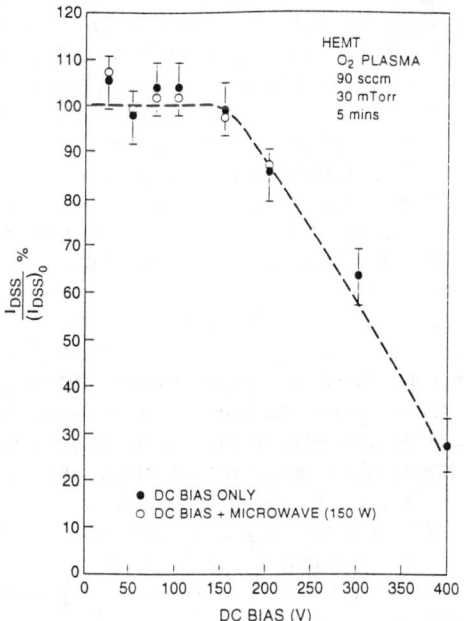

Figure 2. Percentage change in I$_{DSS}$ of HEMT structures exposed to a 30 mTorr, 90 sccm O$_2$ plasma for 5 mins at DC biases from -25 to -400 V. The GaAs cap layer was in place during the plasma exposure.

In this case, the depth of damage introduction cannot be explained without invoking rapid diffusion of the point defects into the structure. The projected range of even 400 eV O$^+$ ions in GaAs is only 22Å with a straggle of 11Å (see Figure 3). We cannot rule out a contribution to the depth of damage introduction from channelling of the implanted oxygen, but this alone cannot account for our results.

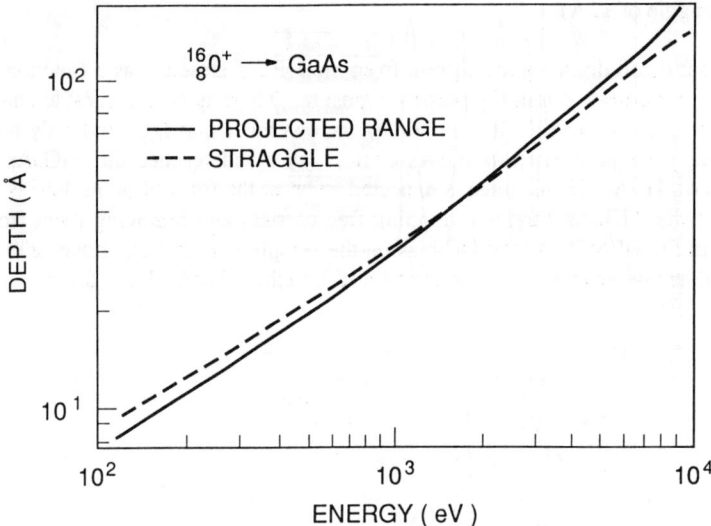

Figure 3. Monte-Carlo calculations of projected range and straggle of 100 eV-10 keV O^+ ions in GaAs.

Typical photoluminescence (PL) spectra of the GaAs-AlGaAs MQW structure before and after exposure to a H_2 plasma with 300 V applied to the cathode are shown in Fig. 4. From spectra of this type we plotted the intensities for the quantum wells in the damaged regions, normalized to the intensity of the 100 Å well buried beyond any feasible damage depth, as a function of sample depth. This is shown in Fig. 5. In our structure we were able to observe damage effects for relatively large self-biases (≥ 200 V) and in many cases there was no degradation of the PL intensity from the 30 Å quantum well. The lines in Fig. 5 therefore represent worse cases, since for the Ar 200 and 300 V and H 200 V curves, the damage may have penetrated significantly less than 1000 Å. For hydrogen we observed deeper, but less severe damage relative to argon. This is expected since the lighter ion should create less nuclear stopping damage resulting in atomic displacements. Once again channelling alone cannot account for the damage depths observed with Ar^+ ions. Migration of point defects created near the surface must occur to depths of ~1000 Å during the plasma exposure. We rule out thermally-enhanced point defect motion, since the temperature rise of our samples was less than 25 °C.

Conclusions

The fact that ion bombarded or dry etched III-V samples display damage depths far in excess of the expected projected range of the impinging ions is a common phenomenon. In some cases channelling of these ions once they enter the lattice can explain the damage depths, particularly for light ions such as H^+, but this cannot account for the results obtained with heavier species such as O^+ and Ar^+. Since thermal motion can be ruled out, one needs to invoke some other mechanism such as recombination-enhanced defect motion. This is feasible in the highly ionized environment encountered during ion beam or plasma processing.

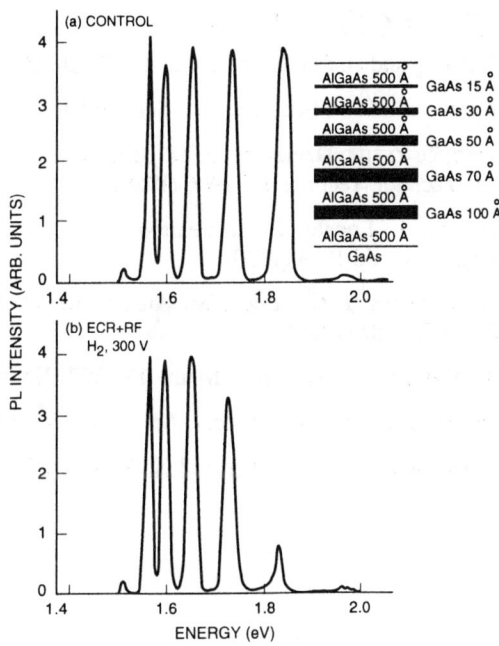

Figure 4. Low-temperature (5K) PL spectra from GaAs-AlGaAs MQW structure before (top) and after (bottom) exposure for 2 min to a 150 W (microwave), 300 V bias H$_2$ plasma.

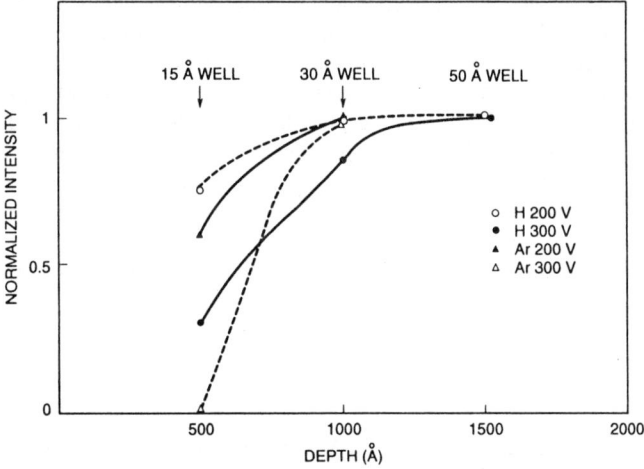

Figure 5. Normalized PL intensity from individual wells in the MQW structure, corresponding to different depths in the sample, after H or Ar plasma exposure with 200 V or 300 V bias on the sample.

REFERENCES

1. S. W. Pang, J. Electrochem. Soc. 133 784 (1986).
2. C. M. Knoedler, L. Osterling and H. Shkitman, J. Vac. Sci. Technol. B6 1573 (1988).
3. A. Scherer, H. G. Craiglead and E. D. Beebe, J. Vac. Sci. Technol. B5 1599 (1987).
4. H. F. Wong, D. L. Green, T. Y. Liu, D. G. Lishan, M. Bellis, E. L. Hu, P. M. Petroff, P. O. Holtz and J. L. Merz, J. Vac. Sci. Technol. B6 1906 (1988).
5. W. C. Dautremont-Smith, J. Lopata, S. J. Pearton, L. A. Kozi, M. Stavola and V. Swaminathan, J. Appl. Phys. 66 1993 (1989).
6. J. Chevallier, A. Jalil, B. Theys, J. C. Pesant, M. Aucorturier, B. Rose and A. Mircea, Semicond. Sci. and Technol. 4 87 (1989).
7. J. P. Biersack and L. G. Haggmack, Nucl. Instr. Meth. 174 257 (1980).
8. S. J. Pearton, U. K. Chakrabarti and F. A. Baiocchi, Appl. Phys. Lett. 55 1633 (1989).
9. S. J. Pearton, U. K. Chakrabarti, A. P. Perley and K. S. Jones, J. Appl. Phys. 68 2760 (1990).

PHOTOLUMINESCENCE CHARACTERISATION OF THE SILICON SURFACE EXPOSED TO PLASMA TREATMENT.

A. HENRY[1], B. MONEMAR[1], J.L. LINDSTRÖM[2], T.D. BESTWICK[3] AND G.S. OEHRLEIN[3]

[1] Department of Physics and Measurement Technology, Linköping University, S-581 83 Linköping, SWEDEN.
[2] National Defence Research Establishment, P.O. Box 1165, S-581 11 Linköping, SWEDEN.
[3] IBM Research Division, T.J. Watson Research Center, Yorktown Heights, N.Y. 10598, USA.

ABSTRACT

A photoluminescence (PL) study on silicon exposed to various plasmas with the conditions of reactive-ion-etching is reported. Depending on the composition of the gas used, typical PL spectra are observed after plasma treatment. Distinct kinds of recombination processes can occur giving rise to (i) sharp PL lines characteristic of transitions at deep neutral defects known from studies of defects produced by high energy irradiation, (ii) broad PL lines tentatively associated with recombination of an exciton bound at a complex defect or (iii) broad PL bands assumed to arise from recombination processes around extended defects. The excitation power dependence as well as the temperature dependence of the intensity of these PL lines and bands are presented.

1. Introduction

Dry etching processes have rapidly replaced wet etching techniques in the fabrication of small dimension semiconductor devices[1]. However dry etching such as reactive-ion-etching (RIE), plasma etching or ion-beam etching, can introduce damage and contamination effects in the exposed materials[2,3]. Although low-temperature photoluminescence (PL) spectroscopy is a powerful technique for the study of defects in bulk semiconductors, it does not yet appear to be extensively used for the characterisation of plasma exposed surfaces, and the number of related publications is very small[4-8]. Therefore we report here a PL study of n- and p-type silicon exposed to various plasmas with the conditions of RIE treatment. Depending of the composition of the gas used, typical PL spectra due to distinct types of recombination processes are observed after plasma treatment.

2. Experimental procedure

Boron (10 Ωcm) and phosphorus (5 Ωcm) doped, <100> oriented Czochralski (Cz) grown silicon wafers with an initial concentration of 3.9×10^{17} and 9×10^{16} oxygen atoms per cm^3 and 3.8×10^{17} and 5×10^{15} carbon atoms per cm^3, respectively, have been used in this study.

Before being loaded into the plasma chamber the silicon wafers were given a 30 seconds dip in dilute HF. They were positioned on the water cooled electrode of a diode reactor to which 225 W of 13.56 MHz rf was fed. The plasma pressure was 25 mTorr, the exposure time 10 minutes and the gas flow rate 100 sccm. The plasmas used were argon (Ar), deuterium (D_2), carbon-tetrafluoride (CF_4), a mixture of 50% argon-50% deuterium (Ar-D_2) or a mixture of 80% helium-20% hydrogen-bromide (He-HBr). The maximum sample temperature reached during plasma treatment was 70-100°C.

Photoluminescence measurements were done at temperatures between 2 K and 120 K using the 514.5 nm line of an Ar^+ ion laser for excitation. The luminescence was dispersed with a SPEX 1404 0.85 m double grating monochromator fitted with two 600 grooves/mm gratings blazed at 1.6 μm. A liquid nitrogen cooled North Coast E0817 Ge detector was used with a mechanical chopper and a conventional lock-in technique to recover the PL signal.

3. Experimental results

Figure 1 shows typical PL spectra at 2 K of p-type silicon samples after various plasma treatments. They all contained the phonon replica of the boron bound exciton (BE) (transverse acoustic B^{TA},

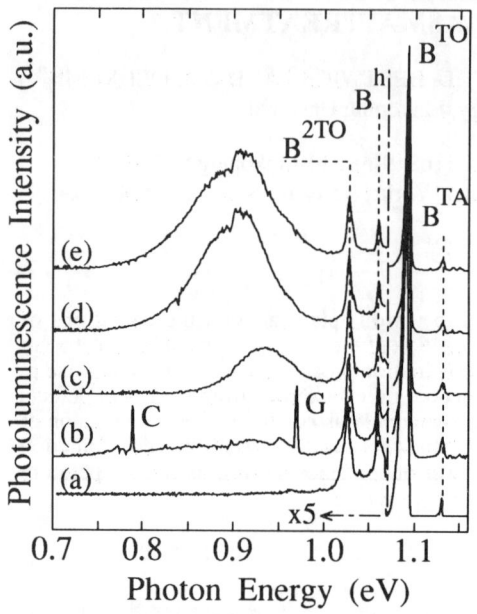
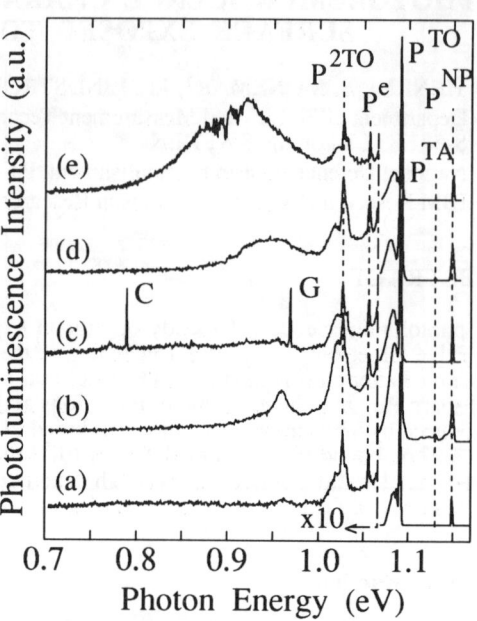

1. PL spectra at 2K of p-type Cz silicon after plasma exposure using as a plasma He-HBr (a), CF$_4$ (b), Ar (c), Ar-D$_2$ (d) and D$_2$ (e).

2. PL spectra at 2K of n-type Cz silicon before (a) and after plasma treatment using as a plasma He-HBr (b), CF$_4$ (c), Ar (d) and D$_2$ (e).

transverse optical B^{TO}, two TO phonon transition B^{2TO} and two-hole TO transition B^{h})[9]. No change in the PL spectrum was observed before and after He-HBr plasma treatment (Fig.1.a), whereas additional PL lines or bands appear after plasma exposure using other gases. The main effect after CF$_4$ plasma treatment is the observation of the G and C lines (Fig.1.b) whereas Ar and D$_2$ (or Ar-D$_2$) plasmas induce a broad PL band located at 935 meV with a halfwidth of 76 meV (Fig.1.c) and at 900 meV (80 meV halfwidth) (Fig.1.d-e), respectively. The structure observed around 900 meV is due to water absorption and the parts of the spectra indicated by the arrow are magnified by the factor specified in the figure.

A similar behaviour is observed for the n-type material except for the He-HBr treatment. On all the spectra the phosphorus BE lines are observed (Fig.2) and are denoted in a similar way as the boron BE lines (P^{NP} and P^e denote the no-phonon line and the two-electron TO transition, respectively). The G and C lines are also observed after CF$_4$ plasma treatment (Fig.2.c) whereas broad bands at 945 meV (halfwidth 76 meV) and 903 meV (80 meV halfwidth) emerged from the PL spectra after Ar (Fig.2.d) and D$_2$ or Ar-D$_2$ (Fig.2.e) plasma exposure, repectively. Moreover after He-HBr plasma exposure two broad PL lines with a halfwidth of 13 meV were observed at 1020 and 958.6 meV (Fig.2.b), respectively, the first (1020 meV) being superimposed on the P^{2TO} line.

Resulting from the plasma exposure, shifts in energy position of the boron BE lines and the electron-hole-droplet (EHD) band were observed to occur towards the low energy side with a maximum of 2-3 meV depending of the plasma characteristics. In any of the n-type plasma exposed material no shifts of the phosphorus BE lines and EHD band were observed.

The dependence on the laser excitation power was measured at 2K for various lines or bands observed on the PL spectra, as shown in Fig.3 in the case of the phosphorus doped sample. The PL intensity of the phosphorus BE line is increased linearly with power until saturation was attained at a high enough excitation power. However the PL intensity of the other emissions such as the G and C lines after CF$_4$ plasma, the broad PL line (958.6 meV) observed after He-HBr plasma on n-type silicon and of the broad PL band observed after Ar, D$_2$ or Ar-D$_2$ plasma on any material, is increased with a sublinear dependence of the excitation power due to the competition between the various

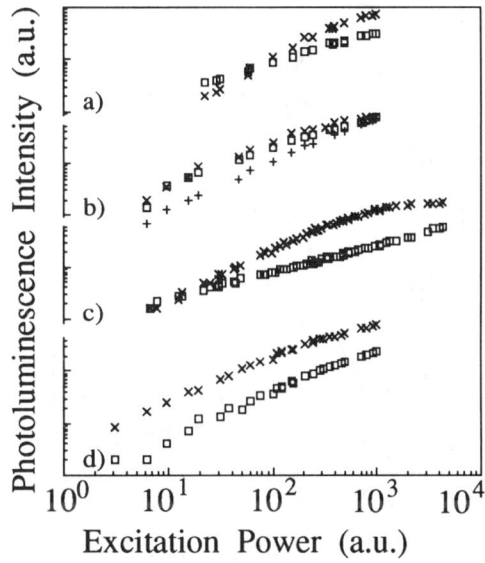
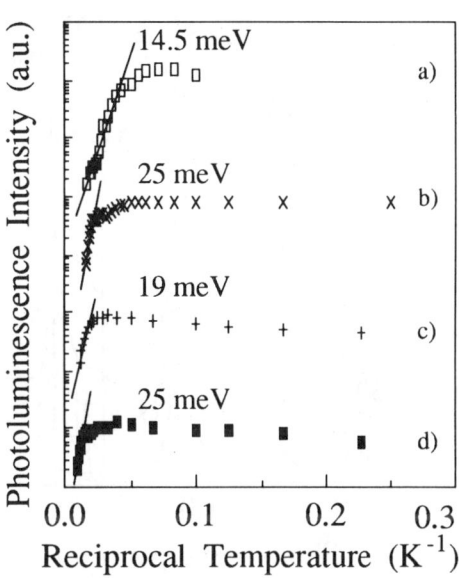

3. PL intensity as a function of the excitation power of x P^{TO} line and □ 945 meV band after Ar plasma (a), x P^{TO}, + G and □ C lines after CF_4 plasma (b), x P^{TO} line and □ 956.8 meV broad line after He-HBr plasma (c) and x P^{TO} line and □ 903 meV broad band after Ar-D_2 plasma (d).

4. Temperature dependence of the integrated □ FE line (a), x He-HBr broad lines (b), + Ar induced band (c) and ■ Ar-D_2 related band (d) intensities on the n-type plasma exposed samples.

recombination channels. Neither energy shifts of the observed PL line and broad PL bands nor any new PL bands are observed as the excitation power is increased.

The temperature dependence from 2K up to 120K of some of the RIE induced PL lines or bands has been investigated, and figure 4 shows their integrated PL intensity as a function of the reciprocal experimental temperature. As expected, at temperature higher than 4K the free exciton (FE) was observed. Its temperature behaviour is plotted in Fig.4.a. as a reference and gives a deactivation energy of 14.5 meV[9], as determined by the slope of the straight line. The 1020 and 958.6 meV broad PL lines observed on He-HBr exposed n-type material show similar temperature behaviour, no energy shifts of their position and no additional lines are observed as the temperature increased (Fig.5.a). They remain rather constant up to 20K with a slight increase in intensity due to the quenching of competing shallow recombination channels, and are quenched above 20K with a deactivation energy of 25 meV (Fig.4.b). The 945 meV PL band observed for the n-type sample after Ar plasma exposure shifts in energy position from 945 meV at 10K to 905 meV at 80K. Its intensity increases with increasing temperature up to 20K and above 30K is decreased with a thermal deactivation energy of ≈19 meV (Fig.4.c). Similar behaviour is observed for the 903 meV (at 2K) PL band on Ar-D_2 exposed n-type material, which shifts to 842 meV at 110K. Some recorded spectra at selected temperatures are shown in Fig.5.b. As the sample temperature increases this broad PL band increases in intensity from 4K to 30K, thereafter the PL intensity decreases with a deactivation energy of 25 meV (Fig.4.d). Figure 6 shows the variation of the energy position of these PL bands as a function of temperature, together with the expected variation of the silicon band gap energy.

4. Discussion

It is now well established that reactive-ion-etching introduces damage into the semiconductor surface. This modification of the silicon near-surface is generally investigated by various surface analysis techniques such as Auger electron spectroscopy, in-situ x-ray photoelectron spectroscopy (XPS) or transmission electron microscopy (TEM). Although photoluminescence is a powerful tool for the

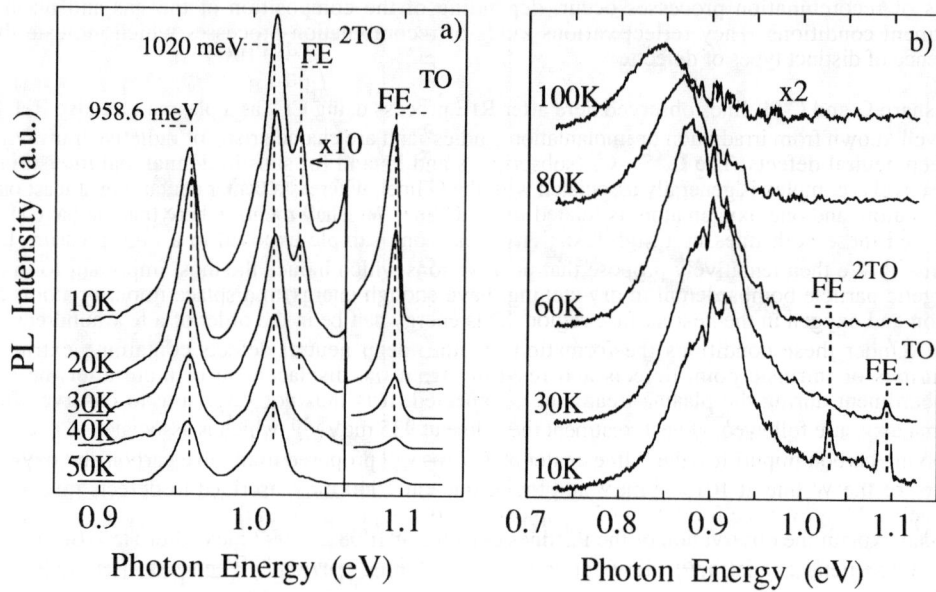

5. PL spectra recorded at the indicated experimental temperature in the He-HBr (a) and in the Ar-D_2 (b) plasma exposed n-type sample.

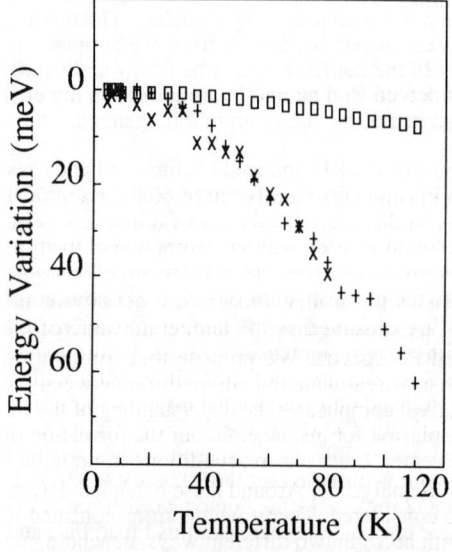

6. Variation of the energy position of the Ar-plasma induced broad PL band (945 meV at 2K) (x), Ar-D_2 related band (903 meV at 2K) (+) and the expected silicon band gap energy (□) as a function of the sample temperature.

characterisation of defects in bulk material its application to the study of near-surface semiconductor properties can give useful information, however. The penetration depth of the exciting 514.5 nm radiation is about 1 μm at 2K in bulk silicon. However, the diffusion depth of photo-generated carriers is much larger, it can be of the order of few tens of micrometers depending of the concentration of defects present in the crystal. Laser excitation of silicon produces a large non-equilibrium concentration of electrons and holes which form excitons at low temperature. These excitons can recombine either directly giving rise to the free exciton (FE) luminescence, or indirectly after various relaxation processes as interaction with defects or localisation at different centres. Radiative decay of excitons bound to impurities or defects gives rise to specific luminescence features, which often allow to extract electronic properties of the respective centers.

The experimental results presented above clearly indicate distinct electron-hole recombination processes occuring in our plasma etched induced samples. There is of course luminescence related to the dopant impurity of the material as the dominant feature in the PL spectra at low temperature, and as expected the FE recombination was also observed at sample temperatures between about 8K and 50K. However in our plasma exposed samples other kinds of recombination processes occur, depending of the composition of the gas and plasma treatment conditions. They reflect various kinds of recombination processes which indicate the presence of distinct types of defects.

The sharp G and C PL lines observed here after RIE process using CF_4 as a plasma (see also Ref.4) are well known from irradiation or implantation studies[9] and are characteristic of radiative transitions at deep neutral defects. The C_s-Si_i-C_s (subscripts s and i stand for substitutionnal and interstitial, respectively) complex is generally associated with the G line, whereas a center containing at least one carbon atom and one oxygen atom is related to the C line. We shall mention here that we have also observed these both lines in a high resistivity float-zone sample exposed to a high pressure D_2 plasma[10]. We then tentatively propose that positive ions which induce the most important form of energetic particle bombardemnt in dry etching, have enough energy to displace impurity atoms as carbon and oxygen in the near surface region. This energy can be in the order of a few hundreds of eV[11]. Under these conditions the formation of other deep neutral defects containing extrinsic impurities or intrinsic point defects and resulting from the displacement of atom after the ion bombardment during the plasma treatment is expected. It is thus not surprising to observe after plasma exposure followed by heat treatment the T line at 935 meV[8,10], which is associated to a defect involving carbon impurities, the P line center at 767 meV[10] proposed to involve carbon and oxygen atoms, or the W line at 1019[7,8] meV and the X line[8] at 1040 meV ascribed to defects involving vacancies. Moreover the formation of complexes involving atoms from the plasma is also possible and can explain the observation of the PL lines detected at 1008 and 997 meV after He-HBr plasma and subsequent annealing[8], these latter two lines being previously reported after hydrogen implantation[12].

We previously reported a detailed PL study of silicon exposed to He-HBr gas as the plasma, including the effect of subsequent annealing[8]. We report then here a brief report of our previous discussion regarding the effect of this treatment on the PL spectra. We suggested that the He-HBr induced broad lines located at 1020 and 958.6 meV, respectively, were associated to the NP and TO phonon replica recombinations of the same exciton which is bound to a complex defect. The binding energy of the loosely-bound particle in the complex is estimated as 25 meV from the temperature dependence of the broad PL lines (Fig.4.b). The identity of the complex cannot be firmly established, however we proposed an involvment of intrinsic point defects and atoms (hydrogen) from the etch plasma. The broadening feature of these lines was assumed to be due to an inhomogeneous strain field around this complex defect.

A comparison of the effect of RIE treatment using as a plasma D_2 or a mixture of 50% Ar and 50% D_2 shows the same PL behaviour whereas after Ar plasma the relative intensity of the observed PL band is less intense and the energy position is higher (about 40 meV), without distinction of the type of the silicon sample. The Ar plasma exposure seems thus to induce less defects or defects with lower luminescence efficiency than when D_2 is added to the plasma. Additionally, the conditions of the plasma treatment, mainly pressure of the plasma[7,10], exposure time[6,10] and temperature of the sample[7], have an important influence on the observed PL spectra. We propose that competition between the formation of defects occurs during the plasma treatment and competition between the various recombination channels (radiative or non-readiative) complicates the understanding of the PL data. If some plasma conditions, as using CF_4 gas as a plasma for instance, favour the formation of point defects or simple defects, such as the G or C line center, other plasma conditions could induce more complex or extended defects as observed with TEM analysis[14]. Around these extended defects local modification of the crystalline potential could be considered. Therefore electrons confined in wells surrounding these complexes could recombine with holes in two different ways, depending on the considered model. Firstly recombination with holes from the valence band and localised far away from these extended defects could be envisaged, if we consider binding to the conduction and valence band without band-gap narrowing as the model proposed to explain the various properties of the so-called new-donors in annealed Cz silicon[15,16]. Secondly if band-gap narrowing occurs as previously proposed[17,18] after plasma exposure, recombination of the electron and hole both confined near the extended defect could be possible. In any case both these models can explain the optical properties of the broad PL bands observed here after plasma treatment, such as the temperature dependence with the shift in energy position. Moreover we believe that the luminescence from these types of recombination is very efficient.

5. Summary

In summary we have performed a photoluminescence (PL) study of optical defects introduced after plasma exposure in the near surface silicon crystal. Ar, D_2, CF_4, mixture of 50% Ar and 50% D_2 and mixture of 80%He and 20% HBr were used as a plasma for the RIE treatment on n- and p-type Cz silicon and are shown to introduce damages in the near surface silicon material. The conclusions from this study are :
(i) Various kinds of defects are introduced in the silicon crystal by RIE treatment. They firstly comprise point defects or simple defects such as carbon and/or oxygen related centers or vacancy related centres giving rise to sharp PL lines known from studies of defects introduced by high energy irradiation. Broad PL lines ascribed to recombination of excitons bound at complex defect are also observed. Finally very broad PL bands are tentatively associated to recombination around extended defects. Competition between the formation of these various defects is shown to depend on the plasma conditions and sample characteristics.
(ii) Competition between the various recombination processes is revealed to play an important role on the understanding of the PL spectra.

REFERENCES

1. see e.g. J.W. Coburn, *Plasma Etching and Reactive Ion Etching*, ed. N. Rey Whetten (American Vacuum Society Monograph Series, New-York, 1982).
2. G.S. Oehrlein, Material Science and Engineering B4, 441 (1989).
3. S.J. Fonash, J. Electrochem. Soc. 137, 3885 (1990).
4. J. Weber and M. Singh, Appl. Phys. Lett. 49, 1617 (1986).
5. G.A. Northrop and G.S. Oehrlein, Mat. Sci. Forum 10-12, 1253 (1986).
6. I.-W. Wu, R.A. Street and J.C. Mikkelsen, Jr, J. Appl. Phys. 63, 1628 (1988).
7. H. Weman, J.L. Lindström, G.S. Oehrlein and B.G. Svensson, J. Appl. Phys. 67, 1013 (1990).
8. A. Henry, O.O. Awadelkarim, B. Monemar, J.L. Lindström, T.D. Bestwick, and G.S. Oehrlein, J. Electrochem. Soc. 138, 1138 (1991).
9. See, for example the review by G. Davies, Rep. Phys. 176, 83 (1989) and references therein.
10. A. Henry, B. Monemar, J.L. Lindström, T.D. Bestwick, and G.S. Oehrlein, submitted to J. Appl. Phys.
11. P.W. May, D. Field, D.F. Klemperer and Y.P. Song, *Plasma Processing*, eds. G.S. Mathad and D.W. Hess (The Electrochem Soc. Inc, Proc. Vol. 90-14, Pennington, NJ, 1990) p.136.
13. V.I. Obodnikov, L.N. Safronov and L.S. Smirnov, Sov. Phys. Semicond. Vol 10 (7), 814 (1976).
14. S. J. Jeng, G.S. Oehrlein and G.J. Scilla, Appl. Phys. Lett. 53, 1735 (1988).
15. A. Henry, J.L. Pautrat, P. Vendange and K. Saminadayar, Appl. Phys. Lett. 49, 1266 (1986).
16. P. Vendange, A. Henry, K. Saminadayar, N. Magnea and J.L. Pautrat, *Defects in Semiconductors*, ed. H.J. von Bardeleben, Mat. Sc. Forum, Vol 10-12 (1986) p.991.
17. M. Singh, J.Weber, T. Zundel, M. Konuma and H. Cerva, *Proc. 15th Int. Conf. on Defects in Semiconductors*, ed. G. Ferenczi, Materials Science Forum, Vol. 38-41 (1989) p.1033.
18. H. Weman, B. Monemar, G.S. Oehrlein and J.S. Jeng, Phys. Rev. B42, 3109 (1990).

AN ANALYSIS OF POINT DEFECT FLUXES DURING SiO$_2$ PRECIPITATION IN SILICON

William J. Taylor, Teh Y. Tan and Ulrich M. Gösele
School of Engineering, Duke University
Durham, NC 27706 USA

ABSTRACT

From studies of diffusion behavior of dopants and neutral impurities, it has been concluded that vacancies (V) and self-interstitials (I) coexist in silicon at elevated temperatures. Adopting this conclusion, we investigate the influences of these point defects on SiO$_2$ precipitate growth over a wide temperature range (400-1200°C). We assume that these point defects play the essential role of providing the strain relief during SiO$_2$ precipitate growth. In this context, we model how the point defect concentrations change in the vicinity of the precipitate, and show that these local variations affect the flux of the oxygen atoms through the oxygen/point defect reaction at the precipitate surface. In the two extreme cases studied, (V-only and I-only) the model predicts an oxygen flux lower than that limited by oxygen diffusion alone (oxygen-diffusion-limited precipitation has been experimentally verified). For the case of V-only, this difference is orders of magnitude, indicating that a V-only model is not capable of fitting experimental data. However, the model for I-only predicts a flux only 30% lower than that limited by oxygen diffusion at temperatures as low as 400°C, which is well within the error range of the oxygen precipitation data. We conclude that, while vacancies may contribute to some extent, self-interstitials play the predominant role in providing strain relief during SiO$_2$ precipitate growth.

1. Introduction

Intrinsic point defects are important in many diffusion and precipitation processes in silicon. In early times, vacancies (V) were the only point defect considered. In 1968, Seeger and Chik[1] proposed that self-interstitials (I) also play an important role, touching off a V vs. I debate which continues to this day.[2] In the last decade, however, studies of impurity diffusion under non-equilibrium conditions has established that V and I both contribute. This has allowed calculation of the fractional vacancy and interstitialcy contributions to diffusion of many substitutionally dissolved elements.[3,4]

While the V/I components of substitutional dopant diffusion has been extensively investigated, the same is not true for precipitation. It is generally agreed that point defects are necessary for relieving the strain associated with the volume increase during SiO$_2$ precipitation, but the fractional V and I contributions to this process have not been examined. In this paper, we address this problem.

2. Approach

We begin by noting two types of experimental observations. First, several studies have shown that the oxygen precipitate growth rate is limited by oxygen diffusion.[5-7] Secondly, the formation of SiO$_2$ in silicon involves a volume expansion of more than 100%, which creates a strain in the surrounding matrix. Hu[8] has shown that without strain relief, this strain is so large that, at most temperatures, it would completely prevent precipitate growth. Therefore, for SiO$_2$ precipitate growth, there must be some means of strain relief. Based upon the second observation, we assume, as others have,[9-18] that the strain relief is provided by point defects: either V absorption or I emission. Since experiments showed that oxygen precipitation is predominantly limited by oxygen diffusion, we conclude that *the flux of the point defects involved in strain relief must be similar in magnitude to the oxygen flux expected for simple oxygen limited precipitation.*

Another conclusion we draw from the above observations is that, during growth, the reaction at the precipitate surface between oxygen and point defects is proceeding faster than the influx of oxygen atoms. From this we assume a local equilibrium between the species, which allows the use of the mass-action law at the precipitate/matrix interface. At the precipitate surface, point defect concentrations affect the oxygen concentrations, and hence change the oxygen flux. This means that *point defect under/supersaturations at the precipitate surface affect the oxygen precipitate growth rate.*

Based upon the above observations and assumptions, we examine a flux balance for a one precipitate case. The fluxes of the species (oxygen and one point defect species) are controlled by the supersaturations in the bulk and at the precipitate surface. The bulk supersaturations are known (oxygen is measurable, and the point defect concentration is assumed to be at equilibrium) and are independent. In contrast, the supersaturations at the precipitate surface are *not* independent, since the local equilibrium is in effect. Therefore, there is only one set of supersaturations at the precipitate surface which both satisfies the local equilibrium *and* allows the fluxes to balance. Upon finding these supersaturations, we can compare the resulting oxygen flux to experimental data, and draw conclusions about the viability of the system.

It is likely that strain relief involves both V and *I*. We wish to determine the relative contribution of each. To do this we study two extreme cases: V-only and *I*-only.

3. Vacancies Only

If vacancies are the predominant intrinsic point defects, the precipitate is entirely dependent upon the matrix to provide the strain relief. The V must arrive at a rate high enough to match the oxygen flux. The integrated oxygen flux J_{ox} into the precipitate is given by

$$J_{ox} = 4\pi r D_{ox} C_{ox}^{eq} \left(\frac{C_{ox}(\infty)}{C_{ox}^{eq}} - \frac{C_{ox}(r)}{C_{ox}^{eq}} \right) , \qquad (1)$$

where we have used the standard equation for the integrated flux into/out of a spherical point sink/source. Here D_{ox} is the oxygen diffusivity, C_{ox}^{eq} is the oxygen thermal equilibrium concentration, and $C_{ox}(r)$, $C_{ox}(\infty)$ are the oxygen concentrations at the precipitate surface and in the bulk, respectively. By changing the prefactor $4\pi r$ in Eq. 1 it is possible to model the flux for precipitates of other shapes. However, it will soon be clear that the prefactor plays no role in the subsequent analysis. Thus, although we model here for a spherical precipitate, the conclusions we draw are just as applicable to other precipitate shapes. This versatility is important, since precipitates take on different shapes at different temperatures.[16]

During the initial stages of precipitation, the vacancies in the vicinity of the precipitate are quickly consumed, creating a local V undersaturation. This creates a V concentration gradient between the precipitate ($C_V(r) << C_V^{eq}$) and the bulk ($C_V(\infty) = C_V^{eq}$), allowing a continuous flux of V flowing into the precipitate which balances the incoming oxygen. This flux, J_V, is given by

$$J_V = 4\pi r D_V C_V^{eq} \left(\frac{C_V(\infty)}{C_V^{eq}} - \frac{C_V(r)}{C_V^{eq}} \right) . \qquad (2)$$

Since there is roughly a 100% volume expansion during SiO_2 formation, a convenient approximation is that for every two oxygen atoms incorporated, one V must also be absorbed. Other factors may affect this ratio slightly, but these changes do not dramatically alter the results we present here. (Such factors may include volume expansions different from 100%, or the ability of the matrix to absorb some strain, reducing the demand for V absorption.) We use this 2:1 ratio of oxygen:V to write a flux balance between the two species as

$$\frac{1}{2} D_{ox} C_{ox}^{eq} \left[\frac{C_{ox}(\infty)}{C_{ox}^{eq}} - \frac{C_{ox}(r)}{C_{ox}^{eq}} \right] \approx D_V C_V^{eq} \left[\frac{C_V(\infty)}{C_V^{eq}} - \frac{C_V(r)}{C_V^{eq}} \right], \quad (3)$$

where the $4\pi r$ terms have canceled out. To solve this equation, we use $D_{ox}=0.2\exp(-2.56eV/k_BT)cm^2s^{-1}$, $C_{ox}^{eq}=3.2\times10^{21}\exp(-1.03eV/k_BT)cm^{-3}$, and $D_V C_V^{eq}=3\times10^{22}\exp(-4.03eV/k_BT)cm^{-1}s^{-1}$ from references 19, 20, and 3 respectively. $C_{ox}(\infty)$ can be measured and we assume $C_V(\infty) = C_V^{eq}$. Justification for this last assumption is given in Section 5. This leaves Eq. (3) with two unknowns: $C_{ox}(r)$ and $C_V(r)$. Since the growth stage of precipitation is known to be oxygen-diffusion-limited and not reaction-limited, we assume that the reaction at the surface between the incoming oxygen and V is proceeding comparatively fast. We can then write the reaction's equilibrium equation

$$\frac{C_{ox}(r)}{C_{ox}^{eq}} = \left(\frac{C_V^{eq}}{C_V(r)}\right)^{\frac{1}{2}} \exp\left(\frac{\sigma\Omega}{rk_BT}\right). \quad (4)$$

where Ω is the SiO$_2$ molecular volume, and σ the surface energy density. Using Eq. (4), we can rewrite Eq. (3) in terms of one variable, $C_V(r)/C_V^{eq}$.

$$\frac{1}{2} D_{ox} C_{ox}^{eq} \left[\frac{C_{ox}(\infty)}{C_{ox}^{eq}} - \left(\frac{C_V^{eq}}{C_V(r)}\right)^{\frac{1}{2}} \exp\left(\frac{\sigma\Omega}{rk_BT}\right) \right] \approx D_V C_V^{eq} \left[\frac{C_V(\infty)}{C_V^{eq}} - \frac{C_V(r)}{C_V^{eq}} \right] \quad (5)$$

For any given temperature, the solution $C_V(r)/C_V^{eq}$ provides the undersaturation of V required at the precipitate surface which will allow a sufficient in-flux of V. Using $C_{ox}=1\times10^{18}$ cm^{-3} and $\sigma=0$, we have plotted solutions to this equation in Figure 1.

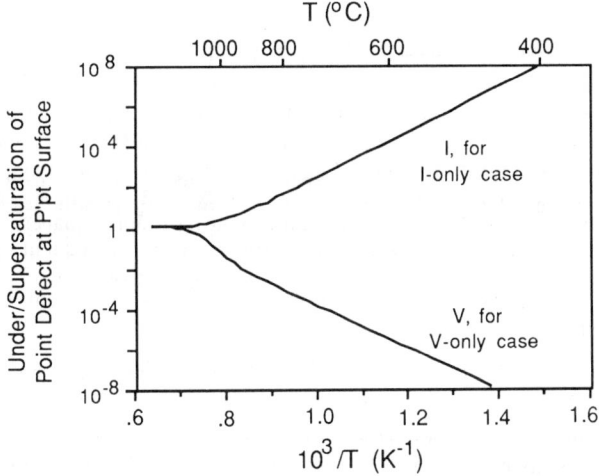

Figure 1: Point defect under/supersaturations required at the precipitate surface in order to obtain a flux balance with oxygen. Assumes initial oxygen concentration of 10^{18} cm^{-3}. From Eq. (5) and Eq. (9).

This large V undersaturation at the precipitate surface mandates, via Eq. (4), a fairly large local supersaturation of oxygen. This decreases the oxygen flux, via Eq. (1). To study the significance of this decrease, we note that SiO$_2$ precipitation data fairly well fits models which ignore the influence of point defect fluxes. If our model is to match the same data, the predicted precipitation rates (oxygen fluxes) predicted by our model should be similar to those obtained from a model which assumes just oxygen-diffusion limited precipitate growth. To test this, we define Q_V, a ratio of the oxygen flux in which V are considered to that in which V are ignored.

$$Q_V = \frac{D_{ox}C_{ox}^{eq}\left[\dfrac{C_{ox}(\infty)}{C_{ox}^{eq}} - \left(\dfrac{C_V^{eq}(r)}{C_V}\right)^{\frac{1}{2}}\exp\dfrac{\sigma\Omega}{rk_BT}\right]}{D_{ox}C_{ox}^{eq}\left[\dfrac{C_{ox}(\infty)}{C_{ox}^{eq}} - \exp\dfrac{\sigma\Omega}{rk_BT}\right]} \quad (6)$$

Values of Q_V much less than 1 would indicate that the process does not match experimental data very well. In Figure 2 we plot Q_V (for $\sigma=0$) as a function of temperature. The extremely low values for Q_V indicate that this vacancy model can not accurately describe published data.[5-7]

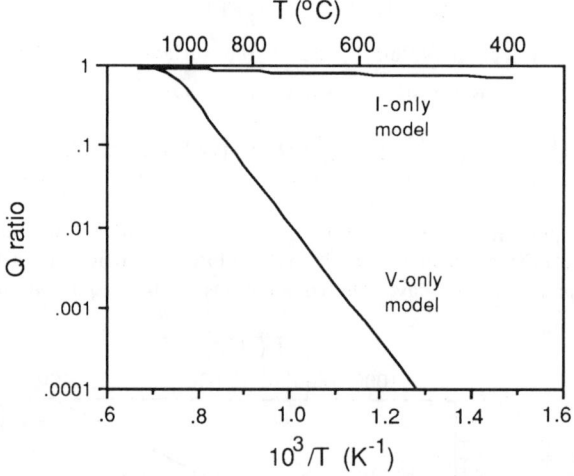

Figure 2: Ratio Q of oxygen flux into precipitate if intrinsic point defects are accounted for, to the case in which point defects are ignored. Values of Q significantly less than 1 indicate a model incapable of fitting experimental data. From Eq. (6) and Eq. (10).

4. Self-Interstitials Only

In this scenario, in contrast to the V-only case, the precipitate has control over the strain relief and point defect flux, since it can emit I whenever necessary. During the early stages of precipitation, this I emission generates a supersaturation of I around the precipitate, forming a gradient which allows continuous out-flux of I during precipitation. The development is similar to the V-only case. The flux balance is

$$\frac{1}{2}D_{ox}C_{ox}^{eq}\left[\frac{C_{ox}(\infty)}{C_{ox}^{eq}} - \frac{C_{ox}(r)}{C_{ox}^{eq}}\right] \approx D_I C_I^{eq}\left[\frac{C_I(r)}{C_I^{eq}} - \frac{C_I(\infty)}{C_I^{eq}}\right] \quad (7)$$

The reaction at the precipitate surface provides

$$\frac{C_{ox}(r)}{C_{ox}^{eq}} = \left(\frac{C_I(r)}{C_I^{eq}}\right)^{\frac{1}{2}}\exp\left(\frac{\sigma\Omega}{rk_BT}\right) \quad (8)$$

Substituting Eq. (8) into Eq. (7) we obtain

$$\frac{1}{2} D_{ox} C_{ox}^{eq} \left[\frac{C_{ox}(\infty)}{C_{ox}^{eq}} - \left(\frac{C_I(r)}{C_I^{eq}} \right)^{\frac{1}{2}} \exp\left(\frac{\sigma\Omega}{rk_BT} \right) \right] \approx D_I C_I^{eq} \left[\frac{C_I(r)}{C_I^{eq}} - \frac{C_I(\infty)}{C_I^{eq}} \right] . \quad (9)$$

Using $D_I C_I^{eq} = 4.57 \times 10^{25} \exp(-4.84 eV/k_BT) cm^{-1}s^{-1}$ from Stolwijk et al.[21], and the assumption that in the bulk the point defect is near equilibrium $(C_I(\infty)/C_I^{eq}=1)$, Eq. (9) is reduced to containing only one unknown: $C_I(r)/C_I^{eq}$. Solutions to Eq. (9) are plotted in Figure 1. These I supersaturations increase the oxygen supersaturation at the precipitate surface, via Eq. (8), thus decreasing the oxygen flux. In analogy to the V case, we define Q_I, a ratio of the oxygen flux in which I are included, to the oxygen flux in which I are ignored:

$$Q_I = \frac{D_{ox} C_{ox}^{eq} \left[\frac{C_{ox}(\infty)}{C_{ox}^{eq}} - \left(\frac{C_I(r)}{C_I^{eq}} \right)^{\frac{1}{2}} \exp \frac{\sigma\Omega}{rk_BT} \right]}{D_{ox} C_{ox}^{eq} \left[\frac{C_{ox}(\infty)}{C_{ox}^{eq}} - \exp \frac{\sigma\Omega}{rk_BT} \right]} \quad (10)$$

We have plotted Q_I in Figure 2 for $\sigma=0$. The slight decrease (roughly 30% at 400°C) in oxygen flux is insignificant compared to the uncertainty in the experimental precipitation data. Although here we have addressed the simple case of $\sigma=0$, we have found that this factor only plays a significant role for very small precipitates (containing less than about 2000 atoms, or r=20Å). We conclude that *self-interstitials, not vacancies, are the predominant point defect during SiO₂ formation.*

5. Additional Notes

An important point to note in this approach is that the results obtained are dependent upon the *products* $D_V C_V^{eq}$ and $D_I C_I^{eq}$, and not simply the diffusivities (D_V and D_I). There appears to be a general agreement on the quantities $D_V C_V^{eq}$[3,22,23] and $D_I C_I^{eq}$[3,21]. However, estimates for diffusivities of point defects span orders of magnitude, so models dependent upon such a quantity may be considered as unreliable.

Note also, that in both cases we assumed that the intrinsic point defect concentrations in the bulk are near their thermal equilibrium values, and derived point defect under/supersaturations at the precipitate accordingly. The other extreme would set point defect concentrations to equilibrium at the precipitate and seek the required super/undersaturations in the bulk. The resulting Q ratios would always be 1, so the model should match data. However, these situations are unrealistic. If large super/undersaturations existed in the bulk, then phenomena such as oxidation- or nitridation-enhanced diffusion,[3,4] which provide enhancements of 2 to 3, would be undetectable. Therefore, we conclude that typically, point defect supersaturations in the bulk are relatively low, and the assumptions made above should hold true.

To further corroborate our findings, we consider the required amount of I or V for precipitating out a large amount of oxygen ($>10^{17}$ cm^{-3}). To relieve the strain, a similar magnitude of point defects must be involved. For the case of I-only, the precipitate has no problem generating this large number of point defects. For the case of V-only, the V must come from the bulk, which is highly unlikely. These V must either exist naturally prior to precipitation (i.e. a large supersaturation of V in the bulk quenched in from the crystal growth process), or must be generated during the precipitation process. We have described above why the first option is unlikely. For the second option (continuous supply of V to the bulk during precipitation), we consider three possible processes, and find that none are viable: 1) in-diffusion of vacancies from the surface, which is much too slow, 2) generation of I-V pairs, which would simultaneously generate large, energetically unfavorable supersaturations of I, and 3) dissolution of V agglomerates, which appears to be insufficient, since Marioton[24] has estimated an upper limit of roughly 10^{15} cm^{-3} for V in agglomerates.

6. Summary

For quite some time, point defects have been assumed to act as strain relief mechanisms during SiO_2 precipitate growth, yet their role in precipitate growth dynamics is still unclear. Precipitation models which ignore point defects can already adequately explain experimental data, which display oxygen-diffusion-limited precipitation behavior. Therefore, by introducing point defects into an SiO_2 precipitation model, we can not make dramatic improvements in data fitting. However, we are able to address which *type* of point defect is involved. We do so by noting that the influx of oxygen must be balanced with a point defect flux of similar magnitude, and by realizing that these fluxes are mediated by the reaction at the precipitate surface. We find that vacancies alone are incapable of providing strain relief at the necessary rates. A model using only self-interstitials, however, can match the data over the temperature range 400-1200°C. We conclude that self-interstitials are the predominant species for strain relief during SiO_2 precipitation.

7. Acknowledgements

We acknowledge support from the Solar Energy Research Corporation, Golden CO (contract XL-8-18097), and the Mobil Foundation.

8. References

1. A. Seeger and K.P. Chik, Phys. Stat. Sol. 29, 455 (1968)
2. J.A. Van Vechten, U. Schmid, and Q.-S. Zhang, J. Electronic Materials 20, 431 (1991)
3. T.Y. Tan and U. Gösele, Appl. Phys. A37, 1 (1985)
4. P. Fahey, R.B. Griffin, and J.D. Plummer, Rev. Mod. Phys. 61, 289 (1989)
5. F.M. Livingston, S. Messoloras, R.C. Newman, B.C. Pike, R.J. Stewart, M.J. Binns, and J.G. Wilkes, J. Physics C: Solid State Phys. 17, 6253, (1984)
6. R.C. Newman, Mat. Res. Soc. Symp. Proc. Vol. 104, 1988, (Mat. Res. Soc., Pittsburgh, PA, 1988) p. 25
7. K. Yasutake, M. Umeno, and H. Kawabe, Phys. Stat. Sol. A83, 207 (1984)
8. S. M. Hu, in *Oxygen, Carbon, Hydrogen and Nitrogen in Crystalline Silicon*, eds. J.C. Mikkelsen, Jr., S. J. Pearton, J. W. Corbett, S. J. Pennycook, Mater. Res. Soc. Symp. Proc. 59 (Mater. Res. Soc., Pittsburgh, PA, 1985), p. 249.
9. S. Mahajan, G.A. Rozgonyi, and D. Brasen, Appl. Phys. Lett. 30, 73 (1977).
10. J.R. Patel, K.N. Jackson, and H. Reiss, J. Appl. Phys. 48, 5279 (1977).
11. H. Takako, J. Osaka, and N. Inoue, Jpn. J. Appl. Phys. 18, Suppl., 18-1, 179 (1978)
12. W. Patrick, E. Hearn, W. Westdorp, and A. Borg, J. Appl. Phys. 50, 7156 (1979)
13. S.M. Hu, Appl. Phys. Lett. 36, 561, (1980)
14. U. Gösele and T.Y. Tan, Appl. Phys. A28, 79 (1982)
15. K. Yasutake, M. Umeno, and H. Kawabe, Phys. Stat. Sol. A83, 207, (1984)
16. J. Vanhellemont and C. Claeys, J. Appl. Phys. 62, 3960 (1987)
17. M. Schrems, P. Pongratz, M. Budil H.W. Pötzl, J. Hage, E. Guerrero and D. Huber, in *Semiconductor Silicon 1990*, ed. H.R. Huff, Vol. 90-7, (The Electrochem. Soc., Pennington, NJ, 1990) p. 144
18. W.B. Rogers and H.Z. Massoud, *Proceedings of the 2nd International Symposium on Process Modeling in Semiconductor Technology*, ed. G.R. Srinivasan, J.D. Plummer, S.T. Pantelides, (The Electrochem. Soc., Pennington, NJ, 1991) p. 495.
19. J.W. Corbett, R.S. McDonald, and G.D. Watkins, J. Phys. Chem. Solids 25, 873 (1973).
20. R.A. Craven, in *Semiconductor Silicon 1981*, eds. H.R. Huff, R.J. Kriegler and Y. Takeishi, (The Electrochem. Soc., Pennington, NJ, 1981) p. 254.
21. N.A. Stolwijk, B. Schuster and J. Hölzl, Appl. Phys. A33, 133 (1984)
22. H. Zimmermann and H. Ryssel, submitted to Appl. Phys. Lett.
23. H. Zimmermann and H. Ryssel, submitted to Phys.Rev.
24. B.P.R. Marioton, Ph.D. Thesis, Duke University, 1988, Durham, North Carolina

DEEP STATES ASSOCIATED WITH COPPER DECORATED OXIDATION INDUCED STACKING FAULTS IN SILICON.

M. KANIEWSKA[1,2], J. KANIEWSKI[1,2], and A.R. PEAKER[1]

[1] Centre for Electronic Materials, University of Manchester, Institute of Science and Technology, PO Box 88, Manchester, M60 1QD, United Kingdom.
[2] Institute of Electron Technology, Al. Lotnikow 32/46, 02-668 Warsaw, Poland

ABSTRACT

In this paper we report for the first time, results of investigations of the deep levels associated with oxidation-induced stacking faults decorated with copper in n-type Si. For this purpose, Deep Level Transient Spectroscopy was used. It is found that the emission behavior of the copper-related states shows similarities with that observed in the case of the stacking faults which were not intentionally contaminated. Their emission properties change depending on metal location on a Frank partial dislocation, and are strongly modified by copper diffusion conditions. It is shown that the electrical activity of the extended defects is crucially important when they are decorated with copper. The results of precise DLTS profiling along the extended defects, depending on their size, are discussed in terms of the gettering effect of metals at the stacking faults.

INTRODUCTION

For manufacturing Si devices various technological steps are necessary in which high temperatures, including oxidation processes, are involved. Different kinds of crystallographic defects, such as dislocations, stacking faults and oxygen precipitations are then generated. Extended defects are considered to be a major factor limiting device performance. For example, Kolbesen et al. [1] established the influence of oxygen-induced stacking faults (OSFs) on increasing leakage current in metal-oxide-semiconductor power devices.
It is a well-known fact that extended defects, especially dislocations, can act as a sink for metallic impurities that themselves are a continuing problem in device technology. This is due to the fact that they introduce levels placed deeply in the energy gap, and that is why they can act as generation-recombination centers disturbing device characteristics.
Moreover, recently it has been demonstrated by Peaker et al. [2, 3] that deep states move to mid-gap positions as a result of common action of metallic impurities and OSFs. Taking into consideration the above facts, we see that extended defects and metals, when present in active parts of devices, become detrimental to their performance [4, 5]. On the other hand, for the same reasons, extended defects, acting as sinks, can play a positive role in eliminating metals when they are outside active junctions. In both cases, investigations of the interaction between the defects and metals seem to be important.
Such studies are also interesting from the point of view of fundamental understanding. Since it has been shown that dislocations can introduce a band of levels, it is still debated whether electrically active states are associated with dislocations or with the metals decorating the dislocations. In this respect it is important to study the electrical activity of extended defects by introducing metal impurities. The investigations may also answer the question of whether extended defects should be eliminated or if their activity can be decreased by excluding metals. It is the purpose of our work to present results of Deep Level Transient Spectroscopy (DLTS) studies of deep levels associated with OSFs decorated with Cu in n-type Si. There has been a considerable amount of work concerning dislocations, including DLTS measurements. However, most of the work has been done on plastically deformed,

and not intentionally contaminated, crystals. The results obtained so far by different authors in n-type Si have been reviewed by Peaker and Sidebotham [6] and Omling et al. [7]. Several groups of deep states, labelled according to original notation from A to D, have been found (for the details see [8]).
In contrast to the extended defects studied previously, OSFs are characterized by unique features when nucleated by surface damage. They have equal lengths and they are present only close to the surface, in a layer whose thickness is determined by their lengths. Thanks to this, we can study electrical activity along the stacking fault, as well as as a function of its size.

EXPERIMENTAL DETAILS

The scheme of our experiments is shown in Fig.1. An Si wafer was cleaved into few groups of slices. The first was polished (with 1µm diamond slurry), carefully cleaned and oxidised in dry O_2 at a constant temperature as a function of time. In this way, OSFs with different final lengths (0.7-6)µm were produced. Next, the slices were contaminated from an infinite source by evaporation of Cu on the back-side of the slices. Diffusion was performed at 750°C for 30 min in N_2 ambient. The slices chosen, containing 3µm long OSFs, were also contaminated with Cu at different temperatures of diffusion (600 - 900)°C. After diffusion, all the slices were cooled in air.
In parallel, reference samples were produced, i.e. samples containing OSFs, not contaminated but annealed under the

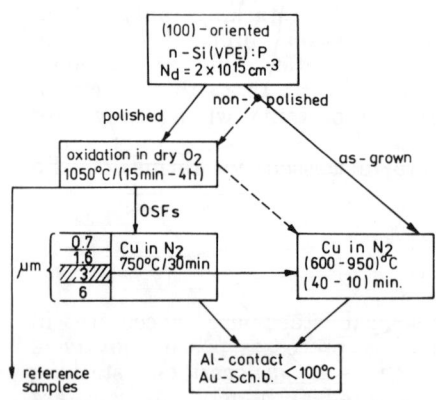

Fig.1 Scheme of experiments.

same conditions as those for which diffusion was performed. Additionally, another kind of reference sample, non-polished but oxidised (without OSFs), and so-called as-grown samples, were prepared. These two groups were also contaminated with Cu as a function of temperature of diffusion. After oxidation and diffusion processes, parts of the slices produced were etched in Y3 [9] and the surfaces were examined by means of optical Nomarski and scanning electron microscopes.
Au-Schottky barriers of 1mm diameter and Al-ohmic contacts were evaporated. DLTS measurements were conducted with a Bio-Rad DL-4600 system. All the DLTS spectra presented were obtained at the rate window $e_n = 200s^{-1}$ and the pulse duration time $t_c = 1ms$.

EXPERIMENTAL RESULTS

From a comparison of the DLTS spectra in contaminated and associated reference samples, it can be concluded that Cu diffusion results in a pronounced peak at about 200 K. Since, no additional levels with high concentrations have been found in the temperature range (80 - 350) K, we attribute the level to Cu. The spectrum in the presence of 3µm long OSFs, contaminated with Cu at 750°C, is given by the largest curve in the back of Fig.2. A broad line is observed when the DLTS signal is integrated through the whole penetration depth of the faults. However, as was observed in the case of so-called "clean" OSFs [8], the line transforms into discrete peaks of point defect-like shapes when the electrical activity is profiled within thin slices along the faults. The measurements were performed by changing the reverse bias V_R while the pulse amplitude was kept constant and equal to 0.5V

Again, the most remarkable feature of the spectra are changes in the characteristic peak temperature and their associated changes in the activation

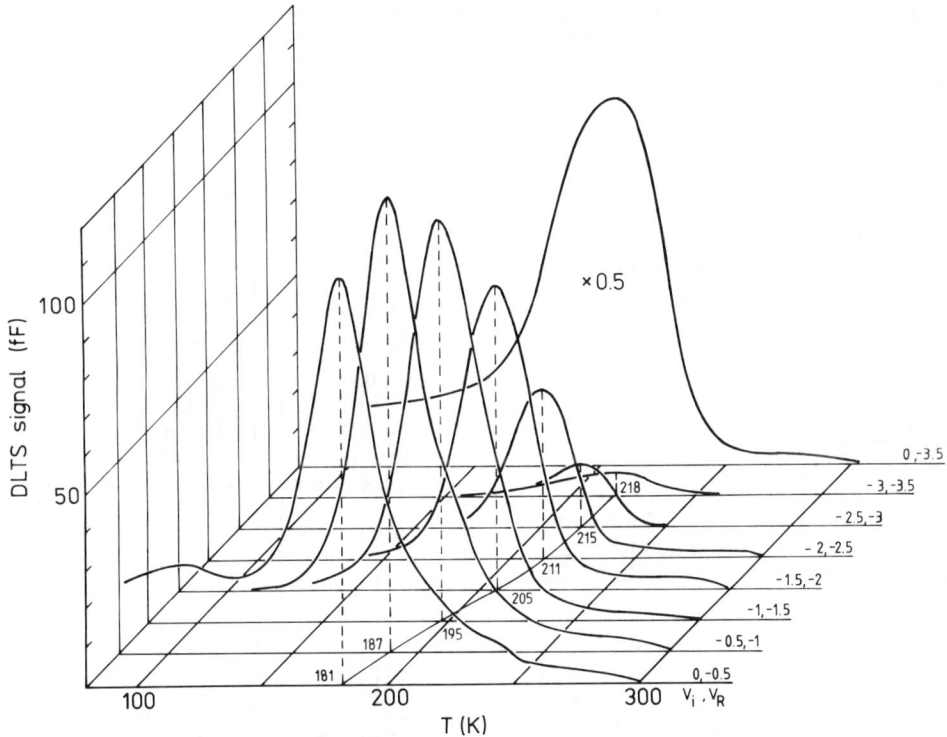

Fig.2 DLTS spectra for n-Si with 3μm long OSFs contaminated with Cu at 750°C/30 min. Each line represents a measurement for different spatial window. The values of reverse bias V_R and voltage levels of pulses V_i are shown at the right hand side. In the back of the figure the DLTS spectrum obtained at $V_R = -3.5V$ with $V_i = 0V$ is shown.

energy. The disappearance of the DLTS peak for observation regions placed far from the surface also suggests a correlation of the Cu-related trap with the presence of the OSFs. In order to check if such a coincidence does not happen accidentially, the spatial distribution of the trap as a function of stacking fault length was determined. Since the penetration depth of the OSFs changes with their length, the spatial distributions were expected to be substantially different. The profiles obtained in the samples containing OSFs with the final lengths equal to from 0.7μm to 6μm are shown in the Fig.3. Diffusion of Cu was performed to all the samples at 750°C. From an inspection of this figure, it can be noticed that the concentration profiles of the Cu-related state follow the penetration depth of our OSFs. In its turn, this means that the metallic impurity is also closely related to the extended defects.

For comparison, an additional profile was measured for the trap associated with Cu in a sample that was not polished, but was oxidised, and contaminated with Cu (750°C/30min). This sample contained extended defects different from OSFs, as was revealed by etching in Y3. In this case, the profile was completely flat and did not show any spatial correlation even up to 2.5μm.

At this point, it should be added that the results in Fig.3 give direct evidence for the gettering effect at OSFs. It is obvious that Cu is eliminated from the regions placed under the OSFs. The efficiency of the effect can be noticed from the dynamics of changes in the trap concentration. From Fig.3, one can notice that the concentration distributions are not homogeneous and the maximum moves deeper into the sample with increasing OSF length. They tell us clearly that the emission behaviour of the trap and its apparent activation energy change according to its location on the FPDs bounding the OSFs.

The emission properties are also strongly dependent on the diffusion parameters. The samples containing the 3μm long OSFs were contaminated with Cu in the temperature range of (600-950)°C. DLTS spectra sampling the whole penetration depth of OSFs, non-typically in a logarithmic scale, are shown in Fig.4. It can be noticed that the peak amplitude increases with increasing temperature of diffusion. At the same time, the peak displaces successively to higher temperatures. Simultaneously, it is less broadened and and nearly exponential when Cu is introduced at the temperature of 950°C. The latter effect is better seen in a linear scale.

Additionally, at this point it is worth underlining that the level of electrical activity of the traps is very low when Si is free from extended defects. The concetration of the traps in the as-grown samples which were not heated before diffusion, and were Cu-diffused even at 950°C, was as low as 5×10^{11} cm^{-3}. On the basis of our results, we can conclude that the Cu-related trap considered seems to be electrically active in the presence of extended defects and, as we noticed, its activity increases substantially when the temperature of diffusion is higher than 600°C.

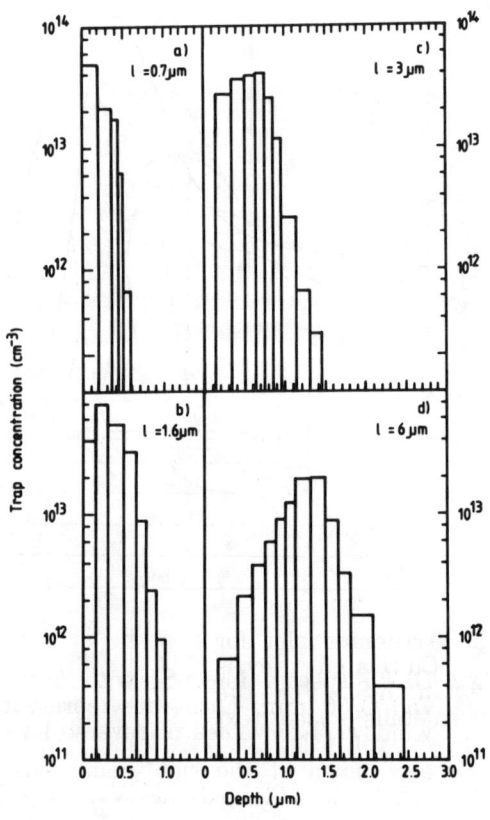

Fig.3 Spatial distribution of the Cu-related trap in the presence of OSFs with different lengths a) l = 0.7μm, b) l = 1.6μm, c) l = 3 μm, d) l = 6 μm. The samples were contaminated with Cu at 750°C/30min.

DISCUSSION OF EXPERIMENTAL RESULTS

The experimental results we obtained so far are summarized in an Arrhenius plot. They are indicated in Fig. 5 by experimental points and solid lines. Levels 1 and 6 are discussed in [8]. For a comparison, data obtained by other authors for plastically deformed n-type Si are also included in the figure - see refs. [3-6] in [8]. Additionally, data for cases when bulk stacking faults are present in Czochralski n-Si are also added (rhombs) [10].

As a result of changeable emission properties, there is a possibility of covering a very wide part of the Arrhenius plot by sampling different regions around OSFs, or by changing conditions of metal diffusion.

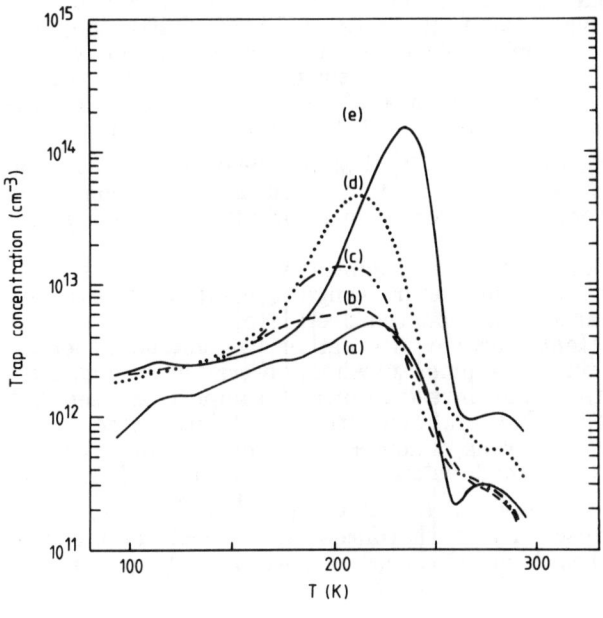

Fig.4 DLTS spectra for n-Si. with 3μm long OSFs contaminated with Cu at:
a) 600°C / 40min,
b) 700°C / 30min,
c) 750°C / 30min,
d) 800°C / 30min,
e) 950°C / 10min.

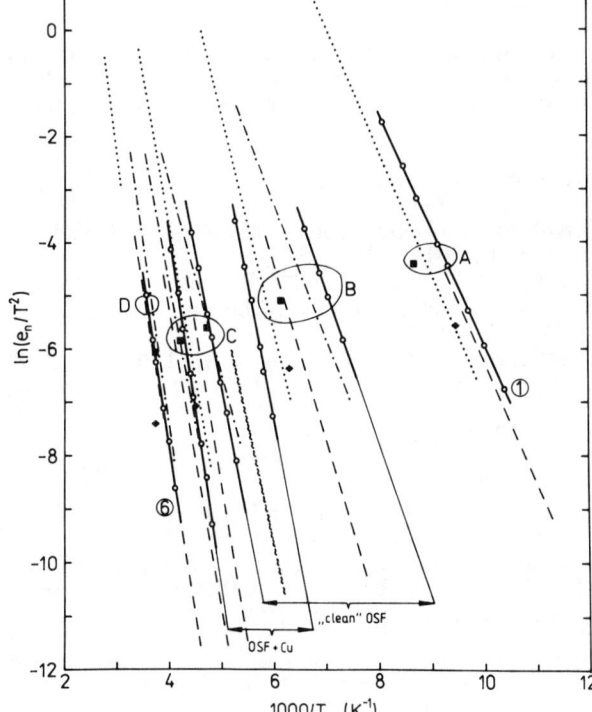

Fig.5 Arrhenius plot for Cu-related levels in n-Si. For the details - see text and ref. [8].

In the Arrhenius plot, results obtained for OSFs decorated with Cu were compared with those found for so-called "clean" OSFs [8]. Because of the very good coincidence of the results in the common region (not shown in the Arrhenius plot), we suppose that also in the case of "clean" OSFs, they are unintentionally decorated with Cu. On the basis of our results, we would thus like to suggest that the whole wide region with activation energy from 0.25eV to 0.56eV is reserved for Cu in the presence of extended defects. In this region, one can notice a Cu-related level at $E_c - 0.4$eV observed by Brotherton [11] when Cu diffusion was performed at 1000°C/30min (waved line). On the basis of our results, it is not surprising that the Cu-related region joins two groups of B and C levels, commonly observed in plastically deformed n-type Si.

It is worth noticing that plastically deformed samples were often annealed at temperatures of (700-900)°C. In this temperature range, decoration of extended defects by Cu results in appearance of pronounced deep levels.

At the end, there is the important question of whether the extended defects themselves are electrically active, or become so when decoration by impurity atoms takes place. From results of our investigations of samples oxidised but not polished, i.e. when extended defects different from OSFs are present, the level of electrical activity can be strongly depressed when non-intentionally contaminated slices are carefully cleaned before a heat treatment. Then, the concentration of traps can be decreased up to below $5 \times 10^{11} \text{cm}^{-3}$. However, in this way the effect cannot be reached at such concentration levels in the case of OSFs, probably because of the high efficiency of gettering of contaminats by stacking faults.

ACKNOWLEDGEMENTS

We are grateful to B.Hamilton and M.R.Brozel of UMIST for discussion, A.Berg and M.Vandini for technical assistance. We would like to thank the UK Science and Engineering Research Council for supporting this work.

REFERENCES

1. B.O.Kolbesen, W.Bergholtz, H.Cerva, F.Gelsdorf, H.Wendt, and G.Zoth, Inst.Conf.Ser. 104 ,421 (1989).
2. G.R.Lahiji, B.Hamilton, and A.R.Peaker, Electron.Lett.24, 1340 (1988).
3. A.R.Peaker, B.Hamilton, G.R.Lahiji, I.E.Ture, and G.Lorimer, Mat.Sci.& Eng.B4, 123 (1989).
4. A.Goetzberger and W.Shockley, J.Appl.Phys. 31, 1821 (1960).
5. P.S.D.Lin, R.B.Marcus, and T.T.Sheng, J.Electrochem.Soc. 130, 1878 (1983).
6. A.R.Peaker and E.C.Sidebotham in Properties of Silicon, ed. C.Hilsum (IEE, London, 1988) p.221.
7. P.Omling, E.R.Weber, L.Montelius, H.Alexander, and J.Michel, Phys.Rev. B32, 6571 (1985).
8. J.Kaniewski, M.Kaniewska, and A.R.Peaker - this conf.
9. K.H.Yang, J.Electrochem.Soc., 131, 1140 (1984).
10. T.Hasegawa and S.Matsumoto, Jpn.J.Appl.Phys. 27, 1906 (1988).
11. S.D.Brotherton, J.R.Ayres, A.H.Gill, H.W. van Kessteren, and F.J.A.M. Greidanus, J.Appl.Phys. 62, 1826 (1987).

ELECTRICAL PROPERTIES OF OXIDATION-INDUCED STACKING FAULTS IN N-TYPE SILICON

J. KANIEWSKI[1,2], M. KANIEWSKA[1,2], and A.R. PEAKER[1]

[1]Centre for Electronic Materials, University of Manchester, Institute of Science and Technology, PO Box 88, Manchester, M60 1QD, UK,
[2]Institute of Electron Technology, Al. Lotnikow 32/46, 02-668 Warsaw, Poland

ABSTRACT

Electrical properties of deep traps in "clean" epitaxial n-type silicon containing oxidation-induced stacking faults have been studied by Deep Level Transient Spectroscopy. Among the various traps detected, one shows a direct correlation with the faults. The deep trap depth distribution proves that electrical activity is associated with a Frank partial dislocation bounding the fault, not the stacking fault plane itself. The Arrhenius plot based similarities of the traps observed, to those detected in plastically deformed n-Si, as well as their connection with impurities, are discussed.

INTRODUCTION

It is well known that extended lattice defects within the electrically active regions are generally harmful to device performance, and hence may cause yield problems - see the refs. in [1]. This is particularly important for silicon technology, where the high temperature treatment, as well as oxidation processes, can give rise to stacking faults along with other associated extended defects. Such defects, especially when decorated with metals, can induce generation-recombination centers. They increase the leakage current and reduce minority carrier recombination and generation lifetime. The main problem is that metals in Si, especially fast-diffusing 3d transiton metals, are difficult to control [2].
Up till now, the electrical properties of extended defects were intensively studied mainly on samples in which structural defects were generated by plastic deformation, followed by annealing at temperatures of up to 900°C [3-6]. It has been proposed by Omling et al. [6] that the deep levels observed by the different authors in plastically deformed n-Si can be divided into four groups on the basis of their Arrhenius plots, as determined by Deep Level Transient Spectroscopy (DLTS). It has been concluded that the B group of levels, with thermal activation energy for emission of electrons (0.27-0.29)eV, and the D levels with energy (0.54-0.68)eV, are related to point defects located within or very close to the dislocation core. The origin of the A levels, with thermal activation energy of about 0.18 eV, has not been suggested. The C groups of levels, C1 (0.37-0.52)eV and C2 (0.48-0.51)eV, observed in DLTS as an unusually broadened peak, has been shown to be produced in direct proportion to the dislocation density, and are believed to be directly related to dislocations.
In contrast to previous studies, we report results of DLTS measurements of the deep levels observed in the presence of oxidation-induced stacking faults (OSFs) in "clean" epitaxial n-type Si. It is well established that the OSFs are {111} extrinsic stacking faults surrounded by a/3 <111> Frank partial dislocations (FPDs), and have semi-elliptical shapes. Their lengths depend on the temperature and time of oxidation. The OSFs nucleated by surface damage show equal lengths because their growth starts from the surface and commences at the same time.

EXPERIMENTAL

The investigations were done on (100)-oriented, n-type Si layers grown by conventional vapor phase epitaxy on n^+-Czochralski substrates. The layers were doped with phosphorous to the level of $2\times10^{15}\text{cm}^{-3}$. In order to produce nucleation centers, the surfaces of the slices were damaged, prior to a heat treatment, by lightly polishing with 1μm diamond slurry (Hyprez Five-Star). The oxidation processes were performed in dry oxygen with a flow of 10 litres/min, at atmospheric pressure and temperature of 1050°C, for 2 hours. The average length (at the surface) and the density of OSFs, as determined from a Nomarski optical micrograph after etching in Y3 [7], were 3.6 μm and $1\times10^7\text{cm}^{-2}$, respectively. Schottky barriers and ohmic contacts were produced by evaporation of Au and Al, respectively. During these processes, the temperature of the samples was kept below 100°C. Electrical properties of the deep traps were studied by DLTS using a BIO-RAD DL-4600 system.

Due to the fact that the OSFs grow on the {111} plane, the technique gives a unique opportunity of profiling electrical activity by controlling the depth of the OSFs.

RESULTS AND DISCUSSION

An example of a typical DLTS spectrum detected in the samples containing 3.6μm long surface OSFs is presented in the back of Fig.1. In order to monitor the nearly whole depth where the OSFs were present, the measurement was performed at high reverse bias, with the filling pulses going to 0V. One can notice two well-resolved peaks at temperatures of about 114 K and 285 K, and a broad band in the (140-250) K temperature range, typical of extended defects. However, this spectrum changes significantly when measurements are performed over a narrow spatial window. The experimental conditions for the measurements are indicated in the Fig.1. The successive spectra presented correspond to deeper spatial windows. It can be noticed, that instead of a broad band, three well-resolved peaks are observed just under the surface, labelled S, 4, and 5. A characteristic feature of the spectra is a displacement of the position of the main peak S to higher temperatures with increasing distance from the surface. When the simple DLTS mode was used, externally applied electric field was similar at different depths, and was about $2\times10^4\text{Vcm}^{-1}$. It was confirmed by the Double Correlation DLTS technique that the displacement of the S peak is very small as the external field changes from $2\times10^4\text{Vcm}^{-1}$ to $7\times10^4\text{Vcm}^{-1}$. That is why a dependence of the thermal emission rate on the electric field cannot explain the effect observed. Therefore, it seems that the emission properties of the defect change along with its location along the stacking faults, (probably FPDs).

The S trap also shows a change in volume concentration with increasing distance from the surface. The profile calculated when the width of the transition region was taken into account exhibits a pronounced continous increase of trap concentration (more than one order of magnitude) up to the depth of about 0.9 μm, and then decreases. At the depth of about 1.5 μm, the trap disappears. The depth coincides very well with the penetration depth of our OSFs, as determined from transmission electron microscopy studies. The distribution of the deep state tells us unumbiguously that the S peak is associated with an FPD, not the stacking fault plane itself.

In the case of trap 1 (with energy $E_c - 0.19\text{eV}$), a slight increase of its concentration was observed up to the depth of about 1μm. This may suggest a weak interaction between the defect and the OSFs. However, this trap also exists in the absence of OSFs, i.e. deeper than the penetration depth of our OSFs.

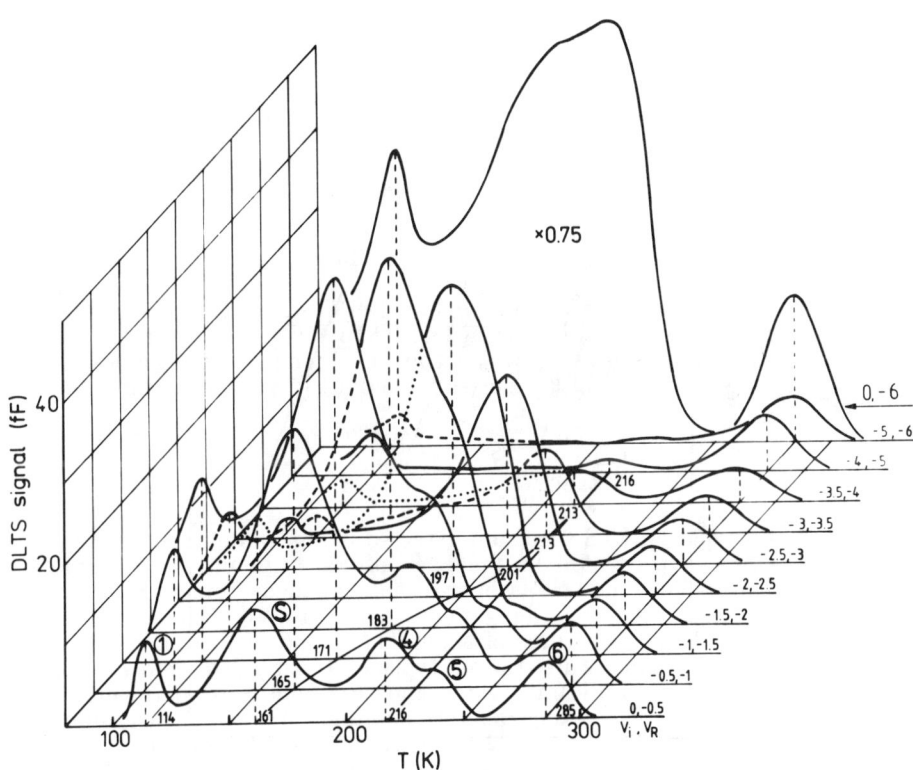

Fig.1 DLTS spectra for n-type Si containing surface OSFs with the final length equal to 3.6μm. The measurements were performed for rate window $e_n = 200s^{-1}$ and pulse duration time $t_c = 1ms$. Each line represents a measurement for a different spatial window. The values of the reverse bias V_R and the voltage level of the filling pulses V_i are indicated at the right hand edge. In the back of the figure, the DLTS spectrum at $V_R = -6V$ and $V_i = 0V$ is shown.

In contrast to traps S and 1, no influence of the OSFs on the trap 6 profile was found. A continous increase of its concentration with increasing depth was observed. The trap 6 ($E_c - 0.54eV$) was also found to be present below the penetration depth of the OSFs.

The characteristic behavior of the main electron trap S has expected consequences for its parameters determined from the Arrhenius plot. It was found that the thermal activation energy and the pre-exponential factor increase with increasing depth, i.e. moving down the stacking fault. In several samples investigated, the thermal activation energy changes in continous manner from $E_c - 0.25$ eV to $E_c - 0.43$ eV. The region of the changes, as well as the limiting cases in the Arrhenius plot, are indicated by the arrows - Fig.2. Since the lowest activation energy was slightly different from sample to sample, the highest value of the energy showed asymptotically a correlation with trap 4. It should be underlined that the values of the energy determined for trap S are apparent values, since the reason for the displacements in the peak temperature positions (Fig.1) has not been established.

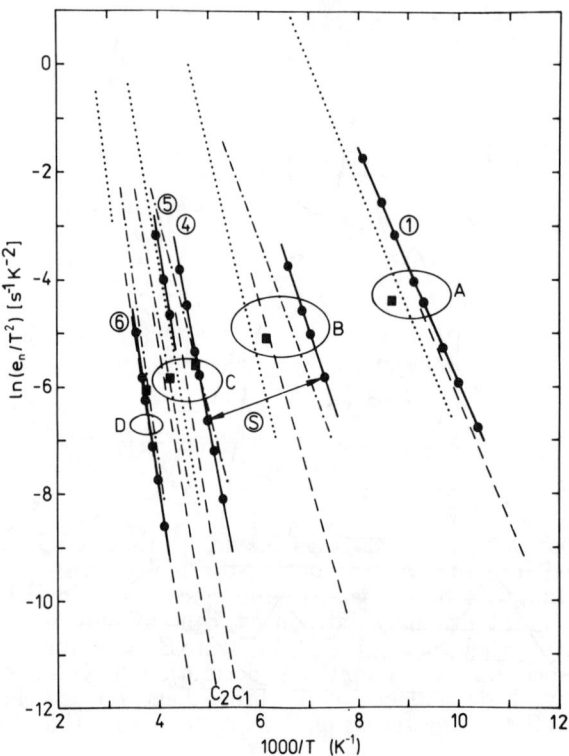

Fig.2 An Arrhenius plot for the DLTS peaks shown in Fig.1. The solid lines with the experimental points - our results. The data obtained in plastically deformed n- Si are also included: squares - after [3], dotted lines - [4], chain lines - [5], dashed lines - [6].

An Arrhenius plot of the traps detected in the samples, containing the OSFs, is closely related to those previously observed in plastically deformed n-Si. The traps 1 and 6 are within the A and B groups, respectively. Due to the specific behavior of the trap S , it shows a correlation with the B as well as C groups. .
In ordering to find the origin of the traps, the samples containing the OSFs characterized by the same final length (3.6µm) were cleaned using a different procedure before the oxidation processes. After standard treatment in hot trichloroethylene, acetone, and methanol, the samples were finally cleaned in three ways. Two of them were based on HF and HF, plus a full RCA procedure [8]. The third group was treated unconventionally. Knowing that the efficiency of a cleaning procedure based on hydrogen peroxide (RCA solution) is time-limited, we utilized this fact in order to gain an uncotrolled contamination effect. The specimens were dipped in a "dirty" RCA solution, i.e. previously contaminated during the washing of many wafers.
The DLTS spectra typical of the three groups of samples are presented in Fig.3. The measurements were performed, when the depletion edge coincided with

the penetration depth of the OSFs. It was found that the DLTS spectra are strongly dependent on the cleaninig procedure. Trap 6 almost disappears after HF plus RCA, and the concentration of trap 1 is also depressed. The effect is not so clear in the case of the broad band associated with OSFs.

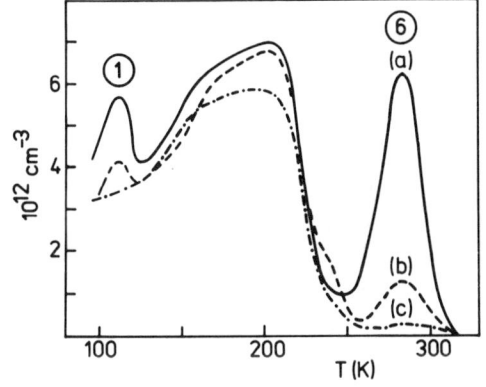

Fig.3 DLTS spectra for n-type Si containing OSFs cleaned in different ways: a) HF + "dirty" RCA, b) HF, c) HF + RCA. The measurements were performed for rate window $e_n = 200s^{-1}$, pulse duration time $t_c = 1ms$, and at $V_R = -6V$, $V_i = 0V$.

The results obtained suggest that traps 6 and 1 are related to the impurities that are introduced into the samples during thermal processes.
We also suppose that the broad band in the central part of DLTS spectrum is associated with non-intentionally introduced contaminants. Specific behavior similar to that presented in Fig.1 for the trap S, was found in the case of the main trap associated with intentional decoration of OSFs with Cu [9]. We suppose that the weak dependence of the broad band on the cleaning procedure used results from efficient gettering of the impurities by the OSFs.

CONCLUSIONS

DLTS spectra reveal several traps in n-type Si containing the OSFs. One of them, characterized by apparent thermal activation energy for electron emission from 0.25eV to 0.43eV, exhibits specific behavior along the FPD, and is closely related to the presence of the OSFs. The trap 1, with thermal activation energy equal to 0.19eV, shows a weak interaction with the OSFs, and can also exist in the absence of extended defects. The trap 6, with energy of 0.54eV, does not exhibit any spatial correlation with the OSFs. We suppose that all the traps obseved are impurity-related.

ACKNOWLEDGEMENTS

We are grateful to B. Hamilton and M.R. Brozel of UMIST for discussions, A. Berg and M. Vandini for technical assistance. We would like to thank to the UK Science and Engineering Research Council for supporting this work.

REFERENCES

1. J.S.Kang and D.K.Schroder, J.Appl.Phys.65, 2974 (1989).
2. E.R.Weber and D.Gilles in Defect Control in Semiconductors, ed. K.Sumino (Elsevier Science Publishers, North Holland, 1990) p. 89.
3. L.C.Kimerling and J.R.Patel, Appl.Phys.Lett.34, 73 (1979).

4. W.Szkielko, O.Breitenstein, and R.Pickenhein, Cryst. Res. Technol. 16, 197 (1981).
5. V.V.Kveder, Yu.A.Osypian, W.Schroter, and G.Zoth, Phys.Stat. Sol. 72, 701 (1982).
6. P.Omling, E.R.Weber, L.Montelius, H.Alexander, and J.Michel, Phys.Rev.B 32, 6571 (1985).
7. K.H.Yang, J.Electrochem.Soc. 131, 1140 (1984).
8. W.Kern and D.A.Pustinen, RCA Review 31, 187 (1970).
9. M.Kaniewska, J.Kaniewski, and A.R.Peaker, this conference.

STUDY OF INTERNAL OXIDE GETTERING FOR CZ SILICON: EFFECTS OF OXIDE PARTICLE SIZE AND NUMBER DENSITY AND ASSESSMENT OF THERMAL STABILITY OF GETTERING FOR COPPER AND NICKEL

Z LACZIK[1], GR BOOKER[1], R FALSTER[2] and P TÖRÖK[1]

[1]Department of Materials, University of Oxford, Parks Road, Oxford OX1 3PHD, UK
[2]MEMC Electronic Materials SpA, Novara, Italy

ABSTRACT

CZ Si wafers containing oxide particles of size ≈ 30 nm were subsequently contaminated with either Ni or Cu, given single- or double-anneals and fast- or slow-cooled. The scanning infra-red microscope (SIRM) was used to determine the number density and size of the active gettering centres. The results showed that the gettering process depended on the type of metal, the oxide particle density and the cooling rate. For the specimens given a double-anneal, the gettering behaviour was often significantly different from the specimens given a single-anneal. The active gettering centres were thermally unstable at high temperatures, indicating that the gettering that occurs during the fabrication of VLSI is a dynamic process. Mechanisms are proposed to explain the results.

1. INTRODUCTION

Due to the increasingly small critical dimensions used by present day VLSI and ULSI technologies, techniques used for the removal of harmful contaminants during device processing are gaining major importance. One such technique is internal oxide gettering (IOG), mainly used for the gettering of metal contamination. For IOG, Czochralski (CZ) silicon wafers containing interstitial oxygen are heat-treated to precipitate the oxygen to produce oxide particles in the bulk, while keeping the surface region of the wafers denuded of oxide particles. During the following device processing steps, the metal species are attracted to the oxide particles and associated crystal defects, and away from the wafer surface where the devices are located.

We have studied IOG in a range of CZ silicon wafers containing oxide particles with different sizes and number densities for a range of fast diffusing metal contaminants [1 to 4]. The haze test [5] as subsequently modified [1] was used to assess the gettering efficiency of the IOG process. We used etching combined with optical microscopy, transmission mode scanning infra-red microscopy (SIRM) [6] and transmission electron microscopy (TEM) to reveal oxide particles, associated dislocations and metal precipitate particles, and in particular to determine particle number densities, particle sizes and 3-D distributions. It was established that both small (10 to 30 nm) and large (40 to 80 nm) oxide particles were efficient in gettering copper and nickel contamination, if present in sufficient number densities within the bulk of the wafers, even though the getting mechanisms were different in the two cases.

After previous studies of specimens closely similar to those used for the present work, the following mechanisms were proposed for the gettering of copper and nickel by small oxide particles [4]. For copper, gettering starts with the precipitation of copper silicide particles at individual oxide particles. The copper silicide particles then form a colony with a dislocation loop bounding the colony. The colony grows by the nucleation of additional silicide particles on the dislocation loop while the dislocation loop extends by climb. The dislocation movement is supported by excess silicon self-interstitials generated by the metal precipitation. For nickel, metal silicide particles precipitate at individual oxide particles and dislocation loops are punched out. Further metal precipitation occurs on the dislocation loops while the loops glide away from the oxide particles and new dislocation loops are generated. In this case dislocation movement is supported by strain caused by the metal precipitation. These mechanisms are similar to results obtained from independent examinations of specimens with metal contamination, but without purposely introduced oxide particles [7,8].

In general, it is advantageous to keep the oxygen precipitation at the lowest level (both for size and number density of oxide particles) at which complete gettering can still be achieved. In the present work we investigated the effect of oxide particle number density variations and the thermal stability of gettering for IOG based on small oxide particles in CZ silicon wafers and for copper and nickel contamination.

2. EXPERIMENTAL

For our present investigation three groups of CZ silicon wafers were given three-step annealing treatments to produce oxide particles in the bulk of the wafers. The as-grown wafers contained $\approx 7 \times 10^{17}$ cm^{-3} oxygen. The first step was used to control the oxide particle number density by dissolving some of the oxide nuclei pre-cursors and this step was either 1100°C for 15 min., 900°C for 15 min. or omitted for the three wafer groups. In the following the groups will be referred to as low, medium and high number density (**L-**, **M-** and **H-**) respectively. The second and third annealing steps were identical for all three groups and were 800°C for 4 hours and 1000°C for 4 hours.

The wafers were then broken into pieces and intentionally contaminated with either copper or nickel. Following the contamination, the specimens were annealed at 1100°C for 10 min. ('drive-in' anneal). Either slow (**-S**) or fast (**-F**) cooling rates of 3 C°/min. or 100 C°/second respectively were used at the end of the drive-in anneal. In the following these specimens will be denoted by, for example, **L-Cu-F** for low oxide particle number density, copper contamination and fast cool after the drive-in anneal. To investigate the thermal stability of the gettering, specimens **M-Cu-S**, **M-Cu-F**, **M-Ni-S** and **M-Ni-F** were given an additional second drive-in anneal and again were cooled at either slow or fast rates. These specimens will be denoted by suffixes such as **-SF**, and **-FS**.

Gettering efficiency was then assessed by the modified haze-test, and the specimens were examined in the SIRM in bright-field plan-view mode, with the SIRM focussed in the middle of the wafers, i.e. $\approx 300\ \mu$m below the front surface. For the SIRM examination, the back surfaces of the specimens were polished mirror flat.

3. RESULTS

Oxide particles
SIRM images from uncontaminated specimens from specimen groups **L-**, **M-** and **H-** showed dark spots $\approx 2\ \mu$m across (the spatial resolution of the SIRM) with very low ($\approx 0.1\%$) contrast corresponding to individual oxide particles. From these SIRM images it was determined that the initial three-step annealing treatments produced oxide particles with number densities $\approx 5 \times 10^6$ cm^{-3} $\approx 6 \times 10^7$ cm^{-3} and $\approx 1 \times 10^9$ cm^{-3} for specimen groups **L-**, **M-** and **H-** respectively and estimated oxide particle size ≈ 30 nm. The latter size was in agreement with TEM results [9] obtained from these specimens.

Gettering efficiency
Following the metal contamination and drive-in anneal, Secco etching of the front surfaces of the specimens (haze test) revealed no surface precipitation, indicating complete gettering of the metal

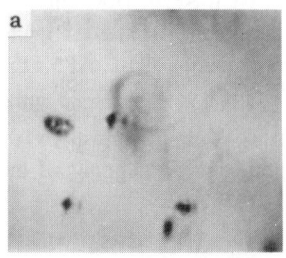

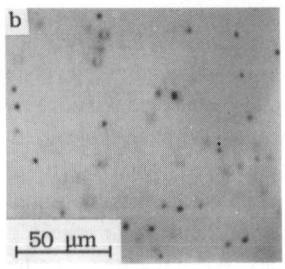

Figure 1 SIRM plan-view images from specimens with Ni contamination and fast-cooling. a) Low and b) medium oxide particle number densities (specimens L-Ni-F and M-Ni-F)

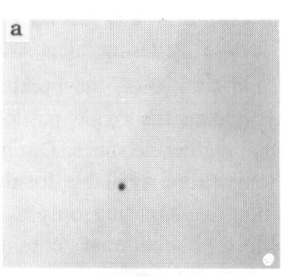

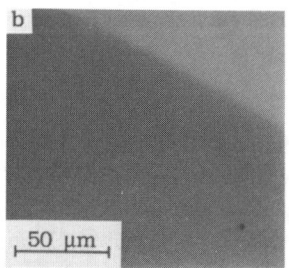

Figure 2 SIRM plan-view images from specimens with Ni contamination and slow-cooling. a) Low and b) medium oxide particle number densities (specimens L-Ni-S and M-Ni-S)

contamination. The only exception was for nickel with the low oxide particle number density when fast-cooled (**L-Ni-F**), for which some surface precipitation was revealed by etching, indicating incomplete gettering.

SIRM images from the contaminated specimens showed dark spots or lines of spots with increased contrast. This increased contrast arises from gettered metal precipitating at the oxide particles in the bulk in the form of metal silicide particles.

Nickel contamination, different oxide particle number densities
For Ni and fast cool, SIRM images (Figure 1) showed spots ≈ 2 μm across with number densities 5×10^6 cm^{-3}, 6×10^7 cm^{-3} and 8×10^7 cm^{-3} for specimens **L-**, **M-** and **H-Ni-F** respectively. The spot contrast progressively decreased on going from the low oxide particle number density specimen to the high number density specimen. For Ni and slow cool, SIRM images (Figure 2) revealed spots ≈ 2 μm across but with significantly lower number densities (2-5×10^6 cm^{-3}) and with higher contrast ($\approx 30\%$) for all three specimens (**L-**, **M-** and **H-Ni-S**).

Nickel contamination, double drive-in anneal
For specimen **M-Ni-FS**, SIRM images (Figure 3a) were similar to images from specimen **M-Ni-S**, showing spots 3-5 μm across with number densities of 2×10^6 cm^{-3} and with 10-20% contrast. For specimen **M-Ni-SF**, SIRM images (Figure 3b) were very similar to images from specimen **M-Ni-F**, showing spots ≈ 2 μm across with 5×10^7 cm^{-3} number density and $\approx 3\%$ contrast.

Copper contamination, different oxide particle number densities
For Cu and fast cool, SIRM images (Figure 4) showed either large star shaped features 10 to 20 μm across with the points of the stars aligned along <110> directions (specimen **L-Cu-F**) or spots ≈ 2 μm across (specimens **M-** and **H-Cu-F**). The number density was 4×10^6 cm^{-3} for the stars and 5-6×10^7 cm^{-3} for the spots. The contrast decreased from up to 30% for the stars to $\approx 2\%$ for the spots. For Cu and slow cool, SIRM images from specimen **L-Cu-S** (Figure 5a) showed large features up to ≈ 60 μm across with fine structure which included arms aligned along <110> directions. The number density for these large features was 4×10^6 cm^{-3}. For specimen **M-Cu-S**, SIRM images (Figure 5b) showed rows of spots ≈ 2 μm across. The rows were 10 to 25 μm long and were aligned along <110> directions. The number density of the rows was 3×10^7 cm^{-3}. SIRM images from specimen **H-Cu-S** were very similar to images from **M-Cu-F** and **H-Cu-F** showing spots with a number density of 4×10^7 cm^{-3} and with $\approx 1\%$ contrast.

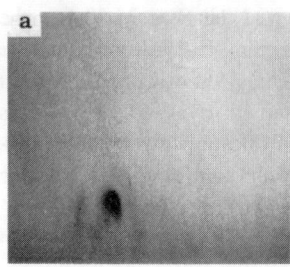

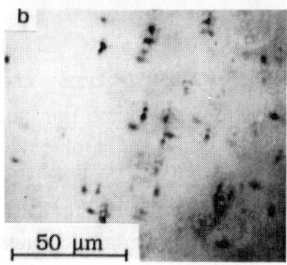

Figure 3 SIRM plan-view images from double-annealed specimens containing oxide particles with medium number densities, and with Ni contamination. a) Fast-cooling followed by slow-cooling b.) Slow-cooling followed by fast-cooling (specimens M-Ni-FS and -SF)

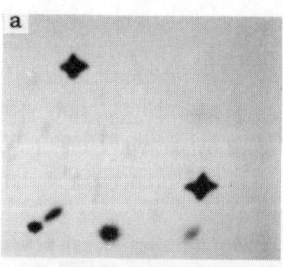

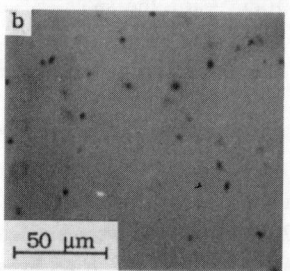

Figure 4 SIRM plan-view images from specimens with Cu contamination and fast-cooling. a) Low and b) medium oxide particle number densities (specimens L-Cu-F and M-Cu-F)

Copper contamination, double drive-in anneal

For both specimens **M-Cu-FS** and **M-Cu-SF**, SIRM images (Figure 6) were similar to images from specimen **M-Cu-S**. showing rows of spots 10 to 25 μm long. For specimen **M-Cu-FS**, the number densities of the rows was 3×10^6 with 1.5-4% contrast while for specimen **M-Cu-SF**, the number density of the rows was 3×10^7 with ≈1% contrast. For **M-Cu-FS**, significant overlapping of the spots occurred within each row.

4. DISCUSSION

Effects of oxide particle number density on the gettering of nickel

For Ni and fast cool, the number densities of active gettering centres (centres showing increased contrast due to metal precipitation) and the number densities of oxide particles were closely the same up to a limit, above which not all the oxide particles acted as active gettering centres (specimen H-Ni-F). This can be explained by considering a diffusion limited gettering mechanism. After the drive-in anneal, on cooling down, metal supersaturation occurs. When reaching a certain level of supersaturation, metal silicide particles nucleate at some of the oxide particles and active gettering centres form through the mechanism described in the introduction. Due to the precipitation of the solute nickel, the supersaturation decreases in a spherical zone around the active gettering site. Since the decreased supersaturation is not sufficient for nucleation at additional oxide particles within the depleted zone, all those additional oxide particles become 'inhibited' and will not act as active nucleation centres. For Ni and fast cool, a saturation in the number density of active gettering sites was observed at $\approx 8 \times 10^7$ cm^{-3}. For nickel and slow cool, the number density of active gettering sites was less than the oxide particle number density for all three oxide particle number densities. During the slow cooling, there is significantly longer time available for the nickel to diffuse while the supersaturation is increasing only at a slow rate, the size of the zones within which all additional gettering centres are inhibited will be much larger than for the fast cool case. This can explain the dramatically decreased number density of active gettering centres.

Thermal stability of gettered nickel

For **M-Ni-FS**, the drastically reduced number density of active gettering sites (2×10^6 cm^{-3} as compered to 6×10^7 cm^{-3} for **M-Ni-F**) can only be explained by considering the dissolution of all or most of the nickel silicide particles during the second drive-in anneal. During the following slow cooling, only a small proportion of the oxide particles will then act as active gettering centres as described above for **M-Ni-S**. In a similar manner, for **M-Ni-SF**, most of the nickel particles dissolve during the second drive-in anneal. During the following fast cool, most oxide particles will initiate active gettering

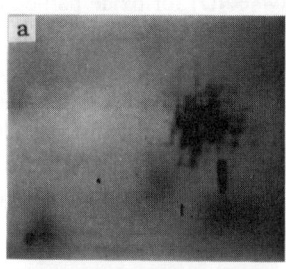
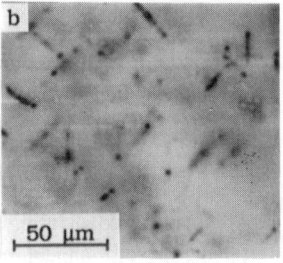

Figure 5 SIRM plan-view images from specimens with Cu contamination and slow-cooling. a) Low and b) medium oxide particle number densities (specimen L-Cu-S and M-Cu-S)

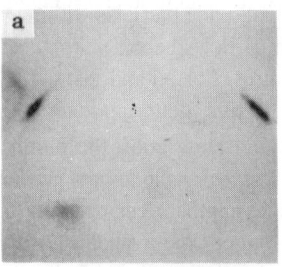
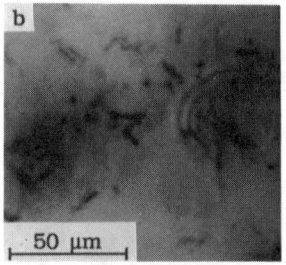

Figure 6 SIRM plan-view images from double-annealed specimens containing oxide particles with medium number densities, and with Cu contamination. a) Fast-cooling followed by slow-cooling b) Slow-cooling followed by fast-cooling (specimen M-Cu-FS and -SF)

centres, resulting in a number density of 5×10^7 cm^{-3} similar to the number density of active centres for **M-Ni-F**. For **M-Ni-FS**, the increased size of the active centres (3-5 μm across) as compared to **M-Ni-F** or **-S** (less than 2 μm) can be explained by considering that during the second drive-in anneal, although most of the metal is dissolved, not all of the punched-out dislocation loops anneal out. Consequently, during the following slow cooling, metal precipitation will start not only at the oxide particles, but also on the remaining loops and more readily generate additional new loops, resulting in large gettering centres. When cooled fast, the remaining loops have less effect, and the resulting gettering centres will be small.

Effect of oxide particle number density on the gettering of copper

For Cu and fast cool, a saturation level ($\approx 5 \times 10^7$ cm^{-3}) in the number density of the active gettering centres can also be observed and can be explained by the diffusion limited gettering mechanism. However, for Cu and slow cool, the number densities of the active centres are higher than what would be expected for the diffusion limited process. The increased number densities can be explained by considering the rate at which the copper silicide colonies can grow as the limiting factor. The growth rate of the colonies can be relatively low since the expansion of the colonies involves dislocation climb and the bounding dislocations can be pinned by the silicide particles.

Thermal stability of gettered copper

For **M-Cu-FS**, the formation of large active gettering centres with low number densities (as compared to **M-Cu-F** and **-S**) can be explained by considering that all or most of the copper silicide particles are dissolved during the second drive-in anneal. Since the colonies produced by the first drive-in anneal/fast cool were relatively small, the dislocation loops bounding the colonies are annealed out or significantly reduced in size during the second drive-in anneal. During the following slow cool nucleation starts at a small number of remaining dislocation loops, resulting in a reduced number density of active gettering centres. However, for **M-Cu-SF**, the first drive-in anneal/slow cool produced large colonies. Although all or most of the metal is dissolved during the second drive-in anneal, most of the bounding dislocations will only reduce in size and will not anneal out completely. During the following cooling down, all of these remaining loops will act as active gettering centres, resulting in number densities similar to the number density of active gettering centres for **M-Cu-S**.

5. SUMMARY AND CONCLUSIONS

The results show that when CZ Si wafers containing oxide particles of size ≈ 30 nm are subsequently contaminated with either Ni or Cu, annealed at 1100 C° and cooled, the gettering behaviour depends on the particular metal, the oxide particle number density and the cooling rate. For example, there is a saturation number density (SND) for oxide particles, above which the additional oxide particles are no longer active in the gettering process. The SND was shown to be 8×10^7 cm^{-3} for Ni and 5×10^7 cm^{-3} for Cu. If such gettered wafers are given a second gettering treatment, the behaviour that then occurs is often significantly different from the earlier behaviour. For example, the gettering can be modified due to the presence of dislocations associated with the individual oxide particles that remained from the first gettering treatment. The results demonstrate that the active gettering centres are not stable at high temperatures, and indicate that the gettering that occurs during the fabrication of VLSI is a dynamic process.

ACKNOWLEDGEMENTS

The authors wish to thank G. Ferrero for his help with some aspects of the work and to acknowledge support by SERC, UK.

REFERENCES

[1]   Falster, R. and Bergholz, W.: J. Electrochem. Soc. 1990 137 1548
[2]   Laczik, Z., Booker, G.R. and Falster, R.: Solid State Phenomenon, 1989, 6, 395
[3]   Laczik, Z., Falster, R. and Booker, G.R.: GADEST 91, Berlin, October 1991, to be published in Solid State Phenomenon
[4]   Bhatti, A.R., Falster, R. and Booker, G.R.: GADEST 91, Berlin, October 1991, to be published in Solid State Phenomenon
[5]   Graff, K., Hefner, H.A. and Pieper, H.: Proc. Matetial Res. Soc. Meeting, Pittsburg, USA, 1985, (Eds. R.D. Fair, C.W. Pearce and J. Washburn), p.19.
[6]   Booker, G.R., Laczik, Z. and Kidd, P.: Proc. of Conference 'Defect Recognition and Imaging in Semiconductors Before and After Processing (DRIP 4)', Wilmslow, UK, 1991, Semiconductor Science and Technology (in press)
[7]   Seibt, M. and Graff, K.: J. Appl. Phys., 1988, 63, 4444
[8]   Ourmazd, A. and Schröter, W.: Appl. Phys. Letts., 1984, 45, 781
[9]   Bhatti, A.R.: private communication

MORPHOLOGY CHANGE OF OXYGEN PRECIPITATES IN CZ-SI WAFERS DURING TWO-STEP HEAT-TREATMENT

MASAMI HASEBE[1], JAMES W. CORBETT[1] and KAZUTO KAWAKAMI[2]

[1]The Institute for the Study of Defects in Solids,
Department of Physics, The State University of New York at Albany,
Albany, New York 12222, U.S.A.

[2]Electronics Research Laboratories, Nippon Steel Corporation
1618 Ida, Nakahara, Kawasaki, Kanagawa 211, Japan

ABSTRACT

We have studied the morphology change of oxygen precipitates during two-step heat-treatments in Czochralski silicon (CZ-Si) wafers by transmission electron microscopy (TEM), high resolution electron microscopy (HREM), and Fourier-transform infrared absorption spectroscopy (FT-IR). We have found that the morphology change during low and high temperature two-step annealing is not necessarily continuous; thermal donors (TDs) dissolve into solute oxygen, and platelets shrink out and simultaneously octahedra are newly nucleated in the second high-temperature annealing. The precipitate morphology is determined by the relaxation process of the lattice stress due to the volume difference of the oxide and the matrix silicon, *i.e.*, the platelet is a favorable shape to reduce the strain at lower temperatures and the octahedron is easily formed with the stress release by self-interstitial emission at more than 1000°C.

Introduction

Oxygen in CZ-Si wafers is well known as a useful impurity to improve the wafer strength [1, 2] and the purification of device region by the so-called intrinsic gettering technique [3, 4]. However, the precise control of oxygen precipitation is needed to obtain the appropriate wafers with the both ability of strengthening and gettering in multiple heat-treatments of the device manufacture. In order to provide such wafers, it is important that the successive precipitation mechanism be clear in the heat-treatments. Although many authors have reported the oxygen precipitation behavior (*e.g.*, Bourret [5] for a review), the detailed mechanism, for example, the question on the continuity of the precipitation process from TDs to octahedra, is not clarified yet. And sometimes we can see peculiar problems of the precipitate distribution in association with the Si-wafer thermal history [6, 7]. The problem still remaining to be solved is how the thermal history in the crystal growth to device fabrication process effects the precipitate formation and distribution. As one of the first steps to know the precipitation mechanism during consecutive heat-treatments, we have studied the morphology change and the formation behavior of the secondary defects such as dislocation loops and stacking faults by TEM, FT-IR, and resistivity measurement.

Experimental procedure

Specimens were n-type (phosphorous doped, 1.0Ω·cm), <110> orientation, CZ-Si wafers. The oxygen and carbon concentrations were between $9.5\text{-}10.0 \times 10^{17}/cm^3$ and less than $10^{16}/cm^3$, respectively. The oxygen concentration was measured at room temperature by FT-IR with a 1106cm^{-1} absorption line and a conversion factor of 3.03 x

$10^{17}/cm^2$. The change in carrier density was examined by eddy-current resistivity-measurement. TEM (Hitachi H800) and HREM (JEOL 4000EX) observations were operated at 200kV to avoid the electron-irradiation damage observed significantly in 400 kV operation. The TEM specimens were cut from the same center positions of the wafers to prevent the influence of oxygen-concentration fluctuation. Heat treatments were performed in quartz furnaces with purified nitrogen ambient gas, following a common wafer cleaning procedure for the device fabrication.

Experimental results

1. Single heat-treatment

The morphology change of the precipitates was observed by TEM with the specimens subjected to single heat-treatments between 400-1100°C. Although no precipitates or no other defects were obtained by TEM in the samples treated at 400 and 450°C for up to 312 hours, TD formation was confirmed by the increase in carrier density from 5.0×10^{15} (as-prepared level) to $11.2 \times 10^{15}/cm^3$ at 450°C for 120 hours. Rod-like defects elongate in <110> were observed between 550-750°C. Platelets lying on {100} planes were formed between 650-1000°C. Octahedra bounded by eight {111} planes but usually truncated by {100} planes were seen at 1100°C. Typical micrographs are shown in fig. 1. As a specific result, at 900 and 1000°C, each precipitate

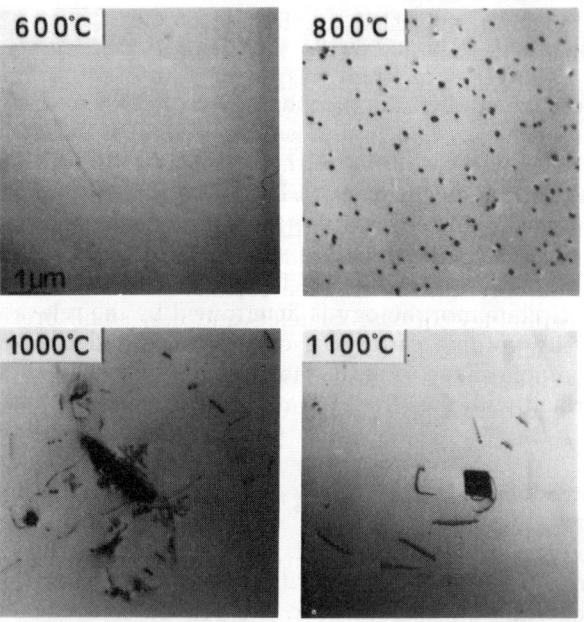

Fig. 1: TEM micrographs in the specimens after single annealing at 600, 800, 1000, and 1100°C for 120 hours.

is indeed nearly square in shape, bounded by a dislocation loop; however, it is not one perfect plate but an oxide colony, which looks like a dendritic or ameba shape. Perfect dislocation loops were observed in 900-1100°C treatments. On the contrary, no stacking faults could be found in every specimen after single-step heating even at 1100°C.

2. Two-step heat-treatment

A. 450°C + second annealing

Using pre-annealed wafers at 450°C for 120 hours, we investigated TD disappearance-behavior during the second annealing at 650, 800, 1000, and 1100°C for 1-16 hours. The increased carrier-density by TD formation in the pre-anneal was recovered to the as-prepared level and thus TDs vanished completely after the second treatments at all above temperatures. Corresponding to TD disappearance, the solute-oxygen concentration (fig. 2) also increased up to the as-prepared level for the first one hour. It indicates that TDs are broken and dissolve into solute oxygen. Afterward the concentration decreased again at 650-1000°C but almost unchanged at 1100°C. This is caused by new precipitate nucleation and growth.

B. 800°C + second annealing

The platelets formed at 800°C also dissolved during the second annealing at 1000 and 1100°C. As an example, a set of TEM micrographs taken from the specimens subjected to 800 + 1100°C treatment is shown in fig. 3. After half an hour, many of the platelets have shrunk out but some still remain, and also stacking faults and new small precipitates are simultaneously generated. After four hour treatment, the platelets have entirely disappeared, and only the stacking faults and the small precipitates can be seen. The

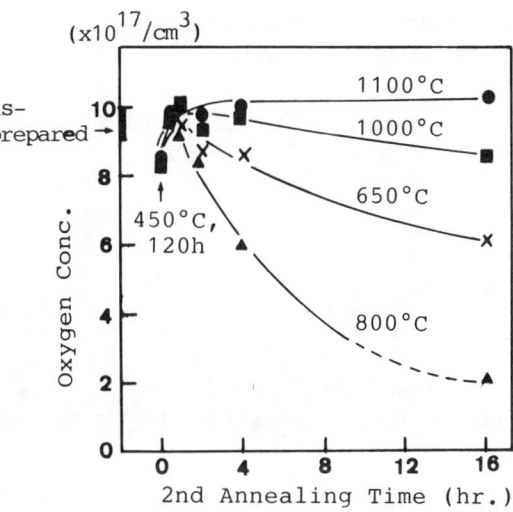

Fig. 2: The change in solute-oxygen concentration during 450 + 650, 800, 1000, or 1100°C two step annealing.

morphology of the platelets and new small precipitates was determined by HREM, as each typical image is shown in figs. 4(a) and (b). The form of the new small precipitates is a truncated octahedron. The octahedral shape but not in perfect was also observed at 1000°C (fig. 6 (b)). The dissolution of platelets and the formation of

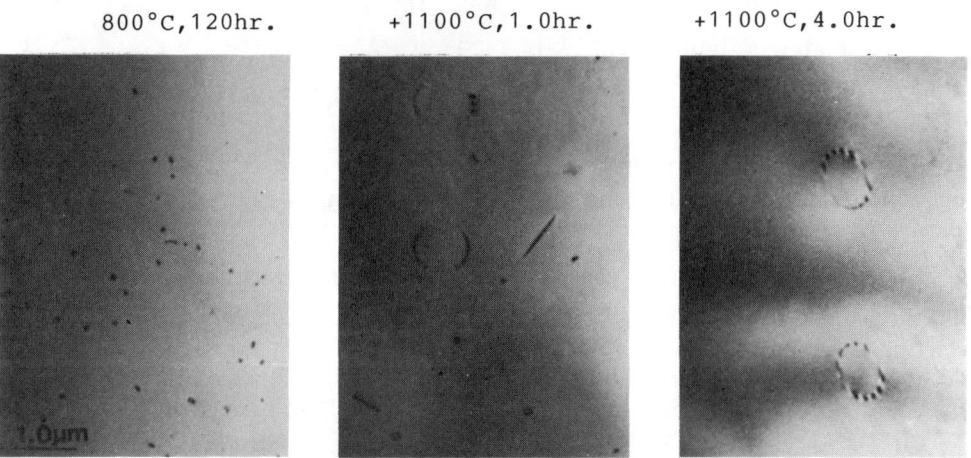

Fig. 3: A set of TEM micrographs showing the morphology change of the precipitates and the secondary defects in the 2nd anneal at 1100°C after 800°C pre-annealing.

octahedra were confirmed by FT-IR (fig. 5(a)) and oxide number-density (fig. 5(b)) measurements. The solute-oxygen concentration is regained over the solubility limit right after the second heating is started. This result describes that the plate-like shape is no longer stable and dissolve at 1000 and 1100°C. Moreover the subsequent re-decrease of the concentration and the increase of the number-density mean that the octahedral shape is more stable so that it is newly nucleated and grow. However, occasionally, we found the thick center parts of relatively large platelets could grow and change continuously into octahedra, which Bergholz et. al. [8] called fins, as shown in figs. 6(a) and (b).

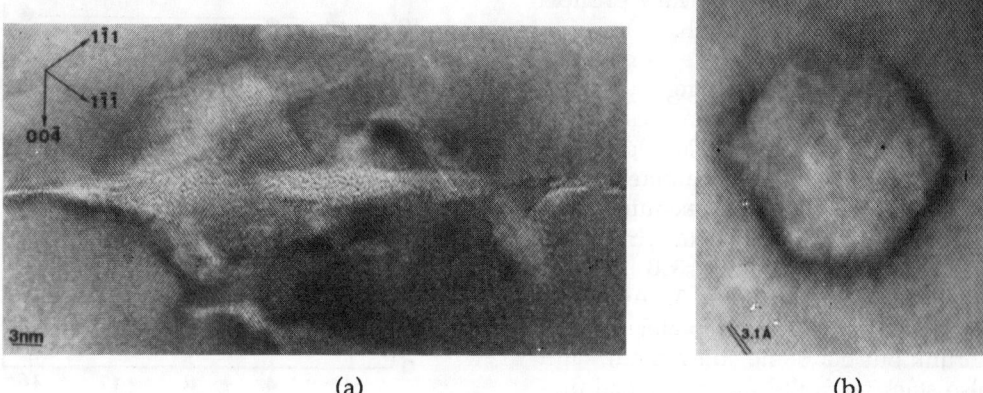

(a)　　　　　　　　　　　　　　　　(b)

Fig. 4: HREM images of a platelet formed after 800°C, 120h (a), and of a small precipitate formed after 800°C, 120h + 1100°C, 4h (b).

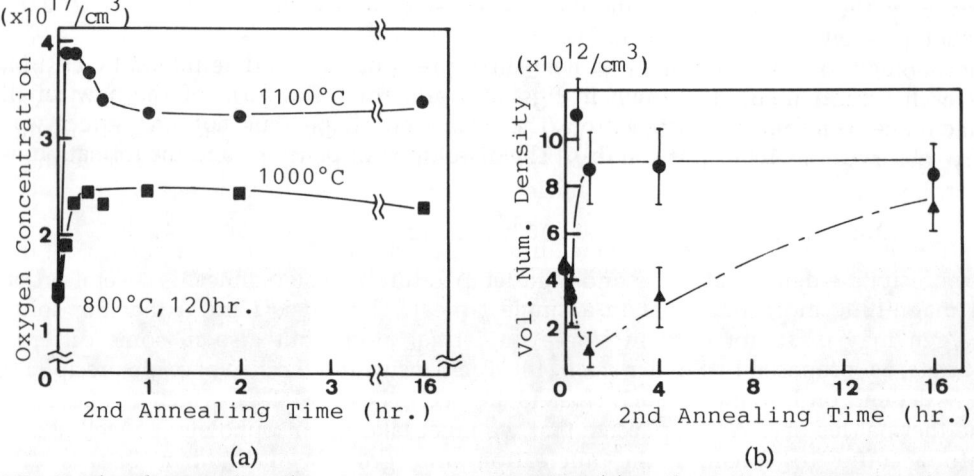

(a)　　　　　　　　　　　　　　　　(b)

Fig. 5: The changes in solute-oxygen concentration (a) and precipitate number-density (b) during the 2nd treatment at 1000 and 1100°C after pre-annealing at 800°C for 120h.

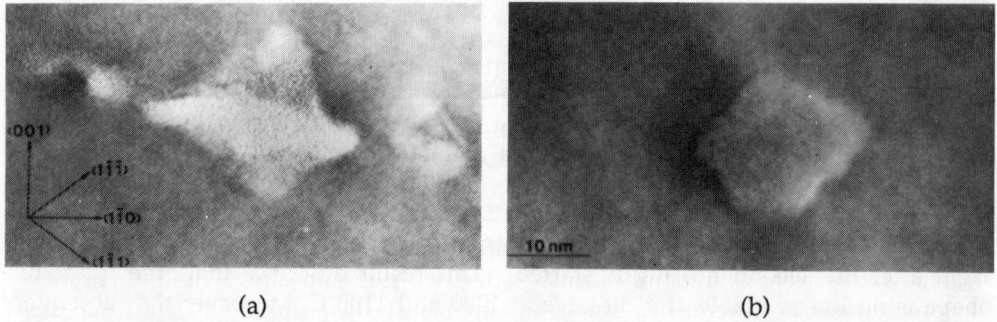

(a)　　　　　　　　　　　　　　　　(b)

Fig. 6: HREM images showing the morphology change of the platelet precipitate. Note that the thick center part changes into an octahedron with dissolution of the other parts. (a): 800°C, 120hr. + 1000°C, 1.0hr.　(b): 800°C, 120hr. + 1000°C, 16hr.

Discussion

1. Morphology in single heat-treatment

The morphology of oxygen precipitates formed in single heat-treatments was small platelets at 650-800°C, large platelets at 900-1000°C, and octahedra at 1100°C. Although these are basically the same results as the previous papers by many authors [e.g. 5, 8], we have found more detailed structure of the large platelets; remarkably at 1000°C, each plate is bounded by a perfect dislocation to form a square but consists of a colony of small particles with dendritic or ameba shape. The volume number-density and the mean size were around $10^9/cm^3$ and a few micrometers, respectively. In that density without new nucleation, since we did not recognize significant density-increase with respect to annealing time, the platelets may extend their size in a dendritic shape on their planes. Why they prefer growing in such a dendritic way is a new problem for the future study.

2. Dissolution and new formation of the precipitates

We observed the dissolution of lower temperature precipitates and the formation of different shape precipitates in the second higher temperature annealing. TDs dissolve into solute oxygen and therefore they cannot become the new nuclei formed at more than 650°C. The platelet shape is fundamentally unstable and shrink out at higher than 1000°C, although some platelets change their shape to an octahedron in the case that their center parts are thick enough to grow. We consider that the pre-formed oxygen precipitates are disintegrated and dissolve unless they are stable in terms of morphology as well as size for the nuclei in the second heat-treatments.

On the other hand, we know the fact, from experience and other reporters [e.g. 9], that the precipitate density in a low and high temperature two-step anneal is several orders of magnitude more than that in a single anneal. Thus, it is natural that the solute oxygen may exist not only in single interstitial atoms but also in some different clusters, as proposed by Snyder et. al. [10]. The density and the size of oxide precipitates may depend on how many such clusters are stable in the wafers, which can be called the thermal history. We emphasize that we must take the morphology as well as size into account to consider the nucleation and growth mechanism because the morphology change is not always continuous.

3. Secondary defect formation and precipitate morphology

The secondary defect formation-mechanisms have been already proposed for the dislocation loops by Tan et. al. [11] and for the stacking faults by Patel [12] and Mahajan et. al. [13]. Both mechanisms are based on the relaxation process of the strain due to the volume difference of the precipitates and the matrix silicon. Dislocations are punched out and stacking faults are formed by self-interstitials emission and agglomeration.

In this study, perfect dislocation loops were mainly observed in single annealing but stacking faults were dominantly formed after two-step annealing. The morphology of the precipitates is different in single treatments at 1000 and 1100°C but the size is similar and very large, more than 500nm. With such a big volume, the mechanical compression-stress around the precipitates is so high that the dislocations are easily introduced during cooling processes in the heat-treatments. However, the concentration of the precipitated oxygen is so small (less than $10^{17}/cm^3$) that the self-interstitial emission is insufficient to form stacking faults. In the two-step treatments, many small octahedral precipitates (less than 50nm in size, 10^{12}-$10^{13}/cm^3$ in density, $5 \times 10^{17}/cm^3$ in amount of precipitated oxygen) occur with the platelet dissolution. The growth of such many octahedra can generate much more self-interstitials than those consumed by the dissolution. Besides, the octahedra can become the stacking-fault formation centers. Therefore, only could we observe the stacking faults after two-step treatments.

From these results, and also considering that no stacking faults are formed at 800°C in spite of a lot of precipitates, we support the idea, suggested by Hu [14] and Tiller [15], that the self-interstitial emission is dominant as a stress relaxation process at more than 1000°C and thus octahedral shape can easily grow. On the contrary, at lower temperatures, the emission may be infrequent so that plate-like morphology is favorable to reduce the strain. Conclusively, the formation of the secondary defects are mutually and strongly related to the precipitate morphology.

Acknowledgment

The authors wish to thank Dr. I. Verner for discussion and Dr. A. J. Yencha for interest and encouragement. The experimental results were mainly obtained by one of the present authors (M.H.) at Electronics R&D Laboratories., Nippon Steel Corp. in Japan.

References

1. J. R. Patel and A. R. Chandhuri, J. Appl. Phys. 33 , 2233 (1962)
2. K. Sumino, I. Yonenaga and A. Yusa, Jpn. J. Appl. Phys. 19 , L763 (1980)
3. T. Y. Tan, E. E. Gardner and W. K. Tice, Appl. Phys. Lett. 30 , 175 (1977)
4. N. Inoue, K. Wada and J. Osaka in *Semiconductor Silicon 1981*, eds. H.R.Huff, R. J. Kriegler and Y. Takeishi (Electrochem. Soc., Pennington, 1981) p.282
5. A. Bourret, Mat. Res. Soc. Symp. Proc. 59 , 223 (1986)
6. T. Y. Tan and C. Y. Kung, J. Appl. Phys. 59 , 917 (1986)
7. M. Hasebe, Y. Takeoka, S. Shinoyama and S. Naito, in *Defect Control in Semiconductors*, ed. K. Sumino (North-Holland, Amsterdam, 1990) p.157
8. W.Bergholz, M.J.Binns, G.R.Booker, J. C.Hutchison, S.H.Kinders, S. Messoloras, R.C.Newman, R.J.Stewart and J.G.Wilkes, Phil. Mag. B59 , 499 (1989)
9. H. Tanaka, J. Osaka and N. Inoue, Jpn. J. Appl. Phys. Suppl. 18-1 , 179 (1979)
10. L. C. Snyder and J. W. Corbett, Mat. Res. Soc. Symp. Proc. 59 , 207(1986)
11. T. Y. Tan and W. K. Tice, Phil. Mag. 13 , 615 (1976)
12. J. R. Patel, in *Semiconductor Silicon 1977*, eds. H.R.Huff and E.Sirtl (Electrochem. Soc., Pennington, 1977) p.521
13. S. Mahajan, G. A. Rozgonyi and D. Brasen, Appl. Phys. Lett. 30 , 73 (1977)
14. S. M. Hu, Mat. Res. Soc. Symp. Proc. 59 , 249 (1986)
15. W. A. Tiller, S. Hann, and F. A. Ponce, J. Appl. Phys. 59 , 3255 (1986)

ANNEALING OF DAMAGE IN GaAs AND InP AFTER IMPLANTATION OF Cd AND In

W. Pfeiffer[1], M. Deicher[1], R. Kalish[2], R. Keller[1], R. Magerle[1], N. Moriya[2], P. Pross[1], H. Skudlik[1], Th. Wichert[3], H. Wolf[3] and
ISOLDE Collaboration[4]
[1] Fakultät für Physik, Universität Konstanz, D-7750 Konstanz, FRG
[2] Physics Department and Solid State Institute, Technion, Haifa 32000, Israel
[3] Technische Physik, Universität des Saarlandes, D-6600 Saarbrücken, FRG
[4] CERN, CH-1211 Geneva 23, Switzerland

ABSTRACT

The removal of damage and the electrical activation after heavy ion implantation of ^{111m}Cd and ^{111}In was investigated using the perturbed angular correlation technique (PAC) and Hall measurements. After implantation at 90 K and subsequent annealing the removal of structural disorder in the vicinity of the probe atom ^{111}In was observed around 300 K in GaAs and InP. The annealing behavior in the high temperature regime (500 K to 1100 K) of GaAs implanted with ^{111m}Cd and ^{111}In was investigated as a function of total implantation dose. After annealing at 600 K part of the Cd probe atoms are located in a slightly perturbed environment, the remainder in a heavily perturbed one. For Cd annealing above 900 K leads to outdiffusion of Cd located in heavily perturbed sites and electrical activation occurs. In contrast to Cd all In probe atoms are located in a slightly perturbed environment and no In is lost by outdiffusion. The differences and similarities of results obtained after Cd and In implantation are discussed in terms of extended defects and their interactions with the probe atoms.

1. Introduction

Most of the investigations concerning the recovery of III-V compound semiconductors on a microscopical scale were performed after irradiation of these materials with electrons, protons [1], or neutrons [2]. The situation after heavy ion implantation is different and up to now not well understood. Ion implantation is leading to a much higher defect concentration, creating amorphous regions in the material already at small doses [3]. Studies of high temperature annealing of implantation damage above 600 K in III-V materials mostly look at the electrical activation of dopants [4] and supply no direct information on the annealing mechanisms. The basic defect reactions were investigated at lower temperature by Rutherford backscattering [4], emission channeling (EC) [5], positron annihilation [2], and Mößbauer measurements [6]. As it was already shown, PAC is also able to supply information on the annealing behavior of implanted dopants [7,8,9]. The extension to different probe atoms (^{111}In and ^{111m}Cd) and the variation of implantation temperature and dose allow a more detailed discussion of the obtained PAC data in this work.

2. Method

The PAC technique is sensitive to electric field gradients (efg) present at the site of the probe atom, in our case ^{111}In ($t_{1/2}$ = 2.8 days) or ^{111m}Cd ($t_{1/2}$ = 48 min), both decaying via the same intermediate nuclear state. The interaction of the efg with the quadrupole moment Q of this state is detected via the modulation of the anisotropy in the angular correlation of the two consecutively emitted γ rays.

	f_u	f_{sp}	f_{hp}	f_{ad}
$\Delta\nu_Q$ (MHz)	< 0.1	0.1 - 10	> 10	100
ν_Q (MHz)	0	0	0	≈ 250

Table 1: *Classification of unperturbed (f_u), slightly perturbed (f_{sp}), heavily perturbed (f_{hp}) environments and strong perturbation involving an associated defect (f_{ad}) with respect to the mean coupling constant ν_Q and its distribution width $\Delta\nu_Q$.*

The recorded coincidence signal R(t) can be described by the following equation with A = -0.13 for ^{111}In and A = 0.13 for ^{111m}Cd:

$$R(t) = A \sum_i f_i \left(\sum_{n=0}^{3} e^{-n\Delta\nu_{Qi}t} S_n \cos\left(n\frac{3\pi}{10}\nu_{Qi}t\right) \right) \qquad (1)$$

A fit of equation (1) to the experimental R(t) spectra yields the fractions f_i of probe atoms in different environments, characterized by their mean hyperfine quadrupole coupling constants ν_{Qi} and the widths of the corresponding distributions $\Delta\nu_{Qi}$. The presence of a single unique efg ($\nu_Q > 0$ MHz, $\Delta\nu_Q = 0$ MHz) leads to a periodic modulation of the anisotropy tagged by the coupling constant $\nu_Q = eQV_{zz}/h$, where V_{zz} is the main component of the traceless efg tensor. A distribution of efg centered at zero results in a continuous decay of anisotropy (Fig. 1B). The width of the efg distribution is equivalent to the mean strength of perturbation and a very narrow distribution ($\Delta\nu_Q < 0.1$ MHz) around zero is regarded as an unperturbed environment. A faster relaxation of anisotropy, equivalent to a broader distribution of efgs, is characteristic for a slightly or heavily perturbed environment of the probe atom. A distribution centered at higher ν_Q values, what is reflected by a damped periodic modulation in the spectrum (Fig 1A), is interpreted as a defect in the nearest neighborhood of the probe atom superimposed by the perturbations of additional more distant defects. In Table 1 the classification of different environments as used in this work is summarized. A more detailed discussion of the method is given elsewhere [10].

3. Experimental Details

In our experiments we have used LEC grown <100> cut undoped semiinsulating GaAs and InP samples which were implanted with stable As, Ga, and Cd and radioactive ^{111}In or ^{111m}Cd under conditions which are summarized in Table 2. In addition to the used energies and doses the resulting maximum concentrations, the depths and the widths of the implantation profiles for GaAs are given as determined by TRIM calculations [11].

The ^{111m}Cd implantation was performed at the on-line isotope separator ISOLDE at CERN. After implantation the samples were annealed under flowing N_2 in a rapid thermal annealing setup for 20 s (GaAs) or 10 s (InP). In order to minimize surface decomposition the samples were covered by a face to face proximity cap, consisting of the respective material. Determination of the sample

Implant	Energy (keV)	Dose (10^{12} cm^{-2})	Depth(width) (nm)	Maximum conc. (10^{17} cm^{-3})
^{111m}Cd	60	< 1	25 (22)	<1
^{111}In (+ Cd)	60	5	25 (22)	15
Cd	60	13	25 (22)	40
Cd - triple	150, 320, 640	2, 10, 15	130 (200)	10
As	45	100	22 (22)	320
Ga	45	100	25 (26)	300

Table 2: *Energies and doses used for different implanted species. Additional information on the resulting implantation profiles calculated by TRIM [11] is given for the case of GaAs.*

Materials Science Forum vols. 83-87

activity before and after annealing allowed to measure the outdiffusion of radioactive dopants with an absolute accuracy of about 10 %. Samples dedicated to electrical measurements were triple implanted with Cd (Table 2) and Au contacts were evaporated on Be implanted corners in order to perform Hall measurements in standard Van der Pauw geometry at liquid nitrogen and ambient temperature.

4. Results and Discussion

In Fig. 1 the R(t) spectra recorded after ^{111}In implantation into GaAs at 90 K (A) and at 300 K (B) are displayed. The cold implantation results in a distribution of strong efgs around 290 MHz, as indicated by the dip in the time spectra between 0 and 30 ns. The large width of the distribution is leading to a strong damping of the R(t) signal, therefore only one period of the modulation is visible. After implantation at 300 K the situation is different. A distribution of efg around zero is observed, as seen in a continuous nearly exponential decay of anisotropy in Fig. 1B.

The R(t) spectra recorded during an isochronal annealing program between 160 K and 380 K were analyzed using equation (1). The results in Fig. 2 show that up to annealing at 250 K v_Q is unchanged. The corresponding strong efgs indicate that defects are located in the direct vicinity of the probe atom, most probably in the first neighbor shell. The observed broad distribution ($\Delta v_Q = 130$ MHz) is attributed to the existence of additional more distant defects leading to a superposition of the strong efg with weaker efgs. At annealing temperatures above 250 K the mean v_Q value of the efg distribution shifts to lower values, indicating a reduction of disorder in the direct vicinity of the probe atom. After annealing at 380 K the next neighbor shell has recovered, as indicated by the disappearance of strong efgs around 290 MHz, leaving the probe atoms in a still

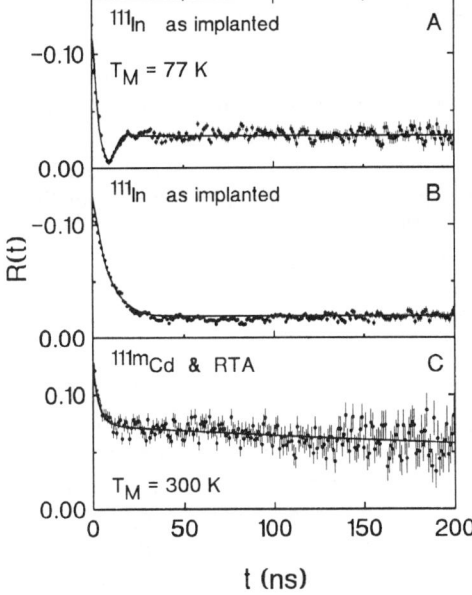

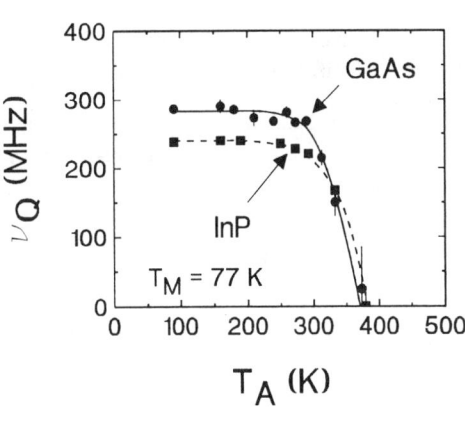

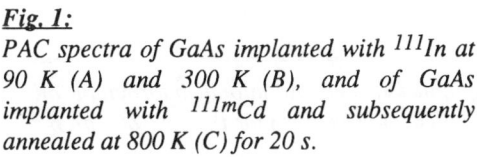

Fig. 1:
PAC spectra of GaAs implanted with ^{111}In at 90 K (A) and 300 K (B), and of GaAs implanted with ^{111m}Cd and subsequently annealed at 800 K (C) for 20 s.

Fig. 2:
Average coupling constant v_Q of the efg distribution in GaAs and InP implanted with ^{111}In at 90 K as a function of annealing temperature. The samples were annealed in an ethanol bath for 10 min.

heavily perturbed environment, however without nearby defects, characterized by $v_Q = 0$ MHz and $\Delta v_Q = 60$ MHz. This remaining perturbation is attributed to a large concentration of point defects in the lattice not directly associated to the probe atoms.

Emission channeling measurements (EC) observed an annealing stage between 200 K and 350 K after [112]In implantation at 100 K, which was attributed to a common recovery of the lattice allowing channeling of electrons emitted from [112]In on substitutional sites [5]. Below 250 K the probe atoms location with respect to the total lattice is not well defined. Since EC and PAC results were obtained under comparable conditions we attribute the above reported annealing stage (Fig. 2) to the recovery of the direct neigborhood of the probe atom, accompanied by the incorporation of In on well defined substitutional Ga sites. Mößbauer experiments reveal also an anealing stage around 300 K, attributed to the incorporation of In on substitutional sites and to the annealing of close defects [12].

The situation after implantation of ^{111}In at 90 K into InP is similar to that in GaAs. A broad efg distribution centered at 260 MHz is observed. According to the annealing data in Fig. 2 the recovery stage seems to occur at the same temperature. But in contrast to GaAs in InP an implantation well above room temperature is necessary to anneal the associated defects during the implantation, visible from the disappearance of the dip in the time spectrum. Therefore the annealing occurs at slightly higher temperature in InP. This is in agreement with the results obtained with EC measurements [13].

The recovery of GaAs at higher temperature was investigated after implantation of [111m]Cd and [111]In. For Cd the heavily perturbed fraction, which is observed after implantation at ambient temperature, splits in two different fractions f_{hp} and f_{sp} after annealing above 500 K. Fig. 1C shows a typical R(t) spectrum after [111m]Cd implantation at ambient temperature and subsequent annealing at 800 K. The initial fast decay of anisotropy corresponds to the 50 % fraction f_{hp} of probe atoms located in a heavily perturbed environment, the remainder f_{sp} is located in a slightly perturbed environment. Whereas the population of both fractions exhibit no significant change below 900 K, the width Δv_Q of f_{sp} continuously decreases (Fig. 3 triangles) until it reaches the value of an unperturbed fraction ($\Delta v_Q \approx 0.1$ MHz).

The annealing behavior of [111m]Cd implanted in virgin material was also investigated in samples preimplanted with Cd (Fig.3, filled square) or As (Fig. 3, open square). With increasing total dose, i.e. an increasing number of defects created in the implanted layer, the decline of the damping

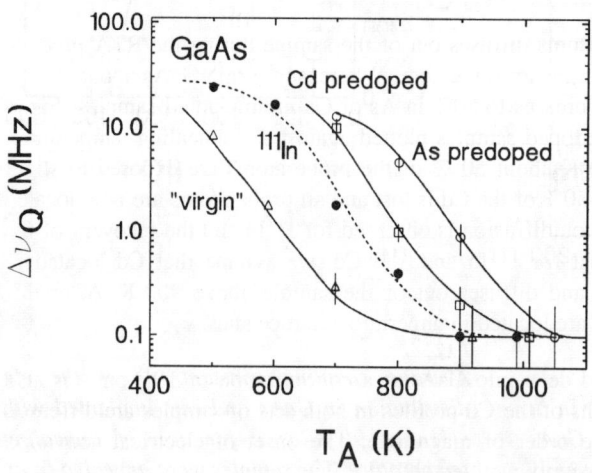

Fig. 3:
Width Δv_Q of the efg distribution of f_{sp} in differently doped GaAs samples as a function of annealing temperature (RTA, 20s). The different curves correspond to different total implantation doses (Table 2). The continuous curves should guide the eye.

parameter Δv_Q is shifted to higher temperature. After ^{111}In implantation and subsequent annealing above 500 K no probe atoms are located in a heavily perturbed environment (f_{sp} = 100 %), nevertheless the associated damping parameter Δv_Q (Fig. 3, filled circle) fits nicely to the dose dependence of Δv_Q in ^{111m}Cd implanted samples. Since the decrease of Δv_Q indicates the removal of distant defects one can conclude from this correlation that the ^{111m}Cd atoms located in slightly perturbed environment and the ^{111}In atoms observe the same annealing process. The slow decline of Δv_Q and its strong dose dependence implies a complex annealing mechanism involving also extended defects. We assign the continuous growth of extended defects known to occure above 500 K to be responsible for the observed reduction of the slight perturbation at the probe atoms site. This growth leads to a reduction of defect concentration, thereby increasing the average distance between probe atoms and defects, and to a reduction of strain.

The environment of the heavily perturbed ^{111m}Cd sites is characterized by a broad efg distribution. With rising annealing temperature Δv_Q increases, the resulting efg distribution reaches to $v_Q > 400$ MHz. The incorporation of ^{111m}Cd in extended defects on a variety of differently perturbed lattice sites can explain the observed behavior of f_{hp}. Since for ^{111}In f_{hp} is zero, In is not incorporated into these defects and an attractive interaction of Cd atoms with the extended defects is deduced.

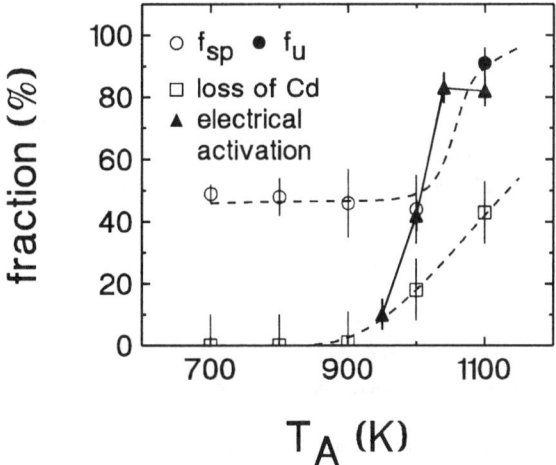

Fig. 4:
Comparison of the fraction f_{sp} of ^{111m}Cd (implanted in Cd predoped GaAs) located in slightly perturbed environment with the relative loss of Cd as a function of annealing temperature. In addition the achieved fraction of electrically activated Cd atoms measured in triple implanted GaAs samples is shown.

Above 900 K part of the implanted Cd atoms diffuses out of the sample during the RTA process. The amount of activity loss increases with the absolute dose. Whereas in virgin GaAs about 30% is lost at 1050 K, this value dramatically increases to 80% in As or Ga preimplanted samples. Fig. 4 shows f_{sp} and the activity loss of Cd predoped samples plotted against the annealing temperature. Up to an annealing temperature of 1000 K about 50 % of the probe atoms are exposed to slight perturbations. After annealing at 1100 K 50% of the Cd is lost and all probe atoms are now located in an unperturbed environment. Since no outdiffusion is observed for ^{111}In and the recovery of the slightly perturbed fraction f_{sp} is identical for ^{111}In and ^{111m}Cd, we assume that Cd located in extended defects (f_{hp}) is highly mobile and diffuses out of the sample above 900 K. After this outdiffusion all remaining ^{111m}Cd atoms are located in unperturbed lattice sites.

The electrical activation of Cd implanted deep into GaAs at a concentration of 10^{18} cm^{-3} is also displayed in Fig. 4. The depths and widths of the Cd profiles in both sets of samples are different, but the Cd concentration is of the same order of magnitude. The onset of electrical activation coincides with the stage of outdiffusion of radioactive ^{111m}Cd. The conflict between the electrical activation of 80% acchieved for deep Cd implantation and the 50% activity loss observed for shallow implantation is easily explained by the different depth profiles. According to the above given interpretation the Cd outdiffusion might reflect the onset of mobility of the extended defects. This process removes extended defects, which can act as efficient carrier traps, from the dopant profile leading thereby to electrical activation. The given interpretation suffers from the problems to identify defects by PAC which are not characterized by an unique efg. It might be helpful to combine the results obtained by PAC after heavy ion implantation with TEM and positron annihilation studies in order to obtain more information on the nature of defects, especially extended defects, present during annealing.

5. Summary

The annealing of implantation damage after heavy ion implantation of Cd and In in GaAs and InP was investigated using the microscopically sensitive PAC method. Two different annealing stages were observed. The first at about 280 K is assigned to the annealing of associated defects in direct neighborhood of the probe atom ^{111}In and the second at high annealing temperature above 500 K to the formation, continuous growth and finally removal of extended defects. The latter stage ends up with the electrical activation of implanted Cd. Increasing Cd outdiffusion was observed for rising total implantation dose.

6. Acknowledgement

This work was supported by the German-Israeli Foundation for Scientific Research & Development (contract No. I-1599-303.1/87) and by the Bundesminister für Forschung und Technologie. We gratefully acknowledge the helpful discussions with S. Winter.

7. References

(1) K. Saarinen, P. Hautojärvi, J. Keinonen, E. Rauhala, J. Räisänen and C. Corbel, Phys. Rev. B 43, 4249 (1991).
(2) A. Goltzene, B. Meyer, C. Schwab, J.P. David and A. Roizes, in *Defects in Semiconductors*, ed. H.J. von Bardeleben (The Metallurgical Society of AIME, Warrendale, USA, 1986) p. 1057.
(3) D.K. Sadana, Nucl. Instr. Meth. B7/8, 375 (1985).
(4) S.J. Pearton, J.S. Williams, K.T. Short, S.T. Johnson, D.C. Jacobson, J.M. Poate, J.M. Gibson and D.O. Boerma, J. Appl. Phys. 65, 1089 (1989).
(5) S. Winter, S. Blässer, H. Hofsäss, S. Jahn, G. Lindner, U. Wahl and E. Recknagel, Nucl. Instr. Meth. B48, 211 (1990).
(6) G. Weyer, S. Damgaard, J.W. Petersen and J. Heinemeier, J. Phys. C: Solid State Phys. 13 L181 (1980).
(7) F. Schneider and S. Unterricker, phys. stat. sol. (a) 85, 211 (1984).
(8) K. Bonde-Nielsen, H. Grann, H. Haas, F.T. Pedersen and G. Weyer, in *13th International Conference on Defects in Semiconductors*, ed. L.C. Kimmerling and J.M. Parsey (Trans Tech Publications Ltd, Aedermannsdorf, CH, 1986) p. 1065.
(9) W. Pfeiffer, M. Deicher, R. Keller, R. Magerle, P. Pross, H. Skudlik, Th. Wichert, H. Wolf, D. Forkel, N. Moriya and R. Kalish, Appl. Surf. Sci. 50, 154 (1991).
(10) Th. Wichert, M. Deicher, G. Grübel, R. Keller, N. Schulz and H. Skudlik, Appl. Phys. A 48 , 59 (1989).
(11) J.P. Biersack and L.G. Haggmark, Nucl. Instr. Meth. 174, 257 (1980).
(12) G. Weyer and S. Winter, priv. communication, to be published.
(13) H. Hofsäß, S. Winter, S. Jahn, U. Wahl and E. Recknagel, EMRS 1991 Spring Meeting Strasbourg, in press.

ION IMPLANTATION INDUCED SHEET STRESS DUE TO DEFECTS IN THIN (100) SILICON FILMS

JIANZHONG YUAN, ANDREW J. YENCHA AND JAMES W. CORBETT
The Institute for the Study of Defects in Solids, Department of Physics, University at Albany-SUNY, Albany, NY 12222, USA

ABSTRACT

Detailed measurements of sheet stress induced by Ar^+ ion implantation were carried out on thin (100) Si films at doses ranging from 10^9 to 10^{16} cm^{-2} at room temperature. The stress is compressive and increases with dose up to of the order of 2×10^{14} cm^{-2} for 100 keV Ar^+ implanted at 0.1 μA/cm^2. After which it decreases and finally attains a constant value, which is approximately 35% of the peak stress value. The peak stress over unit depth decreases as the implantation energy increases, while the dose corresponding to this peak stress increases as the implantation energy increases. RBS/channeling measurements verified that this dose value is less than the implantation critical dose. A model has been established, and a second-order differential equation has been derived that reproduces the stress curve and the well-established amorphization curve of ion implantation.

INTRODUCTION

Ion implantation process is one of the most important VLSI fabrication techniques. The effects and formation of amorphous layer by implanted ions in crystal solids have been extensively studied for decades[1]. However, few works have reported to study the stress associated with this process [2-5]. The study of the stress process can also lead to an understanding of the fundamental physics of ion implantation-induced process in solids. In technical aspect, massive production failures happened due to the lack of understanding of the stress effect of implanted ions. Many subsidiary effects can be caused by the stress induced by ion implantation, including substrate bending and cracking. This process may be detrimental in subsequent processing and device performance.

EerNisse et al [2] used an in situ capacitance technique to measure the integrated surface stress-induced by 10 to 100 keV He^+ implanted into Mo, Nb and Al at room temperature. They found that the low-fluence results provide value for the induced volume expansion per implanted He atom, while the high-fluence results demonstrate that blistering, in Mo and Nb, is directly related to relief of stress. A model describing the blistering phenomenon was also developed in this work. Madakson et al [3] measured the stresses induced by 28 keV Ar^+ and 30 keV Ti^+ ion implantations in <111> Si by using X-ray diffraction method. He found that the dose corresponding to a turning point for stress reduction is about 3.5×10^{15} cm^{-2} for 28 keV Ar^+ implanted while RBS/channeling studies show radiation damage to increase linearly with ion dose and saturate at dose at about 5×10^{14} cm^{-2}. The stress reduction and the amorphization process were not related in this work. An in situ stress measurement of MeV ion implantation in Si was carried out by Volkert[4] by using a reflected laser beam to measure the bending caused by the stress. The accuracy is fair and no keV implantation performed. A model was proposed that described the behavior in terms of the expansion of crystalline silicon by the creation of defects and the flow of amorphous material under the ion beam. Yuan et al [6] used thin (100) Si films as substrate materials to perform the stress measurements of keV implantations for different ions. A dose-rate dependence of the stress curve has been measured[7] for heavy ions such as Ar^+.

Several models have been proposed for the crystalline to amorphous transformation in ion-implanted silicon[8]. The overlap-damage model proposed by Gibbons[1] was used to determine the critical dose for the amorphous transformation. The critical-energy-density (CED) model[9,10] was in good agreement with the measured critical dose. This model assumed that a region would become

---

*Supported in part by the Solar Energy Research Institute, IBM Corporation, and Mobil Foundation.

amorphous if the energy density deposited into atomic processes by the ions exceeded the critical energy density of 6×10^{23} eV/cm^3. Out-diffusion models[11] can explain the temperature dependence. However, little work has been reported to quantitatively model the stress behavior of silicon under ion irradiation. Burnett et al [12] proposed a model to explain the generation and relief of the near-surface stress by ion implantation into sapphire. He proposed that the stress reduction is due to the change of the thickness of the amorphous layer, and thus of the still-crystalline but damaged layer. But he assumed that the variation of the integrated stress with dose to be linear prior to amorphization. Further more, thermal-effects, dose-rate effects and electronic effect have been neglected. This model can only roughly explain the stress reduction. The stress behavior prior to amorphization was still not well understood.

In the present work, stresses induced by 20 to 120 keV Ar$^+$ ion implantation in (100) Si at dose-rate of 0.1 µA/cm^2 have been measured at different dose ranges. Integrated sheet stress curves have been obtained. RBS/channeling measurements were used to measure the amorphization for different doses. A model is proposed. This model can explain both the measured stress curve and amorphous transformation curve.

EXPERIMENT AND RESULTS

Integrated lateral stress within the implanted region can be calculated by measuring the bending of the implanted substrate. The experimental set up was described elsewhere[6]. The stress, σ_s, is given by the modified Stoney's equation[13]:

$$\sigma_s = \frac{E_s t_s^2}{6(1-\nu)t_l} \times \frac{x}{2dL} \qquad (1)$$

where E_s and ν are the Young's modulus and Poisson's ratio for the substrate, respectively; t_s is the substrate thickness, t_l is the thickness of the implanted region ($t_s \gg t_l$), x is beam spot displacement, d is the implanted length, and L is the distance between sample and screen. Samples were cut 0.2-1.5mm × 8mm from two-side polished commercially available thin (100) silicon wafer with thickness of about 8 mm. Ion implantation of 100 keV Ar$^+$ was performed at doses ranging from 10^9 to 10^{16} cm^{-2} at room temperature, using dose-rate at 0.1 µA/cm^2. Figure 1 shows the average integrated compressive surface stress, which was induced by 20 to 120 keV Ar$^+$ implantation at

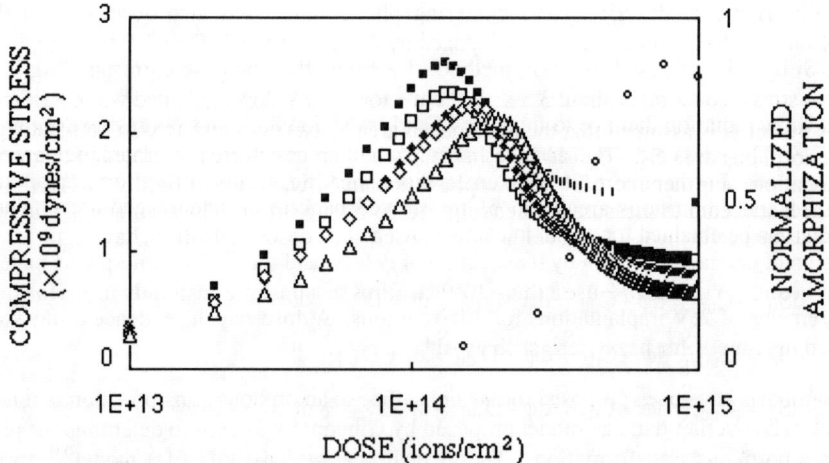

Figure 1. Stresses induced by Ar$^+$ implantation at 20(■), 40(□), 60(♦), 80(◊), 100(▲) and 120(△) keV at doses ranging from 10^{13} to 10^{15} cm^{-2}, and channeling yield (o) along <100> direction.

room temperature, versus dose. For doses below about 2×10^{14} cm^{-2}, the surface stress increases quickly as dose increases, although the increase slows down, thus is not linear, and quickly reaches its peak value. After the stress reaches its peak, it reduces until finally it attains a constant value. This stress curve depends on the implantation energy. In particular, the stress reaches its peak value, which is 2.3×10^9 dynes/cm^2, at dose at 2×10^{14} cm^{-2} for 100 keV at 0.1 µA/cm^2 implantation. The remain stress is 0.74×10^9 dynes/cm^2. The stress curves shift toward higher dose and smaller stress value as the implantation energy increases. RBS/channeling result of the amount of amorphization data is shown on the right axis in Figure 1.

ION IMPLANTATION INDUCED DEFECTS

Energetic ions penetrate the surface of the silicon material, transfer energy to the crystal and come to rest in an approximately Gaussian distribution. The implanted ions lose energy by two mechanisms prior stopping. Inelastic collisions result in the displacement of silicon atoms from the site. These displaced atoms may further proceed to displace other nearby atoms until the energies for both incident ions and the recoiling silicon atoms are insufficient to produce further displacement. Fast moving ions may also lose energy by electronic excitation of the silicon atoms. These process can cause the weakening of the host atom bonding. The analysis of experimental data shows that ion-implantation-induced process is very complicated[14]. In general, one should take into consideration of the formation, dissociation and reconstruction of various complexes; annihilation and migration of defects; capture of the mobile defects by sinks; ionization produced by incident particles; influence of impurities; initial defects, etc. Dvurechenskii et al [15]. summarized three main mechanisms for this process:
1) The evolution of a sequence of defects and the loss of point defects at different imperfections in crystal, such as surface and interface;
2) The accumulation of point defects up to the critical concentration, which arises from the overlapping of individual disordered regions produced by single ions;
3) The formation of regions of amorphous phase produced in the damage created by an individual ion or by the multiple overlap of the damage regions created by individual ions.

There is evidence to show that the creation of complexes such as interstitial atoms, vacancies, divancacies and tetravacancies, etc., can cause volume expansion and thus produce large stress in the crystal[16]. The amorphization process of the crystal is also related to this stress directly. In fact, our results indicate that the crystal experiences a big stress before amorphization. The accumulation of those defect complexes causing the stress in the region increase. After it reaches a critical concentration, the crystal loses its long range order and becomes amorphized, and the stress reduced.

MODEL

In this model, it is assumed that the undamaged area S_u can be transferred to both damaged area S_d and amorphous area S_a. The damaged area, S_a is then transferred to amorphous area by further implantation. Further more, it is assumed that annealing occurs during ion implantation. Both damaged and amorphous areas can be transferred back to an undamaged area. Thus a set of relations can be obtained for a total implanted area S:

$$dS_u = -C_{ud}\frac{S_u}{S}d\varphi - C_{ua}\frac{S_u}{S}d\varphi + D'_{du}\frac{S_d}{S}dt + D'_{au}\frac{S_a}{S}dt \qquad (2a)$$

$$dS_d = C_{ud}\frac{S_u}{S}d\varphi - C_{da}\frac{S_d}{S}d\varphi - D'_{du}\frac{S_d}{S}dt + D'_{ad}\frac{S_a}{S}dt \qquad (2b)$$

$$dS_a = C_{ua}\frac{S_u}{S}d\varphi + C_{da}\frac{S_u}{S}d\varphi - D'_{ad}\frac{S_d}{S}dt - D'_{au}\frac{S_a}{S}dt \qquad (2c)$$

where C_{ud} is an area converted per ion from undamaged to damaged area; C_{ua} is an area converted per ion from undamaged to amorphous area; C_{da} is an area converted per ion from damaged to amorphous area; D'_{du} is annealed from unit damaged area to undamaged area per second; D'_{au} is annealed from unit amorphous area to undamaged area per second; and D'_{ad} is annealed from unit amorphous area to damaged area per second, finally, φ is the dose. For a constant dose-rate i, we have $d\varphi = i \cdot dt$. Let $D_{du} = i \cdot D'_{du}$, $D_{au} = i \cdot D'_{au}$ and $D_{ad} = i \cdot D'_{ad}$, we then get:

$$\frac{dS_u}{d\varphi} = -C_{ud}\frac{S_u}{S} - C_{ua}\frac{S_u}{S} + D_{du}\frac{S_d}{S} + D_{au}\frac{S_a}{S} \tag{3a}$$

$$\frac{dS_d}{d\varphi} = C_{ud}\frac{S_u}{S} - C_{da}\frac{S_d}{S} - D_{du}\frac{S_d}{S} + D_{ad}\frac{S_a}{S} \tag{3b}$$

$$\frac{dS_a}{d\varphi} = C_{ua}\frac{S_u}{S} + C_{da}\frac{S_d}{S} - D_{ad}\frac{S_a}{S} - D_{au}\frac{S_a}{S} . \tag{3c}$$

Equations for S_d and S_a from (3):

$$\frac{d^2S_d}{d\varphi^2} + a_1\frac{dS_d}{d\varphi} + a_2 S_d = a_{d3} S \tag{4}$$

$$\frac{d^2S_a}{d\varphi^2} + a_1\frac{dS_a}{d\varphi} + a_2 S_a = a_{a3} S \tag{5}$$

where
$a_1 = C_{ud}+C_{da}+C_{ua}+D_{au}+D_{ad}+D_{du}$
$a_2 = C_{ud}C_{da}+C_{ua}C_{da}+C_{ua}D_{ad}+C_{ua}D_{du}+C_{ud}D_{ad}+C_{ud}D_{au}+C_{da}D_{au}+D_{ad}D_{du}+D_{au}D_{du}$
$a_{d3} = C_{ud}D_{ad}+C_{ud}D_{au}+C_{ua}D_{ad}$
$a_{a3} = C_{ud}C_{da}+C_{ua}C_{da}+C_{ua}D_{du}$.

Solutions for (4) and (5) can be obtained:

$$S_d = \frac{C_{ud} - \frac{a_{d3}}{a_2}s_2 S}{s_1 - s_2}e^{s_1\varphi} + \frac{C_{ud} - \frac{a_{d3}}{a_2}s_1 S}{s_2 - s_1}e^{s_2\varphi} + \frac{a_{d3}}{a_2}S \tag{6}$$

$$S_a = \frac{C_{ua} - \frac{a_{a3}}{a_2}s_2 S}{s_1 - s_2}e^{s_1\varphi} + \frac{C_{ua} - \frac{a_{a3}}{a_2}s_1 S}{s_2 - s_1}e^{s_2\varphi} + \frac{a_{a3}}{a_2}S \tag{7}$$

where
$$s_{1,2} = \frac{1}{2}(-a_1 \pm \sqrt{a_1^2 - 4a_2}) .$$

The thickness of the damaged area and amorphous area are changing during the implantation[17]. Experimental results indicate that at moderate dose levels, the damage density can be described by a Gaussian distribution along depth x[18,19]:

$$D(x) = D_0 e^{-(x-R_D)^2/2(\Delta R_D)^2} \tag{8}$$

where R_D is the depth of the damage peak and $\sqrt{2}\Delta R_D$ is the half-width of the damage distribution. The peak damage density, D_0, can be expressed in terms of the dose:

$$D_0 = 1 - e^{-C_i\varphi} \tag{9}$$

where C_i is the surface projection of the cross section of the idealized cylindrical damaged zone produced by an implanted ion. By this definition, $C_i \equiv C_{ud}$. Assume the crystal is amorphized if $D(x) \geq D_{ac}$ and the crystal can be considered damaged if $D(x) \geq D_{dc}$, respectively. Then the thickness of the damaged layer t_{dt} and amorphized layer t_a can be calculated:

$$x_d = R_D \pm \Delta R_D\sqrt{2\ln(D_0/D_{dc})} \tag{10}$$

$$x_a = R_D \pm \Delta R_D \sqrt{2\ln(D_0/D_{ac})} \tag{11}$$

$$t_{dt.} = \begin{cases} \dfrac{2|x_d - R_D|}{x_d} & \text{if } |x_d - R_D| < R_D \\ & \text{other} \end{cases} \tag{12}$$

$$t_{a.} = \begin{cases} \dfrac{2|x_a - R_D|}{x_a} & \text{if } |x_a - R_D| < R_D \\ & \text{other} \end{cases} \tag{13}$$

where x_d and x_a are the positions for $D(x) = D_{dc}$ and $D(x) = D_{ac}$, respectively. The thickness of the damaged but not amorphized layer can be given as $t_d = t_{dt} - t_a$. According to EerNisse[20]:

$$\sigma = (\sigma_d V_d + \sigma_a V_a)/V_0 \tag{14}$$

where σ_d is the stress per unit volume if all implanted volume only be damaged and σ_a is the stress per unit volume if all implanted volume amorphized, respectively. Thus we have the expression for the unit surface stress:

$$\sigma = (\sigma_d S_d t_a + \sigma_a S_a t_a + \sigma_d S\ t_d)/S\ t_0 \tag{15}$$

and amorphization:

$$r = S_a t_a / S\ t_0 \tag{16}$$

where t_0 is the depth of the implanted layer which be chosen as $1.5 R_D \approx R_P$. R_P is the projected range and can be calculated by TRIM. Figure 2 shows the calculated stress and amorphization curves compare to the measured curves.

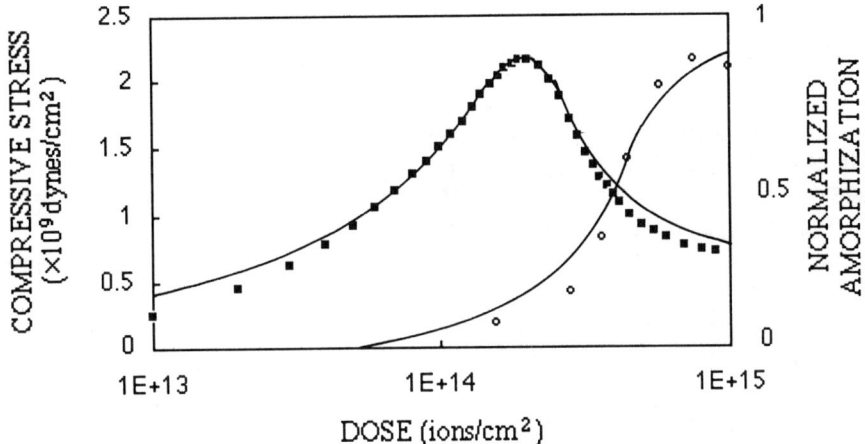

Figure 2. Comparison of the stress (■) and amorphization (o) from Ar+ 100 keV implantation with calculated curves from this model.

DISCUSSION

The results clearly established that the stress reduction and the amorphization are strongly related. The surface stress originates from the volume expansion caused by the implanted ions. In the heavy ion case, the crystal can be amorphized by a single incident ion with damaged areas around this amorphized region. However, for a light ion, the crystal may only be amorphized by accumulation of the damage created by incident ions. No single ion can produce an amorphous region in this case. During the implantation, annealing may occur in the damaged and amorphized region. In particular, in the damaged region, two kinds of annealing may exits: self-annealing and regular annealing of the damages. The first annealing process is clearly related to dose-rate effect, which was studied elsewhere[6]. Because of the size of the present paper, we can not include the study of the parameters in detail. Perimeters C_{ud}, C_{da} and C_{ua} are strongly depend on implanted ions and energy, and we should have $C_{da}>C_{ua}$. Perimeters D_{au}, C_{ad} and C_{du} are strongly depend on implantation dose-rate and sample temperature, and we should have $D_{du}>D_{au}$. These together with the initial conditions of S_d and S_a provide the condition for the solution of eq. (13) and (14). The thicknesses calculated in the above model are average values. The stresses are also integrated values. The depth distribution of the stress has been neglected.

SUMMARY

In summary, stresses induced by 20 to 120 KeV Ar^+ ion implantation in (100) Si at dose-rate of 0.1 $\mu A/cm^2$ have been measured at room temperature. The stresses are compressive and increase nonlinearly with dose up to of the order of 2×10^{14} cm^{-2}. After they saturate, they decrease and finally attain constants. RBS/channeling measurements verified that this dose value is less than the implantation critical dose. A model has been established, and a second-order differential equation has been derived. After considering the thickness change of the damaged and amorphized layer, a stress expression is obtained that reproduces the stress curve and the well-established amorphization curve of ion implantation.

ACKNOWLEDGMENT

The authors would like to thank Dr. Igor V. Verner for useful discussions and Mr. Arther W. Haberl and Mr. Robert P. Lentlie for technical assistance.

REFERENCES

[1] J.F. Gibbons, Proc. IEEE 60, 1062 (1972).
[2] E.P. EerNisse and S. T. Picraux, J. Appl. Phys. 48, 9 (1977).
[3] P. B. Madakson and J. Angilello, J. Appl. Phys. 62(5), 1688 (1987).
[4] C.A. Volkert, Mat. Res. Soc. Symp. Vol. 157, 635 (1990).
[5] D.I. Tetel'baum, Sov. Phys. Semicond. 17(6), 658 (1983).
[6] J. Yuan and J.W. Corbett, Proc. 22nd NATO Int'l School of Mat. Sci. and Tech., in process.
[7] J. Yuan and J.W. Corbett, to be published.
[8] John R. Dennis and Edward B. Hale, J. Appl. Phys. 49(3), 1119 (1978).
[9] H. Muller, K. Schmid, H. Ryssel, and I. Ruge, in *Ion Implantation in Semiconductors and Other Materials*, eds.B.L. Crowder (Plenum, New York, 1973) p.203.
[10] F.L. Vook, in *Radiation Damage and Defects in Semiconductors*, eds. J.E. Whitehouse (Institute of Physics, London, 1972) p. 60.
[11] F.F. Morehead and B.L. Crowder, Radiat. Eff. 6, 27 (1970).
[12] P.J. Burnett and T.F. Page, J. Materials Science, 20, 4624 (1985).
[13] R.W. Hoffman, in *Physics of Thin Films*, ed., G. Hass and R.E. Thun (Academic Press, New York, 1966), 3, 211.
[14] S.T. Picraux and F.L. Vook, Radiat. Eff. 11, 1968 (1969).
[15] A.V. Dvurechenskii and L. S. Smirnov, Rad. Eff., 37, 173 (1978).
[16] G.B. Krefft and E.P. EerNisse, J. Appl. Phys. 49(5), 2725 (1978).
[17] P.J. Burnett and T.F. Page, J. Mater. Sci. 19, 845 (1984).
[18] J.W. Mayer, L. Eriksson and J.A. Davies, in *Ion Implantation in Semiconductors Silicon and Germanium*, (Academic Press, New York and London, 1979).
[19] M. Miyap, T. Miyazaki and Tokuyama, Japan. J. Appl. Phys. 17, 995 (1978).
[20] E.P. EerNisse, J. Appl. Phys., 45, 167 (1974).

OBSERVATION OF A TRIVALENT Ge DEFECT
IN OXYGEN IMPLANTED SiGe ALLOYS

M.E. ZVANUT*, W.E. CARLOS, M.E. TWIGG, R.E. STAHLBUSH, and
D. J. GODBEY
Naval Research Laboratory
Washington, D.C. 20375, USA

ABSTRACT

We report the study of a defect with $g_\| = 1.9998$ and $g_\perp = 2.0260$ in annealed O-implanted SiGe alloys. The material studied is damaged crystalline SiGe and the defect is shown to exhibit the angular dependence of (111) symmetry. We deduce that the electron is in an orbital of a trivalent Ge atom backbonded to Ge or Si atoms.

Introduction

Research has recently revealed the potential of SiGe based alloys and superlattices in microelectronic and optoelectronic applications [1,2]. Furthermore, oxygen implantation into Si, Ge, and SiGe heterostructures has received attention because of the promising characteristics of buried oxides for device isolation [3]. While most of the work has been on silicon based systems, here we report results of Electron Paramagnetic Resonance (EPR) studies on oxygen implanted SiGe alloys.

There has been little work on defects in crystalline SiGe alloys, but several reports discuss the characteristics of point defects in a-SiGe:H [4,5]. The authors identify two defects: the Si-dangling bond, 5G wide at g= 2.0055, and a Ge-dangling bond 30G wide at g=2.017. Although the defects were studied thoroughly as a function of alloy composition and anneal temperature, spectroscopic analysis was thwarted by the lack of hyperfine lines and anisotropy. However, studies of a-Si:H and a-Ge:H [6] do reveal that the shift of the Ge and Si g-values from the free electron value (g_e) scales with the spin orbit coupling parameters as expected.

Also pertinent to our investigation are the defects observed in O-implanted Si and Ge. EPR measurements of Si substrates implanted under conditions similar to those used here, reveal the presence of the Pb center, an electron on an interfacial Si back-bonded to three substrate Si's [7]. The defect, which exhibits the (111) symmetry of crystalline Si, was first identified at the Si/SiO$_2$ interface by Nishi [8] and later fully characterized by Poindexter [9]. In the studies of O-implanted Ge no analogous Ge-dangling bond defect was reported although the Ge E' center was identified [10].

Here we report the study of a Ge dangling bond resonance in annealed O-implanted SiGe alloys. Unlike in the investigation of a-SiGe:H, the material studied in our work is damaged crystalline SiGe and the defect is shown to exhibit the angular dependence of (111) symmetry with $g_\| = 1.9998$ and $g_\perp = 2.0260$. When compared to the Si dangling bond resonance, the value for $g_\perp - g_e$ scales according to the ratio of the Si and Ge spin-orbit coupling constants.

Experimental

Samples used in the current work are 800 nm thick $Si_{.9}Ge_{.1}$ layers grown by Molecular Beam Epitaxy (MBE) on (100) 3" or 4" Si wafers. The 4" wafer was oxygen implanted to a dose of $1.8 \times 10^{18} cm^{-2}$ at a substrate temperature of 600° C. The 3" wafer was implanted with $6 \times 10^{17} cm^{-2}$ O with the substrate heated only by the implantation process (estimated temperature is approximately 100° C). Also implanted to $1.8 \times 10^{18} cm^{-2}$ at 600° C was a plain Si substrate to be used as a control. Following implantation, the wafers were sawed into $0.23 \times 1.5 cm^2$ pieces with the long edge along a (110) direction. The pieces were annealed at 900°C in argon.

Electron Paramagnetic Resonance spectra were obtained at 4K using 0.2 mW incident power and 0.5 G modulation amplitude. Since the alloy samples contained only 10% Ge, the data obtained from the control sample were used as background for the SiGe spectra. The g-tensor for the alloy (fig. 2) was extracted from a spectra obtained by integrating the EPR spectra of both the SiGe and the Si samples, normalizing the spectra to the central line of the Si Pb center, and subtracting. In some cases, where resolution of two lines was poor, the g value was extrapolated directly from the EPR spectra of the alloy alone.

Cross-sectional transmission electron microscopy of the 600° C implanted sample shows an amorphous SiO_2 layer buried between regions of heavily damaged Si or SiGe. Such a structure is typical of the SIMOX (Separation by the IMplantation of Oxygen) process, a potential isolation technology for microelectronics. Extensive studies of Si SIMOX substrates indicate that the damaged layers also contain oxide precipitates. Furthermore, these studies show that a 5 hour anneal at a temperature greater than 1200° C is required to remove the damage and oxide precipitates [11,12]; thus, it is expected that after a one hour 900° C anneal, most of the oxide precipitates remain. For our purposes, the anneal was necessary to remove the broad a-SiGe signal from our spectrum and reveal the narrower lines characteristic of the crystalline defect studied here.

Energy dispersive X-ray analysis (EDXS) on $1.8 \times 10^{18} cm^{-2}$ implanted SiGe sample, reveals that Ge is present in the Si layers both above and below the buried oxide. Calibrating the system using a SiGe layer grown in a manner similar to those used for the implantation, the Ge concentration was shown to be approximately 5% in the upper layer and 7% in the layer immediately below the buried oxide. The Ge concentration in the oxide layer was only 2%.

Results and Discussion

Figure 1a shows the EPR trace obtained from the $1.8 \times 10^{18} cm^{-2}$ implanted SiGe layer aligned with the (100) axis parallel to the magnetic field. Two resonances are apparent: one with a zero crossing at g=2.0063, the other with the zero crossing at g=2.018. The data of annealed a-SiGe:H obtained by Finger and co-workers [4] reveals two lines similar to those seen in O-implanted material; however, the lines are broader in the amorphous material. In the oxygen implanted sample, the linewidth of the resonance at g=2.0063 is approximately 3 G and the g=2.018 linewidth is estimated to be

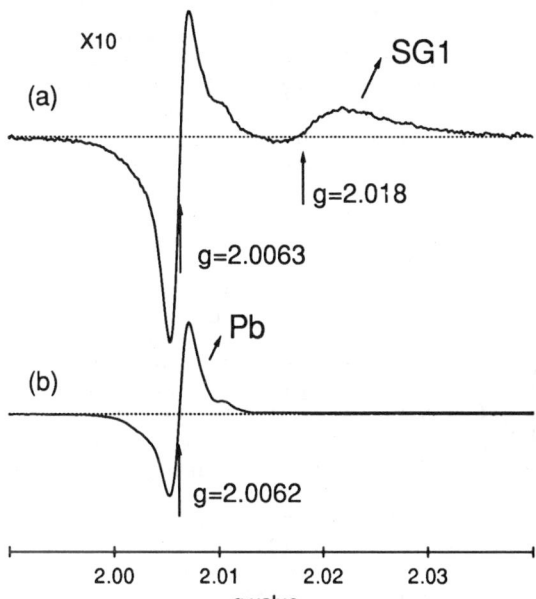

Figure 1 EPR Spectra of O-implanted SiGe (a) and O-implanted Si (b)

10 G. The corresponding linewidths for the a-SiGe:H are 5G and 30G, respectively. It is also apparent, as is shown in figure 2, that the defect in the implanted sample is anisotropic. No such behavior is reported for the annealed a-SiGe:H. The anisotropy indicates that the resonances observed in O-implanted SiGe arise from the crystalline region of the structure rather than from the amorphous buried oxide layer.

The spectra of figure 1b was obtained from the oxygen implanted Si sample with the magnetic field aligned along a (100) axis. The line at g=2.0062 is the Pb center, a SiO_2/Si interfacial defect consisting of an electron on a Si backbonded to three substrate Si atoms. In an earlier report of this center in O-implanted Si it was determined that nearly 90% of the centers were located in the upper Si layer [7]. Thus, it was reasoned that the interface involved was that between Si and the oxide precipitates known to exist in the layer. The signal at g=2.0063 observed in the implanted SiGe material exhibits the same anisotropy and the same hyperfine structure as that observed for the Pb center in O-implanted Si substrates. Therefore, we attribute this line to the Pb center at oxide precipitates in the damaged SiGe layer.

It should be noted here that compared to the Pb center in silicon, the Pb center observed in the O-implanted SiGe material is asymmetrically broadened. Such a broadening could be an indication of the random strains introduced by the presence of Ge in the silicon lattice. That the asymmetry occurs on the side of a higher g value, typical of Ge, further supports this assertion. It might be thought that implantation induced local disorder would account for the broadening of the signal; however, this is unlikely for the following reason. The width of the Pb center in the O-implanted Si is the same as that seen for the Pb centers in thermal oxides. Certainly if the broadening were due to disorder, it would be more apparent in a comparison between a thermally grown oxide film and one formed by 200 keV oxygen implantation than in a comparison of oxygen implanted Si and SiGe.

The angular dependence of the g-tensor for both the Si Pb line and the additional line at g=2.018, hereafter referred to as SG1, is shown in figure 2. The sharpness of the line for the (111) branch allowed for a reasonable determination of the anisotropy of SG1 despite its proximity to the Si Pb line. On the other hand, the large width of the signals for the (1̄11) and (1̄1̄1) branch made difficult the determination of the values of these two branches. The values shown were approximated from two largely overlapping but

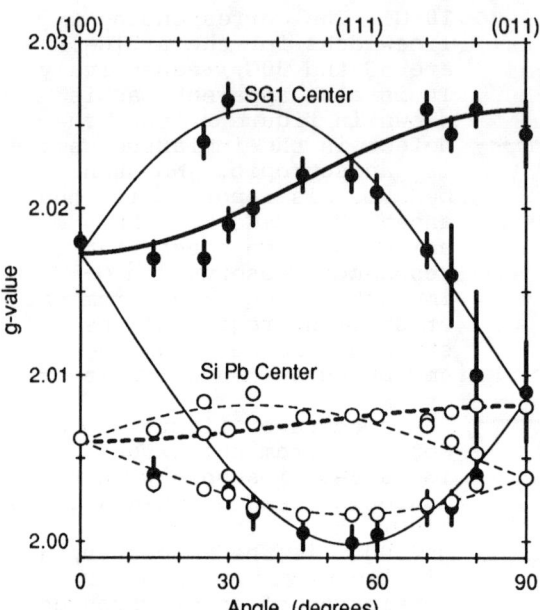

Figure 2 Angular dependence of g value for SG1 (closed) and Si Pb (open) centers. Lines: upper thin, 1$\bar{1}$1; lower thin, 111; bold, doubly degenerate $\bar{1}$11 and 11$\bar{1}$

resolvable lines. Given the limited number of angles at which all three components could be resolved, we did not attempt a fit of the data to obtain both the defect symmetry and magnitude of the g-tensor. Rather, we calculated the lines shown assuming (111) symmetry and adjusting $g_\parallel$ and $g_\perp$ until reasonable agreement with the data was obtained. The initial guesses for $g_\parallel$ and $g_\perp$ we extracted from the high symmetry orientations. The agreement between the data points and the line calculated using $g_\parallel$ =1.9998 and $g_\perp$ =2.0260 is sufficient to support the assumption that the defect exhibits (111) symmetry as expected for a dangling bond associated with the crystalline Si structure.

SG1 appears in the spectra of figure 1a but not in the spectra of figure 1b, which suggests that the defect is related to Ge. That the spin-orbit interaction derived from the g-tensor scales reasonably with that derived from the well-studied Si dangling bond resonance supports this hypothesis. The ratio of the spin orbit coupling constants for Ge (940 cm^{-1}) and Si (142 cm^{-1}) is 7 [13], which, given the simplification of the theory, is a reasonable approximation of the ratio of the shift in $g_\perp$, 4, for the two defects. Furthermore, the width of SG1 is larger than that of the Si Pb signal as is expected for a Ge-related center.

The above discussion suggests that SG1 is associated with Ge and from the observed anisotropy (fig. 2) we conclude that the defect is a dangling bond with the symmetry of the Si lattice. Furthermore, the Ge is most likely backbonded to three other Ge's or a combination of Ge and Si atoms. If oxygen were at two or three of these bonding sites, the zero crossing would most likely be shifted considerably closer to or even below the free electron value. Indeed, $g_\perp$ for the Ge E' center, a dangling bond on a Ge backbonded to three oxygens, is 2.000 and 2.002 in O-implanted Ge [10]. Although replacing one of the Si or Ge backbonds of a Ge dangling bond with oxygen might have little effect on the magnitude of the g-tensor, the deviation from (111) axial symmetry would be apparent in the data of figure 2.

Finally, we discuss the immediate environment of the defect. The unpaired electron could be a site in the damaged SiGe lattice or a broken bond at the interface between Ge and the oxide precipitates. Comparison of the spectra of two different O doses favors the latter possibility. The intensities of the 0.6x10^{18} and 1.8x10^{18} cm^{-2} implant samples indicate that the concentration of the

SG1 center increases with oxygen dose. It is doubtful that the signal increase is due to an increase in implantation-induced disorder since at these high dose levels, the degree of disorder for the two samples should be about the same. Furthermore, the higher dose sample was implanted at a temperature above that of the lower dose samples, thus the degree of disorder may actually be lower in the 1.8×10^{18} cm^2 sample. We suggest, therefore, that the increased signal intensity is related to the increased oxygen content. This is consistent with the notion that the Ge signal arises from the same silicon/oxide precipitate interface as the Si Pb center. Confirmation of such a claim awaits further experiments involving a wider range of O, and perhaps, Ge concentrations.

In summary, we have observed a dangling bond defect with the (111) symmetry of the Si lattice in O-implanted SiGe alloys. From the measured g values, $g_{\parallel} = 1.9998$ and $g_{\perp} = 2.0260$, and observed (111) symmetry, we deduce that the electron is in an orbital of a trivalent Ge atom backbonded to three Ge or Si atoms.

*NRC Postdoctoral Fellow

Acknowledgements: The authors would like to thank W. Jenkins of NRL and M. Liu of Honeywell Inc. for supplying the high dose O implant samples. The low dose O implant sample was fabricated at NRL by H. Dietrich and all the MBE layers were grown by N. Green.

REFERENCES

1. P.J. Wang, B.S. Meyerson, F.F. Fang, J. Nocera, and B. Parker, Appl. Phys. Lett. 55, 2333 (1989).
2. R. People, IEEE J. Quantum Electron. QE-22, 1696 (1986).
3. J.P. Zhang, Y.S. Tang, A.K. Robinson, U. Bussmann, P.L.F. Hemment, B.J. Sealy, S.M. Newstead, A.R. Powell, T.E. Whall, and E.H.C. Parker, Appl. Phys. Lett. 57, 890 (1990).
4. F. Finger, W. Fuhs, G. Beck, and R. Carius, J. Non-Cryst. Solid 97&98, 1015 (1987).
5. M. Stutzmann, J. Stuke, and H. Dersch, Phys. Stat. Sol. B115, 141 (1983).
6. M. Stutzmann, C.C. Tsai, and R.A. Street, J. Non-Cryst. Solid 98&98, 1011 (1987).
7. W.E. Carlos, Appl. Phys. Lett. 50, 1450 (1987).
8. Y. Nishi, Jpn. J. Appl. Phys. 10, 52 (1971).
9. P.J. Caplan, E.H. Poindexter, B.E. Deal, and R.R. Razouk, J. Appl. Phys. 50, 5847 (1979).
10. N.M. Ravindra, T. Fink, W. Savin, T.P. Sjoreen, R.L. Pfeffer, L.G. Yerke, R.T. Lareau, J.G. Gualtiere, R. Lux, and C. Wrenn, Nucl. Instrum. Methods B46, 409 (1990).
11. C.G. Tuppen, M.R. Taylor, P.L.F. Hemment, and R.P. Arrowsmith, Appl. Phys. Lett. 45, 57 (1984).
12. B.-Y Mao, P.-H. Chang, H.W. Lam, B.W. Shen, and J.A. Keenan, Appl. Phys. Lett. 48, 794 (1986).
13. W. Gordy in Techniqes of Chemistry, ed. W. West (John Wiley & Sons, New York, 1980) p. 604.

COMPARISON BETWEEN DEFECTS INTRODUCED DURING ELECTRON BEAM EVAPORATION OF Pt and Ti on n–GaAs

F. D. AURET, G. MYBURG, L. J. BREDELL,
W. O. BARNARD AND H. W. KUNERT

Physics Department, University of Pretoria
Pretoria 0002   SOUTH AFRICA

ABSTRACT

Schottky barrier diodes (SBDs) were formed by electron beam evaporation of Pt and Ti at various rates with and without the GaAs substrates being screened from stray electrons during metallization. Using DLTS it was found that some of the defects formed in the GaAs during metallization have discrete levels while others exhibit a continuous energy distribution in the bandgap. Standard I–V measurements were used to demonstrate the adverse effects of these E–beam defects on the SBD characteristics.

1. Introduction

Electron beam (E–beam) evaporation, utilizing electrons with energies up to 20 keV, is a popular technique to evaporate, amongst others, metals with high melting points. Whereas it has been reported that stray electrons originating at the filament during evaporation cause adverse effects in metal – oxide – semiconductor (MOS) devices[1], not much is known about the influence of these electrons on the characteristics of Schottky barrier diodes (SBDs). Although the introduction of defects with discrete energy levels in the bandgap of semiconductors has been reported after low energy electron irradiation[2], the existence of E–beam induced defects with continuously distributed energy levels in the bandgap has not yet been reported. Such defects can strongly influence the properties of SBDs[3].

In this paper we report for the first time the detection by energy resolved deep level transient spectroscopy (ER–DLTS)[4] of defects with continuously distributed energy levels introduced in GaAs during E–beam evaporation of Pt and Ti SBDs. Further, the combined influence of the E–beam induced defects with discrete and with continuous energy levels on the properties of these SBDs is evaluated by current – voltage (I–V) measurements.

2. Experimental Procedure

Undoped n–type GaAs epilayers with a free carrier concentration $4 \times 10^{14}/cm^3$ and a thickness of about 6 μm, grown on n^+– GaAs substrates by organo–metal vapor phase epitaxy (OMVPE), were used for SBD fabrication. Ni–AuGe–Au ohmic contacts were formed on the n^+ – backsides of the substrates prior to SBD fabrication. After chemical etching[5], circular Pt and Ti contacts, 0.73 mm in diameter and 500 Å thick, were deposited onto the GaAs at rates of 0.1, 1.0 and 10 Å/s through a metal contact mask by E–beam evaporation. This was done without screening the GaAs from stray electrons originating at the filament of the E–beam evaporator. Control diodes were fabricated at a rate of 1 Å/s, but with a metal shield placed in such a position as to prevent stray electrons originating at the filament from reaching the GaAs. The SBDs were electrically characterized by I–V measurements, while DLTS measurements between 15 and 400 K were used to characterize the E–beam induced defects.

3. Results and Discussion

Under the assumption that the dominant current transport mechanism is thermionic emission, the I–V measurements on control Pt and Ti SBDs yielded barrier heights of φ_b = 0.99 eV and 0.83 eV, which are in good agreement with the values reported in the literature for these metals. The ideality factors of the two sets of control diodes were n = 1.04 and 1.01, respectively. When metallization is performed without shielding the substrates from stray electrons during evaporation, it was found that φ_b was lowered and n was increased for both metals, as shown in Table I. From this table it is clear that Pt SBDs show larger deviations from ideality and that a decreased deposition rate resulted in an increased degree of non–ideality. The reason for this appears to be that, for a given deposition rate, the substrates are exposed to a higher electron dose during Pt (Table I) than during the Ti deposition, which is a result of the higher E–beam intensity required to melt Pt than Ti.

TABLE I

	Platinum SBDs					Titanium SBDs				
Rate (A/s)	Dose (e/cm^2)	$(\Delta C/C)_{cl}$	$(\Delta C/C)_{dl}$	φ_b (eV)	n	Dose (e/cm^2)	$(\Delta C/C)_{cl}$	$(\Delta C/C)_{dl}$	φ_b (eV)	n
0.1	3E16	90E–4	40E–4	0.88	1.19	3E15	12E–4	4.4E–4	0.82	1.04
1.0	5E15	26E–4	26E–4	0.95	1.10	1E15	6E–4	1.4E–4	0.83	1.02
10	9E14	3E–4	0.5E–4	0.98	1.05	2E14	0.5E–4	0.2E–4	0.83	1.02
1.0	control	1E–4	1E–4	0.99	1.03	control	– –	– –	0.83	1.01

The conventional DLTS spectra in Fig. 1 (curves (a) – (c) for Ti and (d) – (f) for Pt) exhibit two features. Firstly, there are well defined peaks e.g. EPt3 in curve (f), and ETi3 in curve (c), which do not move if scans are recorded at different DLTS filling pulse voltages, V_f. These peaks are characteristic of discrete level defects (DLDs), i.e. defects with discrete energy levels in the bandgap[2]. From Fig. 1 and Table I (where $(\Delta C/C)_{dl}$ was taken as the height of the EPt3 and ETi3 peaks) it is clear that the concentration of the DLDs increases with increasing deposition rate.

Secondly, the spectra contain peaks that become broader and move to lower temperatures when V_f is increased, e.g. EPt8 in curves (d) – (f) and ETi8 in curves (a) – (c), and also show a baseline elevation of a part of the spectra. This second feature is the result of the presence of continuous level defects (CLDs), i.e. defects with a continuous energy distribution in the bandgap. Fig. 2 and Table I show that the concentration of the CLDs (proportional to $(\Delta C/C)_{cl}$ in Table I) also increases as the deposition rate is decreased.

In order to estimate the energy distribution in the bandgap of the CLDs which result in this second feature, energy resolved (ER-) DLTS[4] measurements were made by recording spectra with small differences in the filling pulse voltage and then subtracting these spectra from each other. The information contained in such spectra originates from two regions in the SBD. The first region is located at

$$x_f = [2\epsilon(V_{bi}-V_f)/qN_d]^{1/2} - [2\epsilon(E_t-E_f)/qN_d]^{1/2} \qquad (1)$$

below the interface and contains information primarily about the DLDs which may extend up to several microns below the interface. The second region, which is of interest for this paper, is the semiconductor immediately adjacent to the interface. Using ER–DLTS this region may be probed for CLDs. These defects will be detectable provided they are in equilibrium with the Fermi level of the semiconductor and not with that of the metal. CLDs at the interface, however, are in equilibrium with the

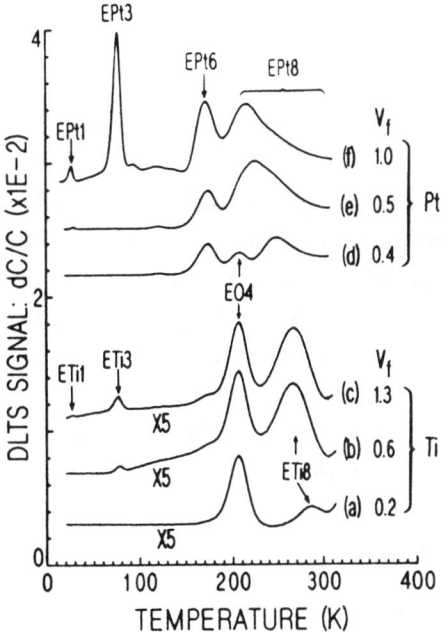

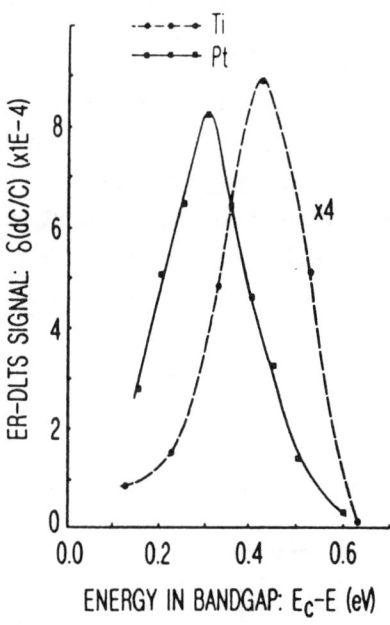

Fig. 1: DLTS spectra for E-beam deposited Ti (curves (a) – (c)) and Pt ((d) – (f)) SBDs. All curves were recorded with a lock-in amplifier frequency of 10 Hz and a reverse bias of $V_r = 1$ V. The filling pulses V_f are given in the figure.

Fig. 2: ER–DLTS signal of Pt and Ti SBDs as function of position in the bandgap. The concentration of continuous level defects, N_{cl}, is proportional to the ER–DLTS signal.

metal Fermi level because the thin interfacial oxide left after chemical etching is not thick enough to prevent carriers to tunnel from them to the metal and hence these defects are not detectable by capacitive methods[6]. When a forward bias filling voltage V_f is applied to the SBD and the semiconductor Fermi level is not pinned at the interface, then ER–DLTS yields information pertaining to defects at $E = \varphi_b - V_f$ below the conduction band. The concentration of CLDs in an energy interval δE at $E_c - E$ that contribute to the ER–DLTS signal may be approximately estimated from

$$N_{cl}(E, \delta E) = [\epsilon q N_d / C \delta E] \cdot \delta(\Delta C / C), \qquad (2)$$

where $\delta(\Delta C/C)$ is the ER–DLTS signal magnitude for a filling pulse increment of $\delta V = \delta E/q$, and the other symbols have their conventional meanings.

ER–DLTS measurements for which $0 < V_f < 1.1$ V and $\delta V = 0.02$ V were performed on Pt as well as Ti SBDs. The results showed that the CLD concentration, N_{cl} in SBDs, which is proportional to the ER–DLTS signal in Fig. 2, varies continuously with energy in a region of the bandgap but exhibits a rather sharp maximum as shown in Fig. 2. Whereas the N_{cl} peak in Pt SBDs is at about $E_c - 0.30$ eV, it is located at about $E_c - 0.42$ eV for Ti. For both metals the peak maximum was found to increase as the deposition rate was decreased. However, the peak concentration for Ti SBDs deposited at 1 A/s is about 4 times lower than for Pt SBDs fabricated at the same deposition rate.

4. Conclusions

The ER–DLTS measurements indicated that apart from the discrete level defects previously reported[2], E–beam metallization also causes the introduction of defects with a continuous energy distribution. This distribution peaks at 0.30 and 0.42 eV below the GaAs conduction band for Pt and Ti SBDs, respectively. The concentration of the discrete and continuous level defects increased if the deposition rate was decreased. Ti SBDs formed by E–beam deposition were found to be of higher quality than Pt SBDs formed at the same conditions. The ER–DLTS results presented here support the view that because less power is required to evaporate Ti, fewer defects are introduced during the deposition of Ti than of Pt, thereby yielding higher quality SBDs. In summary, the results presented here have clearly demonstrated that the defects introduced during E–beam metallization result in an increase in the degree of non–ideality of SBDs.

5. References

1. T.H. Ning, J. Appl. Phys. **49**, 4077 (1978).
2. F.D.Auret and M.Nel, Japan. J. Appl. Phys. **28**, 2430 (1989).
3. D.Bauza and G.Pananakakis, J. Appl. Phys. **69**, 3357 (1991).
4. N. M. Johnson, Appl. Phys. Lett. **34**, 802 (1979).
5. F.D.Auret, G.Myburg and L.J.Bredell, S. A. J. Science **87**, 127 (1991)
6. E. Rosencher, S. Delage and F. D'Avitaya, J. Vac. Sci. Technol. B **3**, 762 (1985)

ENHANCED-DIFFUSION IN ELECTRON-BEAM DOPING OF SEMICONDUCTORS

TAKAO WADA, MICHIHIKO TAKEDA* and TOSHIYUKI KONDO
Nagoya Institute of Technology, Showa, Nagoya, 466, Japan
*Nagoya Government Industrial Research Institute, Kita, Nagoya, 462, Japan

ABSTRACT

The mechanism of electron-beam doping (EBD) methods in semiconductors has been investigated. There are two-kinds of enhanced diffusion; one is the recombination enhanced diffusion (mobility-enhanced) and 2nd is the kick-out mechanism (concentration-enhanced).

1. Introduction

Diffusion in semiconductors can occur via a number of different mechanism[1]. The mechanism of enhanced diffusion can be operative in the presence of ionization or of defects in non-equilibrium concentrations produced by energetic particles[1]. The energy-release-type mechanisms appear to be most effective in raising the mobility of directly diffusing defects in Si, Ge and GaAs[2]. The diffusion of dopants is enhanced, since the concentrations of self-interstitials, which serve as diffusion vehicles for the dopants (interstitialcy mechanism), are raised above their thermal equiribrium values[2].

Recently, new methods of electron-beam doping[3,4] (EBD), electron-beam oxidation[5] (EBO), and electron-beam epitaxy[6] (EBE) processes have been reported by the author and other workers. In the present paper we have investigated the mechanism of electron-beam doping methods (superdiffusion) in semiconductors.

2. Experimental procedure

The wafers used in the experiments were (111)-oriented n-Si($t \simeq 0.25$mm), (111)-oriented n-Ge($t \simeq 0.25$mm) and (100)-oriented undoped semi-insulating GaAs($t \simeq 0.5$mm) grown by liquid encapsulated Czochralski(LEC). Two-layer structures(array II, AII) were of overlayer GaAs//substrate GaAs (AII_0), Ge//Si (AII_1), Si//GaAs (AII_2) and Al/GaAs (AII_3), which means Al evapolated layers on GaAs. The other structures were of GaAs(layer 3)//Si(layer 2)//GaAs (layer 1)($AIII_1$), (a Si wafer was sandwiched between two GaAs wafers) Si//Ge//Si($AIII_2$), and GaAs/Al//Al/GaAs (AIV). The surfaces of the overlayers (Ge for AII_1, Al for AII_3, Si for AII_2 and $AIII_2$, and GaAs for AII_0, $AIII_1$ and AIV) in contact with the substrates were irradiated with a total fluence of $\sim(5-72) \times 10^{17}$ electrons·cm^{-2} at 7MeV, and at 50-60°C from an electron linear accelerator with a pulse width of 3.5µs, a 200Hz duty cycle and a peak electron-beam current of $\sim$50mA·cm^{-2}. During irradiation, the samples were put in an isothermal circulating water bath using a thermoregulator.

After irradiation, the secondary-ion-mass spectrometry(SIMS) measurements for AII_1 were performed by using the primary ion (O_2^+) beam (diameter 1mmφ) with an ion energy of 12KeV in a 2.7×10^{-7} Torr vacuum. The GaAs samples of layer 1 for $AIII_1$, and of AII_2 were annealed at 800°C for 20 min with a SiO_2 cap in a conventional furnace. After stripping off the SiO_2 films, photoluminescence (PL) measurements were performed at 77K. A focused 80mW, 514.5nm argon laser beam was used as the excitation source. The introductions of Ge impurity atoms in Si for $AIII_2$ without annealing were measured by Rutherford backscattering spectroscopy (RBS), using a 1.8MeV $^4He^+$ beam. The yield of scattered He ions was studied as a function of the angle of incidence. Tilt and rotation settings on the goniometer used, could be reproduced to within 0.05°. To avoid pile-up effects in the electronics small currents were used, typically 15nA. PL and lifetime measurements for AII$_8$ and irradiated GaAs wafer were performed by Hamamatsu system consisted of a laser diode (λ=670nm) and a streak camera. The Raman spectra were measured in a usual backscattering configuration, using the 514.5nm line of an argon ion laser operating at 20mW CW as the exciting source. In the cases of AII$_3$ and AIV, after irradiation and etching away Al layer, the measurements were carried out.

3. Experimental results

In the experiments of AII$_1$ system, the concentrations N_{Ge} of Ge atoms by SIMS and RBS measurements in Si EB-doped with the same fluence at 60°C are shown in fig.1 at ~50Å from the Si surface as a function of dose rate (dϕ/dt). The total doses are 5×10^{17} electrons·cm^{-2} at dϕ/dt≃0.4-2.1×10^{17} electrons·cm^{-2}·s^{-1} and 10^{18} electrons·cm^{-2} at 2.7-7.7×10^{17} electrons·cm^{-2}·s^{-1}. The value of N_{Ge} increased with increasing dose rate, and became a maximum value at a proper rate. Then, N_{Ge} decreased in higher dose rates. Figures 2a and 2b indicate PL spectra at 77K for the EBD GaAs samples of layer 1 for AIII$_1$ and of AII$_2$ after annealing. There are three emission peaks: a peak attributed to the band gap transition at 1.51eV, a peak attributed to silicon acceptor with isolated Si atoms on As site Si$_{As}$ at 1.48 eV and a peak attributed to the residual copper in Ga site Cu$_{Ga}$ at 1.36eV. The ratio of the emission intensity for the Si$_{As}$ of GaAs for AIII$_1$ to that of GaAs for AII$_2$ was nearly 4 to 1. It is clear that the AIII$_1$ structure is more effective in EBD technique than the AII$_2$ structure.

Figure 3 shows the angular distribution of elastically scattered ^{4}He$^+$ ions around the ⟨110⟩, ⟨111⟩ and ⟨100⟩ directions obtained for Si (layer 1) of AIII$_2$ EBD Ge-doped Si sample without annealing. Also, the energy spectra in the random and aligned conditions for the sample indicated in the ⟨111⟩ direction, where the scattering yields from Ge atoms exhibited a marked increase above Ge random values. The Ge concentration in Si at the surface for the ⟨111⟩ direction was estimated as about 1.2×10^{20}

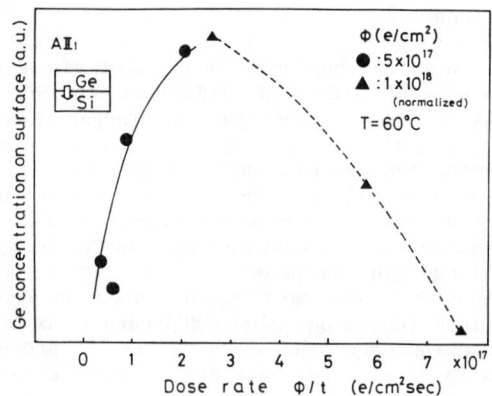

Fig.1 Dose rate dependence of the Ge concentrations at 50Å from the surface of Si before annealing

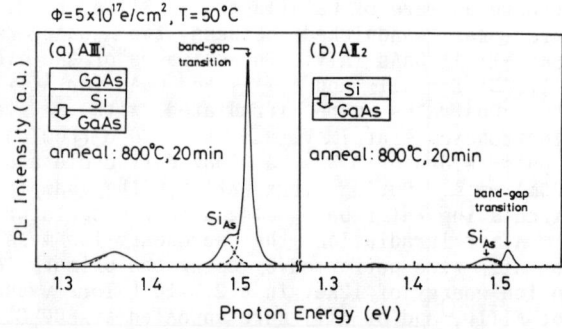

Fig.2 PL spectra at 77K for GaAs irradiated in (a)AIII$_1$ (layer 1) and (b)AII$_2$ after annealing at 800°C for 20 min. Broken lines are Lorenzian curves computed to show the most suitable agreement with spectra.

cm^{-3}. The observed values of the scattering yield(χ_{min}) from the Si atoms for the perfectly aligned direction of ⟨110⟩, ⟨111⟩ and ⟨100⟩ are obtained to be 0.035, 0.05 and 0.04, respectively. Around the ⟨111⟩ direction a narrow peak from the Ge atoms in the mid-channel region was observed, but around the ⟨110⟩ and ⟨100⟩ directions the channeling dips, which were much weaker than the corresponding one for the Si lattice, were measured as χ_{min} = 0.92 and 0.75, respectively. The occurrence of such a peak and dips would be interpreted as being caused by scattering from Ge impurity atoms partially located in both the bond-centered interstitial and a split ⟨111⟩ interstitial in the Si lattice[7]. The typical results of PL measurements and the Raman spectra at 300K for GaAs samples of layers 1 and 2 for AII₈, and of an irradiated substrate only AI before annealing are shown in fig.4. In this PL experiments, a laser diode (λ=670nm) was used as the excitation source. The PL spectra of layers 1 and 2 for AII₈ indicated a clear peak at 1.38eV which is attributed to the Ga antisite defect, Ga_{As}. However, it disappears for AI. The intensity of Raman spectra of the LO phonon at 292 cm^{-1} for GaAs samples of AII₈ is much larger than that of AI. Figure 5 shows Raman spectra (LO phonon) of upper and lower layers of GaAs for AIV and GaAs wafer for AII₃. It is emphasized that all of Raman spectra were observed before annealing. Both intensities of LO phonon in figs.5a and 5b (AIV) are much larger than that in fig.5c (AII₃). In the AIV structures, the solid phase epitaxial layers of $Al_xGa_{1-x}As$ (x≃0.05) were grown in GaAs substrates at 50°C, but for AII₃ no alloying was observed. The characterization was carried out by using an X-ray photoelectron spectroscopy (XPS). They were measured with MgKα radiation at a pressure of ~1×10⁻⁵ Pa. The $Al2p_{1/2}$ core-level binding energies for elemental Al, AlAs and Al_2O_3 are 74, 75.0 and 75.5eV, respectively. The Al2p peak energy for AIV structure is 75eV, but for AII₃ it is 74eV. The composition of EBE $Al_xGa_{1-x}As$ was estimated roughly to be 0.05 by comparing its XPS results with those of MBE $Al_xGa_{1-x}As$ layers for X = 0.3 and 0.1.

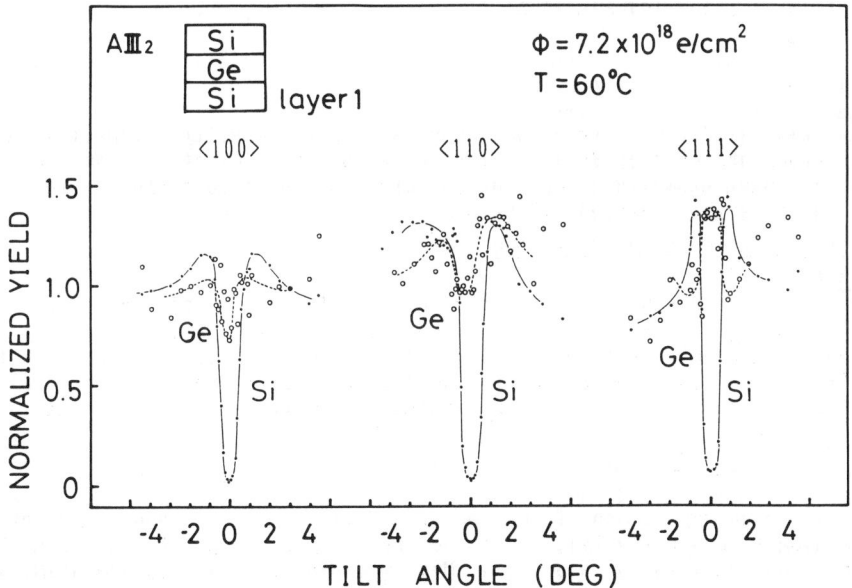

Fig.3 Scattering yield from the Ge atoms (o) and from the Si atoms (·) in Si (layer 1) for AIII₂ as a function of the angle between the incident beam direction and various low-index axes

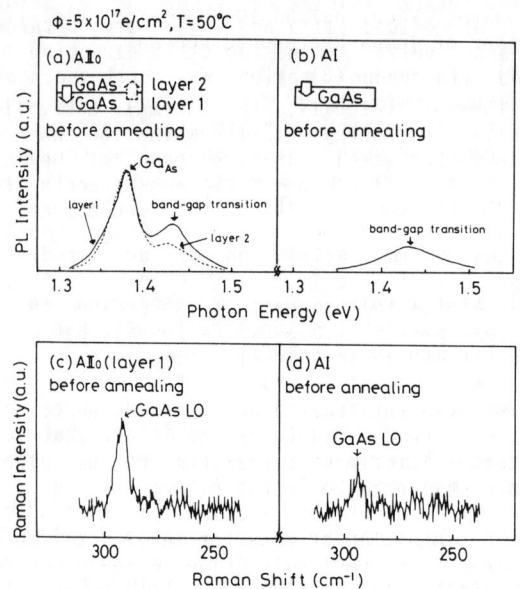
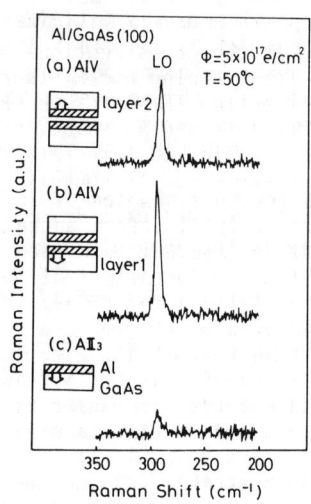

Fig.4 PL and Raman spectra at room temperature for EBD-GaAs before annealing. The samples for PL were (a)layer 1(solid line) and layer 2 (broken line) in AII$_8$, and (b)GaAs in AI. The samples for Raman spectra were (c)layer 1 in AII$_8$ and (d)GaAs in AI.

Fig.5 Raman spectra of EBD-(100) GaAs from the surface before annealing. The samples were (a)upper and (b)lower layers in AIV, and (c)GaAs in AII$_3$

4. Discussion

There are two kinds of migration processes for impurity atoms;(1) dopant impurity atoms are emitted from overlayers by incident electron-beams and (2) introduced dopant atoms into substrates migrate by enhanced diffusion mechanism. Here, we discuss the enhanced diffusion.

4-1 Recombination-enhanced diffusion

a) Diffusivity

The extrapolated ranges of electrons at 7MeV in Si, Ge and GaAs are about 15, 5.6 and 5.65mm, respectively[8]. The wafers are sufficiently thin to allow the irradiating electrons to penetrate into the overlayer and substrate without a significant loss in kinetic energy. The rate of generation G of electron-hole pairs (EHPs) per unit time by an incident electron can be estimated as follows[1],

$$G = \frac{1}{\varepsilon} \cdot \frac{dE}{dx} \cdot \frac{d\phi}{dt} \quad \cdots \cdot (1)$$

where ε is the energy for the formation of EHP (~3.88, 2.79 and 4.63 eV for Si, Ge and GaAs, respectively), dE/dx the energy loss per cm of the path by a fast electron, (3.87, 9.28 and 8.16 MeV·cm^{-1}·electron^{-1}) and $d\phi/dt \simeq 3.1 \times 10^{17}$ electrons·cm^{-2}·S^{-1} during pulse width on electron irradiation. The irradiation results in G$\simeq$3.1, 10.0 and 5.5$\times 10^{23}$ EHPs·cm^{-3}·s^{-1} for Si, Ge and GaAs, respectively. G produces and electron-hole pair concentration of

$$n = G \cdot \tau \quad \cdots \cdot (2)$$

where τ is the carrier lifetime. The value of τ for the irradiated GaAs wafer was obtained to be ~2.7 ns from measurements by a two-dimensional Hamamatsu streak camera coupled with a monochromator. Then, $n \simeq 1.35 \times 10^{15}$ cm^{-3} for GaAs. For an energy-release mechanism[1,9], a number of jumps R are obtained as

$$R = n \cdot \sigma_e \cdot V \cdot \exp\{-(E_c + E_H)/kT\} \quad \cdots \cdots (3)$$

where σ is the cross sections, V the thermal velocity, E_c thermal activation energy for trapping and E_H thermal activation energy of recombination-enhanced defect reaction ($E_H = 0$ for the Bourgoin mechanism[1]). In the case of Ge//Si, $E_c + E_H$ was obtained as ~0.3eV near the surface from the SIMS measurements for various irradiation temperatures. The effective diffusivity for recombination-enhanced diffusion is roughly given by

$$D_{eff} \simeq R \cdot (\Delta \chi)^2 / 4 \quad \cdots \cdots (4)$$

where $\Delta \chi$ is the jump distance. By using their values, D_{eff} is estimated to be about 10^{-15} cm$^2 \cdot$S^{-1}, being roughly in agreement with the experiments near the surface. Assuming $E_c + E_H \simeq 0.17$ eV, D_{eff} becomes 10^{-12} cm$^2 \cdot$S^{-1}.

b) Carrier lifetime dependence

In the case of AII$_1$ experiments for dose rate(dϕ/dt) dependence (see fig.1), the increasing region of N_{Ge} is due to the relationship of G $\propto$ dϕ/dt. The decreasing region is caused by the shorter lifetimes of carriers, because higher concentrations of defects are introduced by larger dose rates.

In the experiments of fig. 2, the ratios of the PL emission intensities for the Si-acceptor (Si$_{As}$) and band-to-band transition of GaAs for AIII$_1$ to those for AII$_2$ were nearly 4 to 1 and 10 to 1, respectively. The time-integrated intensity of the band-to-band transition is proportional to the carrier lifetime[10]. Then, the lifetime of GaAs for AIII$_1$ is much longer than that for AII$_2$. D_{eff} for AIII$_1$ accordingly is much larger than that for AII$_2$. Thus, the Si$_{As}$ concentration for AIII$_1$ is much larger than that for AII$_2$, being in agreement with the experiments.

c) Bourgoin mechanism

In the RBS experiments for AIII$_2$, it is indicated that the Ge interstitial configurations in Si are both a bond-centered interstitial and a split-$\langle 111 \rangle$ interstitial. Weigel et al[11] have suggested that the interstitial in the diamond lattice is a possible example of the Bourgoin mechanism of athermal migration of a defect in the presence of ionizing radiation. In our experiments, the Bourgoin mechanism may supply an athermal mechanism for the interstitial migration via the alternate capture of electrons and holes. The diffusivity of boron in Si for the common diffusion process at elevated temperatures is known to be enhanced by a supersaturation of self-interstitials.

It is suggested that the diffusivity of EBD-Ge in Si is enhanced by a supersaturation of self-interstitials (kick-out mechanism[12]).

4-2 Kick-out mechanism (concentration-enhanced diffusion)

Ion-type intrinsic defects produced by irradiation for GaAs are interstitials As$_i$ or Ga$_i$, or antisites As$_{Ga}$ or Ga$_{As}$. Corbel et al[13]. have showed that 1.5-3 MeV electron irradiation produces negative Ga$_{As}$ antisites and negative V$_{Ga}$ vacancies, and earlier that the arsenic vacancies are involved in the native monovacancy defects while the gallium vacancies are involved in the irradiation-induced monovacancy defect. Spicer et al also have reported that the dominant surface states on GaAs were associated with an As deficit[14].

a)AI (mono-substrate) and AII$_B$

For the both cases of the present electron irradiation and ion implantation[15] for GaAs substrate, Raman intensities similarly decreased with increasing total dose due to damage produced by irradiation (~5×10^{17} electrons·cm^{-2}) and implantation with lower doses, but a peak frequency shift and band width of their spectra did not change. The onset dose at which damage was detectable by Raman measurement was 7.5×10^{12} ions·cm^{-2} in Si^{++} implanted silicon[15].

In the case of AII$_B$, the intensities of Raman (LO phonon) and PL spectra before annealing increase largely compared with that for AI (see figs.4 and 5). Interstitials As$_i$ and Ga$_i$ created by irradiation of the overlayer are emitted from the overlayer and reached to the substrate surfaces. Then, As_i could diffuse into the substrate and fill the As vacancies at or near the surface as follows,

$$As_i + V_{As} \rightleftarrows As_{As} \quad \cdots\cdots(5)$$

The increase of Raman intensity for AII$_a$ may be probably caused by filling the vacancies at the surface.

As the kick-out mechanism may be dominant for EBD experiments, the following reaction for Ga_i may be established via As self-interstitial

$$Ga_i \rightleftarrows Ga_{As} + As_i \quad \cdots\cdots(6)$$

As_i(I) concentration is reduced near the surface, since the surfaces may act as I sinks.

According to eq.6 the formation of Ga_{As} (this formation energy is lower than that of As_{Ga}) requires the generation of I^2. During irradiation for AII$_a$ a supersaturation of I (self-interstitial) is built up. Thus, the PL spectra for AII$_a$ show a large peak attributed to Ga_{As}.

b) Electron-beam epitaxy (EBE)

In the case of EBE-$Al_xGa_{1-x}Sb$, U-shaped diffusion profiles of Al atoms in substrates were observed by SIMS. In a previous paper[4], the U-shaped diffusion profiles of impurity atoms into Si wafers were explained by considering the "kick-out" mechanism. Thermal equilibrium between Al_{Ga} and Al_i may be established via Ga self-interstitials according to the kick-out mechanism[12]

$$Al_{Ga} + Ga_i \rightleftarrows Al_i \quad \cdots\cdots(7)$$

where Al_{Ga} represents an Al atom on a Ga site, Ga_i a Ga atom on an interstitial site. As a supersaturation of Ga_i introduced from the overlayer is built up, alloying may be formed by reaction of eq.7. Further studies on EBE is now in progress.

Acknowledgements

We are grateful to Dr. Y.Shinozuka of Yamaguchi University and Dr. M.Ichimura of Nagoya Institute of Technology for valuable discussions, and to Dr. H.Kan of Hamamatsu photonics KK for measurements of lifetime. This work was partly supported by Nippon Electric Co. Foundation.

References

1. J.C.Bourgoin and J.W.Corbett, Rad.Effects. 36, 157 (1978)
2. W.Frank, A.Seeger and U.Gösele, in Defects in semiconductors, J.Narayan and T.Y.Tan ed. (North-Holland, 1981) p31
3. T.Wada, Nucl.Instr. and Meth. 182/181, 131 (1981)
4. T.Wada and H.Hada, Phys.Rev.B 30(6), 3384 (1984)
5. T.Wada, Appl.Phys.Lett. 52, 1056 (1988)
6. T.Wada and Y.Maeda, Appl.Phys.Lett. 51, 2130 (1987)
7. K.Komaki, Oyo Buturi, 48, 637 (1979) [in Japanese]
8. T.Tabata, R.Itoh and S.Okabe, Nucl.Instr. and Meth. 103, 85 (1972)
9. H.Sumi, Phys.Rev.B 27, 2374 (1983)
10. K.Leo, W.W.Ruhle, P.Norelberg and T.Fujii, J.Appl.Phys. 66(4), 1800 (1989)
11. C.Weigel, D.Peak, J.W.Corbett, G.D.Watkins and R.P.Messmer, Phys.Rev.B 8(6), 2906 (1973)
12. U.Gösele, F.Morehead, W.Frank and A.Seeger, Appl.Phys.Lett. 38, 157 (1981)
13. C.Corbel, F.Pierre, P.Hautojarvi, K.Saarinen and P.Moser, Phys.Rev B 41(15), 10632 (1990)
14. W.E.Spicer, P.W.Chye, P.R.Skeath, C.Y.Su and I.Lindau, J.Vac.Sci.Technol. 16(5), 1422 (1979)
15. K.Mizoguchi, S.Nakashima, A.Fujii, A.Mitsuishi, H.Morimoto, H.Onoda and T.Kato, Jpn.J.Appl.Phys. 26(6), 903 (1987)

HIGH TEMPERATURE DEFECT-FREE RAPID THERMAL ANNEALING OF III-V SUBSTRATES IN METALLORGANIC CONTROLLED AMBIENT

A. Katz,[1] A. Feingold,[1] S. J. Pearton,[1] M. Geva,[2] S. Nakahara[1] and E. Lane[2]

[1] AT&T Bell Laboratories, Murray Hill, NJ 07974, USA
[2] AT&T Bell Laboratories, Solid State Technology Center, Breinigsville, PA 18031, USA

ABSTRACT

High temperature defect-free rapid thermal annealing of III-V substrates was achieved in a load-locked rapid-thermal low-pressure metallorganic chemical-vapor-deposition (RT-LPMOCVD) reactor, under phosphorus and arsenic controlled ambients, using tertiarybutylphosphine (TBP) and tertiarybutylarsine (TBA) metallorganic liquid sources.

Damage-free surfaces of InP and GaAs were obtained for temperatures up to 700°C for InP under TBP ambient, or above 900°C for GaAs under TBA ambient, respectively. Annealing the III-V substrates at low protective ambient pressure (50 mTorr) provided an excellent surface protection through the heating cycle, demonstrating the advantage of using these ambients which are much less toxic than PH_3 or AsH_3, and without resulting in deposition of the group-V elements on the surface and without reducing the efficiency of the process.

I. INTRODUCTION

Efficient ion implantation activation and contact metallization sintering of III-V materials, require high temperature annealing cycles.[1-8] The annealing temperatures are often well above those which lead to surface decomposition due to group-V species incongruent evaporation, and, therefore, protection of the wafer surface is needed to minimize the loss of P or As. The use of rapid thermal annealing (RTA) in III-V technology provides a strong impetus for investigating surface protection methods. In order to achieve a good surface protection, a variety of possible solutions associated with the RTA technique have been tried, but each has deficiencies, as is described below: Dielectric encapsulants (Si_3N_4, SiO_2, or AlN), are commonly used for this purpose. This technique exhibits adhesion problems at the high annealing temperatures and induces considerable stress into the underlying films and substrates, which strongly effects the implanted layer characteristics.[9-10] Second method is using an enclosed graphite cavity for the RTA of both GaAs and InP.[11-13] This method provides a very uniform heating environment for the wafer because of the high emissivity of the graphite, reducing slip formation, and eliminating the need for a guard ring around the circumference of the wafer, which has been suggested earlier as a solution for the temperature nonuniformities.[14-17] The main disadvantage of using the graphite susceptor, however, is the need to recharge the graphite cavity with P prior to each InP wafer annealing, and with As, when annealing GaAs wafers.[18]

In order to eliminate the need for recharging the graphite cavity with the group-V species, we have introduced a modified graphite susceptor with peripheral reservoirs filled with either As or P-containing source material.[19] This arrangement allows for a continuous supply of the group-V element during annealing without the need for recharging. However, while introducing the graphite susceptor solution to a manufacturing line, it presents some complication which are associated with the handling of the susceptor, the need to load the wafer into it, and the actual loading of the entire structure into the annealer. In addition, it provides only an approximate group-V vapor pressure control, which influences the reproducibility and accuracy of the process.

Those techniques were suggested as an improvement to the traditional protective annealing method, the so-called proximity annealing, in which the wafer of interest has been placed face to face with another wafer of the same kind and, thus, an overpressure of the volatile group-V element was created between the two wafers to partially prevent further surface dissociation of the wafer of interest.[20]

As mentioned earlier, all these approaches were suggested in an attempt to get around one of the major deficiencies of the RTA technique, namely, eliminating the III-V semiconductor surface integrity degradation, but yet retaining all the well-reported advantages of this heating method.[21-26]

The most straight forward approach to be taken in order to preserve all the advantages of the RTA technique, while eliminating the surface decomposition, is to provide a global group-V (arsenic or phosphorus) overpressure in the rapid thermal annealer chamber. This concept was suggested by a number of groups in the past,[27-29] but has not been widely used in the III-V technology community due to the very stringent safety requirement associated with the use of the toxic hydride gases. The factors which limited the implementation of AsH_3 or PH_3 overpressure method in routine annealing processes were the acute inhalation toxicity of the hydrides, such as AsH_3 ($LC_{50} = 5 - 40$ ppm) and PH_3 ($LC_{50} = 11 - 50$ ppm), their severe health effects, such as blood hemolysis, the threat of possible catastrophes, the high expenses associated with equipment needed to handle these toxic gases, and in addition the fact that PH_3 has a slow pyrolysis rate, (decomposing by only 25% at 600°C.[30])

During the last three years, new chemicals have been introduced as efficient replacements for the hydrides. These chemicals do not have the high degree of toxicity or the hydrides, and make the possibility of direct protective ambient in the RTA much more attractive. They include organometallic liquid precursors[31] such as tertiarybutylarsine (TBA) and tBP (TBP),[31] which have been successfully used for growing InP and GaAs-based materials and devices.[32-44]

In this paper we describe a high quality RTA process for GaAs and InP substrates, using low pressure TBA and TBP, respectively, in a recently designed, commercially available, RT-LPMOCVD reactor. The surface morphology and metal/semiconductor interdiffusion profiles in the annealed samples were investigated and provide solid evidence for the superiority of this method.

II. EXPERIMENTAL

2" round semi-insulating (SI) InP and undoped GaAs wafers, <100> oriented, were rapid thermally annealed under protective and inert controlled ambients, in order to assess the influence of the ambient on the morphology and surface preservation of the annealed wafers. In addition, some substrates with a WSi_x cap layer (thickness of about 100 nm) were annealed as well in order to evaluate any possible correlation between the evolution of the metal-semiconductor interfacial reactions and the type of ambient gas used to create the group V overpressure in the chamber. Both InP and GaAs wafers were degreased in organic solvents and etched in so-called A-etchant ($1 H_2O : 1 H_2O_2 : 5 H_2SO_4$) at room temperature for 4 min, followed by a de-ionized water rinse and blown dry with N_2 before loading into the reactor. The WSi_x was deposited by rf sputtering.[45,46]

All the annealing were carried out in a prototype of the now commercially available A. G. Associated Heatpulse CVD-800™ system, designed in a joint effort between one of us[46] and A. G. Associates. In brief, this is a load-locked, low pressure, horizontal flow, cold wall chamber, single-wafer rapid thermal processor, capable of processing with inert, hazardous or corrosive ambients. A picture of the system is given in Figure 1. A very detailed description of the system was recently given elsewhere.[46] The temperature of the annealed wafer was monitored by extended range, double-header pyrometer sensing. The optical signal detected from the wafer emission was adjusted to account for the non-wafer radiation sources, such as the light emitted from the quartz isolation tube and the tungsten halogen heating lamps.

The phosphorous and arsenic overpressures were achieved by using the organometallic liquid TBP and TBA, respectively, the vapor of which were pumped directly into the low pressure chamber. Both bubblers were held at 10°C, where their vapor pressure were about 125 Torr and 85 Torr, respectively.[31] Typical flow rates for TBP and TBA were kept constant at 100 and 200 sccm, respectively, using a unit mass flow controller (MFC), to provide a total chamber pressure of about 50 mTorr during annealing cycles at 700-1000°C.

Full area WSi_x layers on InP were patterned with AZ1350J photoresist in a test pattern containing a variety of openings, ranging in a size of 1 to 50 μm. The WSi_x was etched in a pure SF_6

discharge contained within a hybrid Electron Cyclotron Resonance (ECR)/Radio Frequency (RF) system.[48] The pressure was varied from 1 to 35 mTorr, the SF_6 flow rate was held constant at 30 sccm, and the microwave power was fixed at 150W.

Residual gas analysis (RGA) of the chamber ambient through annealing was carried out by means of a Dycor Quadrupole Ametek RGA system. The analytical examinations involved optical microscopy, scanning electron microscopy (SEM), transmission election microscopy (TEM), secondary-ion mass spectrometry (SIMS), and Auger electron spectroscopy (AES). Film sheet resistance was measured by a conventional four-point probe.

Fig. 1   Photograph of the A. G. Associates Heatpulse CVD-800™ system.

III. RESULTS AND DISCUSSION

a. InP Annealing Under TBP and nN2 Ambient

All the heat treatments were carried out in the RT-LPMOCVD reactor chamber, and the gases were always injected into a 5×10^{-6} Torr prepumped atmosphere. Since the system was load-locked, the main chamber was kept relatively dry and clean. All the gases were introduced into the chamber 30 sec prior to initiating the heating cycle, and thus, a controlled ambient and a total constant pressure were achieved prior to turning on the lamps. The gas flow was stopped 25 sec after the lamps were turned off, providing the selected ambient as long as the annealed sample temperature was above 350°C. Figure 2 shows a typical InP annealing parameter plot, for heating under a TBP protective ambient. This plot summarizes all the treatment variables, such as the wafer temperature, condition of each of the mass flow controllers (MFC) through the various annealing steps and the overall chamber pressure, as a function of the process duration. (This graph was produced by the RT-CVD-800™ software in real time, and thus reflects the actual response of the pressure, temperature and the MFC changes.) Figure 2 represents the annealing process for a InP wafer at 700°C for 30 sec under TBP ambient at a total chamber pressure of either 5 mTorr, achieved by holding the TBP bubbler at a constant temperature of 3°C, or 50 mTorr as a result of adjusting the TBP bubbler temperature to 10°C. In both cases, the TBP vapor flow rate was 100 sccm. Subsequent to the controlled ambient anneal, the chamber was purged with N_2 at a total pressure of 5.5 Torr in order to remove residues of the organometallic source prior to unloading the sample through the load-lock.

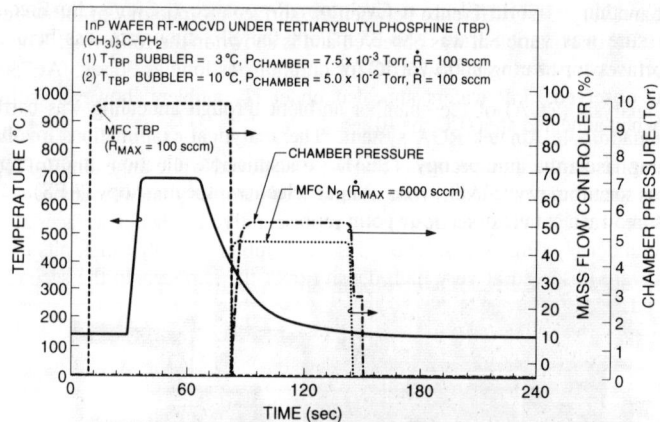

Fig. 2   RT-LPMOCVD process parameter plot of annealing at 700°C for 30 sec under TBP, showing the temperature, total chamber pressure and condition of TBP and N_2 mass flow controllers, as a function of the annealing time.

Other samples were annealed with exactly the same conditions, using a N_2 ambient in place of the TBP. This allowed for a comparison between the process efficiency and wafer surface damage phenomena while annealing under either a protective ambient (TBP) or an inert gas (N_2).

Figure 3 shows the residual gas analyzer spectra during the annealing of InP wafers at 600°C and total pressure of 1×10^{-2} Torr, under N_2 ambient (Fig. 3a) or TBP (Fig. 3b). The existence of both free and carbon-bonded methyl groups, in addition to the large peak of the free phosphorus in the chamber atmosphere when heated under the TBP ambient (see Fig. 3b), may reflect an efficient decomposition of the metallorganic source at this temperature.

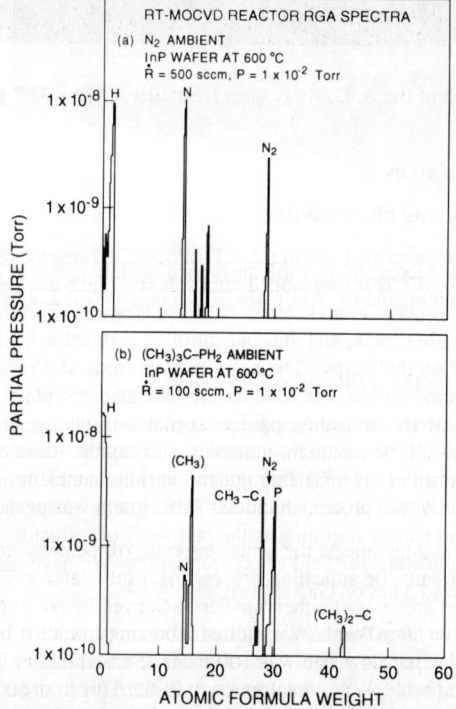

Fig. 3   Residual gas analyzer spectra during low pressure RTA cycles at temperatures of 600°C and pressure of 1×10^{-2} Torr, under (a) N_2, and (b) TBP controlled ambient.

A significant and very clear difference between the surface morphology of InP wafers that were annealed under the TBP or N_2 ambient was observed and is shown in the following. The former gave degradation-free surfaces for heating up to 750°C for durations as long as 30 sec. At 750°C the first surface pitting was observed. InP wafers annealed under N_2 ambient, however, suffered severe deterioration even when annealed at 600°C for 10 sec.

In order to emphasize the improvements that were achieved while annealing InP under a TBP ambient, TEM cross sections were taken from a variety of samples that were annealed under both ambients. Figure 4 shows TEM cross section of two representative InP wafer surfaces after annealing at 700°C for 30 sec under N_2 (Fig. 4a) and TBP (Fig. 4b) ambients. In the first, pits as deep and as wide as 3 to 5 μm were observed at very high density over the surface. In the second, no surface damage was observed.

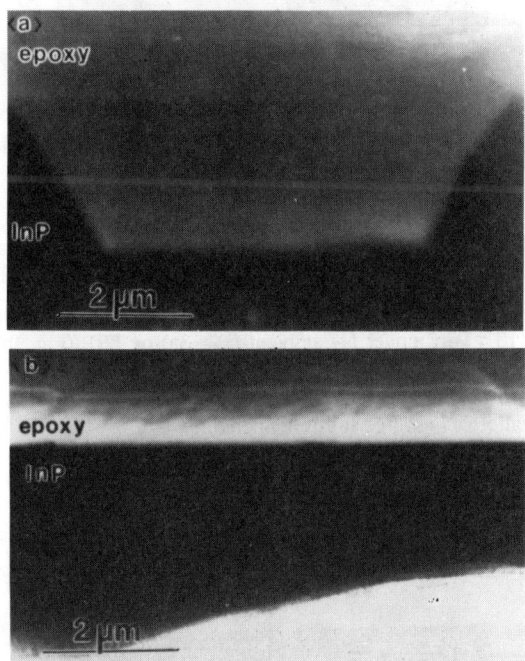

Fig. 4   TEM cross sectional micrographs of InP surfaces after low pressure RTA at (a) 600°C for 30 sec under N_2 ambient, and (b) 700°C for 30 sec under TBP ambient, both at pressure of 50 mTorr.

b. RTP of wSI/InP Contacts Under TBP and nN2 Ambients

One of the technology needs for annealing InP surfaces is the simultaneous activation of implanted dopants and sintering of the metal contacts in self-aligned devices.[49] It is, therefore, important to examine the influence of the annealing under controlled ambient on the contact metal of choice, as well as on the quality of the semiconductor in the neighborhood of the metal contact. Lahav et al.,[7] have reported on extensive pitting at the periphery of refractory gates following RTA of self-aligned GaAs metal-semiconductor field effect transistors. We have observed similar effects during RTA of $WSi_{0.45}$ metallized gates on GaAs substrates.[18-19]

We have rf-diode sputter deposited WSi_x ohmic contacts onto n-InP substrates[49] and subsequently annealed them by means of low pressure RTA under either N_2 or TBP ambient. Figure 5 shows SEM micrographs of WSi_x features on InP after RTA at 600°C for 10 sec in N_2 ambient (Fig. 5a) and at 700°C for 30 sec at TBP ambient (Fig. 5b). As was discussed earlier, severe

InP surface damage was observed by eye for annealing under N_2 ambient at 600°C for 10 sec, reflected in a macropitting and severe degradation of the surface. A controlled protection of the surface, however, was obtained under TBP at pressures as low as 1×10^{-2} Torr. No pitting was observed in the InP around the metallization contact as a result of RTA under TBP ambient at 700°C for 30 sec, and the InP surface preserves its excellent morphology.

In addition, some interaction occurred at the metal contact edges adjacent to the semiconductor, while annealing under a N_2 ambient (see Fig. 5a). This type of reaction was not observed while annealing under a similar condition with TBP ambient, and thus metallurgical analysis was performed in order to inspect the difference in the metal film quality and reactions which were induced by the change in the ambient during the RTA.

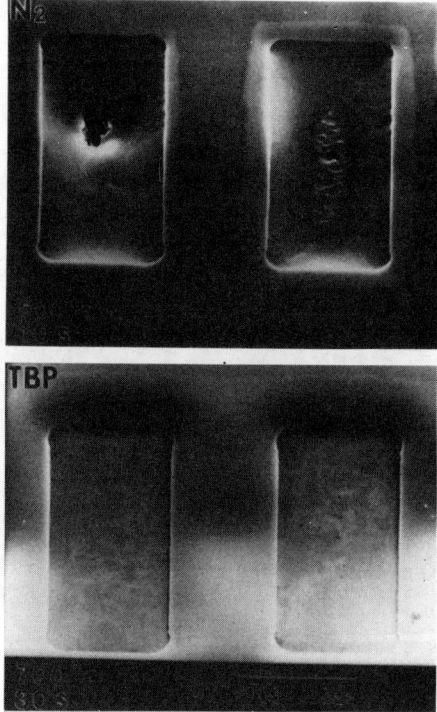

Fig. 5   SEM micrographs of WSi_x contacts on InP after the sample had been annealed at low pressure at (a) 600°C for 10 sec under N_2 ambient, and (b) 700°C for 30 sec under TBP ambient.

Figure 6 shows the sheet resistance of the WSi_x film as a function of the RTA temperature, wafers annealed under either N_2 or TBP ambients. The correlation between the sheet resistance and the film morphology and microstructure evolution in wafers annealed at different temperatures was given elsewhere,[49] However, a negligible influence of the RTP ambient on these characteristics is worth noting. Similar sheet resistance data measured at samples that were annealed for 10 sec under either N_2 or TBP, suggest that in both cases the ambient was inert and did not drive and surface or bulk reactions in the films. One can see, however, that by extending the annealing duration at 650°C to 60 sec under TBP ambient, a decrease of about 20 present in the sheet resistance is achieved. This may result from a metal-semiconductor interaction which occurred under these conditions.

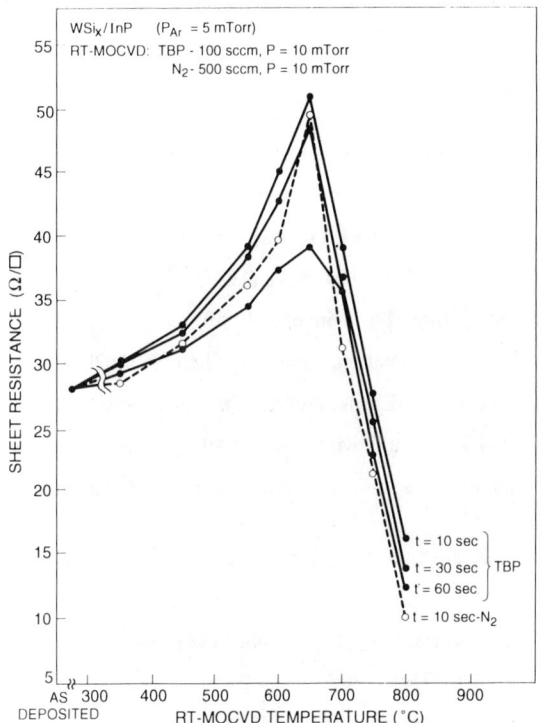

Fig. 6 Sheet resistance of WSi$_x$ contacts on InP after low pressure RTA under N$_2$ and TBP ambients, as function of the annealing temperature.

IV. SUMMARY

The use of the organometallic TBP and TBA as precursors for P and As, within a RT-LPMOCVD reactor, enables high temperature annealing of InP and GaAs with no discernible surface degradation. The TBP and TBA provide the necessary partial pressure for the group V without the need to use the hazardous PH$_3$ or AsH$_3$ gases. This reduces appreciably the risk associated with the process, and indeed enables a much more efficient process due to the higher decomposition of these organometallic sources. In addition to the protective role of TBP during annealing InP, it was found to be inert to WSi$_x$ used as ohmic contacts for self-aligned InP-based devices. This allows compatibility between the semiconductor annealing under a protective ambient and the contact sintering process.

ACKNOWLEDGEMENTS

The authors acknowledge the support and valuable discussions with W. C. Dautremont-Smith, and the AES analysis by E. Lane.

REFERENCES

1. T. N. Jackson, G. Pepper, and J. F. DeGelormo, IEDM Proc. 600, 87 (1987).
2. A. Katz and S. J. Pearton, J. Vac. Sc. Technol. **B8**, 1285 (1990).

3. M. Nishitsuji and F. Hasegawa, Jpn. J. Appl. Phys. **28**, L895 (1989).
4. H. Kanber, R. J. Cipolli, W. B. Henderson, and J. M. Whelan, J. Appl. Phys. **57**, 4732 (1985).
5. C. H. Kang, K. Kondo, J. Lagowski, and H. C. Gatos, J. Electrochem. Soc. **134**, 1261 (1987).
6. A. Katz, P. M. Thomas, S. N. G. Chu, J. W. Lee, and W. C. Dautremont-Smith, J. Appl. Phys. **66**, 2056 (1989).
7. A. Lahav, R. L. Lapinsky, and T. C. Henry, J. Electrochem. Soc. **136**, 1096 (1989).
8. S. G. Liu and S. Y. Narayan, J. Electron. Mater. B, 897 (1984).
9. G. J. Valco and V. J. Kapoor, J. Electrochem. Soc. **134**, 569 (1987).
10. S. J. Pearton and K. D. Cummings, J. Appl. Phys. **58**, 1500 (1985).
11. W. H. Haydl, IEEE Electron. Dev. Lett. **EDL-5**, 78 (1984).
12. S. J. Pearton and R. Caruso, J. Appl. Phys. **66**, 669 (1989).
13. A. Katz, C. R. Abernathy, and S. J. Pearton, Appl. Phys. Lett. **56**, 1028 (1990).
14. R. N. Legge and W. M. Paulson, SPIE Adv. Process. Character. Semicond. III **623**, 163 (1986).
15. M. J. Goff, S. C. Wang, and T.-H. Yu, J. Mater. Res. **3**, 911 (1988).
16. S. J. Pearton, K. D. Cummings, and G. P. Vella-Coleiro, J. Electrochem. Soc. **132**, 2747 (1985).
17. A. Tamura, T. Venoyama, K. Nishi, K. Inoue, and T. Ohuma, J. Appl. Phys. **62**, 1102 (1987).
18. S. J. Pearton, A. Katz and M. Geva, SPIE Proc. 1393, (1990).
19. S. J. Pearton, A. Katz and M. Geva, J. Appl. Phys. **68**, 2482 (1990).
20. B. Molnar, Appl. Phys. Lett. **36**, 927 (1980).
21. W. Wesch and G. Gotz, Phys. Status Solidi **A94**, 745 (1986).
22. S. J. Pearton, J. M. Gibson, D. C. Jacobson, J. M. Poate, J. S. Williams, and D. O. Boerma, Proc. Mater. Res. Soc. Symp. **52**, 198 (1986).
23. A. Katz and W. C. Dautremont-Smith, J. Appl. Phys. **67**, 6237 (1990).
24. A. Katz and Y. Komem, J. Appl. Phys. **63**, 5526 (1988).
25. A. Katz, M. Albin and Y. Komem, J. Vac. Sci. Technol. **B7**, 130 (1989).
26. T. E. Haynes, W. K. Chu, T. L. Aselage, and S. T. Picraux, J. Appl. Phys. **63**, 1168 (1988).
27. C. A. Armiento, L. L. Lehman, F. C. Prince, and S. Zemon, J. Electrochem. Soc. Solid-State Science and Technol. **8**, 2012 (1987).
28. T. Jackson and J. DeGelormo, J. Vac. Sci. Technol. **B3**, 1676 (1985).
29. H. Baratte, T. N. Jackson, P. M. Solomon, D. C. LaTulipe, D. J. Frank, J. S. Moore, Appl. Phys. Lett. **51**, 18 (1987).
30. V. S. Ban and M. Ettenberg, J. Phys. Chem. Solids **34**, 1119 (1973).
31. Cypure™ TBA and TBP source available in electronic grade purity by Electronic Chemicals Department, American Cyanamid Company, Wayne, NJ USA.
32. C. H. Chen, C. A. Larsen, G. B. Stringfellow, D. W. Brown, and A. J. Robertson, J. Crystal Growth **77**, 11 (1986).
33. S. R. Kurtz, J. M. Olsen, and A. Kibbler, private communication (1988).
34. R. R. Saxena, J. E. Foupuet, V. M. Sardi, and R. L. Moon, Appl. Phys. Lett. **53**, 304 (1988).

35. R. Solanki, U. Sudarsan, and J. C. Johnson, Appl. Phys. Lett. **52**, 919 (1988).
36. C. H. Chen, D. S. Cao, G. B. Stringfellow, J. Elect. Mater. **17**, 67 (1988).
37. D. W. Freeman, private communication.
38. G. B. Stringfellow, J. Elect. Mater. **17**, 327 (1988).
39. C. H. Chen, C. A. Larsen, and G. B. Stringfellow, Appl. Phys. Lett. **50**, 218 (1987).
40. R. M. Lum, J. K. Klingert, and M. G. Lamont, Appl. Phys. Lett. **50**, 221 (1987).
41. T. Omstead, P. Lee, P. Van Sickle, and K. F. Jensen, J. Crystal Growth **93**, 20 (1988).
42. P. Lee, T. Omstead, D. McKenna, K. Jensen, J. Crystal Growth, **93**, 134 (1988).
43. C. A. Larsen, N. I. Buchan, S. H. Li, and G. B. Stringfellow, J. Crystal Growth **93**, 15 (1988).
44. S. P. Watkins, G. Haacke, H. Burkhard, M. L. W. Thewalt, S. Charbonneau, J. Appl. Phys. **64**, 3205 (1988).
45. A. Katz, S. J. Pearton and M. Geva, J. Appl. Phys. **68**, 3110 (1990).
46. A. Katz, A. Feingold, S. J. Pearton, U. K. Chakrabarti and K. M. Lee, J. Appl. Phys., to be published (issue of July 1, 1991).
47. A. Katz, unpublished.
48. C. Constantine, D. Johnson, S. J. Pearton, U. K. Chakrabarti, A. B. Emerson, W. S. Hobson, and A. P. Kinsella, J. Vac. Sci. Technol. **B8**, 596 (1990).
49. A. Katz, A. Feingold, S. Nakahara, M. Geva, S. Pearton, and E. Lane, J. Appl. Phys., to be published.

THE PROPERTIES OF INDIVIDUAL Si/SiO$_2$ DEFECTS AND THEIR LINK TO 1/F NOISE

Michael J. Uren
Defence Research Agency, RSRE Malvern, Worcestershire WR14 3PS, England

ABSTRACT

Fluctuations in the occupancy of individual SiO$_2$ defects cause telegraph noise in submicron MOSFETs. The study of this noise has allowed the measurement of defect properties such as cross-section, activation energy for capture and entropy change. It is shown that the noise in large MOSFETs is the result of the superposition of many telegraph signals with the 1/f spectral shape resulting naturally from the distribution of these defect properties. The noise is demonstrated to be due to 'slow' interface states rather than the 'fast' states measured by conventional CV methods.

1. Introduction

This article will consider the origins of 1/f noise in one particularly useful device - the silicon MOSFET. This topic is of importance since these devices, which are usually thought of as digital switches, are increasingly satisfying analogue applications.

The microscopic mechanism which generates 1/f noise has been the subject of intense controversy but without a universally held consensus emerging[1-3]. However, in the case of the MOSFET, the mechanism and origin of the noise has become clear following on from the observation of discrete switching in device conductance due to individual defects in submicron MOSFETs[4,5]. These random telegraph signals (RTSs) arise because the device is sufficiently small that there is only a single defect active in the device. The observation and properties of RTSs in MOSFETs, MIS tunnel diodes and many other systems have been recently reviewed[6-8]. Here I will attempt to summarise our knowledge of the defects' properties and highlight the major unresolved questions. Section 3 will show how our knowledge of the microscopic details of the defects is beginning to allow a more accurate description of the 1/f noise in large devices. I will also show how the defects responsible for the noise are distinct from the defects normally measured by conventional CV techniques on MOS capacitors.

2. RTS Results

2.1 Defect Location

The first observations of RTSs in MOSFETs were made by Ralls et al[4], and in their original paper they observed and described most of the key features of the phenomenon. RTSs had been observed previously in many other systems in the form of burst noise[9] but little progress had been made in their understanding. What made the difference was that the defects in Ralls et al's MOSFET lay in a region of the device that was easily modelled allowing the measurement of their electrical characteristics to be performed for the first time. As an example, figure 1 shows the gate voltage dependence of the change in channel conductivity of an n channel MOSFET measured at room temperature. In this case, the free carriers in the inversion layer are electrons, but similar observations are made on p channel (hole) devices[10-12] As the gate voltage and hence electron concentration was increased, the time in the high conductivity state fell whereas the low state was roughly constant (or increased slightly). This behaviour is simply explained assuming that the

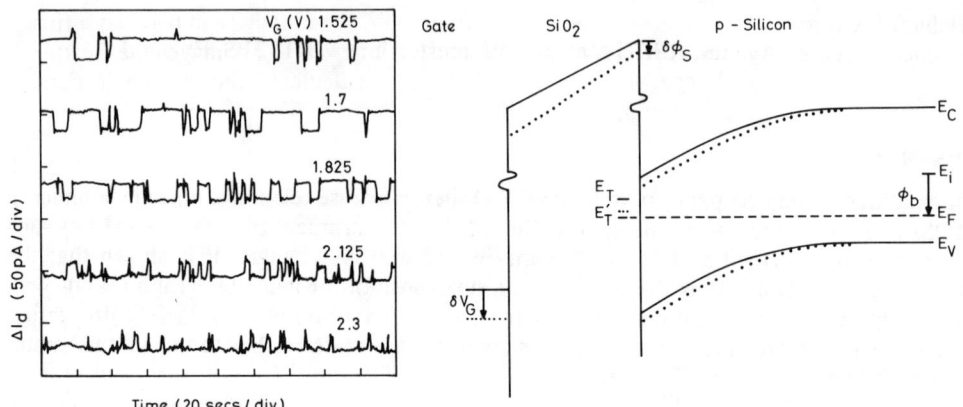

Figure 1. RTSs in 0.5μm×0.75μm MOSFET at the indicated gate voltages. $V_D=4mV$, $T=293K$ (Reference 5).

Figure 2. Band bending in a n channel silicon MOSFET for two values of gate voltage (Reference 6).

defect is located near the Si/SiO$_2$ interface with occupancy level E_T as in figure 2. Increasing the gate voltage changes the band bending and increases the defect occupancy. Hence the capture time of an electron corresponds to the time in the up state in figure 1 and involves a fall in conductivity on capture in this case. (This assignment was elegantly proven via a low temperature experiment where the electron temperature was raised above that of the lattice using the source-drain field[13].) The capture and emission times are found to be exponentially distributed and so the probability of capture is independent of time and the process can be fully described by the mean times τ_C, τ_E[6,14]. Since this is an *equilibrium* measurement, detailed balance requires that:

$$\tau_C/\tau_E = g \exp[(E_T-E_F)/kT] \qquad (1)$$

where g is the electronic degeneracy change on capture. By examining the rate of change of the measured E_T with gate voltage, an estimate can be made of the distance of the trap from the inversion layer: since the deeper the defect is into the oxide, the faster E_T will change with V_G[4,6]. It is found that the traps are located in the oxide typically within 3nm of the Si/SiO$_2$ interface (with some provisos[6]). This gate voltage dependence has also been used to demonstrate that only single electron capture is involved for the simple two level fluctuator (TLF) normally observed[6]. The other main evidence that these traps are oxide defects is that TLFs are seen in the MOS tunnel diodes where the conductivity modulation is weighted towards traps at the middle of the thin tunnel oxide[8]. If the traps were located at the interface or in the silicon, no signal would be observed.

Restle[11] has used the drain voltage dependence of E_T to determine the location of the defect between source and drain. He showed that they are located in the channel and are not associated with the source or drain contacts. Finally, noise measurements as a function of device area have shown that the defects are randomly distributed in the channel and are not associated with the edges of the device[6]. Thus it appears that the defects are characteristic of the channel thermal oxide and we can use simple models of the MOSFET to determine E_F and inversion layer number density n.

2.2 Conductivity Modulation

The first major feature that characterises an RTS is the size of the conductivity step. The RTS gives us the unique ability to study the microscopic effect of changing the charge state of a single defect. In principle, it also gives a microscopic probe into the electron transport in the silicon.

The effect of trapping an electron will be to influence the conductivity in two ways: first, it will lower the local electron concentration and second, it will change the scattering producing a mobility change. Many groups have now calculated the change in number density which leads to an expression of the form[4-6,15]

$$\frac{\Delta I_D}{I_D} = \frac{R}{An} \quad ; \quad R = \frac{\partial n}{\partial N_T} = \frac{\beta qn}{C_{OX} + C_D + C_{IT} + \beta qn} \qquad (2)$$

where $\beta = kT/q$, A is the channel area, n is the inversion layer areal number density, C_{OX}, C_{IT} and C_D are the oxide, interface state and depletion layer capacitances per unit area and N_T is the areal density of active RTS generating traps. R is a measure of the efficiency of coupling of the trap to the inversion layer[16]. This expression for $\Delta I_D/I_D$ has two limits: At high inversion layer concentrations, the trap is fully screened by the inversion layer, so 1 electron is removed from the channel on capture so $\Delta I_D/I_D = 1/(An)$ and $R = 1$. At low concentrations (roughly corresponding to below the threshold voltage), there are insufficient carriers to screen, so the image charge resides on the gate. This means that there is a constant area of the channel affected by the Coulomb field of the defect and hence $\Delta I_D/I_D$ = constant.

This transition from a gate voltage independent amplitude at low n to a 1/n variation at high gate voltage is seen, but superimposed on this all workers observe a wide variation in amplitudes. Figure 3 shows at low concentrations, the amplitude could be nearly a factor of 10 above or below the prediction of equation 2 whereas at high concentration, the scatter is much smaller (figure 3 is a corrected version of figure 15 in reference 6 in which part of the data had a scaling error). At room temperature, the conductivity is always observed to fall on capture indicating that the median effect is described by the number fluctuation expression of equation 2 - the problem is to explain the variation.

Hung et al[17] have considered the effect of the change in scattering on capture in a phenomenological way that has proved useful for the modelling of noise. They noted that the current change will have a correlated mobility change, as well as the number density change already discussed, that is given by

$$\frac{\Delta I_D}{I_D} = \frac{1}{A}\left[\frac{1}{n}\frac{1}{\delta N_T} + \frac{1}{\mu}\frac{\delta \mu}{\delta N_T}\right] \qquad (3)$$

Using Mattheisen's rule they separated the mobility into two parts $1/\mu = 1/\mu_0 + \alpha N_T$ where αN_T is the part limited by oxide charge scattering where α is expected to be a function of gate voltage. Hence:

$$\frac{\Delta I_D}{I_D} = \left[\frac{R}{nA} + \frac{\alpha \mu}{A}\right] \qquad (4)$$

They find values of α that are both positive and negative, corresponding to the observation of amplitudes lying both above and below the value predicted by (2), and have suggested that they can be explained by the presence of both acceptors and donors. However, at room temperature, all traps that have been observed show a fall in conductivity on capture, with a distribution of amplitudes at each value of gate voltage that is roughly log normal[6]. A simple Coulomb model of donors and acceptors would result in distributions that are bimodal or symmetrical in deviation from the value given by (2) and certainly could not explain the largest amplitudes observed.

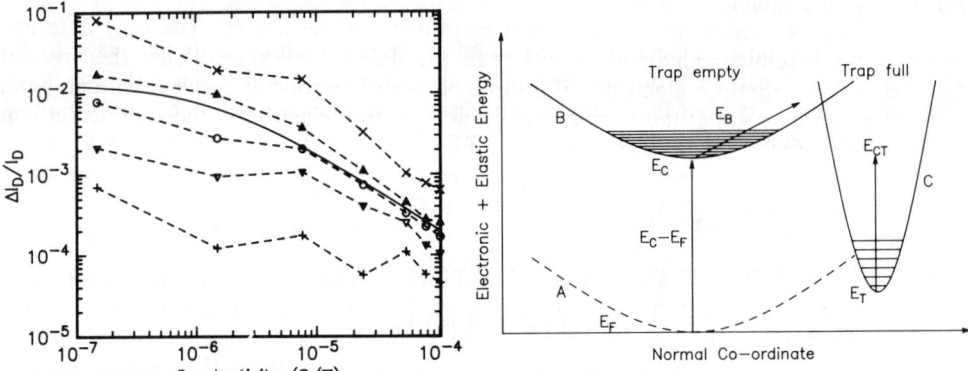

Figure 3. RTS amplitudes in several MOSFETs from the same wafer. The dashed lines correspond to the amplitude below which 0, 25, 50, 75, 100% of RTSs fell. The full line is the fit to (2).

Figure 4. Schematic configuration co-ordinate diagram for a RTS generating defect.

The study of complex RTSs (section 2.3) has provided a pointer to a possible explanation. Some multilevel RTSs exhibit behaviour that is most simply explained by a change in conductivity arising without a change in charge state[20]. The suggestion was made that the distribution in amplitudes as well as the behaviour of these complex defects arose as a result of a local interference mechanism from anisotropic scattering centres[21]. In metals, such a mechanism can cause large scattering cross-sections. At present, a detailed microscopic model to explain the size of the scattering term or its variation from trap to trap does not exist.

At temperatures of 100K or less, a few traps show the reverse sign of the change in conductivity on capture ie the conductivity increases on capture[4,22]. This makes it clear that the scattering is dominant in this regime. At 4.2K, the effect of capturing an electron is to change the quantum interference that occurs over a length of order the inelastic scattering length. This results in an uncertainty in the size of the conductivity change as a result of universal conductance fluctuations, that can be directly observed in multiprobe MOSFETs[23].

2.3 Capture Kinetics

Unlike any other technique, the capture and emission times can be determined simultaneously and with the trap in true equilibrium. This has allowed detailed measurements of the capture kinetics to be made. The original liquid helium measurements of Ralls et al showed that both the capture and emission times were activated. This observation was repeated for defects observable near room temperature[24]. Since the traps were located at distances of only about 2nm from the surface and the activation energies were in the meV range, Ralls et al argued that the capture activation energy could not be due to thermionic emission over a barrier since tunnelling through the barrier should dominate. They suggested that the activation energy is associated with lattice relaxation. (This view is not shared by Karmann and Schulz[25] who argue that there may be special properties of the final state that prevent this tunnel process.) A schematic configuration coordinate diagram for the defect analogous to those developed for deep levels in semiconductors[26] is shown in figure 4, where the electronic free energy + defect elastic energy is plotted against a normal coordinate of the defect. The only difference from the bulk semiconductor diagram is that the defect is displaced from the inversion layer containing the free electrons. Capture consists of starting at curve A with an empty defect and an electron at the Fermi level. Placing a free electron in the conduction band takes us to curve B. At the crossover point of curves B and C, after supplying E_B in lattice energy, the electron can make an off-diagonal non-radiative transition to occupy the defect before relaxing by multi-phonon emission to state C.

The analysis of data has been based on the simplified expression[24,6]:

$$1/\tau_C = nv\sigma = nv\sigma_0 \exp[-E_B/kT] \quad (5)$$

where v is the thermal velocity and σ an activated capture cross-section. Interestingly, we have two independent measures of the cross-section since detailed balance (equation 1) allows us to use τ_E to make an estimate as well.

$$1/\tau_E = T^2 \exp[(E_B+E_{CT})/kT]/(\sigma_0 n\eta) \quad (6)$$

where E_{CT} is the free energy difference between the occupied trap level and the conduction band edge in the inversion layer and η is a constant related to the dimensions of the device and the silicon density of states. Typically these two estimates of σ_0 are found to differ by 1 to 2 orders of magnitude, or equivalently, the trap energy level is incorrectly placed sufficiently far below the Fermi level that the defect would always be full.

The answer to this problem lay in the work of Engstrom and Alm[27] who noted that in thermal experiments, it is the Gibbs free energy changes that are measured and so the trap location relative to the conduction band can be expressed as:

$$E_{CT} = H_{CT} - T\Delta S \quad (7)$$

where H_{CT} is the trap enthalpy difference between the trap level and the conduction band and ΔS is the entropy change on emission of an electon. Thus the slope of the Arrhenius plot of τ_E gives the trap enthalpy change H_{CT} and the combination of the intercepts of τ_E and τ_C give the entropy change on capture.

In measurements on n and p channel devices around room temperature[6,12], every defect was found to be different. Values of E_B normally lay in the range 0.1eV - 0.7eV and σ_0 in the range $10^{-25}m^2$ - $10^{-19}m^2$. There is obviously some correlation between these values via equation (5) since capture and emission times could only be observed in a limited measurement time window of about $10^{-4}s$ - 100s. Hence only a part of the overall distribution of defects can be accessed. These values of σ_0 and E_B are quite consistent with the model of the defects lying in the oxide close to the surface where σ_0 would represent the tunnelling process to the defect at a range of distances and E_B the distortion necessary to ensure cross-over. The wide range of E_B would result from the amorphous environment.

The kinetics related parameters in references 6 and 12 (σ_0, E_B, ΔS) were evaluated at fixed gate voltage with varying T. In practice, these parameters can vary strongly particularly near the MOSFET threshold voltage[6,28].

Perhaps the most surprising observation for the conventional TLF is the size and sign of the entropy change inferred from the temperature dependence of the trap capture and emission times[12]. This is found to be large and positive for emission of electrons to the conduction band in n channel MOSFETs (1.5 - 11.8k), but it is also large and positive for the emission of holes to the valence band in p channel MOSFETs (3.8 - 12.9k). If the same defect occupancy level were involved in both processes, then taking an electron from the valence band to the conduction band via such a trap would result in a total ΔS of 5-20k, far exceeding the electron-hole creation entropy of 2.7k. The most obvious contribution to such an entropy change comes from the change in defect degeneracy due to symmetry changes and spin giving rise to at most $\Delta S = k \log_e(4) = 1.4k$. The next contribution comes from the change in the local phonon density of states which might be expected to be positive for acceptors (donors) communicating with the conduction band (valence band). This would arise since in both cases, emission would change the defect from a charged to a neutral state thus allowing some lattice softening. The size of this effect has been argued to be approximately equal to the entropy change for creation of an electron/hole pair[29]. I am not aware of any calculations for this property in SiO_2, let alone for the complex heterostructure with which we are dealing, but for defects in bulk semiconductors values of 3-6k are found[30].

These contributions are not large enough to account for the measured values of up to 12k (even with a possible systematic error of up to 2k due to uncertainty in the measurement of the band bending). In addition, to explain the sign would require that all the oxide states observed near the conduction band are acceptors, whereas they would all have to be donors near the valence band. Since the defects are actually located within the 9eV gap of the SiO_2, it seems unlikely that these defect energy levels should fortuitously line up with the silicon bandgap, unless the defects are in some way tied to the silicon bandedges. Alternatively, there may be a single defect type and an additional source of entropy change that we have not considered associated with the transfer of an electron (or hole) between the dramatically different mediums of silicon and SiO_2.

2.4 Complex RTSs

The vast majority of signals show the simple behaviour shown in figure 1, but a small proportion show far more complex behaviour (4% in [31]). These include the signal modulated by another RTS, 3 levels, 4 levels, giant steps and others[7,8,14,15,31]. Some signals are most easily interpreted in terms of multielectron capture. It was shown using a simulation, that complex signals could not arise by the Coulombic interaction of randomly distributed defects - the observed incidence was far too high[31]. Since every signal is different, of necessity every signal has required the development of a new model. The signals have been explained in terms of Coulombic or strain mediated coupling within a cluster of defects or by defect re-configuration within a set of metastable configurations of the defects. At present, it has not been clearly demonstrated what the microscopic origin of the effects is.

3. RTS Ensemble Measurements

3.1 1/f Noise Measurements

Flicker noise in MOSFETs has been studied for decades, but despite this intensive effort there has been no universally agreed model that explains all the observations. The primary aspect that has to be explained is the observation of a noise power spectral density that has a power of 1/f over at least 10 decades in frequency. There have been two distinct approaches. The first was based on number fluctuations due to trapping in the near surface

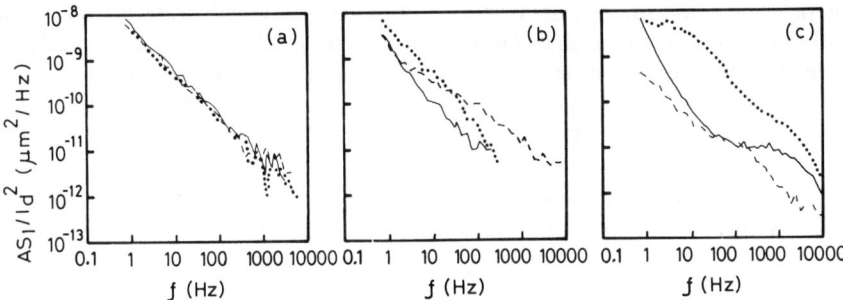

Figure 5. Noise power in n channel MOSFETs for $V_G=2V$ and $T=293K$. (a) Three devices of area $350\mu m^2$. (b) Three devices of area $15\mu m^2$. (c) Three devices of area $0.4\mu m^2$. (Reference 5).

interface states, with the range in time constants resulting from a range in tunnelling distance[32]. The second model holds that the phenomenon is associated with a fluctuation in the bulk mobility via a fluctuation in the phonon population[1].

The availability of submicron MOSFETs was finally able to provide the evidence that the 1/f noise was a defect related phenomenon. This is illustrated in figure 5 which shows how 1/f noise scales with device channel area[5]. In figure 5a, the noise power is plotted against frequency for 3 large MOSFETs and follows a 1/f law with little variation between devices. As the noise is measured in successively smaller devices, the variation from device to device becomes more extreme, culminating in figure 5c which is for $0.5\mu m \times 0.75\mu m$ devices. The noise has been reduced to the sum of 1 or 2 Lorentzian spectra in each device. Since the Lorentzian is the Fourier transform of the RTSs we have discussed in section 2, it is apparent that the noise results from the superposition of many RTSs. Hence the 1/f noise is of trap origin and is not a bulk mobility effect.

Since we have established that the RTSs result in the 1/f noise, the next question that requires examination is what particular property of the defects is it that leads to a 1/f spectrum? To obtain a 1/f spectrum requires a uniform distribution of trap time constants on a log time scale. Hence the number of traps per unit time interval with trap time constant between τ and $\tau + d\tau$ is proportional to $1/\tau$. Since the noise power from a RTS is maximum when the capture time equals the emission time, the time constant for each trap will be determined according to (5). At each value of gate voltage, the trap time constant is determined by both σ_0 and E_B, both of which show a wide distribution of values. Unfortunately, the measurement of the true distributions of σ_0 and E_B requires the detailed analysis of many RTSs and is a task that is not particularly easy to fully automate. The result is that only about a dozen traps have been measured under conditions that allow comparison[6]. Based on this small n channel room temperature dataset, it appears that E_B is fairly uniformly distributed in energy in the range 0.15eV to 0.7eV and $\log(\sigma_0)$ is also uniformly distributed. Such a distribution in either parameter will lead to a uniform distribution of τ in log time and hence 1/f nose. The combination of the two distributions[6] leads to a spectrum that is not 1/f but is in practice indistinguishable from 1/f.

Traditional number fluctuation models of 1/f noise have assumed that the time constant dispersion arises purely from the distribution in σ_0 associated with the tunnelling to traps at a range of distances ie $\sigma_0 \propto \exp(-2\kappa x)$ where κ is related to the height of the oxide barrier and is of order 10nm^{-1} for n channel MOSFETs. To obtain a 1/f spectrum requires that the density of traps in the oxide is uniform with distance on a length scale $>> 1/\kappa$. Summing over all the states, assuming that the density of states $N_T(E_F)$ varies only slowly with E_F, gives the noise power spectral density for the trap fluctuation in occupancy[18]

$$S_N = \frac{N_T(E_F)kTA}{2\kappa f} \qquad (8)$$

As we have seen, the real distribution of time constants is the result of the convolution of two distributions. Hence any estimate of $N_T(E_F)$ based on (8) will be an underestimate of the true value of $N_T(E_F)$.

The next problem is to relate the fluctuation in trap occupation to the noise measured in the current flowing through the MOSFET. Using the measured distributions of amplitude (as well as the measured distributions of σ_0 and E_B), the noise was simulated and gave good agreement with the actual noise measured in a large device[6]. However, although this gives confidence that the approach is valid, it is not particularly useful for interpreting noise measurements where an analytic approach is desired. In the number fluctuation model, equation (8) was related to the current change in the device assuming all the RTSs have the same amplitude at each value of n given by (2). This approach was very successful in predicting and explaining the fact that the noise power in the device current was constant at low n and fell at high concentrations[16]. However, it failed to describe the behaviour accurately above threshold voltage[33].

Recently, Hung et al[18,19] have developed a 'unified model' based on (4) that improves the modelling of the amplitude change by incorporating the mobility in an empirical way. Hence the noise power in the drain current (for low drain bias) is given by

$$S_I = \left[\frac{I_D}{An} (R+\alpha\mu n) \right]^2 S_N \qquad (9)$$

They find that (9) can successfully explain many of the observations that were used to support the mobility fluctuation model even with the approximation that α=constant. From a microscopic point of view, the noise measured in the device results from the sum of all the RTSs, but is weighted towards the largest amplitude signals due to the rough log-normal distribution of amplitudes. Presumably, the extra degree of freedom provided by the parameter $\alpha\mu n$ gives a reasonable description of how the weighted distribution in trap amplitudes changes with gate voltage.

If $\alpha(n)$ were known accurately, then it would be possible to determine the density of 1/f noise generating traps from measurements of the noise as a function of gate voltage. On the assumption that there is not an enormous variation in α and that the size of the mobility related term is not enormous, then an effective trap density can be extracted from the data - essentially using the conventional number density fluctuation model. Based on the measurement of RTS amplitudes shown in figure 3, where the median amplitude follows the number fluctuation prediction, this approach should give the density of states to better than an order of magnitude. It has been found that the inferred density of states varies between device manufacturing processes, often rising rapidly towards the conduction band edge[16,18,19]. This rise can be more than an order of magnitude in size and hence is almost certainly real.

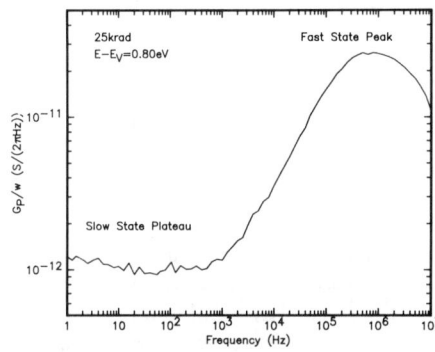

Figure 6. Equivalent silicon parallel conductance for unbiased electron beam irradiated n type MOS capacitor. $A = 6.97 \times 10^{-7} m^2$.

3.2 Relation to Conventional Interface State Measurements

Up to this point, we have discussed the RTS generating traps as if they are conventional Si/SiO$_2$ interface states. Over the years, there has been considerable confusion surrounding this question. At least as far back as 1972, Fu and Sah made a clear distinction between 'slow' states that reside in the oxide and communicate with the silicon by tunnelling and 'fast' states that reside at the interface[34]. But many others have not made this distinction and have compared the densities of interface states inferred from capacitance-voltage or charge pumping techniques with those obtained from 1/f noise. Let us consider the evidence from the CV approach.

Most of the characterisation of interface states has been carried out using capacitors via the high-low frequency CV technique which measures the total number of states through a measurement of the interface state capacitance. More detailed information is provided by the conductance technique which measures the loss (AC conductance) in the capacitor as a function of frequency[35]. For a discrete defect level located at the Fermi level, the silicon conductance (compensating for the oxide capacitance) is given by

$$G_P/\omega = C_T \omega \tau [1 + (\omega \tau)^2]^{-1} \qquad (10)$$

where $1/\tau = \sigma v n$ and C_T is the interface trap capacitance. This expression is similar to the Lorentzian that describes the noise from an RTS with $\tau_C = \tau_E$. As with the noise measurements, a full conductance curve is formed by summing over a distribution of trap energy levels and cross-sections.

Until recently, the conductance data were always found to show states distributed throughout the bandgap but with a single peak as a function of frequency. This implied that all the interface states have a single characteristic capture cross-section. This result is quite incompatible with these states being those responsible for the 1/f noise since they have a wide range of cross-sections. More recent conductance measurements[36] made measurements over a far wider frequency range and an example is shown in figure 6. At high frequency, a peak is found with a single broadened cross-section of around $10^{-19} m^2$ characteristic of the true 'fast' Si/SiO$_2$ interface states. On the low frequency side of the peak are the 'slow' states which had not been previously noticed (similar results have been found from DLTS[37]). The constant conductance as a function of log frequency corresponds

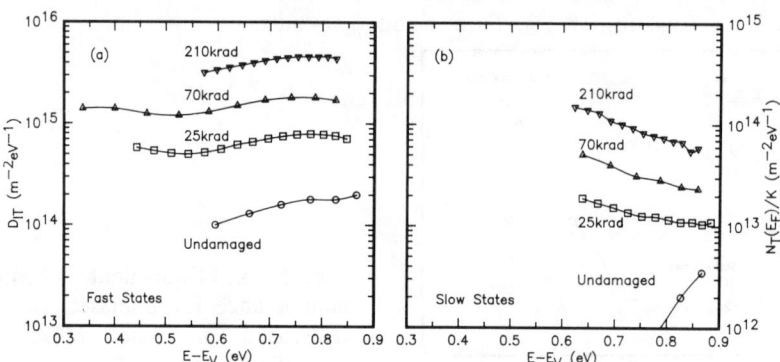

Figure 7. Defect density of states for n type MOS capacitors subjected to the indicated electron beam irradiation with $V_G=0$. (a) 'Fast' state density. (b) 'Slow' state areal density.

to exactly the distribution that results in 1/f noise. The density of 'slow' states $N_T(E_F)$ inferred from the plateau height is also found to be quite consistent with the values of $N_T(E_F)$ recovered from noise measurements[6].

CV measurements principally measure only the fast states, so any correlation between the fast state density D_{IT} and the slow state density is purely coincidental. Figure 7 gives the result of a conductance measurement of both fast and slow state densities before and after radiation damage to a set of capacitors. It is apparent that for both classes of defect the increase in defect density is proportional to the dose, but the energetic distribution is quite different.

4. Conclusions

The individual defects in MOSFETs responsible for the RTSs are located in the oxide in the near interfacial oxide region with each one showing a different set of electrical properties. They all show lattice relaxation and a large entropy change on capture and have a range of capture cross-sections less than or equal to the cross-section of the 'fast' interface states. The size of the entropy change is too large to be accounted for by lattice softening and some other contribution must be present. The other main puzzle is the distribution in size of the conductivity change which cannot be easily explained by simple Coulomb scattering. The actual chemical nature of the defects is unknown, but a clue has been offered by the observation that the pre-irradiation 1/f noise is correlated with post-irradiation fixed oxide charge[38]. Since it seems fairly clear that this positive charge is associated with the E' centres, the 1/f noise and hence the RTSs may be related to the E' precursor - normally taken to be an oxygen vacancy.

The 1/f noise has been clearly shown to be due to the superposition of noise from the 'slow' interface states responsible for RTSs. The understanding of the defect properties that have emerged from the study of RTSs are now beginning to be used to develop more accurate and realistic models of the 1/f noise in MOSFETs. However, a key conclusion of this work is that the form of the measured 1/f spectrum is very insensitive to the details of the distribution of trap properties. Hence the use of noise measurements to carry out the reverse procedure and infer trap properties is fraught with danger.

This article would not have been possible without my collaborators M. J. Kirton, S. Collins and D. H. Cobden. It is a pleasure to acknowledge useful discussions with M. Andersson, O. Engstrom, K. R. Farmer and M. Schulz.

REFERENCES

1. F. N. Hooge, T. G. M. Kleinpenning and L. K. J. Vandamme, Rep. Prog. Phys., 44 479 (1981).
2. M. B. Weissman, Rev. Mod. Phys., 60 537 (1988).
3. C. M. Van Vliet, Sol. State. Elec., 34 1 (1991).
4. K. S. Ralls et al, Phys. Rev. Lett., 52 228 (1984).
5. M. J. Uren, D. J. Day and M. J. Kirton, Appl. Phys. Lett., 47 1195 (1985).
6. M. J. Kirton and M. J. Uren, Adv. in Phys., 38 367 (1989).
7. M. J. Kirton et al, Semicond. Sci. Technol., 4 1116 (1989).
8. K. R. Farmer in *Proc. of INFOS*, eds. W. Eccleston and M. J. Uren, (Adam Hilger, Bristol, UK, 1991).
9. M. J. Buckingham, *Noise in Electronic Devices and Systems*, (Ellis Horwood, Chichester, UK, 1983).
10. A. Karwath and M. Schulz, Appl. Phys. Lett., 52 634 (1988).
11. P. J. Restle, Appl. Phys. Lett., 53 1862 (1988).
12. D. H. Cobden, M. J. Uren and M. J. Kirton, Appl. Phys. Lett., 56 1245 (1990).
13. L. D. Jackel et al in *Proc. of the 17th Int. Conf. on the Physics of Semiconductors*, ed. D. J. Chadi and W. A. Harrison (Springer, New York, USA, 1985).
14. A. Ohata et al, J. Appl. Phys., 68 200 (1990).
15. H. Nakamura et al, Jap. J. Appl. Phys., 28 L2057 (1989).
16. G. Reimbold, IEEE Trans. Elec. Dev., ED31 1190 (1984).
17. K. K. Hung et al, IEEE Elec. Dev. Lett., 11 90 (1990).
18. K. K. Hung et al, IEEE Trans. Elec. Dev., ED37 654 (1990).
19. K. K. Hung et al, IEEE Trans. Elec. Dev., ED37 1323 (1990).
20. M. J. Uren and M. J. Kirton, Appl. Surf. Sci., 39 479 (1989).
21. J. Pelz and J. Clarke, Phys. Rev., B36 4479 (1987).
22. M. Schulz and A. Karmann, Appl. Phys. A, 52 104 (1991).
23. W. Skocpol in *Physics and Fabrication of Microstructures*, ed. M. Kelly and Weissbuch (Springer Verlag, 1986).
24. M. J. Kirton and M. J. Uren, Appl. Phys. Lett., 48 1270 (1986).
25. A. Karmann and M. Schulz, Appl. Surf. Sci., 39 500 (1989).
26. M. Jaros, *Deep Levels in Semiconductors*, (Adam Hilger, Bristol, UK, 1982).
27. O. Engstrom and A. Alm, J. Appl. Phys., 54 5240 (1983).
28. M. J. Kirton, M. J. Uren and S. Collins in *The Physics and Chemistry of SiO_2 and the Si-SiO_2 Interface*, ed. C. R. Helms and B. E. Deal (Plenum, New York, USA, 1988).
29. J. A. Van Vechten and C. D. Thurmond, Phys. Rev., B14 3539 (1976).
30. H. M. Branz and R. S. Crandall, Appl. Phys. Lett., 55 2634 (1989).
31. M. J. Uren, M. J. Kirton and S. Collins, Phys. Rev., B37 8346 (1988).
32. A. L. McWhorter in *Semiconductor Surface Physics*, ed. R. H. Kingston (Univ. of Pennsylvania, Philadelpia, USA, 1957).
33. H. Mikoshiba, IEEE Trans. Elec. Dev., ED29 965 (1982).
34. H-S. Fu and C-T. Sah, IEEE Trans. Elec. Dev., ED19 273 (1972).
35. E. H. Nicollian and J. R. Brews, *MOS Physics and Technology*, (Wiley, New York,

36. M. J. Uren, S. Collins and M. J. Kirton, Appl. Phys. Lett., 54 1448 (1989).
37. H. Lakhdari, D. Vuillaume and J. C. Bourgoin, Phys. Rev. B38 13124 (1988).
38. D. M. Fleetwood and J. H. Scofield, Phys. Rev. Lett., 64 579 (1990).

© British Crown Copyright 1991/MOD
Published with the permission of the Controller of Her Britannic Majesty's Stationery Office

HYDROGEN INDUCED DEFECTS AND DEFECT PASSIVATION IN SILICON SOLAR CELLS

BHUSHAN SOPORI
SERI, 1617 COLE BOULEVARD, GOLDEN, COLORADO 80401

ABSTRACT

Low-energy hydrogen implantation is typically used for passivation of defects and impurities in silicon solar cells. Here we show that this technique of defect/impurity passivation also induces some defects which act as sites of high minority carrier recombination. We describe the nature of such defects and purpose this to be a mechanism that limits the effectiveness of hydrogen for improving solar cell performance. The deleterious effects of hydrogen defects can be particularly important if implantation is done from the junction side of the cell. We have developed a technique for passivating silicon solar cells in which hydrogen is introduced from the back side of the cell. In contrast to the conventional approach of front side hydrogenation, the defects are kept away from the junction side thus minimizing their influence on the cell performance. This technique requires the hydrogen to diffuse through the entire substrate in order to reach the front junction. It is shown that an enhanced diffusion of hydrogen can occur in certain substrates. This feature is applied to achieve a deep diffusion of hydrogen resulting in a significant improvement in the cell performance.

1. Introduction:

Hydrogen in silicon is known to interact with the lattice, with the lattice defects and with the impurities [1,2]. It is determined that hydrogen can occupy interstitial sites in the silicon lattice in several possible configurations. Dopant de-activation due to hydrogen has been studied in some detail and it is believed that acceptor deactivation is due to formation of a donor-like acceptor-hydrogen complex [3,4,5]. In contrast, only a weak deactivation of donors in silicon has been observed and a clear mechanism for donor deactivation is not well understood [5]. Hydrogen interactions with deep level impurities, such as transition metals, are known to result in neutralization of their electrical activity [7]. Hydrogen is also found to enhance oxygen diffusion in silicon presumably by associating with oxygen donor complexes. It is well established that hydrogen can passivate defects such as grain boundaries and dislocations, presumably by tying up the dangling bonds associated with them [8]. Although many of the general features of hydrogen in silicon have been recognized, a detailed knowledge of interactions of hydrogen in silicon is severely lacking.

The ability of hydrogen to passivate defects and impurities is of major interest to silicon solar cells since it can permit production of higher efficiency devices on low-cost substrates. These features of hydrogen are being exploited in the laboratory to improve the efficiency of silicon solar cells. However, the commercial use of hydrogenation for impurity/defect passivation has been limited by somewhat complex behavior of hydrogen in silicon. For example, it has been observed that hydrogen passivation can improve only the low-performance cells and that long process times are needed to diffuse hydrogen deep within the bulk of the substrate. In order to understand these issues it is necessary to develop a coherent model of synergistic effects of hydrogen interactions including dopant deactivation, recombination due to hydrogen induced defects, as well as defect/impurity passivation. In this paper we will discuss the results of our studies on hydrogen-induced defects and the interaction of hydrogen with grown-in defects when hydrogen is introduced by low-energy implantation. It is shown that surface and near-surface defects can be detrimental to the performance of a solar cell. However, the effect of such defects can be minimized if

Figure 1. XTEM micrograph showing surface damage due to a 1.5 Kev implant

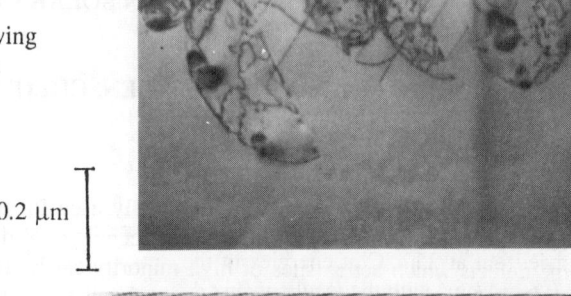

0.2 μm

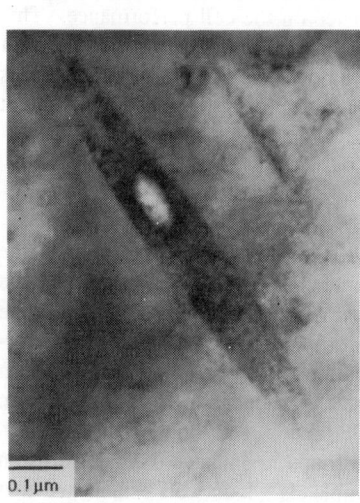

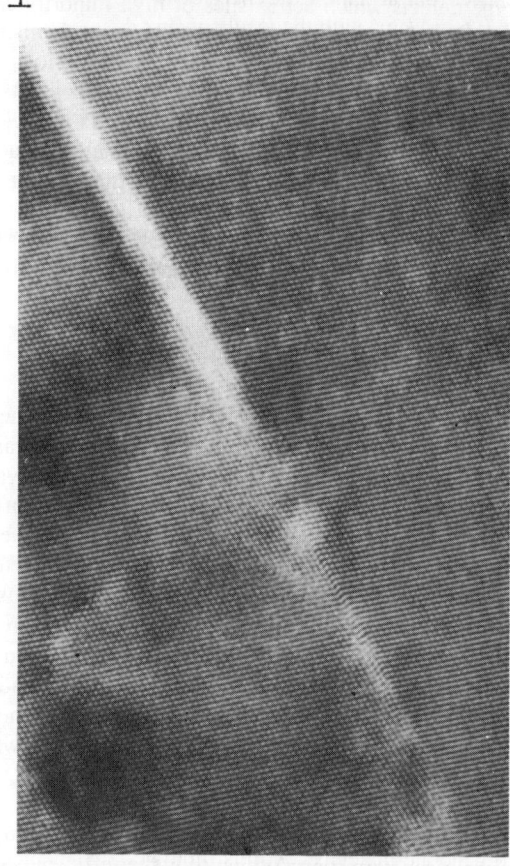

Figure 2a. [011] plan-view TEM micrograph showing a hydrogen platelet

Figure 2b. [011] TEM lattice image of the hydrogen platelet

hydrogen is introduced form the back-side of the cell instead of the junction (front) side. This requires that hydrogen must diffuse through the entire thickness of the solar cell. In our studies on low-energy ion implantation, we have observed an enhanced bulk diffusivity of hydrogen in some polycrystalline silicon substrates. This feature has been applied to develop a back-side hydrogenation technique which circumvents many problems of conventional approach of implanting hydrogen from the junction side of a solar cell.

2. Defects in Silicon Due to Hydrogen Implanted at Low Energies

In our studies hydrogen was implanted by a Kaufman Ion source. Hydrogenation was typically carried out at 250 °C at energies in the range of 0.5 to 2 Kev. The beam currents were limited to 0.6 mA/cm^2 with the resultant flux densities in the range of 5×10^{17}/cm^2. The samples consisted of unprocessed wafers as well as solar cells. The unprocessed wafers included Float Zone, Czochralski, and polycrystalline substrates from different vendors. The solar cells of two different configurations were used. In one case the entire back side of the cell was unmetallized. The second category consisted of completed solar cells with partially open back contact.

Defects due to hydrogen implanted at low energies in silicon may be divided into three categories: surface damage, defects extending into the bulk, and defects due to hydrogen interaction with the extended defects such as dislocations. The near surface damage appears as dislocation loops, stacking faults and entrapment of hydrogen. These defects are believed to be caused by the combined effect of energetic ions and high surface concentrations of hydrogen. Figure 1 shows a cross-sectional TEM (XTEM) micrograph of a sample showing typical structure of the defects due to surface damage. FTIR analyses of hydrogenated samples show absorption peaks around 2100 cm^{-1}. However, after polishing 0.5 µm from the surface i.e after removal of the damaged layer, the absorption peaks are strongly diminished. This analysis also shows that optically active hydrogen is predominantly confined to the near-surface damaged region. The depth of the surface damage clearly depends on the ion energy; the higher ion energy results in deeper damage.

The near surface region also shows a preponderance of "platelets". These defects have been postulated earlier and a limited characterization of these defects has been done. We find that these defects lie in (111) planes and are elongated along {110} directions. Figure 2a shows [001] plan view TEM image of such a platelet. It is seen that such a platelet shows a contrast identifying a core-like structure associated with the defect. Figure 2b shows a lattice image of such a structure indicating a loss of contrast associated with the defect core. We believe that the core represents entrapment of hydrogen and/or an aggregate of vacancies. The trapped hydrogen at the core could be molecular in nature. The platelets are seen to propagate deep into the bulk of the material (deeper than the surface damage described above). The tendency of the platelet formation appears to be related to the impurities in silicon. Low oxygen/low carbon materials have higher tendency to generate platelets. At this time we do not have sufficient data to determine if any orientations are more prone to produce these defects than others.

In addition to producing its own defects, hydrogen interacts with extended defects, such as dislocations, in the material. TEM analysis shows that hydrogen can segregate at dislocation sites. Qualitatively we have observed that hydrogen segregation occurs mainly at dislocation nodes and is more pronounced at "clean" dislocations. However, no segregation is observed if the hydrogen concentration is below 10^{16}/cm^3.

Although we have analyzed several characteristics of hydrogen related defects, here we will only explore those that are directly related to solar cell performance. One such important issue is to identify if the hydrogen defects have a significant effect on the performance of the cell. Our EBIC

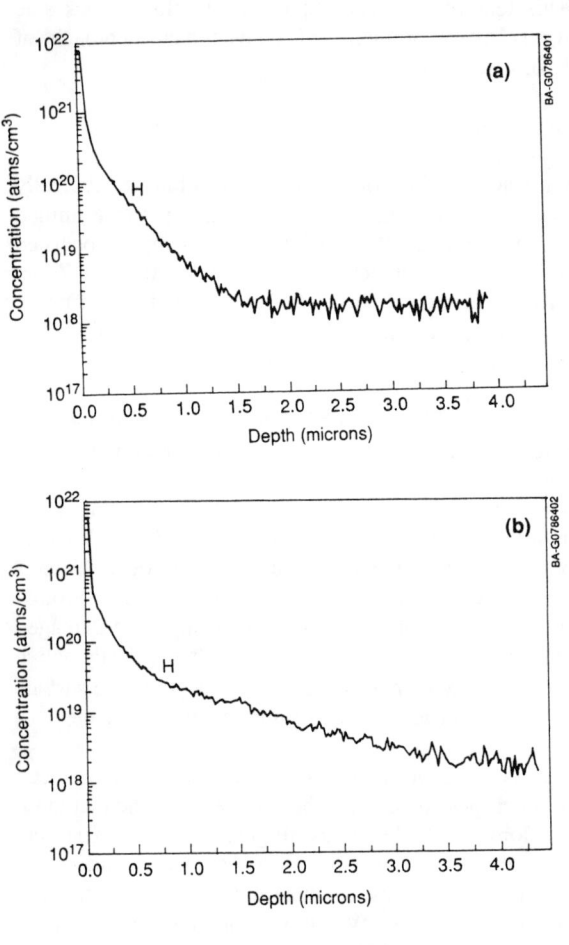

Figure 3. SIMS profile of hydrogen, implanted at 1.5 Kev :
(a) Float Zone wafer,
(b) ribbon

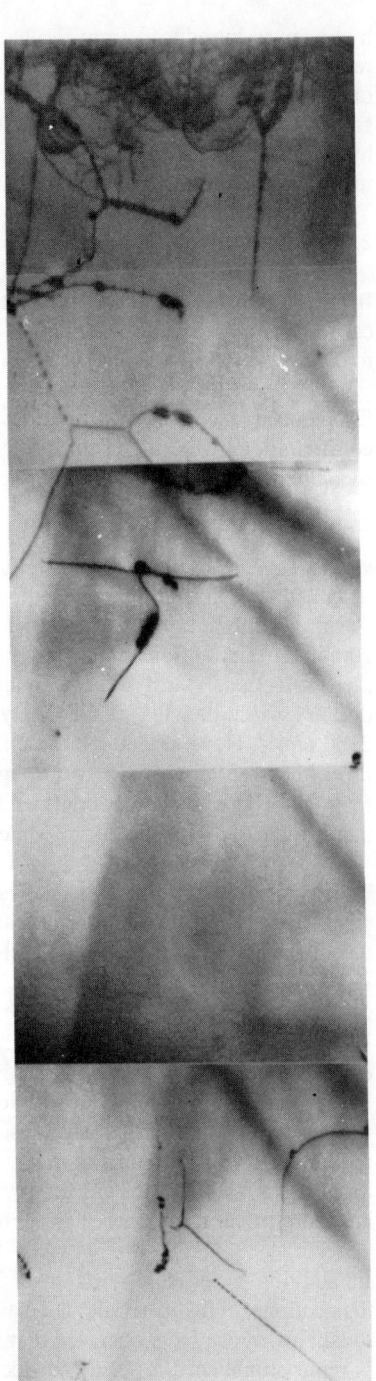

Figure 4. XTEM micrograph montage showing segregation of hydrogen at dislocation sites.

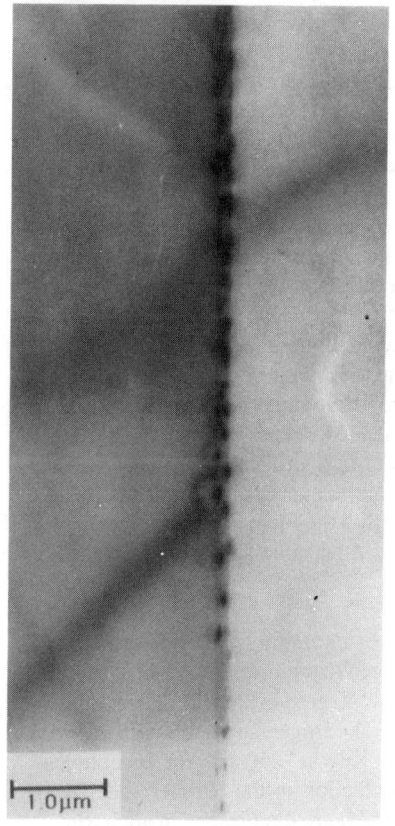

Figure 5. XTEM microgragh showing hydrogen diffusion along a grain boundary (implant conditions are same as for the sample in Fig. 4)

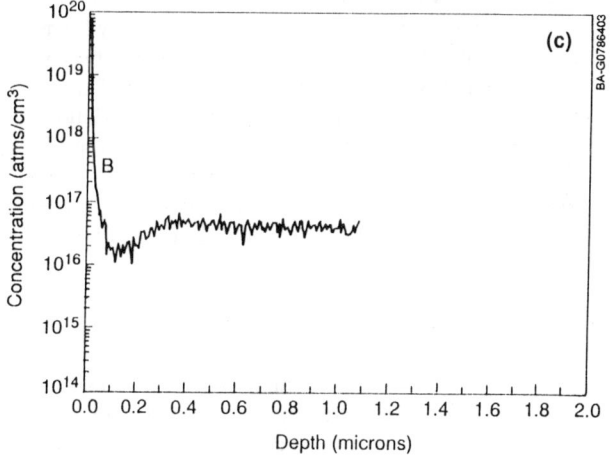

Figure 6. SIMS profile showing out-diffusion of boron

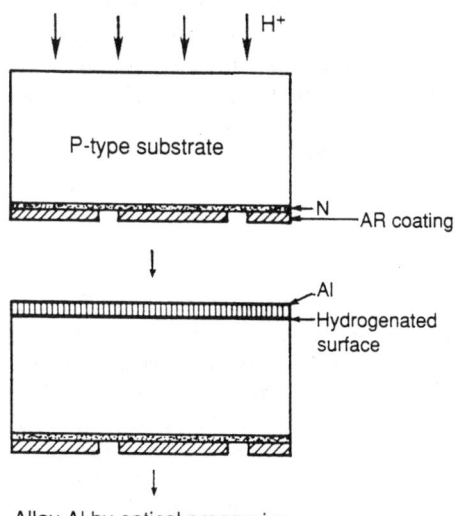

Figure 7a. Schematic of a process for back-side hydrogenation.

studies have shown that the hydrogen induced surface damage results in a high surface recombination velocity indicating that hydrogen does not passivate self-induced defects. Likewise, platelets can be imaged by EBIC contrast thus identifying them to be high carrier recombination regions. We have not been able to determine the effect of hydrogen segregation at the dislocations.

FTIR and resistivity measurements have shown that the majority of hydrogen associated with the defects stays electrically and optically inactive.

3. Hydrogenation diffusion in different silicon substrates:

We have determined the profiles of implanted hydrogen in a variety of different substrates including Float Zone, Czochralski, and polycrystalline substrates obtained from different vendors. Our results show a unique feature that the depth of hydrogen diffusion can be very large in some polycrystalline wafers as compared to that of high quality Float Zone wafers of the same resistivity. Figure 3a shows a SIMS profile of hydrogen implanted at 1.5 Kev in a single crystal Float Zone wafer; the profile is similar to that published in the literature [1]. Figure 3b is a corresponding SIMS profile of a silicon ribbon of the same resistivity and implanted in the same run. It is seen that the diffusion profile in the ribbon is extended deeper into bulk (data taken within one grain). Due to limitations in the SIMS measurement of hydrogen, we have carried out extensive analysis using deuterium implantation. These measurements also showed that deeper diffusion of deuterium occurs in the ribbon samples as compared to the Float Zone and Czochralski wafers. The depth of diffusion within large grains was found to be same as that along the grain boundaries.

The most convincing evidence of deep diffusions was obtained by exploiting hydrogen segregation at the dislocation sites as a semi-quantitative detector for hydrogen depth profile. Selected samples were examined in the cross section with TEM, to determine the degree of hydrogen segregation. Figure 4 is an XTEM micrograph of a ribbon sample showing hydrogen segregation manifested as "bubble" like structures. The sample was implanted at 1.5 Kev at 250 °C for 30 minutes. A decreasing concentration of hydrogen is manifested as a reduction in the number and size of hydrogen "bubbles". The segregation characteristics depicted in Figure 4 represent an approximate concentration of $10^{17}/cm^3$ at a depth of 20 µm. In comparison, a similar concentration of hydrogen is reached at a depth of about 4µm below the surface in Float Zone wafers of the same resistivity. We believe that this is the first observation indicating higher diffusivity of hydrogen than in the Float Zone wafers. Although the exact mechanism of this enhanced diffusion is not well understood at this time, we believe that this is similar to that which causes enhanced diffusion along some grain boundaries. It is important to recognize that unlike grain boundary diffusion, shown in Figure 5, the diffusion seen in Figure 4 is a bulk diffusion. We believe that a vacancy mechanism could be responsible for such deep diffusions. The mechanism of enhanced diffusion is clearly important for solar cell applications since solar cells are bulk devices and we are striving to diffuse hydrogen through the entire thickness of the wafer.

We have shown earlier that the damaged surface exhibits an inversion type of behavior [9]. Such a behavior was attributed to formation of donor type of levels due to the surface damage. Our recent investigations show that out-diffusion of boron can take place during low energy hydrogen implant. Figure 6 shows a SIMS profile of boron following a hydrogen implant at 1.5 Kev. Implantation was done for 30 minutes with substrate held at 250°C. It is believed that boron out-diffusion is enhanced by the surface damage. The mechanism of the enhanced out diffusion will be discussed elsewhere. However, it should be pointed out that such a modification of the dopant profile near a contact can play an important role in determining the conversion efficiency of a solar cell.

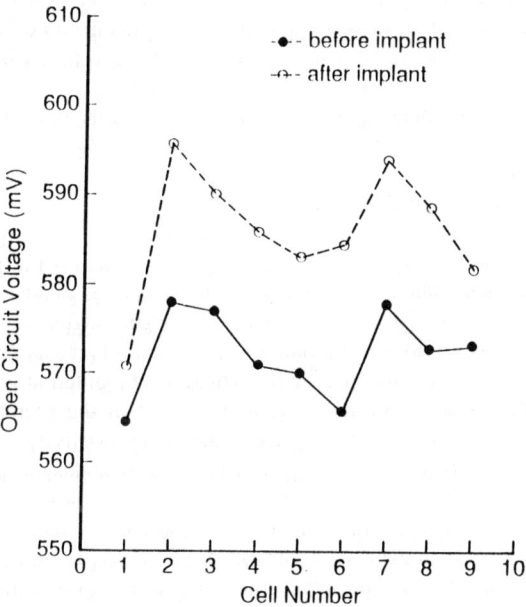

Figure 7b. Plot of the open circuit voltages of a row of solar cells before and after back-side hydrogenation

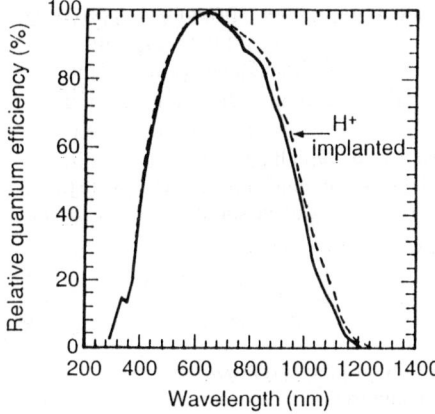

CELL PARAMETERS ARE:

	BEFORE HYDROGENATION	AFTER HYDROGENATION
V_{oc} (mV)	586	587
J_{sc} (mA/cm^2)	25.41	25.60
FF	75.93	76.94
η (%)	11.3	11.7

Figure 8. Spectral responses of a finished solar cell before and after back-side hydrogenation. An increase in the long wavelength response indicates bulk passivation.

4. Backside hydrogenation technique

The surface damage due to hydrogen implantation from the junction side can have deleterious effects on the solar cell performance. The influence of damage can be minimized if hydrogen is implanted from the back side of the cell. However, a back-side hydrogenation requires that hydrogen should diffuse rapidly through the thickness of the cell, typically 300 μm, in order to be effective in improving the junction properties. Such a technique has many other advantages that can make it a production-compatible process[10].

A preferred way to hydrogenate a cell from the back side is prior to making the back-side metallization. Various steps needed for such a process are illustrated in Figure 7a. The solar cell is implanted from the back side and then coated with a thin layer of aluminum, typically about 2000 Å thick. The aluminum is then alloyed in a optically heated furnace, similar to an RTA process. This step serves to drive hydrogen deeper into the cell and also dissolve the damaged region to produce a Si-Al alloy to form an ohmic contact. In addition, it compensates for the out-diffusion of boron discussed in the previous section. Figure 7b shows the effect of such a hydrogenation process on an row of solar cells. The figure shows the open circuit photovoltage of the devices before(solid line) and after (dotted line) the hydrogenation process.

The back-side hydrogenation can also be readily applied to finished solar cells provided the backside metallization is in a grided configuration allowing access for hydrogen to enter silicon through open areas. Figure 8 shows the effect of hydrogenating such a finished cell on the spectral response of the cell; for comparison the other parameters of the cell, before and after the hydrogenation, are also indicated in the figure. From this figure it is seen that the improvement in the cell response is primarily due to increase in the long wavelength response indicating an increase in the minority carrier diffusion length due to the passivating effect of hydrogen. Our experiments have shown that impurity/defect passivation in silicon solar cells is possible even for cells with initial efficiency greater than 12% if the substrate has low oxygen content, typically less than 20 ppma.

Acknowledgement: This work was supported by the U. S. Department of Energy under contract number DE-AC02-83CH10093

5. References

1. S. J. Pearton, J. W. Corbett and T. S. Shi, Appl. Phys. A 43, 153 (1987)
2. See pertinent articles on hydrogen in silicon in Oxygen, carbon, Hydrogen and Nitrogen in Crystalline Silicon, eds. J.C. Mikkelsen, Jr., S. J. Pearton, J. W. Corbett and S. J. Pennycook, MRS Pittsburg, 1986.
3. J. I. Pankov, D. E. Carlson, J. E. Berkeyheiser and R. O. Wance, Phys. Rev. Lett., 51, 2224 (1983).
4. C. T. Sah, J. Y. Sun and J. J. Tzou, Appl. Phys. Lett., 43, 204(1983).
5. M. Stavola, S. J. Pearton, J. Lopata and W. C. Dautremont-Smith, Appl. Phys. Lett., 50, 1086 (1987).
6. N. M. Johnson, C. Herring and D. J. Chadi, Phys. Rev. Lett., 56, 769 (1986).
7. A. J. Tavendale and S, J, Pearton, J. Phys. C16, 1665 (1983).
8. J. C. Muller, Y. Ababou, A. Barhddi, E. Courcelle, S. Unamuno, D. Salles and P. Siffert, Solar Cells, 17, 201 (1986)
9. T. Zhou, Z. Radzimski, B. Patnaik, G. Rozgonyi and B. L. Sopori, Appl. Phys. Lett., 58, 1985 (1991).
10. B. L. Sopori, J. Appl. Phys., 64, 5264 (1988).

A STUDY OF RADIATION INDUCED DEFECTS IN SILICON SOLAR CELLS SHOWING IMPROVED RADIATION RESISTANCE

JAN PETERS[1], TOMAS MARKVART[1], ARTHUR WILLOUGHBY[1]
[1]Engineering Materials, University of Southampton, Highfield, SO9 5NH, U. K.

ABSTRACT

Deep Level Transient Spectroscopy (DLTS)[1] has been used to investigate radiation damage in space solar cells. The radiation induced defects and their role in radiation hard cells which employ a defect gettering region, designed to remove mobile, primary radiation damage, have been studied. The DLTS results are here interpreted in terms of the improved performance characteristics of these cells. These results are compared with conventional Czochralski (CZ) and floatzone (FZ) solar cells with varying oxygen and boron content. A range of defects have been identified; defects that are more prolific in the conventional cells than the denuded zone cells are linked to radiation damage.

1. Introduction

Solar cells for space application are subject to a harsh radiation environment and as a consequence suffer from severe degradation[2]. The degradation in cell performance results from point defects which act as traps or recombination centres for the light generated minority carriers. This work aims to relate the formation of point defects to cell damage and efficiency in addition to the design of radiation resistant solar cells specifically for space use.

A novel solar cell has been designed by Markvart et al[3] that exhibits improved radiation resistance; the cell incorporates an oxygen rich gettering zone deep within the cell (fig 1) which acts as a sink for mobile, primary radiation damage. The type of defect that is most likely to cause significant performance degradation is a mid-bandgap recombination centre with an affinity for both electrons and holes; the radiation tolerance of solar cells has been found to reduce as the doping density increases[4] damage related defects are therefore likely to include the dopant, boron.

2. Experimental

Five batches of CZ solar cells, and four batches of FZ solar cells have been investigated. Two of the CZ batches contain defect gettering zones; the initial oxygen concentration of these two batches was different prior to formation of the denuded zone. The FZ cells are fabricated from wafers of different resistivity, covering more than three orders of magnitude; they were used to correlate the radiation induced defects in the CZ cells. The wafer orientations were not uniform for all the batches due to the limited availability of wafers of specific oxygen or boron content.

The solar cells were fabricated by an n-type phosphorous diffusion into a boron doped substrate to form a junction at $0.1\mu m$; the radiation hard solar cells are fabricated on standard wafers that incorporate a gettering zone in the centre of the wafer, formed by a high - low - high anneal. The cell characteristics, for both CZ and FZ are presented in Table 1 and 2 respectively. The effect of the gettering zone on the solar cell batches C4 and C5 is to modify the band diagram (fig 2) and to modify the cell diffusion length from 150μm to 23μm; the depth of the denuded zone is 17μm and therefore does not affect the efficiency of the cell appreciably.

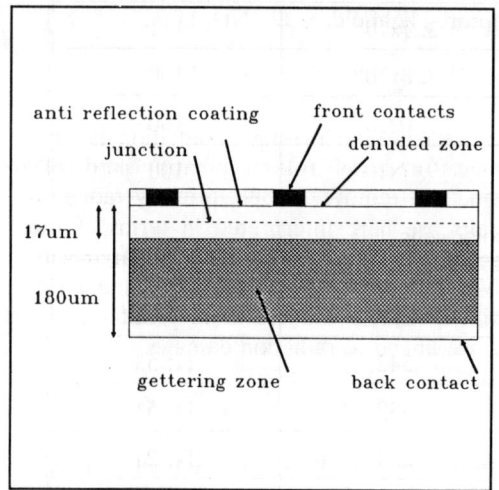

Figure 1 Radiation resistant solar cell with internal gettering zone formed from oxygen precipitates

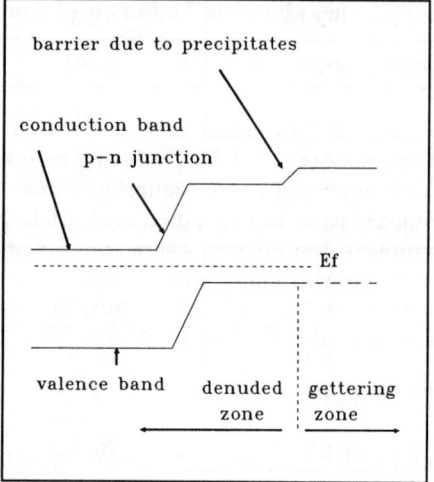

Figure 2 Band diagram of solar cells with gettering zone; the conduction band is modified by ≈0.1 - 0.2μm, which reflects carriers back toward the junction

The CZ cells were irradiated using the Tandem Van de Graaff accelerator at the Central Research Division, BICC, London by 1MeV electrons at a fluence of $1 \times 10^{16} ecm^{-2}$ and the FZ cells were irradiated at $3 \times 10^{14} ecm^{-2}$; the lower irradiation level was used for a previous study. After irradiation, the cells were stabilised under 48 hour AM0 illumination.

Table 1 Parameters of p-type Czochralski material fabricated into n$^+$p solar cells. All these samples were irradiated to 10^{16} electrons/cm^2. * contains a denuded and gettering zone

SAMPLE	RESISTIVITY Ωcm	BORON CONTENT 10^{14}cm^{-3}	OXYGEN CONTENT 10^{17}cm^{-3}	WAFER ORIENT-ATION
C1	17 - 23	17	3.3	111
C2	17 - 23	13	6.9	111
C3	13 - 14	7.6	12.8	100
C4*	11 - 25	5.6	3.8	100
C5*	11 - 25	8.3	2.9	100

The I-V characteristics were obtained using X25 continuous beam simulator which was set to provide AM0 intensity and measured against standard cells at constant temperature. The P_{max} characteristics of the CZ solar cells are presented in Table 3.

3. Results

The I - V characteristics were remeasured after each irradiation up to a total dose of $1 \times 10^{16} ecm^{-2}$, (fig 3). The cells containing the gettering zone demonstrated an improved

Table 2 Average Parameters of the floatzone solar cells

SAMPLE	RESISTIVITY Ωcm	BORON CONTENT cm^{-3}	WAFER ORIENTATION
FZ1	0.3	2.7×10^{16}	111
FZ2	0.94	7.8×10^{15}	100
FZ3	10	7.0×10^{14}	100
FZ4	115	9.7×10^{13}	100

Table 3 Electrical characteristics of the CZ solar cells

SAMPLE	I_{sc} mA/cm²	V_{oc} mV	P_{max} mW/cm²
C1	26.92	542	11.53
C2	27.02	539	11.51
C3	26.72	534	11.21
C4*	21.21	506	8.31
C5*	21.82	502	8.46

relative reduction in P_{max} compared to the conventional cells. The relative degradation of the open circuit voltage, V_{oc}, is much smaller in batches C4 and C5 than in the conventional cells. Contrary to the normal behaviour the open circuit voltage of C4 and C5 degrades more than the short circuit current.

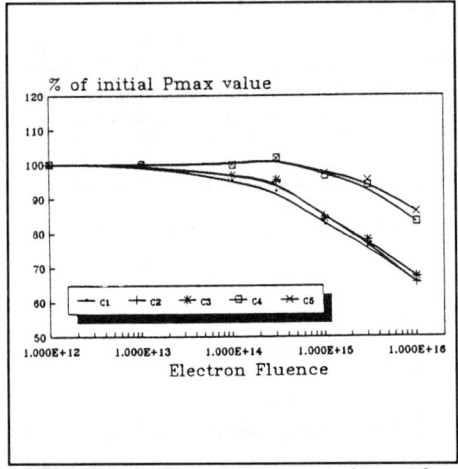

Figure 3 P_{max} characteristics after irradiation up to $1 \times 10^{16} \text{ecm}^{-2}$ on standard solar cells with varying [O] and radiation hard solar cells.

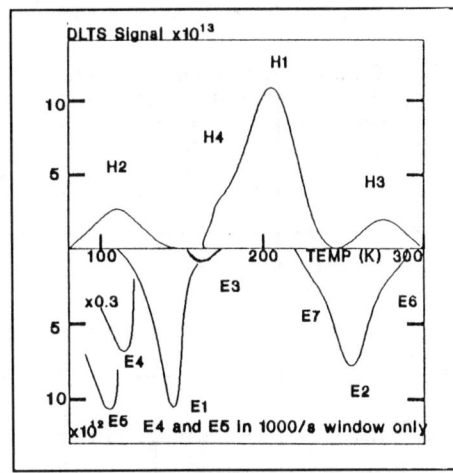

Figure 4 Summary of DLTS peaks observed in CZ solar cells

The results of the DLTS study on the FZ and CZ solar cells are presented in Table 4 and 5. In the CZ samples, four majority carrier traps were observed, labelled H1 to H4; H3 was not detected in some of the samples, and H4 was observed as a shoulder on H1. Seven minority carrier traps were observed in these samples, but only limited information was obtained from them. Two levels were observed in the FZ cells, but due to a lower irradiation level, the concentrations were close to the detection limit of the equipment, i. e., within 10^4 of the dopant concentration. Note: the FZ cells were part of a previous investigation, hence the differing irradiation level. The traps have been compared with levels observed by other workers and correlated with the resistivity and oxygen content of the cells. All temperatures quoted for the traps are observed in the 200/s emission window.

3.1 Czochralski Samples

Level H1 ($E_v+0.33$) is the most prominent trap in all the CZ materials and is now regarded as the $C_i - O_i$ formed by room temperature annealing of C_i^{6b}. H1 demonstrates a sub linear relationship with [O]; Drevinsky reported a saturation of this level at high [O] due to competition from other defects. The formation H1 is suppressed in C4* (which had a higher initial [O] than C5* and hence has a higher concentration of O precipitates) which suggests an increased gettering of C_i during irradiation.

Table 4 The concentrations of majority defect traps in irradiated CZ material ($\times 10^{12} cm^{-3}$). (+ for data marked thus it was not possible to measure the defect concentration for the 200/s emission rate)

LEVEL	H1	H2	H3	H4
TEMP K	230	120	267	200
E_{ACT} meV	320	90-180	470	250
C1	40	6	5	3
C2	65	15	5	+
C3	70	20	-	+
C4*	3.3	1.2	-	+
C5*	40	10	0.1	4

Table 5 Concentrations of the majority-carrier traps in irradiated FZ material ($\times 10^{11} cm^{-3}$), for the 200/s emission rate

LEVEL	[B] cm^{-3}	FH1	FH1$_{intro\ rate}$	FH2	FH2$_{intro\ rate}$
TEMP K		230		200	
E_{ACT} meV		340		280	
FZ1	2.7×10^{16}	90	3.0×10^{-2}	15	5.0×10^{-3}
FZ2	7.8×10^{15}	26	8.7×10^{-3}	6	2.0×10^{-3}
FZ3	7.0×10^{14}	4	1.3×10^{-3}	6	2.0×10^{-3}
FZ4	9.7×10^{13}	2.5	8.3×10^{-4}	7.5	2.5×10^{-3}

Level H2 ($E_v +0.17$eV), is regarded as the divacancy; there is no specific relationship with [B], but a general trend with [O] on a log scale. Metastability has been observed in the divacancy[8].

Level H3 ($E_v+0.47$) is seen sporadically in the CZ samples. This level is likely to be subject to gettering in some form due to the low concentration observed in the denuded zone cells. The position suggests a mid band gap recombination centre which could be related to a heavy metal impurity, the low concentration in C4* and C5* would support this; gettering is normally used remove impurity metals. However, H3 is not present in unirradiated material. The relation of this defect with [O] is also of interest. At $[O] > 10^{18}$ the level is absent: it is also absent from C4* which had a higher initial [O] than C5*. This level may be the same as a level observed in a similar position[5] at $E_v+0.48$eV, but only in 0.3 ohm /cm material, and at $E_v+0.4$eV[9].

Level H4 ($E_v+0.25$) appears as a shoulder on H1 at 200K. Only under certain conditions has it been possible to resolve H4 to a significant degree. Drevinsky et al[6] discuss two levels in this region, DH4($E_v+0.29$) and DH5($E_v+0.30$). Mooney[5] identified a level MH($E_v+0.26$) as the B - O - V, largely by speculation; it is now accepted as the $B_i - C_s$.

3.2 Minority Carrier Traps

Seven traps have been observed by the method of optical injection of carriers. E4 and E5 were only observed in the 1000s^{-1} rate window; one is representative of the A centre. The concentration of the observed levels is an order of magnitude lower than the observed majority carrier traps; the traps were verified by observation of the minority carrier transient on a digitising oscilloscope. E1 is the $B_i - O_i$ defect, observed by other workers[5,6], which has partially annealed during the irradiation sequence to form $B_i - C_s$, H4 (0.28), as described in the reaction sequence[14]. The activation energy of the trap E3($E_c-0.21$) is only approximate due to its low concentration; it has been assigned as a level of the divacancy[11,12] in addition to other configurations[11].

3.3 FZ Samples

Defect introduction rates for the two most strongly observed defects have been plotted against [B]. FH1 ($E_v+0.34$) and FH2 ($E_v+0.28$) refer to the trap identifications: FH1 occurs at 230K and FH2 occurs at 200K. Note that the error for these samples is greater than ± 0.03eV, due to noise as a result of low irradiation levels. FH1 ($E_v+0.342$eV) with a capture cross section of 6.8×10^{-17} cm^{-3} exhibits a linear relationship with [B]. FH1 has been compared with a level seen by Drevinsky[6] (DH5), at $E_v+0.30$eV, $B_i - B_s$. The dependence of this defect on [B], suggests that it is possibly involved with cell degradation since degradation shows a strong dependence on [B]; Weinberg and Swartz[13] have also concluded from other evidence (the reduction in diffusion length as the defect anneals in) that the $E_v+0.31$ level is associated with the $E_c-0.27$ level and is responsible for degradation in FZ solar cells. FH2 (E_v 0.28eV), in contrast, does not exhibit a strong relation with [B], perhaps only at higher concentrations; it is possible that the formation of this defect is suppressed at low [B] due to competition by FH1, this level is the $B_i - C_s$. We note that the present samples had under gone illumination during measurement procedures which facilitates the liberation of B_i from $B_i - O_i$ to form $B_i - B_s$ and $B_i - C_s$ which is consistent with the annealing kinetics proposed by Kimerling[14]. It is probable that illumination, post irradiation, assists the movement of these defects producing $B_i - C_s$ which is not normally produced in high [O] Cz material, $C_i - O_i$ is produced preferentially. This is the probable explanation for photon degradation observed in FZ solar cells[9].

4. Conclusion

There is little direct evidence to identify one defect for solar cell degradation and to explain the cell characteristics of the denuded zone cells (fig 3). However, it is possible to rule out H1, which does not show any strong relationship with [B]; similarly H2, the divacancy. H3 is present in the conventional CZ cells, with lower [O], but undetectable in the high [O] cell and the C4 sample, and low concentration in C5. It would therefore appear likely that this level is a major contributor to solar cell degradation, or at least explain the improved resistance to radiation damage exhibited by the denuded zone cells. The key factor that appears to be reflected in the radiation damage of FZ and CZ cells, both with and without denuded zones, is the oxygen species within the cell. FZ material typically contains [O] $\approx 10^{15}$ to $\approx 10^{16}$ cm^{-3}, whereas CZ material contains [O] $\approx 10^{17}$ to 10^{18} cm^{-3}. FZ solar cells exhibit 'the photon effect'; this is degradation that occurs with illumination subsequent to irradiation[9] and we suggest that the formation of B_i - C_s is responsible for this. CZ solar cells do not display this effect most probably due to the higher oxygen content. In denuded zone cells, oxygen is concentrated into a narrow region in a saturated, precipitated form, away from the junction. The defects associated with this region are able to provide a sink for potentially harmful defects. C3 to C5 have demonstrated improved operation after irradiation compared to C1 and C2; H3 may provide an indication of the reason why C4 and C5 show a significant resistance to radiation; further samples with varying depth of denuded zone are in production. Annealing studies are at present in progress in our laboratories to try to control the behaviour of defects produced in a space environment and to effect their removal in situ.

5. Acknowledgements

The authors would like to thank Wacker Chemitronic for the wafers, TST, Heilbronn, for fabrication of the solar cells and Chris Goodbody of DRA Aerospace Division, for useful Farnborough for useful discussions. This work was supported by the Procurement Executive, Ministry of Defence. Copyright © Controller, HMSO, 1991

6. References

1. D.V. Lang, J. Appl. Phys., 45, 3023, (1974).
2. I. Weinberg in *Current Topics in Photovoltaics 4* (ed Coutts and Meakin), Academic Press, p 87, (1990).
3. T. Markvart, A.F. W. Willoughby and A.A. Dollery, *Proc. 19th Photovoltaic Specialists Conf.*, 709, (1987)
4. A.L. Fahrenbruch and R.H. Bube, *Fundamentals of Solar Cells* Academic Press, p279, 1983
5. P.M. Mooney, L.J. Cheng, M. Suli, J.D. Gerson, and J.W. Corbett, Phys. Rev. B, 15, 3836, (1977).
6. a) P.J. Drevinsky, C.E. Caefer, S.P. Tobin, J.C. Mikkelson and L.C. Kimerling, *Mat. Res. Symp. Proc.*, 104, 167, (1988).
   b) P. J. Drevinsky, C. E. Caefer, L. C. Kimerling and J. L. Benton, *Proc. Defect Control in Semiconductors*, 1, 341, (1990).
7. Y.H. Lee, J.W. Corbett and K.L. Brower, Phys. Stat. Sol. (a), 41, 637, (1977).
8. S.K. Bains and P.C. Banbury, J. Phys. C: Sol. St. Phys. 18, L109, (1985).
9. M. Roux, J. Bernard, R. Reulet and R. L. Crabb, J. Appl. Phys., 56, 531, (1984).
10. L.C. Kimerling, IEEE Trans. NS 23, 1497, (1976).
11. S.D. Brotherton, G.J. Parker and A. Gill, J. Appl. Phys., 54, 5112, (1983).
12. Y.H. Lee and J.W. Corbett, Phys. Rev., B13, 2653, (1976).
13. I. Weinberg and C.K.Swartz, Appl. Phys. Lett., , 693, (1980).
14. L.C. Kimerling, M.T. Asom, J.L. Benton, Mat. Sci. Forum 38-41, 141, 1989.

DEFECTS AND SCHOTTKY BARRIER FORMATION:
A POSITIVE PROOF FOR EPITAXIAL Al ON AlGaAs SCHOTTKY DIODES

JERZY M. LANGER [1] AND P. REVVA [2]

[1] Institute of Physics, Polish Academy of Sciences, Al. Lotnikow 32/46, Warsaw, Poland and
[2] The University of Manchester Institute of Science Technology, Sackville Str., P.O.Box 88, Manchester M60 1QD, England

ABSTRACT

We present here the first direct proof that, at least in the epitaxially grown Al/AlGaAs metal-semiconductor junctions, defects play a key role in the formation of the Schottky barrier. This proof comes from a measurement of a temperature dependence of the Schottky barrier heights of metal-semiconductor junctions. They were grown on both n and p type AlGaAs in a whole composition range. We found that $d\Phi(n)/dT \approx dE_{gap}/dT$, while $d\Phi(p)/dT \approx 0$. This result is in direct conflict with prediction based on a concept of a neutrality level. It can, however, be easily understood if the localized defects cause the Fermi level pinning. This is because for localized defects, whose ground state is of a bonding type, the ionization entropy equals the entropy of the valence to conduction band transitions.

1. Fermi level pinning in Schottky barriers

A time ago it was established that the barrier heights of metal-semiconductor (MS) contacts made on covalent semiconductors are almost independent on the metal work function. Such a behavior suggests that it must be some generic mechanism that is responsible for the Fermi-level pinning in Schottky barriers. Since then many theorists and experimentalists have tried to identify the source of the pinning phenomenon [1,2].

One group of models stresses the role of the metal-induced gap (MIG) states [1,3]. According to proponents of this idea, the reason of the effect lies in the very nature of the intimate metal-semiconductor contact. There the evanescent wave function of a metal produces the gap states (MIGs) at the interface. Such states are defined by a band structure of a semiconductor and a pinning energy corresponds to the neutrality level of a semiconductor. Lannoo and his associates argued, that the neutrality level should be associated with the average energy of anion and cation dangling bonds of a semiconductor [4].

This elegant idea ignores completely all chemical reactions occurring at the interface during the formation of a contact. It is known that surface atoms undergo rebonding. Also, metal deposition may cause generation of defects during metallization [2,5-9]. Such defects, if present, also may cause the Fermi level pinning. Unfortunately, in spite of a year long search, no such defect was positively identified, although there are good reasons to believe that anion antisites are prime candidates [6,7,9].

Most of arguments of the proponents of the MIGs model rely on a positive energy correlation of the Schottky barrier heights and the computed semiconductor neutrality levels [3]. From time to time there are reports producing similarly good correlation of the measured barrier heights and the computed energy levels of particular defects (mostly vacancies) [10]. Such a correlation is in most cased unjustified, because of the well known limits of accuracy of these computations.

From all this vigorous debate one must conclude that the only way to solve this long standing puzzle is to look for some higher order effects than simple energy correlations. It became obvious by now that both intrinsic and extrinsic mechanisms must play some role in the formation of the Schottky barrier. Which of them is dominant is still to be determined.

During a study of the properties of the high quality epitaxial Al/AlGaAs contacts we realized that the temperature evolution of the Schottky barriers can provide a missing insight. If the neutrality level (MIGs) model is valid, than the temperature dependence of the barrier should reflect an individual motion of a given semiconductor band (the conduction band for the n-type substrate and the valence band for the p-type substrate). Although there is no direct method to monitor such a temperature evolution, the theorist agree not only about the direction of these changes, but also about their magnitude [11]. If, on the other hand, defects govern the Fermi level pinning, then their ionization entropy would control the temperature evolution of the Schottky barrier.

2. Ionization entropy of defect and band states

There are two major contributions to the ionization entropy of the semiconductor band states. One comes from the lattice expansion and the second, known as the Debye-Waller factor, from the screening of the electronic interactions by the electron-phonon interaction [11].

$$\frac{dE_{gap}}{dT} = \left(\frac{\partial E_{gap}}{\partial V}\right)_T \left(\frac{\partial V}{\partial T}\right) + \left(\frac{\partial E_{gap}}{\partial T}\right)_V^{el-ph} \tag{1}$$

The expansion factor can be easily computed from known lattice thermal expansion, lattice compressibility and the pressure dependence of the gap states. For the energy gap, its pressure dependence may be used, while for the individual band states knowledge of the deformation potentials [12] is required. For GaAs the lattice expansion amounts only to about 20% of a total change of the energy gap with temperature. In AlAs it is even smaller because of the sign inversion of the deformation potential of the X-valley in comparison with the Γ-valley [12]. In both cases the Debye-Waller contribution dominates the temperature change of the energy gap. This contribution was computed in several papers [11,13]. Summing up the dilation and Debye-Waller terms yields the following values of the temperature dependence of the gap states:

c.b.	GaAs (Γ)	-0.19 meV/K	AlAs (X)	-0.06 meV/K
v.b.	GaAs (Γ)	0.27 meV/K	AlAs (Γ)	0.28 meV/K
E_{gap}	GaAs (Γ-Γ)	-0.46 meV/K	AlAs (Γ-X)	-0.34 meV/K

It is clear from this table that it is the motion of the valence band that dominates the temperature dependence of the energy gap. Therefore, if the formation of the Schottky barrier is governed by the MIG states, the temperature dependence of the Schottky barrier height on n and p-type substrate should be close to each other for GaAs and direct gap AlGaAs. In AlAs or indirect gap AlGaAs the cancellation effects between the expansion and Debye-Waller terms for the X-conduction band should result in a much stronger temperature dependence for the p-type than for the n-type substrate.

Similar analysis can, in principle, be made for the defect states. It can be done much easier by following arguments presented by Van Vechten and Thurmond. They noted that a non zero ionization entropy comes from a change of the local vibration frequencies upon ionization (from ω_i to ω_i') [14,15]:

$$\Delta S_{ion} = -k \sum_i \ln \frac{\omega_i'}{\omega_i} \tag{2}$$

This approach is especially useful in comparing ionization entropies of the gap and defect states. It is obvious from Eq.(2) that the ionization entropy for the effective mass impurity states should be very small as their ionization does not change local phonons. Also, a small ionization entropy is expected for the ionization of antibonding defect states. Quite dramatic difference is expected if the ground state of the defect is a bonding type because defect ionization is nothing but a bond breaking, similar to a bond breaking occuring during creation of an electron-hole pair. Because of this similarity, ionization entropy of the localized defect bonding states should be close to the ionization entropy of the energy gap [14]. Some complication may arise if the defect ionization is accompanied by a large lattice relaxation. Here, additional contribution from a change of configuration (configuration entropy) must be added. In semiconductors, this factor must not be very large, however.

We can conclude then, that if the Fermi level is pinned by defects (most likely some midgap states), either Φ_n or Φ_p should weakly change with temperature depending whether the ground state of the defect is antibonding or bonding. In the latter case, the entropy of the Schottky barrier height Φ is expected to be close to the ionization entropy of the energy gap.

3. Experimental results

To test validity of either of the two models we have measured the temperature dependence of the Schottky barrier height of the MBE grown epitaxial Al on p- and n-type AlGaAs MS junctions. The MBE-growth technique allowed us to get very high quality junctions with almost perfect electrical characteristics. All the layers used in this study (i.e., AlGaAs as well as Al metallization) were MBE grown by M. Missous at UMIST on GaAs (100) p or n substrates in a VG-V90 system using solid sources for the group III and V compounds (see Ref. 10 for details of the growth). The Schottky barrier height has been determined either by the internal photoemission technique (PE) - the Fowler plots or from the capacitance versus voltage C(V) characteristics [$(1/C)^2$ vs voltage exhibited excellent linearity]. We used the current versus voltage I(V) characteristics to test the electrical quality of the diodes. For all diodes at temperatures for which the thermionic model of the transport applies, the ideality factor of the diodes was no larger than 1.03. The barrier height deduced from the temperature dependence of the saturation current agreed very well with those determined by more accurate PE or CV methods.

From a study of the temperature dependence of the Schottky barrier heights we conclude that the Schottky barriers on p-type AlGaAs ($0 \leq x \leq 1$) are very weakly temperature dependent (Fig.1a). On the n-type material, the energetic shift of the Fowler plots is evident. It is slightly larger for all direct gap compositions than for the indirect gap substrates, following the temperature shift of the energy gap (Fig.1b.)

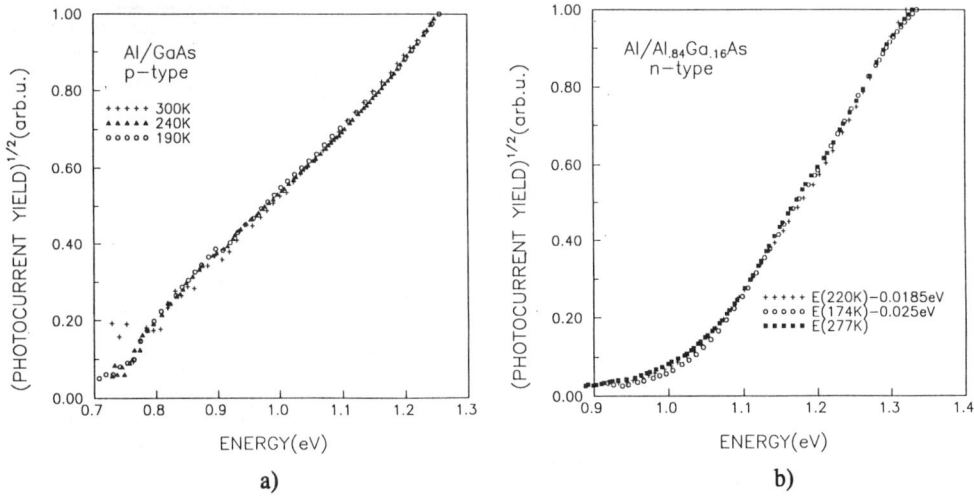

Fig.1 Examples of the Fowler plots for the Al/AlGaAs Schottky barriers on p and n-type substrates. For all p-type AlGaAs compositions the results are as for Al/GaAs(p) as shown in a). For the n-type AlGaAs, the temperature change is larger for the direct gap. To strenghten conclusion about the temperature shift for the n-type AlGaAs, the data in b) were shifted by the amount given in the figure.

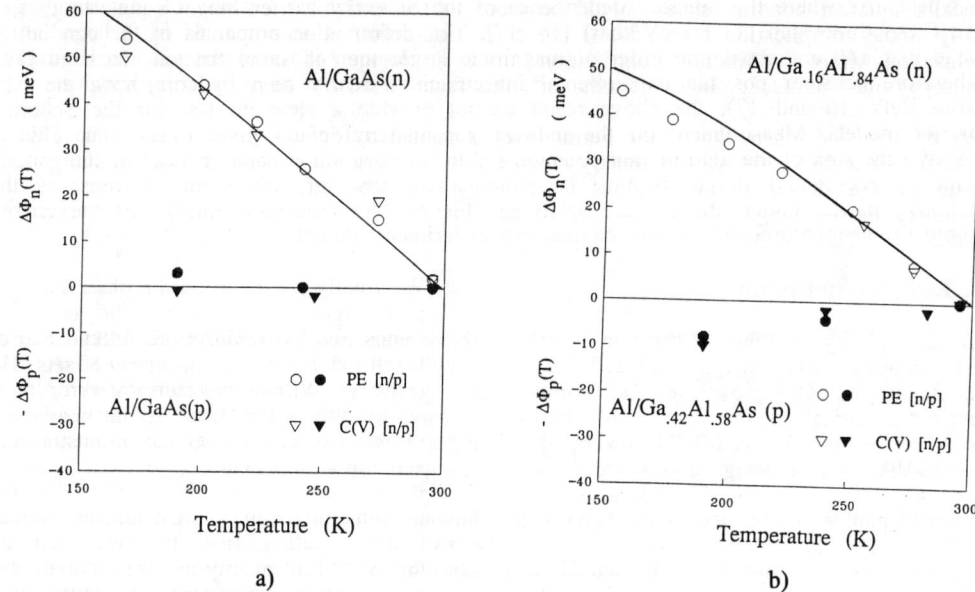

Fig.2 Temperature dependence of the Schottky barrier heights for a) the direct gap GaAs and b) indirect gap AlGaAs. Solid line is the temperature dependence of the energy gap.

4. Conclusions

It is evident from all data gathered in Figs. 1, 2 and 3 that the temperature dependence of the Al/AlGaAs (n-type) follows within the experimental error the temperature dependence of the energy gap (for all AlGaAs compositions, i.e., from GaAs to AlAs). The temperature dependence of the barrier height on the p-type material is much weaker, if any. This result is in direct conflict with predictions based on the neutrality level model.

Following the arguments given in Section 3. it has, however, obvious explanation if the interface defects whose ground state is of the bonding type are responsible for the pinning.

We conclude therefore, that *the Fermi level pinning in the high quality epitaxial Al/AlGaAs MS junctions is indeed due to the interface defects*.

We believe that the reported result is the first positive proof of a dominant role of defects in the formation of the Schottky barrier height. Our measurements do not provide any identification of these defects, however.

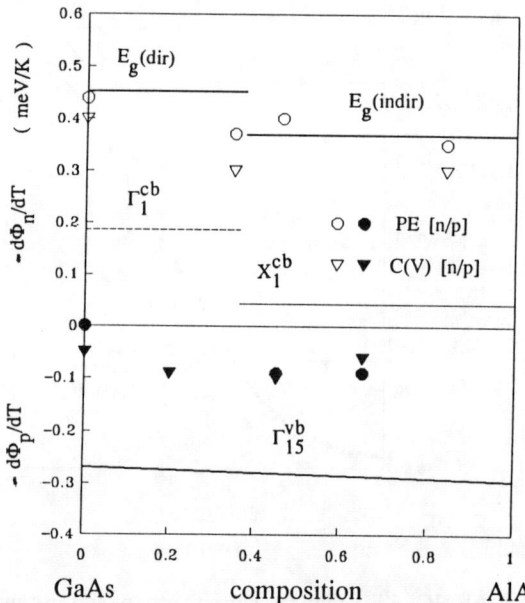

Fig.3 Composition dependence of the Al/AlGaAs Schottky barriers. Note a sign reversal of the temperature dependence. Lines are computed individual band shifts according to the table.

Similar behavior has been observed in the hydrostatic pressure studies of the MS diodes on n-type GaAs where the pressure dependence of the Schottky barrier height equals to that of the GaAs direct gap (11.5 meV/kbar) [16, 17]. Yet, deformation potentials of the conduction band in GaAs are about one order of magnitude larger than those of the valence band [12]. Therefore consider possible experimental inaccuracies (clearly seen by comparing the data from Refs. 16 and 17), the above result cannot provide a clear-cut test for the Schottky barrier models. Measurement on the indirect gap material could give some hope. This is because the sign of the deformation potential for the X-conduction band changes in comparison with a direct gap material. It must be remembered, however, that overall changes of the Schottky barrier under pressure are small and the only reliable technique is the normalized photocurrent measurement, a very challenging experiment, though.

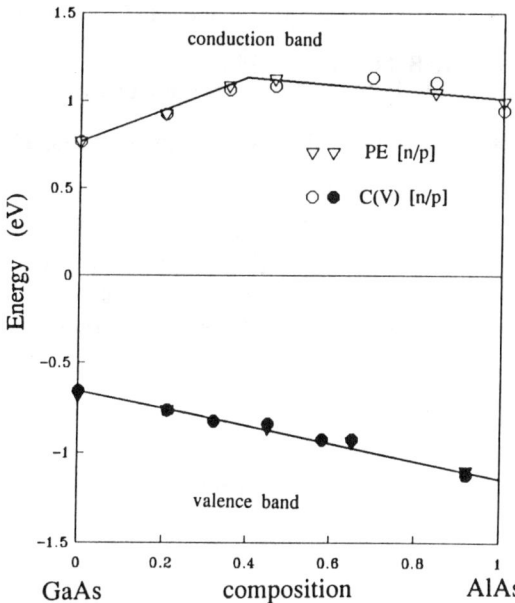

Fig.4 Comparison of the Schottky barrier heights with the band offsets trends. The values of the p and n-type barrier height add well to E_{gap}.

We finally comment on the observed very good agreement between the chemical trends in Al/AlGaAs Schottky barrier heights and GaAs/AlGaAs band offsets [18] (Fig. 4). Within experimental error their composition dependence is practically the same. In the earlier experiments the agreement was worse (see Ref. 19 for a more exhaustive discussion of this problem in a context of a mutual relation between the transition metal energy levels and band offsets), but the diodes were simply evaporated and thus much worse electrically. This result taken alone provides a strong argument for the link between Schottky barriers and heterojunction band-offsets and the idea of a neutrality level based on the screening at the interface [3]. It also justifies the use of the Schottky barriers on AlGaAs as the references for the host band structure in a similar way as it was done recently with the help of transition metal energy levels [3,19].

Analyzing the published data on the composition trends of various defects (not necessary the transition metals) in AlGaAs [20], it becomes obvious that a similar arguments can be used by the proponents of the defect model, as for several point defects, the composition shift of their energy levels is exactly the same as for the band offset or the Schottky barriers. It should be noted, however, that this correlation does not undermine validity of the neutrality level model for heterojunctions, because simply no interface defects down to a very low detection limit were found there.

Acknowledgments

Stimulating and friendly interaction with Professors A. Peaker, K. Singer and E. Rhoderick was most helpful for us while at UMIST. Our most sincere thanks we direct to Dr. M. Missous whose excellence in the MBE growth and numerous discussion on the Schottky barrier problem made this work possible. One of us (JML) would like to thank SERC in U.K. for sponsoring my sabbatical at UMIST and KBN for a financial support in Poland.

REFERENCES

1. F. Flores and C. Tejedor; J. Phys. C, **20**, 145 (1987).
2. I. Lindau and T. Kendelewicz, CRC Crit. Rev.Sol. State Mat. Sci. **13**, 27 (1986).

3. J. Tersoff, Phys. Rev. Lett. **52**, 465 (1984); ibid., **56**, 675 (1986) and Phys. Rev. B **30**, 4874 (1984).
4. M. Lannoo, Revue Phys. Appl. **28**, 789 (1987).
5. J. L. Freouf and J. M. Woodall, Surf. Sci. **168**, 518 (1986).
6. J. M. Langer, Radiation Effects, **72**, 55 (1983).
7. W. E. Spicer, Z. Liliental-Weber, E. Weber, N. Newman, T. Kendelewicz, R. Cao, C. Mc Cants, P. Mahowald, K. Miyano and I. Lindau, J. Vac. Sci. Technol. B **6**, 1245 (1988).
8. R. Ludeke, NATO ASI Series B, **195**, 38 (1989).
9. W. Walukiewicz, Appl. Phys. Lett. **54**, 2094 (1989) and preceding.
10. M. Missous, W. S. Truscott and K. Singer, J. Appl. Phys. **68**, 2239 (1990).
11. M. Cohen and J. Chadi, *Handbook of Semiconductors*, ed. T. Moss, Vol.2, (North-Holland, Amsterdam, 1980) p. 155.
12. Ch. G. Van de Walle, Phys. Rev. B **39**, 1871 (1989)
13 J. Camassel and D. Auvergne, Phys. Rev. B **12**, 3258 (1975); P. Yu and M. Cardona, ibid. B **2**, 3193 (1970).
14. J. Van Vechten and C. D. Thurmond, Phys. Rev. B **14**, 3539 (1976).
15. J. Van Vechten, *Handbook of Semiconductors*, ed. T. Moss, Vol.4, (North-Holland, Amsterdam, 1981) p. 1.
16. W. Shan, M. F. Li, P. Yu, W. L. Hansen and W. Walukiewicz, Appl. Phys. Lett. **53**, 974 (1988).
17. T. Fujisawa and H. Kukimoto, *private communication* (1989).
18. M. Missous, P. Revva and J. M. Langer, in *Proc 20th Int Conf Phys. Semicond.*, ed. E. M. Anastassakis and J. D. Joannopoulous, (World Scientific), p. 359.
19. J. M. Langer, C. Delerue, M. Lannoo and H. Heinrich, Phys. Rev. B **38**, 7723 (1988).
20. D. V. Lang and L. Kimerling, in *Proc. 13-th Conf. Phys. Semicond.*, ed.T. Fumi (Tipografia Marves, Rome, 1976) p. 615; D. V. Lang, R. Logan and L. Kimerling, Phys. Rev. B **15**, 1874 (1977).

RECOMBINATION-ENHANCED DIFFUSION OF Be IN GaAs

MASASHI UEMATSU AND KAZUMI WADA
NTT LSI Laboratories, 3-1, Morinosato Wakamiya, Atsugi-shi, Kanagawa 243-01, JAPAN

ABSTRACT

We have observed recombination-enhanced impurity diffusion (REID) in Be-doped GaAs. Our investigation of current-induced degradation of tunnel diodes reveals that Be diffusion under forward bias is enhanced by a factor of about 10^{15} at room temperature, and the activation energy for the diffusion is reduced from 1.8 eV for thermal diffusion to 0.6 eV for REID. In the REID of Be, the energy related to minority carrier injection at the recombination center enhances the annihilation of the recombination center, in which a point defect that enhances the Be diffusion is generated.

1. Introduction

Device degradation under operation has been widely observed in electron and optical devices such as, Esaki tunnel diodes,[1,2] heterojunction bipolar transistors (HBTs),[3,4] light emitting diodes,[5] and laser diodes.[6] These devices tend to degrade during forward bias operation but not under reverse or zero bias, indicating that minority carrier injection, i.e., the recombination process, plays an essential role in the degradation. However, the detailed mechanism responsible for the degradation must be clarified to obtain an understanding of the basic physics involved and thereby find ways to avoid or suppress it.

This paper describes recombination-enhanced diffusion of Be in GaAs by investigating current-induced degradation of Esaki tunnel diodes. The enhanced Be diffusion coefficients and their decreases with time are obtained from the decay of the tunnel peak current densities. The possible mechanism for the enhanced diffusion is discussed based on the analysis in terms of the kinetics of the decay of the recombination center. These results are evidence that recombination-enhanced defect reaction[7-10] is responsible for the enhanced Be diffusion, which we name recombination-enhanced impurity diffusion (REID).[11]

2. Experimental

The tunnel diodes were fabricated by MBE (Molecular Beam Epitaxy) on (100) n-GaAs substrates, using Be as a p-type dopant and Si as an n-type dopant. The substrate temperature during growth was 550 °C. The carrier concentrations of the p$^+$- and n$^+$- layers of the tunnel junction are 4×10^{19} cm^{-3} and 1×10^{19} cm^{-3}, respectively. Mesas with an area of 5.0×10^{-5} cm^2 were formed by wet etching.

The typical tunnel peak current density (J_p) and the peak voltage (V_p) were measured to be 10 A/cm^2 and 80 meV, respectively, using an HP 4142B Semiconductor Parameter Analyzer. The forward bias operations were carried out under constant current densities in the range 500 - 2000 A/cm^2 over the temperature range 27 - 200 °C. The degradation by thermal annealing was investigated, where the diodes were annealed in nitrogen ambient at 400, 450, and 500 °C under zero-biased condition.

3. Results and discussion
3.1 Current-induced Be diffusion

Degradation of the diodes was observed under forward bias operation. The decreases in peak current densities relative to the initial values with time under current densities of 500 - 2000 A/cm² at 200 °C are shown in Fig. 1. Similar decays were observed at 27, 100, and 150 °C, with the decay rates becoming smaller at lower temperatures.

The peak current density is described by the formula,[12,13]

$$J_p = A \exp(-BW_p), \quad (1)$$

where A and B are constants, and W_p is the width of the space-charge region at small positive bias V_p. The diffusion of the dopants should increase W_p, and hence reduce J_p. Because W_p is on the order of 10 nm, the junction broadening on the order of 0.1 nm should be sufficient to cause a significant decrease in J_p. The broadening of W_p is attributed to the Be diffusion, with the Si donors being immobile, because the rates of the current-induced degradation of HBTs depend critically on the Be doping concentrations.[3] The diffusion coefficient D_{Be} is described in terms of decreasing J_p as[13,14]

$$D_{Be} = -d/dt[\ln\{J_p(t)/J_p(0)\}]W_p B^{-1}(1 + N_A/N_D + e\phi/kT)^{-1} \quad \text{for } t \rightarrow 0, \quad (2)$$

where N_A and N_D are the carrier concentrations of the p- and n-side of the diode, and ϕ the barrier potential at the junction. The B and W_p were calculated from Kane's formula,[12] and the barrier potential ϕ was estimated by considering the forward bias voltage applied during the operation.

The current-induced Be diffusion coefficients were obtained by Eq. (2). Thermal Be diffusion coefficients were also obtained by the same procedure using Eq. (2). These Be diffusion coefficients versus inverse temperature are shown in Fig. 2. Thermal Be diffusion data at high temperatures by Tejwani et al.[15] and Schubert et al.[16] are also shown in the figure, and the present data are consistent with these high-temperature data, which indicates the validity of the analysis noted above.

The current-induced Be diffusion coefficient is given by

$$D_{Be,enhanced} = 8.7 \times 10^{-11} \exp[-(0.59 \pm 0.05) eV/kT] \text{ cm}^2/\text{s} \quad (3)$$

for 2000 A/cm², and the thermal Be diffusion coefficient is

$$D_{Be,thermal} = 8.3 \times 10^{-7} \exp[-(1.8 \pm 0.23) eV/kT] \text{ cm}^2/\text{s}. \quad (4)$$

The exponential prefactors and the activation energies are 2.8×10^{-11} cm²/s, 0.58 ± 0.05 eV for 1500 A/cm², and 1.3×10^{-11}, 0.61 ± 0.05 for 1000 A/cm², respectively. The current-induced Be diffusion coefficient at room temperature is 1.3×10^{-20} cm²/s for 2000 A/cm², and the extrapolated enhancement factor at room temperature is greater than 10^{15}. The difference in the activation energies between the current-induced and the thermal diffusion is about 1.2 eV.

Kimerling and Lang[7,8] have shown that the annealing rate of E3 GaAs defects increases significantly under conditions of minority carrier injection. They concluded that close agreement between the amount of decreasing activation energy for the annealing rate and the E3-valence band transition energy suggests recombination-enhanced annealing. In addition, no degradation was observed under reverse bias with 2000 A/cm² at 200 °C for

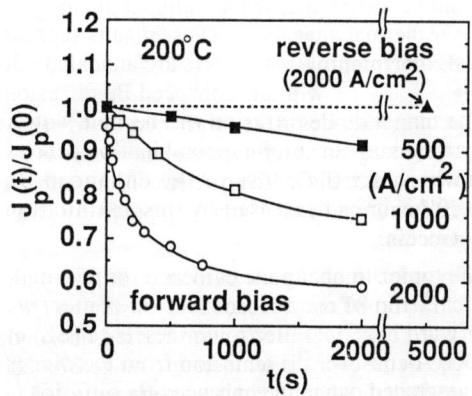

Fig. 1. Decrease in peak current densities relative to initial values with time under forward bias with current densities of 500 to 2000 A/cm² at 200 °C. No degradation was observed under reverse bias with 2000 A/cm² at 200 °C for 5000 s.

5000 sec (Fig. 1), or after a diode had been held for 72 hr at 200 °C with no current flowing, which rules out any possibility of enhancement due to heating by high current density.

These results suggest that the energy difference of 1.2 eV comes from electron-hole recombination events, where the energy related to minority-carrier capture enhances the Be diffusion. The energy level of the recombination center concerned is not specified in the present study. However, by considering the reduced activation energy, the depth of the recombination center is supposed to be approximately 1.2 eV with respect to the conduction or valence band edge.

3.2 Kinetics of enhanced Be diffusion

We investigated the enhanced diffusion from the rates of the decreasing J_p at t = 0 in Section 3.1. In this Section, time-dependence of the enhanced Be diffusion coefficient $D_{Be}(t)$ is investigated and the kinetics of the process is discussed. Equation (2) is also valid for t > 0 if the variation of W_p with time is negligible. In the present case, the decrease in J_p up to $J_p(t)/J_p(0) \sim 0.4$ at 200 °C with 2000 A/cm² for 2000 s is the maximum degree of the degradation. Therefore, the W_p is estimated to increase from the initial value of $W_p(0) = 17.1$ nm to 18.0 nm at most using Kane's formula.[12] On the other hand, $-d/dt[\ln\{J_p(t)/J_p(0)\}]$ decreases about two orders of magnitude as estimated from Fig. 1. Therefore, the variation of W_p with time is negligible.

In Eq. (2), both B and φ are insensitive to the degradation of the diodes. In addition, N_A/N_D may also be insensitive to the degradation because the Be diffusion lengths are so small, as shown below, that the changes in the carrier concentrations are considered negligibly small compared with the decrease in $-d/dt[\ln\{J_p(t)/J_p(0)\}]$. Therefore, Eq. (2) is described as

$$D_{Be}(t) = -d/dt[\ln\{J_p(t)/J_p(0)\}]/D_c, \tag{5}$$

where D_c is a constant. Equation (5) shows the relation between the time dependence of the D_{Be} and the rates of the decreasing J_p.

As can be seen in Fig. 1, the rates of the decreasing J_p were found to decline with elapsed time, and therefore Eq. (5) indicates decreasing D_{Be}. On the other hand, the rates of the decreasing J_p, and hence the Be diffusion coefficients, remain almost unchanged for thermal annealing. These results suggest that recombination centers are annealed out concurrently with the enhanced Be diffusion because the Be diffusion will become slower if the number of the recombination centers which are the cause of the enhanced Be diffusion is decreased by this annihilation process.

In order to obtain the evidence for this annihilation of the recombination center, we investigated the electroluminescence (EL) of the band-to-band emission from the diodes degraded due to the enhanced Be diffusion.[17] The peak intensities of the EL relative to that of no-degraded diode as a function of the degradation time are shown in Fig. 3. The diodes were degraded under forward bias with current density of 2000 A/cm² at 200 °C for 0, 26, 110, 350, and 2000 s. The emission was observed at 77 K with injection

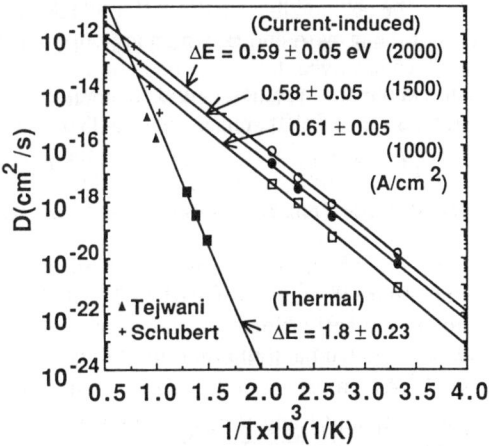

Fig. 2. Current-induced and thermal diffusion coefficients of Be obtained from the decrease in peak current densities. Thermal Be diffusion coefficients at high temperatures by Tejwani et al. (Ref. 15) and Schubert et al. (Ref. 16) are also shown.

current of 120 A/cm² , which induced no further degradation of the diodes during the emission observation.

The EL efficiency was found to increase as the diodes were degraded. In addition, no increase in EL was observed after a diode had been held for 20 hr at 200 °C with no current flowing, which rules out the possibility of the thermal annealing of the recombination center. These findings indicate that upon the enhanced Be diffusion the recombination-enhanced annihilation of the recombination center occurs, in which the energy related to minority-carrier capture enhances the annihilation. The mechanism of the annihilation of the recombination center has been proposed by Kimerling and Lang through multi-phonon emission[7,8] and by Sheinkman via the excited state of the recombination center.[9,10] This EL efficiency increase was simulated based on the minority-carrier lifetime increase due to the single exponential decay of the recombination center using the annihilation rate constant k as a parameter,[17] and we obtained $k = 1.7 \times 10^{-2}$ s^{-1}.

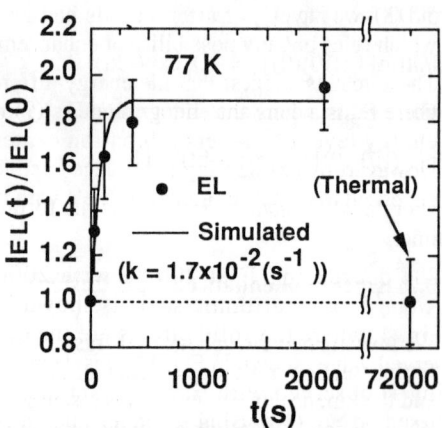

Fig. 3. Peak intensities of electroluminescence of band-to-band emission from diodes relative to that of no-degraded diode as function of degradation time (solid circles). No degradation was observed by thermal annealing at 200 °C for 20 hr. The solid line is the simulated curve based on minority-carrier lifetime with annihilation rate constant $k = 1.7 \times 10^{-2}$ s^{-1}.

In order to account for the decreasing D_{Be}, the relation between the enhancement of the diffusion and the annihilation of the recombination center should be discussed. Following Tan and Gösele,[18] Be diffusion in GaAs is enhanced by the supersaturation of interstitial Ga atoms (I_{Ga}). Therefore, the decreasing D_{Be} indicates the decrease in the point defect which enhances the Be diffusion. Then we propose a model that when the recombination center is annihilated, a certain point defect is generated, which enhances the Be diffusion. As discussed above, I_{Ga} is considered to be one possible candidate for this defect. Furthermore, the recombination-induced formation of the traps A and B, which are only observed in GaAs grown under Ga-rich conditions,[19] has been observed during the operation of AlGaAs light-emitting diodes.[20] Therefore, the supersaturation of I_{Ga} which is generated upon the recombination-enhanced annihilation of the recombination center could enhance the Be diffusion.

We represent these processes by the consecutive reaction scheme as

recombination centers: $A \xrightarrow{k} I_{Ga} \xrightarrow{k'}$ diffusion of I_{Ga} to enhance Be diffusion, (6)

where k and k' are the rate constants for the first and the second step. The first step represents the annihilation of the recombination center to generate I_{Ga} and the second the diffusion of I_{Ga} to enhance the Be diffusion.

We assume that the first step in Eq. (6) is the rate-determining step of the process. Therefore, we take $k \ll k'$, resulting in the familiar steady-state treatment.[21] Then we obtain

$$N(I_{Ga}) = (k/k')N(A) = (k/k')A_0\exp(-kt), \quad (7)$$

where $N(I_{Ga})$ and $N(A)$ indicate the concentrations of I_{Ga} and A, respectively, and A_0 is the initial concentration of A.

Following Tan and Gösele,[18] D_{Be} is proportional to $N(I_{Ga})$. Therefore, Eq. (7) gives

$$D_{Be}(t) = D_0\exp(-kt), \quad (8)$$

where D_0 is the diffusion coefficient at $t = 0$, which we presented in Section 3.1. From Eqs. (5)

and (8), we have

$$d/dt[\ln\{J_p(t)/J_p(0)\}] = -D_R\exp(-kt), \quad (9)$$

where D_R is a constant. Integrating Eq. (9) with $\ln[J_p(t)/J_p(0)]$ being 0 at $t = 0$ gives

$$\ln[J_p(t)/J_p(0)] = a[\exp(-kt) - 1], \quad (10)$$

where a is a constant. Eq. (10) is the formula used for the simulation of the decreasing J_p with time.

The decreasing J_p at 200 °C with 2000 A/cm² in a logarithmic scale is shown in Fig. 4, where the solid circles are the observed values shown in Fig. 1. This decreasing J_p observed with time was simulated based on Eq. (10) using k and a as parameters. The dotted line in Fig. 4 is the simulated curve, and it matched the observed results very closely. This indicates the validity of the model we proposed. Furthermore, the annihilation rate constant $k = 1.3 \times 10^{-2}$ s^{-1} was obtained from the simulation. This rate constant, which was estimated from the enhanced Be diffusion, is close to the annihilation rate constant $k = 1.7 \times 10^{-2}$ s^{-1} obtained from the EL efficiency increase. This strongly supports our proposal that the recombination-enhanced annihilation of the recombination center occurs upon the enhanced Be diffusion and a point defect which enhances the Be diffusion is generated upon this annihilation.

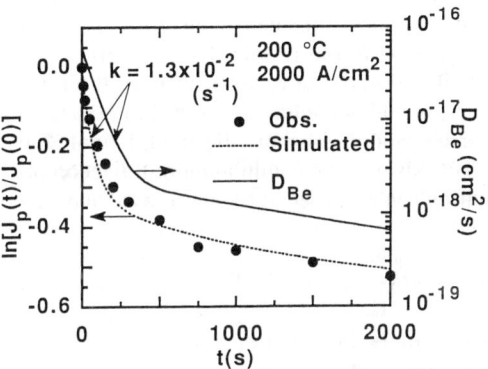

Fig. 4. Decreasing J_p at 200 °C with 2000 A/cm² in a logarithmic scale. The solid circles are the observed values shown in Fig. 1. The dotted line is the simulated curve based on Eq. (10) with $k = 1.3 \times 10^{-2}$ s^{-1}. The solid line is decreasing D_{Be} obtained using the results of the simulation and Eq. (8).

The decreasing D_{Be} at 200 °C with 2000 A/cm² was obtained using Eq. (8), and is shown in Fig. 4 (solid line). The D_{Be} decreases about two orders of magnitude from 0 to 2000 s, and the first decay shows the single exponential decay with $k = 1.3 \times 10^{-2}$ s^{-1}. The Be diffusion length estimated based on the D_{Be} obtained was about 0.9 nm after the elapsed time of 2000 s. This indicates that the change in the carrier concentration is much smaller than the decrease in $-d/dt[\ln\{J_p(t)/J_p(0)\}]$, as mentioned above. In addition, the amount of the Be diffusion estimated from the D_{Be} was found to be within the critical value for Eq. (2) at $t > 0$ being valid. The D_{Be} seems to decay with smaller rate constant than that of the first decay after the elapsed time of about 300 s. The reason is not yet clear; however, one possible factor is that there might be other kinds of recombination centers which have smaller annihilation rate constants than that of the first decay.

The present results lead us to conclude that the recombination-enhanced annihilation of the recombination center occurs, in which a point defect is generated, which enhances the Be diffusion. We name this enhanced Be diffusion recombination-enhanced impurity diffusion (REID). Furthermore, the present study may indicate a novel approach in which the generation kinetics of group III point defects by recombination-enhanced processes can be elucidated from the viewpoint of Be diffusion.

4. Summary

Current-induced degradation of tunnel diodes was investigated using Be as a p-type dopant. The degradation was observed under forward bias operation. On the other hand, no degradation was observed under reverse or zero bias. The diffusion coefficients of Be under forward bias were obtained from the decay of the peak current density. The thermal diffusion coefficient of Be was also obtained by the same procedure. The Be diffusion coefficient under forward bias is enhanced by a factor of about 10^{15} at room temperature, and the activation energy for the diffusion is reduced from 1.8 eV for thermal diffusion to 0.6 eV for REID.

The time-dependence of the REID was investigated based on the analysis in terms of the kinetics of the decay of the recombination center. The increase in the EL from the diodes indicates the recombination-enhanced annihilation of the recombination center upon the REID. The decrease in the peak current density of the diodes was simulated based on the model in which a point defect that enhances the Be diffusion is generated when the recombination center is annihilated. The simulated results matched the observed data very closely, and the annihilation rate constant for the recombination center obtained from the decrease in the peak current density was found to be close to the annihilation rate constant from the EL efficiency increase. The present results suggest that in the REID the energy related to minority carrier injection at the recombination center enhances the annihilation of the recombination center, in which a group III point defect that enhances the Be diffusion is generated .

ACKNOWLEDGMENT

The authors are indebted to O. Nakajima, H.Ito, and T.Ishibashi for valuable discussions, K. Yamada for his collaboration in EL measurement, and K. Hirata and M. Fujimoto for their continuous encouragement.

REFERENCES

1. R. L. Longini, Solid-State Electron. **5**, 127 (1962).
2. R. D. Gold and L. R. Weisberg, Solid-State Electron. **7**, 81 (1964).
3. O. Nakajima, H. Ito, T. Nittono, and K. Nagata, Technical Digest of Int. Electron Device Meeting, San Francisco, Dec. 1990 (IEEE, New York, 1990), pp. 673 - 676.
4. M. E. Hafizi, L. M. Pawlowicz, L. T. Tran, D. K. Umemoto, D. C. Streit, A. K. Oki, M. E. Kim, and K. H. Yen, Technical Digest of GaAs IC Symposium, New Orleans, Oct. 1990 (IEEE, New York, 1990), pp. 329 - 332.
5. H. Kressel and N. E. Byer, Proc. IEEE **57**, 25 (1969).
6. R. L. Hartman, B. Schwartz, and M. Kuhn, Appl. Phys. Lett. **18**, 304 (1971).
7. D. V. Lang and L. C. Kimerling, Phys. Rev. Lett. **33**, 489 (1974).
8. L. C. Kimerling and D. V. Lang, Inst. Phys. Conf. Ser. **23**, 589 (1975).
9. M. K. Sheinkman, JETP Lett. **38**, 330 (1983).
10. M. K. Sheinkman and L. C. Kimerling, in Defect Control in Semiconductors, edited by K. Sumino (North-Holland, Amsterdam, 1990), pp. 97 - 105.
11. M. Uematsu and K. Wada, Appl. Phys. Lett. **58**, 2015 (1991).
12. E. O. Kane, J. Appl. Phys. **32**, 83 (1961).
13. J. H. Buckingham, K. F. Hulme, and J. R. Morgan, Solid-State Electron. **6**, 233 (1963).
14. H. Satoh and T. Imai, Jpn. J. Appl. Phys. **7**, 875 (1968).
15. M. J. Tejwani, H. Kanber, B. M. Paine, and J. M. Whelan, Appl. Phys. Lett. **53**, 2411 (1988).
16. E. F. Schubert, J. M. Kuo, R. F. Kopf, H. S. Luftman, L. C. Hopkins, and N. J. Sauer, J. Appl. Phys. **67**, 1969 (1990).
17. M. Uematsu and K. Wada (submitted for publication).
18. T. Y. Tan and U. Gösele, Mat. Sci. Eng. **B1**, 47 (1988).
19. P. Krispin, J. Appl. Phys. **65**, 3470 (1989).
20. K. Kondo, O. Ueda, S. Isozumi, S. Yamakoshi, K. Akita, and T. Kotani, IEEE Trans. Electron Devices **ED-30**, 321 (1983).
21. K. J. Laidler, Chemical Kinetics (McGraw-Hill, NY, 1950).

ROLE OF THE DIFFUSIVITY OF Be and C IN THE PERFORMANCE OF GaAs/AlGaAs HETEROJUNCTION BIPOLAR TRANSISTORS

F. REN, T. R. FULLOWAN, J. R. LOTHIAN, P. W. WISK, C. R. ABERNATHY, R. F. KOPF, A. B. EMERSON, S. W. DOWNEY AND S. J. PEARTON
AT&T Bell Laboratories, Murray Hill, NJ 07974.

ABSTRACT

GaAs/AlGaAs HBTs with highly Be-doped (4×10^{19} cm^{-3}) base layers show a rapid degradation in current gain during device operation. For example a 2×10 μm^2 device operated at 200°C and a collector current density of 2.5×10^4 A · cm^{-2} shows a decrease in gain from 16 to 1.5 over a period of 12 h. Moreover, both base-emitter and base-collector diode ideality factors worsen dramatically (from 1.33 to 2.39 and 2.01 to 4.51 respectively) during this time. We ascribe this to recombination-enhanced motion of Be interstitials from the base into the adjoining layers. This occurs in both implant-isolated and mesa-etched devices, although the presence of nearby damaged regions does slightly enhance the Be diffusion. By contrast, devices with highly C-doped (7×10^{19} cm^3) bases show no degradation of DC characteristics under the same operating conditions as the Be-doped HBTs. This may be a result of the higher solubility of the carbon, and its occupation of the As sub-lattice.

Introduction

There is currently great interest in the use of GaAs-AlGaAs heterojunction bipolar transistors (HBTs) for a variety of high-speed digital circuit applications. However, a number of authors have reported the degradation of HBTs with conventional p-type dopants (Be or Zn) in the base layer during device operation.[1,2] This involves forward biasing of both base-emitter and base-collector p-n junctions, and it is generally observed that the DC gain of these devices decreases rapidly with time. Uematsu and Wada[3] reported an enhancement by a factor of 10^{15} at room temperature of the Be diffusivity in forward-biased tunnel diodes. This was ascribed to recombination-enhanced diffusion of the Be. Clearly this phenomenon is catastrophic from the viewpoint of stable operation of HBT-based circuits. It is more than a little surprising that this has not been reported earlier given the already relatively long development time of HBTs.

A further problem with the use of Be or Zn as the base dopant is the concentration-dependent diffusivity of these impurities during epitaxial growth, a phenomenon enhanced by the presence of high n-type doping levels in the adjacent emitter and collector layers.[4,5] For this reason, attention has recently been focussed on the use of carbon as the base dopant during MBE,[6] MOMBE[7,8] or MOCVD[9] growth. Carbon has a much lower diffusivity and higher solubility than any of the other p-type dopants, and moreover the diffusion is neither concentration-dependent nor affected by the doping levels in the adjacent layers.

In this paper we report on the current-induced degradation of Be-doped HBTs, and demonstrate that similar C-doped devices do not show this decrease in DC gain under the same conditions. The reason for this stability is most likely the fact that carbon occupies the As sub-lattice and therefore is not affected by Ga$_I$ injection.

Experimental

The Be-doped structures were grown by MBE using Si as the n-type dopant, and the C-doped base structures were grown by MOMBE, with Sn as the n type dopant. The growth temperatures in each case were around 500°C, with the layer structure consisting of a 6000Å n^+ (3×10^{18} cm^{-3}) sub-collector, 4000Å n type (2×10^{16} cm^{-3}) collector, 800Å p^+ (4–7×10^{19} cm^{-3}) base layer, 1000Å n-type (5×10^{17} cm^{-3}) $Al_{0.3}Ga_{0.7}As$ emitter, 2000Å n^+ (1.5×10^{19} cm^{-3}) GaAs emitter cap layer and 300Å n^+ (10^{19} cm^{-3}) InGaAs contact layer. Both small geometry (2×10 μm^2) devices fabricated by a dry-etched, self-aligned process[10] using implant (F^+ and H^+) isolation and large geometry (100 μm diameter) devices fabricated with a wet chemical, mesa-etch process were biased at 200°C and collector current densities of 2.5×10^4 A · cm^{-2} for the small devices and 270Å · cm^{-2} for the large devices for periods up to 12h. The DC gains and junction ideality factors of both Be- and C-doped devices were monitored as a function of time. On some samples, Resonance Ionization Mass Spectrometry (RIMS)[11] measurements of the Be profile in the large structures were made to monitor possible Be motion as a result of either forward biasing of the devices or the implant/anneal cycle used for isolation.

Results and Discussion

Table 1 shows the results of the current-induced changes in the 2×10 μm^2 Be-doped devices. Over the 12h period the DC gain these devices falls from 16 to 1.5, and both base-emitter and base-collector junction ideality factors are severely degraded. While it is not clear as to the physical meaning of the ideality factors being greater than 2, it is clear that the junction characteristics have been greatly compromised. By sharp contrast, there is no significant change in the characteristics of the C-doped devices. We note that this is the case even though the C-doping level is substantially higher than that of the Be. It is possible that under even more extreme levels of minority carrier injection the C-doped devices might show some change in DC properties, but it then becomes an issue as to whether the ohmic contacts remain stable under such conditions. We note that no degradation of the Be-doped HBTs was observed under zero-bias conditions at the same temperature (200°C), implicating the minority carriers in the degradation mechanism and supporting the view that recombination-enhanced diffusion of the Be is the cause.[3] This is also consistent with the decrease in gain and worsening junction ideality factors.

Table 1. Comparison of DC characteristics of 2×10 μm^2 Be-(4×10^{19} cm^{-3}) or C-(7×10^{19} cm^{-3}) doped HBTs operated at 200°C and a collector current density of 2.5×10^4 A · cm^{-2} for 12h

	Be-doped before	Be-doped after 12h	C-doped before	C-doped after 12h
Current Gain	16	1.5	17	16
B-E Ideality Factor	1.3	2.4	1.4	1.4
B-C Ideality Factor	2.0	4.5	1.3	1.3

Figure 1 shows RIMS profiles before and after bias application at 200°C on a large diameter Be-doped HBT. There is a slight, but significant movement of Be from the base into the collector layer, consistent with the mechanism postulated above for the device degradation. This is the first direct measurement of the Be redistribution as a result of forward-biasing in any device structure. We also point out that the Be need only "punch-through" in a few microscopic regions to have an effect on the device operating characteristics and the RIMS data is averaging over quite a large area and so is not as sensitive as the device itself in detecting motion of Be.

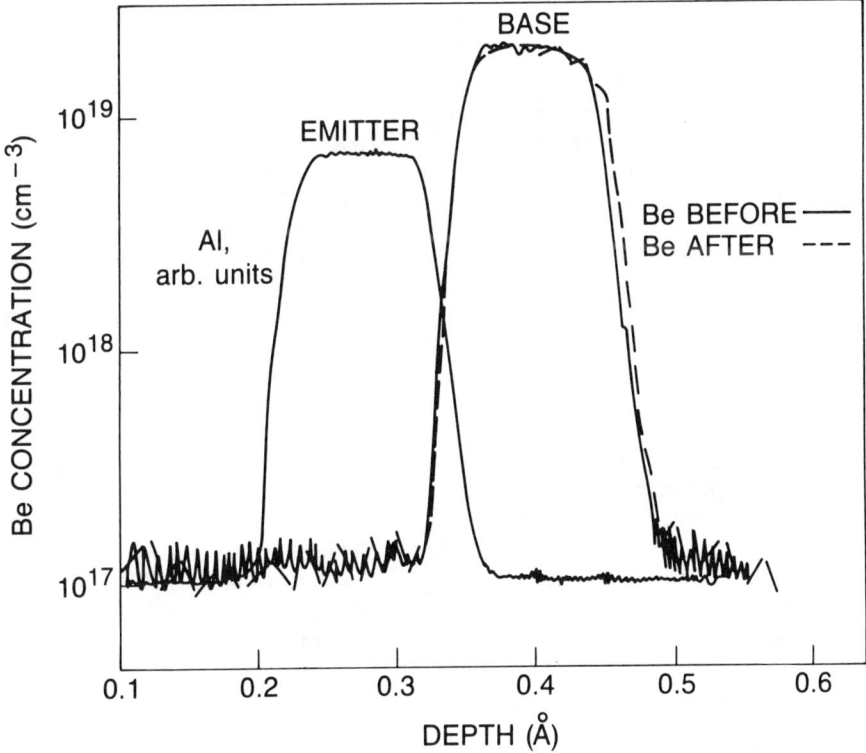

Figure 1. RIMS profiles of Be in an HBT structure before and after bias application at 200°C for 12h.

The role of the nearby implant-damage isolated regions in enhancing the Be diffusion was also examined. Fluorine ions were implanted at doses of 7×10^{13} cm^{-2} (40 keV), 7×10^{13} cm^{-2} (100 keV), 6×10^{12} cm^{-2} (200 keV) and 7×10^{12} cm^{-2} (250 keV) into GaAs p-n junctions (2000Å p = 3×10^{19} cm^{-3}, 3000Å n = 5×10^{16} cm^{-3} and 1000Å n = 4×10^{18} cm^{-3}) grown by MBE with Be and Si dopants and similar samples were also implanted with H$^+$ 40 keV (3×10^{15} cm^{-2}) and 60 keV (3×10^{15} cm^{-2}) ions. Following an anneal at 525°C for 60 sec, RIMS profiles (Figure 2) showed a small amount of Be redistribution, and as stated previously we observed similar rates of device degradation in mesa-etched or implant isolated structures. The slightly greater amount of Be diffusion in the case of F$^+$ implantation is due to the greater amount of damage created relative to fluorine (Figure 3). This additional damage is reflected in the worsened ideality factors from the base collector junction after F^+ or H^+ implantation and annealing (Figure 4).

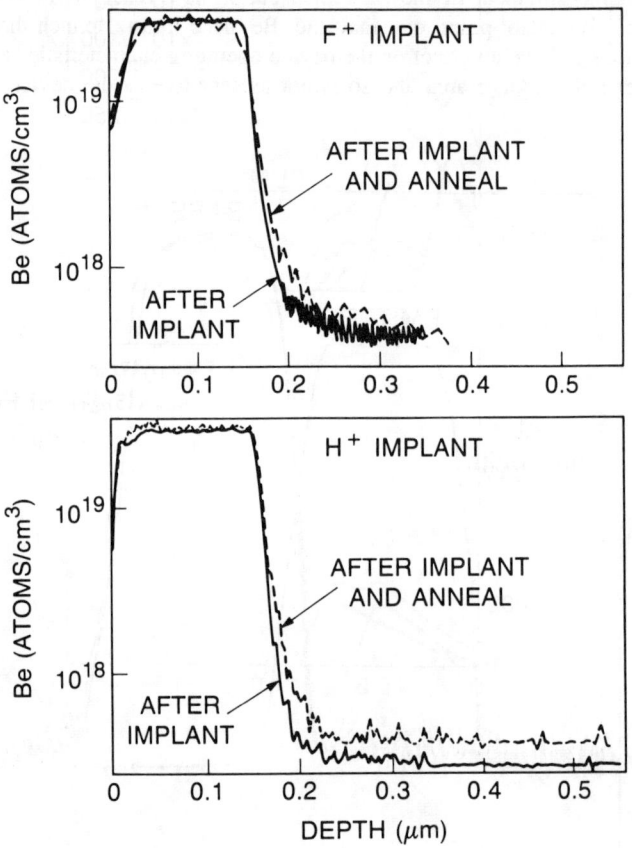

Figure 2. RIMS profiles of Be in HBT structures after implantation with F^+ (top) or H^+ (bottom) and after subsequent annealing to provide isolation.

Conclusions and Summary

The rapid degradation of HBTs with highly Be-doped base layers under elevated temperature bias application has been studied. The DC gain of these devices falls rapidly with time during forward bias application due to an enhanced diffusion of Be out of the base region into the adjoining emitter and collector layers. This appears to be a result of a recombination-enhanced motion of Be interstitials. In sharp contrast, structures with carbon-doped base layers do not show any significant degradation during forward bias operation. We assume that this is due to the As site occupation by the C, so that injection of Ga_I by the high current flow in the devices does not create carbon interstitials. A related fact may be that in the case of MOMBE carbon doping, there is essentially 100% substitutional incorporation of the carbon so that there are few interstitials to begin with. By contrast, the substitutional fraction of Be in MBE-grown base layers falls progressively below unity for doping levels above $\sim 3 \times 10^{19}$ cm^{-3}, so that there is already a supply of Be_I prior to bias application.

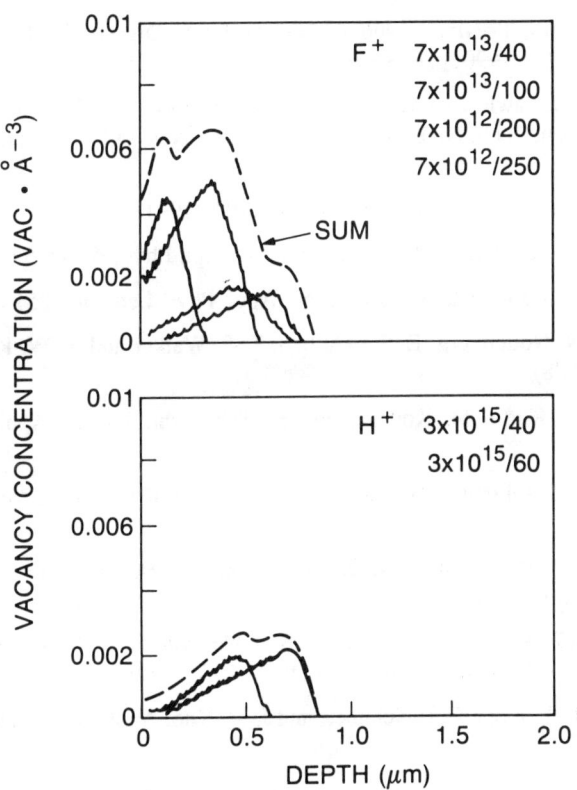

Figure 3. Damage profiles created in GaAs p-n junction structures by multiple F^+ (top) or H^+ (bottom) implantation of the type used in Figure 2.

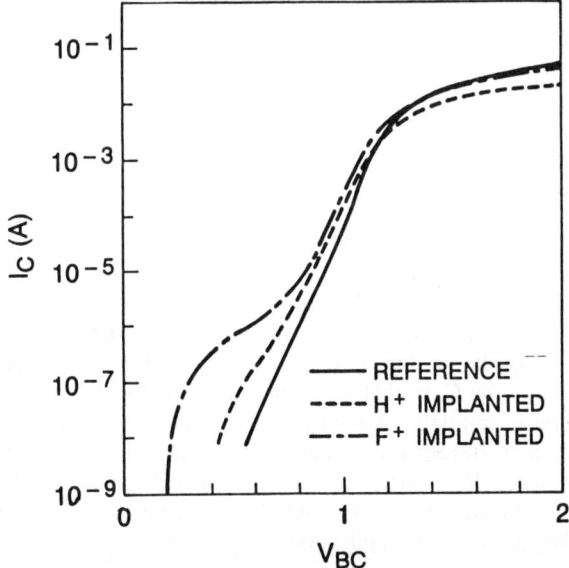

Figure 4. Base-collector I-V characteristics from p-n junction samples implanted with F^+ or H^+ and then annealed.

References

1. O. Nakajima, H. Ito, T. Nittono and K. Nagata, Tech. Digest of IEDM, San Francisco, Dec. 1990 (IEEE, NY 1990) p. 673-676.
2. M. E. Hafizi, L. M. Pawlowicz, L. T. Tran, D. K. Umemoto, D. C. Streit, A. K. Oki, M. E. Kim and K. H. Yen, Tech. Digest of GaAs IC Symp., New Orleans, Oct. 1990 (IEEE, NY 1990) pp. 329-332.
3. M. Uematsu and K. Wada, Appl. Phys. Lett. 58 2015 (1991).
4. P. M. Enquist, J. A. Hutchby and T. J. de Lyon, J. Appl. Phys. 63 4485 (1988).
5. W. S. Hobson, A. S. Jordan and S. J. Pearton, Appl. Phys. Lett. 56 1251 (1990).
6. R. J. Malik, R. N. Nottenburg, E. F. Schubert, T. F. Walker and R. W. Ryan, Appl. Phys. Lett. 54 2661 (1988).
7. E. Tokumitsu, Y. Kudo, M. Konagai and K. Takahashi, Jap. J. Appl. Phys. 24 1189 (1985).
8. C. R. Abernathy, S. J. Pearton, R. Caruso, F. Ren, and J. Kovalchick, Appl. Phys. Lett. 55 1750 (1989).
9. B. T. Cunningham, M. A. Huase, M. J. McCollum, J. E. Baker and G. E. Stillman, Appl. Phy. Lett. 54 1905 (1989).
10. F. Ren, T. R. Fullowan, C. R. Aberanthy, S. J. Pearton, P. R. Smith, R. F. Kopf, E. J. Laskowski and J. R. Lothian, Electronics Lett. 27 1054 (1991).
11. S. W. Downey, R. F. Kopf, E. F. Schubert and J. M. Kuo, Appl. Optics, 29 4938 (1990).

EFFECTS OF THE SUBSTRATE-EPITAXIAL LAYER INTERFACE ON THE DLTS SPECTRA IN MESFET AND HFET DEVICES

M. SPECTOR[1], M. J. GRAY[1], J. D. YODER[1], A. M. SERGENT[2] AND J. C. LICINI[3]

[1] AT&T Bell Laboratories, 2525 N. 12th Street, Reading, PA 19612

[2] AT&T Bell Laboratories, 600 Mountain Avenue, Murray Hill, NJ 07974

[3] Lehigh University, Bethlehem, PA 18015

ABSTRACT

Sidegating and low frequency oscillations are very detrimental to the performance of GaAs MESFETs and AlGaAs/GaAs HFETs. Sidegating is highly enhanced by carbon impurities at the substrate-epitaxy interface. Deep level traps have been associated with low frequency oscillations and light sensitivity of the devices. We have studied the effect of substrate cleaning using ultraviolet-ozone radiation on DLTS spectra in MESFET and HFET devices. The incorporation of a superlattice at the interface as well as variable buffer layer thickness were investigated. Deep level transient spectroscopy was used to identify the traps present in these materials.

The results showed the DLTS peak causing the device light sensitivity can be eliminated by cleaning the interface. This hole trap has an activation energy of 0.5 eV and a capture cross section of $4 \times 10^{-14} cm^2$.

1. Introduction.

Although a considerable number of studies on defects have been performed on GaAs and AlGaAs, their identification and effects on the devices that were fabricated using these materials need further understanding. Additionally, the properties of the substrate-epitaxial layer interface have a major impact on device electrical characteristics, degrading device performance and strongly influencing device reproducibility and yield[1][2][3]. As a result of this, the understanding of the behavior of the deep level occupancy as a function of this interface is essential for the successful fabrication of devices and integrated circuits on GaAs epitaxial wafers.

In this paper, we report the deep levels detected in GaAs MESFETs and HFETs that were fabricated using MBE at AT&T Bell Labs Reading. The possible origin of the deep levels and their effects on device performance were investigated. First, we shall describe the devices used in this study. The conductance DLTS apparatus will be described in detail. The discussion of the results, and finally the conclusions and recommendations are presented in the last section.

2. Experiment.

2.1 Device Structure.

The MESFETs and HFETs used in the experiments were long channel (45 µm) FETs with channel width-to-length (Z/L) ratios of 1/2. Some smaller devices with Z=1.0 µm and L=18 µm were also measured. The large FETs are required for the capacitance DLTS measurements since in a standard FET the capacitance is too small to produce measurable signal levels. The smaller (standard) devices were used to verify that there were no differences with the large FETs spectra due to geometrical or surface potential fluctuations[4].

The MESFET structure consisted of a 1000 Å undoped buffer layer grown on undoped GaAs substrates. An alternating (40 Å GaAs/40 Å AlGaAs) superlattice and another undoped buffer layer (1000 Å) were then deposited. Next the n-GaAs conductive layer was grown with a Si doping concentration of 1×10^{17} cm^{-3}. A heavily (n$^+$) Si doped (2.5×10^{18} cm^{-3}) layer was deposited over this structure to enhance the drain and source Au/Ge/Ni/Au ohmic contacts. Ti/Pt/Au was used for the gate metallization. An oxygen ion implant was used to isolate the devices. The MESFETs were fabricated to assess the influence of several variables on deep levels. We used two thicknesses for the undoped GaAs buffer layer that followed the deposition of the superlattice, standard (1000 Å), and thick (10000 Å). The superlattice was eliminated for some of the wafers. Finally, we studied the effects of two different cleaning procedures on the GaAs substrate surfaces. The standard clean, consists of a 5:2:10 $NH_4OH:H_2O_2:H_2O$ etch. Some wafers received an ultraviolet-ozone clean. Exposure to ultraviolet radiation photochemically oxidizes the carbon on the wafer surface. This oxide is desorbed prior to epitaxial deposition[3].

The HFET structure consisted of a superlattice grown on the GaAs undoped semi-insulating substrate. The growth proceeded by the deposition of an undoped GaAs buffer layer. Undoped spacer layers were grown both above and below the Si-doped AlGaAs donor layer. At the interface between the undoped AlGaAs spacer and the undoped GaAs buffer layer a two dimensional electron gas (2DEG) is formed when the electrons from the donor layer fall into the triangular quantum well created by the band discontinuity between AlGaAs and GaAs. A GaAs cap layer is deposited over the top spacer layer and the other three layers are added to produce the depletion mode structure. The HFET devices were fabricated by using self-aligned tungsten silicide refractory gates. Source and drain n$^+$ ohmic contact regions were formed by Si ion-implantation. The silicon implanted ions were activated by furnace annealing. The ohmic contact metallization was Au/Ge/Ni/Au. An oxygen ion implant was used for device isolation.

2.2 Experimental Technique

Capacitance and conductance DLTS were measured as well as Device I-V characteristics. The capacitance DLTS system used was the original one D. V. Lang designed for his DLTS experiments[5]. The conductance DLTS experiment allows the measurements of DLTS signals from devices which would be ordinarily too small or have an inappropriate geometry for standard capacitance DLTS measurements and it is accomplished as follows: A precision power supply was used to apply a small bias voltage (50 mV) between the source and the drain of the test FET, in order to operate in the ohmic region. The Schottky barrier gate depletion region was changed with a reverse bias pulse from a pulse generator which was triggered through an oscilloscope to a boxcar. The source of the FET was connected to a low noise transimpedance amplifier, and the amplified signal is connected to the input of the gate integrator. The signal was sampled at times t_1 and t_2 during the on/off transient, and the DLTS signal is the difference in decay amplitudes at these two times. The transient-time τ (inverse of the thermal emission constant e) changes with temperature and when this difference is on the order of the window time ($t_2 - t_1$), the boxcar will produce a maximum output signal and therefore a DLTS peak. A DLTS spectrum is generated by sweeping the temperature and plotting the DLTS signal versus temperature. We swept the temperature between -50°C and 150°C in 1°C intervals. The output from the boxcar was connected to a voltmeter. An AT&T PC with a HPIB interface sweeps the oven temperature and reads the temperature and the voltmeter values. The data was plotted using a X-Y plotter connected to the voltmeter and also to the oven with a thermocouple.

3. Results and Discussion

The transient time constant giving rise to the maximum in the boxcar output is the reciprocal of the emission constant of the trap and is given by:

$\tau_{max} = 1/e = (t_1 - t_2) / \ln(t_1/t_2)$

Using the relation[6][7]

$T^2 \tau = (g_c / \sigma <v_{th}> N_c) \exp (\Delta E/kT)$

where $<v_{th}>$ is the mean thermal velocity of the carriers, N_c is the effective density of states in the corresponding band and g_c is the degeneracy of the trap level, the deep level activation energy (E) and the capture cross section (σ) can be determined from the slope and the intercept of the $T^2 \tau$ versus 1000/kT log-linear plot, respectively.

Figure 1a shows the conductance DLTS spectrum with the two traps found for the MESFETs. Their emission temperatures are approximately 0°C and 50°C respectively for 1/10 ms (t_1/t_2) window. Their activation energies were 0.52 eV and 0.83 eV respectively and their capture cross sections were 4.0 x 10^{-14} cm^{-2} and 7 x 10^{-14} cm^{-2}. Figure 1b is the capacitance DLTS spectrum obtained on the same device as represented in Figure 1a showing a hole trap for the first deep level and an electron trap for the second one. The calculated activation energies, obtained by capacitance DLTS were the same as those calculated by conductance DLTS.

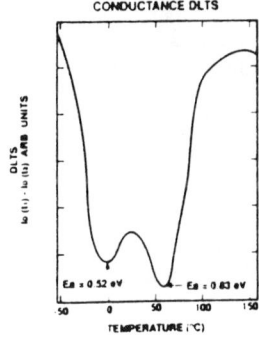
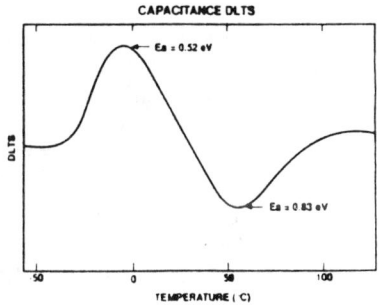

FIGURE 1a
Conductance DLTS Spectrum
showing the two traps
found in the MESFETs

FIGURE 1b
Capacitance DLTS showing a
hole trap for the first peak and
an electron trap for the second one.

The first trap did not appear on the DLTS spectra for MESFETs that received the ozone clean prior to epitaxial growth. From Figure 2 we can see that the first emission trap is not present for these ozone-treated wafers regardless of the undoped GaAs buffer layer thickness used. These same deep levels were measured by DLTS for standard cleaned HFET devices.

MESFET INTERFACE CLEANING EFFECT

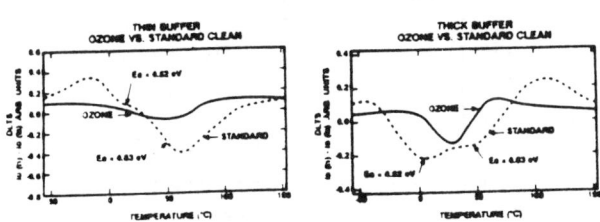
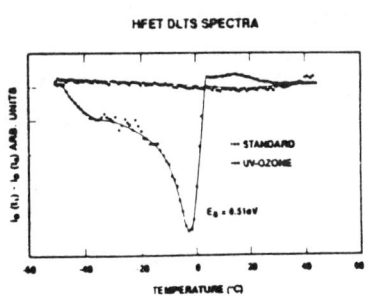

FIGURE 2
The hole trap is not active for either
thin or thick substrates after
UV-ozone cleaning

FIGURE 3
HFET DLTS spectra for UV-clean and
standard clean substrate, where the
first trap doesn't appear for the UV-clean

The 0.52 eV hole trap was also absent for the ozone cleaned HFET-devices (Figure 3). Figure 3[1] contains DLTS spectra of HFET devices fabricated on standard and ozone-cleaned wafers. This trap produces device instabilities when the trap occupancy changes. Figure 4 illustrates a possible explanation. A carbon concentration of 10^{14} cm^{-2} has been measured[1] on the standard clean wafers. This produces a highly p-type interface which pins the Fermi level at the shallow carbon acceptor level and produces a band bending at the substrate-epi layer interface (dashed line). Under these conditions (circles), the hole trap ($E_a = 0.52$ eV) can be populated and detected. When the substrate is UV-ozone cleaned, the carbon concentration is significantly decreased, raising the Fermi level substantially above the 0.52 eV acceptor level. Under these conditions (triangles), the hole trap can not change occupancy, therefore, it is not detected in the DLTS spectrum.

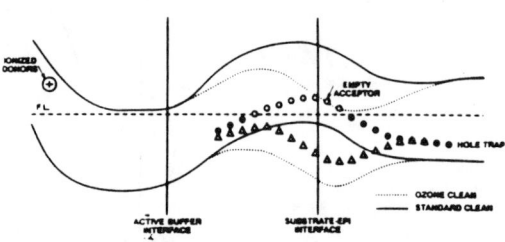

FIGURE 4
The illustration for a possible explanation of the hole trap appearance on the DLTS spectra based on the band bending at the substrate-epi interface

Hole traps in MBE GaAs at growth temperature above 550°C have been associated with the presence of impurities such as Fe, Cu, and Cr. These impurities may be the result of contaminants in the MBE system or due to outdiffusion from the substrate during growth[8]. In order to investigate the signature of the Fe trap, DLTS measurements were performed on intentionally Fe doped substrates (Figure 5).

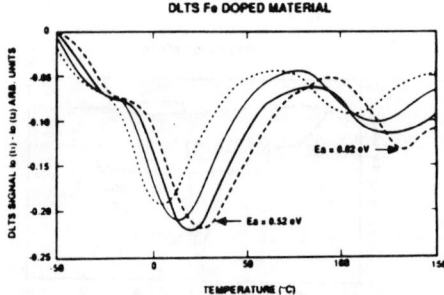

FIGURE 5
DLTS spectra for an intentionally Fe doped ion implanted MESFET.

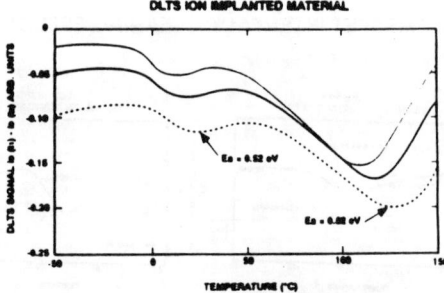

FIGURE 6
Ion implanted MESFET isolated with H$^+$ DLTS spectra. The first trap signal is the Fe and the second one is the EL2.

These MESFET devices were fabricated using ion-implantation technology to form the active and ohmic contact regions. The first trap is due to Fe and the second one is the EL2. The calculated activation energies and capture cross sections of the DLTS spectrum shown in Figure 7 were: Ea_{Fe} = 0.52 eV and σ_{Fe} = 1.3 x 10^{-14} cm^{-2} and E_{EL2} = 0.82 eV and σ_{EL2} = 8 x 10^{-14} cm^{-2}. A thermal activation energy of 0.52 eV was also determined from an Arrhenius plot obtained from a frequency and temperature dependence of the output impedance measurements performed on standard HFETs. A complete description of the output impedance measurements has been reported in a previous paper[1]. The activation energies determined by the output impedance and DLTS measurements were consistent. The capture cross sections were similar also (4.4 x 10^{-14} cm^{-2} by the output impedance method and 4.0 x 10^{-14} cm^{-2} by DLTS measurements). A possible origin for this hole trap may be Fe since the MBE chamber is constructed of stainless steel. The Fe hole trap has been detected in previous studies on MBE grown GaAs[9][10].

The second trap detected by DLTS for the MBE-grown MESFETs and HFETs is an electron trap and it was not influenced by the cleaning procedure. We believe that this electron trap may be responsible for low frequency oscillations that have been measured[11] on Transimpedance Preamplifier Integrated Circuits fabricated on the MESFET structure described in this paper. At room temperature operation these circuits reach 50°C[12] due to the power being dissipated by the circuit under bias. This 50°C operating temperature matches the emission temperature for this second trap. The activation energy and capture cross section determined from the DLTS data were 0.83 eV and 7.4 x 10^{-14} cm^{-2}. From the 1/10 ms window in our DLTS measurements we calculated a frequency of 250 Hz for the trap emission temperature of 50°C. This corresponds closely with the frequency of the oscillation for the circuits, 200 Hz. There is a possibility that this trap may be an oxygen related level. There have been previous reports of an oxygen trap detected for GaAs MESFETs[13][14][15].

A possible origin for this trap may be from the oxygen implant isolation. We have measured DLTS spectra for proton isolated ion-implanted MESFETs, as shown in Figure 6, and we did not detect this trap. The first trap in this DLTS spectrum is the 0.52 eV hole trap possible related to Fe and the second electron trap is the EL2. It should be noticed that both the oxygen-related trap and the EL2 are present on the intentionally doped ion implanted MESFET isolated with oxygen.

Changing the undoped GaAs buffer layer thickness or eliminating the superlattice buffer did not influence the major features of the DLTS spectrum. Both hole and electron traps appeared when the standard clean was used and only the second trap (E_a = 0.83 eV) was observed for the ozone clean wafers.

4. Conclusions and Recommendations.

A 0.52 eV hole trap is present in our standard cleaned MESFETs and HFETs. Its source may be not eliminated but in order for this trap to be electrically inactive the interfacial carbon concentration must be decreased. This can be accomplished by utilizing the UV-ozone wafer surface cleaning. This cleaning is highly recommended since it also suppresses sidegating, and improves electron mobility and electrical device characteristics[3].

The 0.83 eV electron trap needs further investigation. A possible way to eliminate this trap could be by using H^+ for device isolation. Based upon the circuit noise measurements and DLTS measurements it is possible to further evaluate the origin of the low frequency oscillations. A more extensive study of the use of H^+ device isolation may assist in determining the origin of this electron trap. Initial evaluations of sidegating characteristics using this isolation have shown improvement[16].

The conductance DLTS measurements have been shown to be a useful technique to characterize our MESFETs and HFETs. The measurement results are consistent with device output impedance measurements. The DLTS results have demonstrated that the cleaning of the substrate-epi interface is a very important step in the fabrication of high performance GaAs devices.

Acknowledgements.

We would like to acknowledge valuable discussions with D. V. Lang as well as his encouragement and advice. In addition, we want to thank J. M. Parsey for useful discussions, S. H. Wemple for his comments on the manuscript, S. F. Kruse for the fabrication of the devices and E. K. Tobias for the the device assembly.

REFERENCES

[1] M. L. Gray, M. Spector, J. D. Yoder, "MBE Heterostructure Device Instabilities Related to Interfacial Impurities", to be published.
[2] C. P. Lee, "Influence of Substrates on the Electrical Properties of GaAs FET Devices and Integrated Circuits".
[3] M. L. Gray, C. L. Reynolds and J. M. Parsey, Jr., J. Appl. Phys. 68, 169, (1990).
[4] J. P. Harrang, A. Tarseila, M. Rosso, and P. Alnot, "Conductance Transient Spectroscopy of Metal-Semiconductor Field Effect Transistors", J. Appl. Phys. 61(5), March 1987.
[5] D. V. Lang, "Space-Charge Spectroscopy in Semiconductors", Thermally Stimulated Relaxation in Solids, Editor; P. Braunlich, Springer-Verlag, pp 92-129.
[6] J. D. Yoder, "Low Frequency Oscillations in the 1521A GaAs Transimpedance Preamplifier", to be published.
[7] R. S. Miller, T. I. Kamins, J. Wiley & Sons, "Device Electronic for Integrated Circuits", 1977, pp. 156-158.
[8] M. Ilegems, "Properties of III-V Layers", Plenum Press, pp 137, (1987).
[9] W. J. Schaff and L. F. Eastman, B. Van Rees and B. Liles, "Superlattice Buffers for GaAs Power MESFETs Grown by MBE", J. Vac. Sci. Technol. B 2(2), Apr.-June 1984.
[10] A. Mitonneau, G. M. Martin, A. Mircea, "Hole Traps in Bulk and Epitaxial GaAs Crystals", Electron. Lett. 13, 666 (1977).
[11] J. P. Furino, AT&T Bell Labs, private communication.
[12] M. Spector, F. J. James, "Temperature Measurements on GaAs ICs", Thermal Design Forum, MH, October 30, 1986.
[13] M. G. Adlerstein, "Electrical Traps in GaAs MIcrowave F.E.T.s", Electronics Letters, June 10, 1976, Vol. 12 No.12.
[14] G. M. Martin, A. Mitonneau and A. Mircea, Electron. Lett. 13, 191, 1977.
[15] J. Lagowski, D. G. Lin, T. Aoyama, and H. C. Gatos, "Identification of Oxygen-Related Midgap Level in GaAs", Appl. Phys. Lett., Vol. 44, No. 3, February 1, 1984.
[16] A. D. Brotman, AT&T Bell Labs, private communication.

THE STUDY OF INTERFACIAL TRAPS OF InP METAL-INSULATOR-SEMICONDUCTOR STRUCTURES

Lu Liwu, Zhou Jie, Qu Wei and Zhang Shengliang
National Laboratory for Superlattices and Microstructures, Institute of Semiconductors, Academia Sinica, Beijing 100083, P. R. China

The interfacial traps of InP MIS structures with different insulating layers grown by PECVD has been studied using DLTS technique. Experimental results show that the interfacial traps are located in the interface between the insulator and InP, and near the interface in InP. The deep level parameters of the traps indicated that its origin might be due to (1) Part of P atoms evaporate and form P vacancies in InP surface during PECVD; (2) Native defects in InP substrate; (3) Irradiation damage induced by plasma during PECVD.

1. INTRODUCTION

InP is well suited for use in high-frequency and optoelectronic devices by virtue of its high electron mobility and large band gap. Devices require the formation of Schottky contacts. However, most metal contact on InP have low Schottky barrier energy and this gives rise to a large leakage current under reverse bias. Hence, an insulating layer about 100-3000 A thickness is required at metal-semiconductor interface to reduce the leakage current. InP devices based on the MIS structures suffer from current drift and a gradual deterioration in their performance [1]. The interfacial traps of InP MIS structures can play a significant role in device performance [2]. It is the purpose of this work to report the detailed study of the physical behaviour of interfacial traps of InP MIS structures.

2. SAMPLE PREPARATION AND EXPERIMENTAL CONDITIONS

The substrates used here are n-type Sn doped InP LEC crystal with a carrier concentration of $5 \times 10^{17} cm^{-3}$. Then the 2000 A thick SiO_2 layer was deposited on the sample No. 1 and No. 2 by PECVD. And the 1000 A thick SiO_2N and SiO_xN_y layers were deposited on the sample No. 3 and No. 4 by PECVD, respectively. The PECVD system DP-80 is made by Plasma Technology corporation, British. The conditions of PECVD for all samples are as follows: high frequency 13.56 MHz, low pressure 150 mev, work temperature 340 C and power density 0.01 W/cm^2. The Cr-Au was evaporated on the insulating layer to form the metal gate. The ohmic contact was provided by evaporating Au-Ge-Ni on the back surface of samples and heating at 420 C. The

DLTS measurements in the temperature range from 77 to 350 K were carried out using DLS-82E, from SemiLab, Hungary. The typical experimental conditions are reverse bias voltage Vr=-0.2 V, pulse amplitude ΔV = 1.8 V, pulse width tp= 100 μs and frequency f= 23 Hz. The level positions of traps are determined from Arrhenius plots, while capture cross sections are obtained directly from measurements at varying pulse width.

3. EXPERIMENTAL RESULTS

(1) DLTS measurement method

For the DLTS measurement, an MIS diode is biased as illustrated in Fig.1(a). By applying positive-going pulses with V = V_a, surface states and interfacial traps are filled with electrons as shown in Fig.1(b) (the upper part). The diode is then reverse biased at V = Vr. In this period, electrons trapped above the Fermi level are emitted as shown in Fig.1(b) (the lower part). Accordingly, the diode capacitance increses. The diode capacitance also increases by minority carrier generation when Vr is large enough to generate an inversion layer. Therefore, the DLTS signal consists of three components: surface states, interfacial traps and carrier generation. These components can be separated by adjusting Vr and Va. When the surface potential

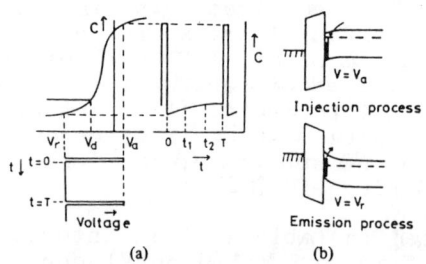

Fig. 1. DLTS measurement.
(a) Shape of bias voltage and capacitance variation.
(b) Energy diagrams during injection and emission.

is adjusted such that no inversion layer is formed during DLTS measurement, the minority-carrier generation is suppressed and DLTS signal represents the emission from interfacial traps and surface states. And both can be separated by measuring shift of DLTS spectra with varying pulse heights V_a, since the peak temperature does not change when traps dominate but it change when states dominate.

(2) DLTS measuring results

The typical DLTS spectra of samples No.1 and No.2 are shown in Fig.2. Seven deep level defects, labeled E1 to E7, which are located at 0.13, 0.32, 0.39, 0.42, 0.15, 0.39 and 0.49ev below E_c, are observed respectively. The deep level parameters defects are listed in Table I. Among above defects, E1(0.13ev) and E5(0.15ev) have the similar deep level parameters to that of trap at 0.16ev of Levinson et al [4]., E2(0.32ev), E3(0.39ev) and E6(0.39ev) have the similar deep level parameters to that of trap at 0.31 and 0.40ev of Yamazoe et al [5]. Although E7(0.49ev) has the similar level position to that of trap at 0.51ev of Lim et al [6]., the difference between both values of capture cross section is much too large (see Table I). The typical DLTS spectra of samples No. 3 and No. 4 are shown in Fig. 3. Six deep level defects, labeled E8 to E13, which are located at 0.21, 0.40, 0.15, 0.36, 0.40 and 0.59ev below E_c, can be observed. Their detailed parameters are also

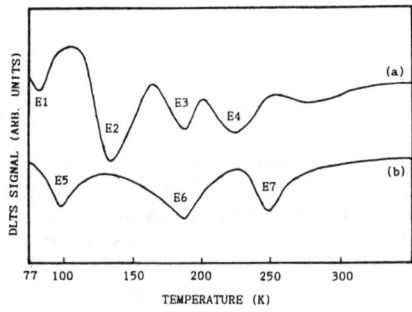

Fig. 2. DLTS spectra of InP MIS structure.
(a) Sample No. 1.
(b) Sample No. 2.
$V_r=-0.2$ V, $\Delta V=1.8$ V, f=23Hz.

Fig. 3. DLTS spectra of InP MIS structure.
(a) Sample No. 3.
(b) Sample No. 4.
$V_r=-0.2$ V, $\Delta V=1.8$ V, f=23Hz.

listed in Table I. It is interesting for us that E8 (0.21ev), E10 (0.15ev) and E11(0.36EV) could be identical with the trap at 0.21, 0.14 and 0.37ev of Levinson et al [4]. And E9 (0.40ev), E12 (0.40ev) and E13 (0.59ev) could be identical to the traps at 0.40 and 0.60ev of Yamazoe et al [5]. Table I summarized the results of previous workers and the results of this work.

Table I. Summary of Experimental Results

Sample No.	Label	Trap notation			Ref.	
		E_c-E_t (ev)	N_t ($\times 10^{14}$cm^{-3})	σ_n ($\times 10^{-16}$cm^2)	E_c-E_t (ev)	σ_n ($\times 10^{-16}$cm^2)
1	E1	0.13	5.0	1.3	0.14	> 2.7 [8]
	E2	0.32	16	0.25	0.31	0.015 [5]
	E3	0.39	12	2.6	0.40	0.33 [5]
	E4	0.42	12	91	0.40	0.33 [5]
2	E5	0.15	20	0.52	0.14	> 2.7 [8]
	E6	0.39	30	3.8	0.40	0.33 [5]
	E7	0.49	30	17	0.51	4×10^4 [6]
3	E8	0.21	5.0	7.7	0.22	> 2.2 [8]
	E9	0.40	8.0	2.4	0.40	0.33 [5]
4	E10	0.15	8.0	0.28	0.14	> 2.7 [8]
	E11	0.36	14	48	0.37	> 1.7 [8]
	E12	0.40	14	1.9	0.40	0.33 [5]
	E13	0.59	8.0	1.6	0.60	5.0 [5]

In all our DLTS measurements, the filling pulse was 1.8 V and thus can be considered large-pulse DLTS. We have also made a series of small pulse DLTS measurements where the filling pulse was only 0.2 V. It can be seen that at the same frequency the same peak consistently appeared indicating that the defects were not interface states but traps, that is, interfacial traps.

4. DISCUSSION

(1) Because of the high vapour pressure of P, P vacancies in InP LEC crystal is easily formed even in low-temperature heat treatment, and Vp is important in the introduction of deep levels in the course of InP device fabrication. The deep levels in InP caused by heat treatment (350 C, 450 C and 550 C) have been systematically investigated by DLTS and PL measurements [5]. As a result of the comparison of DLTS and PL spectra, it has been found that a DLTS peak whose emission activation energy is about 0.40ev strongly correlates the well-known 1.1ev emission peak in PL spectra and origin of this single can be identified as a complex state associated with Vp. According to the defect model proposed by Spicer et al. [7] the level at 0.40ev is associated with a Vp. So the E3 (0.39ev), E6 (0.39ev), E9 (0.40ev) and E12 (0.40ev) may be related to Vp which was formed in the work temperature of PECVD (340 C) due to P evaporation. Although E4 (0.42ev) has the similar level position to trap at 0.40ev of Yamazoe [5], the difference of both values of capture cross section is too large. In view of discrepency, the origin of E4 thus remains unknown.

(2) Yamazoe et al. [5] observed the trap at 0.31 and 0.60ev which might be due to native defects in InP. These traps were annealed out in the N_2 atmosphere and enhanced after heat treatment under excess P pressure. this indicates that they are related to native defects probably associated with P interstitutial or In vacancy. E2 (0.32ev) and E13 (0.59ev) could be identical to traps at 0.31ev and 0.60ev of Yamazoe et al. [5] and may be due to native defects in InP substrates.

(3) Lim et al. [6][9] reported that the trap at 0.51ev is due to oxygen, since it is found in oxygen doped InP and InP MIS fabricated by chemical oxidation. However, its capture cross section is much too large and the peak occurred at too low a temperature for the given emission rate, compared to that of E7 (0.49ev). Thus whether E7 is related to oxygen remains unknown.

(4) Levinson et al. [8] reported the observation of metastable M center in electron irradiated InP, which was shown to exist in either of two configurations, A or B, depending on its charge state. Each configuration is characterized by distinct DLTS spectra. The A configuration is obtained when the sample is cooled to below 160 K without any bias, whereas the B configuration is obtained when a reverse bias is applyed during cooling. The DLTS curve for B configuration showed defect levels at 0.09, 0.14, 0.22 and 0.37ev below E_c. When the defect electronic states are fully occupied, the defect is in A configuration, when three electrons are lost, B configuration is formed according to the following

reaction: $A^0 \rightarrow A^+ + e^- \rightarrow A^{2+} + 2e^- \rightarrow A^{3+} + 3e^- \rightarrow B^{3+} + 3e^-$
In the final state B^{3+}, the defect acts as a trivalent electron trap, and each capture and subsequent emission of electron is responsible for the DLTS peaks. A comparison between our results and Levinson et al. shows that E1(0.13ev), E5(0.15ev) and E10(0.15ev) could be identical with the trap at 0.14ev, E8(0.21ev) could be identical with the trap at 0.22ev, E11(0.36ev) could be identical with the trap at 0.37ev, indicating that E1, E5, E8, E10 and E11 are similar to those found in electron bombarded InP. They may be related to irradiation damage induced by plasma during PECVD.

5. CONCLUSION

We have reported the observation of interfacial traps that are formed in InP MIS structures with different insulating layers. The origin of these traps might be due to (1) Part of P atmos evaporate and form V_p in InP surface during PECVD; (2) Native defects in InP substrate; (3) Irradiation damage induced by plasma during PECVD.

REFERENCES

[1] D. Fritzsche: Inst. Phys. Conf. Ser. 50, 1980, 258.
[2] M. Okamura and T. Kobayashi: Jpn. J. Appl. Phys., 1980, 19, 2143.
[3] T. Katsube, I. Sakata, and T. Ikoma: IEEE Trans. Electron Devices, 1980, ED-27, 1238.
[4] M. Levinson, J. L. Benton, H. Temkin, and L. C. Kimerling: Appl. Phys. Lett., 1982, 40, 990.
[5] Y. Yamazoe, Y. Sasai, N. Nishino, and Y. Hamakawa: Jpn. J. Appl. Phys., 1981, 20, 347.
[6] H. Lim, G. Sagnes, and G. Bastide: J. Appl. Phys., 1982, 53, 7450.
[7] W. E. Spicer, I. Lindau, P. Skeath, and C. Y. Su: J. Vac. Sci. Technol., 1980, 17, 1019.
[8] M. Levinson, M. Stavola, J. L. Benton and L. C. Kimerling: Phys. Rev. B, 1983, 28, 5848.
[9] H. Lim, G. Sagnes, G. Bastide and M. Rouzeyre: J. Appl. Phys., 1982, 53, 3317.

Author index

The order of entries has been determined using the English alphabet after deleting accents. Thus Låßíê is indexed as for Lassie.

Abakumov, V.N.	511
Abe, T.	1327
Abernathy, C.R.	111, 617, 1057, 1063, 1439, 1557
Aboelfotoh, M.O.	179
Achtziger, N.	1097
Ackermann, H.	269
Adamowshi, J.	493
Addinall, R.	1027
Alexander, H.	899, 1315
Allan, G.	865, 1397
Alt, H.Ch.	369
Ammerlaan, C.A.J.	407, 683, 701
Amroun, N.	173
Anderson, F.G.	233, 475
Anderson, G.	33
Anderson, R.A.	63
Andreasen, H.	1003
Ansaldo, E.J.	569
Ardonceau, J.	1201
Asenov, A.	1165
Assali, L.V.C.	143, 221
Aucouturier, M.	45
Auret, F.D.	1499
Azoulay, R.	911
Baba, T.	1409
Baeumler, M.	959
Bagraev, N.T.	1135
Baj, M.	875
Ball, C.A.B.	1147
Ballutaud, D.	45
Banghart, E.K.	499
Bao, X.J.	623
Baranov, P.G.	1207
Baranowski, J.M.	841
Barnard, W.O.	1499
Bates, C.A.	487, 1369
Baumann, F.H.	1339
Baurichter, A.	593
Bauser, E.	853
Bednarek, S.	493
Beeman, J.W.	757
Bennebroek, M.T.	251
Benson, B.W.	339
Benton, J.L.	653, 1433
Benyattou, T.	677

Benz, K.W.	1235
Bergman, J.P.	1363
Bernauer, M.	1165
Bernholc, J.	1351
Bestwick, T.D.	1445
Bishop, S.G.	1195
Blinowski, J.	523
Blöchl, P.	433
Bochkarev, E.P.	1075
Bode M.	1339
Bohling, D.A.	1057
Booker, G.R.	1469
Borenstein, J.T.	51
Borghesi, A.	1069
Bourgoin, J.C.	911
Bourret, E.D.	757
Boutry-Forveille, A.	45
Boyce, J.B.	1
Bozek, R.	729
Bredell, L.J.	1499
Brémond, G.	677, 911
Bretagnon, T.	1021
Brewer, J.H.	569
Briddon, P.R.	457
Broser, I.	1241, 1247
Brozel, M.R.	1027
Brubaker, R.M.	1357
Brunthaler, G.	823
Brüßler, M.	1109
Buchner, R.	1315
Buchwald, W.R.	1153
Bumay, Y.A.	1009
Burkey, B.C.	499
Cai, P.X.	263
Caldas, M.J.	1015
Cannelli, G.	9
Cantelli, R.	9
Canuto, S.	463
Capizzi, M.	9, 599
Car, R.	433
Carlos, W.E.	1493
Chadi, D.J.	447
Chakrabarti, U.K.	617, 1439
Chan, L.Y.	941
Chauvet, O.	1201
Chen, W.M.	251, 333, 356
Chen X.S.	587
Chevallier, J.	539, 629
Cho, K.H.	413
Chow, K.	569, 1115

Christmann, P.	1165
Clark, S.	1285
Claverie, A.	1045
Clerjaud, B.	563
Coffa, S.	203
Colon, J.E.	671
Coluzza, C.	599
Conibear, A.B.	1147
Corbel, C.	923, 979
Corbett, J.W.	27, 51, 57, 395, 1475, 1487
Cordero, F.	9
Cornet, A.	1285
Corradi, G.	279
Cote, D.	563, 729
Cox, S.F.J.	569
Crowder, M.	1421
Dąbrowski, J.	735
Dannefaer, S.	1021
Davies, G.	191, 315
Davis, R.F.	1201
Deák, P.	395
de Mierry, P.	45
Deák, P.	395
de Coteau, M.D.	185
Deicher, M.	593, 1481
DeLeo, G.G.	69
Delerue, C.	659, 787, 865, 911, 965
Demuth, J.E.	1381
DesJardin, W.F.	499
Deubler, S.	593
Diehl, E.	269
Dietrich, H.B.	665
Dmochowski, J.E.	751
Dobaczewski, L.	769
Dodds, S.A.	569, 1115
Doland, C.	33
Dörnen, A.	197, 695, 1213
Downey, S.W.	1557
Drabold, D.	481
Dreszer, P.	841, 875
Drozdova, I.A.	1265
Duarte, A.J.	191
Ducroquet, F.	677
Dunn, J.L.	487, 1369
Dupuy, J.C.	1303
DuVarney, R.C.	569, 1115
Eaglesham, D.J.	653
Ehrhart, P.	947

Elsaesser, D.	671
Emanuelsson, P.	137
Emerson, A.B.	1557
Emiliani, V.	599
Emtsev, V.V.	321
Endrös, A.	1165
Ergezinger, K.-H.	269
Ernst, F.	1315
Eschle, P.	1121
Estle, T.L.	569, 1115
Estreicher, S.K.	27, 63, 119
Evans, K.R.	1051
Ezhevskii, A.A.	701
Falk, A.	823
Falster, R.	185, 1469
Fang, Z.-Q.	991
Farmer, J.W.	817
Fazzio, A.	463
Fedders, P.	481
Feingold, A.	1509
Ferguson, I.T.	1027
Ferrante, R.F.	653
Filo, A.J.	1159
Fischer, B.	269
Fischer, D.W.	533
Fitzgerald, E.A.	653
Fockele, M.	835
Follstaedt, D.M.	81
Ford, W.	881
Forkel, D.	593
Fowler, W.B.	69
Frank, H.-P.	269
Frank, W.	203
Frankl, P.	599
Freitas, Jr., J.A.	1195
Frens. A.M.	251, 356
Fricke, C.	1241
Frova, A.	9, 599
Fujita, S.	75
Fullowan, T.R.	1439, 1557
Gardner, J.A.	971
Gauthier, S.	1177
Gebhard, M.	15, 1097
Geddo, M.	1069
Gehlhoff, W.	137, 719
Gendron, F.	563
Gerasimov, A.B.	517
Geva M.	1509

Ghatneker, S.	209
Giannozzi, P.	611, 635
Gibart, P.	835
Giesekus, A.	257
Gippius, A.A.	1219
Girard, J.C.	1177
Gislason, H.P.	905, 985
Glaser, E.R.	775, 793
Godbey, D.J.	1493
Godlewski, M.	683, 701
Goguenheim, D.	917, 1427
Görger, A.	149
Gorin, S.N.	1075
Gosele, U.M.	1451
Gossard, A.C.	1363, 1375
Grattepain, C.	629
Gray, M.L.	1563
Gregorkiewicz, T.	407, 683, 701
Greulich-Weber, S.	149
Grimmeiss, H.G.	125, 137, 209, 215, 239, 245, 841, 1153, 1235
Grossman, G.	125, 475
Grünebaum, D.	197
Gudmundsson, J.T.	985
Guillot, G.	677, 911, 917
Guitron, J.G.	757
Gürer, E.	339
Gustin, W.	203
Gwilliam, R.M.	251
Hage, J.	401
Hahn, S.	105, 413
Hahn, W.-S.	563
Hallam, L.D.	487
Haller, E.E.	245, 257, 569, 757, 853, 941
Ham, F.S.	233, 475
Han, W.-M.	971
Hansen, W.L.	853
Hara, K.	1141
Harris, C.I.	1363, 1375
Harris, J.S.	605
Harris, T.D.	1063
Hasebe, M.	1475
Hashimoto, K.	227
Häßlein, H.	1109
Hauksson, I.S.	985
Hautojärvi, P.	923, 979
Hawkins, I.D.	769
Heggie, M.I.	457
Heijmink-Liesert, B.J.	407, 683
Heinrich, M.	395
Heiser, T.	161, 173

Heitz, R.	1241, 1247
Hendorfer, G.	959
Hengehold, R.L.	671
Hennel, A.M.	729
Henry, A.	251, 1445
Herms, A.	1285
Herring, C.	33, 605
Heyman, J.N.	257, 757
Higgs, V.	1309
Hillard, R.J.	623
Hirata, M.	309
Hitti, B.	1115
Hobson, W.S.	111, 617, 1063, 1439
Hoffmann, A.	1241, 1247
Hofmann, D.M.	1235
Hoinkis, M.	841
Holtz, P.O.	1363, 1375
Höpner, A.	1315
Horigan, J.	881
Hornauer, U.	1259
Howard, L.K.	1291
Hybertsen, M.S.	1391
Irmscher, K.	137
Irvine, A.C.	997, 1291
ISOLDE Collaboration	1481
Ittermann, B.	269
Jacobson, D.C.	203, 653
Jaklevic, J.	941
Jamila, S.	487
Jantsch, W.	799, 805, 823
Janzén, E.	251, 327, 985
Jeanloz, R.	757
Jia, Y.Q.	965, 1051
Jin, S.X.	587
Johnson, N.M.	33, 605, 1153
Jones, K.S.	1063
Jones, R.	457, 551
Kacman, P.	523
Kaczor, P.	769
Kadono, R.	569
Kalem, S.	629
Kalish, R.	1481
Kalt, H.	1363, 1375
Kaminska, M.	723, 1033
Kaneta, C.	419
Kaniewska, J.	1457, 1463
Kaniewska, M.	1457, 1463

Kasuya, K.	953
Katz, A.	1439, 1509
Kaufmann, B.	197
Kaufmann, U.	793, 959
Kawakami, K.	1475
Keller, H.	1121
Keller, R.	1481
Kennedy, T.A.	775, 1433
Kerr, D.	1021
Khachaturyan, K.	881
Khirunenko, L.I.	425
Kiefl, R.F.	569, 1115
Kightley, P.	1309
Kim, Y.	1339
Kimerling, L.C.	653, 1433
Kirk, P.J.	487
Kisielowski, C.	1171
Klein, J.	1177
Klein, P.B.	665, 1195
Klemm, S.	481
Kleverman, M.	125, 209, 215
Klinger, P.M.	321
Kondo, T.	1503
Kopf, R.	1439, 1557
Kordina, O.	327
Korpas, L.	27
Korsunskaya, N.E.	1265
Korzeniewski, K.	875
Kossut, J.	805
Kozuch, D.M.	111, 617
Krambrock, K.	887
Krause, M.	563
Kreissl, J.	137, 719
Kreitzman, S.R.	569, 1115
Krings, Th.	1259
Krüger, J.	899
Kruger, M.B.	757
Kuech, T.F.	847
Kündig, W.	1121
Kunert, H.W.	1499
Kwok, T.K.	315
Laczik, Z.	1469
Laks, D.B.	1225
Lambert, B.	677
Lamont Schnoes, M.	1063
Lamp, C.D.	1115
Landman, J.I.	1253
Lane, E.	1509
Lang, M.	197, 1097
Langer, J.M.	1545

Lannoo, M.	659, 787, 865, 1397
Latushko, Y.I.	1213
Lavine, J.P.	499, 1159
Le Corre, A.	677
LeBerre, C.	923
Lee, B.Y.	413
LeGoues, F.K.	1271
Leibenzeder, S.	1213
Leitch, A.W.R.	21
Leite, J.R.	143, 221, 707
Lelikov, Yu.	1321
Leon, R.P.	723
Lhomer, C.	677
Li, L.B.	263
Li, M.-F.	853
Li, W.-G.	505
Liang, Z.N.	99
Lichti, R.L.	569, 1115
Licini, J.C.	1563
Lightowlers, E.C.	93, 191, 1309
Liliental-Weber, Z.	723, 1045
Lindström, J.L.	333, 1445
Linnarsson, M.	985
Lino, A.T.	221
Liszkay, L.	923
Liu, L.B.	263
Liu, X.	763
Londos, C.A.	351
Look, D.C.	991
Lopata, J.	617
Lothian, J.	1557
Lu, L.	1569
Lusson, A.	629
Magerle, R.	1481
Mai, F.	269
Maier, K.	1171, 1183
Malakovskaya, V.E.	1009
Manasreh, M.O.	533, 1051
Marchand, J.-J.	677
Maric, Dj.M.	119
Markevich, I.V.	1265
Markvart, T.	1539
Martin, D.	635
Martin, K.R.	69
Martins da Cunha, C.R.	463
Marwick, A.D.	39, 847
Mascher, P.	413
Mashovets, T.V.	321
Masuda, K.	75, 1141
Masuda-Jindo, K.	1333

Mathiot, D.	1303
Matous, G.	533
McAfee, S.R.	1063
McPhail, D.	1027
McQuaid, S.A.	87, 93
Meier, E.	635
Meier, J.	593
Meier, P.F.	119
Melloch, M.R.	1357
Merz, J.L.	1363, 1375
Mesli, A.	161, 173
Metzner, H.	1109
Meyer, A.	1159
Meyer, B.K.	713, 1165, 1235
Michel, J.	291, 653, 1433
Milnes, A.G.	623
Mir, A.	1427
Missous, M.	769
Mizuta, M.	575
Mochizuki, Y.	575
Mohades-Kassai, A.	1027
Monemar, B.	251, 327, 356, 1363, 1375, 1445
Mooney, P.M.	829
Moore, F.G.	665
Morante, J.R.	1285
Morgan, T.N.	859
Morgan-Pond, C.G.	1253
Morier-Genoud, F.	635
Moriya, N.	1481
Moser, P.	979
Müller, H.D.	1183
Murakami, K.	75, 1141
Muto, S.	309
Myburg, G.	1499
Myers, S.M.	81
Myles, C.W.	505
Nakahara, S.	1509
Nakashima, H.	227
Nakata, Y.	1297
Naud, C.	729
Nazaré, M.H.	191
Needels, M.	1391
Nelson, E.T.	499
Neuhalfen, A.J.	689
Neumark, G.F.	1225
Newman, R.C.	87, 93, 1027
Niedermayer, Ch.	569, 1115
Nielsen, J.	1103
Niesen, L.	99
Nishina, Y.	953

Nissen, M.K.	893
Nolte, D.D.	1357
Norman, C.E.	1309
Nylandsted-Larson, A.	273
Öberg, S.	551
Odermatt, W.	1121
Oding, V.	1171
Oehrlein, G.S.	39, 1445
Ogawa, T.	419
Ohno, T.	1403
Omling, P.	137, 713, 1235
Ono, H.	1409
Oshiyama, A.	469
Ostapenko, S.S.	1127
Ostermayer, G.	799, 805, 823
Ott, U.	1259
Ourmazd, A.	1339
Overhof, H.	149, 279, 835
Pajot, B.	539, 581
Pakhomov, A.A.	511
Palmer, D.W.	997, 1291
Pantelides, S.T.	1225
Parker, B.D.	829, 847
Paroskevopoulos, N.G.	1063
Patterson, B.D.	1121
Pavesi, L.	611, 635
Peaker, A.R.	769, 1457, 1463
Peale, R.E.	111, 905
Pearl, E.	1369
Pearton, S.J.	51, 111, 533, 617, 623,
—	1057, 1063, 1439, 1509, 1557
Peiró, F.	1285
Pensl, G.	197, 1097, 1213
Pesant, J.	45
Peters, J.W.	1539
Petersen, J.W.	1003, 1103
Petrov, V.V.	1075
Pettersson, H.	239, 1153
Pfeiffer, W.	1481
Pfiz, T.	569, 1115
Pierre, F.	979
Pillukat, A.	947
Pintanel, R.	707
Pistol, M.-E.	763
Pivac, B.	285, 1069
Poate, J.M.	203, 653
Poindexter, E.H.	1153
Polinger, V.Z.	487

Polovtsev, I.S.	1135
Polyakov, A.Y.	623
Pomrenke, G.S.	671
Ponce, F.	33
Poole, I.	769
Portal, J.C.	911
Porte, C.	563, 581
Prescha, Th.	21, 167
Pressel, K.	695
Pross, P.	1481
Puff, W.	413
Qi, M.W.	263
Qin, G.G.	587
Qu, W.	1569
Que, D.L.	263
Rai-Choudhury, P.	623
Ranz, E.	911
Reade, T.	315
Ready, S.E.	1
Rebane, Yu.	1321
Reeson, K.J.	251
Rehse, U.	137
Reinhart, F.K.	635
Ren, F.	1439, 1557
Rentschler, J.A.	1339
Revva, P.	1545
Richards, P.M.	105
Rigo, S.	1421
Riseman, T.M.	569
Rizk, R.	45
Rodrigues, C.W.	1015
Romano-Rodríguez, A.	303
Romanov, N.G.	1207
Rong, F.C.	935, 1153
Roos, G.	605
Rousset, S.	1177
Rückert, G.	695
Ruvimov, S.S.	1321
Saarinen, K.	923, 979
Sacks, W.	1177
Samuelson, L.	763
Sangster, M.J.L.	1027
Sankey, O.F.	481
Sarto, F.	599
Sassella, A.	1069
Savić, I.M.	1121
Sawyer, W.D.	291

Schaff, W.	1045
Scheffler, M.	735
Scherz, U.	929
Schick, J.T.	1253
Schlesinger, T.E.	623
Schluter, M.	1391
Schmalz, K.	239
Schmidt, J.	251, 356
Schneider, J.	793, 1171, 1183
Schneider, J.W.	569, 1115, 1121
Schönherr, E.	793
Schwab, C.	569
Scolfaro, L.M.R.	707
Seager, C.H.	63
Seghier, D.	677
Seelinger, W.	269
Sergent, A.M.	1563
Shakhovtsov, V.I.	425
Shanabrook, B.V.	793
Sharma, V.K.M.	1027
Sheinkmann, M.K.	787, 1127, 1265
Shi, T.S.	263
Shinkarenko, V.K.	425
Shinozuka, Y.	527
Shlopak, N.V.	1009
Shreter, Yu.	1321
Shulga, E.P.	1265
Siboulet, O.	1177
Sielemann, R.	1109
Sigg, H.	635
Simmler, H.	1121
Sinerius, D.	1235
Singh, M.	251
Sitnikova, A.	1321
Siyanbola, W.O.	997
Skowronski, M.	377
Skudlik, H.	1481
Smargiassi, E.	433
Snyder, L.C.	395
Solomon, I.	1201
Song, C.	581
Sopori, B.L.	1531
Souza, P.L.	1015
Spaeth, J.-M.	149, 835, 887
Spector, M.	1563
Sputz, S.K.	1063
Stahlbush, R.E.	1493
Stam, M.	623
Stathis, J.H.	1421
Stäuble-Pümpin, B.	1121
Stavola, M.	111, 617

Steele, A.G.	191
Stefaniak, M.	1315
Stein, H.J.	81, 105
Stein, R.A.	1213
Stella, A.	1069
Stesmans, A.	1415
Stiévenard, D.	911, 917, 965
Stockmann, H.-J.	269
Stolwjik, N.	197
Stradling, R.A.	751
Straumann, U.	1121
Stutz, C.E.	1051
Stutzmann, M.	9, 629
Su, Z.	817
Suezawa, M.	155, 953
Sugino, O.	469
Sumino, K.	155, 953
Sun, H.-J.	905, 935
Sundaram, M.	1363, 1375
Suttrop, W.	1213
Svensson, B.G.	179, 333
Svensson, J.H.	327
Swaminathan, V.	1439
Swanson, C.C.	1159
Taguchi, A.	641
Tajima, M.	1327
Takahashi, H.	155
Takahei, K.	641
Takeda, M.	1503
Takeda, S.	309
Takeno, H.	1327
Talwar, D.N.	533
Tan, S.S.	263
Tan, T.Y.	1451
Tarhin, D.	1321
Taylor, W.J.	1451
Theis, T.N.	847
Thewalt, M.L.W.	893
Theys, B.	629
Thilderkvist, A.	125, 215
Thonke, K.	695
Thurian, P.	1247
Tilly, L.	239
Tischler, M.A.	829
Tittlebach, K.	239
Tkacheva, T.M.	1075
Török, P.	1469
Toudic, Y.	677
Trequattrini, F.	9
Trimaille, I.	1421

Trombetta, J.	401
Truöl, P.	1121
Trzeciakowski, W.	751
Tucker, J.H.	87
Tuncel, E.	635
Twigg, M.E.	1493
Ueda, O.	1297
Uematsu, M.	1551
Uhrmacher, M.	1097
Ulrici, W.	563, 719
Ulyashin, A.G.	1009
Umerski, A.	551
Uren, M.	1519
Urli, N.B.	285
Van de Walle, C.G.	1225
Van Gisbergen, S.J.C.H.M.	701
Vanhellemont, J.	303
Veloarisoa, I.A.	111
Verner, I.V.	57
Villemaire, A.	893
Vogt, B.	15
von Bardeleben, H.J.	787, 911, 965, 1051
Vuillaume, D.	1427
Wada, K.	1403, 1503, 1551
Wagner, P.	401
Wagner, R.J.	793
Walker, J.	33
Walukiewicz, W.W.	757, 941
Wampler, W.R.	81
Wang, C.	1351
Wang, L.P.	587
Wang, P.D.	751
Wang, Q.N.	1357
Wang, Y.C.	1201
Warren, Jr., W.W.	971
Waterman, J.R.	793
Watkins, G.D.	111, 215, 233, 339, 345, 401, 905, 935
Weber, E.R.	723, 841, 853, 881, 1033, 1045
Weber, J.	21, 167, 1315
Weiner, V.S.	297
Welker, G.	269
Werner, P.	1045
Weronek, K	1315
Wessels, B.W.	689
Wetzel, C.	1165
Weyer, G.	273, 1003
Wichert, Th.	1081, 1259, 1481

Wilamowski, Z.	799, 805, 823
Wilkening, W.	793
Williams, D.M.	689
Williams, P.M.	233
Williams, R.H.	1285
Willoughby, A.F.W.	1539
Wilshaw, P.R.	185
Wilson, R.G.	623
Wisk, P.W.	1557
Witthuhn, W.	15, 593
Wittmer, M.	39
Wolf, H.	1259, 1481
Wolk, J.A.	757
Wysmolek, A.	729
Xie, Y.-H.	653
Yakubenya, S.M.	361
Yamada-Kaneta, H.	419
Yang, B.	985
Yassievich, I.N.	511
Yencha, A.J.	1487
Yeo, Y.K.	671
Yoder, J.D.	1563
Yu, K.M.	723, 941
Yu, P.Y.	853
Yuan, J.	1487
Yuan, M.H.	587
Zach, F.X.	245
Zhan, X.D.	345
Zhang, Q.-M.	1351
Zhang, S.	1569
Zhong, X.-F.	389
Zhou, J.	1569
Zhu, B.	263
Ziegler, C.	929
Zundel, T.	21
Zuppiroli, L.	1201
Zvanut, M.E.	1493

Subject index

Where possible, entries have been grouped together under generic headings. For example, for "Al in Si" see "acceptors in Si — Al".

acceptors in silicon	
— Al	279
— B	269
— Be	255
— see hydrogen in silicon	
acceptors in SiC	
— ODMR data	1207
acceptors in II-VI compounds	
— Li	1259
acceptors in III-V compounds	
— GaAs + radiation, annealing	997
— GaAs:Zn	941
— InP:Zn	941
A-centre in CdTe	1235
A-centre in Si	321, 356
AlGaAs/GaAs — see GaAs/AlGaAs	
amorphous Si bandtails	481
anti-site defects	
— GaP:P_{Ga}-Y_P (Y unknown)	935
— GaAs:Ga_{As}	1171
— in GaAs low temperature growth	1033
— in GaAs + radiation	947, 979
— in T_d semiconductors	447
— see EL2	
As in Ge	297
atomic layer passivation for GaAs:S	1403
Au in amorphous Si	203
B in $Si_{1-x}Ge_x$	1303
Be in GaAs	1551
Be in GaAs/AlGaAs	1557
Be in InAs, InSb	1027
Be in Si	255
bistability	
— of CdF_2:In	493
— of DX in AlGaAs	847
— of Si:Fe-Al	149
— of Si:H-As, Si:H-P	63
carbon	
— in AlGaAs grown by MOMBE	1057
— in magnetic CZ Si	1075
— in Si	285
— in Si:P	339
— in SiC	1201
cascade phonon emission	499, 511
cathodoluminescence of dislocations in Si	1309
charge coupled devices	499

configurational coordinates
— of EL2 917
Cr in semimagnetic semiconductors 523
Cu decoration of oxide stacking faults in Si 1457
defect production by local excitation 511, 517
density functional theory
— and diffusion 433
— of AlAs:Si 635
— of amorphous Si 481
— of C:N 457
— of GaAs:C 551
— of GaAs:EL2 929
— of GaAs:H 611
— of GaAs:S adatoms 1403
— of Si:Al 279
— of Si:As 469
— of Si:B adatoms 1391
— of Si:O+H 551
— of Si:Sb 469
— of ZnSe 1225
deuterium in crystalline Si
— D-Be 69
— diffusion 45
— D-P 75
— D-Sb 99
— effusion 45
— internal surfaces 81
deuterium in III-V's
— in AlGaAs:Si 539
— in AlGaAs - SIMS profiles 617
— in GaAs - SIMS profiles 617
— in GaAs:Zn 539
— in InP - SIMS profiles 617
diffusion in GaAs:Be 1551
diffusion in GaAs/AlAs 1351
diffusion in GaAs/AlGaAs 1339, 1557
dislocations in GaAs
— STM 1177
dislocations in Ge 1333
dislocations in InGaAs 1285, 1291
dislocations in silicon
— 'D' photoluminescence 1309, 1315
— effect of transition metals 1309, 1321
— EFG films 291
— motion 1333
dislocations in Si/Ge 1271
divacancy in Si 321, 1165
DLTS measurements
— digital aquisition 1147
— of AlGaAs:DX 605, 769
— of AlGaAs:Si 605, 817, 829

— of AlGaAs:Te + DX	853
— of AlGaAs/GaAs heterostructures	1063
— of AlGaAS/GaAs HFETs	1563
— of GaAs:EL2	1147
— of GaAs e-beam evaporated	1499
— of GaAs:Er	671
— of GaAs MESFET's	1563
— of GaAs:O	377
— of GaSb:DX	769
— of GaSb + H	623
— of Ge:Ni	245
— of InGaAs	1291
— of InP with Schottky barriers	1569
— of P_b centre	1427
— of Si:Au	1103
— of Si:B + radiation	351
— of Si:C	285
— of Si:C_i-P_s	339
— of Si:Cd-Fe	1097
— of Si:Cu-acceptor pairs	161
— of Si:Cu-radiation damage complexes	179
— of Si:Fe	227
— of Si:Ir	1103
— of Si:Os	1103
— of Si:Pt	1103
— of Si:S	1153
— of Si:Se	1153
— of Si solar cells	1539
— of Si:stacking faults	1457, 1463
— of Si:Ti	227
— of Si:V	227
— temperature dependent field effect	1153
donor-acceptor pair spectra	
— in SiC:Al-N	1195
donors in silicon	
— As	469, 1171
— P	499, 1171
— Sb	273, 469
donors in II-VI's	1265
DX centres	
— density functional theory	735
— in AlGaAs	605, 769, 775, 787, 799, 805,
— —	817, 823, 829, 835, 841, 847, 853
— in AlSb	793
— in GaAs	735, 751, 757, 763, 859
— in GaSb	769
EFG Si	
— carbon related levels	285
— photoluminescence	291
elastic energy loss of Si:B-H	9

electrically detected magnetic resonance
— of Si:Pt   1165
electron beam doping   1503
electron beam induced current
— of dislocations in Si   1309
electronic optical transitions
— of CdF_2:In   493
— of GaAs:Fe   695, 729
— of InP:Fe   695
— of Si:Au   209, 215, 1135
— of Si:Be   255
— of Si:Fe   125
— of Si:Mn   125
— of Si:thermal donors   401
— of Si:Zn   197
— of Si:615 meV line   327
— of SiC:V   1213
electron paramagnetic resonance
— of AlGaAs deep + shallow states   787
— of AlGaAs:Sn + DX   841
— of AlSb:Te   793
— of GaAs   1051
— of GaAs:As_{Ga} 899
— of GaAs:Mn   701
— of GaAs:Tl   959
— of GaAs:Zn + radiation   965
— of GaP:As_{Ga} 899
— of GaP:Fe-S   719
— of GaP:V   701
— of Ge-As   297
— of P_b centre   1415, 1421
— of Si:C_i-P_s 345
— of Si:CrIn   137
— of Si:Fe-acceptors   143, 155
— of Si:Fe-Al   149
— of Si:Fe_4 221
— of Si:H-P   75
— of Si:MnGa   137
— of Si:Mn_4 221
— of Si:Ni^- 233
— of Si:NL8, NL10   407
— of Si:Pd^- 233
— SiC:C sp^2 1201
— of SiC:N   1141, 1183
— of SiC:Ti   1183
— of SiC:V   1183
— of SiGe:Ge   1493
electron spin echo
— of Si:A centre   357
EL2
— density functional theory   735

— in AlGaAs	911
— in GaAs	859, 865, 875, 881, 887,
— —	893, 917, 923, 929, 947, 991, 1033, 1147
F in GaAs	1009
Fano resonances	
— of Si-Be	255
Frenkel pair production in Ge	1109
Fröhlich interaction of CdF_2:In	493
GaAs/AlAs superlattices	
— Al monolayer vibrations	1409
— Zn diffusion	1351
GaAs/AlGaAs superlattices and quantum wells	
— acceptor states	1375
— chemical mapping	1339
— donor states	1369, 1375
— Franz-Keldysh effect	1357
— time-resolved photoluminescence	1363
— Zn diffusion	1351
Ge-O complexes in Si	419
Green's function calculations	
— of GaAs:As_{Ga}	865
— of lattice relaxation	505
— of rare earth inpurities	659
Hartree-Fock calculations	
— C:N	463
— Ge:N	463
— Ge:H-B	119
— Ge:H-C	119
— Ge:H-Si	119
— Si:H-As	63
— Si:H-B	119
— Si:H-H-B	27
— Si:H-H-P	27
— S:H interstitial	119
— Si:H-P	63
— Si:N	463
Heisenburg Hamiltonian	523
hetero-epitaxy	1271 *et seq.*
hydrogen in AlAs	
— in AlAs:Si	635
hydrogen in amorphous Si	1
hydrogen in crystalline Si	
— diffusion	21, 45, 51, 57, 87
— effusion	45
— H-Al	111
— H-As	63
— H-B	9, 93, 111, 119
— H-Be	69
— H-Cd	15
— H-Ga	111
— H-H-B	27

— H-H-P	27
— high concentrations in p$^+$	39
— H-In	15
— H-P	63, 75
— H-Sb	99
— O migration	551
— perturbed angular correlation	1081
— platelets	33
— solubility	93
— surface desorption	45
— thermal donor formation	105
hydrogen in GaAs	
— diffusion	21
— H-C	111, 563
— H-Cd	593
— H-N	581
— H-O	581
— H-Zn	539
— local density theory	611
— passivation of C	551
— photoluminescence	599
— Schottky barriers	587
— SIMS profiles	617
hydrogen in GaP	
— H-Cd	593
— passivation of Zn	575
hydrogen in GaSb	
— diffusion coefficient	623
hydrogen in germanium	
— H-B	119
— H-C	119
— H-Ni	245
— H-Si	119
hydrogen in InAs	
— diffusion	629
— H-Cd	593
hydrogen in InP	
— H-Cd	593
hydrogen in InSb	
— H-Cd	593
hydrogen in Schottky barriers	587
hydrogen in silicon solar cells	1531
hydrogen in III-V alloys	
— GaAlAs:DX	605
— GaAlAs:Si	539, 605
— GaAs/AlGaAs	1357
hydrogenic states in GaAs/AlGaAs	1369
hydrostatic pressure perturbations	
— of GaAs:donors	751
— of GaAs:DX	757
— of GaAs:D*	763

— of GaAs:EL2	875, 929
In in GaAs	1003, 1481
interstitial defects	
— in CdTe	1253
— in Ge	309
— in Si	309, 363, 433, 1433, 1451
— diffusion	527, 1433, 1439
— rôle in electron beam doping	1503
ion channeling	
— in GaAs/AlGaAs ion etched	1439
— of AlGaAs:Te + DX	853
ion implantation	
— stress in Si	1487
isotope effects	
— epr in isotopically modified Ge	297
— in P_b centres	1415, 1421
— radioactive isotopes in Si	1097, 1103
— vibrations of O in ^{74}Ge	425
Jahn-Teller effect	
— of interstitials	527
— of Si:Au-Fe	209
— of Si:vacancy	475
— of Si:Zn	197
— reduction factors	487
junction space charge measurements	
— of Si:Mo	239
— of Si:Ti	239
— of Si:W	239
kinetic modelling of Si:H	51, 57
Li in GaAs	985
Li in II-VI's	1259
linear combinations of atomic orbitals	
— for dislocations in Ge and Si	1333
local mode vibrations	
— dynamical modelling	533
— in C-rich EFG Si	285
— of AlAs:Si	635
— of C:N	457, 463
— of GaAs:Al monolayer	1409
— of GaAs:DX	757
— of GaAs:H-C	111, 563
— of GaAs:H-N	581
— of GaAs:H-O	581
— of GaAs:O,	369, 377
— of ^{74}Ge:O	425
— of InAs:Be	1027
— of InAs:Si	1027
— of InSb:Be	1027
— of InSb:Si	1027
— of Si:H-Al	111
— of Si:H-B	93, 111

— of Si:H-Ga 111
— of Si:N 263
— of Si:O epitaxial growth 1069
— of Si:P + radiation 333
low temperature growth of GaAs 1033, 1045
magnetic circular dichroism
— of GaAs:As$_{Ga}$ 865, 887
— of GaAs irradiated 947
magnetic CZ silicon 1075
magneto-optical transmission
— of GaAs:donors 751
metastability
— and optical nuclear polarisation 1135
— in II-VI's 1265
— of acceptors and donors in T$_d$ crystals 447
— of DX centres 735
— of EL2 735, 911, 917
— of Si:B + radiation 351
— of Si:C$_i$-P$_s$ 339, 345
— of Si:Pt-'X' 167
— of Si:S 251
— of Si:615 meV centre 327
— of transition metal acceptor pairs in silicon 137, 155
Mössbauer spectroscopy
— of GaAs:In 1003
— of Si:H-D, Si:H-Sb 99
— of Si:Sb 273
— review 1081
muonium
— in CuCl 569
— in GaAs 569, 1115, 1121
— in Ge 1115
— in InP 1121
— in Si 1115, 1121
— level crossing 1115
— radiofrequency resonance 1115
negative U
— of AlGaAs:Sn + DX 799
— of GaAs:Au adatoms 1397
— of GaAs:O$_s$ 369, 377, 389
neglect of differential overlaps
— of AlGaAs:Si 1015
neutron activation analysis
— of Au in a-Si 203
nitrogen
— in C 457, 463, 1219
— in GaP 505
— in Ge 463
— in Si 263, 463
— in SiC 1141, 1171, 1183, 1195

nuclear magnetic resonance
— of amorphous Si 1
— of GaAs 971
— of Si:B 269
optically detected cyctron resonance
— of InP:Er 683
— of InP:Yb 683
optically detected electron-nuclear resonance
— of GaP:P_{Ga}-Y_P (Y unknown) 935
— of InP:antisites 905
optically detected magnetic resonance
— of AlAs 775
— of AlGaAs 775
— of AlGaAs:Sn + DX 835
— of CdTe:A centres 1235
— of GaAs:As_{Ga} 887
— of GaP:Mn 713
— of GaP:V 713
— of InP:antisites 905
— of Si:A centre 357
— of Si:S 251
— of SiC:acceptors 1207
optical nuclear polarisation
— of Si:Au 1135
— of SiC:N 1141
out-diffusion of O from Si substrates 1069
oxidation induced stacking faults 1457, 1463
oxygen in C 1219
oxygen in GaAs
— bond-orbital calculations 389
— impurity complexes 377
— substitutional 369, 377
oxygen in germanium
— vibrations in ^{74}Ge 425
oxygen in silicon
— enhanced diffusion from H 87
— epitaxial growth 1069
— Ge-O complexes 419
— magnetic CZ 1075
— nitrogen complexes 263
— positron annhilation 413
— precipitation 1327, 1451, 1475
— surface contamination 1381
— 1/f noise 1519
P in $Si_{1-x}Ge_x$ 1303
P_b centres
— at Si/SiO$_2$ interface 1415, 1421, 1427
— in SiGe 1493
passivation
— of Si solar cells 1531
— see hydrogen in Si *etc.*

perturbed angular correlation
— in Ge 1109
— of Cd in III-V's 593, 1481
— of In in III-V's 1481
— of Li in II-VI's 1259
— of Si:H-Cd 15
— of Si:H-In 15
— review 1081
photoconductivity
— of AlGaAs:Al diodes 1545
— of AlGaAs:Si + DX 799, 829, 823
— of AlGaAs:Sn + DX 835
— of GaAs:EL2 881
photoluminescence
— of AlGaAs grown by MOMBE 1057
— of AlGaAs:Si$_{Ga}$-Si$_{As}$ 1015
— of AlGaAs/GaAs heterostructures 1063, 1363, 1375
— of bound excitons in II-VI's 1241
— of C:N 1219
— of C:Ni 1219
— of C:O 1219
— of GaAs:DX 763
— of GaAs:EL2 893
— of GaAs:Er 665, 671
— of GaAs:F 1009
— of GaAs:H complexes 599
— of GaAs:Li 985
— of GaAs:Si$_{Ga}$-Si$_{As}$ 953
— of GaP:Mn 713
— of GaP:V 713
— of GaP:Zn + H 575
— of GaSb + H 623
— of InAs + H 629
— of InAsP:Yb 689
— of Si:Cu 191
— of Si:dislocations 1309, 1315, 1321, 1327
— of Si:Er 653
— of Si in TD formation 407
— of Si - reactive ion etched 1433, 1445
— of SiC:N-Al 1195
— of SiC:V 1213
— of ZnS:W 1247
— polarised 1127
— quenching in III-V:rare earth 641
photorefractive effect in GaAs/AlGaAs 1357
positron annihilation
— in CZ-Si 413
— in GaAs:EL2 923
— in GaAs + radiation 979
— in InP 1021

radiation damage
— of Ge 309
— of Si 303, 306, 315
— of Si:P 333, 339
— production rate in Si 321
rapid thermal annealing 1509
rare earth impurities
— GaAs:Er 641, 665, 671
— GaP:Nd 641
— InAsP:Yb 689
— InP:Er 677, 683
— InP:Yb 641, 677, 683
— Si:Er 653
— theory 659
reactive ion etching
— of GaAs/AlGaAs 1439
— of Si 1433, 1445
relaxation rates of nmr in GaAs 971
resonance ionization mass spectrospcopy
— of GaAs/AlGaAs:Be 1557
rigid rotator model of Si:D-Be, Si:H-Be 69
Rutherford backscattering
— of Au in a-Si 203
— of EFG Si 291
— of Si:Ar$^+$ implanted 1487
S-Ga surface bonds on GaAs 1403
scanning infra-red microscopy of Si:O 1475
scanning tunneling microscopy
— of GaAs 1177
— of semiconductor surfaces 1381
Schottky barriers
— AlGaAs:Al diodes 1545
— effect of e-beam evaporation 1499
— effect of H 587
— on InP 1569
Se in Si 505
self-diffusion in Si 433
Si in InAs, InSb 1027
Si_{Ga}-Si_{As} in GaAs 953
$Si_{1-x}Ge_x$ epitaxial layers
— dopant diffusion 1303
— strain relief 1271
SIMS measurements
— of AlGaAs grown by MOMBE 1057
— of D in III-V's 617, 629
— of electron-beam doped semiconductors 1503
— of InP:Cu 719
solar cells 1531, 1539
spin marking
— of Si:interstitial 361
spreading resistance of ion etched Si 1433

Steibler-Wronski effect	481
stress-induced dichroism	
— of Si:Ni$^-$	233
— of Si:O	87
— of Si:Pd$^-$	233
superlattices	1339 et seq.
telegraph noise	1519
thermal annealing	
— of GaAs	997, 1051, 1481
— of GaAs:In	1003
— of InP	1481
thermal donors in silicon	
— formation in Si:B, Si:Al	407
— semi-empirical calculations	395
— stress-induced alignment	401
— under H plasma	105
thermally stimulated current spectroscopy	
— of GaAs	991
tight binding approximation	
— of CdTe:interstitials	1253
— of GaAs:Au	1397
Tl in GaAs	959
transition metals in C	1219
transition metals in GaAs	
— Fe	695, 729
— Mn	701
transition metals in GaAsP	
— Cu	707
— Mn	707
transition metals in GaP	
— Fe	719
— Mn	701, 713
transition metals in InP	
— Cu	723
— Fe	729
transition metals in Si	
— Au	215
— Cd-Fe	1097
— Cr-In	137
— Cu	185, 191, 1159
— Cu and 'X' defect	167
— Cu effect on internal gettering	1475
— Cu in Al-doped	167
— Cu in B-doped	161, 167
— Cu in Ga-doped	161, 167
— Cu in In-doped	167
— Cu-radiation damage complexes	179
— dislocation effects	1309, 1315, 1321
— Fe	125, 173, 185, 227, 1159
— Fe-Al	143, 149, 155
— Fe$_i$-Al$_s$	149

— Fe-Au	209
— Fe-B	143
— Fe-Ga	143, 155
— Fe-In	155
— gettering	185
— Mn	125, 227
— Mn-Ga	137
— Mo	239
— neutron activation analysis	1159
— Ni$^-$	233
— Ni effect on internal gettering	1475
— perturbed angular correlation	1081
— Pd$^-$	233
— Pt-'X' formation	167
— surface contamination	1381
— Ti	227, 239
— V	227
— W	239
— 'X' defect	161, 167
— Zn	197
transition metals in SiC	
— Ti	1183
— V	1183, 1213
transmission electron microscopy	
— of AlGaAs/GaAs heterostructures	1063
— of dislocations in Si	1309
— of GaAs low temperature growth	1045
— of In$_{0.54}$Ga$_{0.46}$As	1285, 1297
— of InP:Cu	723
— of InP: rapid thermal annealed	1509
— of irradiated Ge	309
— of irradiated Si	303, 309
— of Si$_{1-x}$Ge$_x$ layers	1271
— of Si oxide precipitates	1475
Two-dimensional electron mobility	
— of In$_{0.54}$Ga$_{0.46}$As	1297
uniaxial stress perturbations	
— of AlGaAs:Te + DX	853
— of GaAs:EL2	893
— of GaAs:H-C	563
— of Si-Au	215
— of Si-Be	255
— of Si:Cu	191
— of Si:Zn	197
— of Si:615 meV line	327
vacancies in InP	1021
vacancies in Si	433, 475, 1327, 1451
V-O centre in Si — see 'A centre'	
Xα calculations	
— on GaAsP:Cu	707
— on GaAsP:Mn	707

— on Si:Fe$_4$	221
— on Si:Fe-acceptors	143
— on Si:Mn$_4$	221
X-ray diffraction	
— of GaAs irradiated	947
— of magnetic CZ Si	1075
X-ray spectroscopy of Si:transition metals	1159
Zeeman effect	
— of GaAs:EL2	893
— of SiC:V	1213
Zn doped GaAs	
— + Li	985
— + radiation	965
Zn doped GaAs/AlGaAs	1351
II-VI compounds	1225 *et seq.*